2nd International Conference on Natural Products and Bioresource Sciences (ICONPROBIOS 2018)

AA006234

IOP Conference Series: Earth and Environmental Science
Volume 251

Tangerang, Indonesia
1 – 2 November 2018

ISBN: 978-1-5108-8552-3
ISSN: 1755-1307

Printed from e-media with permission by:

Curran Associates, Inc.
57 Morehouse Lane
Red Hook, NY 12571

Some format issues inherent in the e-media version may also appear in this print version.

Institute of Physics
Dirac House, Temple Back
Bristol BS1 6BE UK

Phone: 44 1 17 929 7481
Fax: 44 1 17 920 0979

techtracking@iop.org

Additional copies of this publication are available from:

Curran Associates, Inc.
57 Morehouse Lane
Red Hook, NY 12571 USA
Phone: 845-758-0400
Fax: 845-758-2633
Email: curran@proceedings.com
Web: www.proceedings.com

Table of contents

Volume 251

The 2nd International Conference on Natural Products and Bioresource Sciences - 2018

1–2 November 2018, Tangerang, Indonesia

Accepted papers received: 1 March 2019
Published online: 12 April 2019

Preface

Papers

IOP Conf. Series: Earth and Environmental Science **251** (2019) 011001 doi:10.1088/1755-1315/251/1/011001

PREFACE

Thank God, finally the proceedings of the 2nd International Conference on Natural Products and Bioresource Sciences - 2018 (ICONPROBIOS 2018) can be finished. The ICONPROBIOS 2018 organized by Research Unit for Natural Product Technology, Indonesian Institute of Sciences (BPTBA LIPI) and This conference is a continuation of the success of the first ICONPROBIOS 2017 which had been published in IOP Conference Proceedings.

ICONPROBIOS 2018 which was held on November 1 until 2, 2018 at Indonesia Convention Exhibition (ICE) BSD City, Tangerang, Indonesia, with the theme "Natural Product for Sustainable Agroindustrial Production". The ICONPROBIOS 2018 is a part of Indonesian Science Expo (ISE) that is initiated by the Indonesian Institute of Sciences (LIPI). This conference brought together more than 65 participants and presenters from several background such as university lecturer, researcher, student, and industrial practitioners. The 2nd ICONPROBIOS 2018 provides a forum for exchange of information on natural products within all of the related topics as well as aims to build and strengthen scientific cooperation among the research institutions. The topics covered include natural product, bioactive metabolites, microbial-based technology, extraction and separation science, food science and technology, food process engineering, biomass for chemical and biofuel, bioremediation, animal nutrition and feed technology, animal health and production. We hope all of us get lots of benefit from this conference thus it will give us inspiration how to enhance natural product for sustainable agroindustrial production.

Sincerely,

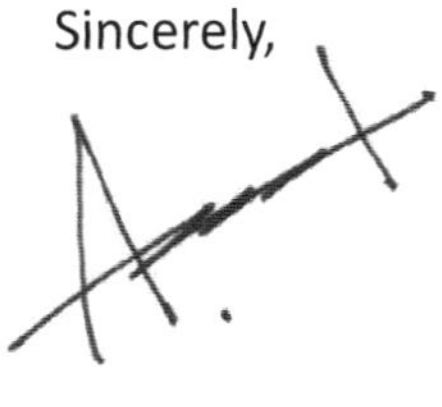

Dr. Asep Nurhikmat

The 2nd ICONPROBIOS 2018 General Chair

IOP Conf. Series: Earth and Environmental Science **251** (2019) 011002 doi:10.1088/1755-1315/251/1/011002

Committee of The 2nd ICONPROBIOS 2018

Host and Organizer: Research Unit for Natural Product Technology, Indonesian Institute of Science (BPTBA LIPI)

Advisory Board:

1. Dr. Laksana Tri Handoko (Chairman of Indonesian Institute of Sciences)
2. Dr. Yan Riyanto (Deputy Chairman for Engineering Sciences of the Indonesian Institute of Sciences)
3. Hardi Julendra, S.Pt, M.Sc. (Director of the Research Unit for Natural Product Technology, Indonesian Institute of Sciences)

Organizing Committee:

Chair person	:	Dr. Asep Nurhikmat
Co-Chairperson	:	Dr. Roni Maryana
Secretary	:	Muslih Anwar, M. Sc.
	:	Betty Hariastuti, SE
Technical Program	:	Dr. Ahmad Sofyan
Committee	:	Dr. Anastasia Wheni Indrianingsih, M.Sc. Eng.
	:	Tri Hadi Jatmiko, M.T.
Registration Web	:	Wahyu Anggo Rizal, S.T.
Portal	:	Wahyu Setyo Prabowo, M.Eng.

IOP Conf. Series: Earth and Environmental Science **251** (2019) 011002 doi:10.1088/1755-1315/251/1/011002

Scientific and Editorial Board of The 2nd ICONPROBIOS 2018

IOP Conf. Series: Earth and Environmental Science **251** (2019) 011002 doi:10.1088/1755-1315/251/1/011002

24. Dr. Achmad Dinoto, M.Sc. (Research Center for Biology, Indonesian Institute of Sciences-Indonesia)
25. Dr. Ainia Herminiati ST., M.Si. (Centre of Appropriate Technology Development, Indonesian Institute of Sciences-Indonesia)
26. Dr. Takahiro Fujiwara (Faculty of Agriculture, Kyushu University, Japan)
27. Hendra Herdian, S.Pt., M.Sc. (Research Unit for Natural Product Technology, Indonesian Institute of Sciences-Indonesia)

IOP Conf. Series: Earth and Environmental Science **251** (2019) 011003 doi:10.1088/1755-1315/251/1/011003

The 2nd ICONPROBIOS 2018

IOP Conf. Series: Earth and Environmental Science **251** (2019) 011004 doi:10.1088/1755-1315/251/1/011004

Peer review statement

All papers published in this volume of *IOP Conference Series: Earth and Environmental Science* have been peer reviewed through processes administered by the proceedings Editors. Reviews were conducted by expert referees to the professional and scientific standards expected of a proceedings journal published by IOP Publishing.

Evaluation of high purity cellulose production from pretreated various agricultural biomass wastes

R Maryana[1], T H Jatmiko[1], D J Prasetyo[1], W A Rizal[1], A Suwanto[1], A S Praharasti[1], A W Indrianingsih[1], M Anwar[1] and A T Rizaluddin[2]

[1] Research Unit for Natural Product Technology, Indonesian Institute of Sciences, Indonesia
[2] Center for Pulp and Paper, Ministry of Industry, Indonesia

Email: roni_yogya@yahoo.co.id

Abstract. Four kinds of non-wood biomass were studied as raw material for high purity cellulose production. Cajuput (*Melaleuca Leucadendron*) twigs, sugarcane (*Saccharum officinarum*) bagasse, oil palm (*Elaeis guineensis*) empty fruit bunch and rice (*Oryza sativa*) straw were treated by NaOH 1 % at boiling temperature and atmospheric pressure for 2 h. The surface structure of these materials and chemical modification of cellulose, hemicellulose, and lignin during the treatment were analyzed. The result showed that the highest cellulose increasing was from 29.6 to 62.7% for rice straw. Meanwhile, the highest and the lowest cellulose content after treatment were 69.7% for sugarcane bagasse and 48.0% for cajuput, respectively. Scanning electron microscopy (SEM) image showed that the smoother surface for all biomass after pre-treatment.

Keyword: cellulose, cajuput, sugarcane bagasse, oil palm, rice straw

1. Introduction

United Nations Environment Programme (UNEP) reported that in 1950 when the world's population was 2.5 billion, the plastic produced was 1.5 million tonnes. Meanwhile, in 2016 when the population over 7 billion, more than 300 million tonnes of plastic were produced. It is predicted the plastic will reach 33 billion tonnes by 2050 if there are no actions taken [1]. One of possible actions is to use biodegradable plastic or bioplastic. Bioplastics can be produced by using renewable sources such as plant-based raw materials, unlike traditional plastics that are produced from refinery of fossil fuel. In addition, other issues such as biodegradability and other environmental hazards associated with traditional plastics have led to interesting research in the field of bioplastics. Bioplastics currently available on the market are mostly amylose-based such as PLA (Poly Lactic Acid) so that it can affect and compete with food supply. Cellulose plastic is a bioplastic that is produced using cellulose derivatives, which are wood-based both hardwood and softwood. However, because of the deforestation, air pollution, and climate change problems, alternative sources such as biomass from agricultural waste are proposed as promising materials.

Cajuput in Indonesia known as "Kayuputih" is mainly planted in Java, Sulawesi, Moluccas and Nusa Tenggara Timur. Cajuput oil is usually used as herbal remedies such as for baby oil. Indonesia produces

650-ton cajuput oil yearly [2]. If the oil yield is 1-2 %, the biomass potency is about 32.500-65000 ton/year. Regarding sugarcane bagasse (SB), it was reported that in 2014, 2.6 million tons of sugar were produced from 0.48 million ha of planted area [3] and sugar production increased to 2.9 million tons in 2015. Indonesian Ministry of Agriculture estimates that sugar production and harvested area will reach around 3.82 million tons by 2019. During milling, sugar cane is destroyed to extract the juice, producing sugarcane bagasse as a by-product. SB potency was about 9.9–11.2 million tons in 2014 [4].

Indonesia produced 29.3 million tons of crude palm oil (CPO), with production increase to 30.8 million tons in 2015 [Ministry of Agriculture Indonesia, 2015]. In the mill, during milling, fresh fruit bunches (FFB) are separated into fruit (about 70%), empty bunches (about 20 wt.%), and other residues (10%). The potential of oil palm empty bunches (EFB) is around 30-40 million tons per year in Indonesia [5].

Meanwhile, another biomass, rice straw (RS) is also very potential. Rice production in 2015 was 75.36 million tons of milled dry grain (in Bahasa "gabah kering giling") or increased by 4.51 million tons (6.37 percent) compared to 2014 [6]. Previous study reported that there is 18, 9 million ha rice field in Indonesia and if the potency 3 ton of rice straw/ha, 56.7 million ton of rice straw will be produced.

Delignification is commonly known as pulping. The first patent dealing with pulping of wood with aqueous solutions of calcium hydrogen sulfite and sulfur dioxide in pressurized systems was granted in 1886 [7]. Nowadays pulping can be separated to alkaline and acid pulping which is the most common methods. Generally, the lignin reaction during kraft pulping is presented in Figure 1.

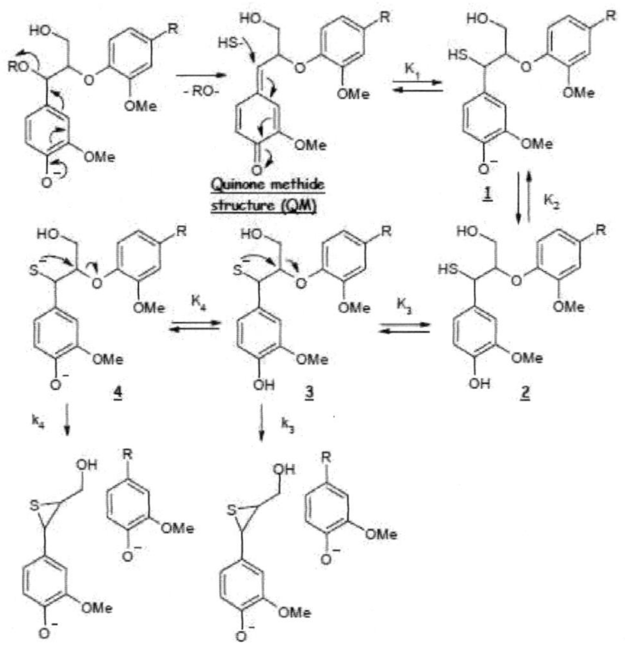

Figure. 1 Cleavage of phenolic b-O-4 linkages during the initial delignification [8].

Cellulose needs to be modified chemically to become bioplastic. Therefore, in this research, isolation, and purification of cellulose from cajuput twigs, sugarcane bagasse, oil palm empty fruit bunch and rice straw as potential cellulose source for bioplastic production, is reported.

2. Materials and Methods

Cajuput twigs (CT) was collected from cajuput oil mill in Gunungkidul region. SB was collected from PG Madukismo, sugar mill in Yogyakarta region. Meanwhile, oil palm EFB was sent from PT Ladang Sawit Mas, Kalimantan Barat and rice straw was collected from rice field in Bantul region, Yogyakarta.

All biomass samples were reduced to 1-2 cm for cooking and further ground to about 40 mesh for chemical analysis. The chemical compositions analysis of raw material and pulp after soda cooking were

IOP Conf. Series: Earth and Environmental Science **251** (2019) 012001 doi:10.1088/1755-1315/251/1/012001

determined. The cellulose, hemicelluloses, lignin and water-soluble compound by using Chesson-Datta method [9]. 150 ml of distilled water was added to about 1 g (oven dry) (a) of each biomass and heated in the oil bath for 1 h at boiling temperature. Furthermore, the mixture was filtered and washed with distillation water. Solid residue then subjected to oven and dried to constant weight (b). Solid residue was further reacted with 1 N sulfuric acid 150 ml and heated in the oil bath 100°C for 1 h. The mixture was again filtered and washed with distillation water until neutral and residue dried in the oven (c). The solid residue was again reacted with 72% of sulfuric acid 10 ml at room temperature for 4 h and then 1N sulfuric acid 150 ml was added and refluxed for 1 h in the oil bath. The mixture was filtered and washed again with distillation water until neutral and residue dried in the oven (d). After that, solid residue was heated in the furnace 400°C and weighted (e). The calculation of % hemicellulose = (c – b)/a x 100%; % cellulose = (d – c)/a x 100%; % lignin = (e – d)/a x 100%.

Soda cooking for all ± 15 g (oven dried) biomass were performed at boiling temperature for 2 h with 1% NaOH dosage and solid to liquid ratio was 1: 10.

3. Result and Discussion

In biomass biorefinery, how to separate cellulose, hemicellulose and lignin effectively are very important. In this study, sodium hydroxide 1% was chosen for delignification at boiling temperature. NaOH is widely used in the pulp and paper industry to separate lignin from cellulose because of its effectiveness and its economic process. Solid to liquid ration was 1:10 in order to make sure that all biomass soaked in the alkaline. The previous study was also reported advantages of NaOH compare to other chemicals [10]. The temperature condition was boiling point at atmospheric pressure. This stage is called delignification-1 which is the first step for a further increment of the reaction temperature and pressure. The chemical composition of various biomass before and after delignification is presented in Table 1.

Table 1. Chemical composition of materials after delignification.

Materials		Water soluble compound	Cellulose	Hemicellulose	Lignin
Rice straw	raw material	24.00	29.55	25.35	9.99
	delignification-1	6.68	62.72	15.68	14.51
Oil palm empty fruit bunch	raw material	21.10	42.56	20.27	15.36
	delignification-1	6.68	62.72	15.68	14.51
Sugarcane bagasse	raw material	16.18	42.31	20.74	18.61
	delignification-1	4.12	69.69	14.73	10.77
Cajuput twigs	raw material	22.16	31.94	17.99	27.51
	delignification-1	6.15	47.98	15.34	30.31

Table 1 shows that the highest cellulose content obtained for sugarcane bagasse that is 69.69% while the lowest is cajuput twigs which are 47.98%. While the highest increasing in cellulose content during delignification was obtained by rice straw which was 111.8%, and the lowest was eucalyptus twigs 50.2%. Lignin content of rice straw and cajuput twigs increased after process can be explained by removal process of lignin, cellulose, and other compounds. For instance, the total percentage of chemical composition of rice straw was 88.9% (about 11.1% was other materials) and after delignification increased to 99.6%. Hence, the dividing factor smaller than the raw material. For comparison, total chemical composition of raw material for SB was 97.8% and 99.3% after delignification. Other chemical compounds in the wood and non-wood biomass usually ethanol or other organic solvents soluble compounds; acid-insoluble lignin; organic acids and ash content.

Overall this cellulose yield is still very low compared to the target of above 90%. Therefore, optimization will be carried out by increasing the process temperature, applying pressure and increasing the NaOH level using will be performed. In addition, bleaching stage will be carried out in the next stage.

The Fourier transform infrared spectra of cajuput twigs before and after treatment with NaOH 1% were shown in Figure 2. The broad peak around 3300-3400 cm^{-1} in all spectra is representative of the stretching of hydroxy (-O-H) groups. The peak around 2800-2900 cm^{-1} and 1970-2200 cm^{-1}, which was observed in all spectra corresponds to the C-H stretching and overtone of aromatic rings of lignin, respectively.

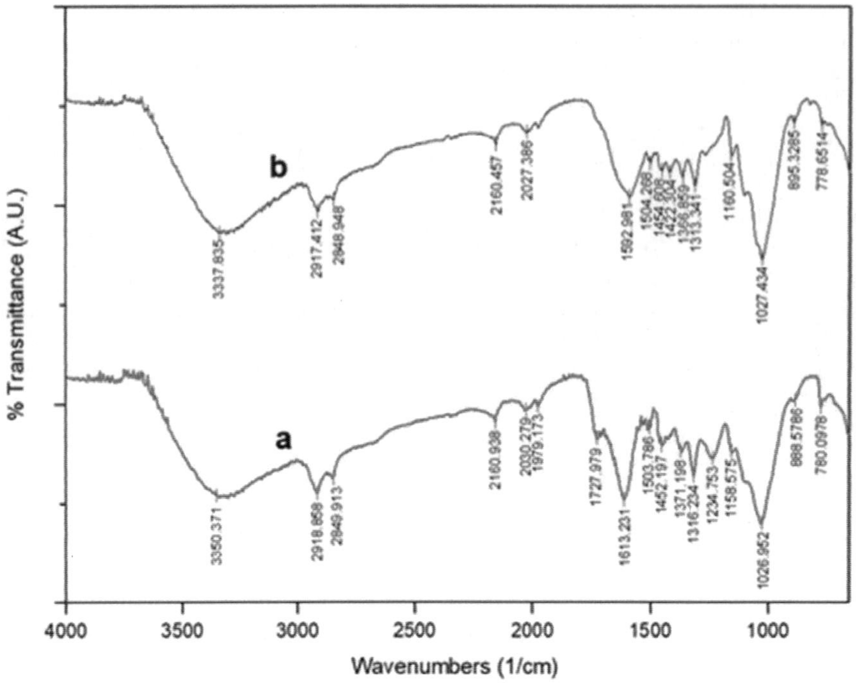

Figure 2. FTIR spectra of raw cajuput twigs (a) and after delignification-1 (b).

After delignification process, one of the samples, namely the Cajuput twigs, analyzed for its spectrum using FTIR to see the effect of the treatment with NaOH. The Fourier transform infrared spectra of eucalyptus trunk before and after treatment with NaOH 1% were shown in Figure 1. The broad peak around 3300-3400 cm^{-1} in all spectra is representative of the stretching of hydroxy (-O-H) groups. The peak around 2800-2900 cm^{-1} and 1970-2200 cm^{-1}, which was observed in all spectra corresponds to the C-H stretching and overtone of aromatic rings of lignin, respectively. A shoulder at 1727 cm^{-1} of the raw eucalyptus trunk exhibit the spectrum of (C=O st) acetyl and ester groups in hemicellulose or carboxylic acid groups in pectin and p-coumaric components of lignin [1]. Alkaline treatment with NaOH 1% could eliminate non-cellulosic materials. Both sets of peaks are assigned the absorption of water (-O-H bend)around 1590-1620 cm^{-1}. The shifting from 1613 to 1592 cm^{-1}represents the weaker bond of -OH bend of water in the product of NaOH treatment. The existing of a peak around 1500 cm^{-1} (C=C aromatic st) in all peaks indicate that alkaline treatment cannot completely eliminate all of the lignin compounds from the eucalyptus trunk. The appearance of a peak around 1422 cm^{-1} is representative of the absorption of -CH_2- symmetric bending of alkaline treatment product. The absorption around 1300-1400 cm^{-1} of both spectra refers to the bending frequency of in-the-plane CH_2 bending. Both peaks around 1160 and 1027 cm^{-1} refer to C-O-C asymmetric and C-C, C-OH, C-H ring, and side vibrations; while and peak at 895 cm^{-1} is representative of COC, CCO, and CCH deformation and stretching [2]. Furthermore, based on the calculation of spectral ratio (3300/1500 cm^{-1}), the ratio of cellulose and hemicellulose to lignin were significantly increased due to alkaline treatment.

Morphology of surface was analyzed by using scanning electron microscopy (SEM). The SEM is SU-3500 Hitachi, Tokyo, Japan with a magnification of 2000x and the SE 5 kV. Figure 3 shows the SEM image of the biomass surface before and after treatment with NaOH. The SEM profile shows that the surface of the material at magnification is 2000 times cleaner than the material. It is suggested that the hemicellulose, some lignin, and other impurities dissolved into black liquor.

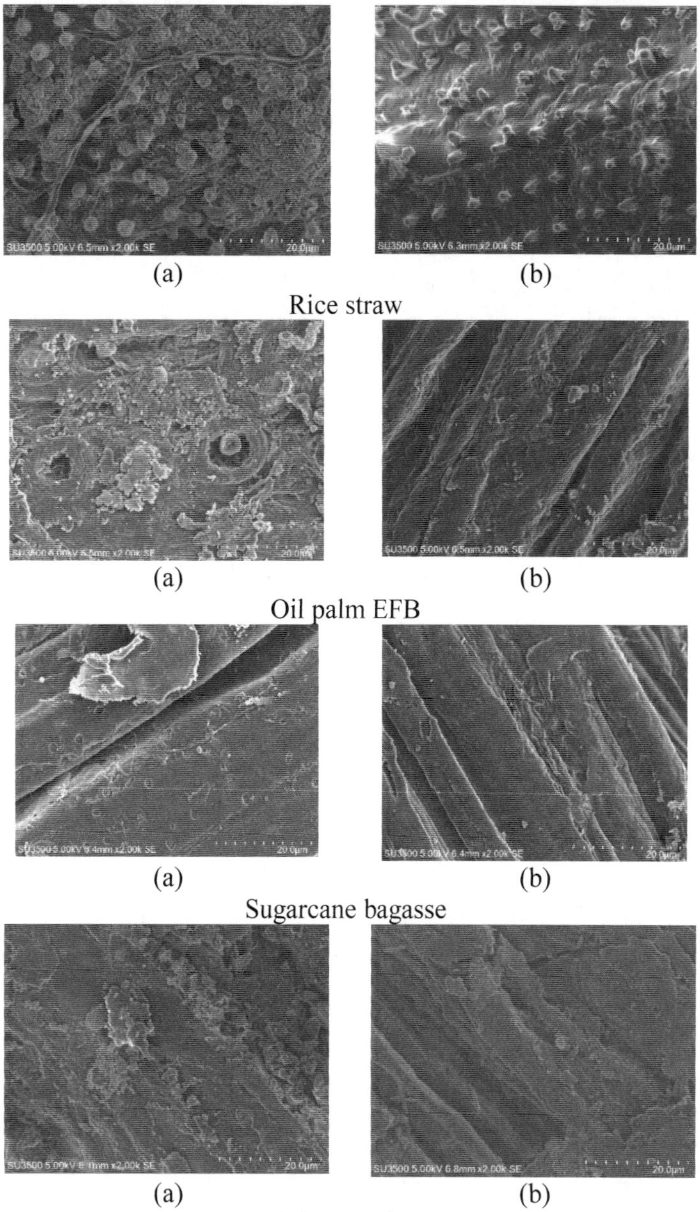

Cajuput twigs

Figure. 3 Scanning Electron Microscopy profile of the materials (a) raw material, (b) after delignification-1

4. Conclusion

The condition of 1% NaOH at boiling temperature and atmospheric pressure for 2 h increased cellulose content 111.8% for the rice straw, the highest among other biomass. In addition, in this condition, the highest cellulose content was 69.7% for sugarcane bagasse.

5. References

[1]. UNEP 2016 *Marine plastic debris* and *microplastics – Global lessons* and *research* to *inspire action* and *guide policy change* Nairobi

[2]. Rimbawanto A, Kartikawati N K and Prastyono 2017 *Minyak Kayu Putih, dari tanaman asli Indonesia untuk Masyarakat Indonesia* (Yogyakarta ISBN 978-979-3666-20-4)
[3]. Statistics Indonesia 2015 *Production of plantation crops by province and crops, Indonesia (thousand tons)* 2012-2014
[4]. Ministry of Agriculture of Indonesia 2014 *Tree crop estate statistics of Indonesia* 2013-2015
[5]. Ministry of Agriculture of Indonesia 2015 *Annual report of Agriculture Development* 2015
[6]. Statistics Indonesia https://www.bps.go.id/pressrelease/2016/03/01/1271/produksi-padi-tahun-2015-naik-6-37-persen.html
[7]. Sjöström, E 1981 *Wood Chemistry-Fundamentals and Applications* (New York : Academic Press, Inc.)
[8]. Tapani Vuorinen, Chemistry of Pulping and Bleaching https://mycourses.aalto.fi/pluginfile.php/140064/mod_resource/content/1/Supporting%20material%201.pdf
[9]. Mahyati, Patong A R, Djide M N and Taba D P 2013 *Int. J. Sci. Technol.* **2** 11 79-82
[10]. Maryana R, Nakagawa-Izumi A, Kajiyama M and Ohi H 2017 Environment-friendly non sulfur cooking and totally chlorine-free bleaching for preparation of sugarcane bagasse cellulose. *J. Fiber Sci. Technol.* **73** 8 182-191.

Optimization of Cellulose Isolation from *Melaleuca leucadendron* Twigs by Box-Behnken Design

D J Prasetyo[1], R Maryana[1], T H Jatmiko[1], Hernawan[1] and S K Wahono[1,2,3]

[1]Research Unit for Natural Product Technology, Indonesian Institute of Sciences, Yogyakarta, Indonesia 55861
[2]Future Industries Institute, University of South Australia, Mawson Lakes 5095, South Australia, Australia
[3]School of Engineering, University of South Australia, Mawson Lakes 5095, South Australia, Australia

E-mail: prasetyo_dwijoko@yahoo.co.id; dwij001@lipi.go.id

Abstract. The production of *Melaleuca leucadendron* essential oil generates a lot of solid waste, since only under 2% take it out as the oil product. The solid waste contains cellulose as an important content which can be isolated by alkaline treatment of sodium hydroxide (NaOH). In this project, optimization of alkaline treatment of *M. leucadendron* twigs (MLT) waste was studied in order to obtain optimum isolated cellulose. The experiments were conducted at atmospheric pressure and boiling temperature. The 3^3 Box-Behnken design (BBD) was used to evaluate the effects of treatment time, solid loading and NaOH concentration. Quadratic model evaluation showed that the model is a prospective to describe the effects of the variables (R^2=0.8994, $p<0.05$). The significant variables were found in linear term of NaOH concentration and quadratic term of treatment time. The optimum result of the experimental variables range was predicted in treatment time, solid loading and NaOH concentrations of 60 minutes, 10%, and 1.83% respectively. In conclusion, BBD is appropriate to optimize cellulose isolation process of MLT.

Keywords: cellulose, sodium hydroxide, *Melaleuca leucadendron*, Box-Behnken Design

1. Introduction

Melaleuca leucadendron (*kayu putih*) is a plant which well-known for its essential oil products [1-2]. It is commonly grown in natural forests or plantations in Indonesia, such as in Gunungkidul, Yogyakarta [2-3]. In this area, *M. leucadendron* leaves and twigs are processed as raw materials for local essential oil plant. Since essential oil content is under 2% [1-2], the extraction process generates a lot of solid waste. The utilization of these waste is not optimum yet, since utilized as wood fuel only such as for tofu production [4]. Since the genus of *Melaleuca* is lignocellulosic biomass material, it has huge prospect as a source of useful materials i.e. bioethanol production from *M. leucadendron* shedding bark [5] and cellulose production from *Melaleuca cajuputi* wood [6]. Nanocellulose is one of cellulose form which has prospective market to develop and evolve [7]. The rise of cellulose demand will trigger the exploration of alternative raw material which *M. leucadendron* waste is one of the prospective one.

Various treatments in order to extract cellulose from biomass have been proposed; alkaline pretreatment is the favorite method among them because of its high efficiency of lignin removal and its milder condition than acid pretreatment [8]. Sodium hydroxide (NaOH) is the most common reagent for alkaline pretreatment of lignocellulosic materials. NaOH treatment has some advantages, i.e. high reaction rate, 50% hemicellulose dissolution and 60-80% delignification [8]. Since some previous study have been successfully conducted for removing lignin from sugarcane bagasse [9-11], palm midrib [12] and sansevieria fiber [13], this study will adopt the method in *M. leucadendron* twigs delignification process.

In order to optimize the process, experimental designs based on response surface method (RSM), such as Doehlert matrix (DM), central composite design (CCD) and Box-Behnken design (BBD) could be applied. Among these methods, BBD and DM are more efficient than CCD for experiment with three factors [14]. In this case, the number of experiments needed for BBD is less than the other methods [14]. Previous studies based on BBD have been successfully conducted for optimizing pretreatment of lignocellulosic such as alkaline pretreatment of rice straw [15] and cellulose extraction of jute fiber [16]. Therefore, the aim of this study is to optimize alkaline pretreatment process of *M. leucadendron* twigs to obtain optimum cellulose content by BBD.

2. Materials and methods

2.1. Materials

Melaleuca leucadendron twigs (MLT) were obtained from local essential oil factory in Gunungkidul, Yogyakarta, Indonesia. The twigs were milled and screened to obtain particles with size 16-20 mesh, then were packed in plastic bag to minimize contact with water vapor in the air.

2.2. Box-Behnken Design

The 3^3 Box-Behnken design with 3 replications at center point was selected as a method to optimize delignification process for cellulose isolation of MLT. The selected variables were time, solid loading, and NaOH concentration. The minimum and maximum levels of three variables are presented in Table 1.

Table 1. Minimum and maximum levels of the variables

Variable	Symbol		Level		
	Code	Real	-1	0	+1
Time (min)	x_1	X_1	30	45	60
Solid loading (%)	x_2	X_2	10	15	20
NaOH concentration (%)	x_3	X_3	1	1.5	2

Second-order polynomial model was used to describe the effects of the three variables. The model was given in Equation 1 below.

$$Y = \beta_0 + \beta_1 x_1 + \beta_2 x_2 + \beta_3 x_3 + \beta_{11} x_1^2 + \beta_{22} x_2^2 + \beta_{33} x_3^2 + \beta_{12} x_1 x_2 + \beta_{13} x_1 x_3 + \beta_{23} x_2 x_3 \quad (1)$$

Y is the predicted response (%cellulose obtained); β_0 is constant; β_1, β_2 and β_3 are linear coefficients; β_{11}, β_{22} and β_{33} are quadratic coefficients; β_{12}, β_{13} and β_{23} are interactions coefficients; x_1, x_2 and x_3 are representing factors. The model evaluation with ANOVA and regression analysis was carried out using Microsoft Excel.

2.3. Experimental

MLT particles at selected solid loading (10, 15, 20%) were dispersed in 100 mL of NaOH solution (1, 1.5, 2%) in a pear-shaped flask. The mixture was heated until reach its boiling point in an oil bath. The reaction was stopped after the selected reaction time (30, 45, 60 min) was reached. The mixture was then filtered to obtain a solid phase. The solid phase was washed with tap water until reach neutral pH and finally washed with aquadest. The solid was dried at temperature of 105 °C until reached its constant weight. Chesson-Datta method [17] was used to analyze % cellulose ($g_{cellulose}/g_{solid}$) before and after treatments.

IOP Conf. Series: Earth and Environmental Science **251** (2019) 012002 doi:10.1088/1755-1315/251/1/012002

3. Results and discussions

3.1. Box-Behnken Design and Model Evaluation

Raw MLT contained 39.20% cellulose and after treatment, its content increased into 42.35-48.07%. The experimental and predicted results (%cellulose) based on Box-Behnken design (BBD) are shown in Table 2.

Table 2. Experimental and predicted results

Run	Code Variables			Real Variables			Response	
	x_1	x_2	x_3	X_1	X_2	X_3	Experimental	Predicted
1	-1	-1	0	30	10	1.5	47.58	47.05
2	1	-1	0	60	10	1.5	48.07	47.69
3	-1	1	0	30	20	1.5	46.15	46.53
4	1	1	0	60	20	1.5	46.38	46.92
5	-1	0	-1	30	15	1	43.65	44.02
6	1	0	-1	60	15	1	45.34	45.57
7	-1	0	1	30	15	2	47.91	47.69
8	1	0	1	60	15	2	47.54	47.16
9	0	-1	-1	45	10	1	42.35	42.51
10	0	1	-1	45	20	1	43.74	42.99
11	0	-1	1	45	10	2	45.51	46.27
12	0	1	1	45	20	2	44.66	44.50
13	0	0	0	45	15	1.5	46.08	45.19
14	0	0	0	45	15	1.5	44.24	45.19
15	0	0	0	45	15	1.5	45.25	45.19

Table 3. ANOVA results for acquired model

Source	Degrees of Freedom	Sum of Squares	Mean Square	F-value	p-value	Characteristics
Model	9	36.90	4.10	4.96	0.046	Significant
Residual	5	4.13	0.83			
Total	14	41.03				

Significant at $p<0.05$ level; R^2=0.8994

Table 4. Regression analysis for acquired model

Source	Coefficients	Standard Error	t Stat	p-value	Characteristics
Constant	45.19	0.52	86.13	0.000	Significant
x_1	0.26	0.32	0.80	0.463	Not significant
x_2	-0.32	0.32	-1.00	0.362	Not significant
x_3	1.32	0.32	4.10	0.009	Significant
x_1^2	1.95	0.47	4.12	0.009	Significant
x_2^2	-0.09	0.47	-0.20	0.850	Not significant
x_3^2	-1.03	0.47	-2.18	0.081	Not significant
x_1x_2	-0.06	0.45	-0.14	0.893	Not significant
x_1x_3	-0.52	0.45	-1.14	0.306	Not significant
x_2x_3	-0.56	0.45	-1.24	0.272	Not significant

The ANOVA and regression analysis of the quadratic models is presented in Table 3 and 4. The coefficient of determination (R^2) value at 0.8994 and p-value of model<0.05, indicated that the model was appropriate to describe the phenomenon. Meanwhile, the result of model evaluation gave relation between %cellulose and three variables as shown in Equation 2. Positive coefficients indicate a linear increase in %cellulose value, while negative ones indicate a linear decrease. Overall, linear term of NaOH concentration and quadratic term of treatment time are found as significant variables as shown in Table 3.

$$Y=45.19+0.26x_1-0.32x_2+1.32x_3+1.95x_1^2-0.09x_2^2-1.03x_3^2-0.06x_1x_2-0.52x_1x_3-0.56x_2x_3 \quad (2)$$

3.2. Effects of Process Variables at Cellulose Content

The effects of each variable at selected cellulose contain were studied in this study. The 3-D response surface graphics are shown in Figure 1, 2 and 3. Each graphic shows the effects of pairing variables at observed range while another variable is fixed in zero levels. The response value was calculated based on Equation 2.

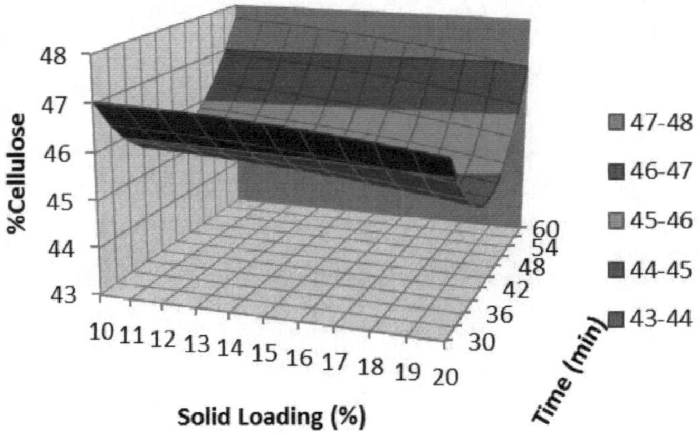

Figure 1. Effect of interaction between solid loading and time

Figure 1 shows the effects of solid loading, time and their interactions with the cellulose content. The cellulose content slightly increased as solid loading decreased. The trend was in line with Equation 2 which the coefficient of solid loading is in a negative value, means has negative effect on cellulose content. Trend of treatment time was different with solid loading. It has a negative peak which means quadratic term has important role. This phenomenon was in line with the regression analysis results in Table 4 which quadratic term of time (x_1^2) has a significant effect. Meanwhile, the interaction between solid loading and time was not significant and was in line with the previous result conducted by Velmurugan and Muthukumar [10].

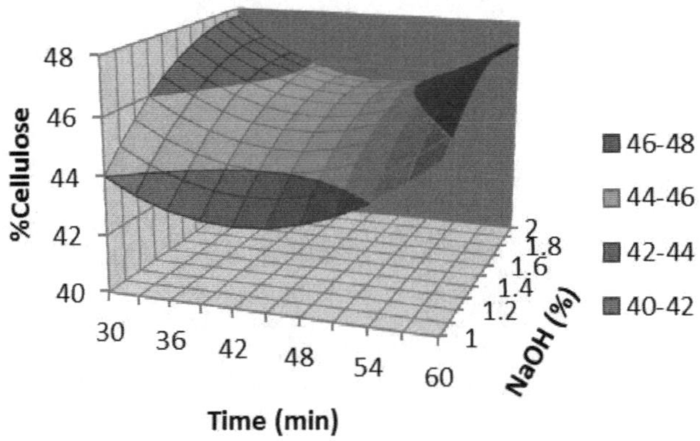

Figure 2. Effect of interaction between time and NaOH concentration

The effects of interaction between time and NaOH concentration is presented in Figure 2. As mentioned before, the time has significant at quadratic term. Meanwhile, NaOH concentration has a positive coefficient which means the increasing of cellulose content increases in line with NaOH concentration increase. This result was supported by the previous studies [12-13]. The insignificant interaction between time and NaOH concentration based on Table 4 is similar with previous results by Velmuragan and Muthukumar [10].

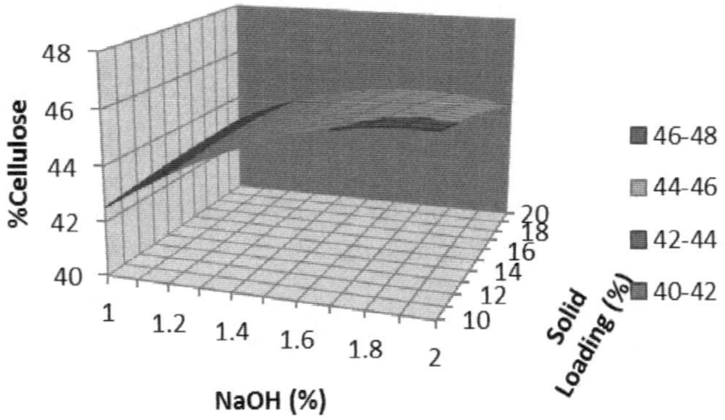

Figure 3. Effect of interaction between NaOH concentration and solid loading

The Interactive effect was found in the interaction between NaOH concentration and solid loading. Figure 3 shows a positive peak in the range of 1.6-2% NaOH concentration and 10-12% solid loading. The low value of solid loading and high value of NaOH concentration means the ratio between NaOH and MLT is high, thus the probability NaOH to react with lignin in MLT become higher. Therefore, decreasing of solid loading and increasing of NaOH concentration was the key factors to optimize cellulose content value.

The prediction of optimum condition for the maximum value of cellulose content was conducted by Solver program in Microsoft Excel. Range value of the optimum condition was set based on experiment value i.e. 30-60 minutes of time, 10-20% of solid loading and 1-2% of NaOH concentration. Maximum predicted the value of cellulose content was 48.14% at a condition of 60 minutes, 10% solid loading and 1.83% NaOH concentration.

4. Conclusion

Utilization of Box-Behnken design (BBD) for optimization of cellulose isolation from *Melaleuca leucadendron* twigs (MLT) has been studied. The Quadratic model has significant conformity between experimental and predicted value with a coefficient of determination value (R^2) at 0.8994 and $p<0.05$. Linear term of NaOH concentration and quadratic term of time was found as significant variables. The predicted optimum condition was found at the condition of 60 minutes, 10% solid loading and 1.83% NaOH concentration with a result of 48.14% cellulose content. Overall, BBD could be adequate to optimize the process of cellulose isolation from MLT.

5. Acknowledgments

This work was funded by Lembaga Ilmu Pengetahuan Indonesia. The authors are grateful to Deputy of Engineering Sciences – Indonesian Institute of Sciences for support to this program.

6. References

[1] Farag R S, Shalaby A S, El-Baroty G A, Ibrahim N A, Ali M A and Hassan E M 2004 *Phytother Res.* **18** 30-5
[2] Pujiarti R, Ohtani Y and Ichiura H 2011 *J. Wood Sci.* **57** 446-51
[3] Widodo, Sutarno, Widoretno S and Sugiyarto 2010 *Biodiversitas* **11** 89-92
[4] Suharto, S K Wahono and H Julendra 2007 *Prosiding Seminar 2007 Pemberdayaan Masyarakat Melalui Pemanfaatan Bahan Baku Lokal (Yogyakarta) vol 1* (Jakarta: LIPI Press) pp 107-12
[5] Ahmed I N, Nguyen P L T, Huynh L H, Ismadji S and Ju Y H 2013 *Bioresource Technol.* **136** 213-21

[6] Junaidi A B and Yunus R 2009 *Jurnal Hutan Tropis Borneo* **10** 284-91
[7] Cowie J, Bilek E M, Wegner T H and Shatkin J A 2014 *Tappi J.* **13** 57-69
[8] Kim J S, Lee Y Y and Kim T H 2016 *Bioresource Technol.* **199** 42-6
[9] Maryana R, Ma'rifatun D, Wheni A, K W Satriyo and Rizal W A 2014 *Energy Procedia* **47** 250-4
[10] Velmurugan R and Muthukumar K 2012 *Bioresource Technol.* **112** 293-9
[11] Hernawan *et al.* 2017 *AIP Conf. Proc.* **1823** 020026
[12] Ferdiansyah M K, Pranoto Y and Marseno D W 2018 *Food Res.* **2** 258-62
[13] Mardiyati, Steven, Rizkiansyah R R, Senoaji A and Suratman A 2016 *AIP Conf. Proc.* **1725** 020043
[14] Ferreira S L C *et al.* 2007 *Anal. Chim. Acta* **597** 179-86
[15] Kim I and Han J I 2012 *Biomass Bioenerg.* **46** 210-7
[16] Manzato L, Takeno M L, Pessoa-Junior W A G, Mariuba L A M and Simonsen J 2018 *Fiber Polym.* **19** 289-96
[17] Datta R 1981 *Biotechnol Bioeng.* **23** 2167-70

IOP Conf. Series: Earth and Environmental Science **251** (2019) 012003 doi:10.1088/1755-1315/251/1/012003

Optimization Ratio of Mixed Metal Soaps and Palm Fatty Acid Distillate (PFAD) on Making Grease

Sukmawati[1], P P Lestari[1], and Safriadi[2]

[1)]Department of Chemical Engineering, Institut Teknologi Medan, Jl. Gedung Arca No. 52, North Sumatera, 20217, Indonesia.
[2)]Department of Chemical Engineering, Universitas Sumatera Utara, Jl. DR.Mansyur No.58,North Sumatera, 20155, Indonesia.

E-mail:sukmawatiuma@yahoo.co.id

Abstract. Palm Fatty Acid Distillate (PFAD) has a high potential to be used as raw material for oleochemical products, one of them is for grease. This is because the fatty acid composition contained in PFAD is not much different from the fatty acid composition contained in palm oil. The purpose of this research is to study the formularization grease from PFAD which has characteristic approaching grease SNI 06-7069-8-2005. The method used in the manufacture of grease consists of two steps. The first step is the process for preparing a mixture of metal soap (Na-Ca) with predetermined variations and analyzing metal alloy soaps. The second step is the manufacture of grease from a mixture of metal soaps (Na-Ca) and PFAD and analyzing the product. Based on the research the optimum solids obtained in accordance with the SNI standard is in composition A2: B1, A2 is the ratio of NaOH: Ca (OH)2 = 80%: 20% and B1 is PFAD ratio: metal soap = 10%: 90% with density analysis 1.108 g/ml, melting point 1100C, titrant value 92 and NLGI 6 consistency.

1. Introduction

Utilization of palm oil in the manufacture of solid lubricants is a substitute for base oil and thickening components. Palm oil is one of alternatives, however, direct use of palm oil as a lubricant remain cause problems namely high density of palm oil leads to imperfect lubrication [1]. The need for lubricating oil made from synthetic materials derived from non-renewable energy sources is increasing, so we must immediately search for alternative ingredients as Lubricants[2]. The type of palm oil used is PFAD. PFAD is a by-product of the process of purifying crude palm oil. Until now the use of PFAD is limited, which is used as a raw material for making low-quality soap [3]. Palm Fatty Acid Distillate (PFAD) oil from vegetable oils family was compared with mineral oil with regard to tribological properties, such as anti-wear, anti-friction, viscosity index, and flash parameter point, using a four-ball tribotester under different temperatures [4]. One type of vegetable oil that has been assessed as a base for lubricants and lubricating oils is Castor oil. Castor oil has a low pour point, resistance to load (wear) and a better viscosity index compared to super-refined mineral oil (SRMO) which is a lubricant base material. This allows castor oil to be used as a variation of the lubricant base material [5]. The characteristics of PFAD are influenced by three basic parameters namely boiling point, specific heat, and latent heat from fatty acids [6].

The solid lubricant is a solid or semi-solid mixture of lubricants with thickener which serves to reduce friction between two surfaces that intersect or rub against each other [7]. Good solid lubricating properties are reducing friction, preventing corrosion, as insulation from dirt or water, preventing leakage, consistency and unchanging structure, not hardening at low temperatures, properties suitable for elastomeric insulation, and having a certain level of pollutant tolerance. Based on previous research, the research team has made grease from Palm Fatty Acid Distillate (PFAD) using Na and Li metal soap. The optimum ratio obtained at the composition of 95:5 for sodium metal and 35:65 for lithium metal 8], but the results of penetration test and melting point are still not optimal if applied to vehicles or machines.

Additives are an additional ingredient that functions as a vitamin for solid lubricants whose uses include, as an anti-corrosion, as anti-wear, as an anti-oxidant, maintaining a thick lubricant viscosity (viscosity index improver). The solid lubricant is composed of several components [9], namely, basic oil (base oil), thickener, additives. According to the latest theory, base thickener and oil always exist together on the surface of the friction and do not justify popular theories that view thickener only as a sponge that releases base oil to the friction surface [10].

The grease is a material that serves to protect several functioning engine components, so that grease can have a positive effect on tools and machines, which can prevent wear due to friction between components with one other component. In addition, grease can also minimize maintenance and repair costs of tools and machinery. Increased development in the industrial and transportation sectors currently increases the use of grease significantly. This means a large amount of grease is needed to meet consumption needs in the industrial and transportation sectors. Generally, many greases found on the market are made from petroleum or petroleum base oil. With the depletion of petroleum reserves, fewer basic raw materials for grease can meet Indonesia's needs. This is predicted to lead to a shortage of grease in the market or more expensive grease due to imports from abroad.

Vegetable oil is one of the raw materials that can be used as basic material of making alternative lubricant for machine. The derivatives of palm oil, especially glycerol and oleic acid have the potential to be developed as a synthetic grease feedstock. In addition, grease can also be synthesized from fatty acid esters derived from vegetable oils. Generally, grease is synthesized from fatty acid esters with carbon chains in the above range, thus having the opportunity to be developed as raw materials for lubricating oils.

Oil development studies have not been conducted in Indonesia and information is still limited. Therefore, this study attempts to develop further the technology of manufacturing grease to assess the extent to which the potential of palm oil and its derivative products can be developed as grease. Palm oil derivative products namely Palm Fatty Acid Distillate (PFAD) is a by-product of cooking oil factory, where its utilization is still limited. In addition, PFAD is only used as a raw material of soap maker with low quality. Meanwhile, PFAD has a high potential for use as a raw material for oleochemical products, one of which is grease. This is because the fatty acid composition contained in PFAD is not much different from the fatty acid composition contained in palm oil. The utilization of byproducts from this cooking oil factory in the manufacture of greases done by processing technique that is mixing between mixed metal soap (sodium and calcium) with palm oil and its derivative is PFAD which is used as base oil.

2. Material and Methods

2.1 Materials

Sodium hydroxide, stearic acid, phenol, glycerol purchased from Bratachem and calcium hydroxide purchased from Rudang Jaya. All of the chemicals were technical grade and were used without any purification. The products were evaluated by the determination of free base content and unsaponifiable matter on metal soap. The result obtained is in the form of the density test, titrant, melting point, and National Lubricating Grease Institute (NLGI) number.

2.2. *Variables and Process Conditions of Preparation of mixed metal soaps*
The functions of mixed metal soap (Na-Ca) is as a thickener in the grease which is produced from the reaction between PFAD and the bases of NaOH and $Ca(OH)_2$. Made by heated up PFAD until melting at 70 °C then added NaOH and $Ca(OH)_2$ 4N solution and stirred at 650 rpm for 30 minutes. The ratio between NaOH and Ca $(OH)_2$ in mixed metals is 90:10 (A1) and 80:20 (A2). The mixed metal soap was evaluated by free base content and non-soaped ingredients.

2.3. *Preparation of Mixed sodium-calcium soaps*
PFAD (100 g) is heated to a temperature of 70 °C, stirred continuously at a speed of 650 rpm for 30 minutes. To the PDAF solution was gradually stirred and added a mixture of 4N sodium and calcium solution according to variations in the manufacture of metal soaps, i.e. the ratio between NaOH and $Ca(OH)_2$ in mixed metals is 90:10 (A1) and 80:20 (A2).

2.4. *Analysis of Free Base Content and unsaponifiable matter*
Analysis of the free base content was carried out by adding 100 ml of 95% alcohol into 5 grams of sample and heated for 30 minutes then titrated with 0.5N HCl. Whereas the analysis of the non-soaped ingredients is carried out by putting in 50 ml of alcoholic KOH 0.5N into a free base analysis solution and heated for 1 hour, then titrated with 0.5N HCl with the Phenolphthalein indicator.

2.5. *Grease Production*
Grease is made by heating PFAD up to 70 °C in the reactor then added mixed metal soap as thickener according to the predetermined variations. The ratio between PFAD and mixed metal soaps is 10:90 (B1) and 20:20 (B2). After that, 5 ml of stearic acid, glycerin and phenol were added to each variations solution and then stirred at a speed of 650 rpm for 4 hours at a temperature of 120 °C until homogeneous.

2.6. *Analyzing of Grease*
The resulting Grease products were analyzed of the density, melting point, penetration, and NLGI analysis parameters. The melting point and NLGI penetration analysis based on ASTM-D-566 and ASTM-D-217, respectively.

2.6.1. *Procedure of ASTM D-217*
The order is to determine the stiffness of a grease. The P0 or unworked and P60 or worked penetration tests followed the ASTM D-217 method for full scale. This is one of the four required tests performed on every single batch of grease at Nye prior to shipping. The penetration test begins with the grease at 25 degrees Celsius, plus or minus 1 degree Celsius, being leveled into a cup. The sample size and cone weight for this test is determined by your application needs. Using a penetrometer, it is dropped into the cup for 5 seconds, creating a hole in the grease. The technician records the depth of a millimeter of this hole. This value is known as the P0 or unworked penetration. The grease is then sheared (or worked) using a mechanically operated device through 60 double strokes, simulating the use of the grease. When this is completed, the technician repeats the P60 value, also known as a worked penetration. Nye technicians can also carry out P10,000 and P100,000 tests, depending on the application's specifications.

The penetration values of P60 are rated using the NLGI grade chart from 000 to 6. These values determine the stiffness of the grease and how the grease will react over time to service and wear. The higher the penetration value, the softer the grease. The lower the penetration value, the stiffer the grease. Grease that is not too easy to pass from the area that needs to be lubricated. Grease that is too stiff may not effectively migrate to those areas that need to be lubricated.

2.6.2. *ASTM-D-566*
Fill a test cup with a sample by pressing the larger opening into the grease to be tested until the cup is filled. Place the corcks on the thermometer depth gage position in the test tube. Replace the depth gage with the grease cup. Suspend the test tube in the oil bath to a depth corresponding to the 76 mm immersion mark on the thermometer. Suspend the second thermometer in the oil bath so that its bulb is at approximately the same level as the bulb of the test tube thermometer. Stir the oil bath and heat. Two determination can be made simultaneously in the same bath

3. Results and Discussion

From the research, it was found that the free base rate for A1 metal soap type was 0.04% and the type of metal soap A2 was 0.03% or slightly above the neutral level (point 0). In addition to the free base content of metal, soaps have also analyzed the characteristics of unsaponifiable matter value, which in the type of metal A1 is 0.88% and metal soap A2 0.84%.

The focus of lubricant production generated in this research is to obtain a grease formula that approximates the characteristics of commercial grease based on ASTM quality standards (American Society for Testing and Materials International). In this research as for the raw materials used in the manufacture of solid grease is PFAD. The metal soaps used in grease formulations are obtained by process of Saponification using Sodium (NaOH) and Calcium ($Ca(OH)_2$) with PFAD raw materials.

Table 1. Test Result Characteristics of Grease

No	Composition (%) (Metal Soap: PFAD)	Density (gr/ml)	Titrant 25 °C ASTM D 217	Dropping point ASTM D566	NLGI Grade ASTM D217
1	A1 : B1	0.827	91	106	6
2	A1 : B2	0.931	88	100	6
3	A1 : B3	1.103	86	96	6
4	A2 : B1	1.108	92	110	6
5	A2 : B2	1.022	87	92	6
6	A2 : B3	1.107	85	85	6

Description :A = Ratio NaOH : $Ca(OH)_2$ (%); B = Ratio Metal Soap : PFAD (%); A1 : B1 = (90 : 10) ; (90 : 10); A1 : B2 = (90 : 10) ; (80 : 20); A1 : B3 = (90 : 10) ; (70 : 30); A2 : B1 = (80 : 20) ; (90 : 10); A2 : B2 = (80 : 20) ; (80 : 20); A2 : B3 = (80 : 20) ; (70 : 30)

3.1 Density Grease

The raw material consisting of base oil and mixed metal soap (Na-Ca) after being physically blended and has become grease and then analyzed its characteristics respectively using ASTM D-566 method for melting point test and ASTM D-217 method for determination titrant and NLGI conducted at LEMIGAS, Oil and Gas Technology Research and Development Centre (PPPTMGB) Laboratory, South Jakarta.

Figure 1 showed that the NaOH ratio with Ca $(OH)_2$ also affects the grease density. This density is affected by the density of each metal, where NaOH has a density of 1.0 g/ml and $Ca(OH)_2$ has a density of 2.2 g/ml [3]. This causes the density of grease with the type of metal soap A2 has a higher density than metal soap A1. Because the metal soap A2 has a composition of 20% $Ca(OH)_2$ compared to 10% metal A1 soap.

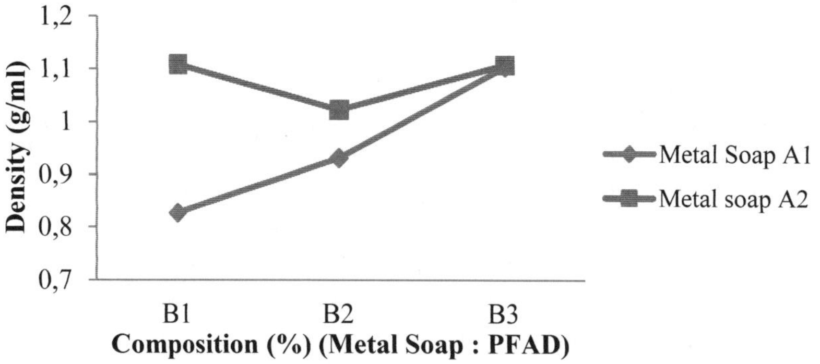

Figure 1. Density grease versus composition of metal soap (%)

The composition of A1: B2 obtained density of 0.931 g / ml of composition A1: B2 obtained density of 0.103 g/ml of composition A1: B3 obtained density of 1.103 g/ml, composition A2: B1 obtained density of 1.108 g/ml, composition A2: B2 obtained density of 1.022 g /ml, and composition A2: B3 obtained density of 1.107 g/ml. This increase in density is due to the decreasing of the metal soap composition and the increased composition of PFAD increasing the density of the grease.

3.2 Titrant

Titrant analysis is conducted with the aim to know the level of hardness or consistency level of the grease that has been made. The results of the analysis that has been done can be seen in Figure 2.

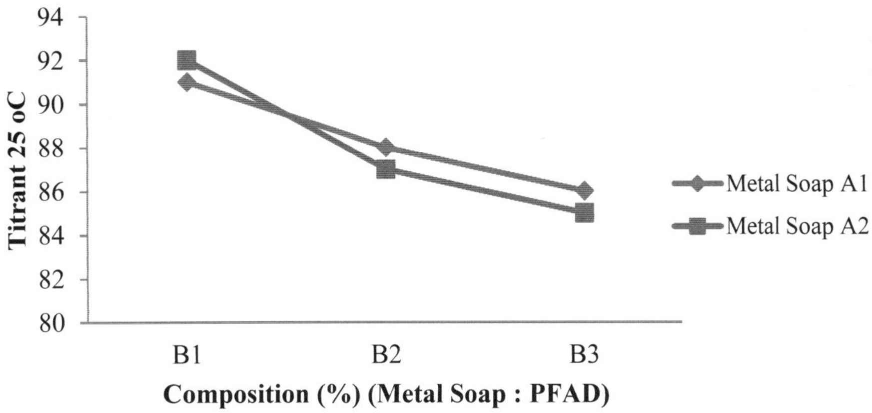

Figure 2. The relationship between % composition (metal soap: raw material) versus the titrant value

The results obtained indicate a decrease in titrant value. Where in composition A1: B1 obtained titrant of 91, composition A1: B2 obtained titrant equal to 88, and at composition A1: B3 titrant equal to 86. In composition A2: B1 obtained titrant equal to 92, composition A2: B2 obtained titrant equal to 87, and on the composition A2: B3 obtained titrant of 85.

Figure 2 showed that the decreasing of the metal soap composition and the increased composition of PFAD resulting in a decrease in the titrant value of the grease. This condition corresponds to the level of NLGI 6 violence.

Dropping point is a critical temperature where the gel structure in grease starts to undergo a phase change to liquid. The melting point represents the highest temperature at which the grease can maintain its structure. The stronger the structure of grease, the more difficult it is to change the phase.

3.3 Dropping point of Grease

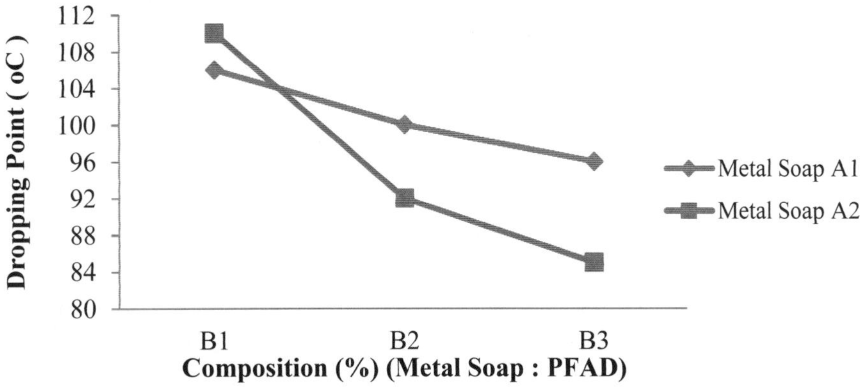

Figure 3. Composition Relation (%) vs. Melting Point (°C)

Where the composition A1: B1 obtained a melting point of 106 °C, composition A1: B2 obtained melting point of 100, and the composition A1: B3 obtained melting point of 96. In composition A2: B1 obtained the melting point of 110, the composition A2: B2 obtained melting point of 92, and on composition A2: B3 obtained melting point of 85.

Figure 3 explained that because of the decreasing of metal soap composition and increasing the composition of PFAD resulting in a decrease in the melting point. This condition is in accordance with the theory that the minimum allowable melting point for solid lubricant is 80 °C where the higher melting point possessed by a solid lubricant the better its resistance in maintaining the solid condition of the solid lubricant, so that the solid lubricant having high melting point will not melt quickly and lifetime increases. This result is in accordance with the specification of solid lubricant performance characteristics and parameters for quality level of NLGI GA, SNI 06- 7069-8-2005.

4. Conclusion

PFAD can be used as raw material for the manufacture of grease based on environment-friendly palm oil and has the characteristic of commercial solid lubricant. Variations of mixed soap metal composition (thickening agent) and PFAD (base oil) in the manufacture of grease affect the results of density analysis, titrant, melting point, and NLGI. The best grease generated according to ASTM standard and SNI 06 - 7069 - 8 - 2005 included in Grade A is found in composition A2: B1, A2 is the ratio of NaOH : $Ca(OH)_2$ = 80%: 20% and B1 is PFAD ratio: metal soap = 10% : 90% with the following criteria: Density: 1.108 g/ml, Titrant: 92 (25 °C), Melting Point: 110 °C, NLGI: 6.

5. Acknowledgment

The authors are grateful to RISTEKDIKTI for financial support.

6. References

[1] Saefudin R, Nazaruddin, Masnur D, 2014 Studi Lubricant Properties (Density) DenganPenambahan Calcium Hypochlorite, (Riau: Universitas Riau)

[2] Hassan A B, Abolarin MS, Nasir A, &Rathel U 2006 Investigation On The Use Of Palm Olein As Lubrication Oil, *LJEPT*, p.1-8

[3] Darnoko, Siahaan D N E and Elishabeth J 2003 TeknologiPengolahanKelapaSawit dan ProdukTurunannya (Pusat PenelitianKelapaSawit, Medan)

[4] Golshokouh, Syahrullail, Nasir Ani F.andMasjuki 2014 Investigation of Palm Fatty Acid Distillate As An Alternative To Petrochemical Based On Lubricants, (Malaysia: Publishing Universitas Teknologi Malaysia, p.25-36

[5] Yanto T and Septiana, PemanfaatanMinyakJarakPagar (Jatropha Curcas L) SebagaiBahan Dasar Pelumas Grease (Banyumas: Publishing Jenderal Soedirman, hal.34-42

[6] Bernandini, E 1985 Oilseeds, Oils and Fats Volume II Oils and Fats Processing B.E. Oil(Roma : Publishing House Via L. Lilio, 19,)

[7] Hartono and Anton J 1991 Lekuk-Liuk-Liuk Pelumas(Penerbit Andi Offset, Yogyakarta)

[8] Sukmawati and T H Jatmiko, 2012 Optimasi Rasio *Palm Fatty Acid Distillate(PFAD)* Dan Sabun Logam Pada Pembuatan Pelumas Padat *(grease)* Biodegradable (Conference Paper: Aceh) Buketetra: 12-13 December 2012 Hal.319-325 ISBN 978-602-17282-0-8

[9] Anonimous 1998 Klasifikasi Grease, in http://www.nlgi.org/grease-glossary/nlgi-grade/

[10] Lansdown A R 2007*Lubrication and Lubricant Selection, A Practical Guide, Third Edition*,(London: Professional Engineering Publishing Limited)

Rapid detection of chlorophyll-a reduction in tropical fruit leaves using violet laser

Isnaeni[1], Masni[2], M Shiddiq[1]

[1]Research Center for Physics, Indonesian Institute of Sciences, Building 442, Puspiptek, South Tangerang, Banten, 15314, Indonesia
[2]Department of Physics, Hasanuddin University, Perintis Kemerdekaan Road, KM.10, Makassar, South Sulawesi, 90245, Indonesia

E-mail: isnaeni@lipi.go.id

Abstract. Chlorophyll is an important substance in photosynthesis process. Chlorophyll turns CO_2 and water into energy required by trees. Detection of chlorophyll in leaves and trees will give us much information, such as chlorophyll quantity and quality. In this paper, we present a simple, yet powerful technique to observe chlorophyll-a degradation in tropical fruit leaves. We conducted experiment on mango, mangosteen, noni, jackfruit and rambutan leaves. We excited fresh leaves using violet (wavelength at 420 nm) laser and detected chlorophyll emission spectra. We found that there were noticeable differences among chlorophyll spectra from 5 leaves. We analyzed chlorophyll peak wavelength and intensity. We also observed time dependence chlorophyll emission upon continuous excitation to study chlorophyll degradation.

Keywords: chlorophyll, leaves

1. Introduction

Photosynthesis is an important process that usually takes place in leaves. Photosynthesis is basically a process when leaves absorb sunlight to produce energy that can be stored in other parts of trees [1]. Many substances play an important role in photosynthesis, one of them is chlorophyll. In photosynthesis, chlorophyll absorbs energy from sunlight to synthesize carbohydrate by converting CO_2 and water [1, 2]. Carbohydrate, which is one of the chemical energies, is then converted into nicotinamide adenine dinucleotide phosphate (NADPH) and adenosine triphosphate (ATP). These energies are stored in the form of sugar and other organic molecules [3, 4].

Chlorophyll, also known as green pigment, gives green color of the leaves. Chlorophyll may reduce during photosynthesis process [3]. Color change, from green to brown, is an obvious sign of chlorophyll reduction in leaves. These aging leaves indicate final life cycle phase of leaves [2, 5]. Chlorophyll reduction in leaves is mostly due to high photosynthesis process and lack of chlorophyll production in leaves [1].

Several detection techniques have been developed to observe chlorophyll in leaves. One of the most common techniques is UV-Vis absorption spectroscopy. This technique can be done in-situ or ex-situ using leave extract [6]. This technique identifies absorbency peak intensity of chlorophyll. However, sometime absorbency spectrum of chlorophyll may overlap with several substances such as chlorophyll-a, chlorophyll-b and beta-carotene [7]. Another technique is then required to specifically distinguish chlorophyll-a substance in leaves [5]. In this paper, we propose photoluminescence

IOP Conf. Series: Earth and Environmental Science **251** (2019) 012004 doi:10.1088/1755-1315/251/1/012004

technique to obtain and study chlorophyll-a reduction in leaves using violet laser as excitation source. We focused on 5 different leaves, which are tropical fruit leaves. They are mango (mangga) leaves, mangosteen (manggis) leaves, noni (mengkudu) leaves, jackfruit (nangka) leaves and rambutan (rambutan) leaves. The purpose of this work is to (1) study chlorophyll content in leaves and (2) to study chlorophyll degradation in leaves.

2. Methodology

This research was done in several steps such as sample preparation, optical setup preparation, and acquisition process and data analysis. Leaves used in this work were mango leaves, mangosteen leaves, noni leaves, jackfruit leaves, and rambutan leaves. Those leaves were collected from trees and freshly measured in the optical laboratory. We studied several parts of parts of the leaves such as center part, leaf tip, and leaf base. All data were measured at least 8 times to confirm repeatability issue.

Photoluminescence setup was comprised of excitation violet laser (continuous wave 420 nm laser at power 60 mW), a laser focusing lens, cuvette stage holder, two lenses acting as emission collector and a fiber-based spectrophotometer (as shown in Figure 1).

The experiment was done based on the following procedure.

- Prepare fresh leaves for photoluminescence experiment.
- Recording chlorophyll-a emission spectra.
- Recording chlorophyll-a reduction (changing of intensity at two specific wavelengths) upon continuous violet laser excitation.
- All experiment was done randomly on entire leaves' surface and repeated at least 8 times per leaf.
- Data fitting analysis using data processing software, including exponential decay fitting for studying chlorophyll-a reduction.

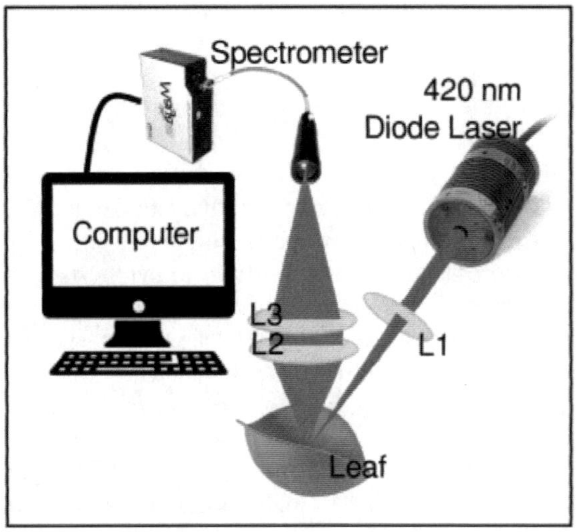

Figure 1. Photoluminescence setup for studying chlorophyll degradation.

3. Result and Discussion

Emission spectra of chlorophyll-a for 5 leaves is given in figure 2. Based on the emission spectrum, we can notice there are two emission peaks, the first peak (P1) exists around 680 nm wavelength and the second peak (P2) occurs around 730 nm wavelength. The spectrum of each leaf was an average spectrum of 8 measurements. In general, we can see that rambutan leaves and mangosteen leaves emit strongest chlorophyll-a emission, indicating abundant chlorophyll-a content in those leaves. On the other hand, mango leaves show the least chlorophyll emission.

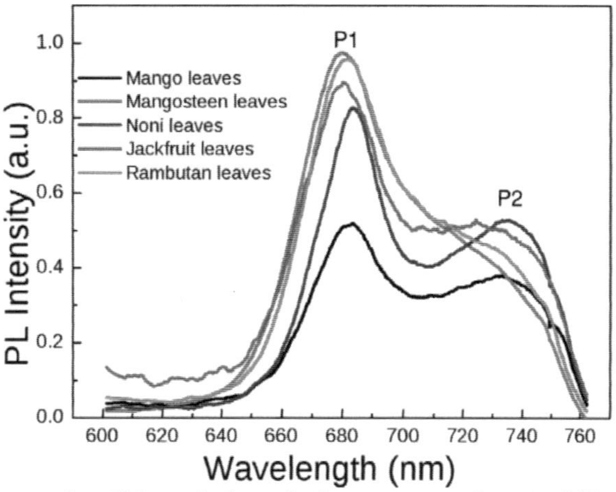

Figure 2. Chlorophyl emission spectra for 5 different leaves.

We notice that peak intensities of P1 and P2 are different. In general, P1 is higher than P2. We also notice that peak wavelength varies. Figure 3a indicates peak wavelength for P1 and P2 for 5 different leaves. We observed for P1; all leaves have similar peak wavelength around 681 nm. The peak wavelength difference among leaves is not obvious. However, P2 show significant difference among leaves. P2 of rambutan leaves has peak wavelength at 723 nm. Meanwhile, P2 of noni leaves has peak wavelength at 733 nm. In the future, this peak wavelength difference can be utilized to determine type of leaves.

In addition, we also notice that peak intensity between P1 and P2 is not always constant for all leaves. As shown in figure 3b, the photoluminescence peak intensity ratio P1 ($I_{\lambda 1}$) and P2 ($I_{\lambda 2}$). Peak ratio varies from 1.3 for mango leaves until 1.9 for jackfruit leaves. In the future, this peak intensity ratio can also be utilized to determine the type of leaves.

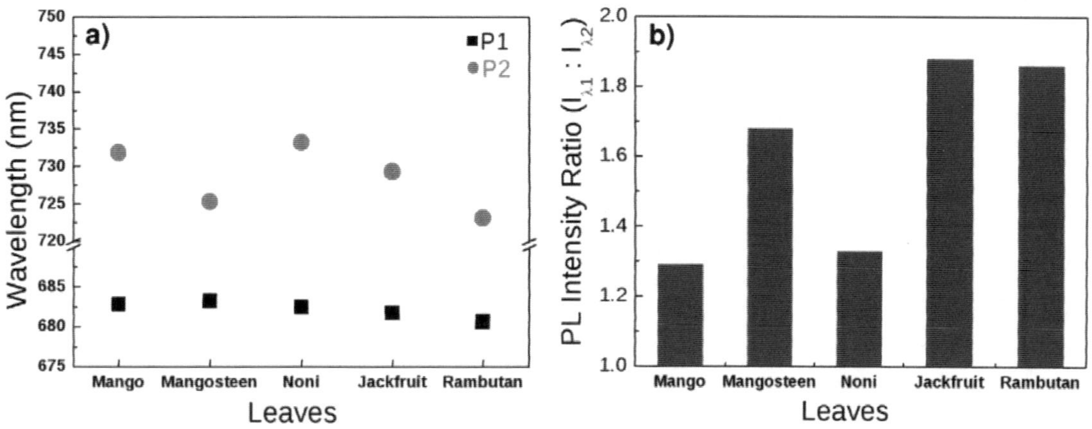

Figure 3. Peak wavelength for P1 and P2 (a) and peak intensity ratio (b) for 5 leaves.

In addition to spectrum analysis of chlorophyll-a emission, we also conducted a study of chlorophyll-a reduction in leaves. Upon continuous excitation of violet laser, we detected intensity of

P1. We recorded peak intensity every 1 second. For all leaves, we found that chlorophyll-a emission intensity significantly reduces.

Time-dependence chlorophyll intensity is then well fitted using single exponential decay. An example of a decay curve of chlorophyll emission is shown in figure 4. Using single exponential fitting, we can obtain decay time for chlorophyll-a reduction. In figure 4, fitted decay time was 9.04 seconds.

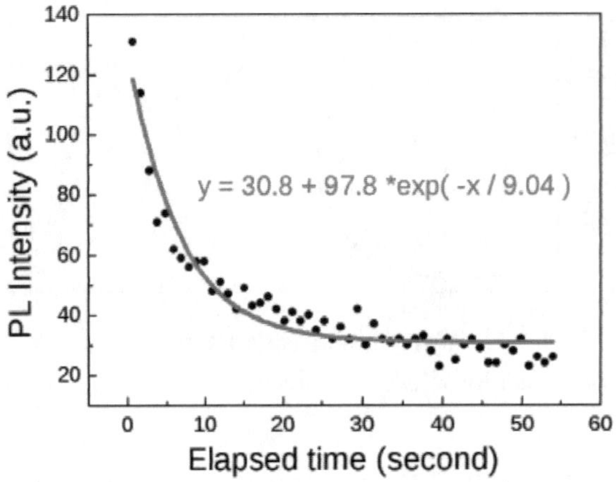

Figure 4. Chlorophyll emission intensity upon continuous violet laser excitation and decay fitting using single exponential fitting.

Figure 5 shows decay time for different leaves. We can see that mangosteen and jackfruit leaves have longer chlorophyll decay time. It means that chlorophyll-a can exist longer in those leaves. Short decay time, such as rambutan leaves indicates that chlorophyll in leaves is easily converted into energy and chlorophyll-a production in leaves was limited since we used fresh leaves picked from the tree.

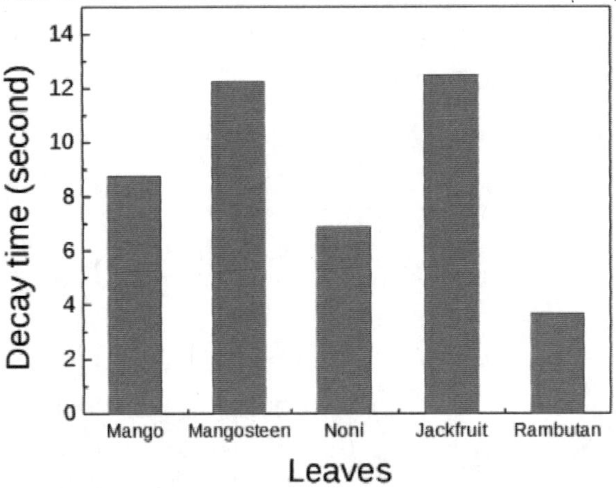

Figure 5. Decay time of chlorophyll emission for 5 different leaves.

The reduction of chlorophyll-a in response to aging which results in natural aging naturally occurs in leaves. Although aging is a genetically controlled program, environmental or external factors can accelerate or suppress aging. External factors cause changes in gene expression, which begin aging. Chlorophyll-a reduction is widely used to describe the loss of green quality in plants such as vegetable crops. Leaf or plant organ detachment is a condition that can accelerate the aging process [8, 9]. Some

IOP Conf. Series: Earth and Environmental Science **251** (2019) 012004 doi:10.1088/1755-1315/251/1/012004

chemical treatments may affect the content of chlorophyll in separate leaves. Although the inherent leaves can recover the contents of chlorophyll-a, chemically contaminated leaves can affect the contents of chlorophyll after leaves are released from the plant [9, 10].

4. Summary

We have shown a simple technique to detect chlorophyll in leaves using photoluminescence by utilizing violet laser. We have found that chlorophyll-a has two emission peaks. Different leaf has different characteristic regarding those two emission peaks. The ratio between P1 and P2 can be used to differentiate the leaves. Furthermore, chlorophyll-a reduction in leave is easily monitored. We found that chlorophyll-a reduction curve follows the exponential decay curve so that chlorophyll-a reduction can be characterized using its decay time. Rambutan leaves have shortest decay time, and jackfruit leaves have longest decay time. Decay time can be correlated to chlorophyll-a content and photosynthesis process.

5. Acknowledgments

This work was supported by Insinas project from Kementrian Riset Teknologi dan Pendidikan Tinggi, Republic of Indonesia No 055/P/RPL-LIPI/INSINAS-1/III/2018.

6. References

[1]. Bryant D A, Frigaard N U 2006 *Trends in Microbiology* **14** 488
[2]. Blankenship R E 2008 *Molecular Mechanisms of Photosynthesis* (Wiley-Blackwell) p 1-10
[3]. Lambers H, Chapin F S, Pons L 2008 *Plant Physiological Ecology* (New York : Springer-Verlag) p 11-99
[4]. Blankenship R E 2008 *Molecular Mechanisms of Photosynthesis* (Wiley-Blackwell) p 157-170
[5]. Baker N R 2008 *Annual Review of Plant Biology* **59** p 89-113
[6]. Cerovic Z G, Ounis A, Cartelat A, Latouche G, Goulas Y, Meyer S, Moya I 2002 *Plant, Cell and Environment* **25** 1663 -1676
[7]. Rehab E M, Salem E S 2016 *Middle East Journal of Agriculture Research* **4** 566 - 571
[8]. Sims D A, Gamon J A 202 *Asian Journal of Agriculture Research* **9** 132 - 138
[9]. Lichtenthaler A R, Rinderle A K 1990 *Radiat Environ Biophys* **43** 329 - 336
[10]. Maharramov A M, Ahmadov I S, Ramazanov M A, Aliyeva S Q, Ramazanli V N 2015 *Journal of Biomaterials and Nanobiotechnology* **6** p 135-143

Enhancement of Cellulase (CMCase) production from marine actinomycetes *Streptomyces* sp. Bse 7-9: Optimization of fermentation medium by Response Surface Methodology

S Ratnakomala*, Fahrurrozi, and Yopi

Research Center for Biotechnology, Indonesian Institute of Sciences
Jl Raya Bogor km 46, Cibinong 16911, Indonesia

E-mail: shanti_ratna01@yahoo.com

Abstract. Optimization of fermentation conditions is important in the development of bioprocesses. Here, we used the Taguchi experimental design and Response Surface Methodology to optimize fermentation medium for enhancing carboxymethyl cellulase (CMCase) production by marine actinomycetes *Streptomyces* sp. Bse 7-9. Optimization using the Taguchi experimental design we found that bagasse has a significant effect on other factors. The Central Composite Design was used to determine the interaction between the factors selected in the fermentation medium to obtain the optimal concentration of ingredients. The experimental results demonstrated that bagasse as a carbon source plays as the key component for the production of CMCase by marine actinomycetes *Streptomyces* sp. Bse 7-9. In conclusion, the optimal fermentation medium for maximum cellulase production by marine actinomycete *Streptomyces* sp. Bse 7-9 was as follows: bagasse concentration (1.84 % w/v), yeast extract (0.5 % w/v) and $CaSO_4$ (0.02 % w/v). The maximum activity of CMCase at this optimum medium was 4.496 U mL^{-1}.

1. Introduction

Cellulase has become an important biocatalyst for widespread applications in the fields of industry, such as in the process of conversion of lignocellulosic biomass. To convert lignocellulose and produce fermentable sugar effectively, microorganisms such as certain cellulolytic bacteria can produce hydrolytic enzymes capable of degrading lignocellulosic substrates. The complete hydrolysis of lignocellulose requires three types of enzymes, i.e., endoglucanases, cellobiohydrolases, and β-glucosidases, which work synergistically to hydrolyzed lignocellulose [1]. Huge of the bacteria are known for the production of hydrolytic enzymes, most of them originating from the fungal cellulases and only a few have been studied from the source of actinomycetes [2].

Actinomycetes are Gram-positive bacteria and living widespread, commonly found in soil, leaf litter, the sediment of freshwater and marine ecosystems [3]. Actinomycetes form an important part of the microbial communities that are responsible for the degradation of cellulose [4], known as one of the cellulose producers. Actinomycetes play a vital role in the degradation of lignocellulose from plant cell walls, thus attracting the interest of research for their potential to generate fermentable sugars from cellulose that can be useful for their growth and also for many industrial applications [5; 6].

Actinomycetes growth to form branching hyphae that are useful to penetrate in lignocellulose degradation. Cellulase of actinomycete is induced extracellular enzymes [7], which can be generated during their growth on a variety of cellulosic materials. A search of new microorganisms that can degrade cellulose is increasing, and a variety of new species have been isolated from a variety of natural habitats. Actinomycetes are rarely explored from the marine habitats, and only limited literature is available on their occurrence, diversity, and enzymatic potential. Therefore, studies on the marine actinomycetes would be of great significance. The marine actinomycetes are largely unexplored for the novel primary and secondary metabolites. In recent years, the possibility of obtaining high yields of useful cellulases from marine origin actinomycetes has attracted the attention of researchers. In this paper, we use marine actinomycetes *Streptomyces* sp. BSE 7-9 of Biotechnology Culture Collection (BTCC). The strains isolated from marine sediments of the seashore on Kuta beach, Bali, Indonesia, for the study of isolation and actinomycetes diversity of coastal environments in Indonesia. The strain can produce cellulase enzyme and examine for the possibility of degradation the cellulose.

Optimization of fermentation conditions is an essential issue in the development of bioprocesses. The interaction between the combinations of components medium with the desired compound production and optimal processes can be developed using procedure experimental design [8; 9]. Effective bioconversion of cellulosic material depends on the properties of the cellulose, the source of cellulolytic enzymes and conditions optimal for the catalytic activity and production of enzymes. The optimization of fermentation media ingredients and environmental factors for the optimum enzymes production is more effective strategies [10].

Statistical experimental designs have been used in various stages of optimization strategies, such as for preliminary screening to find the optimal conditions for a targeted response. One popular choice in applying statistical design for the bioprocessing is Taguchi Experimental design. A Statistical tool of Design of Experiments is the latest trend in the development of the bioprocess and process optimization using response surface methodology (RSM). This tool is searching for the optimum condition through a collection of mathematical and statistical techniques to design an experiment of the factors for the desired response, and evaluate the relative importance of several factors that influence in the presence of complex interactions [11]. RSM is used to determine the various variables under control and optimal process conditions for the production of enzymes, followed by a polynomial equation generation bioprocess models, to estimate the relationship between the variables under control and observe the results [12].

The objective of this research was to study and optimize enzymatic carboxymethyl cellulase (CMCase) production from marine actinomycete *Streptomyces* Bse 7-9 using agricultural waste sugarcane bagasse to provide a high yield of cellulase enzymes.

2. Material and Methods

2.1 Microorganism and culture media

Streptomyces sp. Bse 7-9 used in this study was isolated from the sea sediment in seashore Kuta, Bali Island. Indonesia (E: 115°10'7.8" S: 8°43'5.5"). Approximately 1 g sample was diluted in a serial dilution with artificial seawater (Nihon Pharmaceutical), and 0.2 ml of each dilution was spread on plates of NBRC medium 802 [0.2% (w/v) polypeptone, 0.04% (w/v) yeast extract, 0.02% (w/v) MgSO4 .7H_2O and 1.5% (w/v) agar; pH7.0], The medium was supplemented with 5.0% (w/v) NaCl, 0.002% (w/v) nalidixic acid and 0.005% (w/v) cycloheximide to avoid unwanted growth of Gram-negative bacteria and fungi. After cultivation at 30 °C for one week and single colony isolation twice, strain Bse 7-9 was obtained. The stock culture was maintained on glycerol stock. The medium composition for maintenance comprises of: NBRC medium 802 supplemented with 2.0% (w/v) NaCl, and, 1.5% (w/v) agar; pH 7.0.

2.2 Detection for CMCase producing actinomycete strain

A preliminary analysis for cellulolytic activity was conducted using Congo red dye. The clear zone methodology with Congo red was used to detect the CMCase activity of the strain. *Streptomyces* Bse 7-9 was grown on ISP2 broth without glucose, substituted with 1% bagasse as a carbon source. The medium consists of: bagasse, 1.0% (w/v); yeast extract 0.4% (w/v) and malt extract 0.2% (w/v). The medium was incubated at 30 °C for seven days under a shaking condition at 120 rpm. At the end of the incubation, the medium was centrifuged at 10,000 rpm for 10 minutes. Free cells supernatant were used as the crude CMCase enzyme. Crude enzyme was then concentrated until 40-fold using Amicon. Three µl of the free cell supernatant and 1 µL of 40-fold concentrated free cells supernatant, were spotted onto substrate agar containing 0.5% carboxymethyl cellulose (CMC), and 1.7% agar, diluted in 50 mM Phosphate buffer pH 6.8. The CMC plate was incubated at 30 °C for three days to allow hydrolyzed the CMC substrate. Plates were then flooded with 0.25 % Congo red for 15-20 min, then washed with 1 M NaCl for 15 min. The clear zone formation was indicated hydrolysis of cellulose degradation [13].

2.2.1 SDS-PAGE and zymogram analysis.
The samples supernatant were mixed with the sample buffer containing 1% β-mercaptoethanol. The sample was then heating at 70 °C for 20 minutes and subsequently loaded on 5% stacking gel. For zymogram analysis, CMC with a final concentration of 0.1% (w/v) was incorporated into 12% separating gel as substrate. After electrophoresis was run, SDS gel then soaked in 0.1 M phosphate buffer, pH 6.8 for 15 min with gentle shaking, and then was immersed in 0.1% Triton X-100 for 30 min with gentle shaking. The gel was incubated in the 0.1 M phosphate buffer for 1 h at 35 °C with gently shaking to allow the proteins to refold. The gel was then stained in 0.25% Congo red for 1 h and destained with 1 M NaCl until the clear zone appeared against a red background. Another duplicated SDS gel containing the same loading pattern was stained with Coomassie brilliant blueR-250 and destained with the destaining solution. The positions of the cellulase enzyme on the both SDS gels were compared to a determined molecular mass of the active bands.

2.2.2 Fermentation for cellulase production.
Production of cellulase (CMCase) from marine actinomycete *Streptomyces* Bse 7-9 was analyzed in several media as described in Table 1. The isolate was grown in CYP [14] broth for 24 h under shaking 120 rpm and then inoculated (2%, v/v) into 25 mL fresh of each medium broth

Table 1. Several media for CMCase production by marine actinomycete *Streptomyces* sp Bse 7-9

Medium 1	Medium 2	Medium 3	Medium 4
CMC 10 g/L	KH_2PO_4 1 g/L	Yeast extract 1 g/L	Glucose 0.1 g/L
yeast extract 1 g/L	$MgSO_4.7H_2O$ 0.5 g/L	NaCl 0.5 g/L	K_2HPO_4 0.1 g/L
Peptone 1 g/L	NaCl 1 g/L	Peptone 1 g/L	Yeast extract 0.05 g/L
NaCl 2 g/L	$FeSO_4.7H_2O$ 0.01 g/L	Starch 10 g/L	Tryptone 0.5 g/L
	$MnSO_4.H_2O$ 0.01 g/L		
	NH_4Cl 1 g/L		
	CMC 10 g/L		
	Yeast extract 2 g/L		

Fermentation was conducted at 30 °C under shaking (120 rpm) for four days of incubation. Samples were centrifuged (Tomy Refrigerated Centrifuge MX-307) at 10,000 rpm for 10 min. The cell-free supernatants were considered as crude extracellular enzymes and used as the source of crude cellulase enzymes for assaying cellulase activity.

2.2.3 Enzyme Assay
CMCase activity was determined by monitoring the release of reducing sugars. CMCase activity was measured following the dinitro salicylic acid (DNS) method of Miller [15]. A reaction mixture composed of 0.5 ml of an appropriately diluted enzyme to 0.5 ml of 1 % (w/v) of carboxymethyl cellulose (CMC) in 50 mM Phosphate buffer, pH 6.8 and incubating at 50 °C for 15 min. The amount of reducing sugars released during the reaction was measured using the DNS method, and D-glucose was used as the standard. One unit (U) of carboxymethyl cellulase (CMCase) activity was defined as

the amount of enzyme that liberated 1 μmol of glucose equivalent under the given assay conditions [15].

2.3 Experimental Design and Statistical Analysis

Screening of ingredients components for medium optimization was done by Taguchi experimental design. Experimental design for optimization can make easier to evaluate the effect of several factors in achieving optimum conditions to produces a desirable response, an experiment that uses a small number of the possible [16]. A standard experimental array L_9 (3^4) was used to study the effects of carbon and nitrogen sources, metal ions, and salinity on CMCase production. The levels of the factors investigated and the layout of the L_9 Taguchi's experimental array are shown in Tables 1 and 2. The L and the subscript (9) represent the Latin square and the number of trial runs, respectively. Results obtained from the experiments were then analyzed to determine independent factors that influence the main effect; analysis of variance technique was then applied to determine which factors were statistically significant. The controlling factors are further identified, with the qualified effects and a statistically significant effect. Thus, the optimal conditions are determined by combining the level of the factors that have the highest value of the main effect.

2.4 Optimization of a concentration of the selected medium components using response surface methodology.

Response surface methodology (RSM) was applied to optimize the selected three significant nutrient components: bagasse as a carbon source, yeast extract as a nitrogen source and $CaSO_4$ as a selected metal ion in various concentrations in the fermentation medium. A factorial, central composite design (CCD) for three independent variables were studied at three different levels of the independent variables at the center point, and star points were used in this investigation and shown in Table 5. A centered design consisting of 20 experimental runs (Table 6), including six replicates at the center point was employed. The variables were bagasse (X1), yeast extract (X2), and $CaSO_4$ (X3). The response variable was the CMCase activity (Unit mL^{-1}). The experimental data were regressed according to Equation (1) as a second-order polynomial equation: The quadratic model equation for predicting the optimal point is expressed by Eq. (1) as followed:

$$Y = \beta 0 + \Sigma\, \beta iXi + \Sigma\, \beta iiXi^2 + \Sigma\, \beta iiXiXj + \mathcal{E} \quad \ldots(1)$$

where Y is the predicted response (CMCase activity, $U.mL^{-1}$), β0 is constant coefficient, βi,βii and βij are coefficient for the linear, quadratic and interaction effects, Xi and Xj are factors (independent variables) while Ɛ is the error [17].

A regression model was established, and the experimental verification of the model was validated. Design Expert statistical software (version 9.0.6.2., Stat-Ease Inc., USA) was used for experimental data regression analysis and to plot the response surface.

3. Results

3.1 CMCase producing actinomycetes

Marine actinomycetes strain to produce cellulase has been screened in this study for their potential to hydrolyzed CMC substrate by using the Congo red test. Cellulose in CMC agar was used as the sole carbon source. In a preliminary detection test, marine actinomycete *Streptomyces* sp. Bse 7-9 displayed considerable cellulase-producing ability on carboxymethyl cellulose (CMC) agar plates. This isolate was able to degrade CMC substrate, as indicated by the formation of a clear zone around the dropped crude enzyme (Figure 1a; 1b), measured as the ratio of the size of CMC hydrolysis on dropped enzyme diameter after three days of incubation at 30 °C. The clear zone diameter produced by *Streptomyces* Bse 7-9 was 25 mm for free cells supernatant and 37 mm for 40-fold concentrated free cells supernatant. This result was strong evidence that cellulase was produced in the supernatant of *Streptomyces* Bse 7-9.

The SDS-PAGE of supernatant *Streptomyces* Bse 7-9 revealed a few protein bands after staining (Fig. 1c(i)) Corresponding to the zymogram analysis, supernatant and concentrated free cells supernatant of *Streptomyces* Bse 7-9 showed a few clear bands revealing cellulolytic activity in-gel zymography (Fig. 1c(ii)). Based on the results of this initial detection test, strain Bse 7-9 which showed a substantial cellulase activity and representing from marine ecological habitat was selected for further study.

The denatured of biological mass CMCase was separated by using SDS-PAGE revealed the cellulolytic activity, and shown a molecular weight around 24 kDa (Fig. 1c(i)). In-gel zymography activity contained CMC as substrate suggesting that the enzyme was refolded correctly and had cellulolytic activity after SDS-gel electrophoresis and the following incubation step. The molecular weight of the enzyme estimated from SDS-PAGE was similar to the molecular weight of the bacteria, yeast, beetle, and termites, which are in the range of 29-52 kDa [18; 19; 20; 22; 23; 24].

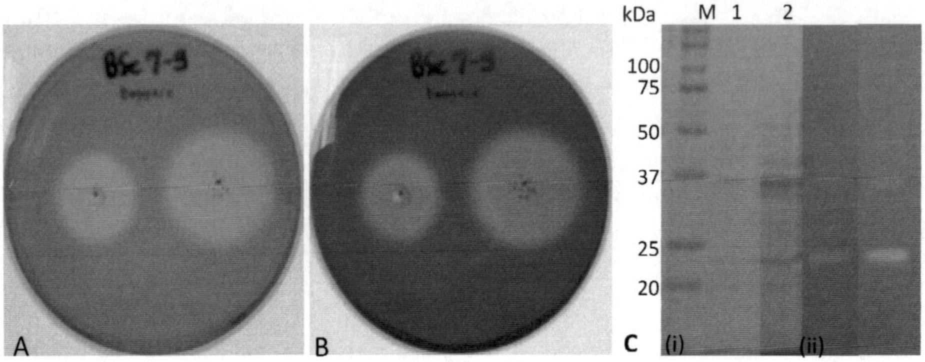

Figure 1. (a) Clear zone by *Streptomyces* sp Bse 7-9 supernatant stained with 0.25% Congo Red, (b) with an addition 5% acetic acid glacial, (c) SDS-PAGE and zymogram analysis of CMCase enzyme from *Streptomyces* sp Bse 7-9. (i) Stained gel with Coomassie brilliant blue R-250 and ((ii) in-gel activity staining or zymography. Lane M: Marker, Lane 1: Crude sample (free cells supernatant), Lane 2: Concentrated 40-fold free cells supernatant

3.2 Medium selection for CMCase production

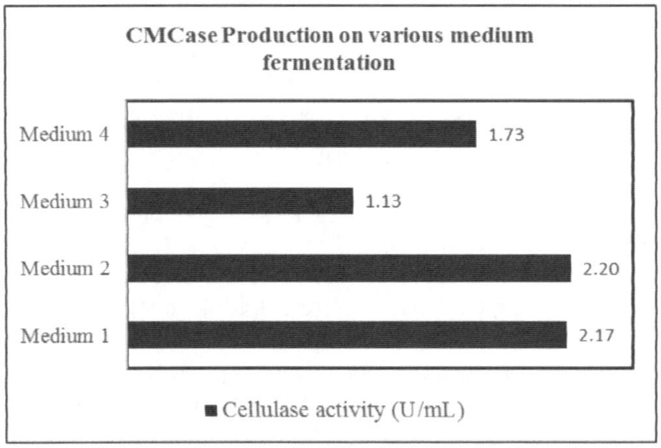

Figure 2: CMCase production by marine actinomycete *Streptomyces* sp Bse 7-9 using several media, fermentation were conducted at 30 °C under shaking 120 rpm for 4 days incubation.

The strain *Streptomyces* sp Bse-7-9 was found to be capable of hydrolyzing cellulose. To optimize culture medium conditions, a set of experiments was performed under different carbon source, nitrogen source, and metal ions. The strain *Streptomyces* sp Bse 7-9 exhibited high cellulase activity. Determination of the CMCase activity of the strain was calculated by estimating the amount of reducing sugar (U mL^{-1}) released by the strain. It was investigated that the highest amount of reducing sugar released due to CMCase activity by the strain *Streptomyces* sp Bse 7-9 was in Medium 2 as presented in Figure 2.

3.3 Selection of significant variables by Taguchi experimental design

Optimization is performed to improve the yield of bioactive substances such as CMCase produced by actinomycetes strain. From the clear zone of hydrolysis, it is observed that strain *Streptomyces* sp Bse 7-9 gives a good hydrolytic activity. Therefore, an attempt has been made to improve the yields of *Streptomyces* sp Bse 7-9 by optimizing biophysical parameters such as salt, metal ions and nutrient concentrations considering at three levels of experiments. Taguchi's experimental design has been applied to find the optimum combination of these four factors for maximum yield. Each factor is taken at three levels in an L^9 experimental array. The variation reflected the importance of the optimization studies for higher cellulase production. The medium was formulated as per the design, and the flask culture experiment was performed. The response was calculated at the rate of enzyme production and expressed as U mL^{-1}. All analyses were performed in duplicate. The average of the rate of enzyme production was considered as the response. The matrix design for the screening of significant variables that have the most profound influence on cellulose production and the corresponding responses are presented in Table 2 and Table 3.

Table 2: The Taguchi experimental design for the screening of medium variables for production of CMCase from marine actinomycete *Streptomyces* sp Bse 7-9

Factors	Level 1	Level 2	Level 3
Carbon source (1%)	Glucose	CMC	Bagasse
Nitrogen source (0.5%)	Peptone	Yeast extract	$(NH_4)_2HPO_4$
Metal ion (0.02%)	Ca_2SO_4	Fe_2SO_4	$MnSO_4$
Salinity (%)	0	1	2

Table 3: L9(3^4) experimental array of Taguchi experimental design and corresponding CMCase, production by marine actinomycete *Streptomyces* Bse 7-9

Run	A: Carbon (1%)	B: Nitrogen (0.5%)	C: Metal ion (0.02%)	Salinity (%)	CMCase (U mL^{-1})
1	2	1	2	3	1.97
2	1	2	2	2	0.48
3	2	3	1	2	0.82
4	3	1	3	2	3.06
5	1	1	1	1	0.55
6	1	3	3	3	0.83
7	3	3	2	1	2.64
8	2	2	3	1	1.60
9	3	2	1	3	3.36

Carbon: 1=Glucose, 2=CMC, 3=Bagasse; Nitrogen source: 1=Peptone, 2=Yeast extract, 3= $(NH_4)_2HPO_4$; Metal ion: 1=Ca_2SO_4, 2=Fe_2SO_4, 3=$MnSO_4$; Salinity: 1=0%, 2=1%, 3=2%.

Among the four ingredients of cultural variables examined based on Taguchi experimental-designed experiments (carbon source, nitrogen source, metal ions, and salinity), one variable Carbon source, A (p-value=0.0011), have maximum influence on the CMCase production (Table 4). Factors showing statistical p- values of less than 0.05 were considered to have a significant effect on the response that is enzyme activity. The model F-value of 26.15 indicates that the model is significant. Only 0.79% chance exists that this large model F-value could be due to noise. The statistical values of

R2=0.9886, Adj R2=0.9581, Pred R2=0.8170, and Adeq Precision = 17.102 indicate the significance of the model and the reliability of the experiments. One of the advantages of the Taguchi experimental design is that it helps to rank the effect of different variables on the measured response independent of its nature either nutritional or physical factor [25].

Table 4: Results of ANOVA obtained for CMCase production by *Streptomyces* sp Bse 7-9 for Taguchi experimental design

Source	Sum of squares	DF	Mean squares	F value	Prob>F	
Model	2.22	2	1.11	26.15	0.0011	significant
A-Carbon source	2.22	2	1.11	26.15	0.0011	
Residual	0.26	6	0.043			
Cor Total	2.48	8				

The results shown in Table 4 suggest that only carbon source (bagasse) is the positive factor in the response, while other variables such as nitrogen source, metal ions, and salinity were found to be relatively unimportant and non-significant variables. Even though marine actinomycetes can grow without salt, they usually showed the lowest growth rate. Na^+ ions are still important for their membrane transport system to maintain high intracellular potassium concentration for the functioning of enzymes [26].

3.4 Optimization of CMCase production by Response Surface Methodology

The three variables for cellulase production identified based on Taguchi experimental-designed experiments were further optimized using RSM, keeping other variables at constant levels. The effective variables were analyzed using the Design- Expert software.

Table 5: Independent variables and their coded levels for the central composite design used for cellulase production by *Streptomyces* sp Bse 7-9

		Code Variable Levels		
Variables	Code	-1	0	1
Bagasse (%)	A	0.5	1	1.5
Yeast extract (%)	B	0.25	0.5	0.75
$CaSO_4$ (%)	C	0.01	0.02	0.03

The experimental values were fed into the response column as presented in Table 6.

Table 6: Central composite design for cellulose production by *Streptomyces* sp. Bse 7-9

Run	Factor A: Bagasse (%)	Factor B: Yeast extract (%)	Factor C: $CaSO_4$ (%)	Response 1:	
				CMCase (U mL^{-1}) actual	CMCase (U mL^{-1}) predicted
1	1.00	0.50	0.02	3.486	3.49
2	0.50	0.25	0.01	2.390	2.09
3	1.50	0.75	0.01	4.026	4.07
4	1.00	0.50	0.02	3.357	3.49
5	1.00	0.50	0.037	2.674	2.91
6	0.16	0.50	0.02	0.631	1.09
7	1.00	0.50	0.02	3.203	3.49
8	1.50	0.75	0.03	4.094	4.11

9	1.50	0.25	0.01	4.190	4.09
10	0.50	0.75	0.01	1.908	1.74
11	1.00	0.50	0.003	2.489	2.66
12	1.00	0.50	0.02	3.517	3.49
13	**1.84**	**0.50**	**0.02**	**4.496**	**4.45**
14	1.00	0.92	0.02	3.580	3.62
15	0.50	0.75	0.03	2.298	2.11
16	1.50	0.25	0.03	4.144	4.02
17	1.00	0.50	0.02	3.381	3.49
18	0.50	0.25	0.03	2.692	2.36
19	1.00	0.50	0.02	4.058	3.49
20	1.00	0.08	0.02	3.473	3.84

The run-13 showed the highest CMCase activity up to 4.496 U mL^{-1} when the experimental culture medium conditions were set to bagasse concentration, 1.84%; yeast extract, 0.5% and $CaSO_4$, 0.02%, respectively. It was noticed that when the experiment was conducted in conditions at temperature, 28 °C; with shaking at 120 rpm, and 4 days incubation, the lowest CMCase activity was attained i.e. 1.09 U mL^{-1}.

Table 7. ANOVA for CMCase production

Source	**Sum of Squares**	**df**	**Mean Square**	**F Value**	**p-value Prob > F**	
Model	15.84	9	1.76	15.07	0.0001	significant
A-Baggase	0.74	1	0.74	6.33	0.0306	
B-Yeast Extract	0.050	1	0.050	0.43	0.5274	
C-$CaSO_4$	0.051	1	0.051	0.43	0.5247	
AB	0.055	1	0.055	0.47	0.5083	
AC	0.056	1	0.056	0.48	0.5033	
BC	0.0051	1	0.0051	0.044	0.8385	
A^2	0.93	1	0.93	7.98	0.0180	
B^2	0.11	1	0.11	0.91	0.3616	
C^2	0.89	1	0.89	7.59	0.0203	
Residual	1.17	10	0.12			
Lack of Fit	0.73	5	0.15	1.69	0.2893	not significant
Pure Error	0.43	5	0.087			
Cor Total	17.01	19				

ANOVA was performed using the experimental design implemented. The Model F-value of 15.07 implies the model is significant (Table 7). There is only a 0.01% chance that an F-value this large could occur due to noise. Values of "Prob > F" less than 0.0500 indicate model terms are significant. In this case, A, A^2, and C^2 are significant model terms. A regression equation for the quadratic model was suggested as:

$$\textbf{CMCase activity} = -60.91 - 81.56A + 1.43B - 1.43C + 0.83AB - 0.84AC + 0.025BC - 25.43A^2$$

In which Y is a function of the variables of bagasse concentration (A), yeast extract (B) and $CaSO_4$ (C).

The values for lack of fit F (1.69) and lack of fit P>F (0.2893) implies that the lack of fit effects is not significant to the pure error. There is a 28.93% chance that a "Lack of Fit F-value" this large could occur due to noise. The fit of the model was also expressed by the coefficient of regression (R2) = 0.9313, indicating that 93.13 % of the confidence level of the model to predict the response. Adj R2=0.8695, Pred R2=0.6367, Adeq Precision=13.928."Adeq Precision" measures the signal to noise ratio. A ratio greater than 4 is desirable. The ratio of 13.928 indicates an adequate signal indicator for correlation between the observed and predicted values and their liability of the experiments

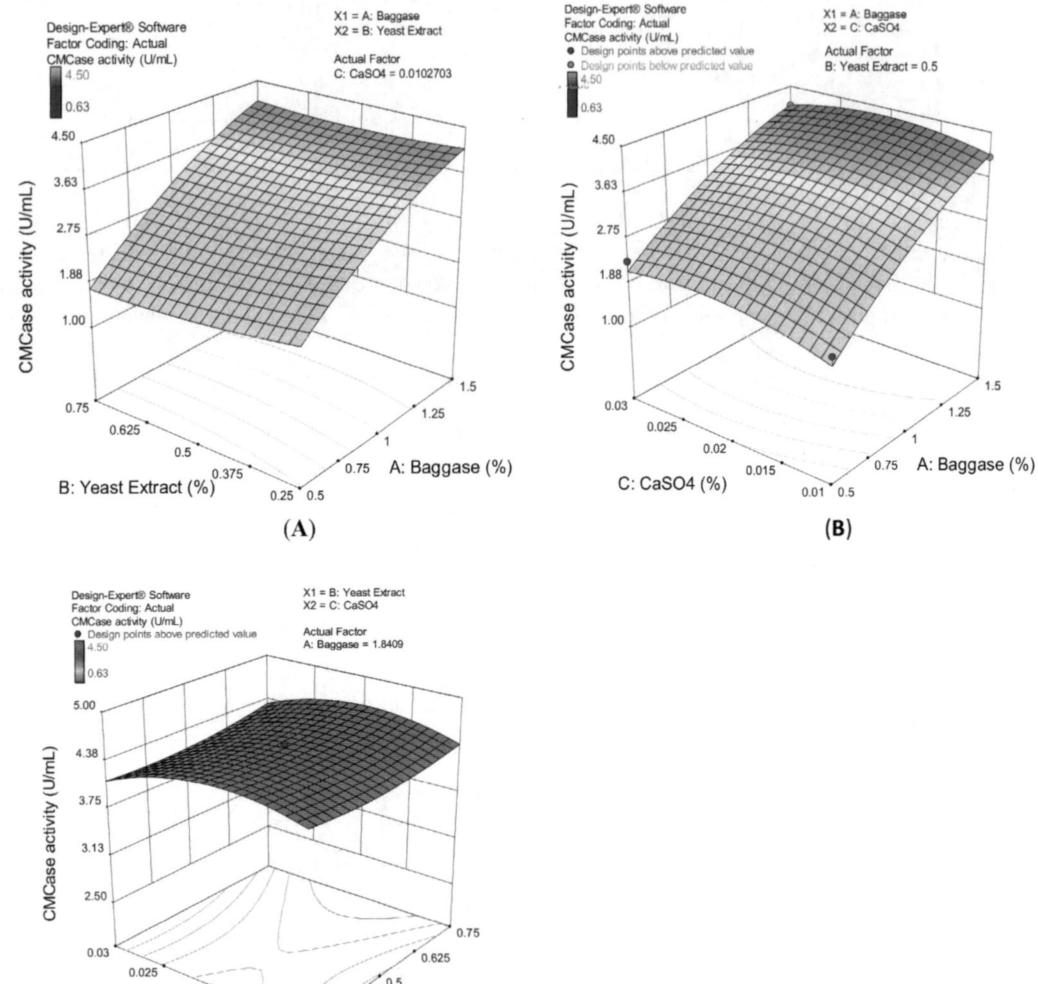

Figure 3. The interaction between (A) Bagasse and Yeast extract concentrations (%); (B) Bagasse and $CaSO_4$ concentration (%); and (C) Yeast extract and $CaSO_4$ concentration (%) on the activity of the cellulase enzyme (U mL^{-1}).

Figure 3 (A) demonstrates the 3D response surface plot for the interaction effects of bagasse versus yeast extract concentration and keeping $CaSO_4$.constant at 1%. By optimizing with a coded factor, the maximum CMCase production observed by the model was 4.496 U mL^{-1} as could be attained when bagasse concentration at 1.84% and yeast extract at the concentration of 0.5%, respectively. The plot revealed that the CMCase production from marine actinomycete *Streptomyces* sp Bse 7-9 was high and increasing bagasse concentration as carbon source resulted in an increasing CMCase production whereas increasing yeast extract concentration resulted in less production of cellulase The CMCase activity was increased when the bagasse concentration was raised from 0.5 to 1.84%.

Figure 3 (B) shows the interaction of bagasse (0.5 – 1.84%) and $CaSO_4$ (0.01 – 0.03%) with the fixed coded value of yeast extract concentration (0.5%). An increasing both bagasse and $CaSO_4$ concentration have initially increased the production of CMCase until they reach optimal CMCase production. Figure 3 (C) reveals the interaction between yeast extract and $CaSO_4$ concentration on CMCase production. When bagasse concentration was fixed at 1.84%, yeast extract concentration and $CaSO_4$ were set to 0.5 – 0.75% and 0.01 – 0.03%. An initial increase is in $CaSO_4$ concentration resulted in increased production of CMCase until reached a point for the optimal production of CMCase, however on increasing in yeast extract concentration; there has no affected in the CMCase production.

By numerical optimization, the quadratic model predicted that the maximum CMCase activity is 4.45 $U.mL^{-1}$ when the optimal values of the factors in the code unit tests that bagasse 1.84%, yeast extract 0.5%, and $CaSO_4$ 0.02% (Table 6). Validation of the results statistically using optimized media was carried out in duplicate. Maximum CMCase activity obtained experimentally found to be 4.496 U mL^{-1}, (Table 6) are in close agreement with the predicted values (4.45 U mL^{-1}). Therefore, this model is considered reliable and accurate to predict the CMCase production by *Streptomyces* sp. Bse 7-9.

4. Discussion

Actinomycetes, particularly *Streptomyces* species are known as the source of thousands of bioactive compounds. These microorganisms have a system of secondary metabolic pathways to produce various compounds that show various impressive and diverse biological activities. Enzymes are one of the important products produced in the metabolic pathway of actinomycetes. However, in comparison with the terrestrial microorganisms, only limited groups of actinomycetes from the aquatic ecosystems are known for the cellulose degradation [6].

Actinomycetes have been found in many unique marine environments, such as marine organic aggregates as the major microbial community [27], seashores and estuaries [28], mangrove soils, sediments and plants [29]. Seashores are highly dynamic ecosystems these ecosystems are home to diverse flora and fauna of marine, freshwater and terrestrial species [30]. Marine microorganisms are encouraged to develop an enzyme system adapted to the complex environment, because of the competition between microorganisms for space and nutrients in the marine environment is one of a significant selective force that can lead to evolution. The constant changes in environmental factors such as tidal and salinity gradients in the mangrove environment can be considered as the driving force to adapt to their metabolic pathway. This situation can lead to the production of unique and unusual metabolites. Therefore, currently has been increased exploitation of microorganisms from the mangrove environment [31]. Microbial fermentation is a process in which the product concentration varies under different conditions. Components of the media play a significant role in metabolism for CMCase production. Various factors of the medium were screened by Taguchi experimental design and then the interaction of the selected factors was studied using CCD.

Taguchi experimental design is the study to identify the effect of individual factors in the system provided by a set of independent variables over a certain level of interest, studying the relationships among variables and also a performance at optimum levels. By studying the major effects of each factor, the trend of influencing factor on the process will be predictable and controlled to provide the desired results. Furthermore, the level of a factor to produce the best results can be predicted [32]. Any individual factor may interact with the other factors creating the possibility of the presence of interactions. From the results of Taguchi experimental design, nitrogen source and metal ions are relatively insignificant coefficients, whereas carbon sources were found significant with positive coefficients.

In this study, to evaluate the accuracy of Taguchi experimental design, a medium of the following composition: temperature 30°C; incubation time 4 days; inoculum size 2% (v/v); agitation speed 120 rpm; bagasse 1%; yeast extract 0.5%, and $CaSO_4$ 0.02 %, which expected to be optimum gives CMCase activity of 3.36 U mL^{-1} which is higher than result obtained from the basal medium (medium no 2) before applying Taguchi experimental design by more than one and half times (2.20 U mL^{-1}).

Bagasse mainly consists of cellulose and hemicellulose, utilized widely by different microorganisms as a carbon and energy source for cellulase production. Several *Streptomyces* strains were isolated from Brazilian soils and grown on sugarcane bagasse in varying concentrations to produce cellulase with the highest production of CMCase (1.11 U mL^{-1}), after 48 hours fermentation [33]. *Streptomyces* sp.LX was shown to secrete a carboxymethyl cellulose (CMC)-liquefying enzyme that cleaves the CMC chains, which has a molecular mass of 9.8 kDa [34]. Jaradat *et al.* [35] also were reported that *Streptomyces* strain (J2) was observed as the highest crude cellulase enzyme activity as 0.432 U mL^{-1} in a medium that was supplemented with 0.5% glucose.

Sugarcane bagasse has a softer structure compare to others that could be easy to break down [36]. In sugarcane bagasse, the cellulose content was reported as high as 32-44% [37]. Indonesia is the large sugar producer in South East Asia region with the production of sugar cane as much as 2.6 million ton; there would be produced solid waste bagasse as much as 13 million ton [38]. Thus, the availability of abundant as agricultural waste from sugar production process can be utilized for the production of CMCase.

In this study, a response surface methodology (RSM) was used to optimize the conditions for CMCase production by *Streptomyces* sp. Bse 7-9. Using RSM based on CCD model, the optimum medium ingredients for CMCase production by *Streptomyces* sp. Bse 7-9 were: bagasse of 1.84% (w / v), yeast extract of 0.5% (w / v), $CaSO_4$ of 0.02% (w / v) and incubation time of 4 days. The maximum CMCase production observed by the models by optimizing medium was 4.496 U mL^{-1}, more than two-fold higher than CMCase activity in basal medium. RSM has been widely employed for the optimization of variables for maximum cellulase as reported in production from *Cellulomonas uda* NCIM 2353 [39] *Bacillus* VITRKHB [40], *Trichoderma reesei* WX-112 [8].

The central composite design revealed the optimum ratio of respective critical media components for CMCase production, while the Taguchi experimental design showed the utmost contribution towards CMCase production to have come from bagasse as carbon sources are known to be the most relevant factors for microbial growth. Response surface methodology (RSM), has successfully been used in the optimization of bioprocesses in evaluating the effects of the concentration of medium ingredients and searching for the optimum conditions of CMCase production for marine actinomycetes *Streptomyces* sp. Bse 7-9.

5. Conclusion

Biodiversity of marine ecosystems provides an abundant natural source of biocatalyst and useful substances. *Streptomyces* sp. Bse 7-9 can efficiently utilize agro waste bagasse from sugar production as a carbon source for growth. CMCase production can be increased by 2-fold by using statistics-based optimization Taguchi experimental design and response surface methodology.

This study focused on the increased production of the CMCase by *Streptomyces* sp. Bse 7-9 at various concentration levels of the media production. The results showed that the optimal level of ingredients in the media for the production of CMCase is a combination of 1.84% bagasse, 0.5% yeast extract 0.02% $CaSO_4$ within four days of incubation. Taguchi experimental design and RSM found to be highly useful in selecting and optimizing medium components in a manageable number of experimental trials. The production improves doubled in CMCase activity. The optimal culture media obtained in this study will be useful for further research to claim the potential of *Streptomyces* sp. Bse 7-9 in the large-scale fermentation for the efficient production of the CMCase for commercial applications.

6. Acknowledgment

This work was conducted collaboratively by The Indonesian Institute of Sciences (LIPI), representing the Government Research Centers (GRC) of the Republic of Indonesia and the Science and

Technology Research Partnership for Sustainable Development (SATREPS), which is a research program structured as a collaboration of the Japan Science and Technology Agency (JST) and the Japan International Cooperation Agency (JICA).

7. References

[1] Lynd L R, Weimer P J, Zyl W H V and Pretorius I S 2002 *Microbiol Mol Biol Rev* **66** 5577
[2] Chellapandi P and Jani H M 2008 *Braz J Microbiol* **39** 122-127
[3] Crawford D L 1986 *FEMS Symposium* pp 34-728
[4] Vinogradova S P and Kushnir S N 2003 *Appl Biochem Microbiol* **39** 573-575
[5] Jang H D and Chen K S 2003 *World J Microbiol Biotechnol* **19** 263–268
[6] Arunachalam R, Wesely E G, George J and Annadurai G 2010 *Curr Res Bacteriol* **3** 15-26
[7] Ibrahim S A, El-Diwany A I 2007 *J Basic Appl. Sci* **1** 473-478
[8] Hao X C, Yu X B and Yan Z L 2006 *Food Technol Biotechnol* **44 (1)**: 89–94
[9] El-Sersy N A, Ibrahim H A H, Abou-Elela, G M 2010 *Curr Res Bacteriol* **3** 1-14
[10] Van Zyl W H 1985 *Biotechnol Bioeng* **27(9)** 1367-1373
[11] Bezerra M A, Santelli R E, Oliveira E P, Villar L S, Escaleira L A 2008 *Talanta* **76** 965-977
[12] Zambare V 2011 *Emir J Food Agric* **23** 37–47
[13] Ariffin H, Abdullah N, Umi Kalsom MS, Shirai Y, and Hassan M A 2006 *Int J Eng Technol* **3** 47-53
[14] Gupta M, Sharma M, Singh S, Gupta P, and Bajaj BK 2015 *Energy Technol* **3** 216-224
[15] Miller G L 1959 *Anal Chem* **31** 426–428
[16] Dong J, Mandenius C F, Lűbberstedt M, Urbaniak T, Nűssler A K N, Knobeloch D, Gerlach J C, Zeilinger K 2008 *Cytotechnology* **57** 251–261
[17] Lai L W, Sarenah S, Yahya M, Nor N M, Sulong M R 2016 *MJAS* **20(1)** 21 – 30
[18] Mawadza C, Hatti-Kaul R, Zvauya R and Mattiasson B 2000 *J Biotechnol* **83** 177-187
[19] Lee S J, Kim S R, Yoon H J, Kim I and Lee K S 2004 *Biochem. Physiol B* **139** 107-116
[20] Inoue T, Moriya S, Ohkuma M and Kudo T 2005 *Gene* **349** 67-75
[21] Thongekkaew J, Ikeda H, Masaki K and Iefuji H 2008 *Protein Expr Purif* **60** 140-146
[22] Wang C Y, Hsieh Y R, Ng C C, Chan H, Lin H T, Tzeng W S, Shyu Y T 2009 *Enzyme Microb Techno* **l44** 373–379
[23] Harshvardhan K, Mishra A and Jha B 2013 *J Mol Catal B Enzymatic* **9**: 51-56
[24] Annamalai N, Rajeswari M V, Elayaraja S and Balasubramanian T 2013 *Carbohydr Polym* **94** 409-415
[25] Youseff G A and Barekaa M M 2009 *Biotechnology* **8(2)** 212-219.
[26] Black J G, and Lewis LM 2005 *Microbiology: principles and explorations* (7th ed.) (New Jersey : John Wiley & Sons) p 156
[27] Lam K S 2006 *Curr Opinion Microbiol* **9 (3)**: 245–251
[28] Dhanasekaran D, Selvamani S, Panneerselvam A, Thajuddin N 2009 *Afr J Biotechnol* **8(17)** 4159-4162
[29] Hong K, Gao A,. Xie Q 2009 *Marine Drugs* **7** 24-44
[30] Jennerjahn T C.and Ittekkot V 2002 *Naturwissenschaten* **89(1)** 23–30
[31] Lee L H, Zainal N, Azman A S, Eng S K, Goh B H, Yin W F, AbMutalib N S, and Chan K G 2014 *The Scientific World Journal* Article ID 698178 14
[32] Chang M Y, Tsai G. and Houng Y Y 2006 *Enzy Microbial Technol* **38** 407-414
[33] Macedo E P, Cerqueira C L O, Souza DAJ, Bispo A S R, Coelho R R R and Nascimento R P 2013 *Braz J Chem Eng* **30(4)** 729 – 735
[34] Li X and Gao P 1997 *J Appl Microbiol* **82** 73-80
[35] Jaradat Z, Dawagreh A, Ababneh Q, Saadoun I 2008 *Jordan J Biol Sci* **(4)** 141-146
[36] Maryana R, Ma'rifatun D, Wheni A, Satriyo K W, Rizal W A 2014 *Procedia* **47** 250 – 254
[37] Supranto S, Tawfiequrrahman A, and Yunanto D 2014 *Somche & Rsce* 1 (013)

[38] Badan Pusat Statistik Indonesia (June 12, 2013) http://www.bps.go.id/tab_sub/view.php?kat=3&tabel=1&daftar=1&id_subyek=54&notab
[39] Kumar D V R R, Chakri S, Sowjanya M, Yugandhar N M and Reddy D S R 2010 *IJBSE* **1(1)** 119-127
[40] Singh K, Richa K, Bose H, Karthik L, Kumar G, Venkata K, Rao B 2014 *3 Biotech* **4** 591–598

Potential Isolates Characterization of Thermophilic Bacteria from Hot Springs and Waste Agricultural Production in Subang District Area

WAgustina[1], T Rahman[1], L Ratnawati[1], Sriharti[1], T Salim[1] and I Nursyahbani[2]

[1]Center for Appropriate Technology Development – Indonesian Institute of Sciences Jl. K.S. Tubun No. 5, Subang, West Java - Indonesia.
[2]State Islamic University (UIN) of Sunan Gunung Djati, Jalan A.H. Nasution No. 105, Bandung, West Java – Indonesia

E-mail: wanwa03@gmail.com

Abstract. The characterization of local thermophilic bacteria originated from hot spring and agricultural waste production have been done. The purposes of this research were to study the morphology colony of isolates, cellulolytic activity, potency to produce ethanol, and other characteristics of thermophilic bacteria. The method used was involved morphology isolates characterization, cellulolytic activity testing, trials ethanol production in an aerobic and anaerobic condition, and characterization of fermented media covered pH, the decreasing of sugar, and titrated acid. The isolates used in this research were coded by B1, B2, B3, C2, C5, N1 and J2. The result of this research was known the isolates characteristic covered form, size, color and margin. Cellulolytic activity test showed that isolate of B2, C2, and C5 was positive when tested by congo red dye and only B2 isolate was positive when tested by I-KI 3% dye. Aerobic and anaerobic fermentation did not show the capabilities of isolates to produce ethanol. The characteristic of anaerobic fermented media shown the decrease of pH number from 7.16 to about 4.43 to 4.49, the decrease of reduced sugar to about 1.02 to 24.43%, the remaining glucose, and xylose in the medium respectively were about 3.45 to 4.07% and 3.40 to 3.99%. The titrated acid as lactic acid is about 0.16 to 0.18%.

Keywords: Thermophilic bacteria, bioethanol, fermentation, cellulolytic.

1. Introduction

Indonesia has a lot of potential thermophilic microorganisms including bacteria. Thermophilic bacteria have an important potential to be developed in various fields of life, especially to support the industry. Thermophilic bacteria can be classified into two groups: obligative and facultative thermophilic bacteria. Obligative thermophilic bacteria can survive at temperatures above 50 °C, while the facultative thermophilic bacteria can grow at a temperature range between 50 – 66 °C or at temperatures lower than 38 °C. Some obligate thermophilic bacteria can grow at a temperature of 77 °C. The habitat of thermophilic bacteria can be found in a high-temperature environment such as hot springs, ground that is always exposed to sunlight and soil fermented compost [1].

The study of the source and the potential of thermophilic bacteria is getting a lot done. Thermophilic Microbes can be used as a source of enzymes and better than animal and plant [2]. Thermophilic bacteria's have the ability to survive in high temperatures caused the bacteria are able to produce enzymes that more stable at high temperatures (thermostable). Nowadays the use of thermostable enzymes become very important as in biotechnology and industrial applications [3].

Increasing the reaction temperature will generally be able to increase the speed of a reaction. Increased reaction temperatures above 10 °C minimum will shift the reaction rate by two times. Additionally, the rising temperatures will inhibit other microbes to live so that the contamination of other microbes can be reduced [4]. Sutiamiharja (2008) also mentions the use of thermostable enzymes that can save the cost of production because this enzyme has a longer shelf life and activity remains optimal at high temperatures [5].

Thermophilic bacteria can be taken or isolated from various potential sources such as hot springs and wastes of industrial or production processes. Microorganisms can be at the hot springs caused by environmental conditions that are supported by biotic and abiotic factors. Biotic conditions in the hot springs can be used as an energy source while abiotic factors also affect the existence of thermophilic bacteria such as alkaline pH caused by high mineral content, causing a high diversity of microorganisms [6]. In the production of waste both liquid and solid wastes generated in addition to indeed have a sufficiently high temperature, normally there will be also a fermentation process that can increase the temperature. The high temperature of the waste will make the microorganisms that grow well adapted.

The purpose of this study was to Determine colony morphology of isolates, cellulolytic activity, the potential to produce ethanol, and other characteristics of thermophilic bacteria originating from the hot springs in the tourist area of Ciater and Batu Kapur and from waste processing production of pineapple and guava in the Subang District, West Java, Indonesia.

2. Experimental method

Materials used in this study were isolated of local thermophilic microorganism obtained from hot springs of Batukapur Subang (B) is (B1, B2, B3), hot springs of Ciater Subang (C) is (C2, C5), waste of guava production (J), i.e. (J2), and waste processing production of pineapple (N), i.e. (N1). Others materials used were tryptone broth, NaCl, yeast extract, KH_2PO_4, $MgSO_4.7H_2O$, glucose, xylose, peptone broth, distilled water, filter paper, and dinitrosalicylic acid. The tools used were the analytical balance, vertical laminar flow, autoclaves, incubators, centrifuges and HPLC Waters e 2695.

2.1. *Morphology Characterization colony*
Isolate scratched in agar media and incubated at 50 °C, and colony growth was seen in the form of shapes, sizes, colors, and edges.

2.2. *Cellulolytic activity*
Bacterial isolates were inoculated on a medium containing substrate of carboxymethyl cellulose (CMC), which had previously been inoculated on media LB. After inoculated on CMC media then incubated for 48 hours at a temperature of 50 °C. Colonies that grew drip with congo red solution or I-KI 3% for ten minutes and rinsed using 1% NaCl. The positive results are shown by the formation of a clear zone in the area around the colony.

2.3. *Fermentation Media*
The medium used is composed of three kinds of media Luria Broth, Seed culture, and fermentation media. In Luria Broth media as much as 0.75 g tryptone broth, 0.375 g NaCl, and 0.375 g yeast extract, dissolved in 75 ml of distilled water, then stir until homogeneous and then poured into a test tube 7 respectively of 10 ml, then sterilized.

Seed culture media made of 0.1875 g yeast extract, 0.075 g KH_2PO_4, 0.0225 g $MgSO_4.7H_2O$, 3.75 g glucose, 3.75 g xylose, and 0.1875 peptone broth dissolved in 75 ml of distilled water, then stir until homogeneous and then poured into a test tube (10 ml), then sterilized.

Fermentation media bolted from 1.875 g yeast extract, 0.75 g KH_2PO_4, 0.225 g $MgSO_4.7H_2O$, 37.5 g of glucose, 37.5 g xylose, and 1,875 peptone broth dissolved in 750 ml of distilled water, stir until homogeneous and then poured 100 ml of the media into 250 ml Erlenmeyer. The sterilization is carried out using an autoclave at 121°C for 15 minutes.

2.4. *Making a thermophilic starter*
Isolate thermophilic bacteria inoculated in Luria broth media, and incubated at 50 °C for 24 hours, after it was taken 4% (v/v) and then centrifuged for 10 minutes at a speed of 3000 rpm, The bacterial pellets were taken and inoculated on the Seed culture medium and incubated at 50 °C. The incubation time is determined by the growth curve of each isolate namely B1, B2, B3, N1 for 24 hours, J2 for 36 hours, C5 for 42 hours and C2 for 54 hours.

2.5. *Fermentation process*
The fermentation process carried out in two ways: aerobic and anaerobic. A total of 10% (v/v) thermophilic starter media were taken and centrifuged for 10 min at 3000 rpm, and the pellet was taken and inoculated into the

IOP Conf. Series: Earth and Environmental Science **251** (2019) 012006 doi:10.1088/1755-1315/251/1/012006

fermentation medium, then incubated at 50 °C. Alcoholic fermentation trials carried out in aerobic conditions for 5 days. In anaerobic fermentation, trials carried out for 7 days. In the aerobic test, sample analysis is stopped until HPLC assay to determine the ability of the alcohol production. While on samples of anaerobic fermentation then carried out a further test which includes parameters of pH, a concentration of reducing sugar, titrated acid total assumed as lactic acid, ethanol concentration, and residues of glucose and xylose. The concentration of ethanol, glucose, and xylose residues were analyzed using HPLC. Total acid for reducing sugar analysis was carried out by the method of Badan Standardisasi Nasional (1992) [7].

3. Result and discussion

3.1. Morphology Characterization colony

Isolates used in this study was isolated that has the ability to grow on the minimum alcohol content of up to 6%. The characteristics of each isolate can be seen in Table 1.

Table 1. Characteristics of isolates from various sources

Isolates	Morphology colonies			
	Shape	Size	Color	Edges
B_1	Round	Large	Beige	Filamentous
B_2	Flower	Medium	Beige	-
B_3	Round	Large	Beige	Undulate
C_2	Round	Medium	Beige	Filamentous
C_5	Irregular	Medium	Beige	Filamentous
N_1	Stem cells	Large	Beige	Stringy
J_2	Round	-	White	-

3.2. Cellulolytic Activity

Testing the ability of cellulolytic activity was carried out to determine the ability of bacterial isolates to degrade cellulose. So that it can be used in the process of lignocellulose hydrolysis. Qualitative test results on cellulolytic activity using two different types of dyes namely congo red dye, and I-KI 3% dye showed that the coloring by red congo contained 3 positive isolates and 3% I-KI staining with one positive isolate.

On staining by congo red positive isolates were isolate of C2, C5, and B2 while the staining by I-KI 3% was isolate of B2. Pictures of the test results can be seen in Figure 1.

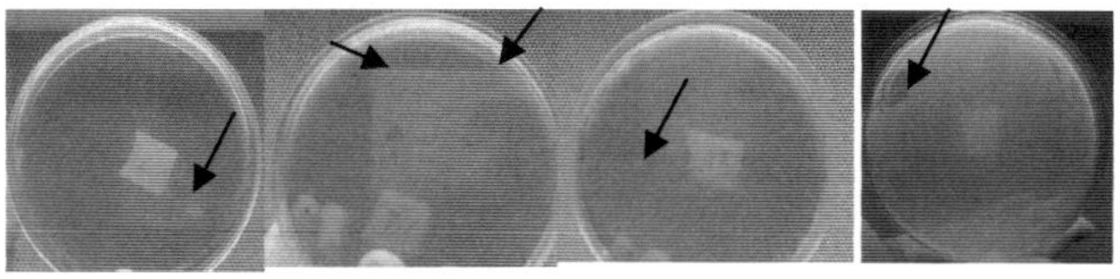

Isolates C_2 Isolates C_5 Isolates B_2 Isolates B_2
by congo red dye by I-KI 3% dye

Figure 1. Isolate thermophilic bacteria capable of producing a clear zone

The formation of a clear zone as shown in figure 1 show that isolates bacteria sourced from hot spring have cellulolytic activity. Establishment of a clear zone around the colony of bacteria means that the isolates of bacteria actively degraded the cellulose derived from carboxymethyl cellulose (CMC) [8].

The existence of cellulolytic activity indicates that the bacteria are able to produce the cellulase enzyme. Previous studies showed that some isolates of thermophilic bacteria capable of producing cellulase enzymes. Isolates of *Bacillus pumilus* isolated from hot springs in the area of Sikumbang-Central Java can produce cellulase enzymes with the optimum temperature of 45 °C, pH 5.0 with an incubation time of 3 days with the substrate concentration of 1.0% (w/v) [9]. Bacterial isolate (S2A) taken from hot springs Kili Ketek Solok which has 97% similarity with *Anoxybacillus flavithermus* AE3 strain capable of producing extracellular cellulase enzymes [10]. Most cellulase enzymes are produced at high temperatures and have optimum activity at high temperatures so that the cellulose enzyme can be produced by thermophilic bacteria [11].

3.3. Bioethanol Ability Test

Bioethanol ability test of bacterial isolates carried out through aerobic and anaerobic fermentation. The fermented medium then examined by using HPLC and the results show the whole isolate was not able to produce bioethanol. The data can be seen in Table 2.

Table 2. Ability Test Bioethanol In Aerobic and Anaerobic

Isolate Thermophilic Bacteria	Bioethanol Production capabilities	
	Aerobic fermentation	Anaerobic fermentation
B_1	-	-
B_2	-	-
B_3	-	-
C_2	-	-
C_5	-	-
N_1	-	-
J_2	-	-

In a previous study conducted, some isolates from Ciater hot springs were tested resistance to ethanol, and the results were all resistant or able to survive on ethanol levels of up to 6 % [12]. However, after the fermentation either aerobic or anaerobic isolates cannot produce ethanol (Table 2). In contrast to research conducted by Martosuyono and Misgiyarta (2000) which states that the thermophilic bacteria (*Bacillus caldoxyloliticus and Geobacillus thermoleovorans*) in aerobic fermentation conditions, can produce ethanol with the amount of ethanol obtained is very small with a concentration of 0.2 to 1.9 g/l, while anaerobic testing shows that bacteria are able to produce higher ethanol concentrations ranging between 2.5 - 3.7 g/l. The highest yield is indicated by the bacterium *Geobacillus thermoleovorans* [13].

The research was supported by a statement of Riyanti (2011) which states thermophilic bacteria can produce ethanol because they have ethanol-producing genes are Alcohol dehydrogenase (ADH) as found in bacteria *Geobacillus thermoglucosidasius, Geobacillus stearothermophilus, Bacillus Sulfolobus and Thermoanaerobacter ethanolic* [14]. Some of the isolates used in this study are supposed there may be capable of producing ethanol when performed on appropriate conditions. This is due to the alcohol fermentation process can be influenced by many factors. The factors that influence the success or failure of ethanol fermentation include inoculum physiological conditions include pH, temperature, growth factor, alcohol. The quality of growth substrate consisting of a carbon source, a source of nitrogen, oxygen, and CO_2 also affect the results obtained as well as the efficiency of fermentation. Inoculum physiological conditions depending on environmental factors, the presence of microbial contaminants will greatly affect the metabolites products produced and inhibited the fermentation process [15].

The primary metabolite of ethanol production is influenced by the growth of microbial cells used. Nutrients are used to the life and growth of cells, including growth factors such as vitamins and minerals. Nutrients needed to form and develop energy cell components. Components containing organic carbon source is used as an energy source for microbes and uses mostly organic components containing protein as a nitrogen source or organic nitrogen source [16].

3.4. Change of pH

The degree of acidity (pH) is one of several important factors that can affect the fermentation process, while the pH test results in fermentation media are as follows :

Table 3. Media Isolates pH levels before and after Fermentation

Media Isolates	pH Before Fermentation	pH After Fermentation
B_1	7.16	4.49
B_2	7.16	4.44
B_3	7.16	4.45
C_2	7.16	4.43
C_5	7.16	4.43
N_1	7.16	4.48
J_2	7.16	4.49

Table 3 showed a significant difference between the pH of media before and after fermentation. From the data, it is known that the pH of media before fermentation was 7.16 and after fermentation pH ranged from 4.43 to 4.49. A decrease in pH caused by the fermentation which can produce acidic metabolites. Some acidic metabolites such as ethanol, acetic acid, levulinic acid, and formic acid by the conditions of the media [17].

In the microbial fermentation required media with optimal pH conditions, the pH value of the media is very influential on the microbial growth, good pH values maximum of about pH 6.5-7.5, while a pH below 5.0 or above 8.5 bacteria will not grow well except acetic acid bacteria [16]. The degree of acidity (pH) optimum for the fermentation process ranges from 4-5, at a pH below 3 fermentation process will decelerate [17].

3.5. Reducing sugar content

Measurement reducing sugar aim to find out how much sugar is in use when the fermentation. These sugars can serve as a source of energy that will be converted into a variety of compounds, such as ethanol, acetic acid, and lactic acid. Sugar is in use on media that is the sugar glucose and xylose sugar. The results in the can are as follows :

Table 4. Total Sugar Reduction

Sample	Sugar concentration (%)		
	DNS	HPLC	
		Glucose	Xylose
Before Fermentation	9.5792	± 5.00	± 5.00
B_1	7.2393	3.7248	3.6660
B_2	9.4817	3.6705	3.6759
B_3	8.4580	4.0767	3.9995
C_2	8.4824	3.4523	3.3975
C_5	8.1411	3.5335	3.5926
N_1	7.6780	3.5341	3.4271
J_2	7.7268	3.5980	3.5690

In Table 4 it can be seen that the initial sugar concentration of the media is tested using the dinitrosalicylic acid (DNS) method that is as much as 9.5793%, which is almost the same as the composition of the sugar added into the media of 5% glucose and 5% xylose. There is a decrease in sugar levels after media fermented, despite the relative decline slightly but this proves that the media added isolates had the fermentation process in which sugars that exist in the media converted into ATP and generate other compounds that make sugar levels lower.

While the sugar levels are measured using HPLC method both glucose and xylose in each medium is 3.45 to 4.07 % and 3.40 to 3.99%, this is possible because the DNS method of measurement carried out on the overall total amount of sugar, whereas the HPLC method measuring total sugars glucose and xylose separately, so the total is divided into two. The decrease in sugar xylose and glucose as xylose and glucose is a monosaccharide included in the group, reducing sugar, wherein the fermentation of xylose and glucose can be converted into hydrogen [18].

Reducing sugars during fermentation will increase after fermentation lasts for three days, after which the reducing sugar will decrease. Fermentation time will affect the content of reducing sugar which on the first day will be increased as much as 7.9% to 16% on the third day [19].

As a pentose sugar xylose being very abundant as hydrolysis products which when fermented lignocellulosic materials can produce bioethanol. While still showing glucose fermentation product of lactic acid as the main products, followed by acetate and ethanol [13].

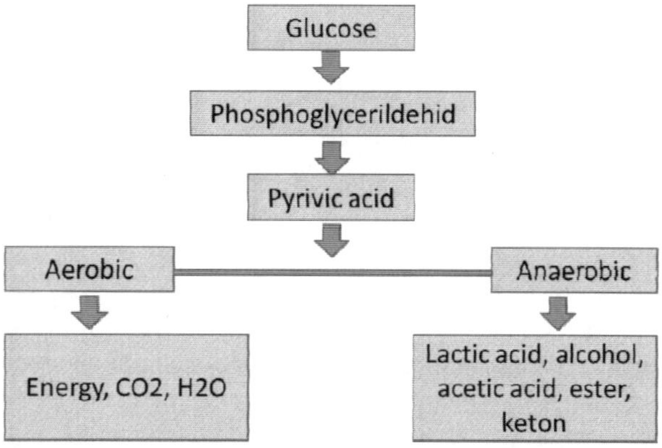

Figure 2. Reaction formation of glucose into other compounds [19].

In contrast to the sample being tested even though it contains xylose and glucose but not producing ethanol, this is possible because the isolate cannot convert the sugars into ethanol but other compounds.

3.6. Test of Titrated Acid (Lactic Acid)

Lactic acid is one of the organic acids that are important in the industry, especially the food industry. In testing the Total titratable acidity can result in the following :

Table 5. Total titratable acidity for lactic acid

Sample	Total acid (%)
Before fermentation	0.1429
B_1	0.1606
B_2	0.1606
B_3	0.1874
C_2	0.1787
C_5	0.1693
J_2	0.1696
N_1	0.1606

Based on Table 5 it can be seen that at the time of fermentation, isolates tested can produce acids that are assumed as lactic acid, the highest total acid produced by B3 isolate as many as 0.1874% and the lowest produced by isolates B1, B2 and N1 as much as 0.16065%. Although the results can be relatively small, this proves that the tested isolates thermophilic bacteria can produce acid and assumed as lactic acid.

The results showed that thermophilic bacteria could produce lactic acid is supported by research conducted by Nurwati and Refdinal [20] which states that the test of glucose fermentation in bacteria thermophilic *Caldicellusiruptor kristjanssonii* produce end products such as acetic acid, lactic acid, ethanol, and formic acid. This was confirmed by research conducted by Rahmawati [21] which states that carbohydrates and proteins in whey fermented by thermophilic bacteria produce lactic acid. The bacteria will use the nutrients contained in the ingredients to turn it into acid.

The acid is fermented to form lactic acid bacteria to convert glucose into lactic acid. According to Hernawati [22] that the lactic acid fermentation of glucose is converted to pyruvic acid and pyruvic acid molecules result glycolysis accept electrons and hydrogen from NADH. Transfer of electrons and hydrogen to produce NAD back. At the same time, pyruvic acid is converted to lactic acid in large quantities.

There are several types of bacteria called lactic acid bacteria are able to convert glucose into lactic acid. *Streptococcus lactis* is one example of the bacteria capable of causing sour milk and break down glucose to lactic acid [22]. The mechanism that is one molecule of glucose is converted to two molecules of pyruvic acid, and pyruvic acid molecules result from glycolysis accept electrons and hydrogen from NADH. Transfer of electrons and hydrogen to produce NAD back and pyruvic acid is converted to lactic acid [22].

4. Conclusion

This research can be concluded that: Isolate tested had an average colony morphology round, medium-sized to large, cream and white colours and had stringly edge line of undulated, filamentous and entire. From all isolates tested, three isolates have cellulolytic activity when tested using a solution of congo red dye, i.e. C2, C5, and B2and one isolate when tested using a solution of I-KI 3 %, i.e. B2. The characteristic of anaerobic fermented media shown the decrease of pH number from 7.16 to about 4.43 to 4.49, the decrease of reduced sugar to about 1.02 to 24.43%, the remaining glucose, and xylose in the medium respectively were about 3.45 to 4.07% and 3.40 to 3.99%. The titrated total acidity as lactic acid about 0,16 to 0,18%.

5. Acknowledgment

This work was supported by Deputy of Science Engineering-Indonesia Institute of Sciences and Head of Center for Appropriate Technology Development are gratefully acknowledged for their facilities supports. The authors would also like to thanks to Mr. Hendris, Prof. Yani Sudiyani, Dedi Sumaryadi and Dini Suciwaty for discussion, experimental assistance and useful suggestion during the course of this project.

6. References

[1] Ginting Y 2009 *Isolasi Bakteri dan Uji Aktivitas Amilase Termofil Kasar dari Sumber Air Panas Semangat Gunung Sumatera Utara* (Medan : Thesis USU Medan)

[2] Yuanita D N S P and Retno P 2014 *Screening* Proteolytic Thermophilic Bacteria from Hot Springs Singgahan Tuban *UNESA Journal of Chemistry* **3** 3

[3] Dewi I M 2008 *Isolasi Bakteri dan Uji Aktivitas Kitinase Termofilik Kasar dari Sumber Air Panas Tinggi Raja Simalungun Sumatera Utara* (Medan : Thesis USU Medan)

[4] Pakpahan R 2009 *Isolasi Bakteri Termofilik dan Uji Aktivias Protese dari Sumber Air Panas Sipoholon Tapanuli Sumatera Utara* (Medan : Thesis USU Medan)

[5] Sutimiharja N 2008 *Isolasi Bakteri dan Uji Aktivitas Amilase Termofil Kasar dari Sumber Air Panas Gurukinayan Karo Sumatera Utara* (Medan : Thesis USU Medan)

[6] Widia F, Agustien A and Febria F A 2015 Karakterisasi Bakteri Termofilik Penghasil Enzim Protease Netral *Jurnal Biologi Universitas Andalas* **4** (1) 9-14

[7] Badan Standardisasi Nasional 1992 *Standar Nasional Indonesia 01-2891-1992 Cara Uji Makanan dan Minuman* (Jakarta: Badan Standardisasi Nasional)

[8] Kanti A 2005 Actinomycetes Selulolitik Dari Tanah Hutan Taman Nasional Bukit Duabelas Jambi. *Biodiversitas* **6** (2) 85-89

[9] Wahyuni V 2001 Aktivitas Selulase *Bacillus Pumilus* Galur 55 yang Diisolasi dari Sumber Air Panas (Bogor : Skripsi IPB Bogor)

[10] Octarya Z 2010 *Skrining dan Identifikasi Bakteri Termofilik Penghasil Selulase dan Amilase dari Sumber Air Panas Bukit Kili Ketek Solok dengan Analisa Gen 16S rRNA Serta Karakterisasi Molekuler Enzim Ekstraselulernya* (Padang : Universitas Andalas Thesis)

[11] Alam M S, Sarjono P R and Aminin A L 2013 Isolasi Dan Karakerisasi Selulase Dari Bakteri Selulotik Termofilik Kompos Pertanian Desa Bayat, Klaten, Jawa Tengah *Jurnal Sains Dan Matematika* **21** (2) 48-53

[12] Ratnawati L, Salim T, Agustina W and Sriharti 2015 Isolation of Thermophilic Microbes Resistant to Ethanol from Ciater Hot Springs, Subang-West Java *Procedia Chemistry* of International Symposium on Applied Chemistry **16** 548-554

[13] Martosuyono P and Misgiyarta 2000 Isolasi Bakteri Termofilik Penghasil Etanol Dari Kompos *Prosiding Seminar Nasional Teknologi Inovatif Pascapanen Untuk Pengembangan Industri Berbasis Pertanian* Balai Besar Penelitian dan Pengembangan Pasca Panen Pertanian

[14] Riyanti E I 2011 Beberapa Gen Pada Bakteri Yang Bertanggung Jawab Terhadap Produksi Bioetanol. *Jurnal Litbang Pertanian* **30** (2) 41-47

[15] Azizah N, Al-Baarri A N and Mulyani S 2012 Pengaruh Lama Fermentasi Terhadap Kadar Alkohol, pH, dan Produksi Gas Pada Proses Fermentasi Bioetanol Dari Whey Dengan Substitusi Kulit Nanas *Jurnal Aplikasi Teknologi Pangan* **1** (2) 72-77

[16] Waluyo L 2004 *Mikrobiologi Umum* (Malang : Malang Press)

[17] Samsuri M, Gozan M, Mardias R, Baiquni M, Hermansyah H, Wijanarko A, Prasetya B and Nasikin M 2007 Pemanfaatan Sellulosa Bagas Untuk Produksi Ethanol Melalui Sakarifikasi Dan Fermentasi Serentak Dengan Enzim Xylanase *Jurnal Makara Teknologi* **11** 1 17-24

[18] Rahmatullah A, Purwanti A and Widjaja A 2013 Pengaruh Variasi pH Dalam Proses Fermentasi Bagas Tebu Menjadi Gas Hidrogen Menggunakan *Enterobacter aerogenes Jurnal Teknik Pomits* **1** (2) 1-5

[19] Suliantri and Winiarti 1991 *Teknologi Fermentasi Biji-bijian dan Umbi-umbian* (Bogor : Departemen P dan K Direktorat Jendral Pendidikan Tinggi Pusat Universitas Pangan dan Gizi IPB)

[20] Nurwati L and Nawfa R 2015 Identifikasi Spesies Isolat Bakteri Galur D dengan Metode Analisa Sekuen Fragmen Gen 16S rDNA *Jurnal Sains dan Seni ITS* **4** (2) 2337-3520

[21] Rahmawati A 2010 Total Bakteri Asam Laktat (Bal), Kadar Laktosa Dan Keasaman Whey yang Difermentasi dengan *Bifidobacterium bifidum* Pada Lama Inkubasi Yang Berbeda (Semarang : Fakultas Perternakan Universitas Diponegoro Skripsi)

[22] Pelczar M J and Chan E C S 1986 *Dasar-Dasar Mikrobiologi* (Jakarta : UI-Press)

IOP Conf. Series: Earth and Environmental Science **251** (2019) 012007 doi:10.1088/1755-1315/251/1/012007

Performance Test And Economic Evaluation Of Pyrolysis Furnace Small Scale-Worth The Business For Producing Liquid Smoke

A Indriati, D D Hidayat, and D A Darmajana[1]

[1] Development Centre of Appropriate Technology Indonesian Institute of Sciences, Jl. KS. Tubun No. 5 Subang West Java, Indonesia, 41213

Email: ashriindriati@lipi.go.id

Abstract. Liquid smoke is water which made of smoke as pyrolysis processing products. It can use in food products as a flavoring, texture, distinctive taste in food products such as meat, fish, and cheese. To produce liquid smoke safely and provide economic value, the effective and efficient a pyrolysis furnace that is indispensable. The study aimed to get the proper scale of the pyrolysis furnace and food grade liquid smoke which feasible for business. The research activities consisted of testing of the pyrolysis furnace, analyzing the content of liquid smoke, determining the effectivity of liquid smoke against various types of spoilage and pathogenic microorganisms and investigating market feasibility. Results of this study showed that the yield of pyrolysis was 42.26%, first distillation was 35.18% and second distillation was 34.33% of the raw materials used. The resulting liquid smoke indicated that no content of Polycyclic Aromatic Hydrocarbon (PAH). The liquid smoke concentration of 20%, 30%, and 40% was able to inhibit the bacteria E. coli, Salmonella, and Staphylococcus aureus; this meant that the liquid smoke was safe for use in food products. The feasibility analysis indicated that the liquid smoke business using the pyrolysis furnace was able to provide benefits and was feasible to run.

1. Introduction

Liquid smoke is a mixture of wood smoke dispersion solution of water produced by the liquid smoke condensation resulting from pyrolysis [1]. Liquid smokes were used extensively in food systems to provide flavor characteristics that were similar to natural smoked food products [2]. The liquid smoke may use to preserve quality and ensure the safety of foods [3,4]. There were many studies done due to liquid smoke. Results of the study showed that liquid smoke had many benefits. As a substitution substance, liquid smoke can replace phosphate in chicken nuggets processing, and replace borax or formalin in the production of beef meatballs [5]. As a pesticide liquid smoke inhibits the development of anthracnose disease, fusarium wilt [6], *Pseudomonas solanaccearum* and *Sclerotium rolfsii* fungi in vitro [7]. In food processing liquid smoke may use for the binder, extends the shelf life of fish meatballs [8], preservative of wet noodles [9], antioxidants and antimicrobials, flavoring and distinctive taste of other food products [10] such as meat, fish, and cheese [11]. Due to the multifunction of liquid smoke, it has a business opportunity to be developed.

The raw materials used to produce a liquid smoke is coconut shell. In Indonesia Coconut is one of Indonesia's plantation potential. Based on the crop estate statistics of Indonesia 2015-2017, the area of coconut plantation was 32,755 Ha with the production of 2,871,280 tons [12]. One of the waste from coconut is coconut shell. To produce an economical product, one of the methods that used is through pyrolysis of raw materials containing hemicellulose, cellulose, and lignin that would be condensed to become liquid smoke. Coconut shell is more suitable for pyrolysis process as it provides lower ash content, and high volatile matter content [13].

The community will undoubtedly require the existence of appropriate technology to process the coconut shell into liquid smoke for both food and non-food purposes. The technology needed is a technology that is safe, easy to operate and can provide benefits. The study aimed to determine the smallest scale of business-worthy pyrolysis furnace and distillation apparatus and to obtain a food grade liquid smoke.

2. Experimental method

2.1. Sites
The study conducted in February and July at the Development Center for Appropriate Technology-Indonesian Institute of Sciences, Subang district-west java province.

2.2. Equipment and materials
Equipment used in liquid smoke production were scales, pyrolysis furnaces, blowers, condensing drums, gas tanks, electrical pumps, plastic dippers, distillation apparatus, graduated cylinders, beaker glasses, and glass jars.

Materials used for liquid smoke production were Kerosine, liquid petroleum gas, and coconut shell. Coconut shells used as raw materials had a moisture content of 7.47% and with an ash content of 0.88%. Kerosene used at the time of initial ignition, and liquid petroleum gas used during the distillation process.

2.3. Research methods
The experiment consisted of testing the pyrolysis furnace and refinement equipment manufactured, and analysis the liquid smoke quality. Parameters observed regarding pyrolysis and distillation process included of equipment specifications, capacity, and temperature of pyrolysis and refinement, as that for the resulting liquid smoke consisted of volume, pH, and chemical content. Determination of liquid smoke chemical content of liquid smoke conducted by Gas Chromatography-Mass Spectrometry (GCMS) analysis and grading of that referred to researched done by Fauzan and M. Ikhhwanus [14].

2.4. The process of Liquid Smoke production

2.4.1. Pyrolysis
Etymologically the word pyrolysis came from Greek, i.e., pyro and lysis. Pyro meaning fire, and lysis meaning to loosen. Pyrolysis, therefore, is the process of thermal decomposition in producing the gas, a liquid, and the char. Pyrolysis usually understood to be thermal decomposition occurred in the oxygen-free atmosphere, but oxidative pyrolysis is nearly always an inherent part of combustion processes [15].

The steps of liquid smoke producing was as follows. A quantity of 26.5 kg of coconut shell imparted to the pyrolysis furnace then initiate the ignition using kerosene. The blower used to drain the hot smoke through the pipe to the condensing drum. The flowing water in the condensing drum will absorb hot smoke in the pipeline. The resulting liquid smoke will come out through the sheltered smoke exit place. The liquid smoke produced from the pyrolysis process classified as grade 3.

2.4.2. Distillation
The first stage distillation conducted to obtain a liquid smoke grade 2. The way to get the liquid smoke grade 2 was to impart the liquid smoke grade 3 into distillation apparatus. The heating process during distillation used LPG fuel.

2.4.3. Redistillation
The second stage distillation carried out to obtain Liquid smoke grade 1. The way to get the liquid smoke grade 1 was to place the liquid smoke grade 2 into distillation apparatus.

Results Analysis

Gas Chromatography-Mass Spectrometry (GCMS) performed To obtain the chemical content of the Liquid smoke, while the financial feasibility analyzed by calculating variable components as follows.

3. Result and Discussion

3.1. Liquid Smoke Tools
The Tool used for the manufacture of liquid smoke comprised of pyrolysis and distillation equipment. The schematic of the liquid smoke making apparatus presented in Figures 1 and 2.

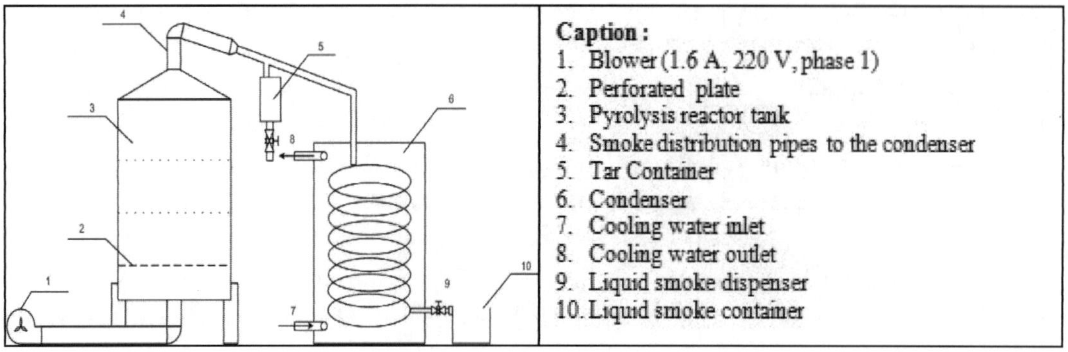

Figures 1. Pyrolysis Tool

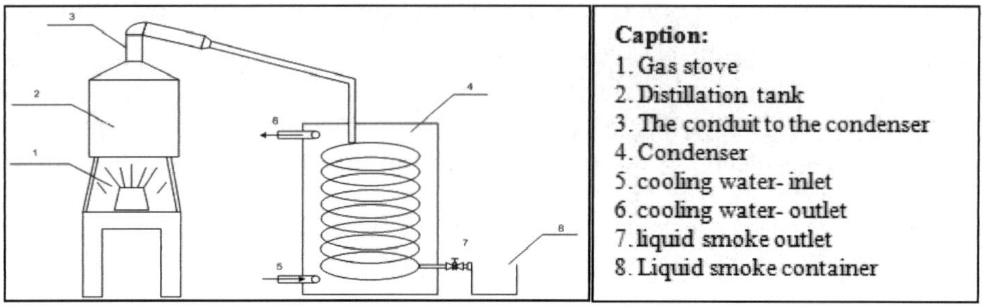

Figures 2. Distillation Tool

In this experiment used a pyrolisis tool which had a capacity of 26.5 kg. The raw material used coconut shell. Results of this experiment were 11.2 liters of liquid smoke that had a pH of 3 and the time required was 2 hours. Assumed that a density of liquid smoke was 1.009 gr/cm^3 [16], the yield of the liquid smoke was 42.26%. The yield was the ratio of liquid smoked amount pyrolyzed from the coconut shell. Yield measurement conducted on the condition of the reactor outside walls temperature of 201 °C and environmental temperature of 27 °C. The temperature measured using an infrared thermometer.

The pyrolysis process produced liquid smoke grade 3. This third-grade liquid smoke then further distilled for 8 hours and produced second grade. The yield was 9.32 lt, i.e., 35.18% with a pH of 2.425. Measurement of result done at a condition of highest temperature outer wall of the distillation apparatus of 81.7 °C, fire temperature of 257 °C and environment temperature of 24 °C.

The last processing stage of liquid smoke was redistillation. The tools used together with previously distillation equipment. Results of redistillation were liquid smoke grade 1. The yield was 9,10 lt, i.e., 34.33 % with a pH of 2.425; the processing time was 4 hours. Measurement of result was conducted at a condition of highest temperature outer wall of the distillation apparatus of 84.4 °C, fire temperature of 258 °C and environment temperature of 26 °C.

3.2. Liquid smoke generated

Liquid smoke obtained from each process, i.e., grade 3, grade 2 and grade resulting from the pyrolysis, distillation and distillation process respectively, then analyzed with Gas Chromatography-Mass Spectrometry (GCMS) The analysis data presented in Table 1.

Table 1.The Liquid Smoke Content

Peak	Name	R. Time	Area	Area (%)
Liquid smoke grade 1				
1	2-Decyne-1-ol	0.021	35474	0.58
2	Boric acid (H_3BO_3)	1.336	237897	3.88
3	2-Propanone	1.395	53934	0.88

Peak	Name	R. Time	Area	Area (%)
4	Methyl acetate	1.429	240895	3.93
5	Acetic acid	1.530	2673810	43.60
6	Propionic acid	1.783	36424	0.59
7	p-Dioxene	2.151	13524	0.22
8	Cyclopentanone	2.321	15840	0.26
9	Allyl n-butanoate	2.625	63179	1.03
10	2-Furancarboxaldehyde	2.661	396524	6.47
11	Acetol acetate	2.940	25002	0.41
12	2-Methyl-2-cyclopentenone	3.396	26174	0.43
13	Phenol	4.626	2035633	33.19
14	Guaiacol	6.144	278252	4.54
Liquid smoke grade 2				
1	Boric acid (H_3BO_3)	1.332	105817	2.46
2	Methyl acetate	1.422	131077	3.05
3	Acetic acid	1.521	2838167	66.00
4	Propionic acid	1.778	17643	0.41
5	1,2-Epoxy-3-propyl acetate	2.151	17056	0.40
6	Allyl n-butanoate	2.626	48658	1.13
7	2-Furancarboxaldehyde	2.665	207828	4.83
8	Acetol acetate	2.940	22298	0.52
9	Phenol	4.603	648333	15.08
10	Phenyl butyrate	4.700	200608	4.67
11	Guaiacol	6.122	62631	1.46
Liquid smoke grade 3				
1	Oct-2-ynoic acid	0.022	7232	0.10
2	Boric acid (H_3BO_3)	1.336	127797	1.81
3	Methyl butan-3-enoate	1.387	62215	0.88
4	Methyl acetate	1.429	49270	0.70
5	Acetic acid	1.529	4468831	63.21
6	Propionic acid	1.780	17505	0.25
7	2-Hydroxyethyl acetate	2.146	84782	1.20
8	Allyl n-butanoate	2.626	80926	1.14
9	2-Furancarboxaldehyde	2.665	220156	3.11
10	3-methyl-hexan-2-one	2.870	29923	0.42
11	Acetol acetate	2.937	83688	1.18
12	2-Methyl-2- cyclopentenone	3.402	9316	0.13
13	Phenol (CAS) Izal	4.609	875559	12.38
14	Tetrahydrofurfuryl alcohol	4.705	851180	12.04
15	Guaiacol	6.145	101437	1.43

Table 2 showed that there was no content of Polycyclic aromatic hydrocarbons (PAH) in liquid smoke produced. PAH was one of the chemical components that might form on the manufacture of coconut shell liquid smoke. Some PAH compounds were carcinogenic compounds. PAH compounds were not present in liquid smoke generated in the pyrolysis temperature below 400 °C. The combustion process occurred in the heat under 400 °C had not formed PAH [8]. Food processing allowed to use liquid smoke with no PAH. Phenol and acid contained in liquid smoke had a function as an antimicrobial [8]. Due to that food preservation can use liquid smoke such as meatballs, noodles, and tofu [14].

3.3. Anti Microbial Test

First-grade liquid smoke is the best liquid quality and produced no harmful compounds to apply to food products [14]. Anti-microbial tests performed only at first-grade level with concentrations of 20%, 30%, and 40%. Determining the level of intensity based on an earlier study conducted by Harun al Rashid that stated the concentration of 30% was the best concentration for preserving the fresh anchovy [17]. Results of other earlier study reported that.

Concentrations level of 0%, 5%, 10% and 15% of liquid smoke tested on fresh fish had no inhibitory effect against Pseudomonas bacteria and 0% and 5% concentrations of that also had not been able to inhibit against E. Coli; inhibitory resistance of liquid smoke just occurred at the level of 5% and 10% against the *Staphylococcus aureus* bacteria [8]. In this study, the antimicrobial test performed by paper diffusion method against bacterial pathogens that contaminated food products such as *E. coli*, *Salmonella*, and *Staphylococcus aureus*. Inhibition zone of each bacteria shown in Figure 3.

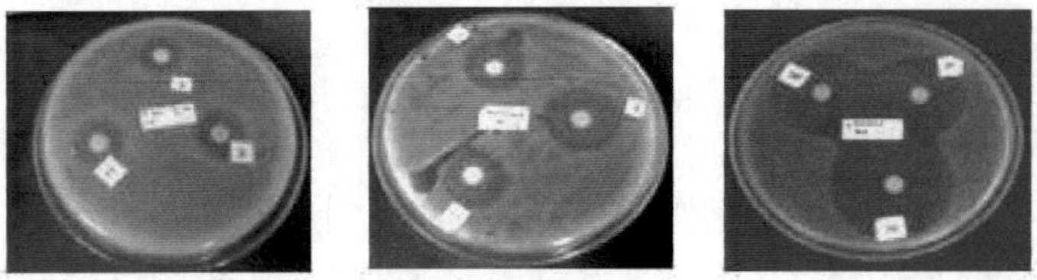

a b c

Figure 3. The inhibition zone against (a) *E. coli* (b) *Salmonella*, and (c) *Staphylococcus aureus*

Figure 3 showed that a concentration of 20 %, 30 %, and 40 % respectively indicated the ability to inhibit *E. coli*, *Salmonella*, and *Staphylococcus aureus*. The results of the research on microbial inhibitory seen in Table 2.

Table 2. The Liquid Smoke Resistance of various concentration level against *E. coli, Salmonella*, and *Staphylococcus aureus*

Concentration (%)	*E.coli*	*Salmonella*	*Staphylococcus aureus*
20	$13.00^{a}\pm2.65$	$14.67^{a}\pm5.03$	$33.00^{a}\pm2.00$
30	$15.00^{ab}\pm1.00$	$17.67^{ab}\pm3.05$	$34.67^{a}\pm2.08$
40	$17.50^{b}\pm0.50$	$24.00^{b}\pm1.00$	$40.00^{b}\pm1.73$

Description: The different notation letters showed a significant
The difference in the significant level (α) of 5%

3.4. *Cost Determinations*

3.4.1. *Investment Costs*

Investment costs represented all expenses incurred either in the form of production equipment and value for supporting apparatus. The calculation of investment costs on the manufacture of liquid smokes showed in Table 3.

Table 3. The equipment of production

Name	Total unit	Unit cost (Rp)	Total cost (Rp)	Economic life (Rp)	Salvage value (Rp)	Depreciation (Rp)
pyrolysis furnace	1	10,000,000,-	10,000,000,-	5	1,000,000,-	1,800.000,-
LPG thank	1	400,000,-	400,000,-	5	40,000,-	72,000,-
electrical pump (120 Watt)	1	350,000,-	350,000,-	5	35,000,-	63,000,-
blower (1,6 A, 220 V, phase 1)	1	350,000,-	350,000,-	5	35,000,-	63,000,-
Liquid smoke container	2	15,000,-	30,000,-	5	3,000,-	5,400,-
Plastic funnel	1	5,000,-	5,000,-	5	500,-	900,-
Plastic dipper	1	10,000,-	10,000,-	5	1,000,-	1,800,-
Total			**11,145,000,-**			**2,006,100,-**

3.4.2. *Operating costs*

Operational cost is the cost used to produce a liquid smoke product using a coconut shell as raw material. Monthly production costs calculated with the assumption of production ten times a month. The annual production costs were the product of the multiplication of monthly production costs by the number of months per year. Operational costs consist of variable costs and fixed costs. The variable costs, fixed costs and total cost presented in Table 4, 5, and 6, respectively.

Table 4. Variables costs

Name	Total unit		Unit cost (Rp)		Cost (1 month) (Rp)	Cost (1 year) (Rp)
Maintenance of assets						500,000,-
- Raw material	26,5	Kg	2,203.39,-	/Kg	583,898.31,-	7,006,779.66,-
- 1-litre plastic bottle	91	pcs	2,000,-	/pcs	182,000,-	2,184,000,-
Pyrolysis process						
- Electric Pump (120 Watts)	2	hours	132,-	/kwh*	2,640,-	31,680,-
- Electric Blower (1,6 A, 220 V, phase 1)	2	hours	387.2,-	/kwh*	7,744,-	92,928,-
Distillation process						
- LPG	1,6	Kg	13,750,-	/kg	220,000,-	2,640,000,-
- Electrical Pump (120 Watts)	8	hours	132,-	/kwh*	10,560,-	105,600,-
redistillation process						
- LPG	1	Kg	13,750,-	/kg	137,500,-	1,650,000,-
- Electrical Pump (120 Watts)	4	hours	132,-	/kwh*	5,280,-	63,360,-
Total					**1,149,622,31,-**	**14,274,347,66,-**

*Electricity Base Tarif Rate B-1

Table 5. Fixed costs

Name	Total unit		Unit cost (Rp)		Cost (1 month) (Rp)	Cost (1 year) (Rp)
Labour	2	persons	500,000,-	/month	1,000,000,-	12,000,000,-

Table 6. Total costs

Total cost (1 year)	
Variable costs	14,274,347.66,-
Fixed costs	12,000,000,-
Total costs	26,274,347.66,-

3.4.3. Cash inflow (Proceed)

The assumption of making liquid smoke used a queuing system; the working hours was 8 hours per day; total consumed time per batch process, i,e, 26 kg of raw materials, was 14 hours (2 days); production frequency was 20 days a month; monthly cash flow was 2,727,000 rupiahs (9,09 x 10 x 30,000 rupiahs). Therefore, based on the existing price of first-grade liquid smoke on the market, the determining annual cash flow presented in Table 7.

Table 7. The annual cash flow

Cash Flow Component	Total
Revenue	Rp, 32,724,000,-
Total cost	Rp, 26,274,347.66,-
Gross profit	Rp, 6,449,652,-
Tax 1%	Rp, 64,496.52,-
Net profit	Rp, 6,385,155.82,-
Depreciation	Rp, 2,006,100,-
Cash inflow	Rp, 4,379,055.82,-

3.4.4. Feasibility Analysis

Business feasibility analysis conducted to determine the financial feasibility of a business. Table 8 showed the results of the feasibility analysis.

The payback period is the period required to recover the cost of an investment. The result of the analysis found that the return period obtained of the liquid smoke making was 2,545 months. The earlier study [18] stated that business defined as feasible if the value of payback periods equaled to that of economic life. Therefore liquid smoke producing in this study was reasonable to run.

The net present value (NPV) is a measurement of profit calculated by subtracting the PV of cash outflows from the PV of cash inflows over a period and. The profitability Index is the proportion of payoff to the investment of a proposed project. Results of the analysis showed that the net present value of 5,454,686.88 which was positive and more than 1; and profitability index of 1,489 indicated that liquid smoke using the existing equipment could provide benefits and was feasible to run.

The internal rate of return (IRR) is used to compare and decide between capital costs. The IRR is the interest will bring a series of cash flows (positive and negative) to a net present value (NPV) of zero (or to the current amount of value of cash invested). Table 9 showed that the analysis result of IRR was 27,81%. It was more than 10 % discount rate. Therefore the business run was acceptable.

Table 8 Feasibility Analysis

Metode	Total
BEP unit	126.18 liter
BEP	Rp, 3,785,298.55,-
PP	2,545 months
NPV	Rp, 5,454,686.88
PI	1,489
IRR	27.81%

4. Conclusion

The maximum capacity of a pyrolysis furnace per batch was 26.5 Kg of coconut shell. The yield of the pyrolysis process, distillation, and redistillation was 42.26%, 35.18 %, and 34.33% respectively. The resulting liquid smoke showed that there was no PAH content. Therefore, it was safe for use on food products. Liquid smoke with a concentration of 30% was the best concentration for inhibiting *E. coli*, and *Salmonella* bacteria and the level of 40% was the best concentration to inhibit against *Staphylococcus aureus*. Feasibility of business tests showed that liquid smoke using the existing equipment could provide benefits and was feasible to run.

5. Acknowledgment

The authors greatly thank Dr. Savitri Dyah WIKR, Taufik Yudi, Dadang G, Teguh Santoso, Neneng Komalasari, Siti Khudaifanny DFAP and those who have supported this research.

6. References

[1] Sholichah E, Agustina W and Desnilasari D 2014 Identifikasi Senyawa Poly Aromatic Hydrocarbon (Pah) Dalam Produk Asap Cair Hasil Samping *Seminar Nasional & Workshop:Peningkatan Inovasi Dalam Menanggulangi Kemiskinan* – LIPI 2013 1–8

[2] Varlet V and Serot T 2010 *Prost C Smoke Flavoring Technology in Seafood Handbook of Seafood and Seafood Products Analysis 1st ed* (Boca Raton, FL: CRC Press)

[3] Schubring R 2008 Use of "filtered smoke" and carbon monoxide with fish *Für Verbraucherschutz Und Leb* **3** 31–44

[4] Martin E M, O'Bryan C A, Lary Jr R Y, Griffis C L, Vaughn K L, Marcy J A, Ricke S C and Crandall P G 2010 Spray application of liquid smoke to reduce or eliminate Listeria monocytogenes surface inoculated on frankfurters *J. Meat Science* **85** 640–644

[5] Abustam E, Likadja J C and Sikapang F 2010 Pemanfaatan Asap Cair sebagai Bahan Pengikat Otot Sapi Bali (The Utilization of Liquid Smoke as a Binder on Meat Balls from Three Different Muscles of Bali Cattle) *Seminar Nasional Teknologi Peternakan Dan Veteriner* 467–73

[6] Aisyah I, Juli N and Pari G 2013 Mengendalikan Cendawan Penyebab Penyakit Antraknosa Dan Layu Fusarium Pada Ketimun (Use of Liquid Smoke from the Destructively-Distilled Coconut Shell to Control the Fungi that Cause Anthracnose and Fusarium Wilt Diseases on Cucumber Plants) *J. Penelit Has Hutan* **31** 170–8

[7] Nurhayati T 2000 Sifat Destilat Hasil Destilasi Kering 4 Jenis Kayu dan Kemungkinan Pemanfaatannya Sebagai Pestisida *Buletin Penelitian Hasil Hutan* **17** 60–8

[8] Budijanto S, Hasbullah R, Prabawati S, Setyadjit, Sukarno and Zuraida I 2008 Identification and safety test on liquid smoke made from coconut shell for food product, *Indones. J. Agric. Postharvest Res.* **5** 32–40

[9] Gumanti F M 2006 *Kajian sistem produksi distilat asap tempurung kelapa dan pemanfaatannya sebagai alternatif bahan pengawet mie basah* (Bogor: Institut Pertanian Bogor)

[10] Karseno, Darmadji P and Rahayu K 2001 Daya Hambat Asap Cair Kayu Karet Terhadap Bakteri Pengkontaminasi Lateks Dan Ribbed Smoke Sheet *Agritech* **21** 10–5

[11] Soldera S, Sebastianutto N and Bortolomeazzi R 2008 Composition of Phenolic Compounds and Antioxidant Activity of Commercial Aqueous Smoke Flavorings *J. Agric. Food Chem.* **56** 2727–34

[12] Direktorat Jenderal Perkebunan 2016 *Statistik Perkebunan Indonesia Komoditas Kelapa 2015-2017*, (Jakarta: Sekretariat Direktorat Jenderal Perkebunan Direktorat Jenderal Perkebunan Kementerian Pertanian)

[13] Sundaram 2009 EGNE Pyrolysis of Coconut Shell: An Experimental Investigation *Eng. Res.* **6** 33–9

[14] Fauzan F and Ikhwanus M 2017 Pemurnian Asap Cair Tempurung Kelapa Melalui Distilasi Dan Filtrasi Menggunakan Zeolit Dan Arang Aktif *Prosiding Semnastek* 1-5

[15] Serio M, Kroo E and Wójtowicz M 2003 Biomass pyrolysis for distributed energy generation *ACS Div. Fuel Chem. Prepr.* **48** 584–9

[16] Wibowo S 2012 Karakteristik Asap Cair Tempurung Nyamplung *J. Penelitian Hasil Hutan* **30** 218–27

[17] Rasyid H A 2010 *Pemanfaatan Asap Cair Tempurung Kelapa sebagai Bahan Pengawet Ikan Teri Nasi (Stolephorus commersanii, Lac) Segar untuk Tujuan Transportasi* (Bogor: Institut Pertanian Bogor)

[18] Kusuma P T W W and Mayasti N K I 2014 Analisa Kelayakan Finansial Pengembangan Usaha Produksi Komoditas Lokal : Mie Berbasis Jagung *Agritech* **34**194–202

IOP Conf. Series: Earth and Environmental Science **251** (2019) 012008 doi:10.1088/1755-1315/251/1/012008

The effect of biomass particle size and chemical structure on the enzymatic hydrolysis reaction of galactomannan from sugar palm fruit by β-mannanase from *Kitasatospora* sp. KY576672

R Pangestu[1], N Rahmani[1], R Palar[1], P Lisdiyanti[1] and Yopi[1]

[1]Research Center for Biotechnology, Indonesian Institute of Sciences, Cibinong, West Java, Indonesia

Email: pangestu.radityo@gmail.com

Abstract. Mannooligosaccharides is essential to support gut health. Unfortunately, obtaining these compounds by chemical synthesis is considered labor extensive. In contrast, production using enzymatic reaction is simpler. Mannan, one of lignocellulose fractions, is a suitable raw material for this purpose. Sugar palm fruit (*kolang-kaling*) contains a high level of galactomannan, which makes it potential to be used as a resource for MOS bio-production. Our previous work revealed that strain which showed highest mannan-degrading activity was *Kitasatospora* sp. KY576672. In this study, the use of β-mannanase enzyme produced by this strain to degrade galactomannan from sugar palm fruit was examined. The effect of biomass particle size and chemical structure of polymer towards reaction were studied by conducting reaction using biomass with three different particle sizes and comparing the hydrolysis profile of sugar palm fruit and locust bean gum. Reaction was performed at 40°C and 190 rpm in sodium phosphate buffer pH 6. Results suggested that reaction rate did not significantly change by altering biomass particle size from 150 µm to 75 µm, but it decreased by the use of 150-300 µm particle-size biomass. In addition, locust bean gum produced a more balanced ratio of hydrolysis products (M1-M6) compared to sugar palm fruit. Mannohaxaose (M6) was the main hydrolysis products from sugar palm fruit and locust bean gum.

1. Introduction

Lignocellulose is a versatile material. Due to its abundance and its potential value as raw material for producing biofuel, chemicals, and food products, lignocellulose is very appealing for industry [1]. Sadly, most of lignocellulose sources are wasted and even burnt. Lignocellulose includes carbohydrate polymer (cellulose and hemicellulose) and non-carbohydrate polymer (lignin) [2]. Agricultural, industrial and forest residuals are prominent biomass sources of lignocellulose [3].

However, about a third of the total components in plants is hemicellulose [4]. In softwoods, such as fruits and plant seeds, hemicellulose is prominently available as mannan [5]. Mannan is able to be hydrolyzed to produce high-value compounds, including mannose and mannooligosaccharides (MOS). Mannose is potentially used as an alternative sweetener, and MOS is known for its prebiotic effect [6,7]. Production of MOS using this approach is beneficial as chemical synthesis of oligosaccharides involves complicated steps, gives poor stereoselectivity and low yields, and faces a problem with scale-up production [8–10].

As one of mega-biodiverse country, Indonesia has myriad of hemicellulose resources which are potential to be explored; sugar palm (*Arenga pinnata*) fruit, or widely known in Indonesia as *kolang-kaling*, is no exception. The cooked endosperm of nut from this fruit is frequently consumed as an additional ingredient for certain drinks [11]. Proximate analysis showed that sugar palm fruit contained about 92% water, 3.42% carbohydrate (mainly galactomannan), 1.52% crude fiber, 1.42% protein and minor constituents [12]. Despite its high content of mannan, the utilization of sugar palm fruit in hemicellulose processing is still limited.

There are enormous mannane-producing microorganisms which had been studied; however, the exploration of Actinomycetes as microbial source is still provoking interest owing to its diversity [13,14]. Several Actinomycetes strains were known to have mannolytic activity [15–18]. Our previous study revealed that various mannan-containing polymers could be hydrolyzed to produce MOS using β-mannanase enzyme from *Kitasatospora* sp. KY576672. This stain exhibited highest mannolytic activity compared to other isolates screened [17]. Present study aimed to investigate the enzymatic hydrolysis of galactomannan from sugar palm fruit using β-mannanase from this strain by comparing its reaction pattern to locust bean gum's, another galactomannan source which had widely been studied, and also consider the influence of physical parameter, focusing on the particle size, which may affect the reaction rate. The information about particle size effect is essential for industry since reducing particle size consumes more energy.

2. Materials and Methods

2.1. Chemicals, reagents, and materials

Cooked endosperm of nut from *A. pinnata* was collected from Caringin, Bogor, West Java, Indonesia. *Kitasatospora* sp. KY576672 was obtained from Research Center for Biotechnology, Indonesian Institute of Sciences. Peptone, malt extract, yeast extract, and agar were purchased from Becton Dickinson & Co (USA). Locust bean gum (LBG) and D-(+)-mannose (M1) were purchased from Sigma-Aldrich Co. (St. Louis, MO, USA). β-D-mannobiose (M2), β-D-mannotriose (M3), β-D-mannotetrose (M4), β-D-mannopentose (M5) and β-D-mannohexaose (M6) were purchased from Megazyme (Wicklow, Ireland). Other chemicals were purchased from Merck (Darmstadt, Germany).

2.2. Production of mannanase enzyme

Kitasatospora sp. (frozen stock) was maintained on ISP2 agar with the following composition (%): yeast extract, 0.4; malt extract, 1.0; glucose, 0.4; and agar, 1.7. It was incubated at 28 °C for 3 days. A single colony was cultivated in ISP2 medium (composition mentioned, without addition of agar) at 28°C and 190 rpm for 3 days. An aliquot of pre-culture was inoculated into fermentation medium with a ratio of 1:9 (v/v). Fermentation medium contained the following compositions (g per 100 mL): locust bean gum (LBG), 1.25; peptone, 0.075; yeast extract, 0.05; $(NH_4)_2SO_4$, 0.14; KH_2PO_4, 0.20; $MgSO_4.7H_2O$, 0.03; $CO(NH_2)_2$, 0.03; $CaCl_2$, 0.039625; $FeSO_4.7H_2O$, 0.0005; $MnSO_4.7H_2O$, 0.0976125; $ZnSO_4.7H_2O$, 0.00014; $CoCl_4$, 0.00036625; and buffer pH 8. Fermentation was conducted in a shaker flask for 4 days at 28 °C and 190 rpm. The enzyme was harvested from culture supernatant by centrifugation (22,540 × g, 4 °C for 30 min.) and filtration (0.2 μm).

2.3. Enzyme activity assay

Mannolytic activity of enzyme obtained was assayed using DNS method [19]. LBG (200 μL, 1%, w/v) was mixed with 100 μL of distilled water and 100 μL of phosphate buffer pH 6 (50 mM). Diluted enzyme (50 μL) was added and the mixture was incubated at 60 °C for 15 min. After reaction, 500 μL of DNS reagent was added. The mixture was heated to 90 °C for 15 min. and, then, cooled by ice. Sugar formed was measured by spectrophotometer at 540 nm (SpectraMax Paradigm, Molecular Devices). The same procedure was also carried out to measure control (without addition of enzyme) and standards (mannose with a concentration of 50, 100, 200 and 300 ppm). Enzyme activity was expressed as unit per mL (U/mL). Unit of enzyme activity was calculated from the amount of sugar released (in μmol) per min.

2.4. Enzymatic hydrolysis reaction

Sugar palm fruit (biomass) was sliced, dried (40-50 °C) and crushed using food processor. The powder was subsequently separated using several sieves (300 μm, 150 μm and 75 μm). Each fraction was diluted by 200 mL of phosphate buffer pH 6 (50 mM) and mixed to 200 mL of crude enzyme. The reaction was conducted in a shaker flask at 40 °C and 190 rpm. Products were sampled at 2, 4, 8, 24 and 32 h. Qualitative physical change of solution was observed during the reaction. Reaction was stopped by heating treatment at 100 °C for 15 min. The products were collected from supernatant after centrifugation at 15,660 × g for 15 min. Similar procedure was applied to locust bean gum (particle size ≤ 75μm).

2.5. Analysis of hydrolysis products

Products were analysed by TLC (qualitative) and HPLC (quantitative). Standards used were mannose (M1), mannobiose (M2), mannotriose (M3), mannotetraose (M4), mannopentaose (M5) and mannohexaose (M6). Qualitative analysis was carried out by spotting samples onto TLC silica gel G60 F_{254} (Merck, Germany). Samples were developed with a mixture of *n*-butanol, glacial acetic acid, and water (2:1:1, v/v/v). Visualization was obtained after the plate was sprayed by DAP reagent (mixture of diphenylamine, aniline, acetone and phosphoric acid, 0.2:0.2:10:1.5, v/w/v/v) and heated at 120°C for 15 min [20,21]. Quantitative analysis was performed using an Agilent Technology 1200 Infinity Series HPLC with refractive index detector (RID) and Hi-PlexCa (Duo) column (300 x 6.5 mm, 8 μm). Before injection, sample was centrifuged (9,700 × g for 15 min.) and filtered (0.2 μm) to remove impurities. Volume of sample injected was 10 μL. Ultrapure water (HPLC grade) was used as mobile phase. Eluent flow rate was 0.60 mL/min. Temperature of column and detector were set at 80°C and 40°C, respectively. Running time of each sample was 30 min. This procedure was the optimization of manufacturer's guideline [22].

3. Results and Discussions

3.1. Production of enzyme and analysis of its activity

Mannooligosaccharides can be generated by degradation of mannan. This process involves breaking β-1,4-linkage of mannose-containing backbone side by hydrolase enzyme, named mannanase [7]. According to our previous study, *Kitasatospora* sp. KY576672 was a prospective microbial source for producing this enzyme [17].

The production of extracellular β-mannanase enzyme was carried out in a 2L Erlenmeyer flask. Locust bean gum was used as a carbon source. Other nutrients, such as nitrogen, water, and salt, were fulfilled by the addition of ingredients mentioned. Concentration of pre-culture inoculated to fermentation media was 10% (v/v) for optimum growth [23]. Considering the mechanical limit of shaker flask, the ratio volume of culture to flask was 1:5. The extracellular enzyme was separated from cell by centrifugation and filtration.

The result of enzyme activity assay is shown in **Table 1**. The activity of enzyme was lower than that of recombinant enzyme from our previous study because the purification process was not conducted [17]. Instead, the crude enzyme was used to figure out its application for industrial process.

Table 1. Activity of crude β-mannanase enzyme produced from *Kitasatospora* sp. KY576672

	Absorbance			Concentration of sugar released (ppm)	Activity (U/mL)
	Sample (S) [a]	Control (C) [a]	S-C		
1	1.2490	0.4714	0.7776	184.14	68.20
2	1.3851	0.5441	0.8410	199.61	73.93
3	1.5120	0.7558	0.7562	178.92	66.27
					69.47 ± 3.98[b]

[a] Values represent the means of duplicate measurements.
[b] Average ± SD.

3.2. Preparation of biomass and enzymatic hydrolysis reaction

Sugar palm fruit (SPF) was collected from *A. pinnata* fruit and prepared in fresh condition to minimize degradation action of microorganism. According to Torio [12], SPF contained about 4% carbohydrate (mainly galactomannan). In other words, the content of galactomannan was about 42% of dried weight of biomass. Thus, SPF was expected to yield a high concentration of MOS. Drying process was an essential key to obtain desired biomass size [24]. Biomass was heated below 60 °C to avoid thermal degradation. Separation process by sieves yielded three different fractions of biomass: SPF-I (particle size < 75 μm), SPF-II (particle size: 75-150 μm) and SPF-III (particle size: 150-300 μm).

The reaction was performed at 40 °C and pH 6 in accordance with the optimization condition of reaction and stability of Actinomycetes mannanase studied by Takahashi *et* al. [25]. Shaking treatment was applied to homogenize the mixture and enhance the reaction rate. The reaction was stopped by high-temperature treatment to denaturalize enzyme. Similar procedure was also conducted to locust bean gum. Hydrolysis products analysis (by TLC) is presented in Figure 1. Generally, all samples produced M1, M2, M3, M4, M5, and M6. Longer unidentified oligosaccharides were also produced since there were several spots with lower Rf values present in all samples. These oligosaccharides might be homo-oligomers containing mannose more than six units and/or hetero-oligomers whose structure possesses galactose unit(s) linked to mannose unit(s) via α-(1→6) linkage.

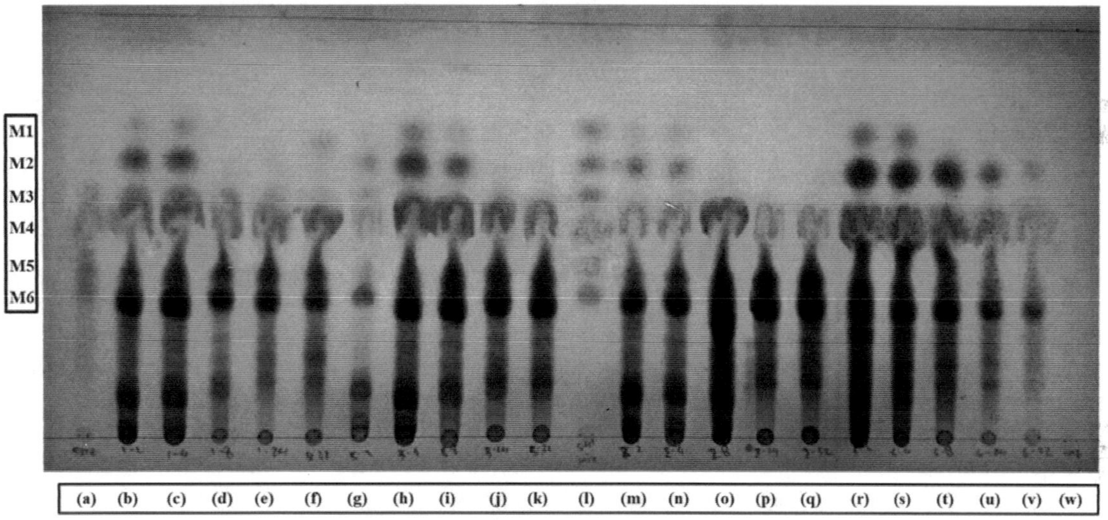

Figure 1. TLC analysis of enzymatic hydrolysis products **(a, w)** crude enzyme (control); **(b-f)** SPF-I at 2, 4, 8, 24 and 32 h; **(g-k)** SPF-III at 2, 4, 8, 24 and 32 h; **(l)** Standard mix M1 (top) - M6 (bottom); **(m-q)** SPF-II at 2, 4, 8, 24 and 32 h; **(r-v)** LBG at 2, 4, 8, 24 and 32 h. See text for details.

Solubility of SPF in sodium phosphate buffer was lower than that of LBG (Figure 2). This was due to the high degree of branching of galactomannan in SPF [12]. However, as enzymatic reaction progressed, the solubility and homogeneity of both SPF and LBG increased. The mixtures became less viscous at 32 h. It seemed that SPF-I and SPF-II became completely soluble and only dirt floated, yet SPF-III was still rather viscous at 32 h. LBG is completely soluble without any dirt left as we use pure galactomannan for this experiment. The observation proved that the enzyme actively hydrolyzed galactomannan (long chain polymer) from biomass to shorter chain of mannooligosaccharides as simpler sugars are more soluble in water.

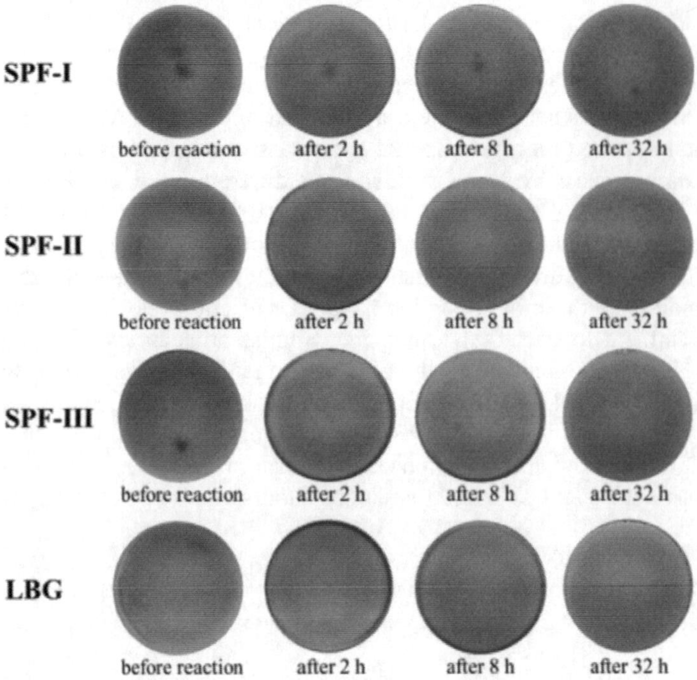

Figure 2. Physical characteristic change of biomass solution during enzymatic hydrolysis reaction. Picture was taken from above flask.

3.3. The effect of biomass particle size

The effect of biomass particle size was studied through the hydrolysis profile of three different particle size fractions of SPF. Figure 3(a)-(c) shows the chromatograms of biomass hydrolysis products at the various time and Figure 4(a)-(c) depicts the concentration change of each MOS.

Generally, M6 was the predominant product from all samples while other MOS were only minors. During reaction, concentration of M6 was 10 times higher than other MOS in all three samples. Concentration of these oligosaccharides remained below 1,000 mg/L, except for M4 in SPF-I at 8 h and 32 h. Further degradation of unidentified oligosaccharides to M6 was observed as its peak area became smaller but the opposite trend occurred to M6. Peak splitting of unidentified oligosaccharides was observed (all samples) at 8 h indicating that there might be more than one unidentified compounds. The concentration of M5 could not be calculated due to peak overlapping with M6, but the spots on TLC analysis (Figure 1) gave more convincing interpretation that M5 was produced.

Meanwhile, the concentration of M1, M2, M3, and M4 showed downward trends after 4 h and 8 h. This interpretation observed in M1 whose peak even disappeared at 4 h and afterward. These drops were probably caused by trans-glycosylation since some of β-mannanase in GH5, and GH113 exhibited this activity [26]. This prediction needs further confirmation. However, M1 peak appeared again in SPF-I at 32 h (observed both in TLC and HPLC analysis), and small increases in M2 and M4 were possibly the products of further-hydrolysis reaction of longer oligosaccharides.

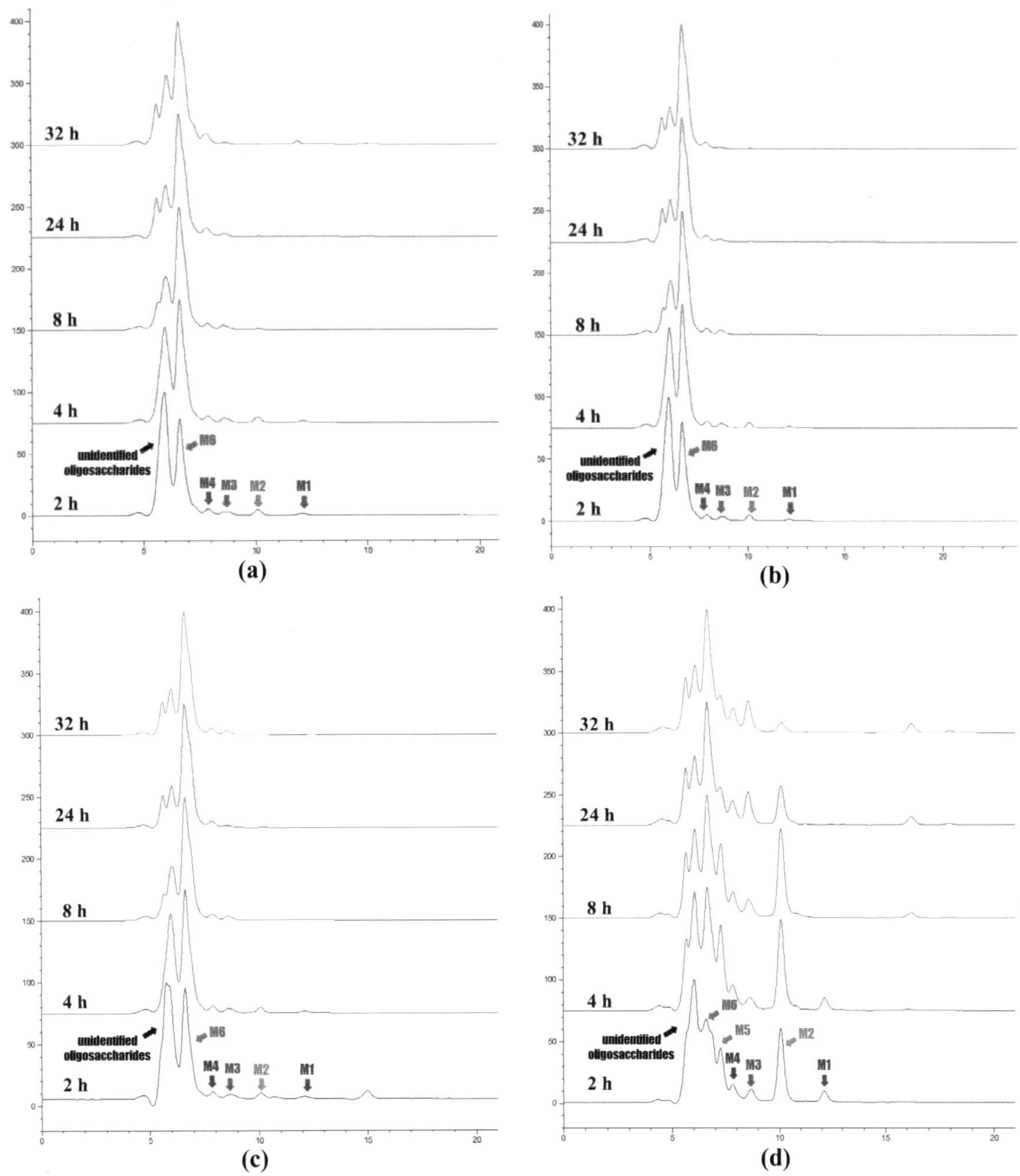

Figure 3. HPLC analysis of hydrolysis products of **(a)** SPF-I; **(b)** SPF-II; **(c)** SPF-III; **(d)** LBG; by β-mannanase enzyme (U = 69.47 $\pm$ 3.98). See text for details.

The effect of biomass particle size was figured out by comparing the concentration of each MOS. Based on theory of reaction kinetics, reduction of particle size can increase the reaction rate. However, the treatment did not affect SPF-I (particle size $\leq$ 75 µm) and SPF-II (particle size: 75-150 µm) as the concentration of MOS from those two did not differ significantly. Nonetheless, the effect was obviously observable in SPF-III (particle size: 150-300 µm). In general, concentrations of hydrolysis products from SPF-III were lower than SPF-I and SPF-II indicating a decrease in reaction rate.

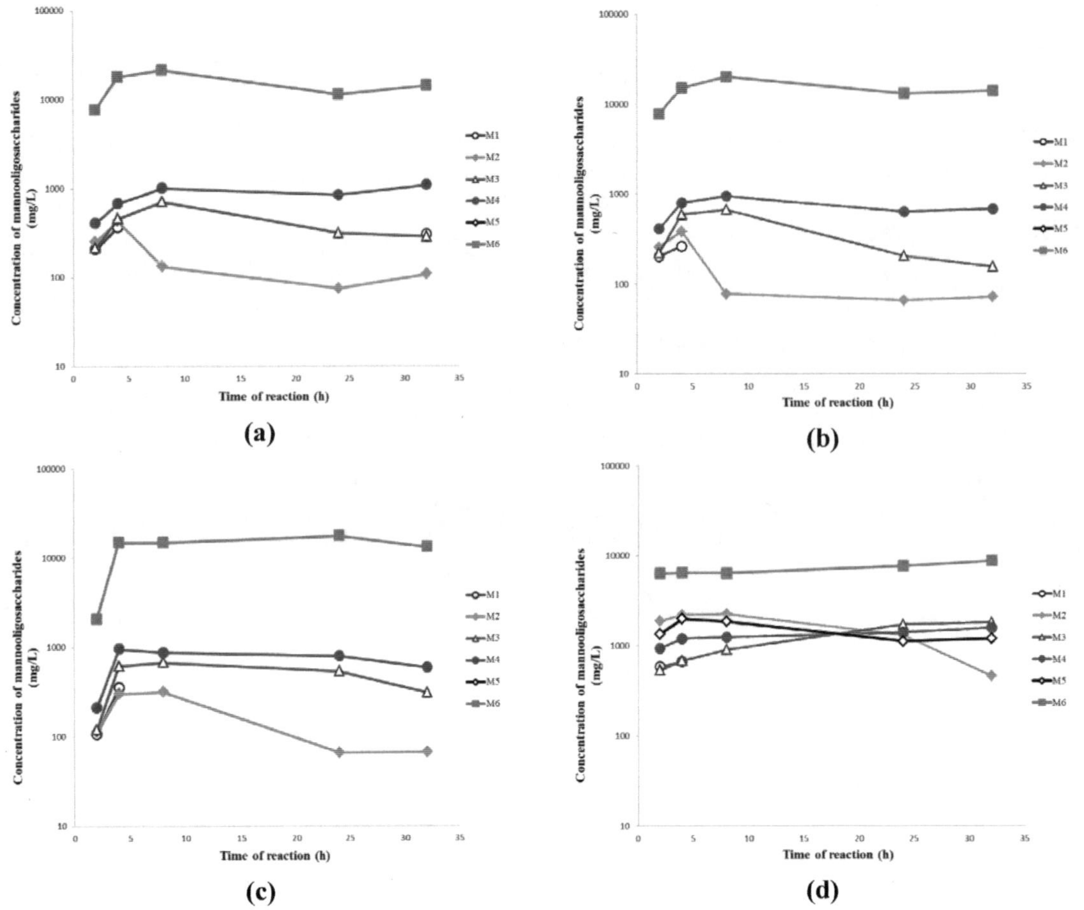

Figure 4. Comparison of concentration of several mannooligosaccharides produced from enzymatic hydrolysis of **(a)** SPF-I; **(b)** SPF-II; **(c)** SPF-III; **(d)** LBG; by β-mannanase enzyme (U = 69.47 $\pm$ 3.98). See text for details.

3.4. The effect of galactomannan chemical structure

The chemical structure of galactomannan varies depending on its source [7]. This diversity may affect the enzymatic hydrolysis pattern of particular biomass. Hydrolysis profile of SPF and LBG could be studied from their chromatograms (Figure 3(a) and (d)) and the comparison of their hydrolysis products (Figure 4(a) and (d)).

In LBG, M6 was also the major hydrolysis product. Interestingly, the quantity of M6 produced from SPF almost doubled and even tripled (at 4 h and 8 h) the concentration of M6 from LBG. Conversely, the concentration of M1, M2, M3, M4, and M5 produced from LBG were considerably higher than that from SPF. Concentration of M2 (at 2 h – 24 h), M3 (at 24 h and 32 h), M4 (at 4h – 32 h) and M5 (at 2h – 32 h) in LBG were higher than 1,000 mg/mL. This could be explained by the difference of mannose/galactose (M/G) ratio of galactomannan. M/G ratio of SPG and LBG were 2:1 and 4:1, respectively. The presence of galactose as branch unit hinder the enzymatic cleavage of backbone chain [27]. Because chemical structure of SPF contained more galactose units, longer-chain hydrolysis products were more favorable.

4. Conclusion

Galactomannan from sugar palm fruit could be hydrolyzed by β-mannanase from *Kitasatospora* sp. KY576672 to produce simpler sugars including mannose, mannobiose, mannotriose, mannotetraose, mannopentaose, mannohexaose, and longer oligosaccharides. Sugar palm fruits with a particle size $\leq$ 75 μm and 75-150 μm were predicted to have relatively similar reaction rate and that with particle size

150-300 μm exhibited a lower reaction rate. Also, in SFP, shorter chain mannooligosaccharides, such as M1, M2, M3, and M4, were produced in a low concentration due to higher concentration of galactose present in its polymer structure. M6 is the major biodegradation product of both SPF (all biomass particle sizes) and LBG.

5. Acknowledgment

The authors gratefully acknowledge the financial from DIPA (National Priority for Health) - Research Center for Biotechnology LIPI and Incubation Program of PUSINOV LIPI.

6. References

[1] Kuhad R C and Singh A 1993 *Crit. Rev. Biotechnol.* **13** 151–72
[2] Anwar Z, Gulfraz M and Irshad M 2014 *J. Radiat. Res. Appl. Sci.* **7** 163–73
[3] Kumar R, Singh S and Singh O V 2008 *J. Ind. Microbiol. Biotechnol.* **35** 377–91
[4] Chauhan P S, Puri N, Sharma P and Gupta N 2012 *Appl. Microbiol. Biotechnol.* **93** 1817–30
[5] Scheller H V and Ulvskov P 2010 *Annu. Rev. Plant Biol.* **61** 263–89
[6] Hu X, Shi Y, Zhang P, Miao M, Zhang T and Jiang B 2016 *Comprehensive Reviews in Food Science and Food Safety* **15** 773–85
[7] Soni H and Naveen K 2013 *Advances in Enzyme Biotechnology* ed. Shukla P and Pletschke (New Delhi: Springer) 41–56
[8] Rastall R A and Bucke C 1992 *Biotechnol. Genet. Eng. Rev. ISSN* **10** 253–82
[9] Monsan P and Paul F 1995 *FEMS Microbiol. Rev.* **16** 187–92
[10] Muthana S, Cao H and Chen X 2011 *Curr. Opin. Chem Biol.* **13** 573–81
[11] Sayuti K, Yenrina R and Anggraini T 2017 *Pakistan J. Nutr.* **16** 69–76
[12] Torio M A O, Saez J and Merca F F 2006 *Philipp. J. Sci.* **135** 19–30
[13] Priyadharsini P and Dhanasekaran D 2015 *J. Saudi Soc. Agric. Sci.* **14** 54–60
[14] Saini A, Aggarwal N K, Sharma A and Yadav A 2015 *Enzyme Research.* **279381** 15 pages
[15] Montiel M D, Rodríguez J, Pérez-Leblic M I, Hernández M, Arias M E and Copa-Patiño J L 1999 *Appl. Microbiol. Biotechnol.* **52** 240–45
[16] Ratnakomala S, Suhartono M T, Meryandini A and Prasetya B 2014 *Annales Bogorienses* **18** 1–12
[17] Rahmani N, Kashiwagi N, Lee J, Nakamura N, Matsumoto H, Kahar P, Lisdiyanti P, Yopi, Prasetya B, Ogino C and Kondo A 2017 *AMB Express* **7** 2–11
[18] Arcand N, Kluepfel D, Paradis F W, Morosoli R and Shareck F 1993 *Biochem. J.* **290** 857–63
[19] Miller G L 1959 *Anal. Chem.* **31** 426–8
[20] Hansen S A 1975 *J. Chromatogr.* **107** 224–6
[21] Humbel R and Collart M 1975 *Clin. Chim. Acta* **60** 143–5
[22] Ball S, Bullock S, Lloyd L and Mapp K 2011 *Analysis of Carbohydrates, Alcohols, and Organic Acids by Ion-Exchange Chromatography - Agilent Hi-Plex Columns Applications Compendium* ed. A Ewen (Pirmasens, Germany: Agilent Technologies) 3
[23] Abulencia J P and Theodore L 2015 *Open-Ended Problems - A Future Chemical Engineering Education Approach.* (Massachusetts: Scrivener Publishing LLC) 394
[24] Jung H, Lee Y J and Yoon W B 2018 *Processes* **6** 1–16
[25] Takahashi R, Kusakabe I, Maekawa A, Suzuki T and Murakami K 1983 *Japan. J. Trop. Agr.*, **27** 140–8
[26] Rosengren A, Reddy S K, Sjöberg J S, Aurelius O, Logan D T, Kolenová K and Stålbrand H 2014 *Appl. Microbiol. Biotechnol.* **98** 10091–104
[27] Mahammad S A, Conford S, Kelly D A and Khan R M 2007 *Biomacromolecules* **8** 949–56

Transformation of Mordenite-Clinoptilolite Natural Zeolite at Different Calcination Temperatures

S K Wahono[1,2,3,*], D J Prasetyo[3], T H Jatmiko[3], A Suwanto[3], D Pratiwi[3], Hernawan[3,4], and K Vasilev[1,2]

[1] Future Industries Institute, University of South Australia, Mawson Lakes 5095, South Australia, Australia
[2] School of Engineering, University of South Australia, Mawson Lakes 5095, South Australia, Australia
[3] Research Unit for Natural Product Technology, Indonesian Institutes of Sciences, Gunungkidul 55861, Yogyakarta, Indonesia
[4] Chemistry Department, Universitas Gadjah Mada, Sleman, 55281, Yogyakarta, Indonesia

*Email: satriyo.krido.wahono@lipi.go.id; satriyo.wahono@mymail.unisa.edu.au; satriyo.krido.wahono@gmail.com

Abstract. Natural zeolite is one of prospective material with various applications. The material has low-cost provisioning and abundant availability in the world. Mordenite-clinoptilolite is one of the common types of natural zeolite; one of the deposits are located in Gunungkidul, Yogyakarta, Indonesia. For optimizing the utilization, some treatment is needed, calcination being one of the possible treatments. The dealuminated natural zeolite was used as the base material for understanding the transformation effect, then calcinated the zeolite in various temperature range between 400 °C and 1000 °C. Some characterization was conducted after the treatment, such as XRD, SEM-EDX, and BET to explore the structure, morphology, composition, and specific surface area. Drastic structure changes were demonstrated at the calcination of 1000 °C which eliminating the zeolite structure into cristobalite as one of the silica structures. The specific surface area decreased after 600 °C, even elapsed after the structure change. Calcination has the ability to transform the zeolite structures and also affected the specific surface area properties. This transformation provides critical information for further application of natural zeolite in the various treatment of calcination.

Keywords: natural zeolite, mordenite-clinoptilolite, calcination, crystal structure, specific surface area

1. Introduction

Zeolites are one of the essential materials used in various applications such as catalyst, adsorbent and porous carrier material. It is available as a natural resource and synthetic structures from manufacturing of alumina and silica. The synthetic products provide uniform structures and purity of phases but have some drawback of complex processing, waste production, and high-cost. In comparison, natural zeolites are available in an abundant quantity which spread out over the world [1-3] and low-cost production [4]. Utilization of the natural zeolite resources is needed to overcome the drawback of synthetic product.

Indonesia has abundant natural zeolites with the high silica concentration of around 60% [5], which are found in various areas, one specific location being Gunungkidul, Yogyakarta [6, 7]. Indonesian natural zeolite is categorized as mordenite and clinoptilolite [3, 8, 9]. By activation and modification, this natural zeolite has good properties of Si/Al ratio, surface area, crystallinity, and functional groups [10-12]. It provides several prospective applications such as in catalyst [5, 13-15], wastewater treatment [16-19], radioactive waste treatment [20, 21], drug delivery [22], bio-ethanol dehydration [10, 11, 23], and biogas purification [7, 24-26].

The various applications of natural zeolites provide room for improvement by activation and modification. One of the standard treatments for the natural zeolite is calcination which is conducted to remove water and impurity gases that are trapped in the crystal pores. This treatment also affects the natural zeolite properties of crystallinity, surface area and porosity [21, 27-31]. Therefore, this study is conducted to understand the characteristic transformation of mordenite-clinoptilolite natural zeolite at different temperatures of calcination.

2. Methods

2.1. Natural Zeolite Modification

CV Mountain Stones supplied Mordenite-Clinoptilolite Natural Zeolite from Gunungkidul, Yogyakarta, Indonesia. It was supplied in powder size of under 40 mesh (425 microns) which is assigned as A0 zeolite in further. A0 zeolite was dealuminated by submerging in 3N HCl (Merck with purity of 37-38%) with a volume ratio of 1:1 for 24 hours which assigned in further as CT1-0 zeolite as the base material. The material was washed and calcinated at a different temperature of 400, 600, 800, and 1000 °C for 3 hours which assigned further as CT1, CT2, CT3, and CT4 zeolite, respectively.

2.2. Natural Zeolite Characterization

For understanding the structural transformation caused by these treatments, the initial characterization of the crystal structure is the XRD analysis. The supporting study of the specific surface area, Si/Al ratio and morphology were conducted to complement the structure transformation results. All of the investigations were performed on all of the modified and pristine natural zeolites. PANalytical Empyrean XRD determined the crystal structure analysis in high angle analysis of 5° – 80° and the scan rate of 0.02 °/sec. The type of crystal was assessed using the High Score Plus Software which complementing the XRD instrument by the Search-Match method.

Micromeritics ASAP 2420 analyzer was utilized to determine the specific surface area (Brunauer–Emmett–Teller/BET Model) and porosity (Barrett-Joyner-Halenda/BJH Model) which carried out in isotherm nitrogen adsorption-desorption at -196 °C. For pre-analysis treatment, the samples were degassed at 350 °C for 24 hours for cleaning the pores from water and impurities [32, 33]. Particularly for non-calcination treatment, the degassing treatment was conducted at 120 °C for removing the water only.

The SEM-EDX analysis was carried out on the powder material analysis for morphology observation and Si/Al ratio determination by the Hitachi SU3500 which equipped with EDX analyzer by EDAX Element. As sample preparation, the sample was dried then a small quantity of powder was splashed on the carbon tape surface. For EDX analysis, the minimum voltage acceleration requirement was set to 3 kV then adjusted with the material conductivity.

3. Results and Discussions

3.1. Mordenite-Clinoptilolite Natural Zeolite from Gunungkidul, Yogyakarta-Indonesia

First, characterizations of crystal structure, surface area and porosity, morphology and Si/Al ratio was carried out to the pristine natural zeolite. The crystal structure of this natural zeolite is implied as mordenite-clinoptilolite and is supported in published studies [3, 8, 9, 26]. Figure 1 shows the calcite peak between 30-60°, whereas the mordenite-clinoptilolite or the zeolite structure between 10-30° which is supported by the search match analysis shown in table 1. The specific surface area of the pristine natural zeolite is 25.95 m^2/g as shown in Table 2. It is consistent with published studies of natural zeolite which report surface area between 14-30 m^2/g [10, 34-36]. The morphology of the pristine natural zeolite is shown in figure 2.a which indicates as the porous material. The Si/Al ratio of the pristine natural zeolite was 4.4 as shown in table 3.

3.2. Crystal structure transformation of mordenite-clinoptilolite natural zeolite

Figure 1 and table 1 show the results of XRD analysis of the natural zeolite crystal structure transformation as a result of high-temperature calcination. The 400 °C was selected as the starting calcination temperature since the organic compound decomposition still occurs under that temperature [32, 37]. Moreover, this temperature was referred to the previous study of the calcination treatment of this natural zeolite which using 400 °C for biogas purification [7, 38].

Figure 1 shows the graph of XRD transformation of the pristine natural zeolite into the dealuminated natural zeolite, then into various high temperatures treatment of calcination. The calcite is visible in the spectrum of the pristine material (A0) and showed a lower peak intensity between 30°-60° in the spectrum of the dealuminated natural zeolite (CT1-0). After calcination of the CT1-CT4, the calcite elapsed and transformed into anorthite (calcium silicate structure) which indicates an interaction between calcium and silica during the treatment.

On the other hand, the zeolite structure remains without transformation. The substantial change of zeolite occurs at CT4 which shows the appearance of a high peak at 21° and the loss of peaks between 10-30° as the zeolite structure indicator in the previous material treatment. The combination of various peaks transformation indicates the shifting of zeolite structure into cristobalite structure as the solid silica structure.

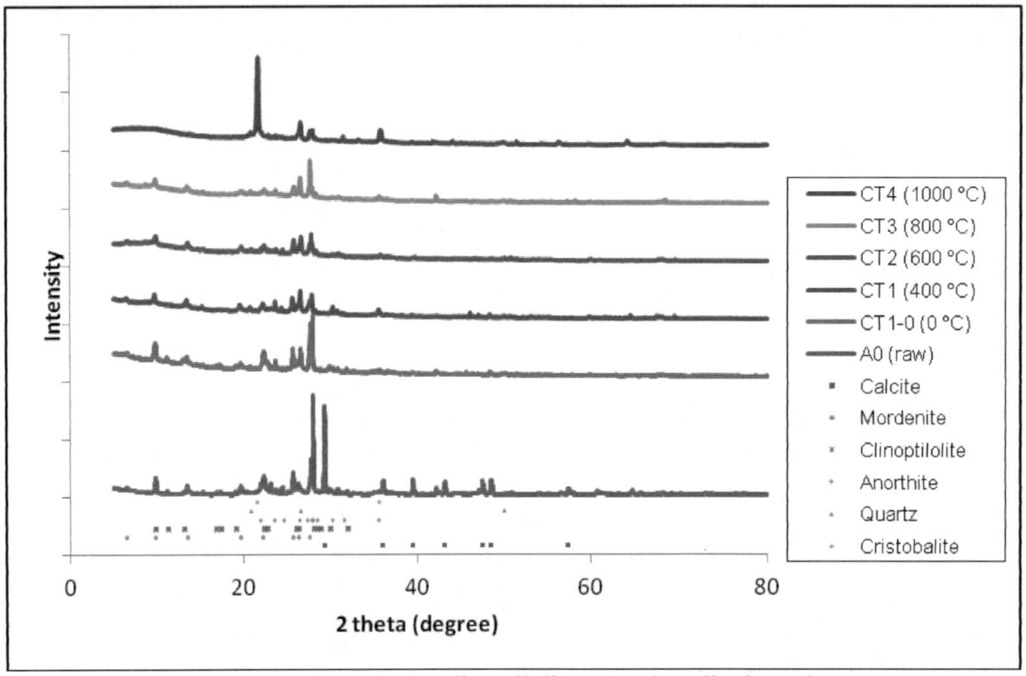

Figure 1. XRD graph of mordenite-clinoptilolite natural zeolite in various treatments

The structural changes are also supported by further analysis using the search-match method from High Score plus software shown in table 1. The match score of the crystal structure in table 1 shows the calcite and zeolite structure in the pristine natural zeolite. The calcium crystal changes after calcination in CT1-CT4, while a zeolite structure change indicates only CT4.

Detail information can be obtained from this search-match analysis about the existence of the quartz structure and the decreasing match score of clinoptilolite and anorthite. The quartz structure pulls out after dealumination treatment and in the further calcination treatment, even increase the match score after calcination. This result indicates that the treatments remove some mineral interaction with silica on the natural structure since the acid reaction or the high-temperature heating, then purifying the silica or quartz structure. CT1-CT4 also shows the decreasing match score of the clinoptilolite and anorthite. It indicates that the high temperature destructs both crystal structure. Moreover, the results also show the mordenite and quartz structure do not affect by high-temperature treatment.

Table 1. Crystal structure match score of mordenite-clinoptilolite natural zeolite in various treatments

Code	A0	CT1-0	CT1	CT2	CT3	CT4
Temp. (°C)	0	0	400	600	800	1000
Structure			**Match Score**			
Calcite	63	56	-	-	-	-
Mordenite	48	51	52	54	52	-
Clinoptilolite	31	26	21	21	17	-
Anorthite	-	-	37	30	25	25
Quartz	-	41	51	59	53	55
Cristobalite	-	-	-	-	-	51

3.3. Surface area and porosity transformation of mordenite-clinoptilolite natural zeolite

Table 2 shows the change in surface area and porosity properties of the various high-temperature calcination treatment of mordenite-clinoptilolite natural zeolite. Calcination treatments effect on the surface area and porosity as one of the aims of calcination, mainly to clean the pores from the impurities [21, 27-31]. This function is appointed at CT1 and CT2 zeolite which increase the surface area until four times from the pristine. Although it is less than the surface area of CT1-0 as the dealumination treatment only, the anorthite structure which presented after calcination is more stable in heating temperature than calcite as mentioned in previous discussion part.

However, the high-temperature calcination does not always produce improvement in the surface area and porosity, especially at too high temperatures. CT3 and CT4 displayed these conditions; even the CT4 shows the lost porosity. It indicates that the material converts from porous material into a solid or compact material which destructs the pore [29, 31]. This condition is supported by the published studies of zeolite calcination treatment above 500 °C which state that the zeolite change into siliceous phase (quartz or cristobalite) or a compact aluminosilicate (labradorite) or amorphous crystal [39-44].

Table 2. Surface area and porosity of mordenite-clinoptilolite natural zeolite in various treatment

Code	Calcination Temperature (°C)	Specific Surface Area (m^2/g)	Pore Volume ($x\ 10^{-2}\ cm^3/g$)	Average Pore Size (nm)
A0	0	25.95	5.31	8.18
CT1-0	0	146.91	10.97	2.99
CT1	400	103.82	8.87	3.42
CT2	600	107.95	9.40	3.48
CT3	800	35.27	6.22	7.05
CT4	1000	0.61	0.18	11.52

3.4. Morphology and Si/Al ratio transformation of mordenite-clinoptilolite natural zeolite

The morphology and Si/Al ratio were determined by SEM-EDX which shown the results in figure 2 and table 3 respectively. For the morphology, the various treatments of mordenite-clinoptilolite natural zeolite do not provide the extraordinary change from the SEM results, except for the figure 2.f (CT4 zeolite) as the most considerable transformation. The morphology of CT4 zeolite shows the melting structure configures the compact structure which is different from others morphology. This condition is supported by the XRD result on the previous section about CT4 properties that the new crystal structure of cristobalite constructs and the loss of surface area after the calcination of 1000 °C.

For the Si/Al ratio, the different value finds in the pristine natural zeolite only (A0 zeolite). After the dealumination, Si/Al ratio increases up to 7.1 for the base material (CT1-0 zeolite) which indicates the effectiveness of Al removal. The calcination in various high temperatures provides the similar Si/Al ratio value between 6.3-6.9. These ratio values close to the ratio of the base material and indicate no effect of calcination which supported the previous studies [45, 46].

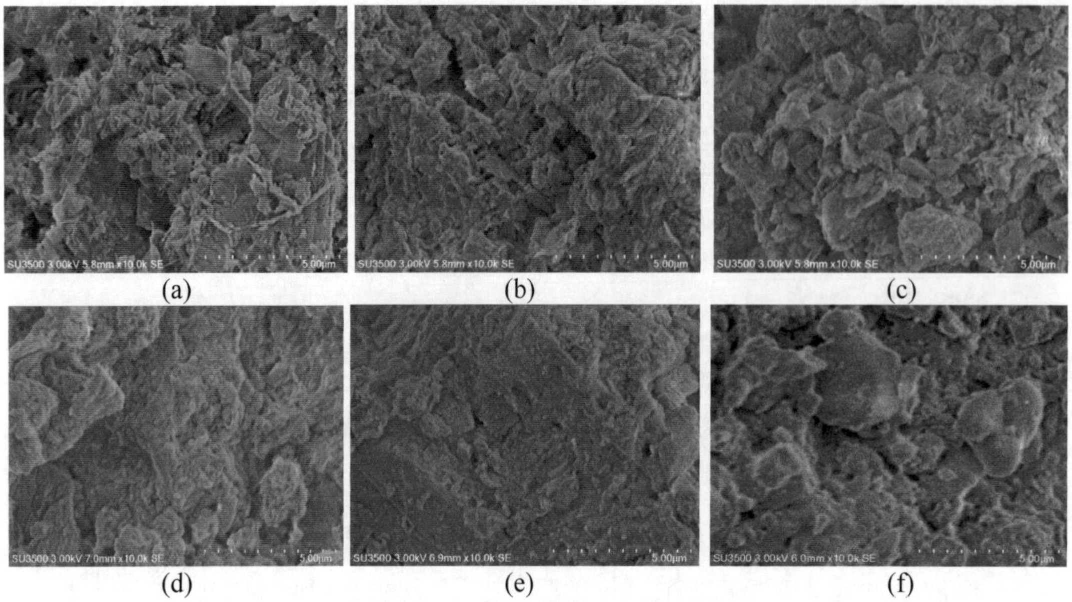

Figure 2. The morphology of mordenite-clinoptilolite natural zeolite in various treatments: (a) A0 zeolite; (b) CT1-0 zeolite; (c) CT1 zeolite; (d) CT2 zeolite; (e) CT3 zeolite; (f) CT4 zeolite

Table 3. Si/Al ratio of mordenite-clinoptilolite natural zeolite in various treatments

Code	A0	CT1-0	CT1	CT2	CT3	CT4
Si/Al ratio	4.6	7.1	6.4	6.3	6.9	6.7

4. Conclusions

In this work, transformational changes in the mordenite-clinoptilolite zeolite, as one of the typical structure of natural zeolite, were carried out. The characterization of mordenite-clinoptilolite natural zeolite at different temperatures of calcination demonstrates the primary effect on the crystal structure and surface area, particularly in the high temperature of above 600 °C. The calcium crystal transforms from calcite into anorthite, while the zeolite crystal transforms from mordenite-clinoptilolite into cristobalite. The crystal transformation into cristobalite destructs the pores of the structure which is also detected in the morphological characterization. However, the Si/Al ratio did not change much with an increase of the calcination temperature. The results of this study guide the appropriate calcination temperature treatments of mordenite-clinoptilolite natural zeolite to enhance its properties for applications in various fields.

5. Acknowledgment

The authors would like to acknowledge the scholarship support from RISET-PRO Project, Ministry of Research, Technology and Higher Education, Republic of Indonesia; also for the research facilities support from Future Industries Institute-School of Engineering, Div. ITEE, University of South Australia and BPTBA LIPI-Indonesia.

6. References

[1] Wang S and Peng Y 2010 Natural zeolites as effective adsorbents in water and wastewater treatment *Chemical Engineering Journal* **156** 11-24

[2] Ozaydin S, Kocer G and Hepbasli A 2006 Natural Zeolites in Energy Applications *Energy Sources, Part A: Recovery, Utilization, and Environmental Effects* **28** 1425-31

[3] Marantos I, Christidis G E and Ulmanu M 2012 *Handbook of Natural Zeolites,* ed V J Inglezakis and A A Zorpas: Bentham Science Publishers) pp 28-51

[4] Misaelides P 2011 Application of natural zeolites in environmental remediation: A short review *Microporous and Mesoporous Materials* **144** 15-8

[5] Widayat, Roesyadi A and Rachimoellah M 2009 The Effect of Time dealumination and Solvent Concentration in Synthesis of Zeolite Catalyst and Catalytic Test for DiEthyl Ether Production Process *AIP Conference Proceedings* **1169** 106-12

[6] Idrus A, Titisari A D, Sudiyo R and Soekrisno R 2008 Geology, characterization, quality improvement and recommended utilization of natural zeolite (Zeolitic Tuff) deposits from gunung kidul, yogyakarta special teritory, Indonesia. In: *In "2nd IASME / WSEAS International Conference on Geology and Seismology (GES'08),* (Cambridge, UK: ISSN 1790-2769; ISBN 978-960-6766-39-8) pp 21-5

[7] Wahono S K and Rizal W A 2014 Biogas Filter Based on Local Natural Zeolite Materials *International Journal of Renewable Energy Development* **3** 1-5

[8] Kogel J E 2006 *Industrial minerals & rocks: commodities, markets, and uses*: SME)

[9] Lestari D Y 2010 Kajian Modifikasi dan Karakterisasi Zeolit Alam dari Berbagai Negara [Review of Natural Zeolites Modification and Characterization from Various Countries]. In: *National Seminar of Chemistry and Education of Chemistry,* (Yogyakarta - Indonesia: Education of Chemistry Department, Yogyakarta State University)

[10] Hernawan, Wahono S K, Maryana R and Pratiwi D 2015 Modification of Gunungkidul Natural Zeolite as Bioethanol Dehydrating Agents *Energy Procedia* **65** 116-20

[11] Wahono S K, Hernawan, Kristiani A, Tursiloadi S and Abimanyu H 2014 Characterization and Utilization of Gunungkidul Natural Zeolite for Bioethanol Dehydration *Energy Procedia* **47** 263-7

[12] Widayat, Mustafa, Roesyadi A and Rachimoellah M 2008 Pengaruh Konsentrasi HCl dan Jenis Reaktan dalam Pembuatan Katalis Zeolit untuk Proses Dehidrasi dari Zeolit Alam [Effect of HCl Concentration and Reactant Type in The Zeolite Catalysts Production for Dehydration Process of Natural Zeolite]. In: *National Seminar on Chemical Engineering and Processes,* (Semarang - Indonesia: Chemical Engineering Department, Diponegoro University) pp H091 1-9

[13] Widayat W, Achmad R and Muhammad R 2013 Diethyl Ether Production Process with Various Catalyst Type *International Journal of Science and Engineering* **4** 6-10

[14] Kristiani A, Sudiyarmanto S, Aulia F, Nurul Hidayati L and Abimanyu H 2017 Metal supported on natural zeolite as catalysts for conversion of ethanol to gasoline *MATEC Web Conf.* **101** 01001

[15] Widayat W, Susanto H and Satriadi H 2014 Optimization Process of H-Zeolite Catalyst Preparation with Surface Response Methods. In: *Green Chemistry Proceeding Of 9th Joint Conference on Chemistry*: Department of Chemistry Diponegoro University) pp 124-9

[16] Kok T 2010 Characterization of Natural Zeolite from Gunung Kidul, Yogyakarta, Indonesia and Its Modification, and Its Use to Reduce the Contamination of Zinc in Waste Water *Journal of Materials Science and Engineering* **4** 64-8

[17] Pich B, Warmada I W, Hendrayana H and Yoneda T 2015 Modified Natural Zeolite and Bentonite as Adsorbents of Heavy Metal Ions from Polluted Groundwater in Yogyakarta Urban Area, Indonesia *2015* **2**

[18] Retnaningrum E and Wilopo W 2017 Removal of Sulphate and Manganese on Synthetic Wastewater in Sulphate Reducing Bioreactor Using Indonesian Natural Zeolite *2017* **17**

[19] Wilopo W, Haryono S N, Putra D P E, Warmada I W and Hirajima T 2010 Copper (Cu2+) removal from water using natural zeolite from Gedangsari, Gunungkidul, Yogyakarta *Journal of Applied Geology* **2**

[20] Rosita W, Ferdiansjah A W Y P and Prihatin T J 2016 Indonesia's Local Material Effect in Clay-Based Ceramic Filter Fabrication as an Alternative for Liquid Radioactive Waste Processing Material *Materials Sciences and Applications* **7** 371

[21] Subagiono R and Biyantoro D 2006 Kinetika Reaksi Proses Adsorpsi Campuran Uranium dan Molibdenum dalam Zeolit [Reaction Kinetics of The Sorption Process of Uranium and Molibdenum Mixture in The Zeolite]. In: *Conference and Science Presentation of Basic Research and Technology of Nuclear,* (Yogyakarta, Indonesia: National Atomic Energy Agency of Indonesia) pp 219-25

[22] Ainurofiq A and Choiri S 2015 Drug release mechanism of slightly soluble drug from nanocomposite matrix formulated with zeolite/hydrotalcite as drug carrier *Tropical Journal of Pharmaceutical Research* **14** 1129-35

[23] Laksmono J, Sudibandriyo M, Saputra A and Haryono A 2017 Development of porous structured polyvinyl alcohol/zeolite/carbon composites as adsorbent. In: *IOP Conference Series: Materials Science and Engineering*: IOP Publishing) p 012006

[24] Kismurtono M, K.W S, M R and A. R d W 2015 Development of a New Phase Change Biogas for Renewable Energy Storage System *2015* 4

[25] Wahono S K, Maryana R and Kismurtono M 2009 Biogas Purification Process to Increase Gen—Set Efficiency *AIP Conference Proceedings* **1169** 185-9

[26] Wahono S K, Skinner W, Addai-Mensah J and Vasilev K 2016 Preliminary study: Combination of amine and natural zeolite as new adsorbent material for CO2 capture. In: *Chemeca 2016: Chemical Engineering-Regeneration, Recovery and Reinvention,* pp 900-10

[27] Lu J, Zhao Z, Xu C, Duan A and Zhang P 2005 Effects of Calcination Temperature on the Acidity and Catalytic Performances of HZSM-5 Zeolite Catalysts for the Catalytic Cracking of n-Butane *Journal of Natural Gas Chemistry* **14** 213–20

[28] Giudici R 1999 Synthesis, dealumination, and adsorption behavior of mordenite extrudates: diss. submitted for the degree of doctor of technical science. (Zurich, Swiss: Swiss Federal Institute of Technology Zurich)

[29] Wang S and Zhu Z H 2006 Characterisation and environmental application of an Australian natural zeolite for basic dye removal from aqueous solution *Journal of Hazardous Materials* **136** 946-52

[30] Widhiyanuriyawan D and Hamidi N 2013 Variasi Temperatur Pemanasan Zeolite alam-NaOH Untuk Pemurnian Biogas [Heating Temperature Variation of Natural Zeolites-NaOH for Biogas Purification] *Jurnal Energi Dan Manufaktur* **6** 53-63

[31] Cobzaru C 2012 *Handbook of Natural Zeolites,* ed V J Inglezakis and A A Zorpas pp 185-213

[32] Figini-Albisetti A, Velasco L F, Parra J B and Ania C O 2010 Effect of outgassing temperature on the performance of porous materials *Applied Surface Science* **256** 5182-6

[33] Thommes M, Kaneko K, Neimark Alexander V, Olivier James P, Rodriguez-Reinoso F, Rouquerol J and Sing Kenneth S W 2015 Physisorption of gases, with special reference to the evaluation of surface area and pore size distribution (IUPAC Technical Report). In: *Pure and Applied Chemistry,* p 1051

[34] Favvas E P, Tsanaktsidis C G, Sapalidis A A, Tzilantonis G T, Papageorgiou S K and Mitropoulos A C 2016 Clinoptilolite, a natural zeolite material: Structural characterization and performance evaluation on its dehydration properties of hydrocarbon-based fuels *Microporous and Mesoporous Materials* **225** 385-91

[35] Dziedzicka A, Sulikowski B and Ruggiero-Mikołajczyk M 2016 Catalytic and physicochemical properties of modified natural clinoptilolite *Catalysis Today* **259** 50-8

[36] Korkuna O, Leboda R, Skubiszewska-Zie¸ba J, Vrublevs'ka T, Gun'ko V M and Ryczkowski J 2006 Structural and physicochemical properties of natural zeolites: clinoptilolite and mordenite *Microporous and Mesoporous Materials* **87** 243-54

[37] Haydar S, Moreno-Castilla C, Ferro-García M A, Carrasco-Marín F, Rivera-Utrilla J, Perrard A and Joly J P 2000 Regularities in the temperature-programmed desorption spectra of CO2 and CO from activated carbons *Carbon* **38** 1297-308

[38] Wahono S K, Maryana R, Kismurtono M, Nisa K, Suharto, Poeloengasih C D, Febrisiantosa A and Herdian H 2010 Metoda dan Alat untuk Meningkatkan Kadar Methana dalam Biogas [Methods and Tools for Increasing Methane Contents of Biogas]. (Indonesia

[39] Roque-Malherbe R 2001 *Handbook of Surfaces and Interfaces of Materials,* ed H S Nalwa: Academic Press) pp 495-522

[40] Roque-Malherbe R, Diaz-Aguila C, Reguera-Ruiz E, Fundora-Lliteras J, López-Colado L and Hernández-Vélez M 1990 The state of iron in natural zeolites: A Mössbauer study *Zeolites* **10** 685-9

[41] Breck D W 1984 *Zeolite molecular sieves: structure, chemistry, and use*: R.E. Krieger)

[42] Hernandez-Velez M, Raymond-Herrera O, Alvarado-Martin A, Jacas-Rodriguez A and Roque-Malherbe R 1995 New materials obtained from high temperature phase transformations of natural zeolites *J Mater Sci Lett* **14** 1653-6

[43] Arcoya A, González J A, Llabre G, Seoane X L and Travieso N 1996 Role of the countercations on the molecular sieve properties of a clinoptilolite *Microporous Materials* **7** 1-13

[44] Mimura H and Kanno T 1980 Processing of Radioactive Waste Solution with Zeolites (I): Thermal-Transformations of Na, Cs and Sr Forms of Zeolites *Science reports of the Research Institutes, Tohoku University. Ser. A, Physics, chemistry and metallurgy* **29** 102-11

[45] Lemoine G 2013 Comparison of different types of zeolites used as solid acid catalysts in the transesterification reaction of Jatropha-type oil for biodiesel production. Worcester Polytechnic Institute)

[46] Windati W, Syah Y and Widati A A 2013 Impregnasi Zeolit Alam dengan TiO2 untuk Degradasi Jingga Metil Secara Fotokatalitik [Natural Zeolites Impregnation by TiO2 for Photocatalytic Degradation of Methyl Orange] *Journal Universitas Airlangga* **1**

IOP Conf. Series: Earth and Environmental Science **251** (2019) 012010 doi:10.1088/1755-1315/251/1/012010

Optimization Study of Charcoal Observation With Scanning Electron Microscope In Various Operating Conditions

E Prasetyo[1], T. H Jatmiko[1], M. F. Karimy[1]

[1]Research Unit for Natural Product Technology, Indonesian Institute of Sciences

Email: trih011@lipi.go.id

Abstract. Scanning electron microscopy (SEM) has been widely used in the development of science, especially in the field of materials science. The development of SEM as a tool of material analysis has also experienced very rapid development. However, in practice, the use of SEM for analysis requires setting parameters that cannot be generalized between materials. This study conducted to analysis charcoal morphology on various parameters of SEM operation. From the result of the study, it is known that some SEM operation parameters such as voltage accelerate, spot intensity, working distance and objective aperture influence the results of charcoal morphology analysis with SEM.

Keywords: SEM, accelerating voltage, aperture, working distance, spot intensity, charcoal.

1. Introduction

Scanning Electron Microscope (SEM) is a microscope that observes a high magnification's object using electrons. SEM scans objects by scanning with high-speed electrons. SEM has a magnification of 10 - 300000 times and the resolution reaches below 1 nm even smaller depending on the type of SEM. The primary functions of SEM can be used to find out topography, morphology, composition of a material. The combination of high magnification, excellent resolution, ability to analyze the composition and crystallographic information make SEM widely used for research and industrial purposes.

SEM working principle is the electron beam produced by a filament on an electron gun fired at the test sample. The commonly used electron gun is a filament of tungsten hairpin which functions as a cathode. A high voltage is applied to the filament which results in the release of electrons in the filament. The released electrons will move towards the anode where the test sample is placed with speed depending on the given rating. In SEM there is a condenser lens that focused the electron beam to a point on the surface of the sample. Condenser lenses focused on the electron beam in a very small diameter, which is about 10-20 nm. When an accelerated electron interacts with there will be an interaction of electrons with the sample. Electron interactions with samples produced two new types of electrons which are secondary electrons and backscattered electrons. Secondary electrons have appeared when electrons from a filament hit an atom at the surface of the sample. Electrons in the certain atomic skin get enough energy to escape from the atom. The loose electron is called a secondary electron which is then captured by the SE detector. Secondary electrons are used to analyze surface information in the form of topography and morphology. The second type of electron is backscattered electron (BSE). BSE is an electron from the filament source which is reflected after the electron hits the sample atom. BSE can be used to analyze the composition differences in a sample because each composition (molecule or atom) will reflect electrons with different intensity depending on the type of the composition of the sample itself. The electron interaction with the material causes secondary electron and scattering electron (BSE) from the surface of the sample. SE and BSE will be detected by the detector and displayed in the form of image images on a computer screen.

The attainment of high-quality SEM images requires testing samples with good conductivity, thermal stability, and a high secondary electrical yield[1,2]. In general, SEM image capturing requires the use of detectors that collect secondary electrons and backscattered electron signals[2-5]. The production of excess electrons or accumulation of free particles on a low conductive sample surface can result in a charge-discharge phenomenon and sensitive to radiation damage[6]. Charge-discharge is a phenomenon that the surface of the sample with poor conductivity force to produce a certain electrostatic charge accumulation under the irradiation of electron beam, which ultimately affects the electronic signal transmission and can generate image distortion problems such as image deformation, charged, electron drifting[2,7]. Gold plating with the surface of the test sample is carried out to improve non-conductive sample conductivity. The gold layer can reduce the accumulation of electrons and reduce damage to the test sample. However, if the gold plating is too thick, it can cause detailed changes in the sample topographic information, dimensional changes, and differences in sample composition which makes it difficult to observe [2,8]. The SEM operation process to produce high-quality images requires several parameter settings that are determined based on the type of sample tested, magnification, and information to be obtained. These parameters are the voltage of electron acceleration, Spot intensity, working distance, and Aperture.

A study of parameter optimization and sample characteristics in the SEM test has previously been carried out, but studies using charcoal have never been done. Previous research, Stein (1974) conducted a study on the effect of accelerating voltage in SEM on a biology sample, then Pretorius (2010) also conduct an investigate effect accelerating voltage with blood platelets samples, then the latest Dan he (2018) conducted study of FE SEM optimization with low conductive samples, namely hydroxyapatite, modified poly (vinylidene fluoride) and zinc oxide nanopillar. Based on a review of a previous study, this study discusses parameters that influence the operating condition of SEM to get the best test results on charcoal micrograph. Also, this study can also be used to reference SEM testing in other samples by determining the characteristics of the sample.

2. Materials and Methods

Charcoal obtained from Hitachi High-Tech at Tokyo Solution Laboratory (Kawasaki City, Kanagawa Prefecture, Japan). The surface morphology of charcoal obtained using SEM Hitachi SU3500 series with the different operating condition including spot intensity, voltage accelerates, working distance and apparture. In this experiment used a secondary electron detector, because the focus of this study is on the morphology of charcoal which only can be analyzed with a secondary electron detector.

3. Results and discussion

3.1. Effect spot intensity

Spot intensity was showing probe current intensity electron beam from filament electron gun. Spot intensity was also showing the size of the electron beam spot. Spot intensity determined based on condenser lens current from highest to lowest [9]. A higher current intensity, resulting in a strong electron beam signal and low image noise. However, resolution of the image decreases, and there was distortion that was attributed to the charge which was the result of electron accumulated on the sample surface [2]. The result of various spot intensity shown in Figure 1.

Based on the observations of charcoal the sample was known if the spot intensity is higher then the image quality is lower. High image quality gets if using a low-intensity spot. However, the use of low spot intensity will be difficult in determining focus because Low-intensity spot causes high noise and number of electron decreases.

IOP Conf. Series: Earth and Environmental Science **251** (2019) 012010 doi:10.1088/1755-1315/251/1/012010

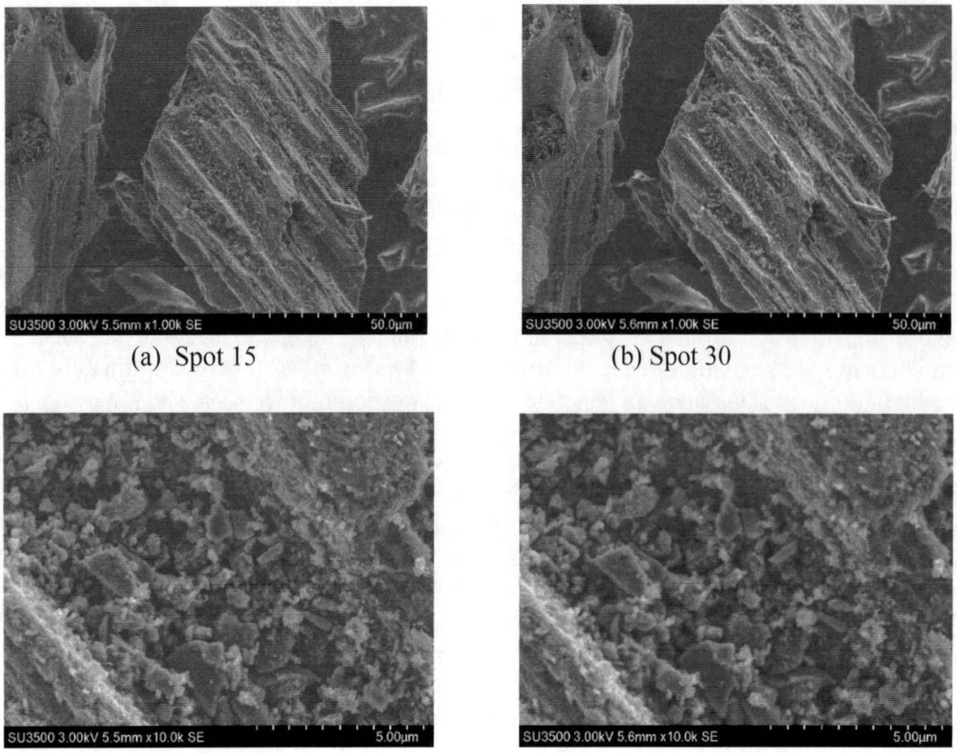

(a) Spot 15 (b) Spot 30

(c) Spot 15 (d) Spot 30

Figure 1. SEM image of charcoal at the different spot intensity under 3 kV accelerating voltage and working distance 5 mm

3.2. Effect Accelerating voltage

Accelerating voltage (Vacc) is the voltage difference between the electron source and the anode which accelerates the electron beam towards the anode. The accelerating voltage of a typical SEM ranges from 0 to 30kV. Changing the accelerating voltage alters the incident electron energy to a specimen. This entails changes area of electron scattering in the specimen, secondary electron emission yield, production charging up and depth penetration. In consequence, various image quality changes as shown in figure 2.

The most important thing about changing the accelerating voltage is the penetration depth. Penetration depth is a measure of the depth of the electron penetrating the sample. To get the high-quality image, it requires penetration depth with a certain value to get maximum sample surface information and resolution [10]. Penetration depth is determined based on the formula of kanayaokoyama.

$$R = \frac{0.0276\, A\, {E_0}^{n}}{(Z^{0.89})}$$

R -- Depth Penetration
A -- Atomic Weight (g/mole)
n -- A constant
E_0 -- Beam Energy (KV)
Z -- Atomic number
ρ -- Density $(g/cm)^2$

Based on the formula above the penetration depth is determined by Vacc, density, and type of atom. Atoms with large or large density atomic numbers will need a large Vacc to get the penetration depth compared to the density and smaller atomic numbers. In this study the sample used was charcoal. Test results with various Vacc variations are shown in Figure 2. charcoal is a sample that is included in a sample of small atomic numbers, and the density is also not too high so that the Vacc that is not too large has obtained a good image. Tests with high Vacc (15kv and 30 kV) have some morphological images that look transparent and light-emitting due to the charging up phenomenon. This happened because the depth penetration was too deep so that the secondary electron not only came from the surface of the sample but also the inside.

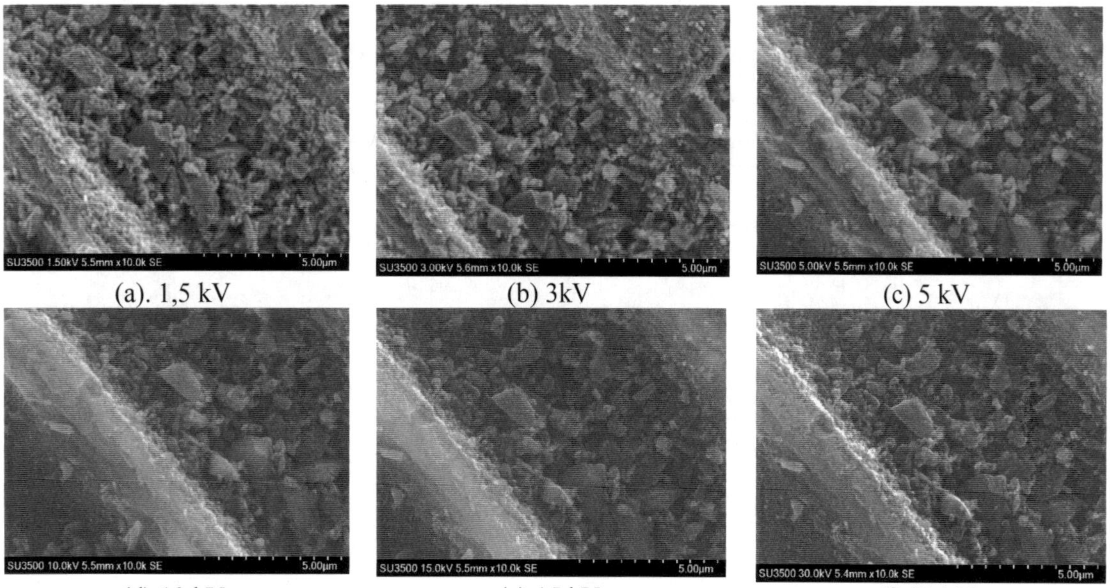

(a). 1,5 kV (b) 3kV (c) 5 kV

(d) 10 kV (e) 15 kV (f) 30 kV

Figure 2. SEM image of charcoal at the different accelerating voltage under working distance 5 mm and Spot intensity 30

Images with Vacc 15 kV and 30 kV shows light exposure due to the charging up phenomenon. This charging phenomenon shows that there is a buildup of electron charges on the surface of this charcoal sample. The parameter that affects this phenomenon is the conductivity value of the sample. Conductivity values indicate the level of a sample flowing electric current (electron). The greater the conductivity, the greater the ability of the ocean to flow electric current. Carbon has a conductivity value of 3 Mohm. The higher the conductivity value, the smaller he is at risk of charging when the Vacc is enlarged. In the charcoal samples tested, the results with Vacc below 10 kV showed the charging phenomenon was decreasing and the best picture on the picture with 5kVVacc. Besides, the higher the Vacc, the risk of the sample being damaged when interacting with high-speed electrons. Understanding the type of sample is very important to determine the Vacc as needed.

3.3. Aperture effect

The objective lens aperture controls the divergence angle of the electron beam passing through it and affects the electron beam diameter, the probe current (number of an electron), focal depth. The aperture at Sem Hitachi SU3500 has five apertures with aperture number 4 having the smallest diameter aperture lens and other hands the aperture number 0 has the largest diameter aperture lens. Probe electron shows the number of secondary electrons generated. If the aperture lens diameter gets smaller, this causes the number of secondary electrons produced decrease. Focal depth D refers to a distance range in the direction of height where an image appears to be focused even when the focal point is deviated due to the surface unevenness of the image. This depth is determined by the

IOP Conf. Series: Earth and Environmental Science **251** (2019) 012010 doi:10.1088/1755-1315/251/1/012010

divergence angle. This signifies that the focal depth of SEM is longer when the size of the objective lens aperture is smaller.

SEM test results on charcoal with various Aperture parameters are shown in Figure 3. The image shows that the smaller aperture value of the resulting image the lower the quality. Theoretically, if the aperture gets smaller, this causes a large divergence angle resulting in shallow focal depth, and then the beam diameter gets bigger. Based on the resulting image shows the aperture number 4 produces the best image. The disadvantage of using a small aperture is that the current probe gets smaller so that the observation and finding focus process will be more difficult.

(a). 0

(b) 1

(c) 2

(d). 3

(e) 4

Figure 3. SEM image of charcoal at the different Aperture under 3 kV accelerating voltage, working distance 5 mm and spot intensity 30.

3.4. Working Distance effect

Working distance (WD) is the space between the bottom of the SEM column and the top of the sample. In general, the shorter the working distance will result in a higher resolution image and present a clearer surface detail. Changing the working distance bring about changes in the beam diameter, the focal length of the objective lens and the divergence angle of the electron beam. Extending the working distance enlarges the diameter of the electron beam but increase the focal depth. Also, when the working distance is extended, the scanning area of SEM widens and magnification decreases.

Test results with working distance variations are shown in figure 4. Based on working distance resulting image at a small magnification, it does not have too much effect. However, at large magnifications, working distance affects where the small working distance values focus better than the large working distance. The high working distance will make the diameter beam bigger, and therefore high magnification will be blurred. Working distance low contrast will make the small diameter beam, and that means the focus will be better.

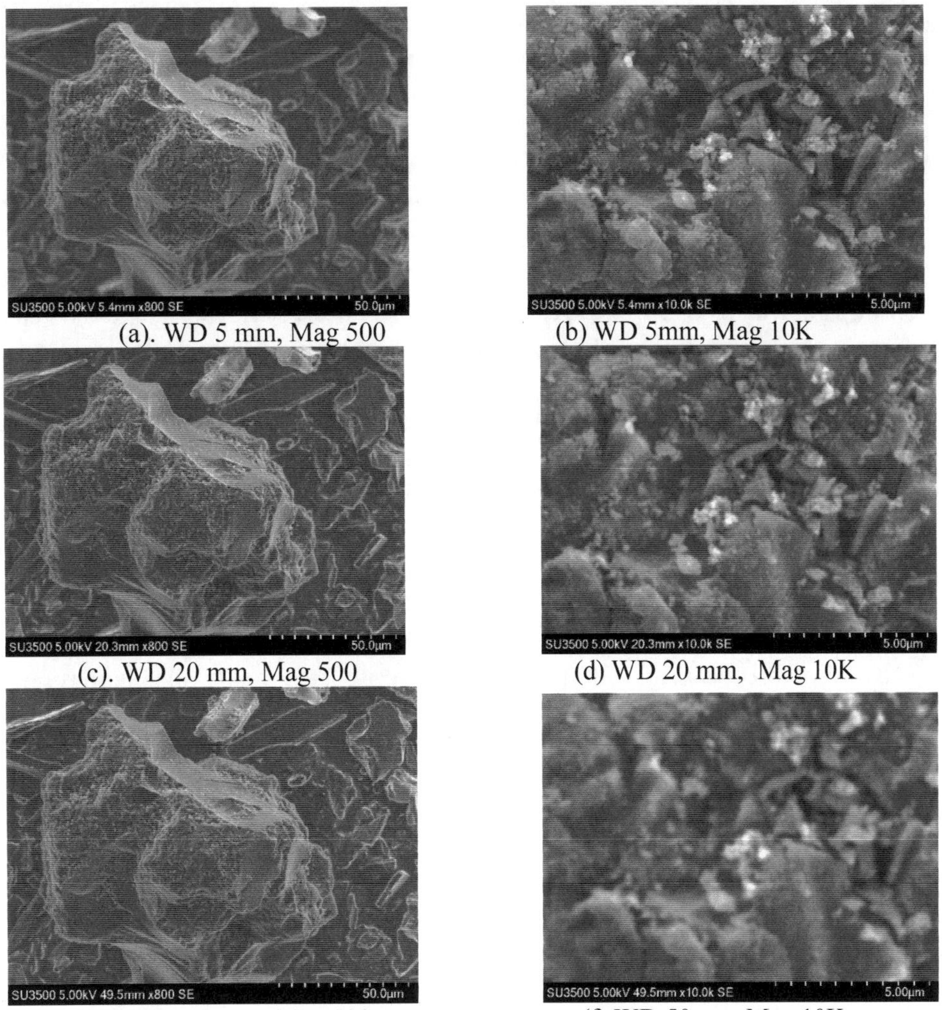

(a). WD 5 mm, Mag 500 (b) WD 5mm, Mag 10K

(c). WD 20 mm, Mag 500 (d) WD 20 mm, Mag 10K

(e). WD 50 mm, Mag 500 (f) WD 50 mm,Mag 10K

Figure 4. SEM image of charcoal at the different working distance under accelerating voltage 3kV, and spot intensity 30.

4. Conclusion

The present study directly characterized a representative sample of charcoal using SU 3500 scanning electron microscope. The optimum conditions for these morphology materials observations were obtained following changes in the different working accelerating voltage, electron beam spot intensity, aperture, and working distance. Also, the application SEM on the direct morphology observation was systematically examined. The optimized study generated a clear improvement in the observation accuracy and detection efficiency of the sample.

5. Acknowledgment

We would show gratitude to Hitachi High-Tech at Tokyo Solution Laboratory and PT. Fajar Mas Murni for supporting this study.

IOP Conf. Series: Earth and Environmental Science **251** (2019) 012010 doi:10.1088/1755-1315/251/1/012010

6. References

[1] Liu J 2000 High-resolution and low-voltage FE-SEM imaging and microanalysis in materials characterization *Mater. Charact.* **44** (4) 353–363

[2] He Dan Fu Cheng and XueZhigang 2018 Optimization study of direct morphology observation by cold field emission SEM without gold coating *Micron.* **109** 53-57

[3] Gomati MME and Walker CGH 2009 Recent developments in the understanding and application of backscattered and secondary electrons in the SEM *Proc. Spie.* **7378** 73780Z-73780Z-8

[4] Elgomati MM Walker C Bonet C Tear SP and Matthew J 2008 Secondary, scattered and low energy loss electrons in the SEM: quantification for nano *Microsc Microanal.* **14** (S2) 908–909

[5] Raynald G Paula H and Eric L 2003 On the contrast and resolution of secondary and backscattered electron images in a FE-SEM *Microsc. Microanal.* **9** (S02) 970–971

[6] Pretorius E 2010 Influence of acceleration voltage on scanning electron microscopy of human blood platelets. *Microsc. Res. and Tech.* **73** 225-228

[7] Kato T Matsukawa T Koyama H and Fujikawa K 1975 scanning electron microscopy of charging effect on silicon. *J. Appl. Phys.* **46** (5) 2288–2292

[8] Bazzana S Dumrul S Warzywoda J and Sacco Jr. A 2006 Low-voltage high-resolution scanning electron microscope imaging of the uncoated and cr-coated zeolite Microporous *Mesoporous Mater* **92** (1–3) 165–172

[9] Gavrilenko VP Novikov YA and Todua PA 2008 Measurement of the Parameters of the electron Beam of a scanning electron microscope *Proc. Spie.* 70420C-70420C1

[10] Stein OL and Gay JL 1974 Effect of accelerating voltage in scanning electron microscopy *Trans. Br. Mycol. Sac.* **63** (2) 404-408

Nutritional Evaluation of *Ulva* sp. from Sepanjang Coast, Gunungkidul, Indonesia

T H Jatmiko[1], D J Prasetyo[1], C. D. Poeloengasih[1], Hernawan[1], Y Khasanah[1]

[1]Research Unit for Natural Product Technology, Indonesian Institute of Sciences, Gunungkidul, Indonesia

Email: trih011@lipi.go.id

Abstract. Seaweed is a food ingredient that abundant availability and high nutritional content. The chemical composition of seaweed differs depending on geographical and seasons and the environmental factors. This present study aim is to evaluate the nutritional composition, including proximate and mineral analysis, of green algae *Ulva* sp. from Sepanjang coast for utilization in the future for functional food. The results of the nutritional analysis of *Ulva* sp. show that the dominant element in the proximate analysis of *Ulva* sp. was carbohydrate (49.09%). Moisture, ash, fat and protein content were 9.89%, 31.4%, 0.38%, and 9.24% respectively, while crude fiber was 3.68%. The sodium and magnesium were dominant minerals than other minerals in *Ulva* sp. According to the results of this study, *Ulva* sp. from Sepanjang coast, Gunungkidul has the potential as an alternative source of an ingredient food for health benefits in the future.

Keywords— *Ulva* sp., proximate, mineral, Gunungkidul.

1. Introduction

Edible macroalgae, also known as seaweeds, is an attractive source of nutrition. Seaweed is a food ingredient commonly consumed in Asia and some Pacific countries because of its abundant availability and high nutritional content [1–4]. The highest consumption of seaweed is in Asia, especially in Japan, China, and Korea, consisting of green algae (5%), brown algae (66.5%) and red algae (33%). However, now the demand for seaweed also extends to America and Europe [5].

Although seaweed has been widely consumed and is an essential food ingredient for diets, the nutritional properties of seaweed have not been fully known, and they are usually estimated only for its chemical composition [6]. There are very few studies on the nutritional composition of seaweed, and most of the information available is only related to traditional Japanese seaweed. The chemical composition of seaweed differs depending on species, geographical and seasons and the environmental factors such as salinity, light, water temperature, nutrients, and minerals availability [1,4,6,7].

Seaweed is known to have low-calorie content, rich in polysaccharides, vitamins, proteins, minerals, and dietary fiber that makes it attracted for commercial purposes. Seaweed or macroalgae has the potential as a value-added food because seaweed is claimed to have health benefits and some types of seaweed have been shown to have antibacterial and anticoagulant activities [8].

Ulva is one of the genera in edible green algae found along the coast in the world, and its nutritional content has the potential to be used as functional or health [7,9,10]. *Ulva* has been regarded as a good source of dietary fiber and a potential source of prebiotics since it cannot be digested in the human gastrointestinal tract [7,10]. There are only a few species of *Ulva* that have been investigated for their utilization in food manufacturing [7]. There is less information about chemical and nutrient composition of *Ulva* sp. from Gunungkidul coast, especially from Sepanjang coast. This study aims to determine the proximate composition as well as the mineral content of *Ulva* sp. from Sepanjang coast, Indonesia.

2. Materials and Methods

2.1. Materials
Fresh *Ulva* sp. obtained from the coast of Sepanjang, Gunungkidul and then dried using Memmert oven at 50°C until constant weight.

2.2. Proximate and minerals analysis
The moisture content, ash, fat, protein was determined according to AOAC [11]. The carbohydrate was estimated by difference: 100 – (moisture + ash + protein + fat) %. The moisture content was determined by drying the *Ulva* in an oven at 105 °C until a constant weight was obtained. Ash content determined by heating *Ulva* sp. in a muffle furnace at 550°C until turned white and free of carbon. Then it is cooled in the desiccator to room temperature and immediately weighed again. Soxhlet extraction used to determine the total fat content, 2 g *Ulva* sp. wrapped with a filter paper and submerged in 120 mL of petroleum ether then heated and allowed to reflux for five hours. The total crude protein content was calculated by multiplying the nitrogen content, which was determined according to the standard AOAC method by a factor of 6.25. All analytical determinations were conducted at least in triplicate. The value expressed as the mean ± standard deviation. Mineral content determined using SEM-EDX of Hitachi SU3500 equipped with EDAX Element for EDX analysis.

3. Results and discussion
The proximate analysis result of *Ulva* sp. collected from the Sepanjang coast, Gunungkidul, Indonesia is shown in Table 1. Moisture content is an essential factor in determining the storage time and quality of the material. High moisture content was causing the growth of microorganisms that will damage the material [12]. The moisture content of *Ulva* sp. in this study was 9.89%. This result agreed with the previous study by Khairy and El-Shafay (2013) that moisture content of *Ulva lactuca* range 8.9–14.57% [13]. But another study report higher moisture content for *Ulva* like Yaich et al. (2011) 14.94% and Rasyid (2017) 16.9%, but Rohani (2012) report lower moisture content for *Ulva* at 6.8%.

Table 1. Nutrient content of *Ulva* sp.

Component	Value (% db)
Moisture content	9.89 ± 3.37
Ash	31.40 ± 0.62
Protein	9.24 ± 0.31
Fat	0.38 ± 0.24
Crude fiber	3.68 ± 0.16
Carbohydrate	49.09 ± 3.99

Ash content shows the total mineral content in a sample. Ash content of *Ulva* sp. in this study was 31.4% which was higher than the ash content of some previous studies by Rasyid (2017) which ranged from 11-18%, Khairy and El-Shafay (2013) range from 17.56- 23.19%, and Yaich et al. (2011) 19.59%.

The fat content (0.38%) of the green algae *Ulva* sp. studied in this work agreed with the literature that in seaweeds, in general, the content of lipids is less than 4% [10]. Ortiz et al. (2006) study found that the fat content of *Ulva lactuca* was 0.3 g/100 g dry weight. Another study by Rasyid (2017) shows that the fat content of *Ulva lactuca* 0.19%, Khairy and El-Shafay (2013) 3.6%, Yaich et al. (2011) 7.87%.

The protein content in green algae is high, ranging from 10% to 30% [8]. In this study protein content of *Ulva* sp. approximately close to a range of another study. Khairy and El-Shafay (2013) reported that the protein content of *Ulva lactuca* range between 16.78–20.12%, Yaich et al. (2011) 8.46%.

Carbohydrate is an essential nutrition for metabolic processes. (Khairy and El-Shafay, 2013). The proximate composition of *Ulva* sp. examined in this study dominate by carbohydrate; approximately 49.09%. This result is higher than reported by Khairy and El-Shafay (2013) range between 42.09–46.42% [13] but lower than reported by Rohani (2012) for *Ulva lactuca* was 59.1% [12] and 58.1% by A Rasyid (2017) 59.1% [4].

The different content of nutrient in the *Ulva* sp from Sepanjang coast, Gunungkidul with another study prove that geographical location and local environmental conditions can influence the proximate composition of seaweeds [4,6,9,10,12,13].

Table 2. The result of EDX analysis of *Ulva* sp.

Element	Weight %	Atomic %
Na	6.86	6.07
Mg	7.92	6.63
S	11.05	7.01
Cl	11.58	6.64
K	4.46	2.32
Ca	1.95	0.99
I	0.27	0.04
Fe	0.86	0.31

The sodium and magnesium were dominant minerals than other minerals in *Ulva* sp. The study by Rohani (2012) shows that dominant mineral from *Ulva lactuca* was Potassium (515.6 mg/100 g) and magnesium (79.1 mg/100 g), Rasyid (2017) was Calcium (1828 mg/100 g) and Potassium (467 mg/100 g). The heavy metal content not detected by EDX sensor and its show that the heavy metal content was lower than the limit of the quality criteria applied to edible seaweeds sold in Indonesia.

4. Conclusion
Green alga *Ulva* sp. from Sepanjang coast, Gunungkidul has great potential for cultivation as a functional food raw material because it has high levels of carbohydrates, minerals, and proteins but low in fat. Its nutritional content is very potential for food and health products.

5. Acknowledgment
This work financially supported by The Ministries of Research, Technology, and Higher Education through INSINAS project no. 06/INS-1/PPK/E4/2018. The authors would also like to thanks to Andri Suwanto for the support during research.

6. References
[1] Lorenzo J, Agregán R, Munekata P, Franco D, Carballo J, Şahin S, Lacomba R and Barba F 2017 Proximate Composition and Nutritional Value of Three Macroalgae: Ascophyllum nodosum, Fucus vesiculosus and Bifurcaria bifurcata *Mar. Drugs* **15** 360
[2] Nagappan T and Vairappan C S 2014 Nutritional and bioactive properties of three edible species of green algae, genus Caulerpa (Caulerpaceae) *J. Appl. Phycol.* **26** 1019–27
[3] Paul N A, Neveux N, Magnusson M and de Nys R 2013 Comparative production and nutritional value of "sea grapes" — the tropical green seaweeds Caulerpa lentillifera and C. racemosa *J. Appl. Phycol.*
[4] Rasyid A 2017 Evaluation of Nutritional Composition of The Dried Seaweed *Ulva lactuca* from Pameungpeuk Waters, Indonesia *Trop. Life Sci. Res.* **28** 119–25

IOP Conf. Series: Earth and Environmental Science **251** (2019) 012011 doi:10.1088/1755-1315/251/1/012011

[5] Marinho-Soriano E, Fonseca P C, Carneiro M A A and Moreira W S C 2006 Seasonal variation in the chemical composition of two tropical seaweeds *Bioresour. Technol.* **97** 2402–6

[6] Yaich H, Garna H, Besbes S, Paquot M, Blecker C and Attia H 2011 Chemical composition and functional properties of *Ulva lactuca* seaweed collected in Tunisia *Food Chem.* **128** 895–901

[7] Almeida AP V L 2013 The Marine Macroalgae of the Genus *Ulva*: Chemistry, Biological Activities and Potential Applications *Oceanogr. Open Access* **01**

[8] Rodrigues D, Freitas A C, Pereira L, Rocha-Santos T A P, Vasconcelos M W, Roriz M, Rodríguez-Alcalá L M, Gomes A M P and Duarte A C 2015 Chemical composition of red, brown and green macroalgae from Buarcos bay in Central West Coast of Portugal *Food Chem.* **183** 197–207

[9] Peña-Rodríguez A, Mawhinney T P, Ricque-Marie D and Cruz-Suárez L E 2011 Chemical composition of cultivated seaweed *Ulva* clathrata (Roth) C. Agardh *Food Chem.* **129** 491–8

[10] Ortiz J, Romero N, Robert P, Araya J, Lopez-Hernández J, Bozzo C, Navarrete E, Osorio A and Rios A 2006 Dietary fiber, amino acid, fatty acid and tocopherol contents of the edible seaweeds *Ulva lactuca* and Durvillaea antarctica *Food Chem.* **99** 98–104

[11] AOAC 1995 *Official methods of analysis of the Association of Official Analytical Chemists* ed P Cunniff (Washington, DC USA.: Association of Official Analytical Chemist)

[12] Rohani-Ghadikolaei K, Abdulalian E and Ng W-K 2012 Evaluation of the proximate, fatty acid and mineral composition of representative green, brown and red seaweeds from the Persian Gulf of Iran as potential food and feed resources *J. Food Sci. Technol.* **49** 774–80

[13] Khairy H M and El-Shafay S M 2013 Seasonal variations in the biochemical composition of some common seaweed species from the coast of Abu Qir Bay, Alexandria, Egypt *Oceanologia* **55** 435–52

Adsorption of Methylene Blue dye using Marine algae *Ulva lactuca*

D Pratiwi, D J Prasetyo, C D Poeloengasih

Research Unit for Natural Product Technology, Indonesian Institute of Science, Yogyakarta, Indonesia 55861

Corresponding author: dee.diahpratiwi@gmail.com

Abstract. Contamination of dyes causes an aesthetic problem, severe public health concern and adds several serious environmental issues as a result of their resistance in nature and nonbiodegradable characteristic. Adsorption of methylene blue (MB) has been conducted by utilizing marine algae *U. lactuca* as biosorbent. In this study, the batch adsorption of Methylene blue dye using marine algae *U. lactuca* was examined. The ability of the *U. lactuca* in removing the dye color was dependent on contact time, algae biomass, dye concentration, and pH. The optimum adsorption was found at around pH 8; contact time 110 min; adsorbent dose 1,25 g/L; initial concentration 25 mg/L. The maximum percentage dye removal value was 91,92 %. This work confirms the potential use of *U. lactuca* for removing Methylene blue dyes solutions.
Keywords: Methylene blue, *Ulva lactuca*, adsorption, biosorbent, removal of dyes

1. Introduction

A dye is a substance that used to impart color to product or material i.e.paper paper, leather, textile, rubber, plastic, and cosmetic. Colour added to the product or material to create an attractive appearance and give added value to an aesthetic and also for a social reason. Coloring compounds can be made from a natural compound or human-made (synthetic). Synthetic dyes are aromatic water soluble and dispersible organic colorants [1,2]. The dyes are having potential application in various industry and have become a major commodity in trade.

The textile industry in one of the major application for synthetic dyes. These industries produce colored wastewater which is heavily polluted with dyes, textile auxiliaries, chemicals used in the process production [3]. Visual pollution is a major problem in water quality because it is not easy to accept red or browns rivers [4]. Contamination of dyes causes not only an aesthetic problem and severe public health concern but also adds several serious environmental issues as a result of their resistance in nature and nonbiodegradable characteristics [5-6]. The textile industrial wastewater could lead to a decline in the quality of the environment. The level of pollutant even in very low concentration is highly visible, undesirable, carcinogenic and will affect aquatic life and food chain. They are generally stable to light, an oxidizing agent and resistant to aerobic digestion, which is raising difficulties in treating colored wastewater and difficult to be degraded

Colour can be removed from wastewater effluent by chemical, physical, and biology methods [7]. Some of the applied methods in the treatment of effluent containing dyes are flocculation, coagulation, precipitation, adsorption, filtration, and ion exchange. All these methods have shortcomings and limitations, sometimes the process is not always effective and economical [1,8]. This problem has prompted the employment of various material to develop a cheaper alternative method by utilizing many varieties of biosorbent which has high sorption capability and an unlimited amount of availability. Marine alga is an alternative adsorbent from biological resources which are obtainable in massive quantities in the world. Bioconversion and biosorption could utilize by using marine algae as an adsorbent. Through bioconversion, some algae will break down the dyes into additional easy compounds. Biosorption process is typically fast and in theory appropriate for extraction of color and metal ion from massive volumes of water. Hence, biosorption is thought of as a promising technique involved with the uptake of unwanted ions from liquid solutions using biological materials [2,8].

U. lactuca is a small genus of marine algae which are available in large quantities in the world. This marine alga is an edible green alga as known by the common name sea lettuce. *U. Lactuca* was approved as usable biomaterials for biological treatment for dye effluent [2,5,6,8,9] and toxic heavy metal from the solution [10,11]. Thus this study aims to investigate the adsorption capability of commonly available green algae *U. lactuca* on from solution and effects of initial dyes concentration, contact time, the concentration of algae biomass, and pH.

2. Material and Methods

2.1. Material

2.1.1. Preparation of reagents

Methylene blue C.I.52015 (Merck) is used as a dye source. The dye stock solution was prepared by dissolving accurately weighted dye in the distilled water to the different initial concentration. Dye concentration determined was performed colorimetrically using a Halo UV-VIS Spectrophotometer RB-10. The absorbance of colors was read at 663 nm. The initial pH was adjusted with concentrated Hydrochloric Acid (Merck) or Sodium Hydroxide (Merck).

2.1.2. Preparation of adsorbent

U. lactuca (green algae) was collected from Sepanjang Beach, Kemadang, Tanjungsari, Gunungkidul Regency, Special Region of Yogyakarta during November 2017. The collected algae were washed with sea water, tap water, and then distilled water many times. The washed algae were oven dried at 100 °C for 2 h. The dried algae were ground as a powder using an electric mill and sieved to uniform particle size 310 μm.

2.2. Methods

Adsorption experiment was carried out in batch condition. A series of 250 ml Erlenmeyer flask containing 200 ml dye solution of a known initial concentration of 25 mg/L were prepared at room temperature (25 ± 2 °C). Weighted amounts (1.25 g/L) of dry algae biomass were added to each flask and agitated continuously using an orbital shaker. The pH of the mixture was kept without measurement. Equilibrium process is directly correlated with time. The sample was drawn at suitable time intervals 5, 10, 20, 30, 40, 50, 60, 70, 80, 90, 100, 110, and 120 min, and then centrifuged for 10 min at 2000 rpm. The left out the concentration of dye in supernatants were analyzed using the spectrophotometer by observation the absorbance changes at a wavelength of 663 nm.

Effect of pH was tested at pH 5, 6, 7 and 8. To verify the effect adsorbent amount, different weight of algae biomass amounts was variated 0.5; 1; 1.25; 1.5; and 2 g/L. The dye removal from aqueous solution depends on the initial MB concentration that carried at different concentration (15, 20, 25, 30, and 35 mg/L.

Biosorption capacity (qe), the number of dye adsorbed per gram of biosorbent (mg/g) can be calculated as follows:

$$q_e = (C_o - C_e)\,V/m$$

Where C_o is initial dye concentration (mg/L), C_e is the equilibrium concentration of dye (mg/L), V is the volume of solution (L), and m is the mass of biosorbent (g). Percentage of dye removal can also be displayed by the percentage of dye removal as follow:

$$Dye\ removal\ (\%) = 100\ (C_o - C_e)C_o$$

3. Results and Discussion

3.1. Effect of contact time

Contact time is the most important issue affecting the efficiency of biosorption. The removal of dye color by *U. lactuca* as monitored at various contact times that shown in figure 1. The adsorption efficiency increases with a rise in contact time up to 110 min, after which it is less constant. Majority of adsorption onto adsorbents was achieved after the first 110 min. The percentage removal of dyes from methylene blue solution increase in contact time and is supported in published studied [2,5,6,9]. As contact time increases, percentage removal also increases initially, but then gradually approaches a constant value. These changes in the rate of removal may be because initial adsorbent sites are vacant and a solute concentration gradient is high [2]. The higher biosorption at the initial contact time may be associated with the driving force of dye into the surface of *U. lactuca* [9].

3.2. Effect of adsorbent dosage

The dye removal values at the variation of *U. lactuca* dosage shown in figure 1. A rise in adsorbent quantity led to a decrease in dye removal and is supported in revealed studied [9]. However, this study is not in agreement with results obtained that supported in publishing studied [5,6] where the sorption capability increased with increasing adsorbent dosage. High biosorbent amounts are known to cause cell agglomeration and subsequent reduction in intercellular distance and produce "screen effect" among a dense layer of cells, resulting to the "protection" of binding sites from metal ions. The obtained knowledge was agreement that supported in publish studied [9,12], the lower biosorbent metal percentage at high adsorbent concentrations [10].

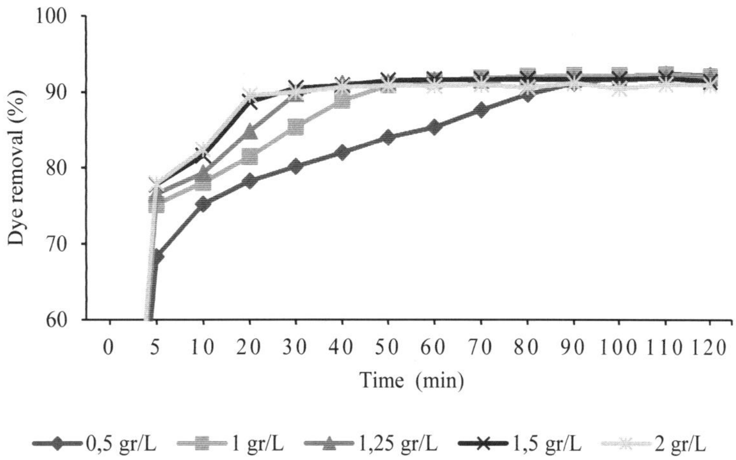

Figure 1. Effect of contact time and adsorbent amount on the percentage dye removal

3.3. Effect of initial concentration of dye

The effect of dye initial concentration on adsorption of MB was investigated over a spread of pH from 15 to 35 mg/L under constant parameter, pH 7, adsorbent dose 1.25 gr/L, contact time 110 min, and temperature 25 °C ± 2 °C. The solution was agitated continuously using an orbital shaker. The effect of variation of dye initial concentration at dye removal percentages is represented in Figure 2. The result is in accordance with research conducted that supported in publish studied [5], that rate of dye removal decrease with increase in initial concentration. This is because of the formation of the monolayer at the lower initial concentration of dye over the surface of adsorbent [5].

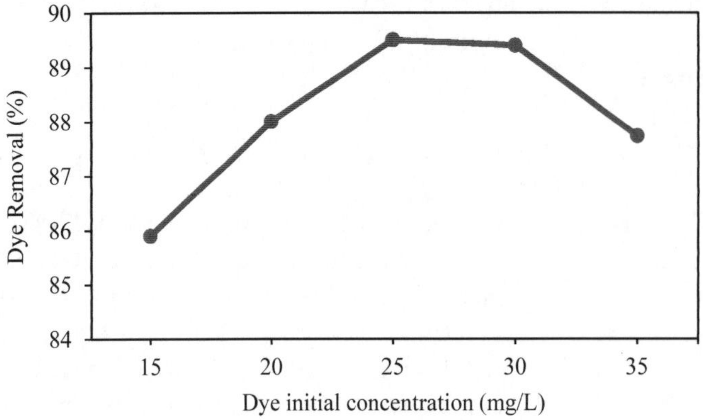

Figure 2. Effect of dye initial concentration on the percentage dye removal

3.4. Effect of initial pH

The effect of initial pH on adsorption of MB was investigated over a variety of pH scale from 5 to 8 under constant parameter, MB concentration 25 mg/L, adsorbent dose 1.25 gr/L, contact time 120 min, and temperature 25 °C ± 2 °C. The pH was adjusted using 0.1 M HCl and 0.1 M NaOH. The solution was agitated continuously using an orbital shaker. The effect of variation PH at dye removal percentages are represented in Figure 3. These results are in accordance with research conducted that supported in publish studied [5,8,9], where an increasing percentage dye removal gradually increased with pH value. The adsorption of dye on biomass surface is controlled by ionic attraction [8]. This Biosorption alga has principally been attributed to the cell wall properties whatever both electrostatic attraction and complexion will play a task. Algae cell wall often consisting of protein and carbohydrates give a functional group like amino, hydroxyl, carboxyl, sulfate, etc., which can act as binding sites for metals [5,6,13]. At low pH value, a fewer anionic adsorption site on the dried *U. lactuca* was generated, and sorption was unfavorable, probably because excess H^+ competing with dye molecule for sorption site on dried *U. lactuca.* The surface algae cell all that are liable for binding of dye molecules [5].

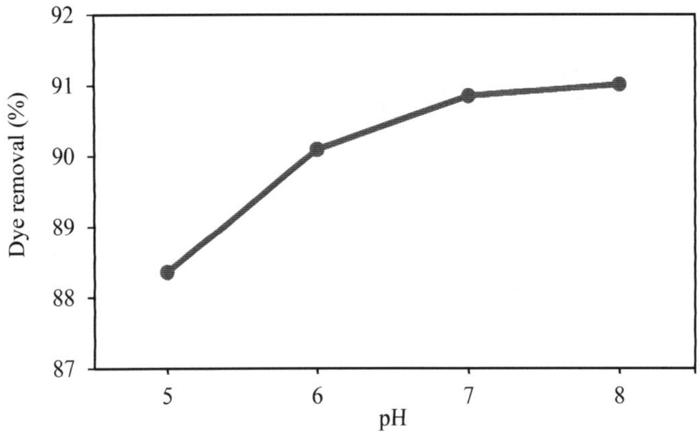

Figure 3. Effect of pH on the adsorption of percentage dye removal

4. Conclusion

The experiment conducted in this study focused on the adsorption dye into *U. lactuca*. The ability of the *U. lactuca* in removing the dye color was dependent on contact time, alga biomass, dye concentration, and pH. The optimum adsorption was found to occur at around pH 8; contact time 110 min; adsorbent dose 1,25 g/L; initial concentration 25 mg/L. The maximum percentage dye removal value was 91,92 %. The result showed that green algae *U. lactuca* could be used as biosorbent material for dyes removal from aqueous solution.

5. References

[1] Oladipo M A, Bello I A, Adeoye D O, Abdulsalam K A and Giwa A A 2013 Sorptive Removal of Dyes from Aqueous Solution: A Review *Adv. Environ. Biol. J.* **7** 3311–27

[2] El Nemr A, Abdelwahab O, Khaled A and El Sikaily A 2006 Biosorption of Direct Yellow 12 from aqueous solution using green alga *Ulva lactuca* *Chem. Ecol.* **22** 253–66

[3] Stefanakis A, Akratos C S and Tsihrintzis V A 2014 Treatment of Special Wastewaters in VFCWs *Vertical Flow Constructed Wetlands* (Elsevier) 145–64

[4] Attia A A, Rashwan W E and Khedr S A 2006 Capacity of activated carbon in the removal of acid dyes subsequent to its thermal treatment *Dyes Pigments* **69** 128–36

[5] Mane R and Sabale A Biosorption of Methylene Blue and Malachite Green From Binary Solution onto Ulva lactuca *Int. J. Curr. Microbiol. Appl. Sci.* **3** 295–304

[6] Tahir H, Sultan M and Jahanzeb Q 2008 Removal of basic dye methylene blue by using bioabsorbents Ulva lactuca and Sargassum *Afr. J. Biotechnol.* **7** 2649–55

[7] Pratiwi D, Indrianingsih A W, Darsih C and Hernawan 2017 Decolorization and Degradation of Batik Dye Effluent using *Ganoderma lucidum* *IOP Conf. Ser. Earth Environ. Sci.* **101** 012034

[8] Dahlia M E M 2013 Evaluation of non-viable biomass of Laurencia papillosa for decolorization of dye waste water *Afr. J. Biotechnol.* **12** 2215–23

[9] El Sikaily A, Khaled A, Nemr A E and Abdelwahab O 2006 Removal of Methylene Blue from aqueous solution by marine green alga *Ulva lactuca* *Chem. Ecol.* **22** 149–57

[10] Ghoneim M M, El-Desoky H S, El-Moselhy K M, Amer A, Abou El-Naga E H, Mohamedein L I and Al-Prol A E 2014 Removal of cadmium from aqueous solution using marine green algae, Ulva lactuca *Egypt. J. Aquat. Res.* **40** 235–42

[11] Ibrahim W M, Hassan A F and Azab Y A 2016 Biosorption of toxic heavy metals from aqueous solution by Ulva lactuca activated carbon *Egypt. J. Basic Appl. Sci.* **3** 241–9

[12] Esposito A, Pagnanelli F, Lodi A, Solisio C and Vegliò F 2001 Biosorption of heavy metals by Sphaerotilus natans: an equilibrium study at different pH and biomass concentrations *Hydrometallurgy* **60** 129–41

[13] Srinivasan A and Viraraghavan T 2010 Decolorization of dye wastewaters by biosorbents: A review *J. Environ. Manage.* **91** 1915–29

Production of xylose, glucose, and other products from tropical lignocellulose biomass by using maleic acid pretreatment

M Oktaviani [1,2], E Hermiati [2], A Thontowi [3], R P B Laksana[2], L N Kholida[3], A Andriani[3], Yopi[3], W Mangunwardoyo[4]

[1]Magister Programme in Biology, Department of Biology, Faculty of Mathematics and Natural Sciences, Universitas of Indonesia, Kampus Baru, UI. Depok, Indonesia 16424.
[2]Research Center for Biomaterials, Indonesian Institute of Sciences (LIPI), Jl. Raya Bogor Km. 46 Cibinong, Bogor, Indonesia. 16911.
[3]Research Center for Biotechnology, Indonesian Institute of Sciences (LIPI), Jl. Raya Bogor Km. 46 Cibinong, Bogor, Indonesia. 16911.
[4]Department of Biology, Faculty of Mathematics and Natural Sciences, Universitas of Indonesia, Kampus Baru, UI, Depok, Indonesia 16424.

E-mail: maul009@lipi.go.id; wibowo.mangun@ui.ac.id

Abstract. Efforts to utilize hemicellulose in the biorefinery of bioethanol has been continuously grown in recent years. The utilization of hemicellulosic hydrolysate after pre-treatment was still limited due to its complex composition of sugars (xylose, glucose, arabinose), sugar degradation products (HMF and furfural), lignin degradation products (phenolic compounds), and organic acids (acetic acid also formic acid). Thus, characterization of hemicellulose hydrolysate components after pre-treatment became important. In this study, the effect of maleic acid concentrations on the production of sugars and by-products from three tropical lignocellulosic biomass: sugarcane trash, corn cob, and sweet sorghum bagasse, has been investigated. The biomass was pre-treated using various concentrations of maleic acid (1, 1.5, 2, 2.5, and 3% v/v) at 121°C for 60 minutes. The results showed the presence of xylose, glucose, acetic acid, lactic acid, HMF and furfural in all biomass hydrolysates after pre-treatment with maleic acid. SEM analysis was also conducted to identify the morphological changes of the biomass caused by maleic acid hydrolysis.

1. Introduction

Lignocellulosic biomass including agro-industrial and forest residues, grasses and woody materials, kitchen and paper waste has been considered as a promising feedstock for bioethanol and other chemical products [1,2]. Lignocellulosic biomass is composed of cellulose, hemicellulose, and lignin, as well as water and other minor components. These components have specific properties to be converted into different products [3,4]. Similar with cellulose, hemicellulose can be broken down into monosaccharides, such as xylose, glucose, and arabinose, through acid or enzymatic hydrolysis [5]. However, utilization of hemicellulose hydrolysate was still limited due to its complex by-products, such as weak organic acids, phenolic compounds, and sugar degradation compounds, which produced during hemicellulose hydrolysis.

There are mainly four processes involved in the conversion of lignocellulosic biomass to ethanol, which are pretreatment, hydrolysis by chemical or enzymatic, fermentation of monosaccharides, and purification of products [6]. To release fermentable sugars from lignocellulosic biomass, that can be converted into ethanol and other chemical products, various pretreatment methods have been investigated. Physical treatment by milling and grinding [7], chemical treatment using acids [8,9], alkali [10–12], or organic solvents [13], physio-chemical treatment using liquid hot water [14] or steam explosion [15], and biological treatment using fungi [16–18], have been reported along with their advantages and disadvantages.

Sulfuric acid was the common and superior catalyst to hydrolyze hemicellulosic fraction of lignocellulosic biomass. However, pretreatment with sulfuric acid generates additional lignocellulose-derived compounds such as sugar degradation compounds and phenolic compounds, which are toxic to microorganisms. These by-products have inhibitory effects on the fermentation process by decreasing yeast performance [2]. In order to solve this problem, an organic acid catalyst such as maleic acid, oxalic acid, and formic acid, was used in the pretreatment process. As reported by many researchers, maleic acid produces fewer inhibitors in hydrolysate than sulfuric acid [19]. Thus, in this study, maleic acid was used to hydrolyze lignocellulose components of three tropical biomass, sugarcane trash, corn cob, and sweet sorghum bagasse. The main objective of this study was to investigate sugars production and the formation of degradation products in the hydrolysates obtained after maleic acid pretreatment. Morphological examination of the original and pretreated biomass was also conducted through Scanning Electron Microscopy (SEM) to support the analysis.

2. Experimental

2.1. Materials

Sugarcane (*Saccharum officinarum* L.) trash used in this study was collected from Sugarcane Factory Rajawali II, Subang, West Java, Indonesia. Corn cob and sweet sorghum bagasse were obtained from a local farm in Sukabumi and Cibinong, West Java, Indonesia, respectively. Prior to use, the raw materials were ground into small particles using a drum chipper, ring flaker, and hammer mill, then sieved to obtain particle size (powder) smaller than 80 mesh. The moisture content (MC) of sugarcane trash (ST) was 7.7%, while the MC of corn cob (CC) and sorghum bagasse (SB) was 5.7% and 5.9%. The materials were stored in sealed plastic bags at room temperature. Maleic acid (Wako Pure Chemical Industries, Ltd., Osaka, Japan) and other chemicals (from Merck) used in this study were analytical grade and used without further purification.

2.2. Methods

2.2.1 Maleic Acid Pretreatment of Raw Materials.

Pretreatment was done with the OFAT method to evaluate the effect of maleic acid concentration. Three grams (oven-dry weight) of biomass were immersed in various concentrations (1; 1.5; 2; 2.5 and 3%; v/v) of MA in laboratory glass bottles (Duran) with solid to liquid ratio of 1:10, then were heated in an autoclave at 121°C for 60 min. The slurries were filtered with an Advantec filter paper (Toyo Roshi Kaisha, Ltd., Japan) using vacuum filtration (GAST DOA-P504-BN, Cole-Parmer, Vernon Hills, IL, USA). The liquid fractions were separated and kept in a refrigerator until further analysis of glucose, xylose, acetic acid, 5-HMF, and furfural using HPLC. The pH of the liquid fractions was also measured using pH meter. The residual solid fractions (pulp) were separated, washed with distilled water until the pH was neutral, and stored in sealed plastic bags in the refrigerator for further experiments.

2.2.2. Chemical components analyses of liquid fraction.

Xylose, glucose, glycerol, acetic acid, 5-HMF, and furfural in the liquid fraction of the pre-treated biomass were analyzed by high performance liquid chromatography/HPLC (Shimadzu LC-20AB, Kyoto, Japan) equipped with a refractive index detector and Bio-Rad (Hercules, CA) Aminex HPX-87H column (300-7.8 mm) at 60 °C; 0.01 N H_2SO_4 as eluent, at a flow rate of 0.6 mL/min) and injection volume of 10 μL. The samples were centrifugated and filtered through the 0.22 μm membrane prior to injection.

2.2.3. Scanning electron microscopy (SEM) analysis.
In order to support the analysis, the surface morphology of the original and pre-treated sugarcane trash was characterized using scanning electron microscopy (SEM) JEOL JSM-5310LV (Jeol Ltd., Japan). Dried samples were placed on a specimen stub (10 mm in diameter). The surfaces of the samples then were coated with a thin layer (approximately 400-500 Å thick) of gold using coating unit Ion Coater iB2. SEM images of the samples were recorded using an accelerating voltage of 20 kV at 500-1500 magnification.

3. Results and Discussions

3.1. Sugars in the hydrolysate of maleic acid-pretreated biomass
Pre-treatment is an essential step to achieve effective hydrolysis of substrates into simple sugars, which can be used directly or fermented by microorganisms [20]. To investigate the effects of maleic acid (MA) pre-treatment on the formation of sugars and by-products in the hydrolysates, experiments were performed at MA concentrations ranging from 1 up to 3%. Results showed that xylose and glucose were formed in the hydrolysates during pre-treatment, while other monosaccharides, such as arabinose and mannose, were not detected in the hydrolysates. Figure 1 showed the effects of MA concentrations on xylose and glucose concentrations in the hydrolysates after pre-treatment. For all biomass, the concentration of xylose was increased with increasing of maleic acid concentrations up to 2.5%, and it was decreased when the hydrolysis was conducted at 3% concentration of MA.

Among all biomass, corn cob produced the highest xylose concentration (14.2 g/L) when treated by 2.5% of MA at 121 °C for 60 minutes. The previous study on MA-pretreatment of corn cob reported by Lee and Jeffries [9] showed that 50 g/L of xylose was achieved when corn cob was hydrolyzed with 15.33 g/L of MA at 170 °C for 18 minutes. Xylose concentration achieved in our study was higher than xylose concentration obtained by Mardawati et al. [21], which was only 5.5 g/L after the corn cob was treated by 6% sulfuric acid at 120 °C for 15 minutes. However, xylose concentration obtained in our study was still lower than other results reported by Hatano et al. [22], which showed that 29.7 g/L of xylose was obtained when corn cob was treated by 2% (w/v) sulfuric acid at 121 °C for 60 minutes. Another study by Misra et al. [23] also showed that 25 g/L of xylose was successfully obtained from corn cob after treatment with 1% sulfuric acid at 121 °C for 30 minutes.

Industrially, sugarcane trash, which consists of sugarcane tops and leaves, was left on the field or burnt [24]. Unlike sugarcane bagasse, utilization of sugarcane trash as an alternative energy source was still limited. As reported, sugarcane leaves consist of 40-44% cellulose, 23-29% hemicellulose, and 10% lignin [25,26]. Therefore, the utilization of sugarcane trash to produce value-added products becomes important. In this research, besides corn cob and sweet sorghum bagasse, we used sugarcane trash and hydrolyzed it using MA. As much as 11.2 g/L of xylose was produced in sugarcane trash hydrolysate during pre-treatment with 2.5% MA for 60 minutes. This xylose concentration was higher than xylose concentration reported by Jutakanoke [26], which was only 9.7 g/L after the sugarcane trash was treated by 1.5% of sulfuric acid at 121 °C for 30 minutes.

Glucose was another mono-sugar obtained in the hydrolysate during acid pre-treatment. For all biomass, the concentrations of glucose were increased with increasing of maleic acid concentrations. Maleic acid concentration of 3% gave the highest glucose concentration. Among the three biomass investigated, sweet sorghum bagasse produced the highest glucose concentration (24.7 g/L) in the hydrolysate, while the lowest concentration of glucose was obtained in sugarcane trash hydrolysate (22.5 g/L). Okur and Eken [27] reported that xylose and arabinose released in the hydrolysate were obtained from hemicellulose, while glucose and cellobiose were obtained from hydrolysis of cellulose. Moreover, Aguilar et al. [28] noted that the presence of glucose in the hydrolysate during acid pre-treatment might also come from heteropolymers of hemicellulose fraction. In addition, according to Li et al. [29], glucose obtained in corn cob hydrolysate during pretreatment may come from glucuronoarabinoxylans and xyloglucans.

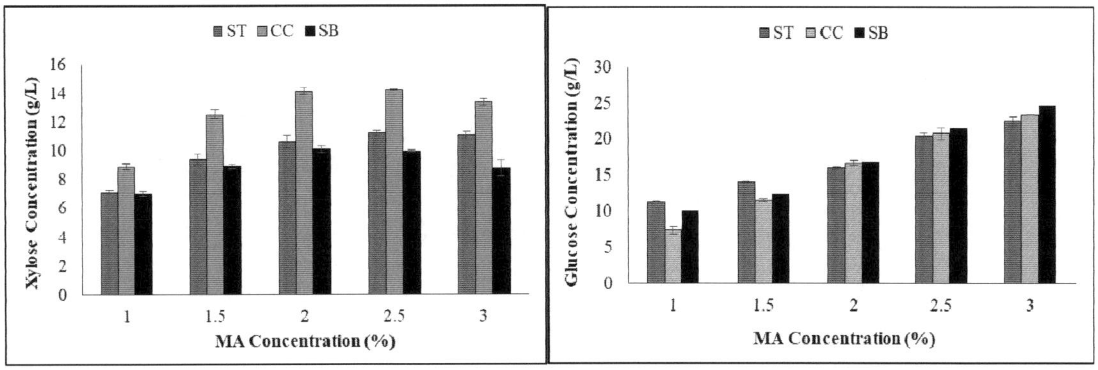

Figure 1. Xylose (A) and glucose (B) concentration in hydrolysate during maleic acid (MA) pre-treatment of sugarcane trash (ST), corn cob (CC) and sorghum bagasse (SB)

The interesting part of this study was the glucose concentration in the hydrolysate. Particle size reduction caused by milling and grinding probably has a positive influence in high glucose released in the hydrolysate. In this study, we used the biomass particle which is smaller than 80 mesh. Banerji et al. [30] reported that the reduction of particle size decreased the cellulose crystallinity of sweet sorghum bagasse. Besides the physical disruption, another possibility was the reaction time. This study used a longer reaction time (60 minutes) to hydrolyze hemicellulose compared to another study. However, a study by Mardawati et al. [21] also reported that both glucose (4.07 g/L) and xylose (5.5 g/L) were also produced in corn cob hydrolysate after treated by 6% sulfuric acid at 121 °C for 15 minutes. Therefore, comprehensive research is necessary to assess these phenomena.

3.2 Formation of degradation compounds in the hydrolysate

During acid hydrolysis, degradation compounds including sugar degradation compounds (HMF and furfural), lignin degradation products (phenolic compounds), and organic acid (acetic acid, formic acid, were produced in the hydrolysates. Figure 2 showed the final pH of hydrolysates after treated with MA. The pH of all biomass decreased as the concentration of maleic acid rose. The changes in the pH of the hydrolysates after treated with maleic acid was related to utilization of some hydrogen ions from the acid for hydrolysis process, which increased the pH of the hydrolysates. Moreover, some hydrogen ions were also formed during pretreatment as degradation products of the carbohydrates, such as acetic acid or formic acid [3]. In addition, some authors noted that during acid pretreatment, a part of the acetyl groups in the hemicellulose was cleaved and produced acetic acid [29,31]. As expected, HPLC results showed that acetic acid, ranging from 0.33 to 2.23 g/L, were formed in all of the biomass hydrolysates (Table 2).

As shown in Table 2, acetic acid and lactic acid were organic acids formed in hydrolysates during pretreatment. The other organic acids, such as levulinic, malonic, and formic acids, were not detected in hydrolysates. The concentrations of acetic acid in hydrolysates were steadily increased with increasing acid concentrations. The highest concentration of acetic acid was found in the corn cob pretreated with 2.5% MA. As was mentioned before, the formation of organic acids in the hydrolysate can inhibit the performance of fermenting microorganisms, although some microorganisms also have tolerance mechanism to avoid an inhibitory effect of the toxic hydrolysates. Weak organic acid in the hydrolysates, such as acetic and lactic acids, can lead to an influx into cytosol of the microorganisms. It caused improper ion transportation and finally inhibit the growth and productivity of the microorganisms [2]. Moreover, microorganism tolerance toward toxic degradation products in hydrolysates was strongly dependant on the physiology and nature of the microorganism itself [32]. Oberoi et al. [33] reported that 1% of acetic acid formed in rice straw hydrolysate could inhibit the growth of *Pichia kudriavzevii*. Another study by Delgenes et al. [32] showed that the growth of

Zymomonas mobilis was inhibited by the presence of 5 g/L acetic acid. The growth of *Saccharomyces cerevisiae* was also inhibited with the presence of 10.2 g/L acetic acid in hydrolysate liquor [34].

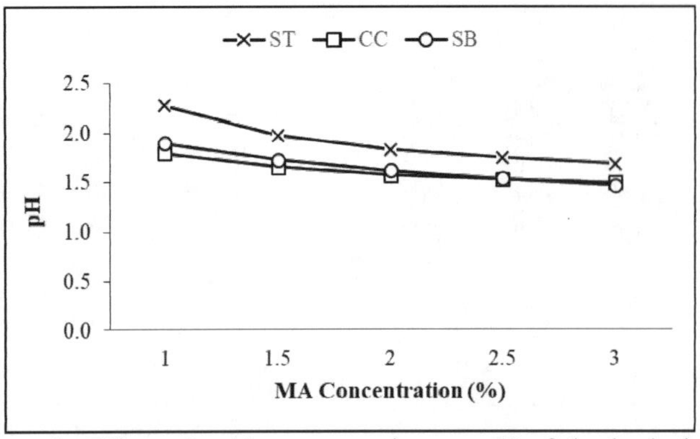

Figure 2. Effect of acid concentration on pH of the hydrolysates after MA pretreatment of sugarcane trash (ST), corn cob (CC), and sorghum bagasse (SB).

Lactic acid was also formed in all biomass hydrolysates due to acid pretreatment (Table 1). The formation of lactic acid was related to depolymerization of hemicellulose or xylan to xylose by an acid catalyst. The xylose then was converted to glycolaldehyde and glyceraldehyde through a retro-aldol condensation reaction, thus released lactic acid as a final product [35]. In this study, the highest concentration of lactic acid was produced in sugarcane trash hydrolysate (99 mg/L) after treated by 1.5% MA. However, this concentration was still low. The study by Favaro et al. [36] reported that *S. cerevisiae* fm17 have a tolerance to the presence of lactic acid, ranging from 1-5 g/L, in the fermentation hydrolysate.

Table 1. Acetic and lactic acid concentrations in hydrolysates after maleic acid (MA) pretreatment of sugarcane trash (ST), corn cob (CC) and sorghum bagasse (SB)

MA	Acetic acid (g/L)			Lactic Acid (mg/L)		
Conc. (%)	ST	CC	SB	ST	CC	SB
1	0.33 ± 0.07	1.09 ± 0.04	0.58 ± 0.49	21 ± 0.2	8 ± 0.6	20.7 ± 0.0
1.5	0.65 ± 0.17	1.35 ± 0.01	1.61 ± 0.45	99 ± 0.7	15 ± 0.2	21.4 ± 0.0
2	0.84 ± 0.11	1.95 ± 0.09	1.47 ± 0.39	48 ± 0.2	39 ± 0.3	21.5 ± 0.0
2.5	1.11 ± 0.15	2.23 ± 0.14	1.14 ± 0.24	25 ± 0.2	19 ± 0.02	21.4 ± 0.0
3	1.22 ± 0.14	2.04 ± 0.11	2.08 ± 0.42	48 ± 0.2	18 ± 0.003	20.9 ± 0.0

In this study, we used maleic acid in order to avoid degradation compounds production in the hydrolysates that can inhibit the fermenting microorganisms. Pretreatment with organic acids, such as maleic acid, produces fewer inhibitors compared with sulfuric acid [19]. Table 2 shows the formation of hydroxymethylfurfural (HMF) and furfural in pretreated hydrolysates, respectively. For all biomass, the concentration of HMF and furfural were smaller than 0.01 g/L. The highest concentration of HMF (7.3 mg/L) was produced in sorghum bagasse hydrolysate after treated with 2.5% MA. Meanwhile, the highest concentration of furfural (13.1 mg/L) was produced in corn cob hydrolysate after treated with 3% MA. These concentrations were lower than those resulted by a study on pretreatment of sweet sorghum bagasse by 0.5-5% sulfuric acid conducted by Banerji et. al. [30] at analogous process, which produced 3.1-3.4 g/L of acetic acid, 1.1 g/L of furfural, and 2.95 g/L of levulinic acid (degradation of HMF). Misra et al. [23] also reported that 1.31 g/L of acetic acid and 0.21 g/L of furfural were formed in corn cob hydrolysate after treated with 1% sulfuric acid at 121 °C for 30 minutes.

Table 2. Hydroxymethylfurfural (HMF) and furfural concentrations in the hydrolysates during maleic acid (MA) pretreatment of sugarcane trash (ST), corn cob (CC) and sorghum bagasse (SB)

MA	HMF (mg/L)			Furfural (mg/L)		
Conc. (%)	ST	CC	SB	ST	CC	SB
1	0.9 ± 0.3	1.1 ± 0.1	0.6 ± 0.1	4.7 ± 0.8	4.4 ± 0.0	5.7 ± 0.0
1.5	0.5 ± 0.3	0.5 ± 0.1	n.d.	4.8 ± 0.6	5.4 ± 0.1	n.d.
2	0.4 ± 0.1	0.7 ± 0.0	4.6 ± 0.4	5.3 ± 0.8	10.3 ± 0.2	4.6 ± 0.0
2.5	0.7 ± 0.0	5.2 ± 0.6	7.3 ± 0.0	9.9 ± 0.6	8.2 ± 0.3	7.3 ± 0.0
3	n.d.	1.6 ± 0.1	5.3 ± 0.5	5.9 ± 1.6	13.1 ± 0.2	5.3 ± 0.5

* n.d. = not detected

3.3. Scanning Electron Microscopy (SEM)

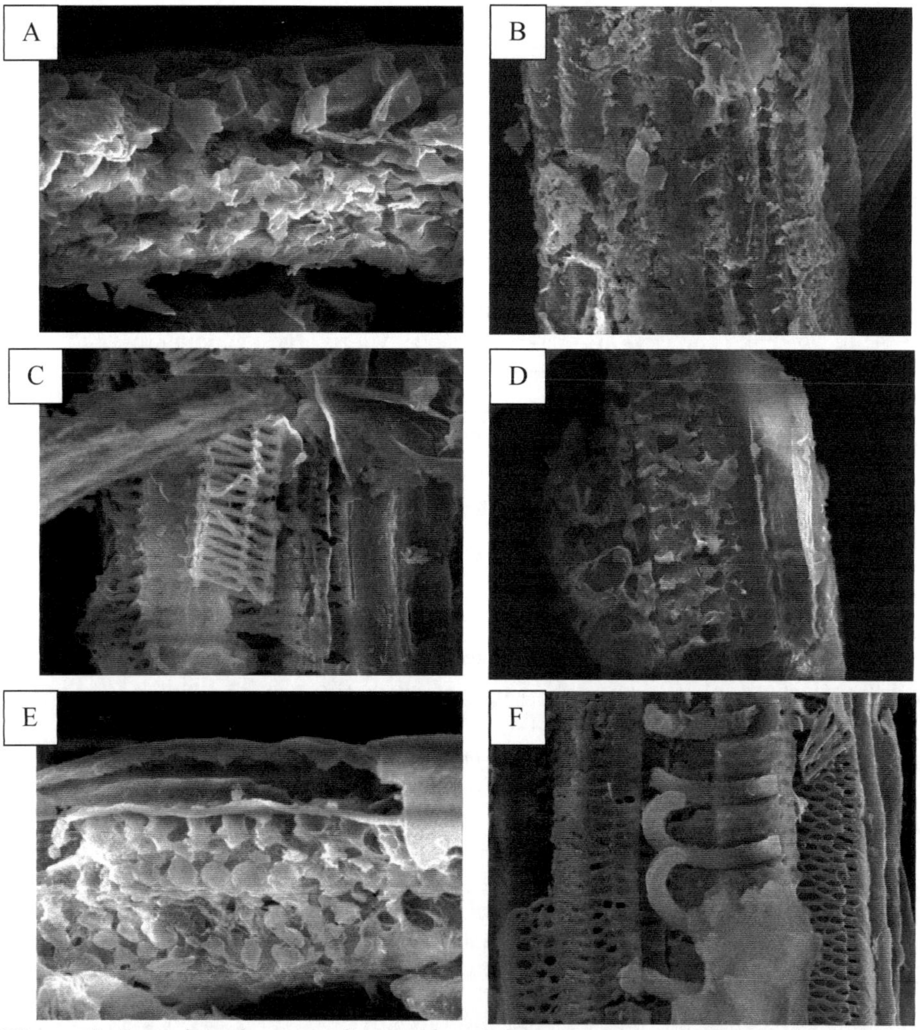

Figure 3. Scanning electron micrographs of the untreated and pretreated sugarcane trash: (A) untreated, (B) pretreated with 1% MA, (C) pretreated with 1.5% MA, (D) pretreated with 2% MA, (E) pretreated with 2.5% MA, and (F) pretreated with 3% MA. Magnification: 500x (B), 1000x (A, D), 1500x (C, E, F)

Scanning electron microscopy was carried out to compare the morphology of the untreated, and maleic acid pretreated biomass. Figure 3A is an image of the untreated sugarcane trash. Figure 3B, 3C, 3D, 3E, and 3F are images of sugarcane trash pretreated with 1, 1.5, 2, 2.5, and 3% MA, respectively. As shown in Figure 3A, the structure of untreated sugarcane trash is compact and rigid, but the surface was clearly degraded probably due to physical treatment. The sugarcane trash became more degraded as increasing of acid concentrations. Numerous porous and holes became clearly apparent as shown in Figure 3C-3F. According to Chen et al. [37], this disruption was probably caused by the removal of reactive amorphous cellulose on the biomass surface. This disruption changes the morphology of the biomass, then increases its porosity and surface area.

4. Conclusions

This study showed the quantitative and qualitative characteristics of sugarcane trash, corn cob, and sweet sorghum bagasse hydrolysates during maleic acid pretreatment. The disruption effect of acid pretreatment was also explored. Xylose and glucose were the main simple sugars produced in the hydrolysates. Among all biomass, corn cob produced the highest xylose concentration (14.2 g/L) after treated by 2.5% of MA at 121 °C for 60 minutes. The highest glucose concentration (24.7 g/L) was produced by sweet sorghum bagasse after treated by 3% of MA at 121 °C for 60 minutes. Acetic acid, lactic acid, HMF, and furfural were also produced as degradation compounds in hydrolysates.

5. Acknowledgments

This work was supported by Insentif Riset Sistem Inovasi Nasional (INSINAS) project 'Produksi Xilitol dari Biomassa Lignoselulosa sebagai Pemanis Alami Rendah Kalori Pencegah Obesitas' of Research Center for Biotechnology, LIPI (Indonesian Institute of Sciences), in the FY of 2018.

6. References

[1] Wang L, Yang M, Fan X, Zhu X, Xu T and Yuan Q 2011 An environmentally friendly and efficient method for xylitol bioconversion with high-temperature-steaming corncob hydrolysate by adapted Candida tropicalis *Process Biochemistry* **46** 1619–26

[2] Kim D 2018 Physico-Chemical Conversion of Lignocellulose : Inhibitor Effects and Detoxification Strategies : A Mini Review *Molecules* **23** (2) 309

[3] Harmsen P, Huijgen W, López L and Bakker R 2010 Literature review of physical and chemical pretreatment processes for lignocellulosic biomass (No. 1184) *Wageningen UR-Food & Biobased Research*

[4] Deng A, Ren J, Wang W, Li H, Lin Q, Yan Y, Sun R and Liu G 2016 Production of xylo-sugars from corncob by oxalic acid-assisted ball milling and microwave-induced hydrothermal treatments *Industrial Crops and Products* **79** 137–45

[5] Brodeur G, Yau E, Badal K, Collier J, Ramachandran K B and Ramakrishnan S 2011 Chemical and Physicochemical Pretreatment of Lignocellulosic Biomass : A Review **2011**

[6] Taherzadeh M J and Karimi K 2008 Pretreatment of lignocellulosic wastes to improve ethanol and biogas production: A review *International journal of molecular sciences* **9** (9) 1621-1651

[7] Norsyarahah S, Kamarludin C, Jainal M S, Azizan A, Sharliza N and Safaai M 2015 Mechanical Pretreatment of Lignocellulosic Biomass for Biofuel Production Mechanical Pretreatment of Lignocellulosic Biomass For Biofuel Production *Applied Mechanics and Materials* **625** 838–41

[8] Sindhu R, Kuttiraja M, Binod P, Sukumaran R K and Pandey A 2014 Bioethanol production from dilute acid pretreated Indian bamboo variety *(Dendrocalamus sp.)* by separate hydrolysis and fermentation *Industrial Crops and Products* **52** 169–76

[9] Lee J-W and Jeffries T W 2011 Efficiencies of acid catalysts in the hydrolysis of lignocellulosic biomass over a range of combined severity factors *Bioresource Technology* **102** 5884–90

[10] Akhtar N, Goyal D and Goyal A 2016 Characterization of microwave-alkali-acid pre-treated rice straw for optimization of ethanol production via simultaneous saccharification and fermentation (SSF) *Energy Conversion and Management* **141** 133-144

[11] Khuong L D, Kondo R, De Leon R, Kim Anh T, Shimizu K and Kamei I 2014 Bioethanol

production from alkaline-pretreated sugarcane bagasse by consolidated bioprocessing using Phlebia sp. MG-60 *International Biodeterioration and Biodegradation* **88** 62–8

[12] Krishnan C, Sousa C, Jin M, Chang L, Dale B E and Balan V 2010 Alkali-Based AFEX Pretreatment for the Conversion of Sugarcane Bagasse and Cane Leaf residues to Ethanol *Biotechnology and Bioengineering* **107** (3) 441-450

[13] Zhao X, Cheng K and Liu D 2009 Organosolv pretreatment of lignocellulosic biomass for enzymatic hydrolysis *Applied Microbiology and Biotechnology* **82** 815–27

[14] Wang W, Zhuang X, Yuan Z, Yu Q, Qi W, Wang Q and Tan X 2012 Effect of structural changes on enzymatic hydrolysis of eucalyptus, sweet sorghum bagasse, and sugarcane bagasse after liquid hot water pretreatment *BioResources* **7** 2469–82

[15] Ballesteros M, Oliva J M, Negro M J, Manzanares P and Ballesteros I 2004 Ethanol from lignocellulosic materials by a simultaneous saccharification and fermentation process (SFS) with Kluyveromyces marxianus CECT 10875 *Process Biochemistry* **39** 1843–8

[16] Ishola M M, Isroi and Taherzadeh M J 2014 Effect of fungal and phosphoric acid pretreatment on ethanol production from oil palm empty fruit bunches (OPEFB) *Bioresource Technology* **165** 9–12

[17] Fatriasari W, Syafii W, Wistara N J, Syamsu K and Prasetya B 2014 Digestibility of Betung Bamboo Fiber Following Fungal Pretreatment *Makara Journal of Technology* **18** 51–8

[18] Kamat S, Khot M, Zinjarde S, RaviKumar A and Gade W N 2013 Coupled production of single cell oil as biodiesel feedstock, xylitol and xylanase from sugarcane bagasse in a biorefinery concept using fungi from the tropical mangrove wetlands *Bioresource Technology* **135** 246–53

[19] Kim S, Park J M, Seo J W and Kim C H 2012 Sequential acid-/alkali-pretreatment of empty palm fruit bunch fiber *Bioresource Technology* **109** 229–33

[20] Triwahyuni E, Muryanto, Sudiyani Y and Abimanyu H 2015 The effect of substrate loading on simultaneous saccharification and fermentation process for bioethanol production from oil palm empty fruit bunches *Energy Procedia* **68** 138–46

[21] Mardawati E, Andoyo R, Syukra K A, Kresnowati M T A P and Bindar Y 2018 Production of xylitol from corn cob hydrolysate through acid and enzymatic hydrolysis by yeast Production of xylitol from corn cob hydrolysate through acid and enzymatic hydrolysis by yeast *IOP Conf. Ser.: Earth Environ. Sci 141 012019* pp 1–19

[22] Hatano K ichi, Aoyagi N, Miyakawa T, Tanokura M and Kubota K 2013 Evaluation of nonionic adsorbent resins for removal of inhibitory compounds from corncob hydrolysate for ethanol fermentation *Bioresource Technology* **149** 541–6

[23] Misra S, Raghuwanshi S and Saxena R K 2013 Evaluation of corncob hemicellulosic hydrolysate for xylitol production by adapted strain of Candida tropicalis *Carbohydrate Polymers* **92** 1596–601

[24] Misran E 2005 Industri Tebu Menuju Zero Waste Industry *Jurnal Teknologi Proses* **4** 6–10

[25] Moodley P and Kana E B G 2015 Optimization of xylose and glucose production from sugarcane leaves (Saccharum officinarum) using hybrid pretreatment techniques and assessment for hydrogen generation at semi-pilot scale *International Journal of Hydrogen Energy* **40** 3859–67

[26] Jutakanoke R, Tolieng V, Tanasupawat S and Akaracharanya A 2017 Ethanol Production from Sugarcane Leaves by Kluyveromyces marxianus S1.17, a Genome-Shuffling Mediated Transformant *BioResources* **12** 1636–46

[27] Telli-Okur M and Eken-Saraçoğlu N 2008 Fermentation of sunflower seed hull hydrolysate to ethanol by Pichia stipitis *Bioresource Technology* **99** 2162–9

[28] Aguilar-Reynosa A, Romanï A, Ma. Rodrïguez-Jasso R, Aguilar C N, Garrote G and Ruiz H A 2017 Microwave heating processing as alternative of pretreatment in second-generation biorefinery: An overview *Energy Conversion and Management* **136** 50–65

[29] Li H, Chen X, Ren J, Deng H, Peng F and Sun R 2015 Functional relationship of furfural yields and the hemicellulose - derived sugars in the hydrolysates from corncob by microwave -

assisted hydrothermal pretreatment *Biotechnology for Biofuels* 1–12

[30] Banerji A, Balakrishnan M and Kishore V V N 2013 Low severity dilute-acid hydrolysis of sweet sorghum bagasse *Applied Energy* **104** 197–206

[31] Hsu T C, Guo G L, Chen W H and Hwang W S 2010 Effect of dilute acid pretreatment of rice straw on structural properties and enzymatic hydrolysis *Bioresource Technology* **101** 4907–13

[32] Delgenes J P, Moletta R and Navarro J M 1996 Effects of lignocellulose degradation products on ethanol fermentations of glucose and xylose by Saccharomyces cerevisiae, Zymomonas mobilis, Pichia stipitis, and Candida shehatae *Enzyme and Microbial Technology* **19** 220–5

[33] Oberoi H S, Babbar N, Sandhu S K, Dhaliwal S S, Kaur U, Chadha B S and Bhargav V K 2012 Ethanol production from alkali-treated rice straw via simultaneous saccharification and fermentation using newly isolated thermotolerant Pichia kudriavzevii HOP-1 *Journal of Industrial Microbiology and Biotechnology* **39** 557–66

[34] Nielsen F, Tomás-Pejó E, Olsson L and Wallberg O 2015 Short-term adaptation during propagation improves the performance of xylose-fermenting Saccharomyces cerevisiae in simultaneous saccharification and co-fermentation *Biotechnology for Biofuels* **8** 1–15

[35] He T, Jiang Z, Wu P, Yi J, Li J and Hu C 2016 Fractionation for further conversion: From raw corn stover to lactic acid *Scientific Reports* **6** 1–11

[36] Favaro L, Basaglia M, Trento A, Van Rensburg E, García-Aparicio M, Van Zyl W H and Casella S 2013 Exploring grape marc as trove for new thermotolerant and inhibitor-tolerant Saccharomyces cerevisiae strains for second-generation bioethanol production *Biotechnology for Biofuels* **6** 1–14

[37] Chen W H, Tu Y J and Sheen H K 2011 Disruption of sugarcane bagasse lignocellulosic structure by means of dilute sulfuric acid pretreatment with microwave-assisted heating *Applied Energy* **88** 2726–34

The effectiveness of anaerobic digestion process by thermal pre-treatment on food waste as a substrate

F N A Saragih[1], C R Priadi[1]*, S Adityosulindro[1], A Abdillah[1], B B Islami[1]

[1]Environmental Engineering, Department of Civil Engineering, Universitas Indonesia, Depok, Jawa Barat, 16424, Indonesia

* email: cindy.priadi@eng.ui.ac.id

Abstract. Thermal pre-treatment is necessary for the feedstock of Anaerobic Digester (AD) to improve the degradation process. This study aims to investigate the effect of thermal pre-treatment on Food Waste (FW) as a substrate through Biochemical Methane Potential (BMP) test followed by pilot scale using a continuous stirred tank reactor (CSTR) digester. TS, VS, COD, SCOD, and solubilization samples were measured during on CSTR reactor. FW was pre-treated by heating at 70 °C, 90 °C, and 120 °C for 45 minutes. Results showed that thermal pre-treatment at 70 °C had the most effective digestion process with TS, VS, COD reduction of 8.26%; 2.54%; and 14% and increase SCOD and solubilization of 10.2% and 24.7%, respectively compared to thermal pre-treatment at 90 °C and 120 °C ($p<0.05$). Therefore, thermal pre-treatment at 70 °C was chosen for AD pilot scale in this study. Meanwhile, pilot scale experiment showed the results that thermal pre-treatment at 70 °C was obtained significant differences of 31% TS; 38% VS; 42% COD; 23% SCOD; and 30% solubilization compare with without thermal pre-treatment ($p<0.05$). Pre-treatment with thermal on FW as a substrate may help in increasing the degradation process at the hydrolysis phase in order to increase the effectiveness of the process in AD reactor.

1. Introduction

Food Waste (FW) is a good source of substrates for Anaerobic Digestion (AD) process because it contains high organic matter, sufficient moisture content, and high biodegradability [9]. AD consists of a phase of a chemical reaction which is hydrolysis, acidogenesis, acetogenesis, and methanogenesis. The substrate in hydrolysis phase was solubilized, and polymers were converted into simple monomers while acidogenesis phase was converted into volatile fatty acid (VFA). Meanwhile, acetogenesis phase was converted VFAs into acetic acid; carbon dioxide; and hydrogen, and methanogenesis phase was converted acetates into biogas [19]. However, the optimization of AD process is necessary due to the long retention time.

Pre-treatment methods have rapidly developed in recent years because it would help to enhance soluble organic material's accessibility and could enhance the biogas production and accelerate the digestion process especially hydrolysis phase which takes the longest time amongst another phase [12]. There were many pre-treatment methods to increase the optimization of AD processes such as mechanical treatment, thermal treatment, and thermo-chemical treatment. In this study, thermal pre-treatment was chosen because it can increase solubilization of organic materials effectively and increase total soluble chemical oxygen demand (SCOD) to enhance digestion process by anaerobic microorganisms compare to another pre-treatment [13]. SCOD represents an available source of energy for the microorganisms to do the activities and to show the condition of the microorganism itself in the whole process. Thus, by providing a pre-treated substrate to AD reactor, it will increase the efficiency of AD process within a shorter retention time [12].

Heating an amount of anaerobic digester's feedstock at 120 °C for 30 minutes could increase the COD solubilization of feedstock from the initial condition of 7.9% to 43.7% while Bougrier *et al.* (2008) reported that thermal pre-treatment for anaerobic digester's feedstock at 50 °C did not increase the COD solubilization, but could increase solubilization 10 – 20% at temperature of 100 °C – 120 °C.

The temperature for thermal pre-treatment can be varied from 50 °C – 250 °C which divided into two types. The temperature below 110 °C was indicated as low thermal pretreatment, and temperature above 110 °C was indicated as high thermal pre-treatment. Li *et al.*, (2016) show that thermal pre-treatment with low temperature (55 – 90 °C) did not show a significant enhancement of COD solubilization, but thermal pre-treatment at 70 °C produced the highest solubilization. The other research from Rafique *et al.* (2010) reported different results using temperature from 25 °C – 150 °C and feedstock with thermal pre-treatment at 100 °C produced the highest solubilization. Meanwhile, feedstock with a heating temperature of 130 °C and 150 °C showed the lowest solubilization.

The main effect of thermal pre-treatment of food waste as AD substrates is the disintegration of the cell membrane which produces organic material solubilization so it will make easier for microorganisms to digest the feedstock within a shorter time [2]. As a result, the enhancement of COD solubilization will make a higher efficiency of the anaerobic digestion process [3].

The aims of this study were to investigate the effects of thermal pre-treatment on food waste as a substrate for AD process. Specifically, the study focused on determining the thermal temperature effects at 70 °C, 90 °C, and 120 °C on Total Solid (TS), Volatile Solid (VS), COD, SCOD, and solubilization. The study was done by laboratory scale first to choose the temperature which has the highest effect on reduction and solubilization process then pilot scale by using continuous stirred tank reactor (CSTR) digester. Highest reduction and solubilization could accelerate the digestion process in a pilot scale [16].

2. Materials and Methods

2.1. Inoculum and Substrate

The inoculum used in this study was taken from an active AD plant treating organic waste at Petamburan Market, Grogol, Jakarta. 25% of inoculum was mixed with 75% of FW, and it was acclimatized and degasified for 30 days in order to adapt the bacteria to the substrate. The inoculum was monitored every day to keep the pH not lower than 6. Sodium bicarbonate ($NaHCO_3$) was added as a buffer solution with 3 gr/L inoculum dosage.

FW as a substrate was used in this study from Canteen in Engineering Faculty, Universitas Indonesia, Depok, Indonesia. The sample collection used quartering method by Wellinger *et al.*, (2013) to get only the organic waste and to avoid a few kinds of FW which have acidic content or hard texture because it would be difficult to process for bacteria. The FW shredded into 2 – 3 mm based on Basaria and Priadi (2016) to make the size of the FW was homogenous and smaller to enhance the digestion process. The FW and CM were taken once as needed and stored in the freezer to be taken at feeding.

2.2. Preparation of BMP Test

BMP process was used to analyze the most effective temperature of substrate pre-treatment based on the TS, VS, COD reduction, an increase of SCOD and solubilization. In addition, it used to determine the highest biogas production from the samples. The method was used for BMP process from Holliger *et al.*, (2016) and carried out in triplicates.

FW as a main substrate for the BMP process was varied by thermal pre-treatment with a temperature of 70 °C, 90 °C, and 120 °C. These variations were added with 25% of inoculum which has been adapting first to the substrate, and it was analyzed to determine their initial characteristics by measure the pH, TS, VS, COD, SCOD, and solubilization by using Indonesia Nasional Standard while flowmeter was used to measure biogas production. The pH of each substrate must be in pH between 7

– 8.5 for the optimum pH range of AD process. Each substrate that had pH lower than 7 was added 3 gr/L sodium bicarbonate ($NaHCO_3$) as a buffer. During the experiment, there were also positive control samples, control sample, blank samples, and experiment control samples as a comparison to the variation samples.

The BMP substrates were mixed from 75% of FW which has been treated by thermal pre-treatment and 25% of inoculum which has been adapting first. 10 gr VS/L of BMP substrates were stored in vial bottles with 100 mL volume which 50 mL was filled with the substrate, and 50 mL was filled for headspace. Headspace provided to prevent high pressure from the biogas produced that will affect the BMP process (Koch, Fernandez, & Drewes, 2015). Each bottle was sealed with rubber and aluminum cap to avoid the gas leak. Then, each bottle was purged using nitrogen gas for 3 minutes to make sure that the condition of each sample was anaerobic to provide a suitable condition for anaerobic microorganisms to grow. All BMP substrate was stored in the incubator with a temperature of 37 °C for 50 days in this research or until the daily production of gas was lowered.

During the incubation period, BMP bottles were shaken once a day to prevent scum forming which could inhibit biogas production. Biogas volume was measured weekly in the first 3 weeks and days until the end of BMP process using flowmeter. BMP process was terminated when the biogas production in three days in a row was lower than 1% of the biogas accumulation volume [10].

2.3. Analytical Methods

Parameters of this study were biogas, Total Solid (TS), Volatile Solid (VS), Chemical Oxygen Demand (COD), Soluble Chemical Oxygen Demand (SCOD), and solubilization which dividing COD to SCOD. Each parameter was measured periodically with the Indonesia National Standard. However, measurement of SCOD in this study using COD method, but beforehand there was a filtration of the sample using 0.40 μm filter.

Comparison of characteristics of BMP substrate before and after was analyzed to determine which temperature has the highest removal of TS, VS, COD, increasing of SCOD and solubilization, and the highest biogas volume. This information was used to determine the most effective thermal pre-treatment temperature to produce biogas and treated the FW in AD reactor. All measurements were carried out in triplicates. One way ANOVA tests and Tukey's multiple comparison tests were performed by using SPSS software version 16.0. The level of significance was 0.05 for all the tests to determine the most effective thermal pre-treatment for AD process and investigate the effect of thermal pre-treatment on AD process.

2.4. Pilot-scale anaerobic digester reactor

The type of AD reactor was a continuous stirred tank reactor (CSTR) which has operational of 60 rpm for 4 hours per day based on our previous study. Dry AD system was conducted in this study as explained in Wijayanti *et al.,*(2018). The most effective thermal pre-treatment temperature was applied to the pilot scale AD reactor which obtained during BMP experiment.

The amount of 90 Kg of the substrate was prepared for 2 weeks worth of feeding. The sample collection used the quartering method by Wellinger *et al.*, (2013) to get only the organic waste and to avoid a few kinds of FW which have acidic content or hard texture. The substrate was heated in a stainless steel container and stirred manually for 45 – 60 minutes. The pre-treated substrate was cooled down to room temperature and stored in the freezer with –20 °C to prevent any change of substrate characteristics before the feeding process. The substrate was taken out from the freezer one day before feeding day.

3. Results and Discussion

3.1. Initial Characteristic

The aims of the experiment investigated the effect of thermal pre-treatment on Food Waste (FW) as a substrate through Biochemical Methane Potential (BMP) test. Initial characteristic of FW was used as a substrate and inoculum has been determining and show in Table 1.

Based on Table 1, FW had an optimum TS concentration with a range of 20 – 50% [15]. In addition, the test was also carried out to determine the initial characteristics of the inoculum. The results showed that TS was still in the optimum range of around 2 – 10% [21].

Table 1. Initial characteristic of FW and inoculum

Sample	TS (%)	VS (%)
Food Waste	23.7 ± 0.15	22.8 ± 0.10
Inoculum	6.06 ± 0.07	3.35 ± 0.05

On the other hand, Table 1 also shows VS concentration. Based on Table 1, VS concentration of FW was 96 ± 0.20% and VS concentration of Inoculum was 55 ± 0.08%. Biogas production can be estimated based on VS content [21]. Tsunatu *et al.*, (2014) showed that the percentage of VS content from TS content must be in a range of 80 – 90% to produce the optimum biogas. The TS and VS content of substrates affects the performances of the anaerobic digestion process. Change in total solids content will lead to change in microbial morphology of AD systems. Thus, the FW in this study was suitable to be used as the substrates for AD process.

3.2. BMP Process Results

The effect of the thermal pre-treatment on the substrate was carried out using BMP testing adapted to the previous study conducted by Wellinger *et al.,* (2013) to obtain the most optimum temperature for the pre-treatment. The results are shown in Table 2.

Table 2. The result of BMP Process

Sample	Efficiency of TS Reduction	Efficiency of VS Reduction	Efficiency of COD Reduction	Efficiency of Increasing SCOD	Efficiency of Increasing Solubilization	Biogas Production
70°C	8.26%	2.54%	14.0%	10.32%	24.7%	16.7 ± 1.68 L
90°C	4.04%	2.32%	15.0%	9.09%	23.2%	15.0 ± 0.74 L
120°C	7.97%	15.32%	14.2%	12.30%	24.4%	14.5 ± 0.47 L
Control	4.89%	0.42%	10.5%	5.89%	14.4%	14.5 ± 0.38 L

As shown in Table 2, the highest efficiency of TS reduction was obtained at a temperature of 70 °C and followed by 120 °C. The results of the statistical estimation showed that there was a great significant between the control sample and pre-treatment, as well as the thermal pre-treatment resulted in thermal pre-treatment at the temperature of 70 °C had the capacity to enhance TS reduction compared to the temperature at 90 °C and 120 °C ($p<0.05$).

VS concentration was also evaluated to identify biogas potential produced. The statistics result showed that there was no significant change in VS content before and after BMP process ($p>0.05$). The thermal pre-treatment also showed a weak correlation between the change of VS content ($p>0.05$). However, the correlation test between VS destruction efficiency and thermal pre-treatment showed a strong correlation based on VS destruction efficiency by up to 15.32% at thermal pre-treatment of 120 °C ($P<0.05$). This was also supported by the previous study showed that the increase of temperature on thermal pre-treatment could help the solid organic decomposition. Thus, thermal pre-treatment at 120 °C showed the highest decreasing solid mass ratio [11].

COD parameter was also evaluated in this study. The result showed that there was a reduction in COD concentration as shown in Table 2. The reduction occurred because of the degradation of organic matters in samples. On the other hand, the thermal pre-treatment and the COD reduction showed a strong correlation up to 0.88. Therefore, the higher temperature on thermal pre-treatment will enhance COD reduction. However, thermal pre-treatment with the temperature above 90 °C will affect the stability of digestion, so organic acids will increase, and pH will decrease. This will affect the metabolism of the microorganisms on AD process.

In this study, SCOD was also evaluated to further study about the level of hydrolysis on substrates. The result showed that there was a significant difference in SCOD efficiency on all samples ($p<0.05$). The temperature of 120 °C will enhance the highest efficiency of SCOD by up to 12.30%. In another hand, the correlation test between the thermal pre-treatment and the efficiency of SCOD conducted a strong correlation with a value of 0.94.

COD and SCOD obtained will be used to understand the percentage of solubilization of organic matters. The results showed that there were a significant enhancement on solubilization level on all samples and a significant difference between thermal pre-treatment ($p<0.05$). The increasing of solubilization value was affected by the increasing of SCOD due to mass transfer of COD or substrate solids in microorganisms. Thus, the higher solubility will help the microorganism on degrade COD [4].

Biogas production statistically proved to give a significant result on biogas production on all samples as shown in Table 2 ($p<0.05$). The temperature of 70 °C was the most optimum condition in producing the highest biogas, as well as a high solubility [18]. Therefore, the temperature of 70 °C will be used on the pilot-scale experiment as the thermal pre-treatment.

3.3. Pilot Scale Results

After a laboratory test conducted by BMP test, the pilot-scale experiment was performed to investigate the effect of thermal pre-treatment on AD process. The initial characteristics of pre-treated and un-pre-treated were shown in Table 3.

Table 3. Feedstock characteristics before and after thermal pre-treatment at 70 °C

Parameter	Unit	Un-pre-treated	Pre-treated
TS	%	25.7 ± 0.15	25.4 ± 0.74
VS	%	22.8 ± 0.10	19.0 ± 0.82

As shown in table 3, TS and VS concentration could be reduced by up to 1.16 ± 0.02% and 16.7 ± 0.21%. Additionally, thermal pre-treatment statistically proven will not give a significant difference in TS concentration ($p>0.05$). In contrast, there was a significant difference in VS concentration ($p<0.05$). VS concentration was reduced after heated by thermal pre-treatment. Thermal pretreatment was more efficient to break long chained or otherwise complex compound into smaller chained molecules for easier digestion. Thus, VS concentration after pre-treatment was lower than before [18].

These results were contradicting with the previous study conducted by [17] which showed that there was an increase in TS and VS concentration by up to 5.4% and 4.5% using thermal pre-treatment. Another study conducted by Dlabaja & Malat'ak [6] was also contradicting with this study by enhancing TS concentration up to 4.9%. The differences might occur due to the different chemical composition of the materials used and the pre-treatment condition (temperature and holding time) [25]. The possible reason for the low concentration of TS and VS on thermal pre-treatment substrates occurred due to the high temperatures entailing the loss of volatile substances and thus the decline in final VS content [8]

Table 4. Comparison of Effluent AD Using un-pretreated And Pre-treated Substrate at 70 °C

Parameter	Unit	Feeding Using Non-Pretreated Substrate	Feeding Using Pretreated Substrate	Increase
TS	%	15.5 ± 0.18	10.7 ± 0.04	30.9%
VS	%	12.2 ± 0.27	7.60 ± 0.05	37.7%
COD	gr/L	264 ± 26.20	154 ± 11.50	41.6%
SCOD	gr/L	114 ± 3.60	87.7 ± 13.00	23.0%
Solubilization	%	43.3 ± 2.97	56.5 ± 4.49	30.2%

From the 15 days of the pilot-scale experiment, this research analyzed the effects of thermal pre-treatment on substrate and digestate of AD pilot reactor. VS of pre-treated substrate reduced significantly to 19%. It happened because there was an organic material's cell membrane disintegration [7].

VS concentration of digestate after 15 days on the substrate pre-treatment operational also showed a significant reduction to 7.6% besides due to the pre-treated substrate and was already sufficient amount of microorganisms that decomposed substrate inside the reactor.

On the other hand, COD also showed a reduction of 41%. This result was higher compared to Naran *et al.,* [20] with the efficiency of removal COD was 5% by using thermal pretreatment. The possible reason for the higher reduction of COD in this study due to the breakage of chemical bonds, including VFA's, polysaccharides, and proteins, by providing external energy in the form of heat [8].

SCOD has been analyzed because it can be used as a performance indicator of the digestion process in AD. It reflects the amount of soluble organic matter present in the substrates in the form of dissolved organic matter [8]. SCOD in the pilot scale experiment showed an enhancement of SCOD up to 43.3% compared to un-pre-treated. The solubilization also increased up to 56.5%. The increasing solubilization could enhance the capability of microorganisms to treat the substrate effectively in order to reduce COD. Furthermore, It might be increased in the production of digestate due to the transformation of solid to liquid in FW.

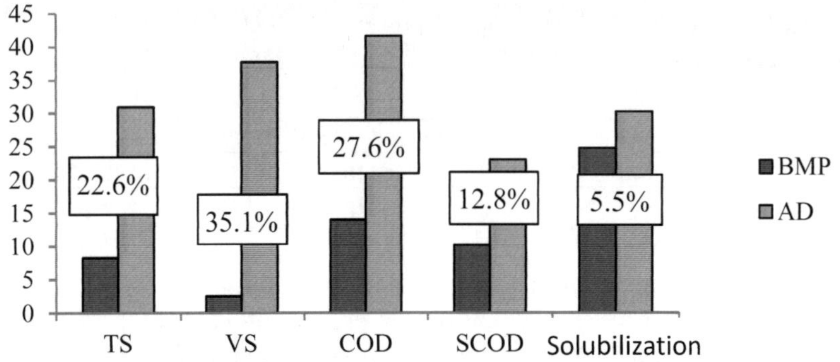

Figure 1. Results Comparison between BMP Process and AD Process

Lastly, the comparison between BMP Process and AD Process in Figure 1. Overall, AD reactor had more effective substrate treatment than BMP. It is shown by the higher value of all parameters between BMP and AD reactor. The differences were also significant as shown in the white box in Figure 1. The greatest difference was the VS reduction that indicated the degradation in AD reactor was more effective than in the BMP process. The efficiency of substrate treatment in the AD reactor improved which could be mainly due to the presence of the continuous stirrer in the pilot-scale reactor which helped maximum contact between microorganisms to treat the substrate.

4. Conclusion

The BMP process showed that a significant result on thermal pre-treatment on the substrates to be treated in AD reactor based on examination of TS, VS, COD, SCOD, solubilization, and biogas production ($p<0.05$). The most effective temperature in biogas production was 70 °C, as well as resulting in the highest solubilization. It will break long chained or otherwise complex compound into smaller chained molecules for easier digestion. This will help the microorganism to degrade COD compare to thermal pre-treatment with a temperature of 90 °C and 120 °C. In addition, thermal pre-treatment at the temperature of 70 °C will also enhance biogas production in BMP process. Thus, thermal pre-treatment at 70 °C can be used to obtain a higher solubilization rate and biogas production.

5. Acknowledgment

Authors would like to thank University of Indonesia because this research was financially supported by *Grant of Indexed International Publications for Student Final Project* (Hibah PITTA), No. 2408/UN2.R3.1/HKP.05.00/2018.

6. References

[1] Basaria P and Priadi C R 2016 Influence of Organic Fraction of Municipal Solid Waste Particle Size on Biogas Production *International Journal of Technology* **7** (8) 1431 – 1437

[2] Bien J B, Malina G, Bien J D and Wolny L 2004 Enhancing anaerobic fermentation of sewage sludge for increasing biogas generation *Journal of Environmental Science and Health. Part A, Toxic/hazardous Substances & Environmental Engineering* **39** (4) 939–949

[3] Bougrier C, Delgenès J P, Carrère H 2008 Effects of thermal treatments on five different waste activated sludge samples solubilization, physical properties and anaerobic digestion *Chemical Engineering Journal* 139 236-244

[4] Carr H 2006 Effect of ultrasonic, thermal and ozone pre-treatments on waste activated sludge solubilisation and anaerobic biodegradability **45** 711–718

[5] D Y Tsunatu, I C Azuaga, J Agabison 2014 Evaluation of the Effect of Total Solids Concentration on Biogas Yields of Agricultural Wastes *International Research Journal of Environment Sciences* **3** (2) 70-75

[6] Dlabaja T and Malaťák J 2013 Optimization of anaerobic fermentation of kitchen waste *Research in Agricultural Engineering* **59** 1–8

[7] Ferrer I, Ponsá S, Vázquez F, and Font X 2008 Increasing biogas production by thermal (70 °C) sludge pre-treatment prior to thermophilic anaerobic digestion *Biochemical Engineering Journal* **42** (2) 186–192

[8] Gandhi P, Paritosh K, Pareek N, Mathur S, Lizasoain J, Gronauer A, Vivekanand V 2018 Multicriteria Decision Model and Thermal Pretreatment of Hotel Food Waste for Robust Output to Biogas: Case Study from City of Jaipur, India *BioMed Research International* 1–13

[9] González, Martín L, Colturato L F, Font X and Vicent T 2010 Anaerobic co-digestion of the organic fraction of municipal solid waste with FOG waste from a sewage treatment plant: Recovering a wasted methane potential and enhancing the biogas yield *Waste Management* **30** (10) 1854 – 1859

[10] Holliger C, Alves M, Andrade D, Angelidaki I, Astals S, Baier U, Wierinck I 2016 Towards a standardization of biomethane potential tests *Water Science and Technology* **74**(11) 2515–2522

[11] Jin Y, Li Y, and Li J 2016 Influence of thermal pretreatment on physical and chemical properties of kitchen waste and the efficiency of anaerobic digestion *Journal of Environmental Management* **180** 291–300

[12] Junoh H, CH Yip, P Kumaran 2016 Effect on $Ca(OH)_2$ pretreatment to enhance biogas production of organic food waste *IOP Conf. Ser.: Earth Environ. Sci.* 32012013

[13] Kavitha S, Rajesh Banu J, Subitha G, Ushani U, and Yeom I T 2016 Impact of thermo-chemo-sonic pretreatment in solubilizing waste activated sludge for biogas production: Energetic analysis and economic assessment *Bioresource Technology* **219** 479–486

[14] Koch K, Fernandez Y, and Drewes J 2015 Influence of headspace flushing methane production in biocemical methane potential (BMP) tests *Waste Management* **186** 173–178

[15] Kothari R, A K Pandey, S Kumar, V V Tyagi and S K Tyagi 2014 Different aspects of dry anaerobic digestion for bio-energy: An overview *Renewable and Sustainable Energy Reviews* 174-195

[16] Li Y, Jin Y, Li J, Li H, and Yu Z 2016 Effects of thermal pretreatment on the biomethane yield and hydrolysis rate of kitchen waste *Applied Energy* **172** 47–58

[17] Ma J, Duong T H, Smits M, Verstraete W and Carballa M 2011 Enhanced biomethanation of

kitchen waste by different pre-treatments *Bioresource Technology* **102** (2) 592–599

[18] Menon A, Ren F, Wang J.-Y and Giannis A 2015 Effect of pretreatment techniques on food waste solubilization and biogas production during thermophilic batch anaerobic digestion *Journal of Material Cycles and Waste Management* **18** (2) 222–230

[19] Metcalf and Eddy I 2004 *Wastewater Engineering: Treatment and* Reuse (4th Ed.) (New York: McGraw Hill Company)

[20] Naran E, Toor U A and Kim D-J 2016 Effect of pretreatment and anaerobic co-digestion of food waste and waste activated sludge on stabilization and methane production *International Biodeterioration & Biodegradation* **113** 17–21

[21] Orhorhoro E K, Ebunilo P O and Sadjere E G 2017 Experimental Determination of Effect of Total Solid (TS) and Volatile Solid (VS) on Biogas Yield *American Journal of Modern Energy* **3** (6) 131-135

[22] Rafique, Rashad, Tjalfe Gorm Poulsen, Abdul-Sattar Nizami, Zaki-ul-Zaman Asam, Jerry D Murphy, Gerard Kiely 2010 Effect of thermal, chemical and thermo-chemical pre-treatments to enhance methane production *Energy* 6 – 10

[23] Wellinger A, Murphy J, and Baxter D 2013 *The Biogas Handbook : science, production and applications* (New Delhi: Woodhead Pub Limited)

[24] Wijayanti D F, Suwartha N and Priadi C R 2018 Effect of the Addition of Fat Oil and Grease (FOG) on the Performance of a Dry Anaerobic Digestion Food Waste Reactor *International Journal Of Technology* **9** (2) 267-274

[25] Wordofa G 2014 Effect of thermal pretreatment on chemical composition and biogas production from kitchen waste, *Thesis*, University of Jyväskylä Department of Biological and Environmental Science.

Antioxidant activity of methanol extracts from *Ganoderma lucidum* Karst. Mycelia

C Darsih[1], A W Indrianingsih[1], W Apriyana[1], S Nur Hayati[1], V T Rosyida[1], K Nisa[1], D Ratih, N Indirayati

[1]Research Unit for Natural Products Technology, Indonesian Institute of Sciences, Gading, Playen, Gunungkidul, Yogyakarta, 55581, Indonesia

Corresponding author: cici001@lipi.go.id

Abstract. The study aimed to evaluate the effect of media cultivation on antioxidant activity of methanol extracts from *Ganoderma lucidum* mycelia. The effect of media was investigated in the submerged flask. Potatoes Dextrose Broth (PDB) and sweet corn were used as media, and cultivation incubation was carried out at 30°C. The results showed that the mean of mycelial dry weight, cultivated in different media were significantly different ($p<0.05$). The mycelia dry weight from sweet corn (0.221±0.004g) was higher than PDB media (0.199±0.008 g). Meanwhile to the metabolites content on the mycelia, the mycelia methanol extracts differed significantly ($p<0.05$). The total polyphenols of mycelia methanol extracts from PDB (13.152±1.89 GAE mg/g) was higher than mycelia methanol extracts from sweet corn (8.177±0.44 GAE mg/g). In line with total triterpenoids of mycelia methanol extracts from PDB and sweet corn with values of 33.085±2.95 mg/g and 11.199±2.74 mg/g, respectively. The scavenging to DPPH radical of methanol extract of mycelia which cultivated on PDB (22.140±3.25 %) higher than sweet corn media (12.840±2.03 %) at 0.2 mg/mL. The scavenging activity was suggested related to the presence of polyphenols and triterpenoids.

Keywords : *Ganoderma lucidum, sweet corn, antioxidant, mycelia, total polyphenol, total triterpenoids*

1. Introduction

Ganoderma lucidum Karts. was known as medicinal mushroom has pharmaceutical and nutraceuticals effects [1]. This mushroom showed an anti-diabetic effect, neuroprotective effects, immunomodulatory, antitumor, antioxidant, hepatoprotective, anti-hypertensive, anticancer, and antimicrobial [2-4]. Triterpenoids, polysaccharides, proteins, sterol, lectins are metabolites in this mushroom that gave biological activities [5, 6].

Generally, this mushroom was cultivated on solid-state fermentation and take six months to form the fruiting body, and its quality is fluctuating. It is encouraging researcher to develop an efficient and effective method for enhancing the mycelia growth and its metabolites. One of the methods is a submerged culture. Most previous research reported that cultivating conditions affected the growth of *G. lucidum* and its metabolites such as polysaccharides and ganoderic acids [7-10]. Previously, our study obtained that the enhancement of water-soluble polysaccharides in the submerged culture was affected by temperature, pH, and media [11]. The pH and temperature for mycelia growing depend on the media. Modification of composition media and control of environmental conditions are important to increase mycelia production and its metabolites.

Previous research reported that corn had been used as a medium culture of *Neurospora* sp [12] and *Penicillium roqueforti* [13]. In this study, cultivation of *G. lucidum* on sweet corn which contains minerals and PDB media at temperature incubation (30 °C) was conducted to evaluate the effect of media cultivation toward the dry weight, total polyphenols, and triterpenoids. The antioxidant activity of methanol mycelia extracts on different media was also investigated.

2. Materials and methods

2.1. Materials

The mushroom was collected from Gubug Jamur Sitiram from Sleman, Yogyakarta. The *G. lucidum* was determined taxonomically at Pharmaceutical Biology Division, Faculty of Pharmacy, Universitas Gadjah Mada. The chemicals and microbiological materials that were used Potato Dextrose Agar (PDA), Potato Dextrose Broth (PDB), sweet corn, potassium dihydrogen phosphate, magnesium sulfate, sodium hydrogen phosphate, methanol, Folin-Ciocalteu reagent, sodium carbonate, distilled water, gallic acid, vanillin, perchloric acid, and ursolic acid.

2.2. Instrumentation

Antioxidant activity of methanol extracts was measured by Elisa reader (Thermo Scintific). Total polyphenols and triterpenoids content were analyzed by UV-Vis spectrophotometer (Dynamica type RB 10). Oven (Memert) was used for drying process of mycelia.

2.3. Cultivation and extraction

The mushroom *G. lucidum* was grown on PDA for seven days at 30 °C and then cultivated into 100 mL Erlenmeyer flasks, each flask containing 25 mL of media and incubated for 14 days. The effect of medium on antioxidant activity, total polyphenols, and triterpenoids of mycelia extracts was studied by growing fungal on PDB and sweet corn media at incubator temperature (30 °C). The sweet corn media contain 5% of sweet corn extract, 0.2% potassium dihydrogen phosphate, 0.05% magnesium sulfate, and 0.01% sodium hydrogen phosphate. Furthermore, the culture and mycelia were separated by filtration, and then the mycelia were dried in an oven at 50 °C until constant weight and ground to be pulverized mycelia. The powder of mycelia was extracted with 50 mL of methanol by sonication for one hour. The extracts were filtered through Whatman No 1 filter paper. The solvent was dried to obtain methanol extracts, MPDB (methanol mycelial extract from *G. lucidum* which cultivated in PDB) and MSC (methanol mycelial extract from *G. lucidum* which cultivated in sweet corn).

2.4. Total of Phenolic Content (TPC)

The TPC of extracts were determined using Folin-Ciocalteu reagent. The reaction mixture contained 500 μL (1 mg/mL) of extract, 500 μL of the Folin-Ciocalteu reagent, and 1.5 mL of 20% sodium carbonate. The final volume was made up to 10 mL with pure water. After 2h of reaction, the absorbance of samples were measured at 765 nm and gallic acid was used as standard [14].

2.5. Total Triterpenoids

Determination of triterpenoids on extracts was carried out using Lin et al. method with slight modification [15]. The mixture of 0.2 mL sample (1 mg/mL), 0.4 mL of vanillin 5% and 1 mL of perchloric acid solution were mixed and then incubated on a water bath at 60 °C for 45 minutes. The samples were then incubated at room temperature for 15 minutes and added with 5 mL acetic acid solution. The absorbance of samples was analyzed by UV-Vis spectrophotometer at 548 nm. The ursolic acid solution was used as a standard.

2.6. *DPPH free radical scavenging assay*

The samples were dissolved in methanol at various concentrations (0.1 mg/mL-0.8 mg/mL) and treated with DPPH (1 mM in methanol) and incubated for 30 min at room temperature in the dark. The absorbance was measured at 517 nm using Elisa reader. The ability of the samples to scavenge the DPPH radical was calculated using the following equation :

$$\% \text{ inhibition} = [(A0-A1)/A0] \times 100 \quad (1)$$

where A0 is the absorbance of the control and A1 is absorbance in the presence of the samples [16].

2.7. Statistical Analysis

Total polyphenols, triterpenoids, the weight of mycelia, and antioxidant activity were expressed as the mean (n=3). Data of total polyphenols, triterpenoids, and mycelia weight were analysed using independent t-Test.

3. Results and discussions

The statistical analysis result was suggested that the mean dry weight of *G. lucidum* mycelia cultivated in different media was significantly different. The mean of mycelia dry weight from sweet corn was higher than the mean of dry weight mycelia from commercial PDB media (Figure 1). The results indicated that sweet corn media which contain 5% of sweet corn extract, 0.2% potassium dihydrogen phosphate, 0.05% magnesium sulfate, and 0.01% sodium hydrogen phosphate was favorable for the cell growth than commercial PDB. The mushroom cell growth was better when a complex nutrient source added in media [17]. Sweet corn, also known as sugar corn, is a special low-starch variety (28%) with a higher sugar content (18%), most of which is sucrose and the protein content ranges from 10-15% [18]. Minerals in the culture media affected mushroom growth [19,20]. Similar results were obtained in the study of Curvetto et al. [21], the addition of limiting mineral components promoted mycelia growth rate of *P. ostreatus* up to 25%. Jonathan et al. reported that magnesium as the best element for enhancing the growth of mushroom. K_2HPO_4 could improve productivity through its buffering action, being favorable for mycelia growth [22].

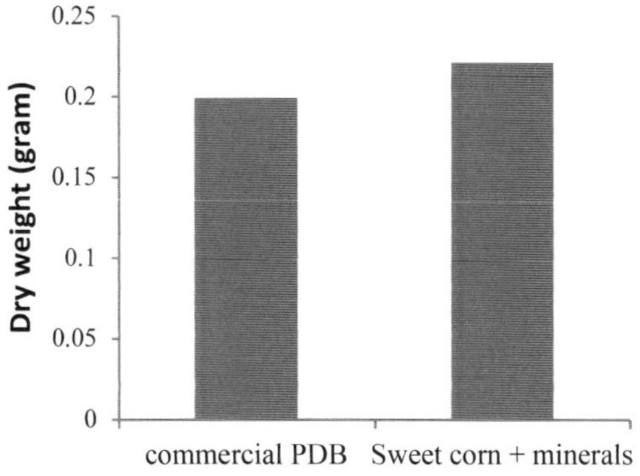

Figure 1. Dry weight (gram) of *G. lucidum* mycelia

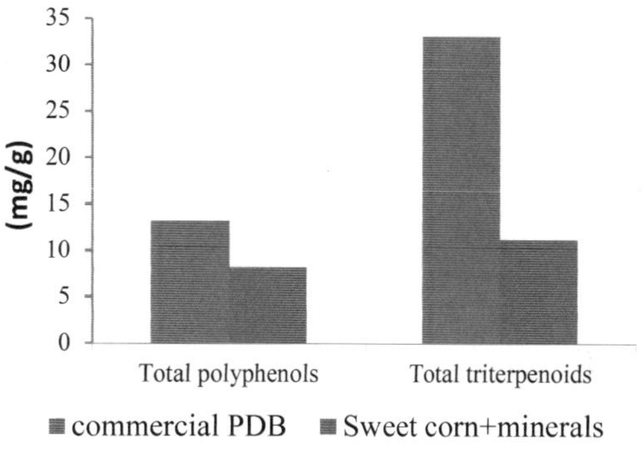

Figure 2. Total polyphenols (GAE, mg/g) and total triterpenoids (mg/g) of *G.lucidum* extracts values of dry weight, total polyphenols, and total triterpenoids are expressed as mean of three parallel measurements

On the contrary secondary metabolites from the extract of mycelia cultivated on sweet corn media lower than mycelia extract from commercial PDB (Figure 2). The mycelia extracts differed significantly ($p<0.05$) in their mean of total polyphenols and triterpenoids. Commercial PDB was commonly used for preparing a liquid broth culture. The composition of commercial PDB was still better for polyphenols and triterpenoids production than sweet corn media.

This study showed that the scavenging activity increased with the concentrations of MPDB and MSC extracts (Figure 3). The DPPH radical scavenging effect of MPDB extract higher than MSC extract. The antioxidant activity of methanol extracts was suggested related to the presence of total polyphenols and triterpenoids on the mycelia. This result was in good agreement with those reported in previous research, the scavenging to DPPH radical correlates directly with the different polyphenols content [5, 23].

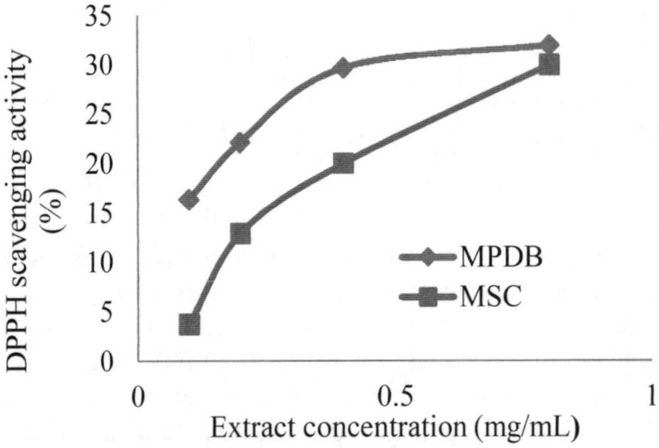

Figure 3. Scavenging effect on the DPPH radical test of methanol extracts from *G. lucidum* mycelia. The data represent the percentage of inhibition induced by increasing concentrations of MPDB and MSC

4. Conclusion

Cultivation media in the submerged flask affected to the antioxidant activity of *G. lucidum* extracts. The sweet corn media which contain minerals was positively effected on cell growth of mushroom, contrary to the secondary metabolites and antioxidant activity. The optimum accumulation of metabolites and its antioxidant activity occurred when *G. lucidum* was cultivated on PDB media. The scavenging to DPPH radical was suggested related to the polyphenols and triterpenoids content on mycelia.

5. Acknowledgments

This work is supported by the Indonesian Institute of Science through the program of Penguatan Kompetensi Inti Satuan Kerja of BPTBA LIPI in 2017.

6. References

[1] Bishop K S, Kao C H, Xu Y, Glucina M P, Paterson R R, and Ferguson L R 2015 *J. Phytochem.* **114** 56-65
[2] Nie S, Zhang H, Li W and Xie M 2013 *J. Bioac. Carbohydr. Dietary Fibre.* **1(1)** 10-20
[3] Ferreira I C, Heleno S A, Reis F S, Stojkovic D, Queiroz M J, Vasconcelos M H and Sokovic M *J. Phytochem.* **114** 38-55
[4] Li Y -b, Wang J -l and Zhong J -J 2013 *J. Process Biochem.* **48(2)** 331-339
[5] Saltarelli R, Ceccaroli P, Iotti M, Zambonelli A, Buffalini M, Casadei L, Vallorani L, and Stocchi V 2009 *J. Food Chem.* **116(1)** 143-151
[6] Batra P, Sharma A K and Khajuria R. 2013 *Int J. Med Mushroom* **15(2)** 127-143
[7] Yang F -C, Ke Y -F and Kuo S -S 2000 *J. Enzyme and Microb. Technol.* **27(3)** 295-301
[8] Yang F-C and Liau C -B 1998 *J. Process Biochem.* **33(5)** 547-553
[9] Tang Y -J, Zang W, Liu R -S, Zu, L –W and Zhong, J-J 2011 *J.Process Biochem.* **46(1)** 404-408
[10] Tang Y -J and Zhong, J -J 2002 *J. Enzyme Microb. Technol.* **31(1)** 20-28
[11] Rosyida, V T, Hayati S N, Apriyana W, Darsih C and Poeloengasih C D 2017 *in IOP Conference Series: Earth and Environmental Science.* **101(1)** 012008
[12] Chutrtong J 2015 *Procedia Social and Behavioral Science* **197** 797-800
[13] Chang S -C, Wei Y -H, Wei D-L, Chen Y-Y, and Jong S-C 1991 *Appl Environ Microbiol.* **57(9)** 2581-2585
[14] Yu L, Haley S, Perret J, Harris M, Wilson J, and Qian M 2002 *J. Agric. Food Chem.* **50(6)** 1619-1624
[15] Lin M, Yu Z, Wang B, Wang, C H, Weng Y, and Koo MALCOLM. 2015. *Sains Malays*, **44** 1685-

1691
[16] Indrianingsih A W, Tachibana S and Itoh K. 2015. *Procedia Environ Sci.* **28** 639-648
[17] Wagner R, Mitchell D A, Lanzi S G, Lopes, A A M A and Berovic M 2003 *Food technol. biotechnol.* **41(4**) 371-382
[18] Panzeri D, Cesari V, Toschi I and Pilu R 2011 *Seed Calorific Value in Different Maize Genotypes, Energy Sources, Part A: Recovery, Utilization, and Environmental Effects*, **33(18)** 1700-1705
[19] Bellettini M B, Fiorda F A, Maieves H A, Teixeira G L, Avila S, Hornung P S, Junior A M and Ribani H R 2016 *Saudi J. Biol Sci.* 1 - 14
[20] Jonathan S G and Fasidi I O 2001 *J. Food Chem.* **72(4)** 479-483
[21] Curvetto N R, Figlas D, Devalis R and Delmastro S 2002 *J. Bioresour Technol.* **84**:171–176
[22] Petre M and Teodorescu A 2009 *J. Ann. For. Res.* **52**: 129-136
[23] Cilerdzic J, Stajic M and Vukojevic J 2016 *J. Curr. Pharm. Biotechnol.* **17(3)** 275-82

IOP Conf. Series: Earth and Environmental Science **251** (2019) 012016 doi:10.1088/1755-1315/251/1/012016

Antibacterial Bioactivity of *n*-Hexane Extract from Mahogany (*Swietenia humilis* Zucc.) Seed and Its Fatty Acid Compound Identification

A P Asmara[1], Hernawan[2], C Nuzlia[1] and R Maryana[2]

[1] Department of Chemistry, Faculty of Science and Technology, Universitas Islam Negeri Ar-Raniry Banda Aceh, Indonesia
[2] Research Unit for Natural Product Technology–Indonesia Science Institute (BPTBA LIPI), Gunungkidul, Yogyakarta, Indonesia

E-mail: anjarpa@ar-raniry.ac.id

Abstract. An investigation of antimicrobial bioactivity followed by GCMS analysis for the fatty acid composition has been conducted to the crude *n*-hexane extract of *Swietenia humilis* Zucc. seed. The goal of this research to determine the inhibition zone diameter for antimicrobial activity and the percentage of fatty acid contained in the extract. The seed of *S. humilis* Zucc. have been extracted by soxhletation with *n*-hexane as the solvent. The extract has been tested through in vitro antimicrobial bioassay toward *Staphylococcus aureus* and *Escherichia coli* by paper disc diffusion method. The fatty acid composition has been identified by fatty acid methyl esters method through gas chromatography coupled with mass spectroscopy. The result shows that the extract is more effective to inhibit the *S. aureus* activity with diameter range from 10 to 17 mm while *E. coli* tends to be resistance. The extract contains palmitic acid 15.49%; palmitoleic acid 0.28%; octadecanoic acid 14.04%; oleic acid 37.02 %; linoleic acid 24.70%; linolenic acid 7.33%; arachidic acid 0.84%; and tricosanoic acid 0.33%. The research confirms that *S. humilis* Zucc. seed oil contains polyunsaturated fatty acid and can be used as an antimicrobial agent.

1. Introduction

Currently, people's interest to consume traditional medicine is increasing, partly may happen because of considering the side effects and high prices of chemical drugs today. Various medicinal plants have been used to prevent or cure a lot of disorders and also used as a natural source of antibacterial agent [1,2]. Furthermore, World Health Organization (WHO) published that medicinal plants are the sources of bioactive compounds against bacterial infections used by as many as 80% of people in developing countries [3]. One variety of medicinal herbs is *Swietenia humilis* Zucc. which is also known as Mahogany or Venadillo from Meliaceae family [4]. The plant's part which contains most of the secondary metabolite compounds is the seed [5].

Antimicrobial constituents from the plants which are safer to the human body with few side effects are unsaturated fatty acids [6]. The mahogany seed contains variant secondary metabolites which are found as essential unsaturated fatty acids (Polyunsaturated Fatty Acid-PUFA) and as potentially antibiotic agent [7]. *S. humilis* Zucc. seed also consists of many PUFA compounds with different composition [4] which are potential as reducing agent of cholesterol and blood triglyceride levels. However, there is a lack of information about its bioactivity against micro bacterium.

IOP Conf. Series: Earth and Environmental Science **251** (2019) 012016 doi:10.1088/1755-1315/251/1/012016

The antimicrobial bioactivity towards *E. coli* and *S. aureus* of acetonic, methanolic, and ethanolic extract of the seed of *S. humilis* Zucc. has already been revealed that all of them are effective for controlling the number of the microorganisms [8]. Furthermore, ethanol–ozonolyzed extract of *S. humilis* Zucc. seed is also capable of inhibiting the growth of these bacteria effectively [9]. While several extracts of the botanical part taken out by polar solvent are confirmed as an antibiotic, further investigation is needed to analyze the ability of non-polar extract regarding bacteria inhibition activity. Therefore, in the context of the fundamental importance of this bioactivity, in this paper, we report the bioactivity of *n*-hexane extract of *S. humilis* Zucc. seed versus *E. coli* and *S. aureus* bacteria using paper disc diffusion. In addition, the fatty acids composition were also investigated by using gas chromatography coupled with mass spectroscopy.

2. Materials and Methods

2.1. *Equipment and Chemicals*

The study was conducted by using a set of soxhlet extraction tools (Borosil, Lucknow), jars, scissors, strainer, separating funnel, blender (Miyako), rotary evaporator (RII0 Buchi, Switzerland) with vacuum pump, oven (GCA Corporation, USA), measuring cylinders (Pyrex), Müller-Hinton agar media (Acumedia, USA), cellulose discs (6 mm sterile diameter Whatman cellulose filters number 5, Germany), Agilent 6890/capillary column HP-5 gas chromatography (30 m × 0.25 mm × 0.25 μm) (Agilent Technologies, Santa Clara, California, US), and UV-Vis spectrophotometer (517 nm on a Cary-50-Bio Varian).

The materials were *n*-hexane (Merck), H_2SO_4 (Merck), $BaCl_2$ (Merck), chloroform (Merck), dimethylsulfoxide/ DMSO (Merck), anhydrous Na_2SO_4 (Merck), *Staphylococcus aureus* ATCC 2921 (USA), *Escherichia coli* ATCC 700609, DPPH free radical solution (2.9 mL, 60 μM, in methanol, Sigma-Aldrich), standard antibiotics Ampicillin as a positive control while DMSO as negative control. All tests were done in duplicate training.

2.2. *Procedure for Sample Preparation and Extraction*

S. humilis Zucc. seeds were collected from Wonosari, Gunungkidul, Yogyakarta Special Region Province, Indonesia. The sample was identified by the Center for Plant Conservation Botanic Gardens–LIPI Bogor with specimen number B-525. The seeds were washed and dried at room temperature and kept away from direct sunlight. The cotyledon then was taken out and cut into 0.5 to 1 cm in diameter size and dried for a few days. The cotyledon pieces were ground into 40–60 mesh then dried at 40 °C for 24 hours. As much as 450 g of sample powder was extracted through soxhletation using *n*-hexane at 60–70 °C for 7–8 hours. The *n*-hexane extract then was dehydrated over anhydrous Na_2SO_4, filtered, and concentrated through the rotary evaporator. The yield (w/w%) was calculated and then stored in a dark-glass bottle at 4 °C.

2.3. *Preparation for McFarland Turbidity Standard*

A total of 9.95 mL of 1% H_2SO_4 was mixed with 0.05 mL $BaCl_2$ 1% in a reaction tube and stirred until homogeneous. This standard suspension is used to determine the suspension concentration of tested bacteria (> 3 × 108 CFU/mL) to be made by comparison of bacterial suspension turbidity with standard suspension McFarland.

2.4. *Preparation of Bacterial Suspension*

S. aureus ATCC 29213 and *E. coli* ATCC 700609 are suspended by adding 0.9% of NaCl solution in different tubes. Turbidity is adjusted to McFarland 0.5 turbidity standard to obtain bacteria with a concentration of 108 CFU/mL. Turbidity is determined by taking a small amount of suspension into a smaller reaction tube, filling it into a hole in the nephelometer, and looking at its turbidity number. If it is less cloudy, the suspension is added to the colony whereas the cloudier one will be added by 0.9% NaCl.

2.5. Preparation for Sample Prime Liquor
Each sample of an extract of *n*-hexane *S. humilis* Zucc. seed was taken as 100 mg then dissolved in 10 mL of solvent (9 ml H_2O + 1 ml DMSO). The prime liquor was diluted to 10, 20, 50, 100, and 1000 μg/mL, then stored in the freezer for subsequent antibacterial trials.

2.6. Preparation of MHA (Müller-Hinton Agar)
A total of 9.5 grams of Müller-Hinton Agar/ MHA (38 g/ L) with medium composition (beef infusion 300 gram, casamino acid 17.5 gram, starch 1.5 gram, and agar) is dissolved in 250 mL distilled water and heated to reach the boiling point and then sterilized in autoclave for 20 minutes in 1 atm of air pressure and 121 °C of temperature.

2.7. Inhibitory Test of n-hexane S. humilis Zucc. *Seed Extract against S. aureus and E. coli*
The antibacterial test was performed by Kirby-Bauer method of paper disc diffusion [10]. The results are based on the measurement of the inhibitory diameter (DDH) of the bacterial growth formed around the disc paper. The *n*-hexane extract of *S. humilis* Zucc. seed was taken as much as 20 μL and dripped onto sterilized disc paper and waited until it becomes saturated. Media for sterile Müller-Hinton was poured into a petri dish with ± 0.5 cm thickness allowed to solidify at room temperature. Then the sterile stick cotton was immersed in the suspension of the tested bacteria and then inoculated in a flattened MHA medium. Wait a few minutes to dry, then place the saturated disc paper with the *n*-hexane extract. Both negative and positive control were also placed on the paper disc for both bacteria. The test was performed with two replications. Furthermore, it was incubated at 35-37 °C for 24 hours. After 24 h, the DDH formed around the disc is observed using SCAN 500® analyzer version 6.3.8.0.

2.8. Analysis of Saturated and Unsaturated Fatty Acids
Fatty acid analysis and fatty acid methylation were carried out by following procedures from AOAC [11] at the Integrated Research and Testing Laboratory-Universitas Gadjah Mada (LPPT-UGM). A sample of 5 mg is added by 10 mL of NaOH 0.5 M; 12 mL of BF_3 14%; and 4 mL of *n*-heptane and then refluxed for 5-10 min at 60 °C. After reaching the room temperature, the mixture is extracted with a saturated NaCl solution to form two layers. The top layer is taken out and then transferred to a reaction tube containing anhydrous Na_2SO_4 then filtered with a glass of fiber. The filtrate is placed into Eppendorf, and a total of 0.1 μL of sample is injected in Agilent 6890 gas-spectroscopy mass chromatography equipped with flame ionization detector and HP-5 silica capillary column (30 m × 0.25 mm × 0.25 μm). The injectors and detectors are set at 250 °C and helium are used as carrier gas (3 mL/ min). The identification of fatty acid compounds is performed by comparing the spectral characteristics of sample components with standard data of C4-C24 methyl esters for 37 fatty acids [4].

3. Results and Discussion

3.1. Antibacterial Activity of n-hexane S. humilis Zucc *Seed Extract*
Antibacterial bioactivity analysis of *n*-hexane extract of *S. humilis* Zucc. seed was tested *in vitro* by paper disc diffusion method through which it was expressed in the inhibitory diameter (mm) as shown in Table 1. The inhibitory test was carried out under varied concentration. It was found that *n*-hexane extract of *S. humilis* Zucc seed inhibited the gram-positive and negative bacteria actively. As shown in the inhibitory diameter, this seed extract is more active against *S. aureus* bacteria with an inhibitory diameter of 10.4 mm at 10 μg/mL, 15.2 mm at 500 μg/mL and 18.3 mm at 1000 μg/mL.

S. aureus bacteria belong to gram-positive bacteria that are sensitive to secondary metabolite compounds from the seed of mahogany species based on several antimicrobial bioactivity assays of various mahogany seed extracts. This matter is regarding the physicochemical factor and the bacterial anatomy that is appropriate to the action mechanism of bioactive compounds from mahogany seed. The gram-positive bacterial cell wall is known to consist of peptidoglycan and teichoic or teichuronic acid. The bioactive compounds of the mahogany seed are predicted to bind the penicillin-binding protein to the transpeptidase enzyme in the bacterial cell wall which causes the inhibition of the transpeptidation reaction. This leads to inactivation and loss of inhibitors of the autolytic enzymes in the bacterial cell wall so that the wall might undergo lysis [12].

Table 1. Bacteria inhibitory of *n*-hexane *S. humilis* Zucc seed extract.

Tested Substance	Concentration (μg/mL)	Bacteria	Inhibitory Diameter (mm)		
			Result 1	Result 2	Average Result [a]
n-hexane extract of *S. humilis Zucc* seed	100	*E. coli*	0.12 ± 0.4	0.22 ± 0.5	0.17 ± 0.2
	500	*E. coli*	2.55 ± 0.7	2.8 ± 0.3	2.68 ± 0.4
	1000	*E. coli*	4.3 ± 0.3	4.02 ± 0.5	4.66 ± 0.6
Positive Control	1000	*E. coli*	28.2 ± 0.7	26.7 ± 0.4	27.45 ± 0.1
Negative Control		*E. coli*	–	–	–
n-hexane extract of *S. humilis Zucc* seed	100	*S. aureus*	10.6 ± 0.2	10.2 ± 0.4	10.4 ± 0.2
	500	*S. aureus*	13.3 ± 0.1	13.1 ± 0.5	13.2 ± 0.4
	1000	*S. aureus*	17.4 ± 0.5	17.2 ± 0.2	17.3 ± 0.2
Positive Control [b]	1000	*S. aureus*	31.5 ± 0.9	31.1 ± 0.3	31.3 ± 0.4
Negative Control [c]		*S. aureus*	–	–	–

[a] Data values determined by mean ± standard deviation with two repetitions (n=2). [b] Ampicillin 1 g/mL. [c] DMSO.

3.2. *Fatty Acid Composition of n-hexane S. humilis* Zucc. *Seed Extract*

The fatty acid composition of the extract was investigated through GC-MS spectroscopy analysis. The result obtained by fatty acid methyl esters (FAME) method is shown in Figure 2. The chromatogram shows eight peaks that appeared based on comparative analysis of the fatty acid retention time in the sample with a mixture of 37 FAME standard C4-C24 under the same conditions. The composition of those eight fatty acid compounds of this analysis as presented in Table 4, sequentially shows the percentage of saturated, monounsaturated, and polyunsaturated fatty acids were 30.98%, 37.02%, and 32.03%. The saturated fatty acid compounds that found at the higher percentage are palmitic acid (C16:0) 15.49% and octadecanoic (stearic) acid (C18:0) 14.04%. Those contents are close to a report of diethyl ether and n-hexane extract of *S. macrophylla* King seed in which the level is 14.62% and 15.47% of palmitic acid, 16.75% and 17.65% of stearic acid and 0.59% and 0.54% of palmitoleic acid [6]. Similar results are also reported that the content of saturated fatty acids in *S. mahagoni* Jacq. are 14.32% of stearic acid and 12.97% of palmitic acid [7].

Consumption of palmitic and stearic acid is reported not to increase the blood fat levels due to its solid state at body temperature (37 °C) and insoluble in blood plasma [4]. Consumption of palmitic acid can increase the levels of high-density lipoprotein (HDL) which may reduce the risk of coronary heart disease as well [13].

Furthermore, the *n*-hexane extract consists of monounsaturated fatty acid (MUFA) that contains 37.02 % of oleic acid (C18:1 cis-9). It means that this is also composed of more oleic acid than the same seed in Mexico which is only 29.27% of the acid [4]. Several studies to the seed of two other types of mahogany, *S. macrophylla* King and *S. mahagoni* Jacq., report that both species contain oleic acid with different amounts depending on their habitat [6,7,14]. As a result, the presence of these compounds indicates that the seed of *S. humilis* Zucc. is potential for a natural herbal remedy to reduce the risk of coronary heart disease [15].

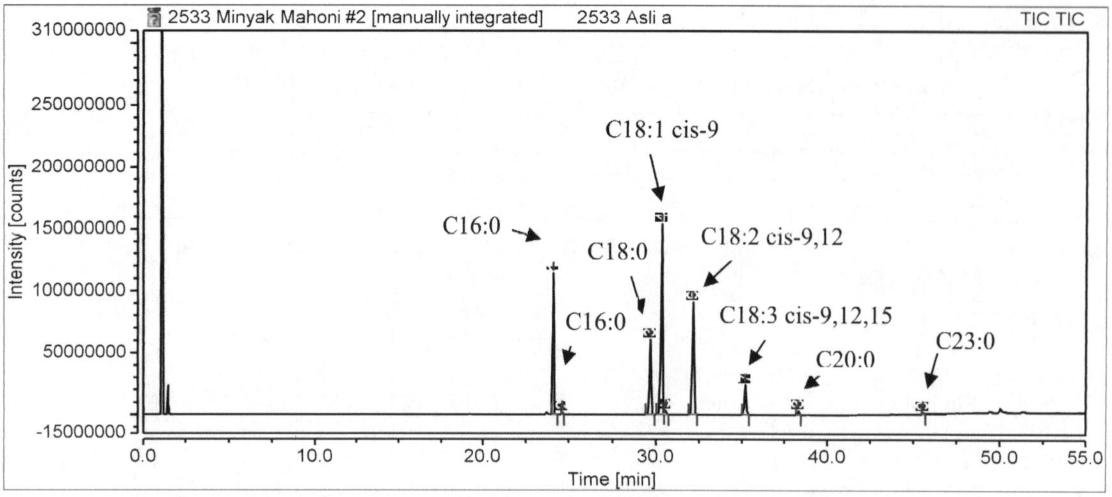

Figure 1. Gases Chromatogram of Fatty Acid Compound from *n*-hexane *S.humilis* Zucc. Seed Extract.

Table 2. Fatty acid composition of *n*-hexane *S. humilis* Zucc seed extract.

Fatty Acid Compound	Type of Fatty Acid	Composition Percentage (%)
Palmitic Acid	Saturated	15.49
Palmitoleic Acid	Saturated	0.28
Octadecanoic Acid	Saturated	14.04
Oleic Acid	Monounsaturated	37.02
Linoleic Acid	Polyunsaturated	24.70
Linolenic Acid	Polyunsaturated	7.33
Arachidic Acid	Saturated	0.84
Tricosanoic Acid	Saturated	0.33

Meanwhile, 32.3 % of PUFA in this extract includes 24.70 % of linoleic acid (C18: 2 cis-9,12) and 7.33% of linolenic acid (C18: 3 cis-9,12,15). These fatty acids are very important for human's health because both of them are precursors for the other essential fatty acids such as DHA (docosahexaenoic acid, C22:6, omega 3) and EPA (eicosapentaenoic acid, C20:5, omega 3) from linolenic acid and arachidonic acid (C20: 4, omega 6) of linoleic acid [4]. Previous research states that both of these compounds are PUFA finding in the seed of the three species of mahogany in the world. Both of them are also associated with the antibacterial bioactivity of *S. macrophylla* seed against *S. aureus, Salmonella typhimurium*, and *Pseudomonas aeruginosa* because the diethyl ether and *n*-hexane extracts of the seed provide a competitive inhibitory zone compared to Streptomycin as its positive control [5].

PUFA compounds in mahogany seed effectively inhibit gram-positive bacteria such as *S. aureus* whereas gram-negative bacteria such as E. coli tend to be more resistant to hydrophobic groups in these fatty acids [2]. In this case, *S. aureus* is inhibited by a termination of the transpeptidase reaction within its cell wall. In addition, the reaction is activated by a transpeptidase enzyme. The enzyme is a receptor for the penicillin drug compound which acts via a bonding interaction between the hydroxyl group in the serine residue and C carbonyl group in the β-lactam ring. Subsequently, it is a nucleophilic addition reaction between a carbonyl group and an ammonium group in the lysine residue which is incorporating carboxylic ions with cations [2]. The bonding of this penicillin establishes a blockade of peptidoglycan that should be bound to the active side (serine and lysine) of the receptor. Based on a literature review about the structure and its activity relationship, the presence of the C=C double bond of the *cis* isomer and the carboxylic group in PUFA is analogous to the penicillin structure so that the action mechanism in the transpeptidase enzyme inhibition may also occur toward PUFA in mahogany [2,16].

4. Conclusion

In summary, we examined the activity of *n*-hexane extract of the seed of *S. humilis* Zucc. which is potential as a natural antibacterial source against *S. aureus*. The analysis of its fatty acid compounds showed that the seed contains 15.49 % of palmitic acid, 0.28 % of palmitoleic acid, 14.04 % of octadecanoic acid, 37.02 % of oleic acid, 24.70 % of linoleic acid, 7.33 % of linolenic acid, 0.84 % of arachidic acid and 0.33 % of tricosanoic acid. Further research is needed to determine the action mechanism of *S. aureus* inhibition activity between unsaturated fatty acid compounds toward transpeptidase enzyme through computational chemical-based molecular modeling.

5. Acknowledgments

The authors are grateful to the Directorate of Islamic Higher Education of the Ministry of Religious Affairs of the Republic of Indonesia which has fully funded this research through the Quality Assurance Scheme of Basic Development Research based on the Decree of Dirjen Pendis number 4653 in 2017 and Unit for Research and Technology of Natural Product of Indonesia Science Institute (BPTBA-LIPI) Gunungkidul which has facilitated the implementation of this research.

6. References

[1] Tadesse M 2010 Antimicrobial natural products from Arctic and sub-Arctic marine invertebrates *Dissertation* Faculty of Biosciences Fisheries and Economics Norwegian College of Fishery Science University of Tromso
[2] Patrick G L 2013 *An introduction of medicinal chemistry fifth edition* Oxford University Press
[3] Alam M K, Mansur F J, Karim M M, and Haque M D 2014 Antimicrobial activity of *Swietenia mahagoni* (seed) against various pathogenic microbes *Indo American Journal of Pharm Research* **4** 5
[4] Perez-Rubio V, Heredia J B, Chaidez-Quiroz C, Valdez-Torres J B, Salazar-Villa E, Allende-Molar R, and Angulo-Escalante M A, 2012 Physicochemical characterization and fatty acid content of 'venadillo' (*Swietenia humilis* Zucc.) seed oil *African Journal of Biotechnology* **11** 22 6138-6142
[5] Suliman M B, Nour A H, Yusoff M M, Nour A H, and Adam M S 2014 Physicochemical characterization and antimicrobial activity of *Swietenia macrophylla* King seed oil. *International Journal of Engineering Research & Technology (IJERT)* **3** 3 1787 – 1792.
[6] Suliman M B, Nour A H, Yusoff M M, Nour A H, Kuppusamy P, Yuvaraj A R, and Adam M S 2013 Fatty acid composition and antibacterial activity of *Swietenia Macrophylla* king seed oil *Afr. J. Plant Sci* **7** 7 300-303
[7] Mostafa M, Jahan I A, Riaz M, Hossain H, Nimmi I, Miah A S, and Chowdhury J U 2011 Comprehensive analysis of the composition of seed cake and its fatty oil from *Swietenia mahagoni* Jacq. growing in Bangladesh *Dhaka Univ. J. Pharm. Sci.* **10** 1 49–52
[8] López-Pantoja Y, Escalante M A, Martínez-Rodríguez C, Soto-Beltrán J, and Chaidez-Quiroz C 2007 Efecto Antimicrobiano de Extractos Crudos de Neem (*Azadirachta Indica* A. Juss) y Venadillo (*Swietenia Humilis* Zucc) contra *E. coli*, S. *aureus* y el bacteriófago P22 *Bioquímica* **32** 117-125
[9] Beltran M S, Jiménez M, Amézquita-López B A, Martínez-Rodriguez C, and Chaidez C 2017 Antibacterial activity of ozonized olive (*olea europaea l.*) And venadillo (*swietenia humilis* zucc.) Oils against *Escherichia coli* and *Staphylococcus aureus. Microbiol Biotech Food Sci.* **6** 3 947-949
[10] Ningsih A P, Nurmiati, and Agustien A 2013 Uji aktivitas antibakteri ekstrak kental tanaman pisang kepok kuning (*Musa paradisiaca* Linn.) terhadap *Staphylococcus aureus* dan *Escherichia coli*, Jurnal Biologi Universitas Andalas (*J. Bio. UA.*) **2** 3 208-209
[11] Association of Official Analytical Chemists 1998 *Official methods of analysis of the*

Association of Official Analytic Chemists (AOAC) 16th ed Gaithersburg MD

[12] Karimi E, Jaafar H J E., Ghasemzadeh A, and Ebrahimi M 2015 Fatty acid composition, antioxidant and antibacterial properties of the microwave aqueous extract of three varieties of *Labisia pumila* Benth. *Biological Research* **48** 9

[13] Tuminah S 2010 Efek perbedaan sumber dan struktur kimia asam lemak jenuh terhadap kesehatan *Bul. Penelit. Kesehat.* **38** 1 43–51

[14] Marpaung H 2003 The analysis of fatty acid compositions in the seeds of *Swietenia mahogany* Jaqc. *Jurnal Sains Kimia* **7** 1 26-27

[15] Kırbaslar F G, Türker G, Özsoy-Günes Z, Ünal M, Dülger B, Ertas E, and Kızılkaya B 2012 Evaluation of fatty acid composition, antioxidant and antimicrobial activity, mineral composition and calorie values of some nuts and seeds from Turkey *Rec. Nat. Prod.* **6** 4 339-349

[16] Thomas G 2003 *Fundamentals of medicinal chemistry* West Sussex: John Wiley & Sons

IOP Conf. Series: Earth and Environmental Science **251** (2019) 012017 doi:10.1088/1755-1315/251/1/012017

Antioxidant properties and toxicity of selected bamboo shoots "iwung" extract: a comparative study

A C Iwansyah*[1], R Kumalasari[1], D A Darmajana[1], and L Ratnawati[1]

[1]Development Center for Appropriate Technology, Indonesia Institute of Sciences. KS Tubun No.5 Subang 41213, West Java – Indonesia

*E-Mail address: chandra.iwansyah@gmail.com

Abstract. This study aims to compare the nutritional, antioxidant properties, and toxicity of various bamboo shoots extract. Physicochemical, total phenolics content, total flavonoids content, antioxidant activity (IC_{50}) and toxicity were analyzed. Folin-Ciocalteau assay and aluminum chloride assay were used to determine the total phenolics and total flavonoid content, respectively. The antioxidant activity determined by DPPH (1,1-diphenyl-2-picrylhydrazyl) scavenging assay. Moreover, the toxicity of bamboo shoots extract was determined by brine shrimp lethality bioassay (LC_{50}). The results showed that bamboo shoots, *viz.*, betung, gombong, buluh, and kuning, are rich in minerals of zinc, iron, and potassium. The ethanolic extract of buluh shoots (BUS) contained the highest total phenolics (27.83 mg GAE/g) followed by KUS > GOS > BES. Bamboo betung shoots extract (BES) is found to have the highest total flavonoids (2.49 mg QE/g) with a ratio of TTF/TP) which is 0.17. Meanwhile, the DPPH assay (IC_{50}) showed that IC_{50} of KUS (347.48 μg/mL) was lowest. Based on these values, KUS is found to have the highest antioxidant activity. Furthermore, the level toxicity of $LC_{50,}$ BES, BUS, and KUS extract is valued as cytotoxic. These results provide useful data about the nutritional, phytochemical, and toxicity of bamboo shoots for functional food purposes.

1. Introduction

In decades, there has been a marked interest in health and diseases related transition of nutrition amongst the urban population. The exploration of functional food or nutraceutical research based on local wisdom and commodities has become a concern. Bamboos are gaining popularity, especially in the utilization of its shoots as healthy and nutritious food [1]. In Southeast Asian cuisines, including Indonesia, there is a long history of bamboo shoots usage as an ingredient in traditional food and medicine. It is usually consumed in various forms as raw, dried, boiled and fermented.

There are a number of bamboo species available in Indonesia, and many of them are used for edible purposes. *Bamboo vulgaris*, *Dendrocalamus asper* (Schult.), *Dendrocalamus longispathus*, *Gigantochloa nigrociliata* are used for an edible purpose in West- and East Java, Indonesia [2,3]. In Java, bamboo shoot is used in a variety of traditional foods such as iwung soup, lumpia, pickles, crackers, and cakes.

Bamboo shoots have good profile minerals, vitamins, dietary fiber, phenol and phytosterol [4]. The consumption of these reported can decrease total serum cholesterol, low-density lipoprotein (LDL), increased fecal volume and bowel movement frequency [5]. However, the bamboo shoots also contain anti-nutrient, cyanogenic glycosides. This compound was responsible for the acridity and peculiar smell in the shoots [6]. Bamboo shoots also reported may interfere with hormone metabolism, reproductive function, and hepatic xenobiotransformation enzymes [7].

Since that edible bamboo shoots species has cyanogenic toxicity, hence it is important to evaluate toxicity and compare of various bamboo shoots extract. Therefore, the aim of this research is to compare the nutritional, antioxidant properties and toxicity of selected bamboo shoots extract for functional food purposes. A relationship between total phenolics content, free radical scavenging activity is also demonstrated.

2. Material and methods

2.1 Materials

Quercetin, gallic acid, Folin-Ciocalteau's reagent, and 1,1-diphenyl 2-picrylhydrazyl (DPPH) were purchased from Sigma-Aldrich, Singapore. *Artemia salina* L. larva (Hobby Artemix) was purchased from Dohse Aquaristik GmbH & CO. Gilsdorf, Germany. Almunium chloride, sodium hydroxide, ethanol, and sodium carbonate were purchased from Merck. All the reagents were analytical grade.

Fresh young bamboo shoots were collected from Rawalele and Dawuan Kaler Village, Dawuan District Subang-West Java, Indonesia, during the month of January – June 2018 (Fig.1). Identification of botanical was performed by "Herbarium Bogoriense", Research Center for Biology, Indonesian Institute of Sciences (No. 957/IPH.1.01/If.07/IV/2018).

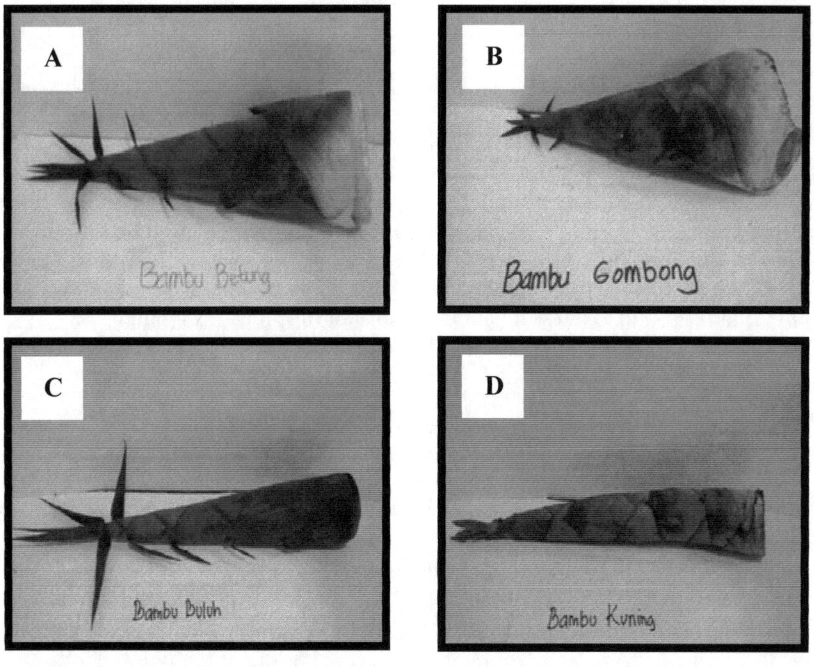

Figure 1. Young bamboo shoots (**A**) *Dendrocalamus asper* (Schult.) Backer ex Heyne; (**B**) *Giganthochla psedoarundinacea* (Steud.) Widjaja; (**C**) *Giganthochla apus;* **(D)** *Bambusa vulgaris* var. *striata*.

2.2 Samples preparation

Fresh bamboo shoots were rinsed, sliced, and steamed for 10 minutes, then dried at 45 °C for 2 days with the tray drier. After those, dried bamboo shoots ground into powder. The powder soaking in ethanol (1:10) for overnight, then extracted for 24 hours (three times) by maceration. The filtrates were consolidated and dried by vacuum evaporator. Accurately weigh of young bamboo shoots extract (0.06 g) put into a centrifuge tube, then added ethanol (10 mL). The sample was shaken for 10 minutes and then centrifuged. The supernatant was then diluted with ethanol until the 10 ml volumetric flask boundary line and mixed for 10 minutes.

2.3 Procedure analysis

2.3.1 Physicochemical properties

Physicochemical properties, *viz.*, yields, water content, and ash content were determined in accordance with [**8**]. Minerals, namely, sodium, potassium, magnesium, iron, zinc, and calcium, were determined by atomic absorption spectrometry methods.

2.3.2 Preliminary phytochemical screening

The powdered of bamboo shoot were tested for the presence of saponin, glycosides, terpenoids, flavonoids, tannins, and alkaloids [9-11].

2.3.3 Total phenolics content

Folin-Ciocalteau assay was used to determine the total phenolics content of young bamboo shoots [12]. Young bamboo shoots extract or blank or gallic acid standard solutions (100 µL) were add sodium carbonate (2 mL, 2%) and distilled water (2.8 mL) then left standing for 4 minutes. After that Folin-Ciocalteu (100 µL) was added into the solutions and left standing for 30 minutes in the dark condition. The measurement was conducted on a spectrophotometer (Shimadzu 1800 UV-Vis, Japan) at λ= 760 nm against the blank. Total phenolics content expressed as mg gallic acid equivalent (GAE) in a gram of dry weight of plant extract and calculated from expression (1). Samples were analyzed in triplicates.

$$\text{Absorbance} = 0.0009 \text{ gallic acid } (\mu g/mL) + 0.0024 \quad (1)$$

2.3.4 Total flavonoids content

The aluminum chloride assay was used to determine the total flavonoids content of young bamboo shoots [13]. Bamboo shoots extract or blank, or quercetin standard solution (1 mL) were add 2 mL aluminum chloride (2%) in methanol solution, mixed, and left standing for 30 minutes. The measurement was conducted on the spectrophotometer (Shimadzu 1800 UV-Vis, Japan) at λ= 415 nm against the blank. Total flavonoids content expressed as mg quercetin equivalent (QE) in a gram of dry weight of plant extract and calculated from expression (2). Samples were analyzed in triplicates.

$$\text{Absorbance} = 0.0145 \text{ quercetin } (\mu g/mL) + 1.558 \quad (2)$$

2.3.5 Antioxidant activity (DPPH) assay

The varying concentration of sample or standard solution or blank (1 mL) was added DPPH methanolic solution (3 mL, 0.004%), then left for 30 minutes in dark condition. The measurement was conducted on a spectrophotometer (Shimadzu 1800 UV-Vis, Japan) at λ= 517 nm against a blank [14]. Data reported as a concentration of antioxidant required for 50% scavenging of DPPH radicals in the specified time period (IC_{50}) and calculated from expression (3). Samples were analyzed in triplicates.

$$\%\text{Inhibition} = \left[\frac{(Ac - As)}{Ac}\right] \quad (3)$$

Where:
Ac= absorbance control or blank,
As= absorbance with sample or standard

2.3.6 Toxicity assay

The brine shrimp lethality (BSLT) method was used to investigate the toxicity of various bamboo shoots extract [15]. Brine shrimp eggs (Hobby Artemix®, Germany), contained salt, in a conical shaped vessel, and filled with sterile distilled water under constant aeration for 48 hours. Ten active nauplii put into each vial containing brine solution (4.5 mL) using a glass capillary. The young bamboo shoots extract (0.5 mL) were added, then left for 24 hours at room temperature. After that surviving larva (a dead and live larvae in each test) were counted. The experiments were conducted along with control. The different concentrations (12.5 - 1000 µg/mL) of each test substances in a set of three tubes per dosage. The probity analysis used to calculate a compound content capable of killing larvae up to 50% (LC_{50}).

2.4 Statistical analysis

Data presented in mean±standard deviation. The differences between treatments were analyzed using ANOVA. Significant differences between mean values were determined using Duncan's Multiple Range Test (α=5%). All statistical analysis was performed using Microsoft Excel 2010 and SPSS ver. 15.00 for Windows.

3. Results and Discussion

3.1. Physicochemical characteristics

Table 1 showed that the yield of various bamboo shoots powder ranged between 5.35% (w/w) and 5.83% (w/w). There is no difference value of water content on various bamboo shoots powder (*P>0.05*). Water content and ash content of various bamboo shoots powder ranged 4.82 - 5.92% and 7.72 – 8.36 %, respectively. According to [Iqbal et al], reported high ash content value in plant material indicates a large amount of inorganic nutrient. The physicochemical characteristics of various bamboo shoots powder were shown in Table 1.

Table 1. Physicochemical and minerals of various bamboo shoots powder

Samples	Bamboo shoots			
	Gombong	Betung	Buluh	Kuning
Yield (%)	5.83±0.39	5.58±0.13	5.38±0.32	5.35±0.32
Water content (%)	5.92±0.66	4.82±0.02	5.61±0.55	5.19±0.31
Ash (%)	7.72±0.70	8.00±0.43	8.36±0.44	7.87±0.48
*Mineral**:				
Sodium (µg/kg)	340.00±80.00	350.00±20.00	350.00±11.0	370.00±14.00
Potassium (mg/kg)	117.19±26.93	121.87±14.56	132.62±26.18	131.21±17.21
Magnesium (µg/kg)	270.00±80.00	260.00±11.0	260.00±8.00	220.00±7.00
Zinc (µg/kg)	710.00±40.00	670.00±40.00	730.00±20.0	690.00±20.00
Iron (µg/kg)	1760.00±89.00	1880.00±70.00	2100.00±63.00	2360.00±76.00
Calcium (µg/kg)	320.00±90.00	320.00±40.00	330.00±10.00	360.00±90.00

Values are mean±SD of triplicate; different alphabetic in the same rows = significant difference (α=0.05)

Minerals are required for metabolic activities of our body. Bamboo shoots are reported rich quantities of useful minerals, *viz*., sodium, iron, potassium, and magnesium [17]. Table 1 showed that bamboo gombong shoots (*Giganthochla psedoarundinacea* (Steud.) Widjaja) powder contained mineral Na (340 µg/kg), K (117 mg/kg), Mg (270 µg/kg), Zn (710 µg/kg), Fe (1760 µg/kg), and Ca (320 µg/kg). Bamboo betung shoots (*Dendrocalamus asper* (Schult.) Backer ex Heyne) powder contained mineral Na (350 µg/kg), K (121 mg/kg), Mg (260 µg/kg), Zn (670 µg/kg), Fe (1880 µg/kg), and Ca (320 µg/kg). Bamboo buluh shoots (*Giganthochla apus*) powder contained mineral Na (350 µg/kg), K (132 mg/kg), Mg (260 µg/kg), Zn (730 µg/kg), Fe (2100 µg/kg), and Ca (330 µg/kg). Moreover, bamboo kuning shoots (*Bambusa vulgaris* var. *striata*) powder contained mineral Na (370 µg/kg), K (131 mg/kg), Mg (220 µg/kg), Zn (690 µg/kg), Fe (2360 µg/kg), and Ca (360 µg/kg) (Table 1). Analysis of variance (ANOVA) showed that there is no significant difference on mineral content, namely., sodium, potassium, magnesium, zinc, iron, and calcium, of several young bamboo shoots powder (*P>0.05*). This results agreement with [Nirmala et al.], reported young bamboo shoots high in potassium and also contain mineral elements such as iron, calcium, chromium, manganese, zinc, and copper. Generally, bamboo shoots are rich in minerals, protein, amino acids, vitamins, and inorganic salts [18].

3.2 Antioxidant properties
Phytochemical screening showed that the ethanolic extract of several of young bamboo shoots is rich in tannins, terpenoids, and glycosides. The results of phytochemical screening are given in Table 2.

Table 2. Phytochemical screening of various bamboo shoots

Constituent	Bamboo shoots			
	Gombong	Betung	Buluh	Kuning
Alkaloids				
Mayer	-	-	-	-
Terpenoids	+	+	+	+
Saponin	-	-	-	-
Tannins	+++	++	++++	+
Flavonoids	+	-	++	-
Glycosides	+	+	+	+

(+) = present; (-) = absent

Table 2 shows the numerous phenolic compounds collected in various extracts of bamboo shoots. The phytochemical screening showed that the ethanolic extract of bamboo shoots did not contain alkaloids, saponin, and some flavonoids. These results were in agreement with [Zhang et al.], which reported that the ethanolic extract of bamboo stem shoots (*P. nigra* var *henonis* (Mitford Rendle)) is rich in triterpenoids compounds.

The quantitative analysis results reinforce the results of phytochemical screening previously, in which the highest total phenolics content found in bamboo buluh shoots extract (BUS) (see Table 3). The total phenolic contents on various bamboo shoots extract depended on the type of ecology, i.e., bamboo species, time of harvesting, climates, and soil. Different of bamboo species and ecology to growth contained different types of phytochemical compounds [20].

Table 3. Antioxidant properties of ethanolic extract of several young bamboo shoots

Samples	Total phenolics (mg GAE/g)	Total Flavonoids (mg QE/g)	TF/TP	Antioxidant activity IC_{50} (μg/mL)
Gombong shoots (GOS)	16.47 ± 0.11^{c}	1.11 ± 0.12^{bc}	0.07	1378.46^{c}
Betung shoots (BES)	14.62 ± 1.12^{d}	2.49 ± 0.07^{a}	0.17	2489.60^{d}
Buluh shoots (BUS)	27.83 ± 0.77^{a}	0.92 ± 0.12^{d}	0.03	418.48^{b}
Kuning shoots (KUS)	26.96 ± 6.42^{ab}	1.11 ± 0.42^{bc}	0.04	347.48^{a}

Values are mean±SD of triplicate; GAE = gallic acid equivalent; QE = quercetin equivalent; a>b>c>d, same alphabetic in the same coloumn = no difference (α=5%).

Table 3 shows that the total phenolic content of ethanolic extract of *bamboo shoots* ranged from 14.62 – 27.83 mg GAE/g dry weight. This indicated that several of bamboo shoots extract has a different phenolic total content. The bamboo buluh shoots extract (BUS) had the highest total phenolic content value (27.83 mg GAE/g) followed by KUS> GOS> BES ($P<0.05$). Meanwhile, The bamboo betung shoots extract (BES) is found to have the highest total flavonoids (2.49 mg QE/g) with a ratio of (TF/TP) which is 0.17, followed by GOS and KUS> BUS ($P<0.05$) (see Table 3). This result is in agreement with [Singhal], which reported that the total phenolics and total flavonoid content of *Bambusa vulgaris* are 29.0 mg GAE/100g and 49.7 mg CE/100 g, respectively.

Antioxidant properties of various bamboo shoots extract were evaluated by DPPH radical assay. Table 3 shows that IC_{50} of the ethanolic extract of bamboo kuning shoots (KUS) was 347.48 μg/mL followed by BUS (418.48 μg/mL)> GOS (1378 μg/mL)> BES (2489 μg/mL). Based on antioxidant analysis (DPPH assay) found that KUS had the highest antioxidant activity, shown by the lowest IC_{50} value ($P<0.05$). However, this activity is still below the standard of gallic acid or ascorbic acid. The phenolics content related to the free radical scavenging activities. According to [Zhang and Hamauzu], the antioxidant activity of phenolic compounds allows scavenging both electrophiles and active oxygen species.

Fig. 2 shows that the total phenolics content has a strong correlation with antioxidant activities of (R^2) 0.725. These results suggest that higher total phenol content will give stronger antioxidant activities.

Fig. 2 also indicates that 72% of the antioxidant properties of various bamboo shoots extract is contributed by the phenolic compounds and the remaining 28% comes from other secondary metabolisms, such as vitamins and others. This result is in agreement with another researcher that reported a close relationship between total phenolics and antioxidant activity on different plants [23-25].

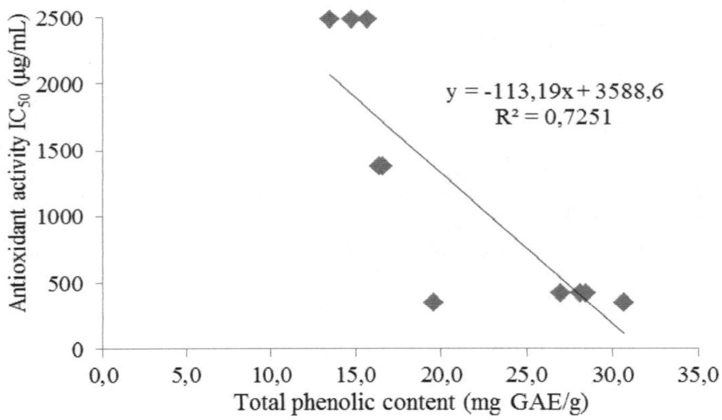

Figure 2. The relationship between total phenolic content and antioxidant activity of ethanolic extract of several young bamboo shoots.

3.3 Toxicity

In this screening, the toxicity of ethanolic extract of various bamboo shoots evaluated by brine shrimp lethality test (LC_{50}) is summarized in Table 4.

Table 4. Toxicity of ethanolic extract of various bamboo shoots

Samples	LC_{50} (µg/mL)	Toxicity[15]
Gombong shoots (GOS)	1088.39	Non-toxic
Betung shoots (BES)	738.08	Toxic
Buluh shoots (BUS)	999.31	Toxic
Kuning shoots (KUS)	773.29	Toxic

Table 4 showed the ethanolic extract of betung shoots, buluh shoots, and kuning shoots having a value of $LC_{50}<1000$ µg/mL (toxic), but not for gombong shoots extract (LC_{50} value more than 1000 µg/mL). The LC_{50} values of ethanolic extract for gombong shoots, betung shoots, buluh shoots, and kuning shoots were 1088.39 µg/mL, 738.08 µg/mL, 999.31 µg/mL, and 773.29 µg/mL, respectively. BES, BUS, and KUS showed lower LC_{50} values and significant toxicity against brine shrimp larvae of *Artemia salina.* It means that BES, BUS, and KUS might have a numerous of phenolic compounds or cyanogenic glycosides. However, this amount compared to the others reported, the ethanolic extract of BES, BUS and KUS were considered high values of toxicity, except bamboo gombong shoots. The diversity of bamboo species and part of the plants available for medicinal use, but some of that may interfere with hormone metabolism and reproductive function [7]. According to [Stewart and Moorthy], the toxic effect of the young bamboo shoots extract appears to be due to free hydrocyanic acid liberated, possibly of one of its cyanogenetic glucosides by a specific enzyme contained in it. These results can be an indicator of potential cytotoxic compounds.

4. Conclusion

Generally, bamboo shoots are rich in minerals, protein, amino acids, vitamins, inorganic salts, and bioactive compounds that provide health benefits. Bamboo shoots powder, viz., buluh, betung, gombong and kuning have a good nutritional characteristics. The ethanolic extract of buluh shoots (BUS) contained the highest total phenolics (27.83 mg GAE/g) followed by KUS > GOS > BES. Bamboo betung shoots extract (BES) is found to have the highest total flavonoids (2.49 mg QE/g) with a ratio of TTF/TP) which is 0.17. Meanwhile, the DPPH assay (IC_{50}) showed that IC_{50} of KUS (347.48 μg/mL) was lowest. Correlation the total phenolics content has a strong with antioxidant activities (R^2: 0.72). Based on these values, KUS is found to have the highest antioxidant activity. Furthermore, the level toxicity of $LC_{50,}$ BES, BUS, and KUS extract is valued as cytotoxic.

5. Acknowledgment

The authors are thankful to the Ministry of Research and Technology and Higher Education, Republic of Indonesia (KEMENRISTEKDIKTI-RI) for financial assistance (No.21/P/RPL-LIPI/INSINAS-1/III/2018) and Indonesian Institute of Sciences (LIPI) for providing access and technical support.

6. References

[1] Nirmala C, David E, and Sharma M L 2007 Changes in nutrient components during ageing of emerging juvenile bamboo shoots *I nt. J. Food Sci. Nut.*, **58** pp. 612-618
[2] Burkill I H 1935 A Dictionary of the Economic Products of the Malay Peninsula. Crown Agents for the Colonies, London, England
[3] Kencana D, Antara N S, and Widia W 2012 Team UNUD-USAID-TPC project. praktek baik budi daya bambu rebeung tabah (*Gigantochloa Nigrociliata* BUSE-KURZ)
[4] Nirmala C, Bisht M S, and Laishram M 2014 Bioactive compounds in bamboo shoots: health benefits and prospects for developing functional foods. *Int. J. of Food Sci and Tech.* **49** pp. 1425–1431
[5] Park E J and Jhon D Y 2009 Effects of bamboo shoot consumption on lipid profiles and bowel function in healthy young women. *Nutrition* **25** pp. 723-728
[6] Rawat K, Nirmala C, and Bisht M S 2015 Processing techniques for reduction of cyanogenic glycosides from bamboo shoots. 10th World Bamboo Congress, Korea
[7] Panee J 2015 Review: Potential medicinal application and toxicity evaluation of extracts from bamboo plants *J. Med. Plants Res.* **9**(23) pp. 681-692.
[8] [AOAC] Association of Official Analytical Chemistry 1990 Official Methods of Analysis. Association of Official Analytical Chemist. Arlington, Virginia (US): AOAC
[9] Harborne J B 1973 *Phytochemical methods* Chapman and Hall, Ltd London. pp. 49-188
[10] Sofowara A 1993 Medicinal plants and traditional medicine in Africa Spectrum Books Ltd, Ibadan, Nigeria p.289
[11] Trease G E and Evans W C 1989 Pharmacognsy. 11th ed. Brailliar Tiridel Can Macmillian publishers
[12] Iwansyah A C and Yusoff M M 2012 Physicochemical, minerals and antioxidant properties of *Labisia pumila* var. *alata* of selected geographic origins. *J. of Agr. Sci. (AGRIVITA)* **34** pp. 94-104
[13] Meda A, Lamien C E, Romito M, Miliogo J, and Nacoulina O G 2005 Determination of the total phenolic, flavonoid, and proline content in burkina fasan money, as well as their radical scavenging activity *Food Chemistry* **91** pp. 571-577
[14] Kumaran R J and Kuranakaran 2006 Antioxidant and free radical scavenging activity of an aqueous extract of *Coleus aromaticus Food Chemistry* **97** pp.109-114
[15] Meyer B, Ferrigni N, Putnam J, Jacobsen L, Nichols D E, and McLaughlin J L 1982 Brine shrimp: a convenient general bioassay for active plant constituents. *J. Medicinal Plant Research*, **45** pp.31-34

[16] Iqbal S, Younas U, Sirajudin, Chan K W, Sarfraz R A, and Uddin M K 2012 Proximate composition and antioxidant potential of leaves from three varieties of mulberry (*Morus. Sp*): a comparative study. *Int. J. Mol. Sci.* **13** pp. 6651-6664

[17] Nongdam P and Tikendra L 2014 Review: the nutritional facts of bamboo shoots and their usage as important traditional foods of north-east India. *International Scholarly Research Notices* pp.1- 17

[18] Goyal A K and Brahma B K 2014 Antioxidant and nutrceutical potential of bamboo: an overview. *International Journal of Fundamental and Applied Sciences* **3**(1) pp. 2-10

[19] Zhang Y, Wu X, Ren Y, Fu J, and Zhang Y 2004 Safety evaluation of a triterpenoid-rich extract from bamboo shavings. *Food Chem. Toxicol* **42** pp. 1867-1875

[20] Maisuthisakul P 2012 Phenolic constituents and antioxidant properties of some thai plants, phytochemicals - a global perspective of their role in nutrition and health, Dr Venketeshwer Rao (ed.)

[21] Singhal P 2016 Effect of processing technique on holistic quality of bamboo shoot (*Bambusa vulgaris*). [Dissertation] Indian Institute of Technology Delhi; India

[22] Zhang D and Hamauzu Y 2004 Phenolics compounds and their antioxidant properties in different tissues of carrots (*Daucus carota* L.). *Food, Agriculture, & Environment* **2**(1) pp. 95-100.

[23] Maisuthisakul P, Pasuk S, and Ritthiruangdej P 2008 Relationship between antioxidant properties and chemical composition of some Thai plants *Journal of Food Composition and Analysis* **21** pp. 229-40

[24] Iwansyah A C, Damanik R M, Kustiyah L, and Hanafi M 2016 Relationship between antioxidant properties and nutritional composition of some galactopoietics herbs used in indonesia: a comparative study *Int. J Pharm Pharm Sci*, **8**(12) pp. 236-243

[25] Yusoff M M, and Iwansyah A C 2011 Comparative evaluation of total phenolics and free radical scavenging activity of aqueous extracts of *Labisia pumila* var. *alata* from Malaysia and Indonesia. International Conference on Biotechnology and Food Science, IEEE, April 1-3, 2011. Bali, Indonesia pp.4-8

[26] Stewart A D and Moorthy F N 1933 Lethal properties of aqueous extract of young bamboo shoots. *The Indian Medial Gazette* pp.320-324

IOP Conf. Series: Earth and Environmental Science **251** (2019) 012018 doi:10.1088/1755-1315/251/1/012018

Antiviral Activity of Isobutyl Gallate to Dengue Virus Serotype 2 *In Vitro*

B E Dewi[1,2,a] M Angelina[3,4] F Nuwwaaridya[5], H Desti[1,2], T M Sudiro[1,2]

[1]Department of Microbiology, Medical Faculty Universitas Indonesia-Cipto Mangunkusumo Hospital Jalan Pegangsaan Timur no 16, Jakarta, Indonesia.
[2]Infectious Disease and Immunology Cluster, Indonesian Medical Education and Research Institute, Jl Salemba no 6 Jakarta 10430, Indonesia
[3]Research Center for Chemistry, Indonesian Institute of Sciences, Kawasan Puspitek Serpong, Indonesia 15416
[4]Doctoral Programme in Biomedical Science Faculty of Medicine, University of Indonesia, Jl. Salemba Raya 6, Jakarta 10430, Indonesia
[5]Undergraduate student, Medical Faculty Universitas Indonesia

Email addresses: betied@yahoo.com

Abstract. Dengue is one of the most prevalent mosquito-borne diseases in Indonesia that lead to a spectrum of disease, ranging from a flu-like illness to a life-threatening dengue shock syndrome. Despite its high incidence and mortality rate, there is currently no antiviral drug available against dengue virus (DENV). Therefore, research regarding the development of anti-viral to DENV is essential. The aim of this study was to evaluate the antiviral activity of isobutyl gallate, a gallic acid derivative, against DENV in vitro. Various concentration of isobutyl gallate was added to DENV-2 strain New Guinea C (NGC) and infected to Vero cell line. Inhibition of DENV replication as aligned with half inhibitory concentration (IC_{50}) value was determined by focus assay. Furthermore, the toxicity of isobutyl gallate that corresponds with cytotoxic concentration (CC_{50}) value was determined by MTT assay. Selectivity index (SI) value was defined as the ratio of CC_{50} to IC_{50}. Isobutyl gallate showed no cytotoxic effects against Huh 7 it-1 cells with CC_{50} values of 167.19 µg/mL and exhibited strong antiviral activity with IC_{50} of 4.45 µg/mL, hence a significant SI value of 25.69. Isobutyl gallate exerts potent antiviral activity against DENV.

1. Introduction

Over the last decades, DENV has emerged as one of the most important mosquito-borne viral diseases in the world. Dengue fever causes a severe flu-like illness as a result of DENV infection. DENV and its vectors have been widely distributed over the last half-century, with more than 125 countries in the world are dengue endemic [1]. According to the data from World Health Organization (WHO), there are up to 100 million cases of dengue fever, and half a million cases of dengue hemorrhagic fever (DHF) require hospitalization with over 20.000 deaths each year, a major percentage of whom are children [2]. Studies suggest that dengue is more prevalent in Indonesia, with proportions ranging between 5.7-7%. Therefore, dengue fever poses a serious risk for Indonesians [3].

Dengue infections are mostly asymptomatic. However, when they are symptomatic, the virus can cause mild dengue fever (DF), or even more severe forms of the disease, which is dengue hemorrhagic fever (DHF) that most likely progress to dengue shock syndrome (DSS) and if left untreated it can be fatal and lead to death [4]. As dengue infection is posing a serious threat, it is important to prevent both the infection and the following complication. Unfortunately, there is currently no specific antiviral available for DENV. One of the compounds that have antiviral DENV properties has been reported was gallic acid [5]. Also reported by others that one the derivative gallic acid, octyl gallic have activity combats DNA and RNA virus. Based on the previous study, our team tries to test a derivative of gallic acid as antiviral DENV [6]. One of the derivatives is isobutyl gallate, a gallic acid derivative that is commonly used in nutraceutical due to its anti-inflammatory and antioxidant properties. Gallic acid itself, a polyphenolic compound whose product of esterification has potentials in inhibiting Hepatitis C virus (HCV), possesses antibiotic, antiviral, and anticancer properties. It has been proven that gallic acid can act as a strong antiviral antioxidant and can exert its antiviral effect on several RNA viruses, including influenza virus and poliovirus [5]. However, research regarding the potential of gallic acid derivatives as DENV antiviral is still limited. Therefore, this research is conducted to evaluate the potential of isobutyl gallate as a candidate for DENV antiviral.

2. Materials and methods

This research was designed to be a laboratory-based experimental study in vitro with triple replications. It was conducted at the Department of Microbiology Faculty of Medicine Universitas Indonesia from August 2017 to May 2018.

2.1. Preparation of Isobutyl Gallate

Isobuthly gallate was synthesized from gallic acid by Dr. Ade Arsianti (Medical Faculty University of Indonesia). To make a testing solution, it was prepared through dilution, diluted in DMSO until the concentration reaches 100 mg/mL. To make isobutyl gallate at the concentration of 160 μg/mL, the diluted compound was homogenized using a vortex, and then 6.08μl of diluted extract was added to 1900μl of the medium. It was further diluted in serial dilution to obtain a concentration of 80, 40, 20, 10, and 5 μg/mL [6].

2.2. Preparation of Dengue Virus

Preparation of dengue virus DENV-2 strain New Guinea C (NGC) used in this study was done by the researcher of the previous study [6].

2.3. Determination of Viral Titer Using Focus Forming Unit Assay

Viral titer was measured using the focus assay method, an immunostaining procedure that uses peroxidase as a marker to count DENV viral titer. Monolayer Huh-7 was seeded in a 96-well plate. After 24 hours of incubation, the cells were treated with a mixture of DENV and various concentrations of isobutyl gallate. The cells were incubated at 37 °C for 2 hours. During the incubation period, the 96-well plate was shaken at 30 minutes intervals. After the cells had been infected with the virus, the remaining DENV was removed from the cells by washing the cells with PBS. Various concentrations of isobutyl gallate were added to the cells before they were incubated at 37 °C for 3 days.

To determine the antiviral activity, the cells were harvested before counting the viral titer using focus assay. The incubated cells that had previously been prepared were added with a serial dilution of viral supernatant and were further incubated at 37 °C with 5% CO_2 for 2 hours. After incubation, the medium was removed, and methylcellulose overlay mediums were added. Fixation was done using PBS with 10% of Formaldehyde at room temperature. The cells were washed thrice with PBS at 5 minutes intervals. Following cell washing, PBS tween 100 μl/well was added and Vero cells were washed with PBS tween after secondary antibody, IgG anti-human labeled as peroxidase had been

IOP Conf. Series: Earth and Environmental Science **251** (2019) 012018 doi:10.1088/1755-1315/251/1/012018

added to the cells. OPD substrate containing H_2O_2 was also added to Vero cell. The infected cell will form brownish foci, originated from one virus. This experiment was done in triplicate [7].

$$\text{Viral titer (FFU/mL)} = \frac{(Total\ foci*Dilution*Volume\ of\ infection)}{1000} \quad (1)$$

2.4. Determination of IC_{50}

The results obtained from focus assay was also used to calculate IC_{50} with the following formula:

$$\text{Infectivity (\%)} = \frac{Viral\ titer}{Average\ titer\ DMSO} * 100\% \quad (2)$$

2.5. Determination of CC_{50} using MTT Assay

Cell that had been prepared was tested using MTT assay. MTT solution was mixed with DMEM and was homogenated using vortex. Three hundred microliter of MTT medium 10% was added to cell in each well plate. The cell was incubated and MTT medium 10% was drained. Two hundred microliter of DMSO 100% medium was added to cell in each well plate. The plate was put on the shaker for 30 minutes. ELISA reader was used to measuring the peak absorbance at 490nm wavelengths [7]. The results were used to calculate CC_{50} value, represented by the percentage of cell viability, with following formula:

$$\text{Cell viability (\%)} = \frac{Absorbance\ of\ sample}{Average\ of\ DMSO\ 0.1\%\ absorbances} * 100\% \quad (3)$$

3. Results and Discussions

Dengue remains one of the infectious diseases that pose a serious risk to health. It is one of the leading causes of death worldwide, especially in endemic areas. There are more than one-third of the world's population who live in endemic areas at high risk for dengue infection and more than 400 million people are infected with dengue every year. Despite the high number of cases, antiviral against DENV is currently underdeveloped. Therefore, research about DENV antiviral is needed to treat dengue infection and prevent the complication, thus can reduce the mortality rate of dengue cases worldwide.

Previous research report activity gallic acid as anti DENV [4]. In this study, the compound tested was isobutyl gallate, one of gallic acid derivatives with the molecular formula of $C_{11}H_{14}O_5$. Gallic acid itself is described as a colorless or yellowish-crystalline compound that can be obtained from tea leaves, tree bark, gallnuts, and berries through hydrolysis of tannins. It is commonly used in nutraceuticals due to its anti-inflammatory and antioxidant properties [6].

3.1. Percentage of Infectivity and IC_{50} value

The antiviral property of Isobutyl gallate against DENV replication was determined by adding various concentrations of Isobutyl gallate to DENV and then infect to the cells. DENV after treated with isobutyl gallate at the concentration of 40 μg/mL or higher showed no DENV that still able to infect the cell (Table 1.) with 0% infectivity. When the concentration decreased to 20 μg/mL, the mean infectivity of the DENV-2 increased to 18.24 ± 9.8 (Table 1.). In addition, the results of DENV-2 infectivity showed a strong correlation between the concentration of the isobutyl gallate and the percentage of DENV-2 infectivity with R^2 value of 0.944 (Figure 1.)

Table 1. Average of percentage infectivity of DENV2 after treated with various concentrations of isobutyl gallate

Concentration (μg/mL)	Average of Infectivity (%) ± SD
160	0.0
80	0.0
40	0.0
20	18.24 ± 9.8
10	44.13 ± 3.74
5	51.78 ± 3.84
DMSO	100 ± 13.21

IC_{50} of isobutyl gallate was determined from the equations obtained from the respective graph (Figure1). Based on the equation of a respected graph, we found that IC_{50} of isobutyl gallate was 4.45 μg/mL.

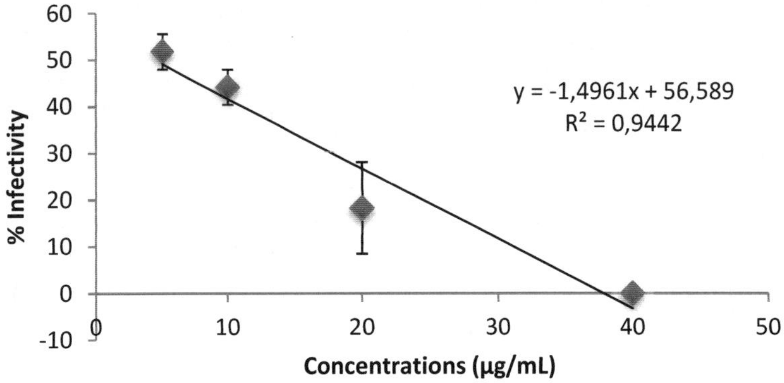

Figure 1. The linear graph with R^2 value and equation to obtain IC_{50} value isobutyl gallate

In this experiment, the virus was treated with various concentrations of isobutyl gallate and infect to the Huh 7 it-1 cell to determine antiviral activity. DMSO at a concentration of 0.1% was used as a negative control of an antiviral experiment of this study. The value of IC_{50} is described as the concentration in which the compound inhibited half of the DENV-2 infection of isobutyl gallate was 4.45 μg/mL. The results suggest that isobutyl gallate can be a good candidate for DENV antiviral because it exerted inhibitory activity against dengue virus at low concentration in vitro. This value is significantly lower compared to *Curcuma longa*, *Psidium guajava*, and *C. papaya,* whose IC_{50} value are 17.91 μg/mL, 7.2 μg/mL, and 6.57 μg/mL, respectively [9,10].

3.2. Toxicity and CC_{50} value of Isobutyl gallate

Potential cytotoxicity of isobutyl gallate towards uninfected cells was determined by measuring the viability of the cell after being treated with various concentrations of isobutyl gallate. MTT assay was used to determine the viability of the treated and control cells.

Table 2. Cell viability of Huh-7it-1 after treated with various isobutyl gallate

Concentration (μg/mL)	Average of Viability (%) ± SD
160	58.43 ± 9.15
80	131.10 ± 15.76
40	149.20 ± 3.08
20	183.70 ± 8.40
10	195.73 ± 10.15
5	212.06 ± 22.82
DMSO	100.10 ± 6.27

The concentration of 5 to 80 μg/mL, showed a great percentage of mean viability (>100%), indicating that the compound was not toxic to the cell at this range. Treated with 160 μg/mL, however, there was a decrease to 58.43% (Table 2). The CC_{50} value was defined as the concentration that resulted in 50% cytotoxicity. To obtain CC_{50} value, the data was plotted on a scattered graph, resulting in a linear regression equation (Figure 2.). By plotting, the value of R^2 was obtained and it revealed a strong correlation between the concentration of isobutyl gallate and the percentage of cell viability, higher concentration of isobutyl gallate resulted in the lower percentage of cell viability. Based on the equation of respected graph (Figure 2.), the CC_{50} value of isobutyl gallate was 167.19 μg/mL.

Moreover, isobutyl gallate has low potential cytotoxicity towards uninfected cells because, at the concentration of 5 to 80 μg/mL, the compound was not toxic to the cell. This is proven by the cell viability results that showed a great percentage of mean viability (>100%) at that range. Potential toxicity was measured by the value of CC_{50}, signifying the concentration at which isobutyl gallate will become toxic to uninfected cells. Isobutyl gallate has a high value of CC_{50} (CC_{50} = 167.19 μg/mL), thus it requires a high concentration of isobutyl gallate in order to result in cytotoxicity towards uninfected cells.

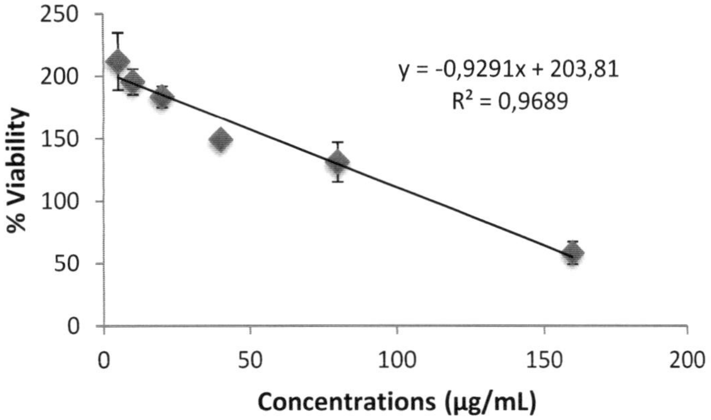

Figure 2. Regression liner of Viability of cell versus concentration of isobutyl gallate to determination correlation and CC_{50} value of isobutyl gallate

3.3. Selectivity Index (SI)

Selectivity Index (SI) is calculated by dividing the value of CC_{50} to the value of IC_{50}. Based on the aforementioned data, the selectivity index of isobutyl gallate was 25.69.

In this study showed that the selectivity index of isobutyl gallate was considered high with the value of 25.69. According to studies conducted by Valdés-Garcia and Quispe, compounds that have SI value of above 10 are considered as selective compounds and SI value of above 10 can be considered as a significant value. It is desirable to have high SI because that means the compound can work selectively towards specific pathogen [8]. In comparison with other plant-derived polyphenolic compounds (bioflavonoids) namely daidzein and naringin, that have significantly lower SI value of 1.03 and 1.3 respectively, isobutyl gallate is more selective towards DENV. In addition, this value is about five times higher than *C. longa* SI value (SI=4.8) and about three times higher than that of quercetin (SI=8.74), which is a phenolic compound. The SI of isobutyl gallate, however, is similar to the SI of *P. guajava,* at the level of 21.28 [10,11].

4. Conclusion

Our results showed that isobutyl gallate exerts a dose-dependent inhibitory to dengue virus in vitro with CC_{50}, IC_{50}, and SI of 167.19 μg/mL, 4.45 μg/mL, and 25.69 respectively. The aforementioned data signifies that isobutyl gallate has low cytotoxicity towards uninfected cells and exhibited anti-viral property at low concentration, thus it can be a good candidate for DENV antiviral in future.

5. Acknowledgments

The authors would like thank to Dr. Ade Arsianti for providing isobutyl gallate. Special thank Chie Aoki Ph.D. for providing Huh7it-1 cell line. This study was supported by grant of Publikasi Terindeks for Tugas Akhir Mahasiswa UI (PITTA) 2018 No: 0588/SK/R/UI/2018

6. References

[1] Murray NEA, Quam MB, and Wilder-Smith A. 2013 Epidemiology of Dengue: Past, Present and Future Prospects *Clin Epidemiol.* **20 5** 299–309

[2] Pozzetto B, Memmi M, and Garraud O 2015 Is Transfussion-Transmitted Dengue Fever a Potential Public Health Threat? *World J Virol.* **12 4(2)** 113–23

[3] Kosasih H, Alisjahbana B, de Mast Q, Rudiman I F, Widjaja S, Antonjaya U, Novriani H, Susanto N H, Jusuf H, van der Ven A and Beckett C G, 2016 *The epidemiology, virology and clinical findings of dengue virus infections in a cohort of Indonesian adults in Western Java. Plos neglected tropical diseases* **10** (2) 0004390

[4] Rajapakse S 2011 Dengue Shock *J. Emerg Trauma Shock* **24 (1)** 120–127

[5] Mahmudah M 2016 *Invitro Potency of Gallic Acid Derivatives as an Antiviral Candidate of Dengue Virus Serotype 2 and Insilico Study on NS5 Protein* Program Studi Magister Ilmu Biomedik 45-50

[6] Misao U, Hisashi Y, Yukiko K, Masanori H, Tomihiko H, and Hajime K 2007 Antiviral Effect of Octyl Gallate Against DNA and RNA Viruses *Antiviral Res.* **73** 85–91

[7] Govea-salas M, Rivas-estilla A M, Rodríguez-herrera R, Lozano-sepulveda S A, Aguilar-Gonzalez C N, Zugasti-Cruz A, et al 2016 *Gallic Acid Decreases Hepatitis C Virus Expression Through its Antioxidant Capacity* Exp Ther Med. **11 (2)** 619–24

[8] Peña-Morán O A, Villarreal M L, Álvarez-Berber L, Meneses-Acosta A, Rodríguez-Lopez V 2016 Cytotoxicity, Post-treatment recovery, and Selectivity Analysis of Naturally Occurring Podophyllotoxins from Bursera fagaroides var. fagaroides on Breast Cancer Cell Lines *Mol Basel Switz.* **4** 21 (8)

[9] Dewi B E, Ridhanya A, Riasanti M, Desti H, Putri D, and Sudiro T 2015 *An Invitro Study of Curcuma longa Activity as Antiviral Against Dengue Virus* Universitas Indonesia 2-6

[10] Saptawati L, Febrinasari R P, Yudhani R D, Yono H, Faza A G, Luthfiani S, Ummiyati H S, Sudiro T M and Dewi B E 2017 In vitro study of eight Indonesian plants extracts as anti Dengue virus *Health Science Journal of Indonesia* **8** (1) 12-18

[11] Zandi K, Teoh B-T, Sam S-S, Wong P-F, Mustafa MR, and Abubakar S 2011 Antiviral Activity of Four Types of Bioflavonoid Against Dengue Virus Type-2 *Virol J.* **28**

Characterization of phytochemical, physicochemical, and microscopic from five selected mangrove associate leaves

M Basyuni[1,2*], D P Sari[3], D N Illian[3], P A Z Hasibuan[3]

[1]Department of Forestry, Faculty of Forestry, Universitas Sumatera Utara, Jl. Tri Dharma Ujung No. 1 Medan, North Sumatera 20155, Indonesia
[2]Mangrove and Bio-Resources Group, Center of Excellence for Natural Resources Based Technology, Universitas Sumatera Utara, Medan, North Sumatera 20155, Indonesia
[3]Department of Pharmacology, Faculty of Pharmacy, Universitas Sumatera Utara, Medan 20155, Indonesia

[*]E-mail: m.basyuni@usu.ac.id

Abstract. This present study reports the analysis of phytochemical, physicochemical, and microscopic of selected mangrove associates leaves, namely *Acacia auriculiformis*, *Hibiscus tiliaceus*, *Pandanus odoratissimus*, *Pometia pinnata*, and *Ricinus communis*. Phytochemical was performed on mangrove leaves based on the developed method. Physicochemical properties of mangrove leaves were determined by simplicial powder that consists of moisture content, water-soluble, ethanol soluble, ash content and the essay extract was carried out to the standard method. Microscopic observation on the simplicial powder was done according to World Health Organization (WHO) method. Phytochemical screening of simplicial powder shows variation among the species, only saponin and triterpenoid/phytosterol were detected in all species. By contrast, the phytochemical analysis in the hexane extract alkaloids, flavonoids, glycosides, saponin, and tannin was not identified, however, only triterpenoid/steroid was observed in all species. Physicochemical properties also depict diversity among the species and significant finding on the water content was under 10% as a condition for the drug. Microscopic screening found a different type of stomata in mangrove associates leaves, *H. tiliaceus* and *P. pinnata* had paracytic stomata-type, *P. odoratissimus* and *R. communis* contained anomocytic, only *A. auriculiformis* was distinct type due to having two kinds of stomata: paracytic and anisocytic. The present study may provide potential biological and medicinal properties from mangrove associates leaves.

1. Introduction

Mangrove plants, including mangrove associates or coastal plants, are an excellent source of secondary metabolites generally derived from isoprenoid alcohol [1-2]. Polyisoprenoid alcohols are undeviating five unit polymers that are in virtually existing all living organisms. Secondary metabolites are useful in their interaction with the environment, in the development molecules valued, and exploited for their phytochemical and other activities [3]. Because of their wide-ranging biological activities, isoprenoids are considered as important as active biologically compound for medicinal properties [4], and mangrove associates have long been practised in traditional medicine [5].

Furthermore selected species, namely *Acacia auriculiformis, Hibiscus tiliaceus, Pandanus odoratissimus, Pometia pinnata,* and *Ricinus communis* are common mangrove associates in North

Sumatra [2]. A number of biological and phytochemical activities of mangrove associates have been described. For example, the antinociceptive and anti-inflammatory activity of phenolic from *H. tiliaceus* and *P. odoratissimus* leaf extracts have been described [6-7]. *R. communis* leaf extracts have been shown to have antimicrobial property [8]. Recently it has been reported the screening of twelve mangrove associate leaves to have significant antibacterial activities that open up the possibility for antimicrobial drug development [9]. In spite of the importance of secondary metabolites from mangrove associates as potential sources of phytochemistry, limited information on the phytochemical and physicochemical of mangrove associate in North Sumatra has not previously been presented. This current study, therefore, aimed to describe the phytochemical, physicochemical, and microscopic analysis of selected mangrove associates leaves to extend our previous works to search isoprenoid and polyisoprenoid compounds from mangrove plants [1-2,10].

2. Materials and method

2.1. Materials

Selected mangrove associates leaves, i.e. *Acacia auriculiformis*, *Hibiscus tiliaceus*, *Pandanus odoratissimus*, *Pometia pinnata*, and *Ricinus communis* were collected from mangrove forest at Lubuk Kertang, Langkat, North Sumatra, Indonesia. The identification of plant species is conducted at Herbarium Medanense, Medan, North Sumatra. The specimen vouchers have been deposited there.

2.2. Phytochemical investigation of simplicial and hexane extract

A phytochemical method in simplicial and hexane extract was performed on mangrove associates leaves to the standard process as earlier described [2]. Phytochemical analysis composed of several primary, secondary metabolites, alkaloids, flavonoids, glycosides, saponins, tannins, triterpenoids/sterols were done as previously reported [11].

2.3. Physicochemical properties

Physicochemical parameters of mangrove associate leaves were analyzed by simplicial powder that involves of moisture content, water-soluble, ethanol soluble, ash content, and the essay extract was implemented to the standard of WHO [12] and the previous study [11].

2.4. Microscopic analysis

Microscopic studies on the simplicial powder of five selected mangrove associate leaves were performed based on WHO method [12]. Concisely, a microscopic investigation of the simplicial powder was conducted by robust the solution of chloral hydrate on the slide, interspersed with simplicial powder, then shielded with a cover glass and observed under a microscope.

3. Results and Discussions

3.1. Phytochemical investigation of selected mangrove associate leaves

The phytochemical analysis of simplicial powder and extracts of n-hexane from chosen mangrove associate leaves, *A. auriculiformis*, *H. tiliaceus*, *P. odoratissimus*, *P. pinnata*, and *R. communis* were performed to get data on the group of secondary metabolites. The results of simplicia powder on the phytochemical investigation and n-hexane extract from five mangroves associate were depicted in Table 1. The classification results are then used as a further emphasis to categorize which groups of compounds have activity against phytopharmacological activities.

Phytochemical investigation of simplicial powder expresses diversity among the species; only saponin and triterpenoid/phytosterol were detected in all species. By contrast, the phytochemical analysis in hexane extract shows no alkaloids, flavonoids, glycosides, saponin, and tannin detected, however, only triterpenoid/phytosterol found in all species. Phytochemical tests displayed that the simplicial powder comprised significant flavonoids, glycosides, saponins, tannins and triterpenoids/steroids with minor of alkaloids (Table 1). The adding of Molisch reagent and focused sulfuric acid formed a purple ring which indicates the occurrence of glycoside compounds (Table 1). The development of stable foam with shaky in water and did not lose with the addition of 2N HCl suggests the existence of saponin compounds [11]. The addition of Mg powder and concentrated hydrochloric acid yields a red solution, and with the accumulation of amyl alcohol, the red color is attracted to the amyl alcohol layer which indicates the presence of flavonoids [11]. The addition of FeCl3 gives a blackish blue color that shows the presence of tannin compounds [8]. The addition of LB reagents forms blue-green, purple and red-purple colors which specify the presence of triterpenoid/steroid compounds [11, 13, 14].

While the addition of Dragendorf, Bouchardat, and Mayer reagent solutions give positive results on *A. auriculiformis* and *R. communis* leaves only, by forming turbid solutions. The content of alkaloid compounds in the others samples of simplicia (*H. tiliaceus, P. odoratissimus,* and *P. pinnata*) was not traced (Table 1). Polyisoprenoids have tested *in vitro* is more useful as an antimicrobial agent and an oxidant in mangrove associate leaves [9]. Recently it has been reported that polyisoprenoids from *Nypa fruticans* leaves inhibited cyclooxygenase 2 expression of WiDr colon cancer [14]. Dolichols from *Avicennia marina* and *A. lanata* has been shown to inhibit WiDr cells proliferation [15].

Table 1. Phytochemical screening in simplicial and hexane extract

No.	Compound	Solvent	Mangrove species	Simplicial	n-hexane
1.	Alkaloids	Dragendorff	*A. auriculiformis*	+	–
		Bouchardat	*H. tiliaceus*	–	–
		Mayer	*P. odoratissimus*	–	–
			P. pinnata	–	–
			R. communis	+	–
2.	Flavonoids	Zn + HCl	*A. auriculiformis*	–	–
			H. tiliaceus	+	–
			P. odoratissimus	+	–
			P. pinnata	+	–
			R. communis	+	–
3.	Glycosides	Molisch	*A. auriculiformis*	–	–
			H. tiliaceus	–	–
			P. odoratissimus	+	–
			P. pinnata	+	–
			R. communis	+	–
4.	Saponin	Hot water + HCL	*A. auriculiformis*	+	–
			H. tiliaceus	+	–
			P. odoratissimus	+	–
			P. pinnata	+	–
			R. communis	+	–
5.	Tannins	$FeCl_3$ 1%	*A. auriculiformis*	–	–
			H. tiliaceus	–	–
			P. odoratissimus	+	–
			P. pinnata	+	–
			R. communis	+	–
6.	Triterpenoid/Steroid	Liebermann-Burchard	*A. auriculiformis*	+	+
			H. tiliaceus	+	+
			P. odoratissimus	+	+
			P. pinnata	+	+
			R. communis	+	+

(+): contain the compound
(–): not contain

3.2. Physical and chemical properties

Table 2 recapitulates the physicochemical of five selected mangrove associate leaves. Description testing leads of simplicial of five mangroves associate *A. auriculiformis, H. tiliaceus, P.odoratissimus, P. pinnata*, and *R. communis* for moisture content 6.92%, 6.46%, 4.73%, 5.76%, and 6.12%, respectively, usually incorporate to the of requirements for drug development $\leq$ 10%. The water content, therefore, contained more than 10% lead the process of decay and damage to the material.

Table 2. Physicochemical characteristic of mangrove associate leaves

No	Species	Water content (%)	Water soluble (%)	Ethanol soluble (%)	Total ash (%)	Ash insoluble acid (%)
1	*A. auriculiformis*	6.92±0.2	28.63±3.6	17.82±2.8	7.24±	1.83±0.2
2	*H. tiliaceus*	6.46±0.3	28.37±3.7	18.57±5.7	3.45±	1.79±0.4
3	*P. odoratatissima*	4.73±0.4	22.98±3.8	21.42±4.2	3.71±	0.97±0.1
4	*P. pinnata*	5.76±0.5	24.17±7.1	20.68±2.5	5.34±	1.73±0.3
5	*R. communis*	6.12±0.1	17.67±6.7	15.16±5.4	4.26±	2.89±0.1

Data are mean of triplicate analyses ± SD (*n*= 3).

Determination of the concentration of the mangrove associate leaves which in water-soluble is carried out to search the materials that are present in the water solvent [14]. Analysis of the content of ethanol soluble indicates the substances that are present in ethanol solvents, which are regarded as the polar and non-polar indication. Determination of the content of soluble in water and ethanol was performed to show the number of polar compounds which are soluble in water and ethanol. This result suggests that more dissolved substances in water solvents occurred. Determination of ash content is carried out to destruct organic compounds and their derivatives [11, 16]. The ash content is not acid soluble to define the levels of inorganic compounds which are not soluble in acids such as silica gel.

The high total ash content depicts the existence of inorganic substances of metals, such as Ca, Mg, Fe, Cd and Pb which may partly derive from contamination [17]. High heavy metal content can endanger health, therefore it is necessary to determine the total ash content and acid insoluble ash content to confirm the simplicial leaf does not enclose certain heavy metals surpassing the definite value because it is toxic to health [17]. The simplisia used as raw material for medicines must meet the requirements documents published by the Ministry of Health, Indonesia.

3.3. Microscopic analysis

The microscopic screening established a diverse type of stomata in mangrove associates leaves in line with [18]. *H. tiliaceus* and *P. pinnata* possed the paracytic stomata-type, while *P. odoratissimus* and *R. communis* contained anomocytic, only *A. auriculiformis* was distinguishable type due to have two type of stomata: paracytic and anisocytic. It has been shown that microscopic study to be an indicator type of vegetation [19].

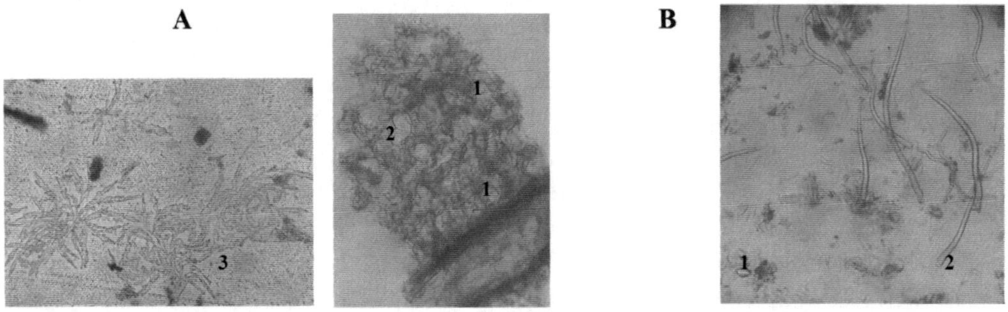

Figure 1. Microscopic analysis on the simplicial powder of *A. auriculiformis* leaves (A), which 1. Paracitic type stomata, 1. Anisocytic type stomata 2. Parenchymal tissue, 3. Fibre. B. Simplicial powder of *H. tiliaceus* leaves, where 1. Paracytic stomata, 2 Hair cover.

Figure 1 represented the microscopic examination of simplicia powder of *A. auriculiformis*, showing the presence of parenchymal tissue and specific fibers. Furthermore, microscopic investigation of simplicia powder of *H. tilliaceus* displayed the presence of specific cover hair.

In this circumstance, microscopic analysis is one of the simplest and cheapest methods to start with for forming the correct identification of the source materials from plants including medicinal plants. Therefore the phytochemical profile along with microscopic data will support the standardisation of clarity and purity of sample proof of identity.

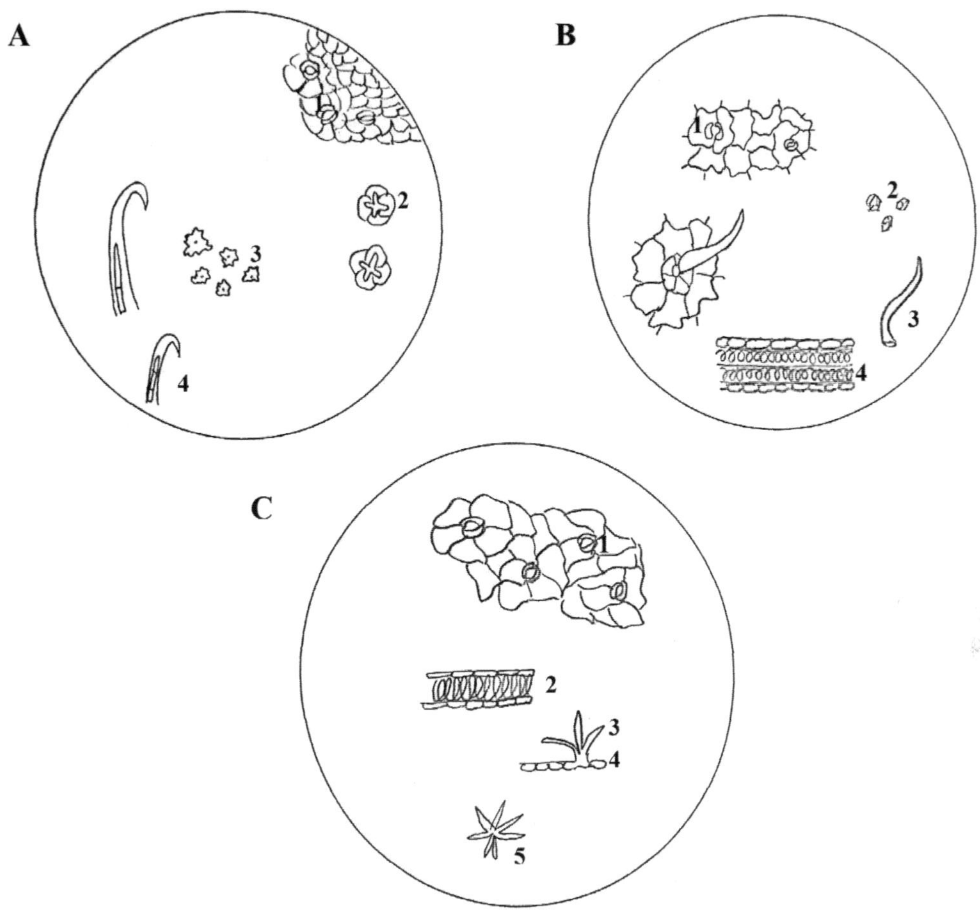

Figure 2. Microscopic analysis of the simplicial powder of *P. odoratissimus* leaves (A), which 1. Anomocytic type stomata, 2. Secretory glands, 3.Druse from oxalate chrystal, 4. Hair cover, B. Simplicial powder of *P. pinnata* leaves, where 1. Paracytic type stomata, 2. Calcium oxalate crystals form prisms, 3.Trichoma, 4. Xylem with thickening of the spiral wall. C. Simplicial powder of *R. communis* leaves, where 1. Anomocytic stomata, 2. Transport files with spiral thickening, 3. Hair cover, 4. Epidermis, 5. Hair cover star shape

4. Conclusions
The current report clarified the physicochemical properties also showed the diversity among the species and essential finding on the water content was fewer than 10% as a requirement for drug development. Our study may provide potential phytopharmacological characteristics from mangrove associate leaves, which may open up another possibility of non-wood mangrove potential.

5. Acknowledgment
The authors are grateful the Ministry of Research, Technology and Higher Education through Excellent Research for Higher Education 2017 (No. 003/SP2H/LT/DRPM/IV/2017).

6. References
[1] Basyuni M, Sagami H, Baba S, Iwasaki H and Oku H 2016 Diversity of polyisoprenoids in ten Okinawan mangroves *Dendrobiology* **75** 167–175
[2] Basyuni M, Wati R, Sagami H, Sumardi, Baba S and Oku H 2018 Diversity and abundance of polyisoprenoid composition in coastal plant species from North Sumatra, Indonesia *Biodiversitas* **19** 1–11

[3] Atanasov A G, Waltenberger B, Pferschy-Wenzig E M, Linder T, Wawrosch C, Uhrin P, Temml V, Wang L, Schwaiger S, Heiss E H and Rollinger J M 2015 Discovery and resupply of pharmacologically active plant-derived natural products: a review *Biotechnol. Adv.* **33** 1582–1614

[4] Ksouri R, Ksouri W M, Jallali I, Debez A, Magné C, Hiroko I and Abdelly C 2012 Medicinal halophytes: potent source of health promoting biomolecules with medical, nutraceutical and food applications *Crit. Rev. Biotechnol.* **32** 289–326

[5] Bandaranayake W M 1998 Traditional and medicinal uses of mangroves Mangroves Salt Marshes **2** 133–148

[6] Maganha E G, da Costa Halmenschlager, R, Rosa R M, Henriques J A P, de Paula Ramos, A L L and Saffi J 2010 Pharmacological evidences for the extracts and secondary metabolites from plants of the genus Hibiscus. *Food Chemistry*, ***118*(1),** 1-10

[7] Londonkar R, Kamble A and Reddy V C 2010 Anti-inflammatory activity of *Pandanus odoratissimus* extract *Int. J.* Pharmacol. **6** 311–314

[8] Naz R and Bano A 2012 Antimicrobial potential of *Ricinus communis* leaf extracts in different solvents against pathogenic bacterial and fungal strains Asian Pac. *J. Trop. Biomed.* **2** 944–947

[9] Sumardi, Basyuni M and Wati R 2018 Antimicrobial activity of polyisoprenoids of sixteen mangrove species from North Sumatra, Indonesia Biodiversitas **19** 1243–1248

[10] Basyuni M and Sumardi 2017 Bioinformatics approach of salt tolerance gene in mangrove plant *Rhizophora stylosa J. Phys.: Conf. Ser.* **801** 012012

[11] Basyuni M, Ginting P Y and Lesmana I 2017 Phytochemical analysis of binahong (*Anredera cordifolia*) leaves extract to inhibit in vitro growth of *Aeromonas hydrophila* AIP Conf. Proceed. **1904** 020072

[12] WHO (World Health Organization) 1998 Quality Control Methods for Medicinal Plant Materials Geneva p. 28–31

[13] Basyuni M and R Wati 2017 Bioinformatics analysis of the oxidosqualene gene and the amino acid sequence in mangrove plants *J. Phys. : Conf. Ser.* **801** 012011

[14] Sari D P, Basyuni M, Hasibuan P A Z and Wati R 2018 The inhibition of polyisoprenoids from Nypa fruticans leaves on cyclooxygenase 2 expression of WiDr colon cancer cells *Asian J. Pharm. Clin. Res.* **11** 154–157

[15] Illian D N, Basyuni M, Wati R and Hasibuan P A Z 2018 Polyisoprenoids from *Avicennia marina* and *Avicennia lanata* inhibit WiDr cells proliferation Pharmacogn. Mag. Doi 10.4103/pm.pm_201_18

[16] Basyuni M, Amri N, Putri LA, Syahputra I and Arifiyanto D 2017 Characteristics of fresh fruit bunch yield and the physicochemical qualities of palm oil during storage in North Sumatra, Indonesia *Indones. J. Chem.* **17** 182–190

[17] Wuana R A and Okieimen F E 2011 Heavy metals in contaminated soils: a review of sources, chemistry, risks and best available strategies for remediation *Isrn Ecology* **24** 2011

[18] Fryns-Claessens E and Van Cotthem W 1973 A new classification of the ontogenetic types of stomata *Bot. Rev.* **39** 71–138

[19] Umbanhowar Jr C E and Mcgrath M J 1998 Experimental production and analysis of microscopic charcoal from wood, leaves and grasses *Holocene* **8** 341–346

Total phenolic contents and antioxidant activity of rice bran fermented with lactic acid bacteria

K Nisa[1], V T Rosyida[1], S Nurhayati[1], A W Indrianingsih[1], C Darsih[1], and W Apriyana[1]

[1] Research Unit for Natural Products Technology, Indonesian Institute of Sciences, Yogyakarta, Indonesia

E-mail: nisa.khoirun @ yahoo.com

ABSTRACT

Rice bran, a by-product of the rice milling process, is rich source of bioactive compounds which have potential nutraceutical effect. Its dietary fiber and phenolic compounds are reported enhancing the functionalities of foodstuffs related to their beneficial properties as antioxidant. The aim of this research is to evaluate the antioxidant activity and total phenolic contents of fermented rice bran using two lactic acid bacteria (LAB), *Lactobacillus lactic* and *Lactobacillus plantarum.* Rice bran was fermented in a solid state with ratio of single culture LAB were 5%, 7.5%, and 10%. The antioxidant activities were found to be highest in 10 % ratio of *Lactobacillus plantarum* as well as its total phenolic contents. The functional properties of rice bran were also investigated at varied time of fermentation (24 h, 36 h, and 48 h). Antioxidant activity of rice bran fermented with *Lactobacillus plantarum* for 48 h showed a almost two-fold increase compared to non-fermented counterpart. The results indicated that the phenolic contents and antioxidant activity of rice bran showed significant improvement upon fermentation with LAB.

Keywords: Rice bran, fermentation, lactic acid bacteria, total phenolic contents, antioxidant activity

1. Introduction

Rice bran, the outer husk of paddy, is one of the most by-products of rice mils production which rich in vitamins, proteins, oil and fiber. It becomes the most abundant agricultural wastes in Indonesia due to the highly demand of rice in this country. Rice bran extract contains several types of polyphenolic compounds as sources of antioxidants, such as phytic acid, ferulic acid and oryzanols [1]. Fermentation process is commonly used for production of value-added products and can be applied in the food, health and cosmetics industries [2]. In the agricultural wastes handling, fermentation using lactic acid bacteria could improve the chemical composition and bioactivity of the substrates by produces enzymes that degrade the cell wall of plants. A number of fermentation studies have been performed on rice bran throughout the years. Fermentation of rice bran produced biologically active metabolites such as lipid, protein, vitamin B, and essential amino acids, phenolic acid, ferulic acid, vanillic acid, protocatechuic acid, γ-oryzanol, phytic acid, and inositol [3]. Several features that beneficially alter the bioactivity to promote health have been evaluated in fermented rice bran with bacterial or fungal agents through previous studies. Fermentation of rice bran with combination of *Lactobacillus rhamnosus* and *Saccharomyces cerevisiae* was reported considerably reduced the melanin synthesis of the resulting extract to B16F1 melanoma cells [4]. Numerous plant phenols, such as ferulic acid, are often biologically unavailable after intake, thus fermentation could assist improvement of the antioxidant potency [5]. This study aimed to investigate the phenolic contents and antioxidant capacity of rice bran fermented with *Lactobacillus plantarum* and *Lactobacillus lactic*.

2. Materials and Methods

2.1. Collection of plant materials
The rice bran was collected from local rice milling company, located in Yogyakarta, Indonesia. Rice bran was packed and stored at room temperature with silica gel until used use in the fermentation process.

2.2. Preparation of lactic acid bacteria
For the starter culture, *Lactobacillus plantarum* (FNCC 0027) and *Lactobacillus lactic* (FNCC 0080) were cultured in MRS broth medium (Merck, Darmstadt, Germany) at 37 ºC for 24 h and kept in the microbiology laboratory.

2.3. Fermentation of rice bran using lactic acid bacteria
Fermentation of rice bran was carried out using *Lactobacillus plantarum* and *Lactobacillus lactic* in MRS broth medium at static condition. Approximately, 30 g of rice bran in Erlenmeyer flask was suspended with 50 mL of distilled water and sterilized at 121 °C for 15 min. After cooling to the room temperature, the starter bacteria were added into each flask at varied ratio 2.5%, 5%, 7.5%, and 10% of total medium, mixed, and incubated at 37 °C for 24 h. Each experiment was performed in triplicate. Non fermented rice bran was used as a control. For time-controlled fermentation experiment, the same procedure was carried out to other rice bran samples. The rice bran sample was inoculated with 10% starter. The fermentation time was varied from 24 h, 36 h, and 48 h [6].

2.4. Extraction of the fermented rice bran
The incubated samples were harvested and dried at 60 °C for 24 h. Dried samples (5 g) were exhaustively extracted with methanol in an ultrasonic bath for 3 x 60 min at room temperature. The supernatant was filtrated and evaporated for analysis.

2.5. Determination of the total phenolic contents
The total phenolic content of fermented rice bran extracts was determined by a Folin-Ciocalteu reaction according to our previous method. An aliquot of 0.5 mL samples from the stock solution was mixed with 7.5 ml distilled water and 0.5 mL of Folin-Ciocalteu reagent. After 8 min of incubation, 1.5 mL of 20% Na_2CO_3 solution was added and further incubated for 2 hours. The absorbance was measured at 765 nm using a UV-Vis spectrophotometer (UV Vis Spectrophotometer Hitachi HALO RB-10) against blank sample. The results were expressed as gallic acid equivalents (mg GAE/g dw) of dry weight and the experiment was conducted in triplicate analysis [7].

2.6. Assay for antioxidant activity
The fermented rice bran extracts were homogeneously dissolved in methanol solvent at a series of concentrations (0.1, 0.2, 0.4, 0.6, and 1.0 mg/mL). The sample extracts (3.2 mL) were added to 1.2 mL of methanolic DPPH (Merck, Darmstadt, Germany) solution (0.2 mM). The reactions were

performed in the dark at room temperature for 30 min. Absolute methanol was used as a blank and gallic acid (Sigma Aldrich, USA) as a positive control. The absorbance was measured at 517 nm and the scavenging activities were expressed as the percent inhibition of DPPH radicals calculated using the formula:

$$\text{Percentage Inhibition } \% = \left[\frac{A_0 - A_1}{A_0}\right] \times 100\ \%$$

where A_0 is the absorbance of the control (absence of sample), and A_1 is the absorbance of the sample [7].

3. Results and Discussion

3.1. Total Phenolic Content

The fermentation of rice bran in this study was performed using two starter lactic acid bacteria, which are *L. lactic* and *L. plantarum*. Fermentation process on rice bran might stimulate the structural breakdown of cell walls, causing the releasing or bio-synthesis of various bioactive compounds [8]. The efficiency of the fermentation process could be studied by its phenolic compounds evaluation. Total phenolic content was determined according to fermentation parameters such as the amount of lactic acid bacteria during the fermentation (LAB ratio) and the fermentation time. The phenolic contents were expressed as gallic acid equivalents (% GAE) of dry weight. The LAB ratios used in this fermentation were 2.5%, 5%, 7.5%, and 10% as shown in the Table 1 below.

Table 1. Total phenolic content of fermented rice bran based on LAB ratio

LAB Rasio	Total Phenolic Contents (mg GAE/g dw)	
	L. lactic	*L. plantarum*
Control	24.922	24.922
2.5%	25.789	25.789
5%	26.222	26.020
7.5%	27.464	29.081
10%	28.966	30.410

Control = non fermented rice bran

The effectiveness of fermentation depends on the starter used. The optimized ratio of the starter varied according to the each bacteria strains. Lactic acid bacteria require media and nutrients that are quite complex to grow in which rice bran can provide them. Microbial activity in LAB fermented bran will increase the amount of organic acids which each LAB has different capabilities in producing organic acids. The higher phenolic content were demonstrated by the increasing of LAB ratio that used in the fermentation of rice bran. However, on the fermented rice bran involving *L. lactic* was not particularly different among each LAB ratio compared by non fermented rice bran (control). Both of *L. lactic* and *L. plantarum* gave the highest total phenolic content of rice bran at the ratio of 10% as 28.966 mg GAE/g dw and 30.410 mg GAE/g dw, respectively. While the phenolic content of fermented rice bran by *L. plantarum* was slighty higher than that of fermented rice bran by *L. lactic*. The effectiveness of fermentation process was not only evaluated based on the starter used, but also the fermentation time of rice bran as shown in the table 2. The optimum fermentation time obtained in this experiment would be the basic of fermentation of rice bran by each starter.

Table 2. Total phenolic content of time-controlled fermented rice bran

LAB Rasio	Total Phenolic Contents (mg GAE/g dw)	
	L. lactic	*L. plantarum*
Kontrol	45.844	45.844
24	56.714	47.293
36	64.534	48.493
48	73.144	50.416

Control = non fermented rice bran

The higher of phenolic content were exhibited by the increasing of fermentation time of rice bran. There is a different capability between *L. lactic* and *L. plantarum* to breakdown the sub-strate by fermentation process. The higher ability to breakdown the substrate was shown by *L. lactic* than L. *plantarum*. In the fermented rice bran involving *L. plantarum*, the phenolic content was not particularly different during fermentation process in 24 h and 36 h. However, in the 48 h, the fermentation achieved the optimum process, by giving the total phenolic value of 50.416 mg GAE/g dw. In contrast, the fermentation of rice bran significantly increased by addition of fermentation time. The optimum phenolic content was reached in the 48 h of fermentation as 73.144 mg GAE/g dw.

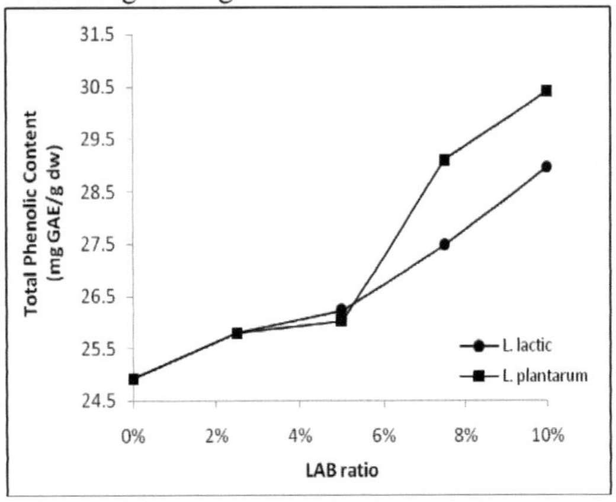

Figure 1. Total phenolic content of fermented rice bran based on LAB ratio

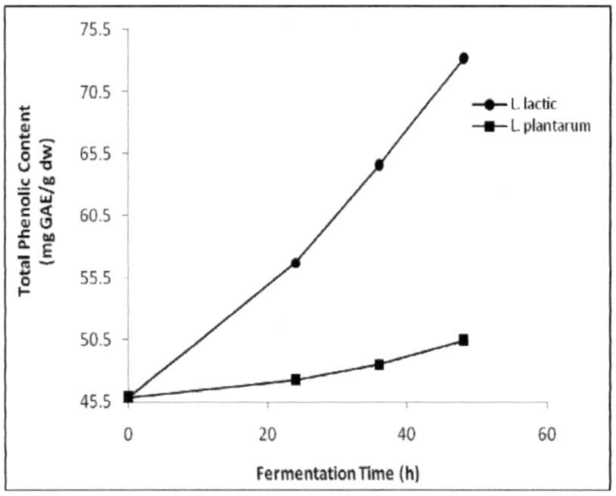

Figure 2. Total phenolic content of time-controlled fermented rice bran

During fermentation, microorganism synthesizes such enzymes which able to break ester links and effect in the releasing bound phenolic acids. Consequently, it may improve the nutraceutical value of cereals and increase the bioavailability due to the improvement of free phenolic acids [9]. An improvement of phenolic contents in plants commonly relates to the action of enzymes produced by microorganism such as β-glucosidase, α-amylase and laccase, together with other enzymes [10]. The increasing of ferulic acid in fermented samples inoculated with *P. acidilactici*, may be due to production of ferulic acid esterase enzyme by *P. acidilactici* [11]. Fermentation process would cleave the bonds between phenolic compounds with other substance and release the monomers of phenolic compounds or antioxidants.

3.2. DPPH radical scavenging assay

Figure 3 shows the inhibition percentages of the DPPH radical by fermented rice bran using *L. lactic* and *L. plantarum*. An antioxidant reduces the DPPH radical to diphenylpicrylhydrazine (a yellow-colored compound), and the level of discoloration is depended on the hydrogen-donating capacity of its antioxidant. The antioxidant activity was investigated in the varied of LAB ratio used in fermentation process and fermentation time. By increasing ratio of *L. lactic* used as starter in fermentation process caused significant improvement of antioxidant capacity compared with the non-fermented rice bran. However, there is no different between 5%, 7.5%, and 10% ratio of *L. plantarum* handling in the fermentation medium.

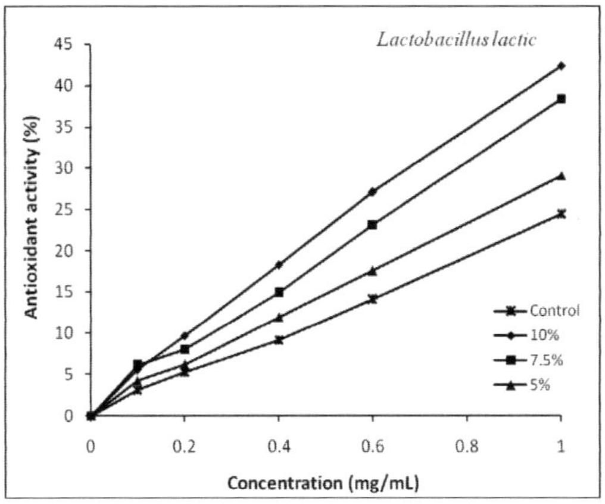

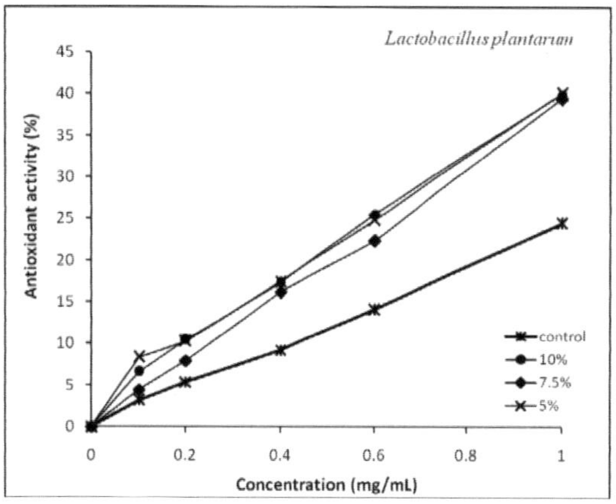

Figure 3. Antioxidant activity of fermented rice bran based on LAB ratio

The highest antioxidant activity of fermented rice bran extract was demonstrated by addition of 10% both *L. lactic* and *L. plantarum* in the fermentation process with shows 42.36 ± 1.08% and 39.94 ± 2.25% of inhibition of DPPH radicals at 1 mg/mL, respectively. The low of activity may be partly due to the fermentation process was conducted inadequate innoculation time.

Table 3. Antioxidant activity of fermented rice bran based on LAB ratio

LAB ratio	Antioxidant activity (1 mg/mL)	
	L. lactic	*L. plantarum*
Control	24.43 ± 0.47	24.43 ± 0.47
10%	42.36 ± 1.08	39.94 ± 2.25
7,5%	38.35 ± 0.55	39.38 ± 0.33
5%	29.03 ± 0.52	40.06 ± 1.83

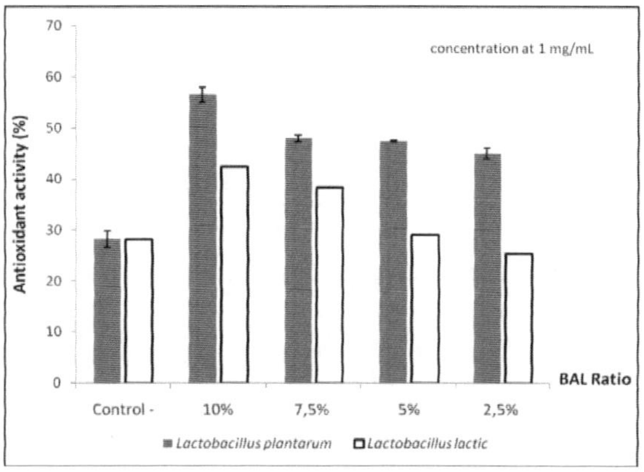

Figure 4. Antioxidant activity of fermented rice bran at concentration 1 mg/mL.

The optimized fermentation process was determined by observation of antioxidant activity within varied fermentation time, which is 24 h, 36 h, and 48 h. The scavenging effect of the DPPH radical in all fermented extracts samples improved through increasing fermentation time. The rice bran fermented with *L. plantarum* evenly showed the higher antioxidant activity than rice bran fermented with *L. Lactic*. This result indicated that *L. plantarum* has ability to breakdown the ester bonds composed in rice bran liberating free phenolic compounds. Phenolic, the major secondary metabolites presence in plant, are commonly established in conjugated forms of the hydroxyl group. Phenolic compounds could act as antioxidants through several different mechanisms, such as free radical scavenging capacity, chelating of metal ions and the inhibitory activity of pro-oxidant enzymes [12]. The increasing of antioxidant activity is related to the phenolic contents produced in fermented rice bran which suggest that it might be attributed to hydrolytic enzymes. These enzymes work through the substrate and turn up the free hydroxyl groups on the phenolic structure, consequently the antioxidant activity of the substrate is increased due to the presence of the content of free phenolic [13]. The obvious difference of antioxidant activity in non fermented and fermented rice bran extracts may be due to the high antioxidant activity of bioactive compounds released during fermentation of rice bran, such as ferulic acid, α-tocopherol and β-oryzanol [14].

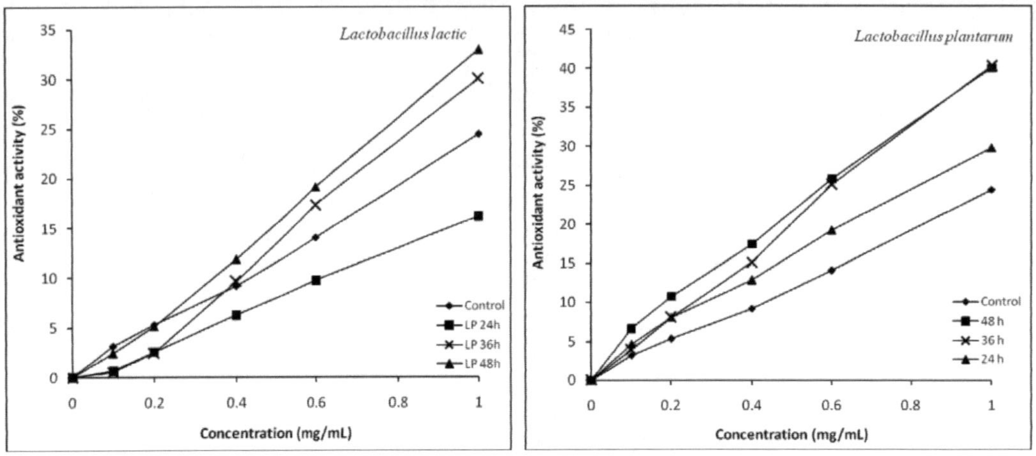

Figure 5. Antioxidant activity of time-controlled fermented rice bran.

The strongest scavenging effects on the DPPH radical were found in rice bran after 48 h fermentation with both starters. The highest scavenging effect at 1 mg/mL was reached by rice bran fermented with *L. Plantarum* (Table 4). Fermentation with LAB had favourable effect on its antioxidant capacity, which is showed by [6], who observed increase of bioactive compound in rice bran resulting in an improved antioxidant activity of rice bran after fermentation with *P. acidilactici, P. pentoseous and L. lactis*.

Table 4. Antioxidant activity of time-controlled fermented rice bran

Fermentation time (h)	Antioxidant activity (1 mg/mL)	
	L. lactic	*L. plantarum*
Control	24.43 ± 0.47	24.43 ± 0.47
24 h	16.19 ± 0.74	29.80 ± 1.35
36 h	30.05 ± 0.76	40.31 ± 0.92
48 h	32.96 ± 0.48	40.08 ± 0.23

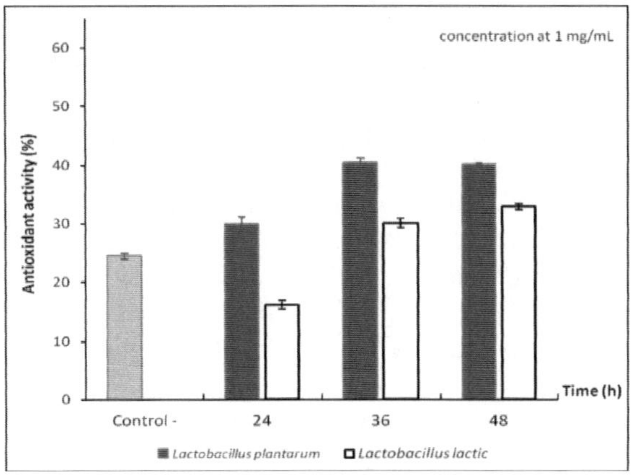

Figure 6. Antioxidant activity of time-controlled fermented rice bran at concentration 1 mg/mL.

Bioprocessing methods through microbial fermentation involving enzymatic reaction have been widely applied to release the chemical bounds, which can improve its free forms and bioavailability [15].

4. Conclusion

This study demonstrates the increasing of the total phenolic contents subsequently the antioxidant capacity of rice bran through microbial fermentation with lactic acid bacteria. The highest total phenolic content of fermented rice bran extract achieved after 48 h fermentation process with 10% of lactic acid bacteria used in medium. The strongest scavenging effects on the DPPH radical were found in rice bran after 48 h fermentation with both starters. The highest scavenging effect at 1 mg/mL was reached by rice bran fermented with *L. Plantarum*. The antioxidant activity of fermented rice bran extracts were positively correlated with their total phenolic contents. It suggested that fermentation of rice bran with lactic acid bacteria, could improve its functional properties. As the result, further study is required to determine the active phytochemicals in the extract of fermented rice bran.

Acknowledgements
This research was supported by Insinas Research Grant, Ministry of Research, Technology and Higher Education of the Republic of Indonesia

References

[1] Iqbal S, Bhanger M I, Anwar F 2005 Antioxidant properties and components of some commercially available varieties of rice bran in Pakistan *Food Chem.* 93 265–272
[2] Pourali O, Asghari F S, Yoshida H 2009 Sub-critical water treatment of rice bran to produce valuable materials *Food Chemistry* 115 1-7
[3] Soi-Ampornkul R, Junnu S, Kanyok S, Liammongkolkul S, Katanyoo W, Umpornsirirat S 2012 Antioxidative and neuroprotective activities of the pre-germinated brown rice extract *Food and Nutrition Sciences* 3 135-140
[4] Kim H Y, Kim J H, Yang S B, Hong S G, Lee S A, Hwang S J, Shin K S, Suh H J, Park M H 2007 A polysaccharide extracted from rice bran fermented with *Lentinus edodes* enhances natural killer cell activity and exhibits anticancer effects *J Med Food* 10 25–31
[5] Sindhu S C, Khetarpaul N 2002 Effect of probiotic fermentation on antinutrients and in vitro protein and starch digestibilities of indigenously developed RWGT food mixture *Nutr Health.*16:173–81

[6] Rashid N Y A, Razak D L A, Jamaluddin A, Sharifuddin S A, Long K 2015 Bioactive compounds and antioxidant activity of rice bran fermented with lactic acid bacteria *Malaysian Journal of Microbiology*. 11 (2) 156-162

[7] Nisa K, Nurhayati S, Apriyana W and Indrianingsih A W 2017 Investigation of total phenolic and flavonoid contents, and evaluation of antimicrobial and antioxidant activities from *Baeckea frutescens* extracts *IOP Conf. Ser.: Earth Environ. Sci.* 101 012002

[8] Katina K, Liukkonen K H, Kaukovirta-Norja A, Adlercreutz H, Heinonen S M, Lampi A M 2007 Fermentation-inducedchanges in the nutritional value of native or germinated rye *J. Cereal Sci*. 46 348–355

[9] Estrada B A A, Gutierrez J A G, Saldivar S O S 2014 Bound phenolics in foods, a review *Food Chemistry* 152 46-52

[10] Zheng Z, Shetty K 2000 Solid-state bioconversion of phenolics from cranberry pomace and role of Lentinus edodes b-glucosidase *J. Agr. Food Chem*. 48 895–900

[11] Anastasia S H, Ida R, Stine G, Stefanie S, Judith N, Stefan S 2012 Improved bioavailability of dietary phenolic acids in whole grain barley and oat groat following fermentation with probiotic *Lactobacillus acidophilus*, *Lactobacillus johnsonii*, and *Lactobacillus reuteri*. *Journal of Agricultural Food Chemistry* 60 6369-6375

[12] Amorati R, Pedulli G F, Gabrini L, Zambonin L, Landi L 2006 Solvent and pH effects on the antioxidant activity of caffeic and other phenolic acids *J. Agric. Food Chem*. 54 (8) 2932–2937

[13] Bhanja T, Kumari A, Banerjee R, 2009 Enrichment of phenolic and free radical scavenging property of wheat koji prepared with two filamentous fungi *Biores. Technol*. 100 2861–2866

[14] Xu Z, Hua N, Godber J S 2001 Antioxidant activity of tocopherols, tocotrienols, and gamma oryzanol components from rice bran against cholesterol oxidation accelerated by 2,20-azobis (2- methylpropionamidine) dihydrochloride *J. Agric. Food Chem*. 49 2077–2081

[15] Sungsopha J, Moongngarm A, Kanesakoo R 2009 Application of germination and enzymatic treatment to improve the concentration of bioactive compounds and antioxidant activity of rice bran *Australian Journal of Basic and Applied Sciences* 3(4) 3653-3662

IOP Conf. Series: Earth and Environmental Science **251** (2019) 012021 doi:10.1088/1755-1315/251/1/012021

Comparisons of antioxidant activities of two varieties of pumpkin (*Cucurbita moschata and Cucurbita maxima*) extracts

A W Indrianingsih[1], V T Rosyida[1], W Apriyana[1], S Nur Hayati[1], K Nisa[1], C Darsih[1], A Kusumaningrum[1], D Ratih, N Indirayati

[1]Research Unit for Natural Product Technology, Gading, Playen, Gunungkidul, Yogyakarta 55861, Indonesia

E-mail: anastasia_wheni_i@yahoo.com / anas004@lipi.go.id

Abstract. Pumpkin is a plant that is widely found in the territory of Indonesia. This study has evaluated the antioxidant activity, morphological, physicochemical and chemical properties of three components of *C. moschata* and *C. maxima* (the pulp, seed, and rinds). The *C. moschata* and *C. maxima* extracts were obtained by the maceration method using ethanol as the solvent. The higher antioxidant activity measured by DPPH assays was found in the rinds of *C. maxima* as of 33.8% of radical scavenging activity, followed by the pulp of *C. maxima* and seed of *C. moschata* as of 29.6% and 25.4% respectively. The higher carbohydrate content was established in the pulp of *C. moschata* with 78.6% and pulp of *C. maxima* with 69.5%. On the other hand, the seed of *C. moschata* and *C. maxima* had higher amounts of fats with 28.5% and 10.4%, respectively, while proteins were 19.2% and 12.3%, respectively. The *C. moschata* pulp, seed, and rinds also had been analysed using Scanning Electron Microscopy (SEM) for their morphological characteristic, The Fourier transform infrared spectroscopy (FTIR) for their functional groups and Color Reader for their color diversity. The result of this study showed that different part of *C. moschata* and *C. maxima* has good antioxidant activity and proximate composition so that appear to be promising comestibles for technological exploitation.

1. Introduction

Free radicals can cause oxidative stress and dangerous to the human body [1]. For that reason, antioxidants are necessary, because it can slow down the oxidation process. However, synthetic antioxidants have negative impact if it accumulates in the body such as cause cancer and other chronic diseases. Meanwhile, antioxidants from nature such as from plants became an interesting topic of research in recent years. Several extracts and phytochemicals have been shown to have an antioxidant ability [2]. Many antioxidants have been isolated from plants such as spices, vegetables, grain, herbs, fruits, roots, and leaves [3].

Pumpkin is a plant that is cultivated and widely found in Indonesia. There is an increasing interest for many variety uses of pumpkin because of its high nutritional value, digestive effects and good sensory characteristics [4,5]. The pulp of pumpkin is usually used in soups, pies, and bakery while the seeds are used as a snack and pharmaceutical. Therefore, pumpkin has high potential application for food, cosmetic and pharmaceutical industries [6]. Due to the good acceptance and industrialization of the derived products from pumpkins, more scientific information about the chemical and physicochemical properties of this fruit is important. The purpose of this study was to compare the antioxidant activities of pumpkin seed, pulp, and rinds and also to study morphology and functional groups of the pumpkin seed, pulp and rinds flour. This work is important in the development of new products from pumpkin.

2. Material and Methods

2.1. Materials and General Instrumentations
The pumpkin was collected from Yogyakarta and Central Java Province, Indonesia. 1,1-diphenyl-2-picrylhydrazyl (DPPH), ascorbic acid, methanol, and ethanol were obtained from Sigma-Aldrich (Japan). Wako (Japan). The equipment used in this study was glassware, digital balance, Elisa Reader, SEM, FTIR, oven, and desiccator.

2.2. Preparation of ethanol extracts
The pumpkin was cut into pieces, peeled the rinds and separated the seeds. After drying process using oven for 3 days at 50 °C, it was made into flour. The ethanol extracts were obtained by macerated the pulp, seeds, and rinds flour with ethanol (1:8 w/w) for 3 days at room temperature.

2.3. Characterization of pumpkin flour morphology and functional group
SEM of the surface of dried pumpkin flour was performed using a Hitachi SU-70 instrument operating at 4 kV. FTIR analysis of the pumpkin flour was performed on a wavelength from 500 cm^{-1} to 4000 cm^{-1}. The proximate analysis procedure uses method that refers to SNI 01-2891-1992 [7].

2.4. Colour test
Colour instrumental parameters were determined using colour reader (CR 20 Color Reader Konica Minolta). Results were expressed in L*, a* and b*, with L (how light or dark the sample is), ranging from black (0) to white (100), a* ranging from green (-60) to red (60) and b* ranging from blue (-60) to yellow (+60). Colour instrumental analyses were performed in three replicates.

2.5. DPPH assay
Several concentrations of *C. moschata* extracts were dissolved in methanol and reacted with DPPH (1.01 mM) at room temperature. It was left to stand in the dark condition for 30 min. Absorbance was recorded at 517 nm using a UV-Vis spectrophotometer [8]. The scavenging activity was calculated using equation 1:

$$\text{DPPH scavenging activity (\%)}=(\frac{A_0 - A_1}{A_0} x100) \quad (1)$$

where A_0 is absorbance of the control and A_1 is absorbance of the sample. The DPPH radical scavenging assays were carried out in three replicates.

3. Results and Discussion

3.1. Proximate compositions of pumpkin flour
Proximate compositions of pumpkin flour (pulp, seed, and rind) of two varieties from *C. moschata* and *C. maxima* are shown in Table 1. The differences in pumpkins composition probably depend on several factors such as soil fertility, the variety, stage of maturity, and climate [9].

Table 1. Chemical compositions of pumpkins flour

Samples	Water (%)	Ash (%)	Fat (%)	Protein (%)	Carbohydrate (%)
Seeds (*C. moschata*)	7.67±0,05	5.18±0.04	28.49±0.4	19.23±0.06	39.51±0.5
Rind (*C. moschata*)	9.64±0.08	4.95±0.07	2.44±0.07	8.05±0.04	75.04±0.8
Pulp (*C. moschata*)	13.01±0.09	4.32±0.05	0.16±0.01	3.99±0.06	78.64±0.9
Seeds (*C. maxima*)	8.15±0.08	5.4±0.08	10.42±0.04	12.33±0.08	63.70±0.7
Rind (*C. maxima*)	9.96±0.05	10.3±0.06	5.59±0.04	12.84±0.06	61.25±0.6
Pulp (*C. maxima*)	16.77±0.05	6.68±0.04	2.14±0.03	4.90±0.06	69.51±0.7

The data indicated that the seeds of *C. moschata*, the seeds of *C. maxima*, and the rinds of *C. maxima* contained a higher percentage of protein, fat, and ash as of 19.23±0.06, 12.33±0.08, and 12.84±0.06 (for protein), respectively; 28.49±0.4, 10.42±0.04, and 5.59±0.04 (for fat), respectively and 5.18±0.04, 5.4±0.08, and 10.3±0.06 (for ash), respectively. In the case of carbohydrate content, the pulp of *C. moschata* and the pulp of *C. maxima* had the highest percentage as of 78.64±0.9 and 69.51±0.7, respectively. These results were in agreement with the literature [6]. It was also reported that because of its low water activity, pumpkin flour has long shelf life and quality [10]. Another study was discovered that there was no gluten was found in pumpkin flour [6].

3.2. Morphology structure of pumpkins flour

Figure 1 shows the SEM image of the surface of the *C. moschata* flour.

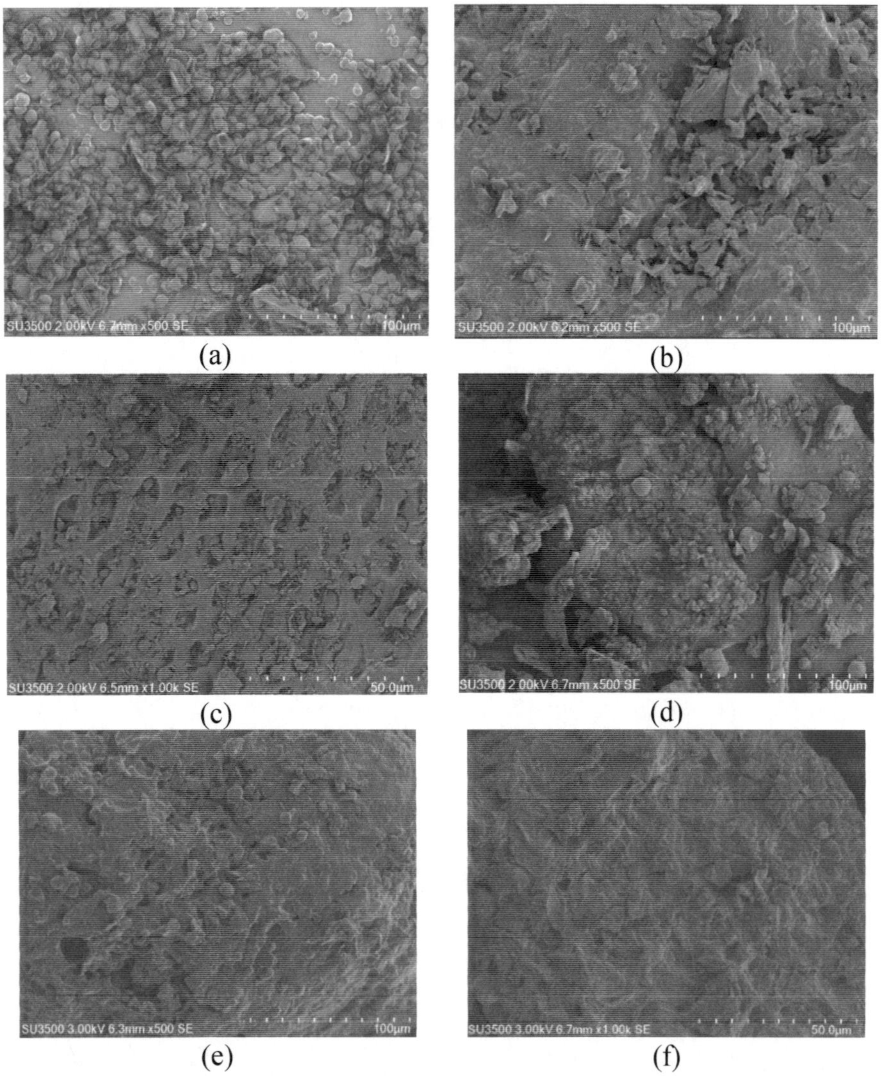

Figure 1. SEM images of the surface morphological structure of pumpkins flour: a. seed of *C. moschata* b. pulp of *C. moschata* c. rind of *C. moschata* d. seed of *C. maxima* e. pulp of *C. maxima* and f. rind of *C. maxima*

The SEM of seed, pulp, and rind of pumpkins flour are shown in Fig. 2a–f. The micrographs of the pumpkins flour show that the surface is rough and compact. The granule size of pumpkins flour is ranged from 2.5–12 nm. Its compact structure might be affected by protein-lipid complexes [11].

3.3. Antioxidant activity

The antioxidant activity of the extract of *C. moschata* and *C. maxima* was evaluated using DPPH assay.

Table 2. Antioxidant activity of ethanol extract of *C. moschata*

Samples	DPPH scavenging activity (%) at 800 µg/ml
Seed (*C. moschata*)	25.42±1.5
Rind (*C. moschata*)	21.58±1.2
Pulp (*C. moschata*)	23.85±1.1
Seed (*C. maxima*)	17.99±0.8
Rind (*C. maxima*)	33.80±2.1
Pulp (*C. maxima*)	29.60±1.8

From Table 1, it can be observed that the scavenging activity of pumpkin flour extract (seed, rind, and pulp) ranged from 17.99 to 33.80% at concentration extract of 800 µg/ml. Ascorbic acid as a positive control in this experiment exhibited an activity of 89.31% at concentration of 200 µg/ml. Among the pumpkins flour extracts, the rind of *C. maxima* shows the highest scavenging activity of 33.80% at concentration extract of 800 µg/ml. It was followed by the pulp of *C. maxima*, seed of *C. moschata*, pulp of *C. moschata*, rind of *C. moschata*, and seed of *C. moschata* as of 29.60%, 25.42%, 23.85%, and 21.58% at concentration extract of 800 µg/ml, respectively. The extract of pumpkins have differences activities in DPPH assay possibly related to the differences compounds in the extract.

Several active compounds were detected on the pumpkin such as morin and catechin [12]. Pumpkin seeds are considered to be wastes, although in several countries they are consumed at a domestic scale [13]. Phenolic compounds and tocopherols which are rich of proteins, antioxidants and lipid were obtained in the seeds of pumpkin [14]. It is usually used in traditional medicine due to its antidiabetic, anti-bacterial, and antineoplastic activities [15]. A study also showed that protein isolated from pumpkin seed exhibited anti-peroxidative properties [16]. In particular, another study showed that diet consumption of pumpkin seed prevents the start of prostate testosterone-induced hyperplasia [17].

3.4. FTIR spectroscopy

FTIR spectroscopy is a technique for evaluating the functional groups as part of a compound in chemical analysis. In the food industries, it was also used as a monitoring quality control because of its rapid screening and quantification of chemical components in samples. Figure 2 shows the spectra of pumpkins flour (pulp, rind, seed) of *C. moschata* and *C. maxima*.

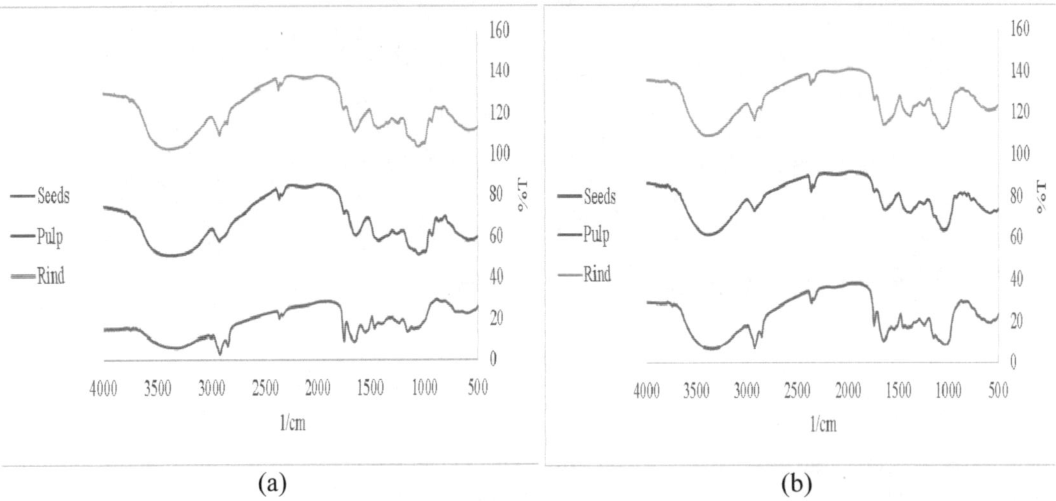

Figure 2. The FTIR spectra of (a) *C. moschata* and (b) *C. maxima*.

As can be seen in Figure 2, the FTIR spectra of pumpkins flour of seed, pulp, and rind of *C. moschata* (a) and *C. maxima* (b) were evaluated in frequency of 4000–500 cm^{-1}. The range of frequency at 2950–2850 cm^{-1} is related to CH absorptions. Generally, a frequency at 3450–3250 cm^{-1} is hydroxyl groups (O-H) from carbohydrate or other compounds such as carboxyl acid and ketone. The absorption around 3120– 3030 cm^{-1} is associated with (N-H) groups from amide. The bands at 1105–1010 cm^{-1} is associated with lipids. These spectra were in accordance with the literature [6].

3.5. Colour test

Table 3 shows the result of colour analysis of seed, pulp, and rind of *C. moschata* flour.

Table 3. Colour parameters of seeds, pulp and rinds of the pumpkin flour from two varieties

Samples	L*	a*	b*
Seeds (C. moschata)	43.5±0.2	10.6±0.2	29.0±0.1
Rind (C. moschata)	67.8±0.5	4.9±0.3	22.9±0.3
Pulp (C. moschata)	75.1±0.6	6.8±0.4	31.9±0.8
Seeds (C. maxima)	66.7±0.7	5.5±0.2	35.9±0.9
Rind (C. maxima)	61.1±0.7	17.0±0.6	44.6±1.0
Pulp (C. maxima)	74.3±0.4	9.8±0.6	43.0±1.3

Highest b* values are related with the yellowish colour of the *C. moschata* flour. The b* values range from 22.9 to 44.6. The *C. maxima* pumpkin have higher b* values than the *C. moschata* pumpkin. The a* values are correlated with the reddish colour of the pumpkin flour. The a* values range from 4.9 to 17.0. The L* are related with the lightness, in this study the L* values range from 43.5 to 75.1.

4. Conclusion

In conclusion, the antioxidant activities of *C. maxima* were higher than those of *C. moschata*. This study showed that both of *C. maxima* and *C. moschata* flour (pulp, rind, and seed) is potential comestible that rich of antioxidants with DPPH radical scavenging activity in the range of 17.99 to 33.80% at concentration of 800 µg/ml.

5. Acknowledgment

This study was supported by Kemenristekdikti (Indonesian Research, Technology and Higher Education Ministry of Indonesia) 2018, through Insinas Project with contract no. 13/E/KPT/2018.

6. References

[1] Ani V, Varadaraj M C and Naidu K A 2006 *Eur. Food Res. Technol.* **224** 109–115
[2] Krishnaiah D, Sarbatly R and Nithyanandam R 2011 *Food Bioprod. Process.* **89** 217–233
[3] Ramarathnam N, Osawa T, Ochi H and Kawakishi S 1995 *Trend Food SciTech.* **6** 75–82
[4] Noelia J V, Mario Roberto M J, José de Jesús Z M and José Alberto G I 2011 *Food Res. Int.* **44** 2587–2593.
[5] Konopacka D, Seroczyńska A, Korzeniewska A, Jesionkowska K, Niemirowicz-Szczytt K and Płocharski W 2010 *LWT - Food Sci. Technol.* **43** 302–309.
[6] Quintana, S E, Marsiglia R M, Machacon D, Torregroza E and García-Zapateiro L A 2018 *Contemp. Eng. Sci.* **11** 1003–1012
[7] National Standardization Agency of Indonesia 1992 *Standar Nasional Indonesia (SNI) 01-2986-1992*
[8] Yen G C and Chen H Y 1995 *J. Agric. Food Chem.* **43** 27–32
[9] Iwasaki K and Matsumura A 1999 *Soil Sci. Plant Nutr.* **45** 909–920
[10] Aziah A A N and Komathi C A 2009 *J. Food Sci.* **74** S328–S333
[11] Nakhon P P S, Jangchud K, Jangchud A and Prinyawiwatkul W 2017, *Int. J. Food Sci. Technol.* **52** 1–9
[12] Rocha-Guzman N E, Gallegos-Infante J A, Delgado-Nieblas C I, Zazueta-Morales J J, Gonzalez-

Laredo R F, Cervantes-Cardoza V, Fernando Martinez-Bustos and Aguilar-Palazuelos E 2012 *Int. J. Food Eng.* **8** 1–13
[13] Patel S 2013 *J. Nutr Metab* **6** 183–189
[14] Jorge N, da Silva A C and Malacrida C R 2015 *Nat. Prod. Res.* **29** 2313–2317
[15] Yadav M, Jain S, Tomar R, Prasad G B K S and Yadav H 2010 *Nutr. Res. Rev.* **23** 184–190
[16] Gossell-Williams M, Davis A and O'Connor N 2006 *J. Med. Food.* **9** 284–286
[17] Nkosi C Z, Opoku A R and Terblanche S E 2006 *Phytother. Res.* **20** 935–940

Pre – Formulation and Evaluation of Jamu Uyup – Uyup (an Indonesian Herbal Galactogogue)

S Nur Hayati[1], W Apriyana[1], V T Rosyida[1], A W Indrianingsih[1], K Nisa[1], D Ratih, N Indirayati

[1]Research Unit for Natural Products Technology, Indonesian Institute of Sciences, Yogyakarta, Indonesia

E-mail: hayanurrizal@gmail.com

Abstract. Jamu Uyup – Uyup has established as the most used natural galactagogue in Indonesia, especially among Javanese people. Jamu Uyup – Uyup is available on the market in the fresh liquid form of a herbal mixture. Liquid forms of herbal preparations have certain limitation related to large dose volumes, difficult packaging, and poor stability. Developing and formulation Jamu Uyup – Uyup into solid dosage form would ensure reproducibility of product quality, accurate doses, improve acceptability and easy compliance with usage directives. The preparation of solid dosage form was carried out using spray drying technique. Microparticle comprising Jamu Uyup – Uyup encapsulated in a lactose matrix have medium yellow color, have distinct typical odor of Zingiberaceae, taste mildly bitter, pungent, slightly sweet, and have an acidic pH (4.7) to keep curcuminoids content stable. The solid dosage form had moisture content of 3.21 ± 1.71% dw and curcuminoids content of 35.86 ± 2.13 mg/g dw. Surface morphology showed that microparticles containing Jamu Uyup – Uyup encapsulated in a lactose matrix have spherical shape with a smooth surface and various sizes (20 – 100 μm) with no apparent cracks or fissures.

Keywords: pre-formulation, Jamu Uyup – Uyup, galactogogue, spray drying, encapsulation

1. Introduction

Galactogogues are pharmaceutical agents, foods, or herbal supplements that are used to support the initiation, continuation, or augmentation of breast milk production [1-7]; hence they are often given when a mother has an inadequate milk supply. There are several botanical species ethno – medically used for the galactagogue by Indonesian people, such as katuk (*Sauropus androgynus* (L.) Merr., family Phyllanthaceae), torbangun (*Coleus amboinicus* Lour, family Lamiaceae), papaya leaf (*Carica papaya*, family Caricaceae), temulawak (*Curcuma xanthorizza*, family Zingiberaceae), fenugreek (*Trigonella foenicum-graecum* L., family Fabaceae), black cumin/habbatussauda (*Nigella sativa* Linn., family Apiaceae), cumin (*Cuminum cyminum*, family Apiaceae), fennel (*Foeniculum vulgare* Miller, family Apiaceae), anise (*Pimpinella anisum*, family Apiaceae), etc. Some of them are ingredient in the traditional galactogogue formulation known as Jamu Uyup – Uyup which has established as the most used natural galactagogue in Indonesia, especially among Javanese people.

Jamu Uyup – Uyup is available on the market in the fresh liquid form of a herbal mixture which may lead to variations in dose each time a treatment is prepared. Furthermore, liquid forms of herbal preparations have certain limitation related to large dose volumes, difficult packaging, and poor stability. Developing and formulation Jamu Uyup – Uyup into other dosage forms than fresh liquid form would ensure reproducibility of product quality, accurate doses, improve acceptability and easy compliance with usage directives. Solid oral dosage forms represent the preferred class of product for orally administered drugs. Solid preparations often have higher stability and are easier to standardize which adds to an increase in their therapeutic acceptance, efficacy and product value [8]. Pre-formulation testing is the first step in the rational development of dosage forms of herbal preparation. The objective of pre-formulation tests is to develop information useful to the formulator in developing a stable bioavailable dosage form [9]. Pre-formulation involves the characterization of physical, chemical and mechanical properties of an herbal preparation alone and when combined with excipients. The present investigation was carried out to perform pre-formulation testing of Jamu Uyup – Uyup in solid dosage form.

2. Materials and Methods

2.1. Preparation of Jamu Uyup – Uyup dried extract

Formula of an herbal mixture was based on formula given by one of producers of Jamu Uyup – Uyup in Yogyakarta, shown at table 1. Fresh rhizomes of some Zingiberaceae family (*C. aeruginosa, C. xanthorriza, C. domestica,* and *Kaempferia galanga)* and pulp of tamarind *(Tamarindus indica)* fruit were pureed and squeezed to collect the juice. The resultant juice was mixed with lemon (*Citrus aurantifolia*) juice and honey (*Mel depuratum*). Beside of its functional constituent, tamarin and lemon were used as natural acidity regulator, while honey was used as a sweetener. The fresh liquid form of a herbal mixture was evaporated to dryness under reduced pressure at 45 °C.

Table 1. Formula of fresh liquid Jamu Uyup – Uyup

Formulae	Dose
Curcumae aeruginosae rhizoma succus	60 mL
Curcumae xanthorrizae rhizoma succus	60 mL
Curcumae domesticae rhizoma succus	60 mL
Kaempferiae galangae rhizoma succus	60 mL
Mel depuratum	60 mL
Citri aurantii fructus succus	20 mL
Tamarindi pulpa succus	10 mL

2.2. Preparation of Jamu Uyup – Uyup powder

The aqueous phase was prepared by dissolving the lactose at a suitably selected concentration (10% w/v) in deionized water. 160 grams of aqueous dried extract was added into solution and stirred vigorously. The suspension was dehydrated using spray drying. The spray dryer (Buchi Mini Spray Dryer B – 290) was operated at an inlet temperature of 110 °C, aspirator of 90%, pump of 5%, and nozzle cleaner of 5. The spray – dried samples were kept at −18 °C for further analysis.

2.3. Physicochemical evaluation

Determination of physical characteristics of extract and powder include organoleptic properties, pH, moisture content, and Scanning Electron Microscopy (SEM). pH of 1 % solution of the formulation was determined by means of a digital pH – meter (Euthech PC 700). Moisture content was measured by moisture analyzer digital (AND MX – 50). The surface morphology of the spray-dried lactose microcapsule loaded with an active compound was examined by means of a Hitachi SU – 3500 microscopes. Photographs were taken at an excitation voltage of 5 kV.

2.4. Determination of the total curcuminoid content

The quantification of curcuminoid was performed using UV – Vis spectrometry method from Pothitirat and Gritsanapan (2006) [10] with slight modification. Calibration curve was made by curcumin added with methanol and adjusted to a final concentration of 2, 3, 4, 5, and 6 ppm. The absorbance was measured at 422 nm. The estimation of total curcuminoid content was carried out in triplicate and the results were averaged.

2.5. Determination of the total terpenoid content

The determination of total triterpenoids was performed according to the colorimetric method of Lin et al. (2015) with slight modification [11] and then expressed as milligram ursolic acid equivalent/gram dry weight (dw). The test samples or ursolic acid standards in tubes were added with 0.4 mL of 5% vanillin/glacial acetic acid (w/v) and 1.0 mL of perchloric acid solution. For the reaction, the tubes were placed in a water bath at 60°C for 45 min. Then the mixed solution was cooled and diluted with 5 mL of acetic acid solution. The absorbance was measured at 548 nm against blank using a spectrophotometer.

3. Results and Discussion

The present study demonstrates pre-formulation testing of Jamu Uyup – Uyup in solid dosage form. The preparation of solid dosage form was carried out using a spray drying technique. A microparticle comprising Jamu Uyup – Uyup encapsulated in a lactose matrix. Lactose is odorless and slightly sweet-tasting; a-lactose is approximately 20% as sweet as sucrose, while b-lactose is 40% as sweet. Lactose is widely used in pharmaceutical formulations as a filler [12]. The carbohydrate (such as lactose) has properties such as low viscosities at high solids contents and good solubility that are desirable in an encapsulating agent [13]. In addition, their diversity, low cost and widespread use in foods and drinks make them the preferred choice for encapsulation. Spray drying technique has been widely used for drying heat-sensitive foods, pharmaceuticals, and other substances, because of the solvent rapid evaporation from the droplets. Although most often considered a dehydration process, spray drying can also be used as an encapsulation method when it entraps active material within a protective matrix, which is essentially inert to the material being encapsulated. Compared to the other conventional microencapsulation techniques, it offers the attractive advantage of producing microcapsules in a relatively simple continuous processing operation. The encapsulation process comprises the steps of forming an aqueous solution, with stirring for an appropriate amount of time and at certain temperature to form a homogeneous solution; incorporating at least one active compound defined above into the previous solution; spray-drying the aqueous solution at certain temperatures to obtain a stable product encapsulated within the active compound. The encapsulation of Jamu Uyup – Uyup may improve their cohesion and bioavailability. The encapsulated Jamu Uyup – Uyup can be used in various applications of oral solid dosage form, such as instant drink powder, capsule, tablet, etc.

3.1. Physicochemical evaluation

Measurable properties that describe a state of a physical system were reported. The changes in the physical properties of a system can be used to describe its alteration between its transient states. Physical properties are often exhibited as observables. Organoleptic are physical characteristics which are easily identified by simple methods as visualization, touch, smell or taste. All these are used for the preliminary identification of the herbal extract [14]. The fresh liquid of Jamu Uyup – Uyup was a suspension since the liquid form was heterogenous system consisting of two phases. Finely divided solid particles were dispersed in a liquid phase and prone to settle out become sediment. Sediment produced easily re-suspended using a moderate amount of shaking. Curcumin content tended to crystallize out of aqueous acidic solutions (pH <7). The curcumin crystals formed were relatively large (10–50 μm), which made them prone to rapid sedimentation [15]. The aqueous dried extract was viscous and sticky, whereas the spray – dried samples were in the form of fine powder. The color of fresh liquid was dark yellow, the aqueous dried extract is brownish yellow, and the spray – dried powder was medium yellow. Yellow color is related to curcuminoid compounds. The fresh liquid and dried extract had strong typical odor of *Zingiberaceae* whiles powder had distinct typical odor of *Zingiberaceae*. The taste of fresh liquid, dried extract, and powder successively as follow: moderately bitter, extremely bitter, and mildly bitter. The three of these were slightly sweet and had an unequal strength of pungent taste. Odor comes from the essential oil or other compounds while taste varies depending on the constituent of material. The bitter taste mainly comes from bitter substances of *Curcuma aeruginosa* and *Curcuma xanthorrhiza*.

Table 2. Physicochemical properties of Jamu Uyup – Uyup

Parameters	Description		
	Fresh liquid	Dried extract	Spray – dried powder
Texture	Suspension, finely divided solid particles are dispersed in a liquid phase and prone to settle out become sediment	Viscous, sticky	Fine powder
Color	Dark yellow	Brownish yellow	Medium yellow
Odor	Typical of *Zingiberaceae*, strong	Typical of *Zingiberaceae*, strong	Typical of *Zingiberaceae*, distinct
Taste	Moderately bitter and pungent, slightly sweet	Extremely bitter and pungent, slightly sweet	Mildly bitter and pungent, slightly sweet
Moisture content	85.71 ± 1.24% DW	11.73 ± 1.54% DW	3.21 ± 1.71% DW
pH	4.2	4.3	4.7

The results of moisture content as shown in table 2 indicates that the fresh liquid had a moisture content of 85.71 ± 1.24%, the aqueous dried extract had 11.73 ± 1.54%, and the powder had 3.21 ± 1.71%. Examination of moisture content help reduce mistakes in the valuation of the real weight of drug material. According to the European Pharmacopoeia [16] and Indonesian Herbal Pharmacopeia [17], weight loss on drying should not pass 10%. The lower moisture content of Jamu Uyup – Uyup powder is always requested since it contains sugar from honey and lactose which can absorb moisture easily or deteriorate quickly in the presence of water. Lactose may develop a brown coloration on storage, the reaction is accelerated by warm, damp conditions [12]. So, the powders must be stored in an airtight container and provided with moisture absorbent hence not to absorb excessive moisture. An excess of water will promote microbial growth (especially mold) and deterioration subsequently hydrolysis. Low moisture of solid dosage form is very crucial to provides better stability against degradation of product.

The term pH refers to the relative amounts of hydrogen in a given chemical environment. It is a measure of the acidity or alkalinity of a given formulation. The pH is important in aqueous drug product formulation, especially since it involves drug solubility, activity, absorption, stability, sorption and patient comfort [18]. The rate of hydrolysis of product may vary depending on the pH of the solution. Determination of the pH will give information as to which excipients could be added to a drug to ensure that the product is stable and can be tolerated physiologically. Curcumin is a terpenoid – related polyphenol constituent, an oil-soluble pigment, practically insoluble in water at acidic and neutral pH, and soluble in alkali, and highly susceptible for pH change. In aqueous systems like water, at neutral and alkaline pH, curcumin is not stable to chemical degradation and gets easily degraded into compounds like vanillin, ferulic acid, etc. since the acidic phenol group in curcumin donates its hydrogen, forming the phenolate ion. In the pH range 1-7, water solubility is very low, and solutions are yellow. At pH>7.5, the color changes to red [15]. It is very important to keep the dosage form contained curcuminoids in acidic pH.

Figure 1. SEM images of the surface morphological structure of lactose monohydrate (Bratachem), voltage 5.00 kV, magnification: (a) 500x; (b) 1,000x; (c) 2,500x; (d) 5,000x

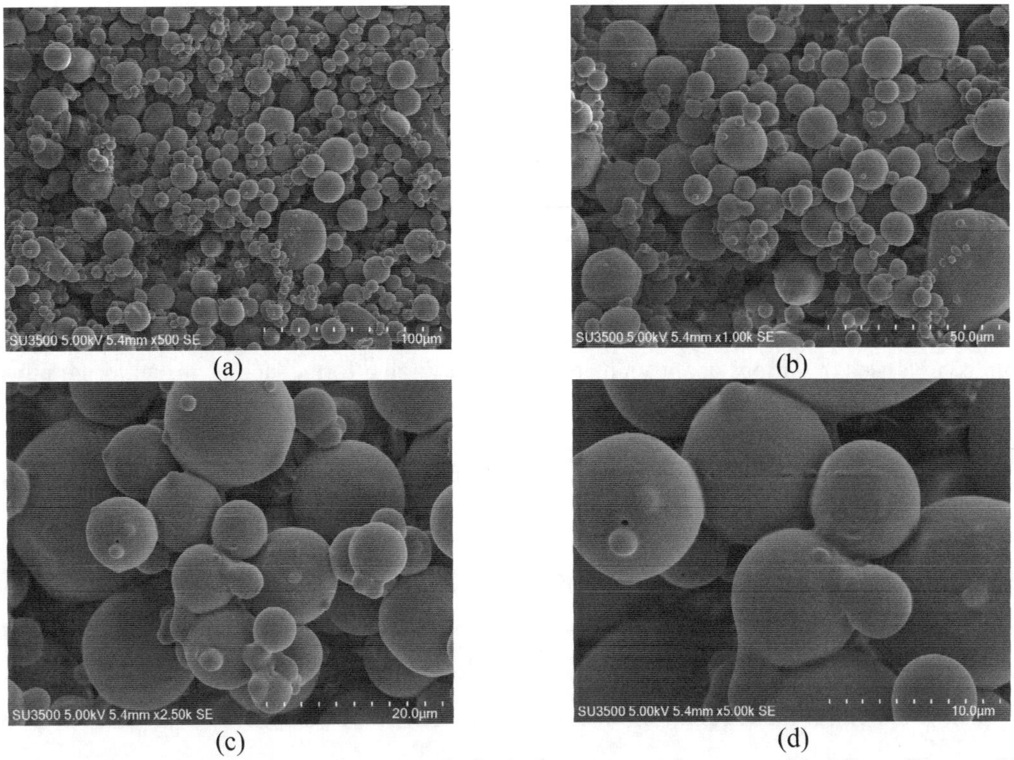

Figure 2. SEM images of the surface morphological structure of spray – dried Jamu Uyup – Uyup microparticles, voltage 5.00 kV, magnification: (a) 500x; (b) 1,000x; (c) 2,500x; (d) 5,000x

SEM pictures of lactose particles are presented in figure 1 and spray-dried active compound-loaded lactose microparticles are presented in figure 2. Observing the external morphology, spray – dried Jamu Uyup – Uyup particles showed a spherical shape, smooth surface, and various sizes (20 – 100 μm) with no apparent cracks or fissures, which is an advantage, since it implies that capsules have lower permeability to gases, increasing protection and retention of the active material. Moreover, the variety in size is a typical characteristic of particles produced by spray drying. According to Ré (1998) [19], imperfections are formed when there is a slow process of film formation during drying of the atomized droplets, associating the presence of surface depressions to the collapse suffered by the droplets during the initial stages of drying. Similar morphological characteristics were found on encapsulation of flaxseed oil by Tonon et al. (2011) [20] and d-limonene by Soottitantawat et al. (2005) [21].

3.2. Phytochemical evaluation

Analyzing the phytochemicals in medicinal plants provides scientists with insight into how effective plants are medicinally and understanding how and why they are effective can lead to the development of new medicines [22]. The main phytochemical in Jamu Uyup – Uyup is curcuminoid compounds which come from *Zingiberaceae* ingredients (*Curcuma aeruginosa, Curcuma xanthorriza, and Curcuma domestica*). The scientific correlation between curcuminoid and galactogogue is still not clearly understood, nevertheless *Zingiberaceae* family has been used empirically as a galactogogue in Indonesia for a long time ago. Although curcumin is relatively stable to heat and can be used in thermally treated, there is a decrease in curcuminoid level of encapsulated powder. Similarly, it occurred with terpenoid level (table 3). Curcumin has antibacterial activity [23] and a very powerful antioxidant effect [21, 22, 23]. So, it may be a preservative of the dosage form as well.

Table 3. Phytochemical properties of Jamu Uyup – Uyup

Parameters	(mg/g dw)	
	Dried extract	Spray – dried powder
Curcuminoid	48.78 ± 1.62	35.86 ± 2.13
Terpenoid	131.95 ± 1.32	113.95 ± 2.16

4. Conclusion

The present investigation demonstrates that the fresh liquid form of Jamu Uyup – Uyup can be formulated into solid dosage form. The preparation of solid dosage form was carried out using spray drying technique. Microparticles containing Jamu Uyup – Uyup encapsulated in a lactose matrix have spherical shape with a smooth surface and contained 35.86 ± 2.13 mg/g dw of curcuminoid. The solid dosage form has an acidic pH (4.7) to keep curcuminoids content stable. The encapsulated Jamu Uyup – Uyup can be used in various applications of oral solid dosage form, such as instant drink powder, capsule, tablet, etc.

5. Acknowledgments

This research was supported by National Priorities Programme 2018 – Indonesian Institute of Sciences.

6. References

[1] Lawrence R A and Lawrence R M 2005 *Breastfeeding: A Guide for the Medical Profession* Philadelphia: Elsevier Mosby
[2] La Leche League International 2004 *The Womanly Art of Breastfeeding* New York: First Plum Printing
[3] Gabay M P 2002 Galactogogues: Medications that induce lactation *J. Hum Lactation* **18**:274–279
[4] Christine M and Betzold N P 2004 Resource for clinician galactagogue *J. Midwifery Womens Health* **49**:151–154

[5] Tiran D 2003 The use of fenugreek for breast feeding women. *Complement Ther Nurs Midwifery* **9**:155–156
[6] Academy of Breastfeeding Medicine Protocol Committee 2011 ABM Clinical Protocol #9: Use of galactogogues in initiatingor augmenting the rate of maternal milk secretion (1st rev). *Breastfeed Med.* **6(1)**:41-49
[7] Gabay M 2002 Galactogogues: medications that induce lactation *J Hum Lact* **18(3)**: 274-279
[8] Qusaj Y, Leng A, Alshihabi F, Krasniqi B and Vandamme T 2012 Development Strategies for Herbal Products Reducing the Influence of Natural Variance in Dry Mass on Tableting Properties and Tablet Characteristics *Pharmaceutics* **4(4)**: 501-516
[9] Shivanand P, Viral D, Manish G, Subhash V and Jaganathan K 2009 Formulation and evaluation of Cedrus deodara loud extract *Int. J. of ChemTech Research* **1(4)**: 1145-1152
[10] Pothitirat W and Gritsanapan W 2006 Variation of bioactive components in Curcuma longa in Thailand *J. Current Science* **91**: 1397 – 1400
[11] Lin M, Yu Z, Wang B, Wang C H., Weng Y and Koo M 2015 *Sains Malays* **44**: 1685-1691
[12] Raymond C R, Sheskey, P J and Owen, S C 2006 *Handbook of pharmaceutical excipients.* American Pharmacists Association. 389 – 395
[13] Desai, K G H and Park H J 2005 *Drying Technol* **23**: 1361
[14] Prakash P R, Rao N R and Chowdary S 2010 Formulation, evaluation and anti-inflammatory activity of topical etoricoxib gel *Asian J. of Pharm. and Clinical Res* **3**: 126
[15] Mahesh K, Zheyuan D, Guodong Z and David J Mc 2017 Physical and Chemical Stability of Curcumin in Aqueous Solutions and Emulsions: Impact of pH, Temperature, and Molecular Environment. *J. Agric. Food Chem.* **65 (8)**:1525–1532
[16] Council of Europe, European Directorate for quality medicines and Healthcare and European Pharmacopoeia Commission 2007 *European pharmacopoeia*
[17] Menteri Kesehatan RI 2009 *Indonesian Herbal Pharmacopoeia*
[18] Allen, L V Jr and Edmond O K 2011 pH and Solubility, Stability, and Absorption, *Int. J. of Pharmaceutical Compounding* Part II, **1(8)**: 3-8
[19] Ré M I 1998 Microencapsulation by spray drying *Drying Technology* **16 (6)**: 1195–1236
[20] Tonon R V, Grosso C R F and Hubinger, M D 2011 Influence of emulsion composition and inlet air temperature on the microencapsulation of flaxseed oil by spray drying *J. Food Research International* **44(1)**: 282–289
[21] Soottitantawat A, Bigeard F, Yoshi H, Furuta T, Ohkawara M and Linko P 2005 Influence of emulsion and powder size on the stability of encapsulated Dlimonene by spray drying *J. Innovative Food Science and Emerging Technologies* **6(1)**: 107–114
[22] Doughari J H 2012 Phytochemicals: Extraction Methods, Basic Structures and Mode of Action as Potential Chemotherapeutic Agents. Department of Microbiology, School of Pure and Applied Sciences, Federal University of Technology, 1-3
[23] Negi P S, Jayaprakash G K, Jagan M R L and Sakariah K K 1999 Antibacerial activity of turmeric oil; a byproduct from curcumin manufactuer. *Journal of agricultural and food chemistry* **47**: 297-300
[24] Ruby A J, Kuttan G, Dinesh B K, Rajasekharan K N and Kuttan R 1995 Antitumor and antioxidant activity of natural curcuminoids *Cancer Letters* **94**: 79–83
[25] Osawa T et al 1995 Antioxidative activity of the tetrahydrocurcuminoids *J. Biosci. Biotechnol Biochem* **59(9)**: 1609-1612
[26] Sreejayan N and Rao M N 1994 Curcuminoids as potent inhibitors of lipid peroxidation *J. Pharm. Pharmacol.* **46**: 1013

Acid-base evaluation of chitosan-ferulic acid conjugate by a free radical grafting method

M Anwar[1a], K Nisa[1] and N Indirayati[1]

[1]Research Unit for Natural Product Technology, Indonesian Institute of Sciences, Gunungkidul, Yogyakarta, Indonesia.

[a]email: muslih.anwar@gmail.com

Abstract. Chitosan-ferulic acid conjugate has been synthesized based on the free radical-mediated grafting reaction. This study investigated the application of potentiometric titration to evaluate the free radical grafting method of ferulic acid on chitosan. The synthesis was done by hydrogen peroxide/ascorbic acid induced free radical reaction to conjugate ferulic acid with chitosan. The conjugate product of ferulic acid-chitosan was characterized by FTIR, UV-Vis spectrophotometry, potentiometric titration, and SEM images. Evaluation of the grating ratio of the ferulic acid-g-chitosan was determined by Folin–Ciocalteu method. The chitosan-ferulic acid conjugate gave 84% in yield with total phenolic showed the value was 92 .53 mg per gram of dry polymers and the conjugate product only soluble at pH around 6. The antioxidant activity of chitosan was increased after functionalized with ferulic acid.

Keywords: Ferulic acid, chitosan, potentiometric titration, free radical grafting, antioxidant activity

1. Introduction

Chitosan (Cs) is a partially composed of D-glucosamine and N-acetyl-D-glucosamine linked by β(1-4) bond which is obtained by deacetylation of chitin with alkaline treatment [1]. Chitin mainly comes from the exoskeleton of shrimp, prawn, lobster, and insects. Chitosan contains an amino (-NH_2) and hydroxyl (-OH) groups in D-glucose backbone, which has a wide range of applications in tissue engineering [2], pharmaceutical [3-4], food preservation [5-6], cosmetics [7], and agriculture [8], due to its nontoxicity, biodegradability and biocompatibility. However, the utilization of chitosan is limited by its poor solubility in neutral or alkaline media, since chitosan is only soluble in acidic condition. Chemical modification to the chitosan functional groups improves its physicochemical and biological properties. Chitosan can be modified by many chemical reactions, such as carboxymethylation, esterification, alkylation, N-quarternization, and graft copolymerization. In order to introduce certain functional molecules into chitosan, graft copolymerization reaction has been most frequently used [9]. Grafting some phenolic compounds such as eugenol [10], gallic acid [11], epigallocatechin gallate [12] onto chitosan remarkably increased its biological activities.

Phenolic compounds are plants secondary metabolites which have precious biological activities such as anticancer, antioxidant and antimicrobial [13-14]. Ferulic acid (FA) (4-hydroxy-3-methoxycinnamic acid) is a natural phenolic acid which commonly isolated from Grain [15]. Because of its low toxicity and has many physiological functions, including anti-inflammatory, antioxidant, anticancer [16], and antimicrobial activity [17], this bioactive compound has been extensively used in pharmaceutical, food, and cosmetic industries. Ferulic acid contains phenolic, methoxy and alkene conjugated to carboxylic acid groups which readily oxidized in the high temperature. The phenolic molecules covalently grafted onto polymers (i.e., cellulose and chitosan) expand their stabilities, physicochemical properties and biological activities (e.g., antioxidant, antimicrobial, antifungal and antitumor) [18-19].

Several studies reported that chitosan-phenolic compounds conjugates formed complex had good solubility in water [20]. Due to the instability of ferulic acid in the high temperature, free radical grafting was preferable than another grafting method. This study aimed to produce chitosan-ferulic acid conjugate by a simple modification with the eco-friendly approach. Ferulic acid was grafted onto chitosan by free radical grafting with hydrogen peroxide/ascorbic acid. The synthesized product was evaluated by using total phenolic contents with Folin-Ciocalteu reagent, potentiometric titration, UV-Vis spectrophotometry, and Fourier-transform infrared (FT-IR) to confirm the conjugation.

2. Materials and Methods

2.1. Materials

Ferulic acid (Sigma-Aldrich), chitosan from Indonesian crab shells (66.5% degree of deacetylation), hydrogen peroxide (Merck), ascorbic acid, 2,2-diphenyl-1-picrylhydrazyl radical (Merck, Darmstadt, Germany), Folin-Ciocalteu reagent (Sigma-Aldrich), acetic acid (99%w/w) (Merck), sodium hydroxide (Merck), ethanol (Merck), and hydrochloric acid (37% w/w) (Merck) were used without any purification. The synthesis products were identified by infrared spectrophotometer (FTIR, Shimadzu-8201PC) over a wavenumber range of 400–4000 cm^{-1} on KBr pellet, UV-Vis spectrophotometer (*Dynamica* Scientific *Halo RB*-10), pH meter (Thermo Scientific Orion Star A214), and Scanning Electron Microscopy (Hitachi SU3500).

2.2. Synthesis of Ferulic Acid Grafted-Chitosan

Free radical grafting method of Chitosan-Ferulic acid was conducted following to Curcio et al. (2009) [21], with some modification. Chitosan (0.25 g) was dissolved in 25 mL of acetic acid solution (2% v/v) in a 100 mL flask tube. In the ice bath, 0.027 g of ascorbic acid in 1 mL hydrogen peroxide 1M was added slowly into chitosan solution. 0.2708 g of ferulic acid in 5 mL ethanol was dropped into the flask, and the mixture was continuous stirring under atmospheric condition at room temperature for 24 h. NaOH 2 M was added into the mixture and pH was adjusted to 10, followed by filtration. The residue was washed with absolute ethanol three times to give a yellow solid with 84% in yield.

2.3. Total phenolic compounds evaluation

The quantification of the total phenolic group on chitosan-ferulic acid conjugate by Folin-Ciocalteu procedure according to the literature with some modifications [22]. Chitosan-ferulic acid conjugates (25 mg) was dissolved in acetate buffer pH 5.6 up to 25 mL in a volumetric flask. Folin-Ciocalteu reagent (0.5 mL) was supplemented to the sample solution (0.5 mL) and was mixed thoroughly, followed by addition of 1,5 mL of sodium carbonate (20%w/v), 7,5 mL distilled water and allowed to stand for 2 h. The concentration of phenolic content was measured at wavelength 765 nm with the chitosan solution under the same conditions as a blank. The concentration of total phenolic groups in the chitosan conjugate product was represented as ferulic acid equivalent concentrations. The calibration curves of ferulic acid were made by plotting the concentrations to 20, 45, 60, 80, 100, 125 and 150 ppm versus absorbance. Total phenolic of ferulic acid-chitosan was expressed as milligrams of ferulic acid per gram conjugate product.

2.4. Potentiometric titration experiments

The degree of substitution of chitosan was estimated by potentiometric titration adapted from Kong (2012) [23]. The weighed chitosan-ferulic acid conjugate (50 mg) was dissolved in 10 mL of aquadest and adjusted to below pH 2 with 0.1 M hydrochloric acid, and NaOH solution (0.1M) was used as the titrant. The figure was obtained based on the sodium hydroxide volume against the corresponding pH and first derivatives of electric potential difference $\Delta E(mV)/\Delta V(mL)$ value recorded.

2.5. Antioxidant activity test by DPPH method

The antioxidant activities of chitosan-ferulic acid conjugate were determined using 2,2-diphenyl-1-picrylhydrazyl (DPPH). Briefly, the stock solution (1000 ppm) was prepared by dissolving a conjugate product in acetate buffer pH 5.6. The series of concentrations of the conjugate product (50, 100, 150, 200, 250, and 500 ppm) were made by adjusted the stock solution with acetate buffer pH 5.6 and methanolic DPPH (0.2 mM). The mixtures were kept in the dark condition for 30 min at room temperature to perform radical scavenging reaction. The same portion of absolute methanol and acetate buffer pH 6 used a blank. The absorbance was determined at 517nm, and the antioxidant activities were defined as inhibition of DPPH radicals calculated as follows :

$$Inhibition\ (\%) = \left[\frac{A_0 - A_1}{A_0}\right] x\ 100\%$$

where A_0 is the absorbance of the standard (mixture methanolic DPPH and acetate buffer pH 6), and A_1 is the absorbance of the sample only. The IC_{50} value denoted as the concentration of chitosan-ferulic acid conjugate that inhibited 50% DPPH radicals [24].

3. Results and Discussion

3.1. Synthesis of Chitosan-Ferulic Acid Conjugate

Chitosan-ferulic acid conjugate was successfully grafted by free radical method using hydrogen peroxide/ascorbic acid. The reaction was carried out temperature for 24 h at ambient with the mass ratio of chitosan, and ferulic acid was 1:1 (w/w). The reaction mechanism divided into two steps, first is the chitosan initiated by the radical reaction and the second step is the grafting of the ferulic acid into chitosan. The reaction started by oxidation of ascorbic acid by hydrogen peroxide in cold temperature to form ascorbate radical ($Asc^{\bullet-}$) instead of hydroxyl radical ($OH^{\bullet}$). The existence of ascorbate radicals abstracts hydrogen atom of chitosan chain to produce dehydroascorbic acid and chitosan radicals [25]. The grafting reaction was occurred by added ferulic acid into chitosan radicals to yield conjugate product. Afterward, pH of the reaction mixture was adjusted into alkaline (pH) condition to dissolve unreacted ferulic acid and to precipitate the conjugate product. The yellow solid product was washed with absolute ethanol to eliminate residue of ferulic acid. This method is simple to produce conjugate of chitosan-ferulic acid without dialysis membrane, by changing the pH condition, to precipitate the desired compounds. The radical reaction can be performed in lower temperatures to prevent the ferulic acid oxidation. Moreover, this reaction does not generate toxic side products.

The reaction product was insoluble in acidic and alkaline condition but fully dissolved at pH around 5-6 (Figure 1). The solubility of the product might be influenced by both chitosan and ferulic acid properties. In acidic condition, chitosan is soluble, but ferulic acid is insoluble. Opposite in the alkaline condition, ferulic acid is soluble, but chitosan is insoluble. The conjugate product only soluble in the pH 5-6 might due to the grafted reaction occurred not in the functional group which responded to the solubility such as amino ($-NH_2$) in chitosan and phenolic (Ph-OH) or carboxylic acid (-COOH) in ferulic acid.

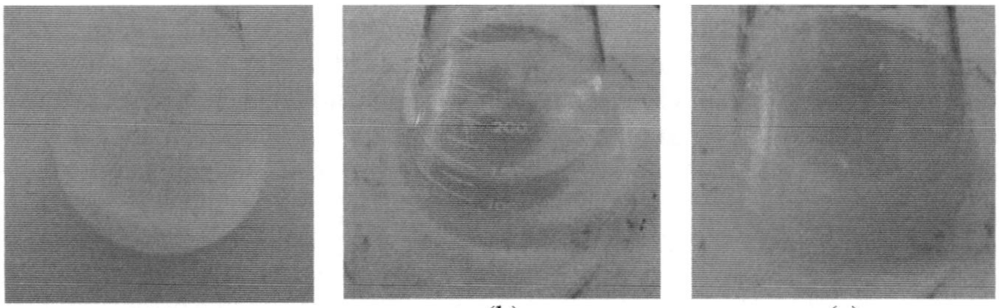

(a) (b) (c)

Figure 1. The solubility of ferulic acid grafted-chitosan in acidic (a), pH 5-6 (b), and alkaline condition (c).

3.2. Characterization of Chitosan-Ferulic acid Conjugates

The conjugate product of free radicals grafted method between chitosan and ferulic acid was evaluated by using Fourier transform infrared, UV-vis spectrophotometry, potentiometric titration, and total phenolic compounds with Folin-Ciocalteu method to determine the degree of substitution.

The Folin-Ciocalteu phenol reagent can be used to estimate the amount of ferulic acid conjugate onto chitosan chain. Ferulic acid as a phenolic compound reacted with phosphomolybdic and phosphotungstic acids existing in the Folin-Ciocalteu reagent to form complex redox mixture. The ferulic acid content in the conjugate product was 92.53 mg/g of conjugate product. It has been reported that radically initiated grafting gallic acid- and catechin-chitosan were 7 and 4 mg/g of the dry product, respectively [21].

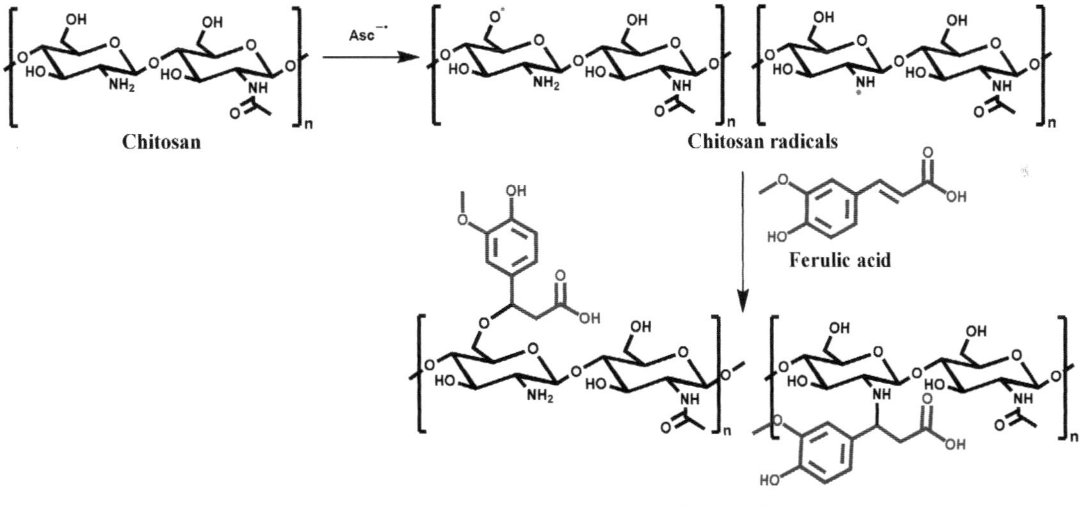

Figure 2. The proposed schematic reaction of ferulic acid grafted into chitosan.

The Fourier transform infrared spectra of chitosan, ferulic acid and conjugate product were shown in Figure 3. Infrared spectra of chitosan exhibited broad peaks at around 3460 cm^{-1} (-O-H st) and 1656 cm^{-1} (C=O amide of residual chitin groups). Ferulic acid showed the appearance of sharp peaks at 3434 cm (-O-H st), alkenes characteristic band around 1690 cm^{-1}. FTIR spectra of the conjugate product showed shifting of C=O stretching of α,β-unsaturated carboxylic acid of ferulic acid at 1690 cm^{-1} to aliphatic carboxylic acid at 1700 cm^{-1}. The absence of a peak at 941 cm^{-1} (C=C-H), overtone at 1822 cm^{-1} and no peak at 1660 cm^{-1} (C=C st) [26] indicate that the alkene group of ferulic acid was converted into a single bond (Figure 2). It implies that covalent bond between ferulic acid and chitosan were formed. Moreover, no peak at 1730 cm^{-1} in the conjugate products, implying no ester bond formed between carboxyl groups of ferulic acid and hydroxyl groups of chitosan, which has been reported by previous researcher [21, 27].

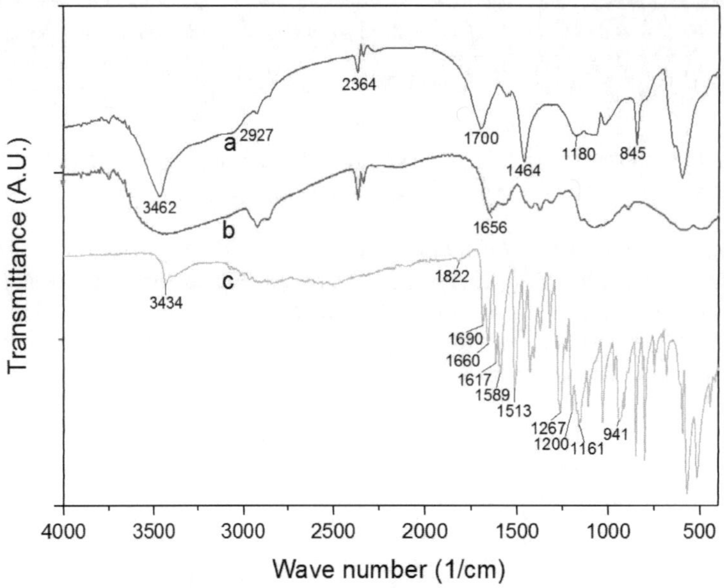

Figure 3. FTIR spectra of Chitosan-ferulic acid conjugate (a), chitosan (b) and ferulic acid (c).

To confirm that grafting between ferulic acid and chitosan was done by comparing UV spectra of chitosan, ferulic acid and chitosan-ferulic acid conjugate (Figure 4) over 190 - 400 nm of wavelength. Chitosan in 2% acetic acid (v/v) revealed no absorption peak between 230 to 400 nm. Ferulic acid in ethanol solution showed a maximum absorption peak around 320 nm. The UV-vis spectra of chitosan-ferulic acid conjugates gave hypsochromic shifting from ferulic acid spectra with maximum absorption around 210 nm. The hypsochromic shifting of maximum absorption band from 320 to 210 nm might due to the breakdown of the alkene group from conjugate carboxylic acid. This spectra supports the FTIR data that insertion of ferulic acid into chitosan was done.

The titration was performed by dissolving chitosan, ferulic acid and the conjugate product with water and adjusted to pH 2 by hydrochloric acid 0.1 M. The excess of HCl was titrated by using NaOH 0.1 M. The titration curve was plotted by volume NaOH (mL) against pH and first derivatives versus $\Delta E(mV)/\Delta V(mL)$. The typical titration curves of chitosan (Figure 5) showed two inflection points as amino and hydroxyl groups which both contribute to acid-base balance. The first inflection point of chitosan at pH 4.81 was represented as hydroxyl (-OH) and hydroxide ($-O^-$) groups balance, whereas the second inflection point at pH 7.15 characterized as the balance of ammonium ($-NH_3^+$) and amino ($-NH_2$) groups. The characteristic potentiometric titration curves of ferulic acid are shown in Figure 6. The curves exhibited three inflection points (pH 3.87, 7.27, and 10.64) which might correspond to acid-base balance of phenolic, α,β-unsaturated and carboxylic acid groups of ferulic acid. The potentiometric titration curves (Figure 7) of the conjugate product of ferulic acid and chitosan appeared only two inflection points (5.65 and 11.56). The reduced of total inflection points of the product from the total inflection points of ferulic acid due to one of the functional groups of ferulic acid was break down and converted to another functional group which was not contributed to acidity. The potentiometric titration curves reinforced the FTIR spectra that unsaturated group of ferulic acid might be changed into a single bond by free radical method.

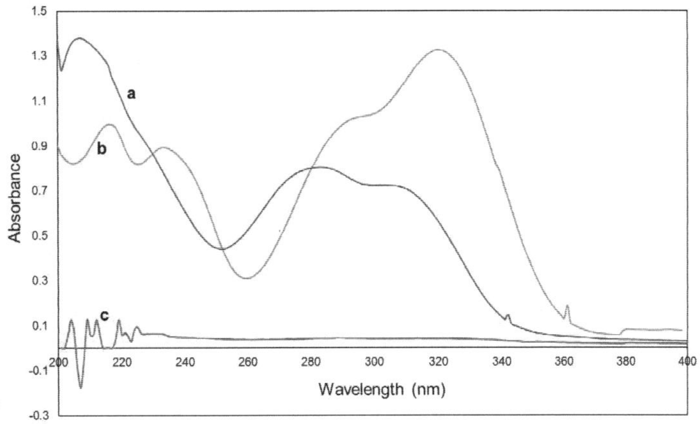

Figure 4. UV-vis spectra of Chitosan-ferulic acid conjugate (a), ferulic acid (b), chitosan (c).

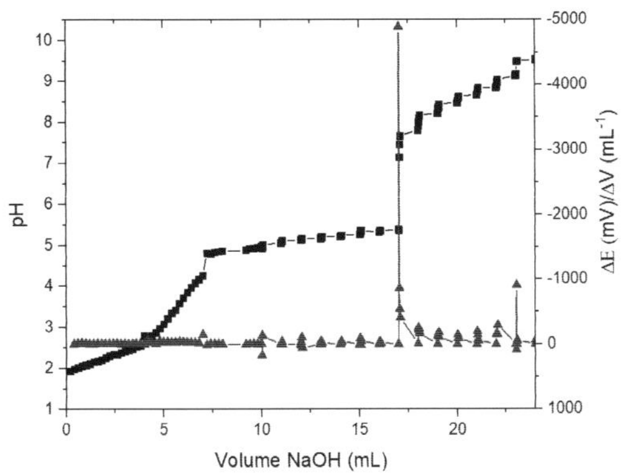

Figure 5. The titration curves of chitosan.

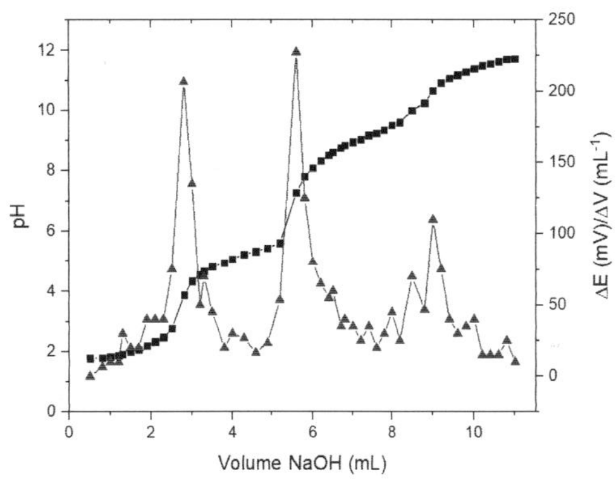

Figure 6. The titration curves of ferulic acid.

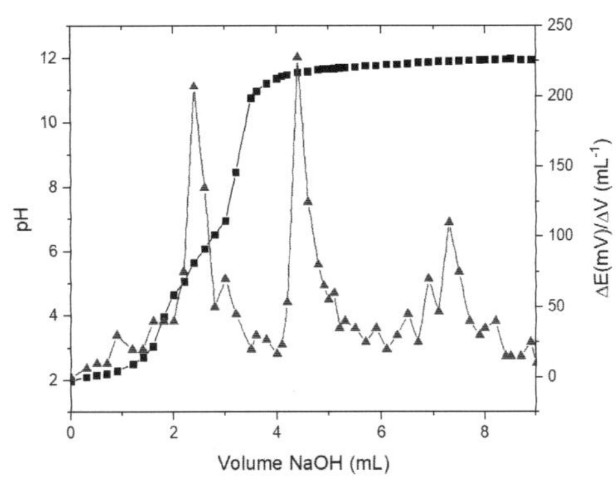

Figure 7. The titration curves of chitosan-ferulic acid conjugate product.

The surface morphology images (Figure 8) were carried out by using scanning electron microscope with voltage acceleration 2 kV and a working distance of 6.5 to 6.8 mm. Grafting reaction of chitosan and ferulic acid by using radical method provided the significant change of morphology between chitosan as reactant (a) and the conjugate of chitosan-ferulic acid product (b). The morphology of chitosan exhibited as a big rolled sheet, whereas the conjugate product appeared the flake-like morphology in the smaller size.

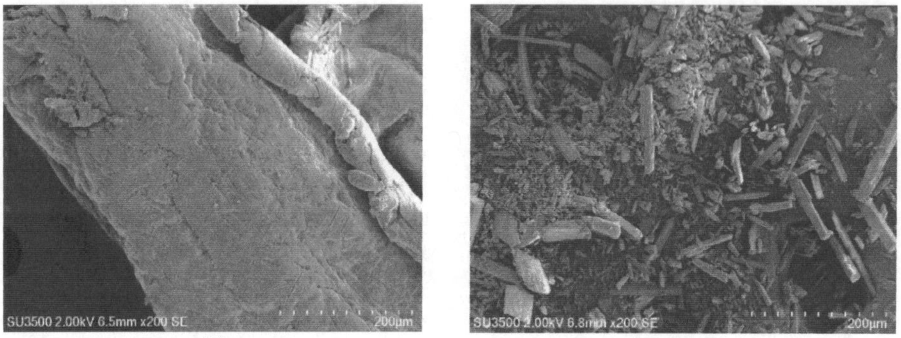

(a) (b)

Figure 8. The morphology images of chitosan (a) and chitosan-ferulic acid conjugate product (b)

3.3. Antioxidant activity of Chitosan-Ferulic acid Conjugates

Radical scavenging capacities of chitosan-ferulic acid conjugates were evaluated by DPPH test with the decrease in absorbance of yellow-colored compound at 517 nm. Reduction of diphenyl-picrylhydrazine radicals is depended on the hydrogen-donating ability of the conjugate product which might be contributed by ferulic acid content in the chitosan chain. The antioxidant activity of ferulic acid is caused by proton transfer from the phenolic and unsaturated carboxylic acid group to DPPH radicals. The DPPH scavenging activities of chitosan-ferulic acid conjugate are shown in Figure 9. The increasing concentration of conjugate product from 0 to 0.6 mg/mL, showed an increased trend of scavenging activity up to 71.19±1.97% at 0.6 mg/mL with IC_{50} = 0.386 mg/mL. This conjugate product has a better antioxidant activity compared with the plain chitosan (24.86% and at 0.2 mg/mL). It means that solubility and antioxidant activity of chitosan were improved by conjugating with ferulic acid. This product has a potential to be used in pharmaceutical, biomaterial and food industries.

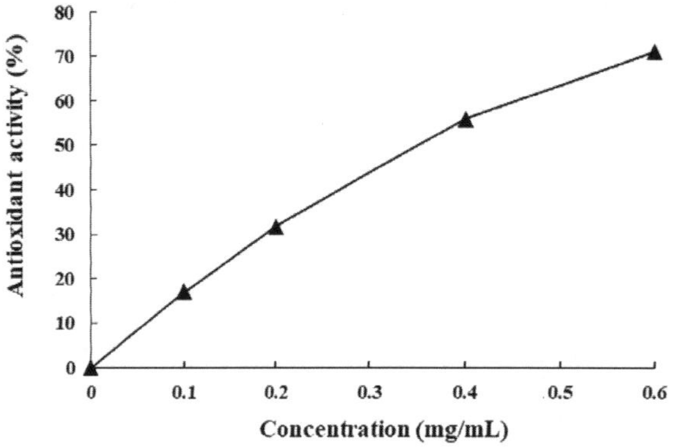

Figure 9. Antioxidant activity of chitosan-ferulic acid conjugate by DPPH method

4. Conclusion

In the present study, chitosan-ferulic acid conjugate was successfully synthesized by using free radical method with hydrogen peroxide/ascorbic acid. The conjugate product was confirmed by FTIR, UV-Vis and potentiometric titration. It suggested that the grafting reaction happened between hydroxyl group of chitosan and α,β-unsaturated carboxylic acid groups of ferulic acid. Furthermore, the product was improved in solubility and radicals scavenging activities against DPPH compared with chitosan. In addition, simple preparation of chitosan-ferulic acid conjugate by free radical and precipitate the product by changing pH condition could be further explored in many applications area.

5. Acknowledgments

This research was funded by ITSF (Indonesia Toray Science Foundation). The authors would like to thank Edi Prasetyo for the SEM Images, Teguh Santosa for the potentiometric titration data, and also for the research facilities support from BPTBA LIPI.

6. References

[1] Anwar M, Anggraeni A S and Al Amin M H 2017 AIP Conference Proceedings **1823** 1 020071
[2] Rodríguez-Vázquez M, Vega-Ruiz B, Ramos-Zúñiga R, Saldaña-Koppel D A and Quiñones-Olvera L F 2015 *Biomed. Res. Int.* **2015** 821279
[3] Ahmed T A and Bader M A 2016 *Drug Des. Devel. Ther.* **10** 483–507
[4] Yadav R and Chauhan M 2017 *Int. J. Pharm. Sci. Res.* **2** 06-11
[5] Higueras L, López-Carballo G, Gavara R and Hernández-Muñoz P 2015 *Food Bioprocess Tech.* **8** 3 526-538
[6] Quesada J, Esther S, Casilda N and Sayas-Barberá E 2016 *Foods* **5** 3 57
[7] Aranaz I, Niuris A, Concepción C, Begoña E, Javier M, Carolina C, Gandía M L and Caballero A H 2018 *Polymers* **10** 2 213
[8] Xing K, Xiao Z, Xue P, and Qin A S, 2015 *Sustain.* Dev. **35** 2 569-588
[9] Thakur V K and Thakur M K, 2014 *ACS Sustain. Chem. Eng.* **2** 2637–2652
[10] Woranuch S and Yoksan R 2013 *Carbohyd. Polym.* **96** 2 495-502
[11] Hu Q, Wang T, Zhou M, Xue J, and Luo Y 2016 *J. Agr. Food Chem.* **64** 29 5893-5900
[12] Moreno-Vásquez M J, Valenzuela-Buitimea E L, Plascencia-Jatomea M, Encinas-Encinas J C, Rodríguez-Félix F, Sánchez-Valdes S, Rosas-Burgos E C, Ocaño-Higuera V M and Graciano-Verdugo A Z 2017 *Carbohyd. Polym.* **155** 117-127
[13] Roleira F M, Tavares-da-Silva E J, Varela C L, Costa S C, Silva T, Garrido J and Borges F 2015 *Food Chem.* **183** 235–258
[14] Shahidi F and Ambigaipalan 2015 *J. Funct. Foods* **18** 820–897
[15] Quinde-Axtell Z and Byung-Kee B 2006 *J. Agric. Food Chem.* **54** 26 9978–9984
[16] Ou S and Kwok K C 2004 *J. Sci. Food Agri.* **84** 1261–1269
[17] Rabea E I, Badawy M E T, Stevens C V, Smagghe G, and Steurbaut W 2003 *Biomacromolecules* **4** 6 1457-1465
[18] Cota-Arriola O, Plascencia-Jatomea M, Lizardi-Mendoza J, Robles-Sánchez R M, Ezquerra-Brauer J M, Ruíz-García J, Vega-Acosta J R and Cortez-Rocha M O 2017 *CyTA-J. Food* **15** 1 65-74
[19] Yakub G, Ignatova M, Manolova N, Rashkov I, Toshkova R, Georgieva A and Markova N 2018 *Int. J. Biol. Macromol* **107** 689-702
[20] Kumar S, Deepak V, Kumari M and Dutta P 2016 *Int. J. Biol. Macromol.* **84** 349-353.
[21] Curcio M, Puoci F, Iemma F, Parisi O I, Cirillo G, Spizzirri U G and Picci N 2009 *J. Agr. Food Chem.* **57** (13) 5933-5938
[22] Chatterjee N S, Panda S, Navitha M, Asha K K, Anan and R, Mathew S 2015 *J. Food Sci. Tech.* **52** 7153-7162

[23] Kong X 2012 *Carbohyd. Polym.* **88** 1 336-341
[24] Ardestani A and Yazdanparast R 2007 *Food Chem.* **104** 21–29
[25] Liu J, Pu H, Chen C, Liu Y, Bai R, Kan J and Jin C 2017 *J. Agr. Food Chem.* ***66*** (1) 279-289
[26] Pretsch E, Bhuhlmann P, and Affolter C 2000 *Structure determination of organic compounds* (Berlin:Springer-Verlag) pp. 272-315
[27] Pasanphan W and Chirachanchai S 2008 *Carbohydr. Polym.* **72** 169–177

Physicochemical Properties of Noni Fruit, Yam Root, Rose Petal, and Betel Leaf Transparent Soap and Their Antimicrobial Activities

V T Rosyida[1], K Nisa[1], S N Hayati[1], W Apriyana[1], C Darsih[1], A W Indrianingsih[1], D Ratih

[1]Research Unit for Natural Products Technology, Indonesian Institute of Sciences, Yogyakarta, Indonesia

E-mail: vitataufikarosyida@yahoo.co.id

Abstract. Natural soap is prepared without an artificial surfactant, and with the addition of functional ingredients such as a natural substance, essential oil, or plant extract. The present study was to perform transparent soap formulation enriched with noni (*Morinda citrifolia* L.) fruit, yam (*Pachyrizus erosus*) root, rose (*Rosa damascena*) petal, and betel (*Piper betle* L.) leaf extract and investigate its antibacterial activity. These herbal transparent soaps were made by hot process method. Physicochemical characteristics tested were moisture content, pH, total fatty matter, free fatty acid content, and chloride content. Antimicrobial activities of herbal transparent soaps were investigated by agar well diffusion method. The result showed that transparency was maintained when the moisture content was greater than 17% but not exceed about 21%. pH value of herbal transparent soap ranged from 9.22 to 9.47, while the free fatty acid content ranged from 0.12% to 2.15%. Yam root transparent soap showed the highest total fatty matter (45.12%). While the lowest total fatty matter was shown by the betel leaf transparent soap (23.20%). The chloride content of noni fruit, yam root, rose petal, and betel leaf transparent soap was 14.38%, 14.17%, 21.81%, and 16.52%, respectively. Noni fruit, yam root, betel leaf transparent soaps had stronger antimicrobial inhibitory than control soap, while the rose petal transparent soap had comparable antimicrobial inhibitory with control. All herbal transparent soap and control at the concentration level of 10% and 20% exhibited antibacterial activity against all microorganism tested (*Staphylococcus aureus*, *Escherichia coli*, *Pseudomonas aeruginosa*, *Candida albicans*) to a certain degree.

Keywords: transparent soap, *Morinda citrifolia* L., *Pachyrizus erosus*, *Rosa damascene*, *Piper betle* L., antibacterial activity

1. Introduction

Nowadays, the harmful effects of synthetic surfactants, the environmental friendliness, and environmental sustainability of large – scale commercial soap making in the factory are being emphasized. On the contrary, natural soap is prepared without an artificial surfactant, with the addition of functional ingredients such as a natural substance, essential oil, or plant extract, and typically run in small scale by home/small industry. Based on the making method, natural soap may be generally divided into a melt – pour soap, a hot process soap, and a cold process soap. The hot process soap is also named a transparent or translucent soap. This soap is made by mixing caustic soda with oil at a high temperature. To make the hot soap, the elements should be heated in a water bath or directly heated for 40 to 50 minutes, and thereafter, the heated elements should be matured in normal temperature for 2 weeks. The soap has good detergency or cleansing power, good moisturizing effects, long-lasting fragrance, and less of an irritant.

Aucubin, L-asperuloside, alizarin, and scopoletin in the Noni (*Morinda citrifolia* L.) fruit are proven, antibacterial agents. These compounds have been shown to fight against infectious bacteria strains such as *Pseudomonas aeruginosa*, *Proteus morgaii*, *Staphylococcus aureus*, *Bacillus subtilis*, *Escherichia coli*, *Salmonella*, and *Shigella*. [1, 2]. Extracts from the ripe noni fruit exhibited moderate antibacterial properties against *Pseudomonas aeruginosa*, *M. pyrogenes* and *Escherichia coli* [3]. Locher reported that selected plants including noni have a history of use in Polynesian traditional medicine for the treatments of infectious disease [4]. Essential oil, absolute and hydrosol of *Rosa damascena* has wide spectrum antimicrobial activities. The in vitro antibacterial activities of essential oil from *Rosa damascena* were exhibited inhibitory and bactericidal activities against gram-positive *Staphylococcus aureus* (ATCC 25923), gram-negative *Escherichia coli* (ATCC 25922), gram-negative *Pseudomonas aeruginosa* (ATCC 27853), and yeast *Candida albicans* (ATCC 14053) [5, 6]. Antibacterial effect of major components of rose oil (citronellol, geraniol, and nerol) was reported [7]. Antibacterial properties of rose absolute could be attributed to its high phenyl ethyl alcohol content [8]. *Piper betle* L. has been demonstrated in vitro to have significant antimicrobial potential against a broad spectrum of microorganisms including bacteria and fungi and has been further described as an effective alternative to reduce biofilm formation in membrane bioreactors caused by both *Pseudomonas aeruginosa* and *Candida albicans* [9, 10, 11, 12]. The antimicrobial potential is most likely due to hydroxychavicol, stearic acid, and palmitic acid. The antimicrobial activity explains the use of betel leaf as an oral care agent in Asian countries [13]. Betel leaf has also displayed the greatest potential value, against both Gram-negative and Gram-positive multidrug-resistant bacteria [14]. Further, some herbal ingredients have been added to the soaps to add its functionality. The present study was to perform transparent soap formulation enriched with this functional herbal extract and investigate its antibacterial activity.

2. Materials and Methods

2.1. Preparation of herbal extract

Selected herbal used in this study were noni (*Morinda citrifolia* L.) fruit, yam (*Pachyrizus erosus*) root, rose (*Rosa damascena*) petal, and betel (*Piper betle* L.) leaf. The ripe noni fruit and yam root extract were pureed and squeezed to collect the juice. Each of fresh juice was evaporated to dryness under reduced pressure at 45 °C. While rose petal and betel leaf extract prepared by collecting the hydrolat (hydrosol, aromatic water) from the steam distillation process. Hydrolat obtained during distillation of aromatic plants. Hydrolat can be collected since a small hydrophilic fraction, oxygenated, odor imparting, water-soluble oil constituents escape into the distillation/condensate water stream.

2.2. The making process of transparent soap

Stearic acid (8.5%), citric acid (0.3%), virgin coconut oil (32%), ricini oil (1%), and alcohol 96% (14%) were mixed while heated until the blend solution was clear at a temperature range between about 50 to 60 °C. Neutralized the batch by slowly adding an aqueous solution of sodium hydroxide while ensuring the temperature does not exceed about 85 °C. After aqueous sodium hydroxide has been added, the mixture was kept between about 80 to 85 °C for a while to ensure the saponification reaction was complete. Added aqueous mixture of sucrose and propylene glycol to the batch and mixed vigorously until solution was clear. While maintained gently agitation, cooled the batch to a temperature about 60 °C and slowly added herbal extract (8%) and mixed until a homogenous mixture was achieved. The composite was then poured into molds and allowed to solidify for 24 hours. The cooled transparent soap removed from the molds and allowed in normal temperature for 2 weeks for aging process.

2.3. Physicochemical properties of soap
Determination of physicochemical properties of soap includes pH, moisture content, total fat matter, free fatty acid content, and chloride content. Standard procedures as described by AOCS [15], were used for the analysis of physicochemical properties with slight modifications. pH of 10 % solution of transparent soap was determined by means of a digital pH – meter (Euthech PC 700). Moisture content was measured by moisture analyzer digital (AND MX – 50). Method for measurement of total fat matter, free fatty acids, and chloride content was gravimetric, volumetric, and argentometric methods, successively.

2.4. Antimicrobial activity
The bacterial strains used to assess the antimicrobial properties of herbal transparent soap included one Gram-positive (*Staphylococcus aureus*) and two Gram-negative bacteria (*Escherichia coli*, *Pseudomonas aeruginosa*), and one fungus (*Candida albicans*). The investigated microbial strains were obtained from the Department of Agricultural Microbiology, Universitas Gadjah Mada, Indonesia. The antimicrobial activities of herbal transparent soaps were evaluated by agar well diffusion [16, 17] using Nutrient Agar medium for the bacterial assay and Potato Dextrose Agar medium for the fungal assay. The microorganism was activated by inoculating a loopful of the bacterial strain in the Nutrient Broth (10 mL) and incubated at temperature 37 °C for 24 hours, while the fungal strain in the Potato Dextrose Broth (10 mL) and incubated at temperature 30 °C. Then 100 µl of bacterial inoculum (inoculum size was 10^8 cells/ml as per McFarland standard) was inoculated into the Nutrient Agar media (20 mL) and fungal inoculum into the Potato Dextrose Agar media (20 mL). After proper homogenization, it was poured into Petri dishes. Wells were made in the seeded plates with the help of a cup-borer (7 mm). 25 µL of test solution and control solution (Lifebuoy® transparent soap) at a level concentration of 1%, 5%, 10%, 20% were introduced into the well. Bacterial plates were incubated at 37 °C for 24 hours and fungal plates were incubated at 30 °C for 48 hours. The experiment was performed 3 times under strict aseptic conditions.

3. Results and Discussion
The soap bar of this study has gloss-like clarity. The fatty acids – stearic acid contribute to lathering and washing properties of the soaps [18]. Coconut oil is a source of lauric acid (12 carbons) which can be made into sodium laurate. This soap is very soluble and will lather easily. Fatty acids with only 10 or fewer carbons are not used in soaps because they irritate the skin and have objectionable odors [19]. The amount of fatty acid to be neutralized with a polyol and or polyol blend. The presence of non-volatile polyols enhances the clarity of the product and prevents shrinkage of the bar during storage and use. The sodium hydroxide was used in bar soap making as alkalis that react with glycerides to provides further neutralizing activity for production of optimum transparency. These are naturally different from soaps made from divalent metals such as calcium, magnesium, iron or aluminum which are not soluble in water [19]. Since the neutralization of the fatty acid is an exothermic reaction, addition of sodium hydroxide must be controlled consequently the temperature will not exceed 85 °C.

3.1. Physicochemical evaluation
The physicochemical characteristic of soap depends on several factors which include the strength and purity of alkali, the kind of oil used and completeness of saponification. Physicochemical characteristics include moisture content, pH, total fatty matter, free fatty acid, and percentage chloride [20]. Good quality soap for cleansing purpose is one that strikes a balance in all the mentioned physicochemical parameters. The results obtained from the analysis of different physicochemical parameters of the different soaps are shown in Table 1.

Water is an important ingredient because the hardness and clarity of the finished soap are strongly dependent on its total moisture content. There are several sources of water in this formulation such as the caustic soda solution and the water generated during the neutralization reaction. The addition of free water to the bar formulation will also influence the final product. Generally, water addition of less than 5% total (not formed in situ or introduced by other ingredients) will usually result in a bar that is too hard and tends to form crystals with associated loss of clarity. Free water addition more than about

15% will usually result in a bar that is too soft. The present study showed that transparency was maintained when the moisture content was greater than 17% but not exceed about 21%. Higher moisture content in soap would lead to a reaction of excess water with unsaponified fat to give free fatty acid and glycerol in a process called hydrolysis of soap on storage.

Table 1. Physicochemical properties of transparent herbal soap

Parameters	Transparent herbal soap				
	Noni fruit	Yam root	Rose petal	Betel leaf	Control
Moisture content	17.69% dw	18.94% dw	20.98% dw	18.91% dw	20.23% dw
pH	9.31	9.22	9.47	9.65	9.18
Total fatty matter	37.75%	45.12%	39.17%	23.20%	36.80%
Free fatty acid	0.65%	2.15%	0.77%	0.12%	0.08%
Chloride content	14.38%	14.17%	21.81%	16.52%	59.12%

A correct pH range and the use of an adjusting agent are critical for achieving transparent soap bars from starting formulations. It has been discovered that adjusting the pH within a range of 9 to 9.5 will result in the desired end products. Obtaining a pH outside the preferred range will opacify the product. The high pH indicates a high percentage amount of unspecified and unsaponifiable matter due to incomplete alkaline hydrolysis. pH and free fatty acid content are critical balance to obtain an acceptable transparent product. Free fatty acid content and pH correlated with objective observations for transparency in the end products. Adding citric acid increased the free fatty acid content but decreased the pH. pH value of herbal transparent soap ranged from 9.22 to 9.65, while the free fatty acid content ranged from 0.12% to 2.15%. The transparency of end products was sustained when the pH did not drop under 9.1 and the free fatty acid content did not exceed 4.0%.

The total fatty matter is one of the most important characteristics describing the quality of soap and it is always specified in commercial soap. It is defined as the total amount of fatty matter, mostly fatty acids. Total fatty matter is usually associated with hardness and quality of soaps [21]. The soap with the higher total fatty matter gives more lather, lasts longer and, more importantly, cleans your skin better and more gently. Yam root transparent soap showed the highest total fatty matter (45.12%). While the lowest total fatty matter was shown by the betel leaf transparent soap (23.20%). These differences in the total fatty matter are possibly due to the imperfection of saponification process. The lower total fatty matter value may be caused by the presence of unreacted NaOH in the mixture [22].

The determination of percentage chloride levels in soap is important by means of excess quantity causes soaps to crack [23]. The chloride content of noni fruit, yam root, rose petal, and betel leaf transparent soap was 14.38%, 14.17%, 21.81%, and 16.52%, respectively. The cause of high chloride content in the soap is due to the use of chlorinated water to dissolve NaOH pellets [23].

3.2. Antimicrobial activity

The bacteria and fungus studied are clinically important ones causing several infections and it is essential to overcome them through some active therapeutic agents. Antimicrobial activities of herbal transparent soap were determined by measuring the diameter of the zone of inhibition and the mean values are presented (Table 2).

All herbal transparent soap and control at a concentration level of 10% and 20% exhibited antibacterial activity against all microorganism tested to a certain degree. However, all herbal transparent soap and control were inactive against *P. aeruginosa* and *C. albicans* at a concentration level of 1% and 5%. Noni fruit transparent soap at a concentration level of 20% exhibited the strongest antimicrobial inhibitory against *S. aureus* dan *P. aeruginosa*. Yam root transparent soap at a concentration level of 10% and 20% exhibited the strongest antimicrobial inhibitory against *C. albicans*, though the inhibition zone was relatively small (9.5 – 11.97 mm). Betel leaf transparent soap at all concentration level exhibited the strongest antimicrobial inhibitory against *E. coli*.

Table 2. Antimicrobial activities of transparent herbal soaps

Soap sample	Concentration (%)	Zone of inhibition (mm)			
		S. aureus	*E. coli*	*P. aeruginosa*	*C. albicans*
Control	1	15.91	13.02	0	0
	5	19.38	17.49	0	0
	10	20.93	18.9	9.9	9.49
	20	22.64	20.27	11.5	11.34
Noni fruit	1	15.68	15.87	0	0
	5	21.07	18.47	0	0
	10	23.72	19.65	10.77	10.09
	20	25.22	22.08	12.81	11.62
Yam root	1	16.95	14	0	0
	5	19.33	18.04	0	0
	10	23.26	19.38	10.8	9.5
	20	24.83	20.74	12.33	11.97
Rose petal	1	14.58	13.29	0	0
	5	19.47	17.7	0	0
	10	21.6	18.58	10.53	9.11
	20	22.5	20.55	10.88	10.38
Betel leaf	1	14.87	15.57	0	0
	5	19.3	20.3	0	0
	10	22.22	20.87	10.98	9.31
	20	24.56	23.9	12.75	10.96

S. aureus was the most susceptible and *C. albicans* was the most resistant among all the microorganism strains investigated in the present work. All herbal transparent soaps were more active against the *S. aureus* than the Gram-negative bacteria and even *C. albicans*. This comes to an agreement with previous reports that plant extracts are more active against Gram-positive bacteria than against Gram-negative bacteria [24, 25].

4. Conclusion

The present investigation demonstrates that noni (*Morinda citrifolia* L.) fruit, yam (*Pachyrizus erosus*) root, rose (*Rosa damascena*) petal, and betel (*Piper betle* L.) can be formulated into a transparent soap to add its functionality. Physicochemical properties and antibacterial activity of herbal transparent soaps were comparable to control soap.

5. Acknowledgments

The authors acknowledge Mr. Zamhari for technical assistance and the Indonesian Institute of Sciences for providing financial support.

6. References

[1] Atkinson N 1956 Antibacterial substances from flowering plants 3 Antibacterial activity of dried Australian plants by a rapid direct plate test *Australian J. Exper Biol* **34**: 17-26

[2] Duncan S H, Flint H J and Stewart C S 1998 Inhibitory activity of gut bacteria against Es cherichia coli 0157 mediated by dietary plant metabolites *FEMS Microbiol Lett* **164**: 283-58

[3] Bushnell O A, Fukuda M and Makinodian T 1950 The antibacterial properties of some plants found in Hawaii. *Pacific Science* **4**: 167-83

[4] Locher C P, Burch M T, Mower H F, Berestecky J, Davis H and Van Poel B et al 1995 Anti-microbiol activity and anti-complement activity of extract obtained from s elected Hawaiian medicinal plants *J. Ethnopharm* **49**: 23-32

[5] Andoğan B C, Baydar H, Kaya S, Demirci M, Özbaşar D and Mumcu E 2008 Antimicrobial activity and chemical composition of some essential oils *Arch Pharm Res* **25**: 860 – 864

[6] Lisin G, Safiyev S and Craker L E 1999 Antimicrobial activity of some essential oils *Acta Horticulturae (ISHS)* **501**: 283–288

[7] Gochev V, Wlcek K, Buchbauer G, Stoyanova A, Dobreva A, and Schmidt E et al 2008 Comparative evaluation of antimicrobial activity and composition of rose oils from various geographic origins, in particular Bulgarian rose oil *Nat Prod Commun* **3**: 1063 – 1068
[8] Etschmann M M W, Bluemke W, Sell D and Schrader 2002 *J.* Biotechnological production of 2-phenylethanol. *Appl Microbiol Biotechnol* **59**: 1 – 8
[9] Ali I, Khan F, Suri K, Gupta B, Satti N, Dutt P, Afrin F, Qazi G and Khan I 2010 In vitro antifungal activity of hydroxychavicol isolated from P. betle L. *Annals of Clinical Microbiology and Antimicrobials* **9(1)**: 7
[10] Siddiqui M, Sakinah M, Ismail A, Matsuura T and Zularisam A 2012 The anti-biofouling effect of *P. betle* extract against Pseudomonas aeruginosa and bacterial consortium. *Desalination J.* **288**: 24-30
[11] Pradhan D, Suri K, Pradhan D and Biswasroy P 2013 Golden heart of the nature: P. betle L. *J. Pharmacognosy and Phytochemistry* **6**: 147-167
[12] Rekha V, Kollipara M, Gupta S, Bharath Y and Pulicherla K 2014 A review on P. betle L: Nature's promising medicinal reservoir. *American Journal of Ethnomedicine* **1(5)**, 276-289
[13] Ramji N, Iyer R and Chandrasekaran S 2002 Phenolic Antibacterials From Piper betle In The Prevention Of Halitosis. *J. of Ethnopharmacology* **83(1)**, 149-152
[14] Valle D, Andrade J, Puzon J, Cabrera E and Rivera W 2015 Antibacterial activities of ethanol extracts of Philippine medicinal plants against multidrug-resistant bacteria *Asian Pacific Journal of Tropical Biomedicine* **5(7)**: 532-540
[15] American Oil Chemists' Society (AOCS). 1997. *Official and Recommended Practices of the AOCS* 7th Ed
[16] Perez C, Paul M and Bazerque P 1990 An antibiotic assay by the agar-well diffusion method. *Acta Biol Med Exp* **15**: 113-115
[17] Nair R and Chanda S 2005 Anticandidal activity of Punica granatum exhibited in different solvents. *Pharm Biol* **43**: 21-25
[18] Chalmers L and Bathe P 1978 *Chemical Specialist, Domestic and Industrial* **2nd** Ed. 1-15
[19] Ainie K, Hamirin K and Peang-Kean L 1996 Chemical and Physical Characteristics of Soap Made from Distilled Fatty Acids of Palm Oil and Palm Kernel
[20] Roila A, Salmiah A and Razmah G 2001 Properties of Sodium Soap Derived from Palm-Based Dihydroxystearic Acid. *Journal of Oil Palm Research* **1**: 33-38
[21] J. David Anneken, Sabine B, Ralf C, Georg F, Udo S and Alfred W 2006 "Fatty Acids" in Ullmann's Encyclopedia of Industrial chemistry, Wiley- VCH, Weinheim
[22] Roila A, Salmiah A and Razmah G 2001 Properties of Sodium Soap Derived from Palm-Based Dihydroxystearic Acid *J. of Oil Palm Research* **1**: 33-38
[23] Taiwo A, Oluwadare, I, Shobo A and Amolegbe S 2008 Physical and Chemical Characteristics of Soap *Scientific Research Essay* **3**: 515-517
[24] Vlietinck A J, van Hoof L and Totte J et al 1995 Screening of hundred Rwandese medicinal plants for antimicrobial and antiviral properties *J. Etyhnopharmacol* **46**: 31-47
[25] Rabe T and Van Staden J 1997 Antibacterial activity of South African plants used for medicinal purposes *J. Ethnopharmacol* **56**: 81-87

IOP Conf. Series: Earth and Environmental Science **251** (2019) 012025 doi:10.1088/1755-1315/251/1/012025

Diversification derivatives product of ripe banana: banana leather

R Ekafitri[1], N K I Mayasti[1] and N Afifah[1]

[1] Development Center for Appropriate Technology, Indonesian Institute of Sciences Subang, Indonesia

E-mail: riyantiekafitri@yahoo.com

Abstract. The banana leather is one of the ripe banana derivatives in sheet form. This product is categorized as a snack food. The purpose of this study was to determine the effect of the addition of egg and cheese on chemical properties and panelist preference of the banana leather product. Three samples of banana leather were prepared, i.e. control treatment, egg addition treatment, and cheese addition treatment. The samples were analyzed to assess the chemical properties, which included the water, ash, fat, protein, carbohydrate, calcium, and phosphorus contents. Then, an organoleptic assessment was done to assess the panelists' preference of the color, aroma, taste, crispness, and overall preference. The addition of egg whites and cheese increased the nutritional value of the banana leather products, particularly in the ash, fat, and protein contents, and the total energy. The ash content of control was 3.49% and then increased to 3.9-5.28% with the addition of egg white and cheese. Meanwhile, the fat content increased from 0.54% to 0.77-12.95%, the protein content rose from 4.11% to 9.52-11.7%, and the total energy rose from 363.42 kcal/100g to 364.97-410.37kcal/100g. The addition of cheese on the banana leather product also increased the calcium and phosphorus content to rech 4578 mg/100g and 5947.98 ppm, respectively. However, it reduced the carbohydrate content to 6.72%. in the organoleptic assessment, the addition of egg whites on the banana leather product increased all parameters, which was aroma, color, crispness, crispness, taste and overall preference, with highest overall preference score of 5.93.

Keywords : banana leather, chemical properties, panelist preference

1. Introduction

Banana (*Musa paradisiaca*) is a horticulture commodity with high potential for food diversification, food security, and agribusiness in Indonesia. Bananas contain carbohydrate, mineral, and fiber that meet the requirements of food commodities and food diet. The number of harvested bananas often excessive during the harvest season. One method to address this issue is by processing banana into flour, which also aims to extend the shelf life of the banana.

The preparation of banana flour from unripe banana has been informed by Rodriguez-Ambriz *et al.* Vatanasuchart *et al.*, Daramola and Osanyinlusi, Anyasi *et al.*, while the observation of banana flour in various ripening stages has been reported by Bugaud *et al.*, Egbebi and Bademosi, Abbas *et al* [1-7]. Meanwhile, the making of ripe banana has been informed by Abbas *et al* [8]. Unripe banana flour has high dietary fiber and resistant starch which have many benefit for human's health [9-10]. The production of ripe banana flour can propose new products with a standardized nutrient composition for domestic and industrial uses. The advantages of ripe banana flour include high sugar content that is suitable for incorporation into food products requiring sweetness, solubility, and high energy content. Ripe banana flour has a good banana flavor compared to the unripe one. Meanwhile, ripe banana flour has low yield because of low starch content, high-cost production, and still limited application. So that required the development of other products based on ripe bananas. Another innovative product base on the ripe banana is banana leather.

The banana leather is one of the ripe banana derivatives in sheet form. This product is categorized as a snack food. The similar product that has been already on the market is made from seaweed that is known as *nori* and other snack food in sheet form that is called crepe [11]. The crepes made from various flours, edible oil, skim milk, and egg flour [12]. In this study, banana leather was made from ripe banana flour mixed with egg and cheese. The addition of egg and cheese aimed to improve the nutritional value of the banana leather as the ripe banana flour has high sugar content but low protein content. The ripe banana flour had 3.15% protein and 12.8% sugar [6]. Tricalcium phosphate, an anti-caking agent, was also added to prevent the formation of lumps in the banana flour due to the high sugar content with hygroscopic nature. The purpose of this study was to determine the effect of the addition of egg and cheese on chemical properties and panelist preference of the banana leather product.

2. Materials and Methods

2.1. Raw materials

The banana leather was formulated with these following ingredients: banana puree, egg white, cheese, and tricalcium phosphate. The banana puree was prepared from ripe banana (Musa paradisiaca sp. AAA group, dessert subgroup that called locally as '*Pisang Ambon Hijau*'). Other ingredients were obtained from the local market. This research use chemical reagents with an analytical grade.

2.2. Banana leather preparation

Three samples of banana leather were prepared viz. control treatment, egg addition treatment, and cheese addition treatment. The control treatment was made by making banana puree using a chopper, mixing banana puree with tricalcium phosphate using a hand mixer (Philips Mixer, HR Model 1538) for 5 min at speed 4 (1200 rpm), pouring the puree dough on a plastic coated sheet, flattening the dough on baking sheet, drying 55 °C for 9 hours, cutting the sheet into 7 x 10 cm samples, second drying for 15 hours in 55 °C, and packaging. The second treatment was made by making egg foam using a hand mixer (Philips Mixer, HR Model 1538) for 5 min at speed 4 (1200 rpm) and making puree of banana with a chopper, mixing banana puree, egg foam, and tricalcium phosphate, then pouring the puree dough on a plastic coated sheet, baking dough on the pan, drying 55 °C for 9 hours, cutting the sheet into 7 x 10 cm samples, second drying for 15 hours in 55 °C, and packaging. The third sample was made by making banana puree using a chopper, mixing banana puree, grated cheese, and tricalcium phosphate using a hand mixer (Philips Mixer, HR Model 1538) for 5 min at speed 4 (1200 rpm), pouring the puree dough on a plastic coated sheet, flattening the dough on baking sheet, drying 55 °C for 9 hours, cutting the sheet into 7 x 10 cm samples, second drying for 15 hours in 55 °C, and packaging. Drying was done using a cabinet dryer. Before analysis, the samples were stored at freeze temperature in sealed aluminum foil bags.

2.3. Analyses of banana leather

2.3.1. Chemical analysis. Chemical analysis in this study including: proximate, energy total, calcium, and phosphorus content. Proximat analysis referred to National Standardization Agency of Indonesia (SNI 01-2891-1992) regarding analytical methods for food and beverage, which includes proximate (moisture, fat, protein, and ash contents), and carbohydrate content was determined using 'by difference' method [13]. Energy total was obtained following Ekafitri *et al.* and determined by calculation using Atwater factors according to the composition of carbohydrates, fats, proteins of these foods [14]. The energy total of banana leather was determined by adding the multiplication results between the Atwater carbohydrate (4 Kcal), protein (4 Kcal), and fat (9 Kcal) factors to the levels of carbohydrates, proteins, and fats in product. Calcium and phosphorus content were determined by using ICP-EOS (Inductive Couples Plasma-Optical Emission Spectrometry) Agilent type 720,USA.

2.3.2. Organoleptic assessment. Organoleptic assessment was performed by scoring method in which the samples were given to 30 panelists to be assessed in particular scores within the range of his or her favorite levels, which comprised of strongly dislike (1), dislike (2), moderately dislike (3), neutral (4), moderately like (5), like (6), and really likes (7). Assessment parameters were five quality criteria of color, aroma, taste, sweetness, and overall preference.

2.3.3. Statistical analysis. This research used a completely randomized design (CRD) to evaluate the effect of egg white and cheese addition on the banana leather products. This study used a factorial design in completely randomized design. Factors tested were the type of filler, which comprised of three levels: control, 20% egg white froth addition, and 20% cheese addition. Each treatment was repeated twice. Variance analysis of data of each treatment was carried out on a statistical package SPSS. The significant difference among mean values was analyzed using one-way analysis of variance (ANOVA), and then used Duncan's test at a significance level of ($P < 0.05$).

3. Result and Discussion

3.1. Proximate analysis

The proximate content of three samples of banana leather includes moisture, ash, protein, fat, carbohydrate contents are displayed in Table 1.

Table 1. The proximate content of three samples of banana leather.

Parameter	Control	Egg addition	Cheese addition
Water content (%)	6.33^{ab}	5.76^{a}	8.32^{b}
Ash content (%)	3.49^{a}	3.96^{b}	5.28^{c}
Total fat (%)	0.54^{a}	0.77^{a}	12.95^{b}
Protein (%)	4.11^{a}	9.52^{b}	11.74^{c}
Total carbohydrate (%)	85.54^{a}	79.88^{b}	61.72^{c}

*values with the same superscript in the same row had no significant difference on 0.05 significance.

The water content of the three banana leather samples ranged from 5.76 to 8.32%, whereas the banana leather samples made with the addition of egg whites were significantly lower than other treatment ($p < 0.05$). The addition of egg whites speeds up the evaporation of water so that in the same drying time, the banana leather sample with the addition of egg whites has lower water content. The water contents of the three samples were less than 10% for dry products, which means that all samples were safe from damage. Winarno stated that the safe water content of food for storage is less than 14%.[15]

The ash content of banana leather products ranged between 3.49-5.28%, and the ash contents of the three products were significantly different at $p < 0.05$. Table 1 shows that the addition of egg white and cheese increased the ash content of banana leather. The addition of egg whites increased the ash content to 3.96%, and cheese treatment increased the ash content to 5.28%. This increase in ash content can be caused by the mineral content in egg white and cheese. Egg whites contain calcium, magnesium, phosphorus, sodium, and zinc and cheese contain minerals such as calcium, iron, magnesium, sodium, phosphorus, and zinc with different compositions depending on the type of milk used in cheese making and cheese ripening process [16-17]. Meanwhile, Wall explained that bananas contain minerals such as calcium, phosphorus, magnesium, sodium, and high concentration of potassium that reaches 2.7% [18-19]. Ash content roughly describes the mineral content contained in a food product. Ash is a carbon-free residue after the food is burned. The higher the ash content of a product, the higher the minerals contained in it. Thus, the addition of egg white and cheese increased the ash content of the product.

The total fat content of banana leather products ranged from 0.54 to 12.95% (Table 1). The fat content of banana leather control (0.54%) and the samples with egg whites addition (0.77%) was not significantly different (p>0.05), but the fat contents of these two treatments were significantly different from the fat content of the sample with cheese addition (12.95%) at p <0.05. The addition of cheese increases the fat content of the product. According to the Food Standard Agency, cheese has a fat content of 32 g/100g of material, which is higher than the fat content of egg white that reaches 11.5% [20]. Hence, the addition of cheese increased fat content higher than the egg white at the product.

The protein content of banana leather products ranged from 4.11 to 11.74% (Table 1). The protein contents of the three samples were significantly different at p <0.05. The samples with egg white and cheese addition had higher protein levels. The addition of egg whites increased the protein level to 9.52%. Egg whites contain 12.8% protein [20]. The addition of cheese raised the protein level to 11.74%. According to the Food Standard Agency cheese contains 25.4% protein [21]. Egg whites and cheese are food sources of protein. In this study, the addition of the two ingredients aimed to increase the protein content of banana leather, and the results showed that the objective had been met.

The total carbohydrate content of banana leather products ranged from 61.72 to 85.54% (Table 1). The carbohydrate content of the control was larger (85.54%) and significantly different from the samples with the egg whites (79.88%) and cheese (61.72%) addition. This is due to the decreased percentage of ripe bananas in the total dough with the addition of egg whites and cheese. Ripe bananas are a source of carbohydrates. According to Sogo *et al.,* ripe bananas contain 24.57% carbohydrates [22]. Even though egg whites and cheese are high-protein foods, they only have a slight amount of carbohydrates. The carbohydrate content of egg whites and cheese are 0.7% and 0.1%, respectively [22-23].

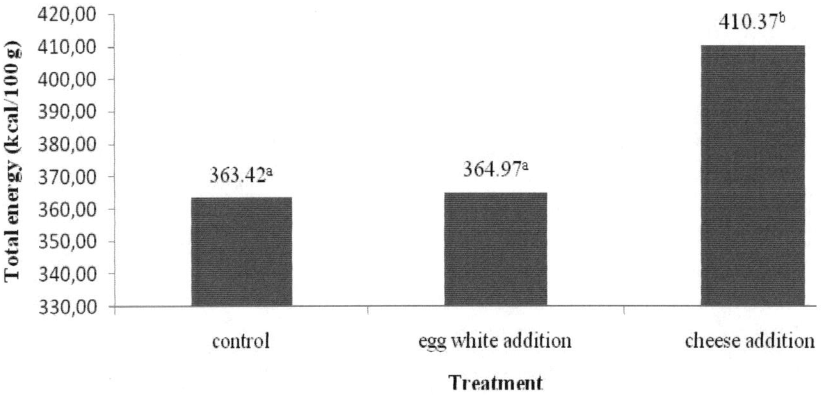

Figure 1. The total energy of the banana leather product.

3.2. Total energy

Based on carbohydrate, protein, and fat contents, the total energy of the banana leather product was then calculated. Total energy was obtained by calculating the value of protein content multiplied by 4 kcal, carbohydrate content multiplied by 4 kcal, and the fat content multiplied by 9 kcal. The results exhibit that the total energy of the banana leather product ranged from 363.42-410.37 kcal/100g (Figure 1). Figure 1 shows that the total energy of banana leather control was not significantly different (p> 0.05) with the sample with egg whites addition and significantly different (p <0.05) with the sample with cheese addition. This could be caused by the fat and protein content of the sample with cheese addition was higher than the other two treatments.

3.3. Mineral content (calcium and phosphorus)

The calcium content of banana leather products ranged from 299.87-457.86 mg/100g (Figure 2). Treatment with cheese addition was significantly higher than control and egg white addition treatment ($p<0.05$). The highest calcium content was obtained in the sample with cheese addition that reached 457.86 mg/100g because the cheese contains calcium of 739 mg/100g, while egg whites only contain 6 mg/100g calcium and banana only has 3.8-9.7 mg/100g calcium [18,20]. The calcium content in the banana leather products also came from the anti-caking agent, which is tricalcium phosphate with 0.25% concentration.

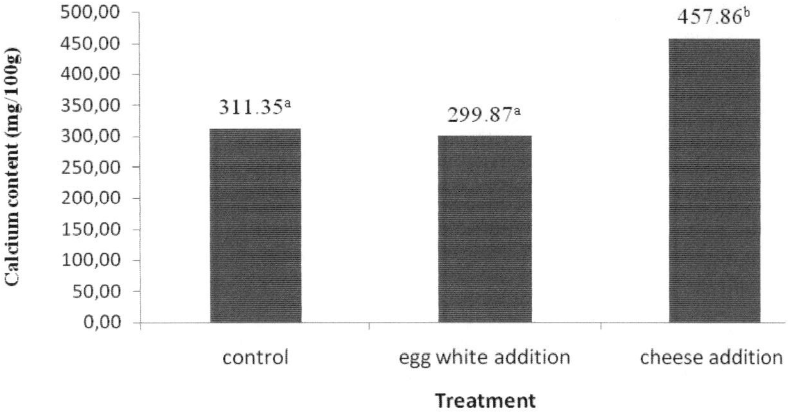

Figure 2. Calcium content of banana leather.

The phosphorus content of banana leather products ranged from 1875.45 – 5947.98 ppm (Figure 3). The phosphorus content of the control and was not significantly different ($p> 0.05$) with the sample with the egg whites addition but significantly different ($p <0.05$) with the sample with cheese addition. The highest phosphorus content was obtained in the sample with cheese addition that reached 5947.98 ppm. Abubakar and Usmiati reported that low-fat white cheese made from modified milk contained phosphorus ranging from 346.62 - 832.15 mg/100g, while egg contains 0.18 mg/100g [23-24].

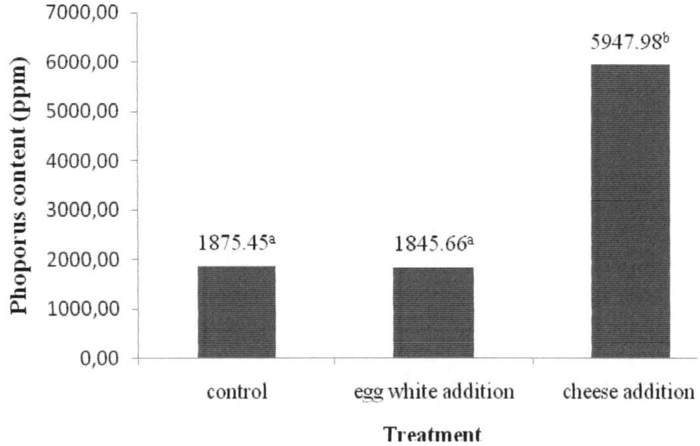

Figure 3. The phosphorus content of banana leather.

3.4. Organoleptic assessment of the banana leather products

The organoleptic assessment of banana leather products was tested by a hedonic rating test of taste, aroma, color, crisp, and overall acceptance parameters with average values are displayed in Figure 4.

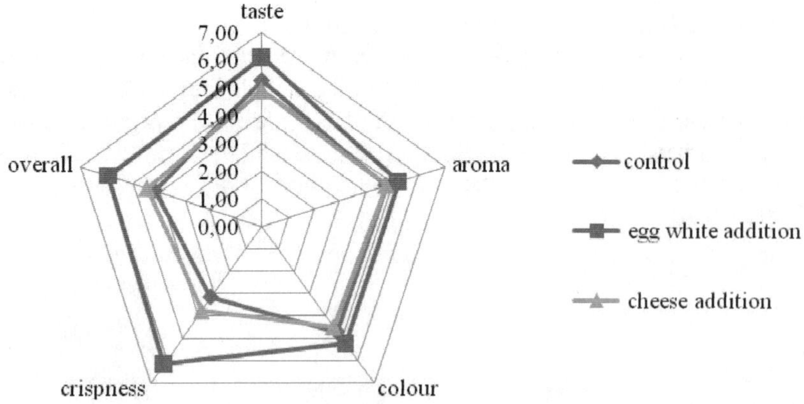

Figure 4. The spider web of the average scores of consumers' preference on various parameters of the banana leather products.

The average of the taste score of banana leather products ranged from 4.90 to 6.13. Figure 1 shows that the sample with egg whites addition had the highest average (6.13) and was significantly different ($p < 0.05$) from the averages of the control (5.27) and the sample with cheese addition (4.90). The sample with cheese addition had the lowest average of the taste score that could be caused by the salty and savory flavor of cheese addition disguised the taste of bananas. The 20% addition of cheese was probably higher than what the panelists favored. Chambers *et al.* explain that cheese had flavors associated with dairy-related attributes (dairy fat, dairy sour, dairy sweet), fundamentally tasted (bitter, salty, sour, and sweet) sharp and astringent, which were found in every type of cheese tested [25] Cheese also has a different flavor depending on the type.

The average of the aroma score of banana leather products ranged from 4.77 to 5.27. The sample with egg whites addition had the highest average (5.23), but it was not significantly different ($p > 0.05$) from the averages of the control (4.77) and the sample with cheese addition (4.77). This suggests that the addition of egg whites and cheese can maintain the banana aroma on the banana leather products. Facundo *et al.* stated that the main factors that affect the banana consumption were aroma and taste [26]. Chemically, the aroma and taste of bananas are caused by the presence of volatile components received by the olfactory receptors. More than 150 volatile components are present in bananas, especially isoamyl and isobutyl ester groups together with 2-pentanone [27].

The average of the color score of banana leather products ranged from 4.50 to 5.27. As shown in Figure 1, the sample with egg whites addition had the highest average (5.27). The sample with cheese addition was significantly lower than control and egg white addition sample. This suggests that the addition of egg whites increased the preference for color while on the contrary, the addition of cheese decreased it. The addition of egg whites produced a brighter banana leather than the controls while the addition of cheese produced banana leathers with a darker color. This could be caused by the higher protein content in cheese so that when drying occurred, protein denaturation resulting the darker color of banana leather products.

The average of the crispness score of banana leather products ranged from 3.17 to 6.17. Figure 1 displays that the sample with egg whites addition had the highest average (6.17), and it was significantly different from the averages of the control (3.17) and the sample with cheese addition (3.77). This suggests that the addition of egg whites increased the preference for crispness. This could be caused by the porous nature due to the addition of egg whites. The drying process with foaming produces better product reconstruction and quality than the dried product without foaming [28]. The drying process with foaming created a product that was more porous and preferred by consumers.

IOP Conf. Series: Earth and Environmental Science **251** (2019) 012025 doi:10.1088/1755-1315/251/1/012025

The average of the overall preference for banana leather products ranged from 4.17 to 5.93. The sample with egg whites addition was significantly higher than control and the sample with cheese addition. This suggests that the addition of egg whites increased the overall preference compared with the other two treatments. The addition of egg whites on the banana leather product increased all parameters of the organoleptic assessment, which was taste, aroma, color, crispness, and overall preference.

4. Conclusion

The addition of egg whites and cheese increased the nutritional value of the banana leather products, particularly in the ash, fat, and protein contents, and the total energy. The ash content of the control was 3.49%, then increased to 3.96-5.28% with the addition of egg whites and cheese. Meanwhile, the fat content increased from 0.54% to 0.77-12.95%, the protein content rose from 4.11% to 9.52-11.74%, and the total energy rose from 363.42 kcal/ 100g to 364.97-410.37 kcal/100g. The addition of cheese on the banana leather product also increased the calcium and phosphorus contents to reach 457.86 mg/100g and 5947.98 ppm, respectively. However, it reduced the carbohydrate content to 61.72%. In the organoleptic assessment, the addition of egg whites on the banana leather product increased all parameters, which were aroma, taste, color, crispness, and overall preference, with the highest overall preference score of 5.93. Therefore, it can be concluded that the addition of cheese improved the nutritional value of the banana leather product, but the sample with the egg whites addition was more preferred by the panelists.

5. References

[1] Rodrı´guez-Ambriz S L, Islas-Herna´ndez J J, Agama-AcevedoE, Tovar J and Bello-Pe´rez L A 2008 Characterization of a fiber-rich powder prepared by liquefaction of unripe banana flour *Food Chem.* **107** 1515–1521

[2] Vatanasuchart N, Boonma N, and Karuna W 2012 Resistant starch content, in vitro starch digestibility and physicochemical properties of flour and starch from Thai bananas. Maejo *Int. J. Sci. Techno.* **6** (2) 259-271

[3] Daramola B, and Osanyinlusi S A 2006 Production, Characterization, and application of banana (Musa spp) flour in whole maize *Afr. J. Biotechnol.* **5** (10) 992-995

[4] Anyasi TA, Afam IOJ and Godwin RAM 2015 Effect of organic acid pretreatment on some physical, functional and antioxidant properties of flour obtained from three unripe banana cultivars. *Food Chem.* **172** 515–522

[5] Bugaud C, Chillet M, Beaute M P, and Ce´cile, S 2006 Physicochemical analysis of mountain bananas from the French West Indies *J. Scienta.* **108** 167-172

[6] Egbebi A O and Bademosi T A 2011 Chemical compositions of ripe and unripe banana and plantain *Int. J. Trop. Med. Public Health* **1** 1-5

[7] Abbas F M A, Saifullah R, and Azhar M E 2011 Differentiation of ripe banana flour using mineral composition and logistic regression model *Int. Food Res. J.* **16** 83-87

[8] Abbas F M A, Saifullah R, and Azhar M E 2009 Assessment of physical properties of ripe banana flour prepared from two varieties: Cavendish and Dream banana *Int. Food Res. J.* **16** 183-189

[9] Faisant N, Gallant D J, Bouchet B, Champ M 1995 Banana starch breakdown in the human small intestine studied by electron microscopy *Eur. J. Clin. Nutr.* **49** 98-104

[10] Juarez G E, Agama-AcevedoE, Sayago-Ayerdi S G, Rodriguez-Ambriz S L, and Bello-Perez L A 2006 Composition, digestibility and application in breadmaking of banana flour *Plant Foods Hum. Nutr.* **61** 131-137

[11] Xin X S, Fang Yong, Yang J, Yang F M, Huqiu H, Pei F, Zhao L, Xin Z, Gu J, Gu W, Gao J 2012 Method for producing instant flavored seaweed sheets by utilizing microwave vacuum drying technology *Chinese Patent* CN 102613611

[12] Lawrence B and Markay M B 2011 Crepe mix *American Patent* CA 2433693 C

[13] The National Standardization Agency of Indonesia 1992 SNI No 01-2891-1992 *Cara uji makanan dan minuman*

[14] Ekafitri R, Novita I, Rima K and Doddy A D 2012 Pengaruh Penambahan Tepung Komposit Pada Pembuatan Mie Instan Jagung Terhadap Nilai Gizi *Jurnal Pangan* **21** (4) 363-374

[15] Winarno F G 2008 *Kimia Pangan* (Jakarta : PT. Gramedia Putaka Utama)

[16] Bologa M, Pop I M and Aida A 2013 Research On Chemical Composition Of Chicken Egg From Different Systems Of Production (Conventional And Organic) *Lucrări Ştiinţifice-Seria Zootehnie* **59** 80-85

[17] Martin I G, Hernadez-Hierro J M, Revilla I, Vivar-Quintana A, Lobos-Ortega I and González-Pérez C 2009 Changes in the Mineral Content in Cheeses of Different Compositions during 6 Months of Ripening *Czech J. Food Sci.* **27**

[18] Wall M M 2006 Ascorbic acid, vitamin A, and mineral composition of banana (Musa sp.) and papaya (Carica papaya) cultivars grown in Hawaii *J. Food Compos. Anal.* **19** 434–445

[19] US Department of Agriculture, US Department of Health and Human Services (USDA/HHS), 2004-2005 *Dietary Guidelines Advisory Committee Report.* /http://www.health.gov/dietaryguidelines/dga2005/reportS

[20] Directorate of Nutrition 1996 *Daftar Komposisi Bahan Makanan* (Jakarta: Bhatara Karya Aksara)

[21] Food Standard Agency 2002 *The Composition of Foods* 6 ed, (Cambridge: Royal Society of Chemistry)

[22] Sogo T C M, Idowu O A, Idowu E 2014 Effect of Biological and Chemical Ripening Agents on the Nutritional and Metal Composition of Banana (Musa spp). *J. Appl. Sci. Environ. Manage.* **18** (2) 243 -246

[23] Abubakar and Usmiati S 2016 Mutu Keju Putih Rendah Lemak Diproduksi Dengan Bahan Baku Susu Modifikasi *Buletin Peternakan* **40** (2) 144-156

[24] Miranda J M, Anton X, Redondo-Valbuena C, Roca-Saavedra P, Rodriguez J A, Lamas A, Franco C M, and Cepeda A 2015 Egg and Egg-Derived Foods: Effects on Human Health and Use as Functional Foods *Nutrients* 7 706-729

[25] Chambers D H, Edgar C I V, and Dalas J 2005 Flavor description and classification of selected natural cheeses *Culinary Art and Science* 641-654.

[26] Facundo H V de V, Deborah S G, Beatriz R C, and Franco M L 2013 Isolation of volatiles compounds in banana by HS-SPME: Optimization for the Whole Fruit and Pulp *Int J Biosci, Biochem and Bioinforma* **3** (2) 110-115

[27] Jordan M J, Kawaljid T, Philip E S, and Kevin L G 2001 Aromatic profile of aqueous banana essence and banana fruit by gas chromatography-mass spectrometry (GC-MS) and gas chromatography-olfactometry (GC-O) *J. Agric. Food Chem.* **49** 4813-4817

[28] Rajkumar P, Kailappan R, Vishwanathan R, and Raghavan G S V 2007 Drying Characteristics of Foamed Alphonso Mango Pulp in a Continuous Type Foam Mat Drier *J Food Eng* **79** (4) 1452- 1459

Evaluation of Physicochemical, Functional and Pasting Properties of Soybean, Mung Bean and Red Kidney Bean Flour as Ingredient in Biscuit

L Ratnawati, D Desnilasari, D N Surahman and R Kumalasari

Development Centre for Appropriate Technology, Indonesian Institute of Sciences, Subang, Indonesia

Email : lia.romeo@gmail.com

Abstract. This present study was aimed to evaluate the physicochemical, functional and pasting properties of soybean, mung bean and red bean flour. The physicochemical (moisture, ash, protein, fat, carbohydrate, energy, starch, total dietary fiber and color), functional properties (swelling power, solubility, water holding capacity, oil holding capacity and bulk density) and pasting properties (peak viscosity, breakdown viscosity, setback viscosity, final viscosity and pasting temperature) of flours were evaluated. Soybean flour had highest values of fat, ash, protein content, solubility, and oil holding capacity while the lowest value for starch, carbohydrate, swelling power and peak, breakdown, final and setback viscosity. Mung bean flour had highest values of swelling power and peak viscosity, while the lowest value for ash content, solubility, water holding capacity, and total dietary fiber. Red bean flour had the highest starch, carbohydrate, water holding capacity, total dietary fiber, breakdown and setback viscosity, while the lowest value for ash, fat and protein content. From the three types of legumes were analyzed in this study, mung bean flour can be the best choice for wheat flour substitution as ingredient biscuit because it has high swelling power and peak viscosity, and low setback, breakdown viscosity and pasting temperature.

Keywords: physicochemical properties, functional properties, pasting properties, soybean flour, mung bean flour, red kidney bean flour

1. Introduction

Legumes are edible seeds and belong to the family Leguminosae and widely cultivated in tropical and subtropical countries such as Indonesia. Legumes seeds are prime importance in human and animal nutrition, due to their high protein content (20-50 %) [1]. Some types of legumes that are widely grown in Indonesia include soybean, mung bean and red bean. Production of soybean, mung bean and red bean in Indonesia includes 538,253 tons in 2017 [2], 271,463 tons in 2015 [3] and 100,316 tons in 2014 [4].

Processing legume into flour is a simple food processing and can be used as a raw material for biscuits. Biscuits are one of the dried food products that are made by baking dough made from flour or its base ingredients, oil or fat with or without the addition of other permitted foods. The minimum protein content of the required biscuits is 5%. One alternative of vegetable ingredients to replace or substitute flour is legume flour because of its relatively high protein content. Before being used as an ingredient in making biscuits, it is important to know the physicochemical, functional and pasting properties of each type of legume flour.

Functional properties are fundamental physicochemical properties that reflect the complex interactions between composition, structure, molecular conformation and physicochemical properties of the components together with the associated environment. Functional characteristics are needed to evaluate and predict how protein, fiber, fat, and carbohydrates play a role in the food system [5]. Several functional properties of food, especially flour may include swelling power, solubility, water or oil absorption capacity and bulk density.

In addition to physicochemical and functional properties, the pasting properties are also important to learn. Testing of pasting properties serves to determine the sensory characteristics of a better product and determine the use in various foods. Pasting property can also be exchanged for anything that might be expected during food processing. Several previous studies related to physicochemical and functional properties of legume flour tests have been conducted by Du et al. [7]. However, previous studies did not discuss the characteristics of soybean flour, and the application of legume flours on the food product. This study aimed to evaluate the physicochemical, functional and pasting properties of soybean, mung bean, and red kidney bean flour as an ingredient in a biscuit.

2. Materials and Methods

2.1. Raw materials
The soybean (*Glycine max*), mung bean (*Vigna radiata* L.) and red kidney bean (*Phaseolus vulgaris* L.) were obtained from a local market in Subang and Lembang, West Java.

2.2. Flour preparation
Soybean, mung bean and red kidney bean were washed and soaked using water at 60-70°C for 3 h. After that, the legumes were dehulled and dried at 50°C, 12 h. Then, the samples were reduced particle size using disk mill and sieved using a 40 mesh sieve. Samples were kept into sealed plastic and stored until further use at 4°C.

2.3. Flour analysis

2.3.1. Physicochemical properties. Physicochemical analysis of legumes flour was evaluated such a proximate, energy, starch, amylose, amylopectin, protein digestibility, total dietary fiber, phytic acid, and color analysis. Proximate analysis was evaluated following Indonesian National Standard [8] procedures and included the following: moisture content, ash content, crude fat by Soxhlet extraction. Protein content was analyzed using DuMaster protein analyzer (DuMaster D-480, Buchi, Switzerland). Total carbohydrate was calculated from the difference (100-(% moisture + % ash + % protein + % fat)). Total dietary fiber content of flour was performed using a combination of enzymatic and gravimetric method [9]. TDF was calculated by equation (1) :

$$TDF\ (\%) = \left[\frac{weight\ residue - weight\ protein - weight\ ash}{weight\ sample}\right] x\ 100 \qquad (1)$$

The color of flours was measured using colorimeter (NH310, China). Each sample was measured three replicates. The result determined were L* value indicates the lightness, 0-100 representing dark to light, a* value indicates the degree of redness to greenness and b* value indicates the degree yellowness to blueness.

2.3.2. Functional properties. Functional properties of legumes flour were analyzed such as swelling power, solubility, water absorption capacity, oil absorption capacity and bulk density. Swelling power and solubility were determined by following the method described by Pranoto et al. [10] with slight modification. A sample (0.2 g) was put in a pre-weighed centrifuge tube, added with 10 mL distilled water and mixed using vortex. The sample was allowed to stand at room temperature (20-25°C) for 5 min and then put in a water bath at 95°C for 30 min. After that, it was cooled at 20-25°C for 10 min. The sample was centrifuged at 3000 rpm for 15 min to separate gel and supernatant. The gel after separating from supernatant was expressed as swelling power. While the supernatant is placed on a plate that has been known to weigh and then dried in an oven to a constant weight. Swelling power (2) and solubility (3) were calculated by using following equation:

$$Swelling\ power\ (g/g) = \frac{W2 - W1}{W0} \qquad (2)$$

Where W_0 = weight of the dry sample (g), W_1 = weight of the dry sample + centrifuge tube (g), W_2 = weight of gel + centrifuge tube (g)

$$Solubility\ (\%) = \frac{W2}{W1}\ x\ 100\% \qquad (3)$$

Where W_1 = weight of the dry sample (g) and W_2 = dry weight of supernatant (g)

Water absorption capacity (WAC) and oil absorption capacity (OAC) were determined by following the method described by Brishti et al. [11] with slight modification. The 0.25 g of flour was mixed with 5 mL distilled water or oil in pre-weighed centrifuge tube for 30 secs using a vortex. Then, sample was allowed to stand at room temperature (20-25°C) for 15 min and centrifuged at 3000 rpm for 15 min. After centrifugation, the supernatant was decanted, and the centrifuge tubes + precipitate were re-weighed. The WAC and OAC were expressed as grams of water/oil absorbed per gram of the sample. The WAC and OAC were calculated by using following equation (4) :

$$WAC\ or\ OAC\ (g/g) = \frac{W2}{W1} \qquad (4)$$

Where W_1 = weight of the dry sample (g) and W_2 = weight of precipitate + centrifuge tube (g)

Bulk density of flour was determined using the procedure of Butt and Batool [12], 10 g of samples were put into a 100 ml measuring cylinder and tapped to a constant volume and the bulk density (gcm-3) calculated using the formula: Bulk density = weight of flour (g) / flour volume (cm^3).

2.3.3. Pasting properties. Pasting properties of flours were analyzed using Rapid Visco Analyzer (RVA-TecMaster, Macquarie Park, Australia). A suspension of 3.5 g (14% wb) of flour in 25 g distilled water was stirred at 50°C (160 rpm) for 1 min, heated from 50 to 95°C for 7.5 min, held at 95°C for 5 min, cooled from 95 to 50°C for 7.5 min and held at 50°C for 2 min. the following data were recorded: peak viscosity, setback viscosity, breakdown viscosity, final viscosity and pasting temperature.

2.4. Statistical Analysis

The data reported in all of the tables are the averages of triplicate analysis. Significant differences for multiple comparisons were determined by one-way analysis of variance (ANOVA) followed by Duncan test by SPSS statistical package.

3. Result and Discussion

3.1. Physicochemical properties

The physicochemical properties of soybean, mung bean and red kidney bean flour are shown in Table 1.

The highest of moisture content was observed for red kidney bean flour (7.60 %) and the lowest for soybean flour (4.66 %). Whereas the difference between red kidney bean flour and other samples i.e. soybean and mung bean flour was significant ($p<0.05$) and between soybean and mung bean flour was not significant. The previous studies were reported that moisture content of mung bean, red kidney bean, and soybean flour are 8.30, 7.16 and 5.4 %, respectively [12; 13; 14].

The ash content of red kidney bean, mung bean, and soybean flour are 2.70, 2.70 and 4.23 %, respectively. There was significant difference ($p<0.05$) between soybean flour with mung bean and red kidney bean flour, but no significant difference between mung bean and red kidney bean flour. The mung bean flour in this study has ash content (2.70 %) similarly with the study was conducted by

Brishti et al. [11] that is 3.02%. While, the raw and boiled red kidney bean flour have ash content, i.e. 4.4 and 2.2 % [6]. The result of this study slightly different from the previous study was reported that the ash content of soybean flour is 3.60% [15]. Ash content can indicate the mineral content in food.

Table 1. Physicochemical properties of soybean, mung bean and red kidney bean flour

Physicochemical Properties	Types of flours		
	Soybean flour	Mung bean flour	Red kidney bean flour
Moisture, % wb	4.66^a	5.63^a	7.60^b
Ash, % wb	4.23^b	2.70^a	2.70^a
Fat, % wb	25.01	1.31	1.11
Protein, % wb	40.94^b	24.99^a	22.53^a
Carbohydrate, %wb	25.32^a	65.41^b	66.50^b
Energy, Kcal/100 g	488.77^a	373.01^b	365.88^c
Starch, % wb	8.66^a	55.18^b	55.46^b
Amylose, % wb	1.79^a	23.45^c	20.14^b
Amylopectin, % wb	6.87^a	31.74^b	35.33^c
Protein Digestibility, %	80.54^c	33.36^a	39.59^b
Total Dietary Fiber, % wb	9.89^b	6.57^a	23.80^c
Phytic Acid, % db	2.92^b	1.79^a	1.58^a
Color			
L (Lightness)	79.07^a	83.37^{ab}	86.74^b
a (Redness)	2.60^b	0.86^a	1.31^a
b (Yellowness)	19.61^c	13.51^b	6.79^a

Different superscript alphabets indicate significant differences among treatments ($p<0.05$)

The fat content of legume flours ranged from 1.11-25.01%. The soybean flour has highest fat content (25.01 %) when compared with other samples like mung bean (1.31%) and red kidney bean (1.11 %). The fat content of mung bean (1.31%) in the present study similarly with the previous study was conducted by Butt and Batool [12] and Brishti et al. [11], reported that the fat content of mung bean is 1.24 and 1.53%. According to Jahreis et al. [15] and Alamu et al. [14], soya flour had the fat content are 22.8 % and 17.56 %, lower than this study. While, based on previous study conducted by Pangastuti et al. [13], stated that the fat content of red bean flour is 8.13% higher than this study.

The results for protein content of legumes flour were evaluated higher amount (40.94 %) in soybean followed by mung bean (24.99 %) and red kidney bean (22.53 %). There was significant difference ($p<0.05$) between soybean flour with mung bean and red kidney bean flour, but no significant difference between mung bean and red kidney bean flour. Previously, Butt and Batool [12], reported that mung bean flour has 25.90 % protein. While, Audu and Aremu [6] reported that raw and boiled kidney bean flour has protein content, 15.3 and 23.6 %. The protein content of soybean flour in this study (40.94 %) is known to be higher than the previous studies (35.60 and 33.53 %) [15; 14]. The three types of legume flour in this study belong to foods that contain high protein. So that all three can be used as ingredients in a biscuit. Legume flour can be an alternative ingredient for making high protein in order to overcome protein deficiency.

Red kidney bean flour had the highest carbohydrate content (66.50%) than soybean (25.32 %) and mung bean flour (65.41 %). The difference between soybean flour and other samples i.e. mung bean and red kidney bean flour were significant ($p<0.05$). The previous study conducted by Dzudie and Hardy [16] reported that the carbohydrate of mung bean flour is 67.1%. Based on previous study conducted by Pangastuti et al. [13] and Audu-Aremu [6], the carbohydrate content of red bean flour was 58.48% and 49%.

The energy value of legumes flour ranged from 365.88-488.77 kcal/100 g. The highest calorie value is soy flour (488.77 kcal /100 g), and the lowest is red beans (365.88 kcal /100 g). There were significant differences ($p<0.05$) between all samples of legume flour. According to Adelakun et al. [17] reported that soybean flour had the energy value 446 kcal/100 g. The difference in chemical composition in this study with the previous studies can be influenced by genetic, variety and growth environment (geographical location and growing session) [18].

Red kidney bean flour contained the highest amount of starch (55.46 %), and lowest was found in soybean (8.66%). There was significant difference ($p<0.05$) between the starch content of soybean flour with mung bean and red kidney bean flour. Starch content of soybean flour based on previous study higher than present study, i.e. 14.85 % [14]. Starch in flours consists of amylose and amylopectin. Amylose and amylopectin ratios affect the pasting properties of flour or starch. Amylose content of legumes flour was ranged between 1.79-23.45%. The results showed that there were significant differences ($p<0.05$) in the amylose content between soybean, mung bean and red kidney bean flour. Amylopectin content was significantly different ($p<0.05$) among the legume flours and the ranged between 6.87-35.33%.

The protein digestibility of the legume flours significantly differed ($p<0.05$) from each other and ranged between 33.36-80.54% with soybean flour having the highest protein digestibility (80.54%) and mung bean flour having the lowest.

The values for total dietary fiber of red kidney bean, soybean, and mungbean flour were 23.80, 9.89 and 6.57 %, respectively. The differences between soybean and mung bean flour were not significant ($p>0.05$). However, there were significant differences between red kidney bean flour and other samples. High levels of dietary fiber in food can be useful in providing bulk to foods to relieve constipation as well as maintaining digestive tract health.

Phytic acid of legume flours was ranged between 1.58 until 2.78% with the highest phytic acid content is soybean flour, and the lowest is red kidney bean flour. Phytic acid content of red bean flour in the study conducted by Pangastuti et al. [13] was 1.75%.

The color value of soybean, mungbean and red kidney bean flour are shown in Table 1. The L* value of legume flours was ranged between 79.07-86.74. The difference between soybean and red kidney bean flour was significant ($p<0.05$). Among the legume flours studied, mungbean showed a higher L* parameter (83.37), indicating its lighter color than soybean (79.07) and red kidney bean (86.74). The* value of mung bean, red kidney bean, and soybean flour were 0.86, 1.31 and 2.60, respectively. The results showed that there were significant differences ($p<0.05$) in the* value between soybean and other samples, i.e., mung bean and red kidney bean flour. The* parameter of mungbean was lowest (0.86) than other samples, indicates that the presence of a slight green color in it. Red kidney bean in this study has a* value 1.31 and b* value 6.79. Based on the study was conducted by Pangastuti et al. [13], stated that the L*, a* and b* values of the red kidney bean flour are 87.42, 1.62 and 7.27, respectively. The b* value of legume flours was ranged between 6.79-19.61 and significant differences ($p<0.05$) value between legume flours.

3.2. Functional properties

The functional properties of three types of bean flour were evaluated. There is such as swelling power, solubility, water absorption capacity, oil absorption capacity, foam stability, and foam capacity. The result of analysis is shown in Table 2.

The swelling power of the samples ranged between 4.82-10.52 g/g. The results showed that there were significant differences ($p<0.05$) in the swelling power between soybean and mung bean flour as well as between soybean and red kidney bean flour. The results of this study indicate that the highest swelling power is mung bean flour (10.52 g/g) with the highest amylose content of 23.45%. While the lowest swelling power is soybean flour (4.82 g/g) with amylose content of only 1.79%. This shows that there is a positive correlation between swelling power and amylose content. Singh [19] states that the ability to swell power is influenced by amylose content. The higher the amylose content of a material indicates that there are more hydrophilic groups so that more water is bound and swelling power increases. In addition to amylose content, swelling power is also influenced by the size of starch granules. High swelling power can indicate that an ingredient can be applied to improve the characteristics of baked products.

Table 2. Functional properties of soybean, mung bean and red bean flour

Functional properties	Types of flours		
	Soybean flour	Mung bean flour	Red bean flour
Swelling Power, g/g	4.82^{a}	10.52^{b}	10.09^{b}
Solubility, %	34.49^{b}	18.80^{a}	24.21^{a}
Water Absorption Capacity, g/g	4.07^{a}	3.18^{b}	4.39^{a}
Oil Absorption Capacity, g/g	3.60^{b}	3.03^{a}	3.40^{ab}
Bulk Density, g/cm^3	0.51^{a}	1.08^{b}	0.53^{a}

Different superscript alphabets indicate significant differences among treatments ($p<0.05$)

Solubility of the samples are within in the range 18.80-34.49 %, with the highest solubility value is soybean flour (34.49%), and the lowest solubility value is mung bean flour (18.80%). %. The results showed that there were significant differences ($p<0.05$) in the solubility between soybean with mung bean and red kidney bean flour. The solubility which related to the presence of soluble molecules [18]. The solubility value of red kidney bean flour in the previous study was 29.97% [13].

Water absorption capacity (WAC) or oil absorption capacity (OAC) is defined as the absorbed amount of water or fat per gram of flour. One of the factors affecting WAC and OAC values is protein content of foodstuffs [12]. Protein has both hydrophilic and hydrophobic properties to interact with water and oil in foods. WAC indicates the hydrophilic capacity of the protein while the OAC can indicate the hydrophobic capacity of the protein. WAC and OAC were used to indicate protein ability in the food material to prevent fluid loss from a product during food storage or processing [20].

The WAC for soybean, mung bean, and red bean are 4.07, 3.18 and 4.39 g/g, respectively. The WAC was significantly different ($p<0.05$) between mung bean flour and other samples, i.e., soybean and red kidney bean flour. WAC value of red bean flour is the highest of the others although the protein content is not as high as soybeans. In addition to being influenced by protein content, WAC is also influenced by carbohydrate content [21]. This also includes the level of dietary fiber in ingredients because dietary fiber has the ability to absorb water. The intrinsic factors that affect water-binding properties of food flours with relatively high protein content relate to amino acid composition, protein conformation and surface polarity [22]. The high-water absorption capacity of flour makes it desirable for use in bakery products [23].

The OAC of samples were ranged from 3.03-3.60 g/g. The OAC of the soybean and mung bean flour was significantly different ($p<0.05$). The ability of flour to absorb and retain water or oil can improve texture and mouthfeel, flavor enhancer and reduce moisture and fat loss of food products like comminuted meats, extenders or analogs and baked dough [24; 25].

Bulk density is the physical properties of materials in dry mixes and is an important parameter in determining product packaging requirements [26]. In this present study, bulk densities for soybean, mung bean, and red bean are 0.51, 1.08 and 0.53 g/cm^3, respectively. The results showed that there were significant differences ($p<0.05$) in the bulk density between mung bean with soybean and red kidney bean flour. Mung bean flour had the highest bulk density than other flours. It indicates that mung bean flour heavier than soybean and red bean flour. The advantage of flour that has high bulk density is that it does not take up too much space when distributing and decreases packaging costs. It can reduce the need for space per unit weight and packaging costs when compared to soybeans and red bean flour. The high flour density indicates that they can function as a good thickener in food products as well as their suitability for use in processed foods. While soybean flour had the lowest bulk density than other samples. The lower bulk densities of soybean flour indication that it is lighter than others.

On the other hand, low bulk density would be an advantage in the formulation of complementary foods [23]. The result of this study similar with previous study was conducted by Pangastuti et al. [13], reported that the bulk density of red kidney bean flour is 0.41 g/cm^3.

3.3. Pasting properties

The pasting properties of soybean, mung bean and red bean flour are shown in Table 3. Pasting properties were evaluated such as peak viscosity, breakdown viscosity, final viscosity, setback viscosity and pasting temperature.

Table 3. Pasting properties of soybean, mung bean and red kidney bean flour

Pasting Properties	Types of flours		
	Soybean flour	Mung bean flour	Red kidney bean flour
Peak Viscosity, cP	19.17^a	909.00^b	827.67^b
Breakdown Viscosity, cP	5.33^a	40.00^a	85.00^b
Final Viscosity, cP	32.83^a	1387.67^b	1606.33^c
Setback Viscosity, cP	19.17^a	518.67^b	863.67^c
Pasting Temperature, °C	ND	77.85	81.73

Different superscript alphabets indicate significant differences among treatments ($p<0.05$)

The peak viscosity of mung bean (909 cP) and red bean flour (827.67 cP) was higher than soybean flour (19.17 cP). The results showed that there were significant differences ($p<0.05$) in the peak viscosity between soybean with mung bean and red kidney bean flour. The peak viscosity of soybean flour is the lowest due to lower starch and amylose content and higher protein content of mung bean and red kidney bean. Protein content in legume flours can inhibit starch granule swelling and reduce viscosity [27]. In addition to proteins that affect starch viscosity, the fat content in flour also affects viscosity. Soybean flour has the highest fat content, so the peak viscosity is low. High-fat content will inhibit the interaction of starch molecules, limiting the swelling of starch molecules so that it affects the viscosity of starch [7].

The breakdown viscosity of the red kidney bean flour significantly differed ($p<0.05$) from soybean and mung bean flour. The breakdown viscosity of legume flours ranged between 5.33 to 85 cP. The higher the breakdown viscosity, the lower of starch ability to heating and shear stress during cooking [25].

The red kidney bean flour showed higher final viscosity, setback viscosity and pasting temperature, 1606.33 cP, 863.67 cP, and 81.73 °C, respectively. While, the soybean flour was lowest final viscosity and setback viscosity, 32.83 cP and 19.17 cP, respectively. The final viscosity also called cold past viscosity is used to define a particular quality of starch and also indicates the stability of cooked starch paste in actual use. The increased final viscosity for samples could be attributed to aggregation of the amylose molecules on cooling [18].

Setback viscosity is the viscosity that occurs due to a decrease in temperature so that the starch molecules retrograde or reconnect. Setback viscosity correlates with the texture of various products. High setback viscosity causes the product to experience syneresis during freezing/thawing cycles. Setback viscosity of red kidney bean flour shows the highest (863.67 cP) from the other samples.

Pasting temperature is the temperature required by starch granules to begin to swell. Pasting temperature for soybean, mung bean, and red bean is not detected (ND), 77.85 and 81.73 °C, respectively. Pasting temperature of soybean not detected (ND) when analyzed using RVA. It may be related to its lower starch content, and the amylose content in soybean starch is only 1.79 %. When analyzed using RVA, soy flour does not undergo a gelatinization process which is characterized by the mixture remaining dilute. While the red kidney bean has higher pasting temperature (81.73°C), it may be related to its higher starch content (55.46%) and amylose content (20.14%). From the results of a study by Hamid et al. [27], stated that higher amylose content in red cowpea than black pea which caused lower pasting temperature of red cowpea. The pasting temperature values of red kidney bean and mung bean flour were 77.9 and 73.2 °C, respectively. The pasting temperature value in this study is higher than the previous study. This is likely due to differences in varieties, flour processing methods, equipment and also analysis methods. Low pasting temperature indicates that minimum temperature is needed to cook an ingredient so that it will reduce the energy cost to prepare a product.

From the results of pasting properties analysis of legume flour, it is known that starch legume is a type C starch, namely starch which has a peak viscosity value that is not so well detected. Starch type C showed limited granular inflating and was relatively constant during cooking. Type C starch is suitable to be applied as raw material for making cookies, biscuits, cakes and other bakery products. Pasting profile of flour or starch influenced by kind of starch, granule size, concentration, amylose/amylopectin ratio, other endogenous materials (particularly lipids and phosphors), added ingredients (salts, sugar, pH modifier etc.) and physical history after pasting (time, temperature and shear stress/shear rate) [28].

4. Conclusion

Soybean flour had highest values of fat, ash, protein content, solubility, and oil holding capacity while the lowest value for starch, carbohydrate, swelling power and peak, breakdown, final and setback viscosity. Mung bean flour had highest values of swelling power and peak viscosity, while the lowest value for ash content, solubility, water holding capacity, and total dietary fiber. Red kidney bean flour had the highest starch, carbohydrate, water holding capacity, total dietary fiber, breakdown and setback viscosity, while the lowest value for ash, fat and protein content. From the three types of legumes were analyzed in this study, mung bean flour can be the best choice for wheat flour substitution as ingredient biscuit because it has high swelling power and peak viscosity, and low setback, breakdown viscosity and pasting temperature.

5. Acknowledgment

The authors are grateful for funding through INSINAS 2018 from Ministry of Research, Technology, and Higher Education and Development Centre for Appropriate Technology for providing the experimental facilities.

6. References

[1] Singh N, Sandhu K S and Kaur M 2004 Characterization of starches from Indian Chickpea (*Cicer arietinum* L.) Cultivars *Journal of Food Engineering* **63** 4 441-449

[2] Yuniartha L, Kartini D 2018 Produksi kedelai nasional per Februari capai 148.918 ton. https://industri.kontan.co.id/news/produksi-kedelai-nasional-per-februari-capai-148918-ton [27 Agustus 2018]

[3] Indonesian Statistics Agency 2015 https://www.bps.go.id/site/resultTab

[4] Badan Pusat Statistik and Direktorat Jenderal Hortikultura 2015 *Produksi, luas panen, dan produktivitas sayuran di Indonesia*, Jakarta

[5] Siddiq M, Nasir M, Ravi R, Dolan K D and Butt M S 2009 Effect of defatted maize germ addition on the functional and textural properties of wheat flour *Int. J. Food Prop.* **12** 860-870

[6] Audu S S and Aremu M O 2011 Effect of processing on chemical composition of red kidney bean (*Phaseolus vulgaris* L.) flour *Pakistan Journal of Nutrition* **10** 11 1069-1075

[7] Du S, Jiang H, Yu X and Jane J 2013 Physicochemical and functional properties of whole legume flour LWT-*Food Science and Technology* xxx 1-6

[8] Badan Standarisasi Nasional (BSN) 1992 SNI 01-2891-1992 *Cara uji makanan dan minuman.* BSN, Jakarta

[9] AOAC 1995 *Official Methods of Analysis, Association of Official Analysis Chemistry.* (Washington D.C : Benyamin Franklin Station)

[10] Pranoto Y, Rahmayuni, Haryadi and Rakshit S K 2014 Physicochemical properties of heat moisture treated sweet potato starches of selected indonesian varieties. *Indonesian Food*

Research Journal **21** 5 2031-2038

[11] Brishti F H, M Zarei S K S, Muhammad M R, Ismail-Futry and Shukri R 2017 Evaluation of the functional properties of mung bean protein isolate for development of textured vegetable protein *Int. Food Research Journal* **24** 4 1595-1605

[12] Butt M S and Batool R 2010 Nutritional and functional properties of some promising legumes protein isolates *Pakistan Journal of Nutrition* **9** 373–379

[13] Pangastuti H A, Affandi D R and Ishartani D 2013 Karakterisasi sifat fisik dan kimia tepung kacang merah (*Phaseolus vulgaris* L.) dengan beberapa perlakuan pendahuluan *Jurnal Teknosains Pangan* **2** 1 20-29

[14] Alamu E O, Therese G, Mdziniso P and Bussie M 2017 Assessment of nutritional characteristics of products developed using soybean (*Glycine max* (L.) Merr.) pipeline and improved varieties *Cogent Food & Agriculture* **3** 1-12

[15] Jahreis G, Brese M, Leiterer M, Schafer U, Bohm V and Jena 2015 Legume flours: nutritionally important sources of protein and dietary fiber *Ernahrungs Umschau* **63** 02 36–42

[16] Dzudie T and Hardy J 1996 Physicochemical and functional properties of flours prepared from common beans and green mung beans *J. Agric. Food Chem* **44** 3029-3032

[17] Adelakun O E, Duodu K G, Buys E and Olanipekun B F 2013 Potential Use of Soybean Flour (Glycine max) in Food Fortification in *Soybean-Bio-Active Compounds* (London : InTech)

[18] Kaur A, Singh N, Ezekiel R and Guraya H 2007 Physicochemical, thermal and pasting properties of starches separated from different potato cultivars grown at different location. *Food Chemistry* **101** 643–651

[19] Singh S, Singh N, Isono N and Noda T 2010 Relationship of granule size distribution and amylopectin structure with pasting, thermal and retrogradation properties in wheat starches *J. Agric. Food Chem.* **58** 1180–1188

[20] Kiosseoglou V and Paraskevopoulou A 2011 Functional and physicochemical properties of pulse proteins. In: Tiwari, K.B., Gowen, A. and McKenna, B., Eds., Pulse *Food: Processing, Quality and Nutraceutical Applications* (London : Elsevier Inc.) 57-90

[21] Adejuyitan J A, Otunola E T, Akande E A, Bolarinwa I F and Oladokun F M 2009 Some physicochemical properties of flour obtained from fermentation of tigernut (*Cyperus esculentus*) sourced from a market in Ogbomoso, Nigeria. *Afr. J. Food Sci.* **3** (2) 51-55

[22] Fekria A M, Isam A M A, Suha O A and Elfadil E B 2012 Nutritional and functional characterization of defatted seed cake flour of two Sudanese groundnut (*Arachis hypogaea*) cultivars *Int. Food Research Journal* **19** 629-637

[23] Appiah F, Asibuo J Y and Kumah P 2011 Physicochemical and functional properties of bean flours of three cowpea (*Vigna unguiculata* L. Walp) varieties in Ghana *African Journal of Food Science* **5** 100–104

[24] Kisambira A, Muyonga J H, Byaruhanga Y B, Tukamuhabwa P, Tumwegamire S and Gruneberg W J 2015 Composition and functional Properties of yam bean (*Pachyrhizus* spp.) seed flour. *Food and Nutrition Sciences* **6** 736-746

[25] Adebowale A, Adeyemi I A and Oshodi A A 2005 Functional and physicochemical properties of flours of six mucuna species *African Journal of Biotechnology* **4** 1461–1468

[26] Mohamed T K, Zhu K, Issoufou A, Fatmata T and Zhou H 2009 Functionality, in vitro digestibility and physicochemical properties of two varieties of defatted foxtail millet protein concentrates *International Journal of Molecular Sciences* **10** 5224-5238

[27] Hamid S, Muzaffar S, Wani I A and Masoodi F A 2015 Physicochemical and functional properties of two cowpea cultivars grown in temperate Indian climate *Cogent Food & Agriculture* **1** 1-11

[28] Shimelis E, Meaza M and Rakishit S 2006 Physicochemical properties, pasting behaviour and functional characteristics of flours and starches from improved bean (*Phaseolus vulgaris* L) varieties grown in East Africa *Agricultural Engineering International* **3** 1–18, FP 05015

IOP Conf. Series: Earth and Environmental Science **251** (2019) 012027 doi:10.1088/1755-1315/251/1/012027

Gluten-free snacks cheese stick based on mocaf (modified cassava) flour: properties and consumer acceptance

A Kusumaningrum, Miftakhussolikhah, E R N Herawati, A Susanto and D Ariani

Research Unit for Natural Product Technology, Indonesian Institute of Sciences, Yogyakarta 55861, Indonesia

E-mail address: annisa.kusumaningrum@lipi.go.id

Abstract. *Mocaf* (modified cassava) flour is the substitution of wheat flour that gluten free and suitable for snacks. Effects of different ratio *mocaf* flour to cheese stick snacks was investigated for chemical, physic (hardness, color) and sensory properties. The *mocaf* and wheat flour were mixed in the ratio of 0/100; 25/75; 50/50; 75/25 and 100/0. Based on the results, chemical properties i.e. moisture, ash, protein, fat and carbohydrate between cheese stick based on 100% *mocaf* and 100% wheat flour were significantly different ($p < 0.05$). Moisture content of cheese stick snacks based on 100% wheat flour and 75/25 (mocaf/wheat) was not significantly different. Fat and carbohydrate in flour blend snacks ratio of 25/75 and 50/50 were not significantly different with 100% wheat flour. Ash content in cheese stick snacks significantly varied among the difference flour ratio. The increasing level of *mocaf* flour substitution has a tendency of decreasing the protein content while the highest fat content in cheese stick snacks was shown by 100% *mocaf* flour (36.37%). In *mocaf* flour substitution, significant variation of color lightness level ranged from 62.84 to 68.44. The hardness of cheese stick snacks based on 100% wheat flour and 100% *mocaf* flour were 5.82 and 4.62 N respectively and non-significantly different. Chemical, hardness and color properties in *mocaf* flour are caused by microorganism activities in fermentation process. The highest sensory value in overall acceptability of cheese stick was *mocaf* flour level of 25% and 100% while texture and flavor showed non-significant variation among the different ratio of flour blend snacks. In the present work reported that gluten-free snacks of cheese stick based on 100% *mocaf* flour resulted in good quality products and acceptable for consumers in market.

1. Introduction

In this digital and technology era, food producers can promote their products easily and popular quickly. Hence products innovation and modification i.e. cooking process, flavor, packaging, storage qualities, additional nutrition and media promotion should be upgrade continually to increase sale value of the product [1]; [2]. Nowadays snacks become favorite food for consumers due to simple, ready to eat and have a long period of shelf life. *Global Industry Analysts, Inc.* predicts that global trend market of snack exceed US$630 billion in 2020 mainly for foods with healthy benefits i.e. functional snack, protein fortification, organic snack and natural snack because of consumers awareness about healthy diets and changing eating habits [3];[4]. Generally, the raw material of snack is wheat flour that have a elasticity characteristic because of gluten protein content on its constituent component [5]. However some people who have allergy to gluten, wheat flour should be avoided [6]. One of the substitution of wheat flour is *mocaf* (modified cassava) flour.

Many studies were conducted about *mocaf* flour production methods and resulted a different characteristics of *mocaf* flour [7]; [8]. The production method of *mocaf* flour that fermented by both *L. plantarum* and *L. acidophilus* resulted an increase in crude protein content whereas there was a decrease in content of crude fat and ash [9]. The *mocaf* flour from Research Unit for Natural Product Technology, Indonesian Institute of Sciences is used as raw material for making various food products like cookies, noodle, meatball and traditional snacks at home industry. The *mocaf* flour contains moisture content (4-5%), ash (0.08-0.1%), crude fat (0.25–0.3%), protein (0.5-0.7%), carbohydrate (90-95%), energy (383 kkal/100g), crude fiber (0.1-0.2%), pH (6- 6.5), HCN (9.5-10 ppm) [8,10]

Cheese stick was originated from onion stick that reformulated with cheese flavor and mostly wheat flour is used as raw material [11]. The aim of this research were to formulate gluten-free snack cheese stick from *mocaf* flour and to evaluate chemical properties, hardness, color, sensory properties to untrained panelists and consumers acceptance to visitors in Yogyakarta Education Park.

2. Materials and methods

2.1. Materials

Mocaf flour size of 80 mesh was produced from Research Unit for Natural Product Technology, Indonesian Institute of Sciences, Yogyakarta. These flour made from cassava chips that fermented with "Starmof" starter [8]. Wheat flour and cheddar cheese were bought from the local market in Yogyakarta, Indonesia.

2.2. Preparation of product

The *mocaf* and wheat flour were mixed in the ratio of 0/100, 25/75, 50/50, 75/25 and 100/0. 50% from *mocaf* weight was mixed with hot water at 70 °C to the gel formed. Gel was then mixed with 18.75% cheddar cheese, 41.25% seasoning and 2.5% chopped of celery then stirring well. Dough was grinded in Oxone 355-AT machine with the thickness of 3 mm and length about 8 cm. After fried at 175 °C for 15 min and drained, the snacks were stored in sealed plastic 0.8 mm at room temperature for further investigation [10]

2.3. Microstructure of mocaf flour

Microstructure of *mocaf* flour was conducted using Scanning Electron Microscope (SEM) Hitachi SU3500. The mocaf flour were fixed on alumunium stubs using double sided adhesive tape and were coated with a thin layer of gold. An Acceleration potential of 3.00 kV was used during micrography.

2.4. Moisture content

Two grams of the samples (W) was dried in oven at 105 °C for 3 h. The dry samples then cooled in the desiccator and finally weighed till constant (W1) [12]

$$water\ content = \frac{W}{W1} x100\% \qquad (1)$$

2.5. Ash content

Two grams samples (W) was placed in porcelain cup (W2) then putted in muffle furnance at 550 °C. The dry samples cooled in desiccator and weighed till constant (W1) [12]

$$ash\ content = \frac{W1-W2}{W} x100\% \qquad (2)$$

2.6. Protein content
0.51 g of the samples was mixed with 2,5 g SeO_2 powder , 100 g K_2SO_4 , 20 g CuSO4.5H_2O and 25 mL concentrated H_2SO_4 in *Kejdahl* flask and heated in hot plate for 2 h. The blend was diluted with water to 100 mL. 5 mL of the solution, 5 mL NaOH 30% and 2 mL PP indicator was distilled for 10 min. Use 10 mL of boric acid 2% as a container. Titration with HCL 0.01 N and determine the blank fix [12]

2.7. Crude Fat content
Crude fat was determined by acid hydrolysis method. Sample of 2 g (W) was putted in closed paper tube then dried in oven at 80 °C for 1 h (W_1). The dry sample was moved in *Soxhlet* instrument and extraction process using n-hexane solvent for 6 h. The samples was distilled from n-hexane and dried in oven at 105 °C. Finally cooled and weighed till constant (W_2) [12]

$$crude\ fat = \frac{W-W_1}{W_2} \qquad (3)$$

2.8. Carbohydrate

% Carbohydrate: 100% - (% Ash + % Moisture + % Protein + % Crude fat) (by differences) (4)

2.9. Hardness
The hardness of cheese stick was measured using *Universal Testing Machine Zwick* SA/0.5. A test speed before reach the sample was 50 mm/min with pre-load of 0.02 N. While the test speed when reach a sample was 10 mm/min.

2.10. Color
The color of cheese stick was determined by *Chromameter Konica Minolta CR-400* using CIE method. The CIE color scale was estimated as:
L* = lightness 0 – 100 is white color
*a** = red color 0 – 60 and green color 0 – (-60)
*b** = yellow color 0 – 60 and blue color 0 – (-60)
The total color difference (ΔE) [13] :

$$\Delta \boldsymbol{E} = \sqrt{(\Delta \boldsymbol{L})^2 + (\Delta \boldsymbol{a})^2 + (\Delta \boldsymbol{b})^2} \qquad (5)$$

Where :
ΔL = (L sample – L standard)
Δ*a* = (*a* sample – *a* standard)
Δ*b* = (*b* sample – *b* standard)
L* *a** *b** standard used parameter of cheese stick snacks color based on 100% wheat flour.

2.11. Sensory evaluation
Sensory evaluation of cheese stick was conducted by 30 untrained panelists. The method using 7 point hedonic scale for color, texture, flavor and overall acceptability with 1 (dislike extremely), 2 (dislike), 3 - 4 (neither like nor dislike), 5 (like) and 6 - 7 (like extremely) [14]. The sensory evaluation was performed in sensory laboratory where it was blocked with a cabinet for each panelist. Fresh water was prepared for neutralizing the attributes of the samples. One out of five samples based on sensory evaluation was selected to evaluate consumer acceptance. The consumer acceptance evaluation was conducted by 100 visitors in

Yogyakarta Education Park. A form which contain about age, *mocaf* knowing, education level, jobs, consumer consideration factors, flavor, texture, color and overall acceptability of cheese stick were filled by visitor.

2.12. Statistical analysis
The data were analyzed using CoStat 6.4 (CoHort software, Monterey, CA, USA) and analysis of variance (ANOVA) with Duncan's means test (significance level of $p < 0.05$). All the data are presented as the mean with standard deviation. Three replications of analysis from different cheese stick snacks were used in all of the quality measurements.

3. Results and discussion

3.1. Microstructure of mocaf flour
The micrograph of *mocaf* flour is presented in Figure 1. The amorphous of *mocaf* granules indicates that fermentation process in *mocaf* flour production. Fermentation in food processing is related to organoleptic quality of the food [4,15]. Microorganism work for one's living by doing chemical reaction process that called by metabolic activities. The metabolic activities release the energy by degrading complex molecule to simple molecule.

In the beginning process, microorganism release an enzyme to destroy cassava surface area and it happens degrading starch [16]. The next process microorganism released an enzyme to break starch linkages into glucose. Through anaerobic path, it converts glucose into lactic acid to get energy for microorganism cell [14,17]. The organics acids produced by microorganism contribute to the refreshing flavor, aroma and texture [4].

3.2. Chemical properties of cheese stick snacks
Table. 1 is the chemical properties of cheese stick snacks. The moisture content between cheese stick snacks based on *mocaf* and wheat flour were significantly different ($p < 0.05$). This result indicated that the structural differences in the amylopectin and amylose of starch in different flour [13]. Moisture content of cheese stick snacks based on flour blends 25/75 increased significantly ($p < 0.05$). Generally cassava and wheat have more amylopectin than amylose. Amylose is linear or slightly branched component of starch that has a strong bond. While amylopectin is highly branched component in starch and the size of amylopectin is larger than amylose. Amylopectin has a shorter chains and a lot of α-(1,6)-branches that easy to bind with water. In flour blends (*mocaf*/wheat flour) ratio of 75/25 the moisture content decreased significantly and changed on 100% mocaf flour ($p < 0.05$) due to difference number of amylopectin and amylose between *mocaf* and wheat flour [18].

Ash content in cheese stick snacks based on difference flour ratio showed that significantly different ($p < 0.05$). Ash content indicates anorganic component or mineral content in the food product [19]. Cheese stick based on 100% wheat flour rich in mineral content than 100% *mocaf* flour. This might be probably due to cassava fermentation process in *mocaf* flour production. Minerals source in cassava used by microorganism for growth cell and electron acceptor in bioenergy reaction [20].

The higher protein content in cheese stick snack was shown by 100% wheat flour. Protein in the wheat flour has a complex structure, called by the gluten complex. Gluten proteins in wheat flour give dough its viscoelastic property necessary in baking process [6,21,22]. The result implies that the increasing level of *mocaf* flour substitution has a tendency of decreasing the protein content ($p < 0.05$).

Fat content changed not significantly in cheese stick snacks based on 100% wheat flour, flour blends (*mocaf*/wheat flour) 25/75 and 50/50. The increase of fat content by increasing the *mocaf* flour ratio 75/25 ($p < 0.05$) and the highest fat content in cheese stick snack was shown by 100% *mocaf* flour (36.37%). Carbohydrate of cheese stick snacks was decreased significantly ($p < 0.05$) in flour blends (*mocaf*/wheat flour) ratio of 50/50, 75/25 and 100/0 respectively. The decrease in carbohydrate and increase in fat content with *mocaf* flour substitution may be due to carbon and nitrogen source in carbohydrate during cassava fermentation process used by microorganism for its growth and to be converted into fatty acid [23].

3.3. Hardness and color of cheese stick snacks

The color of the food product appears from variety and raw material composition then some processing i.e. mixing and frying. In market place consumers usually have a tendency to choose product with natural color and non-synthesis color ingredient of the food product. In the present work, the color of cheese stick snacks referred to cheese stick based on 100% wheat flour (Lightness= 63.09). In *mocaf* flour substitution, the lightness level ranged from 62.84 to 68.44. The lightness of cheese stick snacks based on 100% wheat flour and mocaf flour 75% were not significantly different. The total color difference of cheese stick snacks significantly decreased ($p < 0.05$) with *mocaf* flour substitution ratio of 25/75 and 50/50 and increased on 100/0. Non-significant variation in *mocaf*-wheat flour blend snacks of 50/50;75/25 and 50/50; 100/0. During the frying process has occurred water and cooking oil mass transfer due to difference of concentration between water and cooking oil in surface and inside area of dough. The reaction between sugars and amino acids caused the primary color in cheese stick products [13].

The hardness of cheese stick snacks based on 100% wheat flour (control) and 100% *mocaf* flour were 5.82 and 4.62 N respectively and non-significantly different ($p < 0.05$). The effects of *mocaf* flour substitution ratio to the hardness of cheese stick snacks were changed significantly on 50/50 and 75/25 towards 100% wheat flour. This can be explained by the water binding capacities of starch in *mocaf* flour [24]. Moreover the fermentation cassava in *mocaf* flour production was also affected to the stabilizing of amorphous regions in starch granule and leads to the product texture. However wheat flour has a higher gluten complex that influence the hardness of cheese stick snack.

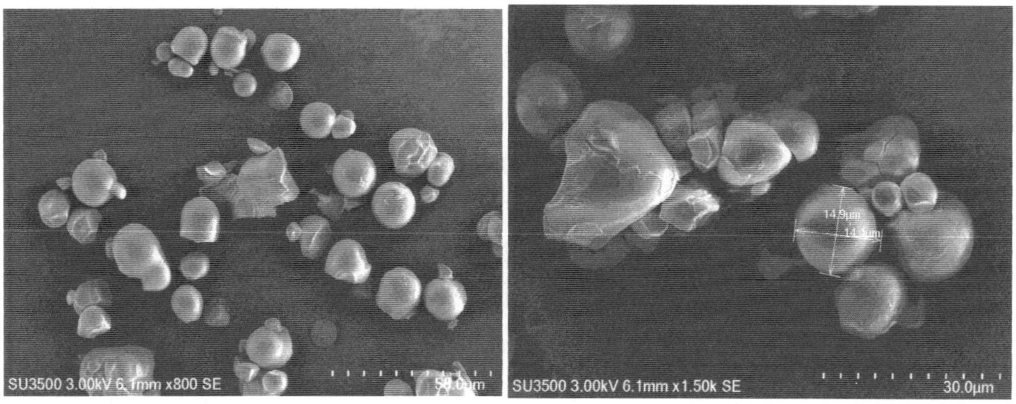

Figure 1. Micrographs of *mocaf* flour

Table 1. Chemical properties of cheese stick snacks

Mocaf/wheat flour ratio	Moisture (%)	Ash (%)	Protein (%)	Fat (%)	Carbohydrate (%)
0/100	4.30±0.06[c]	3.72±0.04[a]	9.99±0.63[a]	25.19±0.53[c]	56.79±0.60[a]
25/75	5.42±0.13[a]	2.88±0.03[e]	8.08±0.39[b]	26.78±1.77[c]	56.85±1.48[a]
50/50	5.39±0.28[a]	3.34±0.01[c]	6.93±0.17[c]	26.72±0.38[c]	57.61±0.65[a]
75/25	4.28±0.13[c]	3.11±0.13[d]	4.99±0.12[d]	32.94±0.31[b]	54.67±0.36[b]
100/0	4.74±0.04[b]	3.49±0.06[b]	4.07±0.12[e]	36.37±0.09[a]	51.33±0.08[c]

Values are expressed as mean ± standard deviation. Means in the same column with different letters were significantly different at $p < 0.05$

Table 2. Color and hardness of cheese stick snacks

***Mocaf*/wheat flour ratio**	**L***	**a***	**b***	**ΔE**	**Hardness (N)**
0/100	63.09±0.42[d]	8.73±0.13[a]	27.95±0.20[a]	-	5.82±0.78[a]
25/75	68.44±0.45[a]	6.60±0.29[c]	25.07±0.37[cd]	6.44±0.64[a]	5.49±0.90[ab]
50/50	65.33±0.33[b]	7.50±0.13[b]	25.83±0.15[b]	3.34±0.19[bc]	4.26±0.30[bc]
75/25	62.84±0.17[d]	7.45±0.15[b]	25.46±0.19[bc]	2.82±0.24[c]	3.95±0.79[c]
100/0	63.95±0.45[c]	7.26±0.06[b]	24.79±0.30[d]	3.61±0.29[b]	4.62±0.14[abc]

Values are expressed as mean ± standard deviation. Means in the same column with different letters were significantly different at $p < 0.05$

3.4. Sensory evaluation and consumer acceptance

The result to sensory evaluation of color, texture and flavor of cheese stick snacks are shown in Table. 3. Sensory evaluation was conducted by 30 untrained panelists to cheese stick based on different flour. Texture and flavor showed non-significant variation among the different ratio of *mocaf*-wheat flour blend snacks. The ratio of *mocaf* flour level showed increasing trend of color value of cheese stick snacks. Significant difference was observed in overall acceptability of cheese stick snacks based on flour blends 25/75 and 0/100, 75/25 with the highest sensory value in *mocaf* flour level of 25 % and 100%. According to sensory value, cheese stick snacks based on 100% *mocaf* flour were selected to observe consumer acceptance in market.

The 100 visitor in Yogyakarta Education Park consists of female 70% and male 30% with age variation were 1 - 25 years = 33%; 26 - 50 years = 62 %, and 51 – 75 years = 5%. The education level was junior high school to Phd and other. While jobs i.e. students, college, entrepreneur, housewife, teachers, soldiers, dealer, private employees, lectures, civil servants and farmers. Visitor knowing about *mocaf* flour products about 43% and 57% did not know. The result implies that food products based on *mocaf* flour do not popular yet in Indonesian community. In United Kingdom, for example, people became more aware of celiac diseases so they will demand healthier eating alternatives that gluten free food. Gluten Free Foods Ltd. was established in 1970 to meet the needs of this demand and introduced gluten free bakery with products such as biscuits, cookies, pasta and wafer bars [25]. The trend market of healthy snacks is predicted rapid growth and there is potential market for processing snacks based on *mocaf* flour. Figure 2 indicated that the flavor (43%) and packaging (22%) were the primary factors for consumer to purchase a food product. Figure 3 showed that a good response from consumer sensory perception of cheese stick snacks based on 100% *mocaf* flour.

Consumers sensory perception i.e. flavor (taste and aroma), color and texture of food attributes plays a vital role in food preference whereas design, graphics, color and material of food packaging influence to consumer purchase decision [26,27]. The principal roles of food packaging are to protect food products from outside influenced, damage and to provide consumer with ingredient and nutritional information [28]. Products quality and unique packaging of cheese stick snacks based on *mocaf* flour should be a concern for commercialization scale. The introduction of snacks based on *mocaf* flour would contribute the diversity of functional food products if the product fortified with micronutrient for enhace of physiological function.

Table 3. Sensory evaluation of cheese stick snacks

Mocaf **/ wheat flour ratio**	**Color**	**Texture**	**Flavor**	**Overall acceptability**
0/100	4.17±1.21^{b}	4.43±1.28^{a}	4.73±1.26^{a}	4.40±1.22^{b}
25/75	5.20±0.76^{a}	5.00±1.02^{a}	4.70±1.09^{a}	5.05±0.86^{a}
50/50	4.80±0.96^{a}	4.93±1.01^{a}	4.73±0.98^{a}	4.79±1.07^{ab}
75/25	4.23±1.28^{b}	4.43±1.25^{a}	4.40±1.07^{a}	4.39±1.15^{b}
100/0	4.87±0.90^{a}	4.87±0.82^{a}	4.80±1.00^{a}	4.84±1.07^{ab}

Values are expressed as mean ± standard deviation. Means in the same column with different letters were significantly different at $p < 0.05$

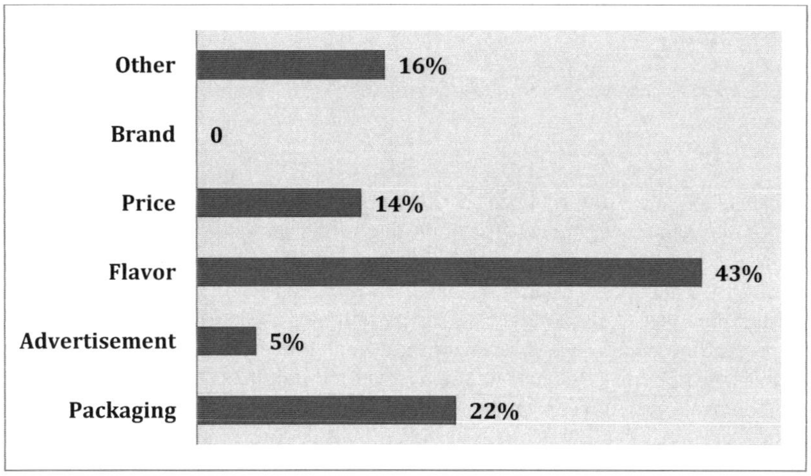

Figure 2. Consumers preference factors to a food product

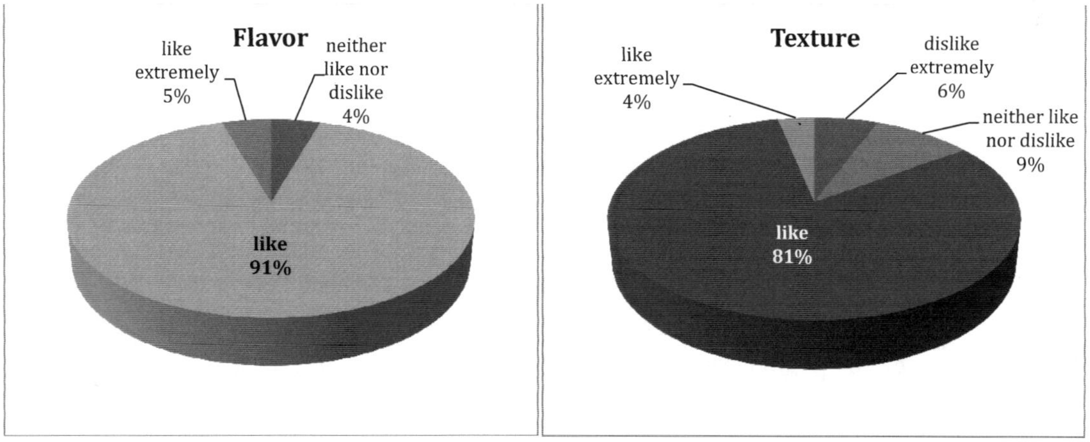

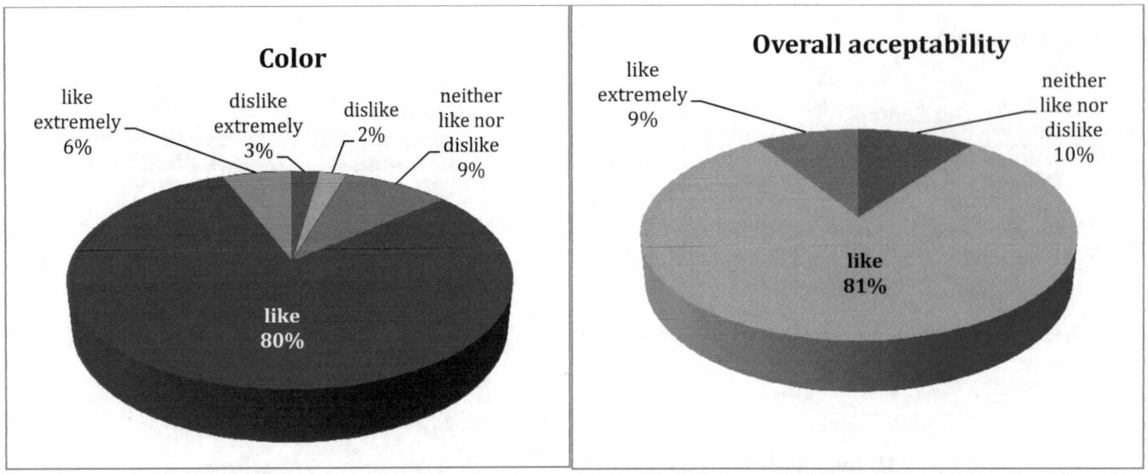

Figure 3. Consumer acceptance to gluten-free snacks "cheese stick" based on 100% *mocaf* flour

4. Conclusion

In this study, *mocaf* flour is the substitution of wheat flour to produce gluten free snack cheese stick. Chemical properties of cheese stick showed the significant variation due to the fermentation process in *mocaf* flour whilst gluten complex in wheat flour. The color and hardness of cheese stick based on *mocaf* and wheat flour have the same qualities. In the sensory evaluation, texture and flavor showed non-significant variation among the different ratio of *mocaf*-wheat flour blend snacks. The highest sensory value of cheese stick snacks was *mocaf* flour level of 25% and 100% in overall acceptability. The consumer acceptance evaluation indicated that the flavor and packaging were the primary factors for consumer to purchase a food product. Cheese stick snacks based on 100% *mocaf* flour showed a good response from consumer sensory perception.

5. Acknowledgements

The authors are thankful to Applied Science and Technology Program in Region (Iptekda) Indonesian Institute of Sciences 2017, Wiwin Widiastuti, Nisa Sepdifa, Sri Endartini and "Mekar" Small and medium enterprises (SME) in Gading district, Playen, Gunungkidul, Yogyakarta, Indonesia for the contribution.

6. References

[1] Rahmawaty U and Maharani Y 2014 Pelestarian Budaya Indonesia melalui Pembangunan Fasilitas Pusat Jajanan Tradisional Jawa Barat *J. Tingkat Sarj. Bid. Senirupa dan Desain* **2**

[2] Adenle A A, Aworh O C, Akromah R and Parayil G 2012 Developing GM super cassava for improved health and food security: Future challenges in Africa *Agric. Food Secur.* **1** 1–15

[3] Global Industry Analysts I 2015 *Surging Popularity of Functional, Protein-Fortified and Natural Snacks Drives the Global Snack Food Market* (USA)

[4] Freire A L, Ramos C L and Schwan R F 2015 Microbiological and chemical parameters during cassava based-substrate fermentation using potential starter cultures of lactic acid bacteria and yeast *Food Res. Int.* **76** 787–95

[5] Ferrari M C, Teresa M, Silva P and Chang Y K 2014 A comparative study among methods used for wheat flour analysis and for measurements of gluten properties using the Wheat Gluten Quality Analyser (WGQA) *Food Sci. Technol.* **2014** 235–42

[6] Waga J 2004 Structure and allergenicity of wheat gluten proteins – a review *Pol. J. Food Nutr. Sci.* **13** 327–38

[7] Kusumaningrum A and Sumardiono S 2016 Perbaikan Sifat Tepung Ubi Kayu Melalui Proses Fermentasi Sawut Ubi Kayu dengan Starter Bakteri Asam Laktat (Artikel Kolom) *J. Apl. Teknol. Pangan* **5** 31–3

[8] Nurhayati R, Khasanah Y, Kurniadi M, Angwar M and Ariani D 2017 *Module of production technology of Mocaf flour and its processed* (Yogyakarta)

[9] Frediansyah A and Kurniadi M 2016 Comparative influence of salinity and temperature on cassava flour product by Lactobacillus plantarum and Lactobacillus acidophilus during single culture fermentation *Nusant. Biosci.* **8** 207–14

[10] Kusumaningrum A, Rahayu NH E, Miftakhussolikhah, Susanto A, Ariani D, Widiastuti W, Endartini S and Sepdifa N 2017 *Module of IPTEKDA LIPI training: Diversification of mocaf flour processed* (Yogyakarta)

[11] Berrykitchen 2015 Resep Cheese Stick Renyah, Gurih Renyah Si Stik Keju *https://www.berrykitchen.com/blogs/detail/189/resep-cheese-stick-renyah-dan-gurih*

[12] SNI (Standar Nasional Indonesia) 1992 *Cara Uji Makanan dan Minuman SNI 01-2891-1992* (Jakarta)

[13] Mir S A, Bosco S J D and Shah M A 2017 Technological and nutritional properties of gluten-free snacks based on brown rice and chestnut flour *J. Saudi Soc. Agric. Sci.*

[14] Adeniran H A and Ajifolokun O M 2015 Microbiological studies and sensory evaluation of breadfruit and cassava co-fermented into gari analogue *Niger. Food J.* **33** 39–47

[15] Bourdichon F, Casaregola S, Farrokh C, Frisvad J C, Gerds M L, Hammes W P, Harnett J, Huys G, Laulund S, Ouwehand A, Powell I B, Prajapati J B, Seto Y, Ter Schure E, Van Boven A, Vankerckhoven V, Zgoda A, Tuijtelaars S and Hansen E B 2012 Food fermentations: Microorganisms with technological beneficial use *Int. J. Food Microbiol.* **154** 87–97

[16] Penido F C L, Piló F B, Sandes S H de C, Nunes Á C, Colen G, Oliveira E de S, Rosa C A and Lacerda I C A 2018 Selection of starter cultures for the production of sour cassava starch in a pilot-scale fermentation process *Brazilian J. Microbiol.* 1–9

[17] Odey E A, Li Z, Zhou X and Yan Y 2018 Optimization of lactic acid fermentation for pathogen inactivation in fecal sludge *Ecotoxicol. Environ. Saf.* **157** 249–54

[18] Bertoft E 2017 Understanding Starch Structure : Recent Progress *Agronomy* **7**

[19] Kusumaningrum A, Ariani D, Khasanah Y and Triwiyono 2017 Effects of storage period to traditional food "jenang saban" characteristics *J. Penelit. Teknol. Ind.* **9** 23–36

[20] Crundwell F K 2003 How do bacteria interact with minerals? *Hydrometallurgy* **71** 75–81

[21] Al-saleh A and Brennan C S 2012 Bread Wheat Quality: Some Physical, Chemical and Rheological Characteristics of Syrian and English Bread *Foods* **1** 3–17

[22] Stiegert K and Blanc J 1997 Japanese Demand for Wheat Protein Quantity and Quality *J. Agric. Resour. Econ.* **22** 104–19

[23] Haryo R, Setiarto B, Widhyastuti N and Saskiawan I 2016 Effect of Lactic Acid Bacteria , Fungi and Yeast Fermentation on Nutritional Quality of Sorghum Flour *AGRITECH* **36** 440–9

[24] Seyhun N, Sumnu G and Sahin S 2005 Effects of different starch types on retardation of staling of microwave-baked cakes *Food Bioprod. Process.* **83** 1–5

[25] FAO and IFAD 2004 Global cassava market study Business opportunities for the use of cassava *Proceedings of the validation forum on the global cassava development strategy*

[26] Bhuiyan F R, Torab A and Rahim M A 2015 Consumer 's sensory perception of food attributes : A survey on flavor *J. food Nutr. Sci.* **3** 157–60

[27] Kingdom U, Farooq S, Habib S and Aslam S 2015 Influence of product packaging on consumer purchase intentions *Int. J. Econ. Commer. Manag.* **III** 538–47

[28] Marsh K and Bugusu B 2007 Food packaging - Roles, materials, and environmental issues: Scientific status summary *J. Food Sci.* **72**

IOP Conf. Series: Earth and Environmental Science **251** (2019) 012028 doi:10.1088/1755-1315/251/1/012028

Formulation Banana Flakes using Different Type and Concentration of Fillers

D N Surahman[1], R P Rahayu[2], D Desnilasari[1], R Ekafitri[1], W Cahyadi[2], T Rahman[1]

[1]Development Center For Appropriate Technology, Indonesian Institute of Sciences, Subang, Indonesia,
[2]Department of Food Technology, Pasundan University, Bandung, Indonesia
Email : diki.lucky@gmail.com

Abstract. The making banana flakes using ripe banana flour had disadvantages on its physical properties, causing the product to be highly hygroscopic. Therefore the formulation needs the addition of other starch to improve the physical properties of banana flakes. The aim of the research was to determine the type and concentration of the fillers to make the best banana flakes. This research conducted using Randomized Block Design (RAB) with two factor which were type of filler (rice flour (A1), corn flour (A2), modified cassava flour (A3)) and concentration of fillers (5g (B1), 10g (B2), 15g (C)) with three repetition. The responses of the research were physical response and moisture content of the product. Physical responses included hardness test, water absorption index (WAI) and water solubility index (WSI). The best product selecting using de Garmo's index of effectiveness, then the product was analyzed by the chemical characteristic. The best product of Banana flakes was addition with 15% rice flour as fillers. This product has characteristic 3,24% of ash, 3,19% of crude fiber, 7,35% of protein, 2,14% of fats and 83,45% of carbohydrate.

Keywords: Flakes, banana flour, fillers

1. Introduction

According to Mathews [1], skipping breakfast time can cause a negative effect on the body. This is due to low blood sugar levels that are automatically lowing the blood pressure and weaken the nerve impulses to the body would limp and cause the decreased passion for work. Therefore, breakfast is somehow needed as the source of calories to increase the levels of the blood sugar.

Breakfast is very important as an energy source, not only for adults but also for children to support the developments of their brain. For the children having breakfast will help them to be able to concentrate and more active in school, so there is a probability that they will become smarter. The benefit of having breakfast will also reduce the risk of micronutrient deficiency problems, especially vitamins and minerals that supported its developments.

Breakfast contributes about 25% of daily energy that consists of about 450-500 calories and about 8-9 grams worth of proteins [2]. A breakfast that contains 25% of nutritional needs a day is part of the compliance the balance of nutrition and also can affect the thinking process and activated of a person throughout the day, even more so in children of growing age. Sukiniarti argued that about 18,05% Indonesian kids always skipping breakfast [3].

Many factors that make it difficult for kids to eat breakfast such as the difficulty of waking up in the morning and the number of parents who works in the morning cause limited time to prepare breakfast. To compromise with the problems that, it is necessary to make a breakfast menu that requires an easy and fast to prepare and also attract children for having them. One of the suitable food products is the *flake-shaped* food products. This is due to the changing lifestyle of the society that demands everything in a fast-paced and practical, no exception to the foods, so the demand for fast-served breakfast such as cereals is increasing.

Today most breakfast foods are made from cereals such as wheat, corn, and rice. Breakfast can also be made from fruit as a source of fiber and vitamins. Materials selection for mixed-formulation (composite) is important to produce good product [1]. One type of fruit that can be used as a breakfast meal is a banana. According to Musita, banana is one of the fruits that were riches with nutrition, carbohydrate, vitamins, and mineral sources [4].

In additions for being a source of carbohydrates, bananas also have a good nutritional content of vitamins (pro-vitamin A, B, and C) and minerals (potassium, magnesium, phosphorus, iron, and calcium) are important for the body [5]. Banana is one of the distinguished fruit that found in Indonesia and becomes one of the fruits that the productivities is always increasing every year. However, people's consumption is not so much.

The growth productivity of banana in Indonesia started since 1980-2013 also tend to increase. If the 1980 production of banana in Indonesia amounted 1.98 million tons, then in 2013 has reached 6.28 million tons. The increase of banana production during this period has reached 3.94% per year, where the growth in Java is slightly higher. In 1980-2013 banana production in Java, reached 61.22% of Indonesia's total banana production, while the total production from outside java was 38.78% [6].

Based on the data from the Survei Sosial Ekonomi Nasional (SUSENAS) that conducted by Badan Pusat Statistik (BPS), the consumption of bananas tends to decrease in the last five years, about 1.80% per year. According to the Ministry of Agriculture data, the consumption of Plantain and Ambon's bananas is generally higher than the other bananas. Ambon's banana consumption in Indonesia 2012 is 1.825 kg/capita per year while in 2013 it decreased to 1.251 kg/capita.

Ambon's banana is one of the fruits that contain many nutrients and has a distinctive flavor and aroma, yet also easily damaged, so it needs to be processed into a durable meal, easy to store, and easy to consume. One way to make the Ambon's banana become a durable material is turning it into a banana flour [7].

The flour from Ambon's bananas can be used as a substitution or even as basic ingredient in the manufacture of various foods. For example, making flakes from Ambon's banana is a way to make food diversification and also can be an alternative for breakfast which also can be a source of nutrition for children that answered the problem about the breakfast preparation for children.

One of the most prominent characteristics of flakes is crispness. Therefore, it is necessary to add a fillers material that can improve the characteristics of the flakes according to Triyono, fillers properties dispersed in water rapidly, high solubility, low hygroscopic, able to form *'body,'* low browning but strongly binding with any main materials [8].

One of the characteristics of cereal products that desired by consumers, in general, is its crispiness so that cereals can last long after the addition of milk. This condition can be achieved by the addition of starch in the form of the flour, whether modified or not [9-10]. According to this, it is needed to reformulated banana flakes using varieties of filler such as rice flour, corn flour, and mocaf flour. The research aimed to determine the type and concentration of the fillers to make the best banana flakes.

2. Materials and Methods

2.1. Materials

The materials used for making Banana Flakes are banana flour from Ambon's banana, sugar, eggs, baking soda and high-calcium skim-milk. The materials that used for chemical analysis are aquadest, Kjeldahl salt, H_2SO_4 (p), NaOH, zinc granule, HCl, PP, and n-hexane indicator. The tools that used in making the Banana Flakes are weighing plastic, digital scale, Philips's mixer with speed at 1, oven deck, baking sheet paper, dough sheeter, two spatulas, scissors, and baking sheets. The tools that used for chemical analysis are Kjeldahl flasks, boiling stones, Bunsen, tripod, pumpkin powder, distillation apparatus, titration apparatus, soxhlet, reflux, desiccator, filter paper, cups, oven, furnace, thimble paper and centrifuge.

2.2. Methods

2.2.1. Methods
This research used the Randomized Group Design which was arranged in a factorial way into two factors; each factor consists of three levels of treatment. The first factor is fillers type with rice flour, corn flour, and modified cassava flour. The second factor is the concentrating the fillers with 5g, 10g, and 15g levels of concentrate. From these two factors, nine treatments were obtained, and each treatment was repeated three times, then 27 tests were obtained each. The data were analyzed using ANOVA with significance at p<0,05

2.2.2. Formulation of Banana Flake
Formulation of Banana Flake is shown in Table 1.

Table 1. Formulation of Banana Flake all Treatment

Ingredient	Treatment								
	A1B1	A1B2	A1B3	A2B1	A2B2	A2B3	A3B1	A3B2	A3B3
Banana flour (g)	50	50	50	50	50	50	50	50	50
Skim Milk (g)	20	20	20	20	20	20	20	20	20
Egg (g)	12.5	12.5	12.5	12.5	12.5	12.5	12.5	12.5	12.5
Baking powder (g)	1	1	1	1	1	1	1	1	1
Sugar (g)	20	20	20	20	20	20	20	20	20
Rice flour (g)	5	10	15	-	-	-	-	-	-
Corn Flour (g)	-	-	-	5	10	15	-	-	-
Modified cassava flour (g)	-	-	-	-	-	-	5	10	15

2.2.3. Analysis Physical and Chemical of Banana Flakes
Physical response of Banana Flakes including hardness of product, water absorption index, and water solubility index. The hardness of banana flakes was analyzed using Texture Analyser. Water absorption index and water solubility index was analyzed according to Triyono [8]. The best sample was selected by using [11] method. The Best banana flakes analyzed for chemical response for ash, crude fiber, protein, fat, and carbohydrate content of the product. Ash, crude fiber, protein, fat, and carbohydrate content analyzed according to Triyono [8].

3. Results and discussion

3.1. Analysis Physical Response of The Banana Flakes

3.1.1. The hardness of the Product
The hardness of the product might be affected by the ration of amylose and amylopectin to the feedstock. Tjokroadikoesoemo explained that amylopectin in starch has high adhesive properties, so the higher amylopectin level in the raw materials used will cause higher compactness of a product [12].

Based on the results of the research, the average hardness with 9 (nine) banana flakes treatment with 3 (three) repetitions are ranged between 247.48gf - 340.50gf. More data can be seen in Table 2.

Table 2. Mean value of hardness responses

Types of fillers	Concentration of fillers		
	B1 (5%)	B2 (10%)	B3 (15%)
A1 (Rice Flour)	247.48a	257.67a	256.41a
A2 (Corn Flour)	292.8a	317.77a	306.36a
A3 (Modified-Cassava Flour)	340.59a	290.51a	300.94a

*Notation same letters in the same column and line indicate no significant difference at significance level of 5%

The results of the hardness test showed that the type of fillers and the concentration of the fillers, then the interaction between the type of fillers and the concentration were not significantly different.

However, from Table 2 we see that the average of hardness on the treatment shows that the trend is rising at each concentration. This can be attributed to the higher concentration of the starch content on fillers. The higher the starch contents in the material means the amylose and amylopectin levels also higher. Furthermore, it would be faster the water absorbent that causes the product to become crusty.

According to [13] amylopectin will turn the food product into a light, porous, crisp, and crunchy one. Amylose tends to produce the crustiness products, and blooming processes occur in a limited way. According to [14] if the amylose levels are high, then the starch will be dry, less adhesive and tend to absorb water (hygroscopic). According to [15] the water content in food also affects the fracture of the flakes. The presence of water in the intercellular cavities of material can decrease the stiffness of the cell so that it will decrease the hardness of the product.

3.1.2. Water absorption index (WAI)

According to [16] Water Absorption Index (WAI) indicates the ability of the material to be able to interact with water. Based on the results of the research, the average water absorption index has been performed with 9 (nine) banana-flakes treatment with 3 (three) repetitions ranging from 1.92 to 2.18 ml/g. The Water Absorption Index (WAI) test indicates that it is not significantly different, whereas for the interaction between fillers type and fillers concentration gives a real or significant different effect. The interaction effect between factor A and B can be seen in Table 3.

Table 3. The Effect of the types of fillers interaction and fillers concentration response to water absorption index banana flakes

Types of fillers	Concentration of fillers		
	B1 (5%)	B2 (10%)	B3 (15%)
A1 (Rice Flour)	$_b2{,}08^A$	$_a2{,}12^A$	$_b2{,}18^A$
A2 (Corn Flour)	$_b2{,}05^A$	$_a2{,}06^A$	$_a2{,}03^A$
A3 (Modified-Cassava Flour)	$_a1{,}92^A$	$_a2{,}08^B$	$_a2{,}05^B$

*different uppercase in the same line and lowercase in the same column indicate significant difference at a significance level of 5%

Table 3 shows that the type of rice flour fillers is not significantly different in the addition of concentrations of 5%, 10%, and 15%. Likewise, cornstarch fillers were not significantly different at the5%, 10% or 15% concentrate. While the modified cassava flour fillers the addition of 10% and 15% concentrate was not significantly different, but in 5% concentrate was significantly different. This can be attributed to the starch content contained in modified cassava flour is slightly ranged from 67.77% compared to rice flour and corn flour [17].

In the 5% concentrate treatments, the rice flour and the corn flour was not significantly different but on modified cassava flour was significantly different. Same as for the 10% concentrate, the addition of rice flour, corn flour, and modified cassava flour were not significantly different. Also in the concentration of 15% corn flour and modified cassava flour was not significantly different, except for the rice flour. The starch content of rice flour can cause this is slightly higher than modified cassava flour and corn flour ranges from 76% to 82% [18].

If we highlighted the average of each concentration, although only slightly, this happened due to the increasing concentration of the added fillers rising the starch content contained in the material that causes more water absorption, hence the more concentration of starch added and the more water were absorbed then the swelling of the starch is even greater causing the product to be developed (puffed) and facilitate the structure of slabs or pieces of the dough and make the product more crunchy.

The higher level of the flakes starch then the water absorption value will increase due to the more gelatinization of starch. The higher the starch content in a product, the water absorption also tends to be higher. This is because the starch granules have a very large water-absorbing capacity because the amount of hydroxyl starch groups is very large [15].

When compared between the three flours (rice flour, corn flour, and modified-cassava flour), the content of starch in the rice is ranging from 76 - 82% [18]. The Amylose content in rice is about 25%, and the amylopectin is about 75%. According to [19] corn flour has a content of 72-73% starch. According to [17] modified-cassava flour has starch of 67.77%. According to [20] maize contains amylose 25-30% and amylopectin 70-75%, while amylopectin content in modified-cassava flour 75% and amylose of 25%. Based on the average data of starch content, modified-cassava flour starch has the lowest content compared to rice flour and corn flour, as well as its amylopectin content so that the water absorption index value of modified-cassava flour is lower than rice and corn flour.

Amylose-amylopectin ratios, molecular weight distributions, and chain lengths, and the degree of branching and conformations determine the swelling power and solubility [21]. Swelling power is the increase of the volume and the maximum weight of starch during the process in the water. Swelling power shows the ability of starch to expand in water. The higher swelling power means that the ability of starch to expand in water is high as well [22]. Swelling is a trait influenced by amylopectin [23]. A high proportion of the branch chain amylopectin has contributed to an increase in swelling value. Besides, there is a negative correlation between swelling power and amylose levels; swelling power decreases with increasing amylose levels [23].

3.1.3. Water solubility index (WSI)
The Water solubility index shows the ability of material to soluble with water that seen by a large number of particles (g) dissolved in a certain amount of water (ml) [19]. Based on the results of the research, the average water solubility index with 9 (nine) treatments of banana flakes with 3 (three) times of repetitions was ranged from 0.223 g / ml - 0.249 g / ml. Water Solubility Index (WSI) test indicates that the type of fillers, fillers concentrations, and the interaction between the type of fillers and the concentration of fillers not significantly affected WSI product. The average value of each treatment can be seen in Table 4.

Table 4. Mean value of the water solubility index in banana flakes

Types of fillers	Concentration of fillers		
	B1 (5%)	B2 (10%)	B3 (15%)
A1 (Rice Flour)	0.233a	0.238a	0.245a
A2 (Corn Flour)	0.223a	0.249a	0.249a
A3 (Modified-Cassava Flour)	0.228a	0.236a	0.248a

*Notation same letters in the same column and line indicate no significant difference at a significance level of 5%

The average value of each concentration shows that the WSI levels were increased after the 5% addition of flour but has no effect after the 10% or 15% addition. The more addition of the filler's concentration, the level of starch is higher as well as the amylose level. By comparing the type of flours, the rice flour has the level of 76% - 82% of starch [19], it is higher than the starch of wheat, corn or modified-cassava flour. The content of amylose in rice is about 25% and about 75% worth of amylopectin.

According to [19] corn has a content of 72-73% starch with amylose 25-30% and amylopectin 70-75% [19] [20], while modified-cassava flour has 67.77% starch with amylopectin 75% and amylose 25% [20]. The higher the starch content added in the ingredients, the value of WSI is also increasing because of the amylose levels contained in the material. According to [14] if the amylose levels are higher, then the starch will be dry, less adhesive and tend to absorb more water (hygroscopic).

Water solubility is associated with the proneness of the water molecules to interact with the granule molecules in starch and replace the hydrogen interactions between the molecules so that the granules will easily absorb water and have a high development. The existence of such development will suppress the granules from the inside so that the granules will break and the starch molecules, especially amylose will come out [24]. According to [25] when the starch molecule is completely hydrated, its molecules begin to spread to the outer media and the first out is the amylose molecules that have short chains. The higher the temperature the more starch molecules will come out of the starch granules. During the heating process, the starch granules will start to break down, and the starch with the higher amylose content will release more amylose

3.2. Analysis Chemical Response of The Banana Flakes

3.2.1. The moisture content of the Product

Moisture content is very important in determining the durability of the food because it affects physical, chemical, microbiological, and enzymatic changes [26]. Based on the result of the research, the average water content on 9 treatments of the Banana Flakes with 3 times repetitions was ranged from 3.82% – 5.37%. More data can be seen in Table 5 below.

Table 5. Mean Value of moisture content of the banana flakes

Types of fillers	Concentration of fillers		
	B1 (5%)	B2 (10%)	B3 (15%)
A1 (Rice Flour)	5.37a	4.66a	3.82a
A2 (Corn Flour)	4.82a	4.77a	4.64a
A3 (Modified-Cassava Flour)	5.36a	4.82a	4.75a

*Notation same letters in the same column indicate no significant difference at a significance level of 5%

The result of moisture contents shows that the type of fillers, the interaction between the type of filler and the concentration of the fillers has no significant effect. On the other hand, the concentration of fillers indicates a significant effect on the moisture content of the banana flakes. The further tested of different concentration of filler by Duncan showed in Table 6 below.

Table 6. The Effect of fillers concentration on moisture content of banana flakes

Concentration of Fillers	Average from different Filler for Moisture Content	Significance level 5%
B3 (Con.of 15%)	4,40	a
B2 (Con. of 10%)	4,75	a
B1 (Con. of 5%)	5,18	a

*Notation same letters in the same column and line indicate no significant difference at a significance level of 5%

The higher concentration of the added ingredients will somehow decrease the water content of the water. This may be due to the differences in starch and protein that contain in the fillers. The higher concentration of the added ingredients, the more starch, and protein contained in the material, the water content of the material will be lower. This is due to the more water that is bound in the material and the free water content will be lower.

Water content is related to the protein content, where proteins will bind water through hydrogen bonds. At the time of cooking, the starch molecules will bind with the proteins through hydrogen bonds. Because of the weakening of these hydrogen bonds, water molecules can infiltrate between protein and starch molecules, so that at the cooling time the reinforcement of hydrogen bonds between hydrogen starch molecules involves strongly bonded water molecules is difficult to be released by evaporation or drying. The free water content can easily disappear when evaporation occurs, whereas bound water is difficult to be free in such a way [15].

3.3. Selected The Best Product of Banana Flakes

3.3.1. Selected sample of Banana Flakes

The products were selected based on [11] method. Response variables included in this method are moisture content, crustiness, water absorption index and water solubility index. Based on the effectiveness test using [11] method, there is a rank of treatment from one to nine as shown in Table 7.

Based on the tables above, we can see that the banana flakes with addition 15% of rice flour's is the most selected banana flakes. The best product of banana flakes then continued with the testing of the ash, crude fiber, protein, fat, and carbohydrate content.

Table 7. Ranks of De Garmo effectiveness test

Rank Order	∑Nh	Sample Code	Details
1	0.315	A1B3	15% of Rice Flour
2	0.271	A1B2	10% of Rice Flour
3	0.257	A2B1	5% of Corn Flour
4	0.231	A1B1	5% of Rice Flour
5	0.225	A3B2	10% of Modified Cassava Flour
6	0.153	A3B3	15% of Modified Cassava Flour
7	0.140	A2B3	15% of Corn Flour
8	0.132	A2B2	10% of Corn Flour
9	0.086	A3B1	5% of Modified Cassava Flour

3.3.2. The Result of the proximate test

The selected banana flakes product was continued with the chemical test for seeing the level of ash, the crude fiber, fat, and carbohydrate content. The following Table 8 is the result of the analysis.

Table 8. Banana flakes chemical quality test with the addition of rice flour 15%

Parameter	Value
Ash (%)	3,24
Crude Fiber (%)	3,19
Protein (%)	7,35
Fat (%)	2,14
Carbohydrate (%)	83,45

4. Conclusion

Formulation banana flakes using different type and concentration of filler has been done. The physical response showed that hardness of the product, the water absorption index, and the water solubility index has not significantly affected by the type and the concentration of the filler, and also the interaction between two of them. The result of the moisture content shows that the type of fillers, the interaction between the type of fillers and the concentration of the fillers has no significant effect. Based on De Garmo effectiveness index method, the banana flakes with 15% of rice flour's addition is the most selected banana flakes that have 3,24% ash, 3,19% crude fiber, 7,35% protein, 2.14% fat, and 83,45% carbohydrate content.

5. Acknowledgment

The authors would like to acknowledge financially, and facility supports from Development Center for Appropriate Technology, Indonesian Institute of Sciences.

6. References

[1] Tegar T 2010 Optimasi Formulasi Breakfast Meal Flakes (Pangan Sarapan) Berbasis Tepung Komposit Talas, Kacang Hijau, dan Pisang *Skripsi* (Bogor : Departemen Ilmu Dan Teknologi Pangan, Fakultas Teknologi Pertanian Institut Pertanian Bogor)

[2] Perdana F 2013 Analisis Jenis, Jumlah dan Mutu Gizi Konsumsi Sarapan Anak Indonesia *Jurnal Gizi Dan Pangan* **8** 39–46

[3] Sukiniarti 2015 Kebiasaan Makan Pagi Pada Anak Usia SD dan Hubungannya dengan Tingkat Kesehatan dan Prestasi Belajar *Jurnal Pendidikan Biologi Indonesia* **1** 315–321

[4] Musita N 2009 Kajian Kandungan dan Karakteristik Pati Resisten Dari Berbagai Varietas Pisang *Jurnal Teknologi Industri Dan Hasil Pertanian* **14** 68 – 79

[5] Abdillah F 2010 Modifikasi Tepung Pisang Tanduk (*Musa paradisiaca Formatypica*) Melalui Proses Fermentasi Spontan dan Pemanasan Otoklaf untuk Meningkatkan Kadar Pati Resisten *Thesis* (Bogor : Program Studi Ilmu Pangan, Sekolah Pasca Sarjana Institut Pertanian Bogor)

[6] Kementrian Pertanian 2014 *Outlook* Komoditi Pisang Pusat Data dan Sistem Informasi Pertanian

[7] Pratomo A 2013 Studi Eksperimen Pembuatan Bolu Kering Subtitusi Tepung Pisang Ambon *Food Science and Culinary Education Journal* **2** 17–30
[8] Triyono A. 2010 Pengaruh Maltodekstrin dan Substitusi Tepung Pisang (*Musa paradisiaca*) Terhadap Karakteristik Flakes *Prosiding Seminar Nasional Teknik Kimia* "Kejuangan" Pengembangan Teknologi Kimia Untuk Pengolahan Sumber Daya Alam Indonesia 1–7
[9] Gaman P M dan KB Sherington 1981 *Ilmu Pangan Pengantar Ilmu dan Nutrisi Pangan, Mikrobiologi* (Yogyakarta : Universitas Gajah Mada Press)
[10] Ekafitri R, D N Surahman dan N Afifah 2015 Pengaruh Penambahan Dekstrin dan Albumen Telur (Putih Telur) Terhadap Mutu Tepung Pisang Matang *Jurnal Litbang Industr*i **6** 13-24
[11] De Garmo E P, W.G. Sullivan., dan C.R. Candra 1984 *Engineering Economi.*7th edition (New York : Mc Millan Publ. Co)
[12] Tjokroadikoesoemo P S 1986 *HFS dan Industri Ubi Kayu Lainnya* (Jakarta : PT. Gramedia)
[13] Muchtadi T dan Sugiyono 1989 *Ilmu Pengetahuan Bahan Pangan* (Bogor : Institut Pertanian Bogor)
[14] Zulaidah A 2013 Modifikasi Ubi Kayu Dengan Kombinasi Proses Penggaraman dan Proses Biologi Untuk Subsitusi Terigu *Dinamika Sains* **11**
[15] Febrianty K 2015 Pengaruh Proporsi Tepung (Ubi Jalar Terfermentasi : Kecambah Kacang Tunggak) dan Lama Perkecambahan Terhadap Kualitas Fisik dan Kimia Flake *Jurnal Pangan Dan Agroindustri* **3** 824–834
[16] Resty D 2008 Sifat Kimia, Fisik dan Mikrobiologi Snack Ekstrusi yang Diperkaya Tepung Putih Telur Sebagai Sumber Protein Selama Penyimpanan.*Skripsi* (Bogor : Program Studi Teknologi Hasil Ternak, Fakultas Peternakan Institut Pertanian Bogor)
[17] Fitriany 2016 Kolerasi Konesntrasi Tepung Olahan Ubi Kayu Terhadap Tepung Untuk Pisang Goreng (*Flour For Banana Fritter)* Menggunakan Regresi Linier Sederhana *Tugas Akhir* Program Studi Teknologi Pangan Fakultas Teknik Universitas Pasundan
[18] Adicandra R M 2016 Beras Analog Dari Ubi Kelapa Putih (*Discorea alata L.*): Kajian Pustaka *Jurnal Pangan Dan Agroindustri* **4** 383–390
[19] Hidayat T 2008 Karakteristik Fisik dan Organoleptik *Tortilla Corn Chips* dengan Penambahan Tepung Putih Telur Sebagai Sumber Protein *Skripsi* (Bogor : Program Studi Teknologi Hasil Ternak, Fakultas Peternakan Institut Pertanian Bogor)
[20] Fitriani A A N 2013 Pengaruh Proporsi Tepung Jagung dan Mocaf Terhadap Kualitas "*Jamof Rice*" Instan Ditinjau Dari Sifat Organoleptik *E-Jurnal Boga Dan Gizi* **2** 34–43
[21] Moorthy S N 2004 Tropical Sources of Starch Di dalam: Ann Charlotte Eliasson (ed) Starch *in Food: Structure, Function, and Application* (Florida : CRC Press)
[22] Suarni 2009 Produk Makanan Ringan (*Flakes*) Berbasis Jagung dan Kacang Hijau Sebagai Sumber Protein Untuk Perbaikan Gizi Anak Usia Tumbuh *Prosiding Seminar Nasional Serealia* 297–306
[23] Li J Y and Yeh A I 2001 Relationship Between Thermal, Rheological Characteristics, and Swelling Power for Various Starches *J. Food Engineering* **50** 141-148
[24] Purnamasari I W 2015 Pengaruh Penambahan Tepung Labu Kuning dan Natrium Bikarbonat Terhadap Karakteristik Flake *Talas Jurnal Pangan Dan Agroindustri* **3** 1375–1385
[25] Fleche G 1985. Chemical Modification and Degradation of Starch in: G M A V Beynum dan J.A Roels (eds.) Starch Conversion Technology (New York : Marcel Dekker, Inc)
[26] Buckle 1987 *Ilmu Pangan Universitas* (Jakarta : Indonesia Press)

IOP Conf. Series: Earth and Environmental Science **251** (2019) 012029 doi:10.1088/1755-1315/251/1/012029

Evaluation of Plasticizer Addition in Composite Edible Coating on Quality of Fresh-Cut Mangoes during Storage

N Afifah, L Ratnawati, and D A Darmajana

Development Center for Appropriate Technology, Indonesian Institute of Sciences Subang, Indonesia

Email: syabiljafa2008@gmail.com

Abstract. A composite edible coating made from carrageenan - beeswax with two plasticizers was used to cover fresh-cut mangoes. The purpose of this research was to develop edible film with plasticizer treatment and investigate the effect of an edible coating on quality changes in fresh-cut mangoes during refrigerated storage. The film-forming solution was poured onto an acrylic plate 20x20 cm^2 and dried at 50 °C for 24 hours in a cabinet dryer. Coated and uncoated mangoes were placed into styrofoam then over-wrapped with cling wrap and stored at 6°C for 6 days. Edible films containing two plasticizers had a higher thickness, water vapor transmission rate (WVTR), and solubility than the single plasticizer film. It was found that edible coatings could preserve a fresh-cut mango quality by reducing weight loss, delaying an increase in total soluble solids, maintaining pH, total acidity and growth of microorganisms. It also indicated no difference in sensory perception between the coated and the uncoated fresh-cut mangoes.

1. Introduction

Mango (*Mangifera indica* L.) is a seasonal and climacteric fruit, which is cultivated most land in Indonesia [1]. Fresh-cut ready to eat fruits have become important, for expanding food industry. Increase in consumer demand for food available in fresh and convenient condition is an opportunity to fresh-cut server mango [2]. Consumer acceptance and important consideration in marketing is determined by internal and external fruit quality, on the basis of freshness and appearance of fruit at the time of purchase [3]. One of the minimal processing of fresh fruit is mechanical operations, such as peeling, slicing, or cutting. However, minimal processing operation can affect the integrity of fruits, giving a negative effect on product quality such as texture weight loss, breakdown, leakage of nutrients, browning, off-flavor development and possibility increase in the presence of microorganisms on fruit surface which may endanger the safety of fresh-cut fruits [4]. Therefore, alternative methods were needed to maintain the quality attributes of fresh-cut mango during handling, distribution and retail sale [3].

Thin layer of edible materials is having in edible film and coating. It is applied to the surface of food products, friendly with an environment that offers substantial advantages for shelf-life enhancement of many food products [5]. Edible coating affects oxygen partial pressure and metabolism of fresh food, help to retain volatile components, control moisture migration between a food product and an environment protecting it from deterioration and improving textural quality, and decrease the risk of pathogen growth [6]. Edible films are an answer to consumer demand for more natural products and the lower contamination of the environment because they are formed of fully biodegradable and biocompatible polymer derived from natural sources including protein, polysaccharides, and lipids. This research becomes importantly in scientific and industrial field interest worldwide [7]. Hence, natural coating materials which can preserve fresh fruits and retain their quality are highly desired.

Generally, the coating made of hydrophilic substances, such as polysaccharides including carrageenan, have a good mechanical property and a good barrier against gas transfer [8]. Carrageenan is a complex mixture of several polysaccharides and is derived from red seaweed [9] that has a high potential as a film-forming. However, hydrocolloid films are characterized by their poor water vapor permeabilities. Hydrophobic properties of lipid, such as carnauba wax and beeswax, can overcome this weakness by exploiting their great water barrier properties [10].

The other substance like plasticizer is added to improve the mechanical properties of coatings. Major plasticizers have been used polyols such as glycerol, polyethylene glycol, and sorbitol, besides, disaccharides such as sucrose and monosaccharides (e.g., fructose, glucose and mannose) [11]. Type and amount of plasticizers affect properties of the coating [12]. Hence, their polymer types and optimum concentration of plasticizers should be determined for the success of its use in a variety of condition.

Strong interactions can occur between plasticizers (plasticizer-plasticizers interactions) when some plasticizers are used in the film matrix; which can improve certain functional properties of the film [**13**]. The starch-based edible film shown better results when film made from a combination of glycerol-sorbitol 1: 1 as a plasticizer than using glycerol or sorbitol separately [14].

A number of studies were interesting and very promising was reported, which showed minimally processed fruits and vegetables and efficient improvement of quality and safety of fresh fruit and vegetables [5]. Some studies reported that applying a chitosan coating on sliced mangoes effectively prolong the quality attributes and extends their shelf life during storage at 6 °C for 7 days [6; 15]. A polysaccharides coating did not consistently improve the quality of cut mango slices, and the addition of calcium ascorbate with citric acid and N-acetyl-L-cysteine could maintain cut mango slices attractiveness in storage by keeping light color [16]. The coating of papaya using carrageenan with the addition of glycerol as the plasticizer and showed that carrageenan of 0.78% (w/v) and glycerol of 0.85% (w/v) were able to delay ripening and prolong their shelf-life [17].

To our knowledge, has not been much scientific literature cover the applying a carrageenan-beeswax coating with the addition of two plasticizers on fresh-cut mangoes to maintain the quality and extend their shelf-life. Therefore, the purpose of this research was to develop edible film with plasticizer treatment and investigate the effect of an edible coating on quality changes in fresh-cut mangoes during refrigerated storage at 6 °C for 6 days.

2. Materials and Methods

2.1. Raw materials

The edible film materials (glycerol, fructose, glucose, and tween 80) were purchased from Brataco-Bandung, carrageenan from Setiaguna-Bogor, and beeswax from Bees Nature-Tangerang. Mangoes (*Mangifera indica L.*) were used for this study obtained from a traditional market in Subang. Uniformity in size, color, shape, and absence of damage became parameters in the selection of samples.

2.2. Edible film and fresh-cut mangoes preparation

Film solution was prepared by dissolving carrageenan powders in distilled water while stirring and heating at 70-80 °C until the solution became clear. The melting wax of 0.1% (w/v) and tween 80 of 0.2% (w/v) were added into solution. Two plasticizers were used, i.e., glycerol of 1% (w/v) and a monosaccharide of 1%. Two types of monosaccharides were chosen as treatments, i.e., glucose and fructose. The film-forming solutions (85ml) was poured onto acrylic plate 20x20 cm^2 and dried at 50 °C for 24 hours in a cabinet dryer [18].

Mangoes were washed in tap water then peeled manually used a stainless-steel knife and cut in pieces sized about 3 x 3 cm. Fresh-cut mangoes were dipped in film solutions for 1 minute. Samples

IOP Conf. Series: Earth and Environmental Science **251** (2019) 012029 doi:10.1088/1755-1315/251/1/012029

allowed standing at room temperature (27 °C) for 15 minutes, placed into styrofoam, and then over-wrapped with cling wrap. Fresh-cut mangoes without coating were used as the control. All samples were stored at 6 °C for 6 days

2.3. Analyses of edible film
Measurement of edible film thickness and water vapor transmission rate based on the method described in [19], while solubility described in [20]. Thickness was measured using a hand micrometer (Mitutoyo) with to the nearest 0.001 mm at five random positions around the film, and the average values were used in the calculation.

Water vapor transmission rate was determined by sealing the top of a cup containing silica gel of 10g. Cup was placed in a desiccator containing saturated magnesium nitrate solution (50% RH) at 25 °C. The cup was weighting every day (24 h) over 6 days. The WVTR was calculated by dividing slop of weight to sample.

Film solubility was evaluated by cutting film (diameter 2 cm). The film was weighed, immersed in 50 mL of distilled water, and then periodically agitated during 24h at 25 °C. The dry mass content of initial and final samples was measured by drying the samples at 105 °C for 24h.

2.4. Analysis of Fresh-Cut Mangoes
In order to determine weight loss, samples were prepared in individual packs. The initial weight of the sample was weighed at the beginning of the experiment and observation on days 2, 4, and 6 of storage. Weight loss was expressed as a percentage loss of the initial total weight [15].

The pH of mango puree was measured using a pH meter (Mettler Toledo FE20), while total soluble solids were measured using a digital refractometer (Hand-Held Refractometer ATAGO PAL-1). Fruits were cut into small pieces and pulped using a hand-held blender. Analysis of total acidity was done by adding 100 ml distilled water into 5 g mango puree. This solution then was titrated using NaOH solution with normality 0.1 N. Total acidity was represented in g of citric acid per 100 g of sample. The equation was used to calculate total acidity as follow [21]:

$$Titratable\ acidity = \frac{V_{NaOH} \times N_{NaOH} \times 0.064}{W_{sample}} \times 100\% \quad (1)$$

Where
W_{sample} : the mass of the fruit sample taken for analysis (g)
V_{NaOH} : the ml of NaOH used for titration (ml)
N_{NaOH} : the normality of NaOH solution (N)
0.064 is the conversion factor for citric acid

For the microbiological analysis, fresh-cut mangoes (5 g) crushed using a blender, and then put into 45 ml Buffered Peptone Water (Merck, Germany) aseptically. Sample homogenized using a vortex and dilution. To calculate the total plate count, the sample has been diluted was transferred into Plate Count Agar (Merck, Germany) using the pour plate method, and then, incubation occurred at 37 °C for 48 h. Three samples in each group were analyzed and reported as log CFU/g (colony forming units per g of sample) [22].

Sensory analysis was conducted by 30 untrained panelists to give the score in term taste, flavor, color, firmness and overall liking using a seven-point hedonic scale (1-dislike very much, 2-dislike moderately, 3-dislike slightly, 4- neither like nor dislike, 5-like slightly, 6-like moderately, 7-like very much). The samples were evaluated at day 0.

2.5. Statistical *Analysis*
The experimental design in this research uses a completely randomized design. When analysis of variance (ANOVA) revealed a significant effect (at the 0.05 level), treatment means were compared using Duncan's multiple range test.

3. Result and Discussion

3.1. Edible film characteristics

An important property for effective food packaging is its ability to prevent or minimize moisture transfer between the food and the surrounding environment. Some parameters of edible film related this function are thickness, water vapor transmission rate (WVTR) and solubility and presented in Table 1.

Table 1. Edible film parameters

Plasticizers	Thickness, mm	WVTR, g/mm^2/day	Solubility, %
Glyerol	0.048^a	12.61^a	58.99^a
Glycerol-fructose	0.078^b	23.27^b	77.96^a
Glycerol-glucose	0.076^b	22.55^b	77.84^a

Same letters within a column indicate no significant differences among treatments ($p > 0.05$)

Edible films with two plasticizers were significantly higher in thickness, and water vapor transmission rate (WVTR) compared to the film with the single plasticizer ($p <0.05$). The addition plasticizer concentration increases the film thickness, due to the increase in total solids in the film solution. The interstitial distance between the polymer chains in the film matrix could be increased because of the dispersion of plasticizer molecules in the film matrix, which could contribute to the increase in film thickness [23]. A film thickness with an inversion sugar plasticizer was between 0.04 mm and 0.12 mm [24]. Edible films prepared from semi-refined kappa-carrageenan (SRC) plasticized with glycerol or sorbitol at 20, 25, and 30% (w/w) had thickness ranged 0.066 – 0.079 mm

The increase in WVTR can be attributed to the hydrophilic nature of the plasticizer. Similar results were reported by some studies [18, 25], increased hydrophilic plasticizer concentrations lead to reorganization of polysaccharide tissues and an increase in free volume and segmental movement, allowing water molecules to diffuse more easily and provide higher WVTR.

The solubility of films were not significant differences between samples ($p >0.05$). Edible film made with two plasticizers generated solubility higher than mono plasticizer film. Edible film prepared with cassava starch of 5% (w/v) and glycerol of 1.5% (w/v) gave solubility of 73% [26]. Higher concentrations plasticizer resulted in more water-soluble films because plasticizer is hydrophilic. The interaction of plasticizer with the film matrix by increasing the space between the chains, facilitates water migration into the film and, consequently, increasing solubility [27].

3.2. Fresh cut-mangoes quality

3.2.1. Weight loss

Fruit weight loss was mainly related to loss of water and other volatile components in the respiration process and transpiration (water release in the form of water vapor) during storage. The result indicated that the composite layer could inhibit weight loss of fresh-cut mango (Figure 1).

All sample treatments showed a continuous weight loss during the storage period. The weight loss of control (uncoated mango slices) was greater than coated mango slices. The edible coating on fruits could make weight loss is relatively low because the edible coating has the ability to prevent the loss of water in the fruits. A significant difference between control and coated samples was observed at the $p < 0.05$ level when fruit was stored for six days at 6°C. However, there was no significant difference between fruit treated with a type of plasticizer ($p>0.05$). At the end of storage, the weight loss of coated samples with plasticizer glycerol, glycerol-fructose, and glycerol-glucose were 14.34%, 12.47%, and 14.81%, respectively, whereas the uncoated sample had 17.15% loss in weight. The results are consistent with previous studies in which coating based chitosan was found to be effective for delaying weight loss in fresh-cut mangoes [3,15,21] revealed that addition beeswax as a component in the composite coating exhibited a higher reduction in the weight loss of fruit.

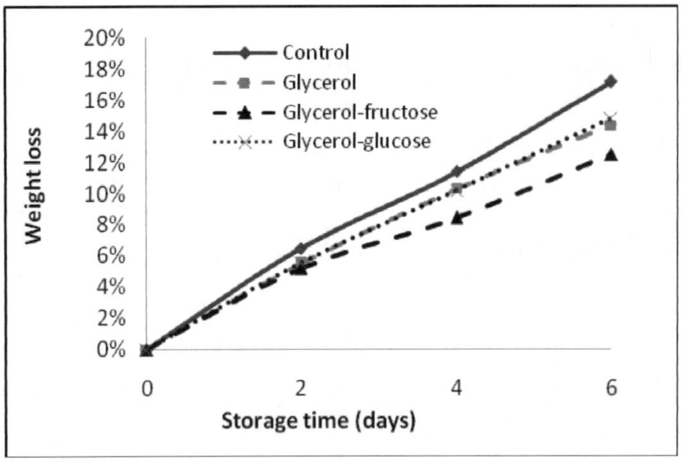

Figure 1. Weight loss value of fresh-cut mangoes stored at 6 °C

3.2.2. pH and total acidity

Generally, some indicators of fruit ripening and tend to increase upon ripening are pH and total acidity. The pH values of uncoated and coated samples during storage were presented in Table 2 and total acidity on figure 2. The initial pH value of an uncoated sample of 3.89 decreased during storage, resulting in a significant difference in storage days 2, 4, and 6, while the coated samples were no significant differences among the pH value of the variously coated samples during the storage period.

Table 2. Change in pH of fresh-cut mangoes during stored at 6 °C

Time (days)	pH			
	Control	*Glycerol*	*Glycerol fructose*	*Glycerol glucose*
0	3.89^{Aa}	3.97^{Aa}	$3{,}99^{Aa}$	3.92^{Aa}
2	3.56^{Ab}	3.97^{Ba}	3.97^{Ba}	3.91^{Ba}
4	3.41^{Ab}	$3{,}97^{Ba}$	3.98^{Ba}	3.94^{Ba}
6	3.38^{Ab}	3.96^{Ba}	3.99^{Ba}	3.93^{Ba}

Same capital letters within a row and same small letters within a column indicate no significant differences among treatments ($p > 0.05$)

After six days of storage, the pH of uncoated sample was 3.38, and for coated sample of 3.96; 3.99; and 3.93 for composite coating with plasticizer glycerol, glycerol and fructose, and glycerol and glucose, respectively. There was no significant difference among the pH value of fresh-cut mango coated with chitosan and dextrin after 6 days storage at 4 °C [28].

The result showed that the total acidity of coated samples no significant differences during storage, while the uncoated samples were significant differences at 2, 4, and 6 days. After 6 days of storage, the total acidity of coated samples was 0.49%, 0.49%, and 0.57%, while the uncoated sample was 0,80%, respectively (figure 2). The total acidity of chitosan-coated and uncoated mango fruit was not significantly different ($p>0.05$) after 7 days of storage, indicated that chitosan coating did not affect the total acidity of samples [15]. During storage, the total acid of fruit increased because of physiological process of the fruit itself and fermentation process resulting in organic acid. By serving as a gas barrier, the edible coating could act to control metabolic reaction and delayed respiration process. The surface coating could increase the resistance of fruit surface to gas permeability, reducing the respiration rate, and creating a modified internal atmosphere [29].

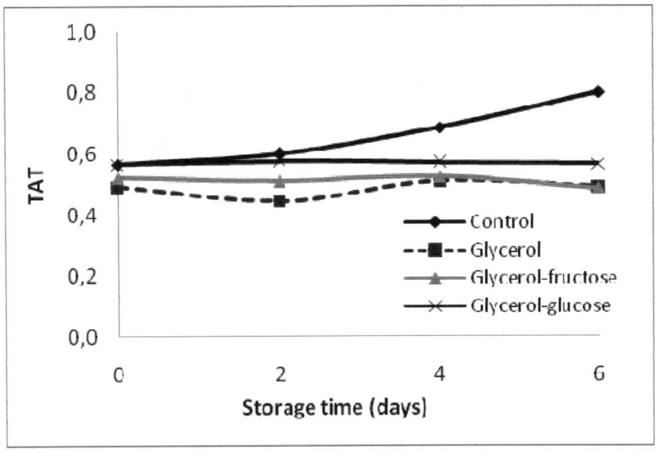

Figure 2. Total acidity value of fresh-cut mangoes stored at 6 °C

3.2.3. Total soluble solids

Changes in the total soluble solids of uncoated and coated fresh-cut mango over the storage period are shown in Figure 3.

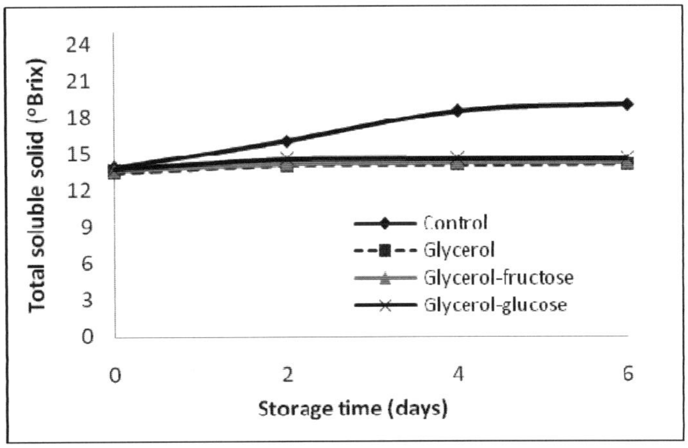

Figure 3. Total soluble solid value of fresh-cut mangoes stored at 6 °C

Fresh-cut mango slices that were not coated contained higher total soluble solids than coated mango slices. The soluble solids content of coated samples remained constant until the sixth day of storage and showed no significant difference ($p>0.05$) among the treatments. Only the soluble solid content of the control sample was higher and differed significantly from the rest of the fruits. At 0 and 2 days storage time, the total soluble solids content of uncoated samples was not significantly different ($p>0.05$), but at 4 and 6 days storage, there was a significant difference. The increase total soluble solid of fruit could be caused a lot of water loss during storage and acid metabolism as a result of fruit ripening still continuous by converting starch and acid to sugar [21; 30]. The metabolic activity involving oxygen from the environment will accelerate the ripening of fruit. The edible coating showed the ability to retain an increase total soluble solids of fresh-cut mango because edible coating could withstand oxygen transfer from the environment.

3.2.4. Microbiological analysis

In the preservation of minimally processed food, one of the most important factors to be considered is microbial safety. The microbiological analysis of coated and uncoated fresh-cut mangoes was presented in Table 3.

Table 3. Change in microbial counts of fresh-cut mangoes during stored at 6 °C

Time (day)	Microbial counts (Log CFU/g)			
	Control	Glycerol	Glycerol fructose	Glycerol glucose
0	2.74^{Aa}	2.54^{Ba}	2.08^{Ba}	2.11^{ABa}
2	5.90^{Aa}	3.11^{Ba}	2.78^{Ba}	5.85^{ABa}
4	5.94^{Aa}	3.64^{Ba}	3.42^{Ba}	6.77^{Aa}
6	7.09^{Ab}	3.30^{Ba}	4.36^{Ba}	6.49^{Ba}

Same capital letters within a row and same small letters within a column indicate no significant differences among treatments ($p > 0.05$)

The growth of microorganisms on control (uncoated samples) significantly increased after storage 6 days ($p \leq 0.05$). The total plate count of control (uncoated samples) at the end storage increased from 2.74 to 7.09 logs CFU/g. The carrageenan-beeswax coating on the fresh-cut mango effectively inhibited the growth of microorganisms. Nevertheless, the type of plasticizer did not further affect the growth of microorganisms ($p>0.05$). Microorganisms need oxygen to grow. The edible coating could withstand oxygen transfer from the environment, so decrease the risk of microorganism growth. The result was in accordance with previous studies [3; 15] which reported that the application of chitosan to sliced and fresh-cut mango respectively, could inhibit the growth of microorganisms.

The Indonesian National Standard (SNI 7388: 2009) concerning the maximum limit of microbial contamination in food for wet fruit sweets gives the total plate count limit (30 °C, 72 hours) of $1x10^5$ colonies/g. The control treatment showed that fresh-cut mangoes had exceeded the permitted standard threshold on the second day, as well as the glycerol-glucose treatment. In the treatment of glycerol plasticizer and glycerol-fructose plasticizers, they were below the maximum threshold for microbial contamination until the sixth day.

3.2.5. Sensory evaluation

Sensory data for fresh-cut mangoes are presented in Table 4. Five characteristics of mango (taste, flavor, color, firmness and overall liking) were analyzed for acceptability.

The odor, the color, and the firmness scores of control and coated mangoes were not significantly different ($p>0.05$). The edible coatings covered the fruits are transparent and thin, so they do not show significant differences. There were significant differences ($p \leq 0.05$) in the taste and the overall liking scores of mango fruit coated by glycerol-fructose with other treatments. The taste, the odor, the firmness, and the overall liking of uncoated fresh-cut mango received the highest score. It could be explained that the possibility of some ingredients of edible coating decreased the sweetness, resulted in a particular odor, and increased the hardness of treated samples. In term of color, coated mango received the better score, related with the glossy coating. The sensory quality deterioration of peeled litchi fruit could be delayed by coating used chitosan [31]. Chitosan coatings could preserve fresh-cut mango quality by maintaining sensory attributes in visual appearance, firmness and overall liking [15].

Table 4. Sensory quality of fresh-cut mangoes during stored at 6 °C

Attribute	Control	Glycerol	Glycerol fructose	Glycerol glucose
Taste	4.77^{b}	4.30^{ab}	4.00^{a}	4.67^{b}
Odor	4.33^{a}	3.90^{a}	3.80^{a}	4.17^{a}
Color	4.53^{a}	4.83^{a}	4.30^{a}	4.40^{a}
Firmness	4.73^{a}	4.27^{a}	4.23^{a}	4.50^{a}
Overall liking	4.57^{b}	4.40^{b}	3.90^{a}	4.43^{b}

a-b means within a column with different letters are significantly different ($p \leq 0.05$)

4. Conclusion

The combination of plasticizers utilized in the carrageenan-based edible films had remarkable effects on increasing thickness, water vapor transmission rate (WVTR), and solubility compared to the film with the single plasticizer. This edible coating could prolong fresh-cut mangoes during storage at 6 °C for 6 days. The edible coating could reduce weight loss, maintain pH and total acidity, delaying an increase of total soluble solids, and retard the growth of microorganisms in fresh-cut mangoes. In addition, there was not different in sensory perception between the coated fruit compared to the uncoated fruit.

5. Acknowledgment

The authors are grateful to The Centre for Appropriate Technology Development for providing the experimental facilities and to the Indonesian Institute of Sciences for financial support

6. Reference

[[1] NO Tridjaja and MS Mahendra 2012 Maturity Indices and Harvesting Practice of Arumanis' Mango Related to The Target Market *J. Teknol. Pertan.***1** 3 pp 17–22

[2] BS de Souza, T J O'Hare, JF Durigan, and PS de Souza 2006 Impact of atmosphere, organic acids, and calcium on quality of fresh-cut 'Kensington' mango *Postharvest Biol. Technol* **42** 2 pp 161–167

[3] PJ Chien, F Sheu, and FH Yang 2007 Effects of edible chitosan coating on quality and shelf life of sliced mango fruit *J. Food Eng* **78** 1 pp 225–229

[4] MA Rojas-Graü, R Soliva-Fortuny, and O Martín-Belloso 2009 Edible Coatings as Tools to Improve Quality and Shelf-Life of Fresh-Cut Fruits *Fresh Prod.* pp 65–72

[5] SA Valencia-Chamorro, L Palou, MA Delŕio, and MB Pérez-Gago 2011 Antimicrobial edible films and coatings for fresh and minimally processed fruits and vegetables: A review," *Crit. Rev. Food Sci. Nutr.* **51** 9 pp 872–900

[6] MA Rojas-Graü, R Soliva-Fortuny, and O Martín-Belloso 2009 Edible coatings to incorporate active ingredients to fresh-cut fruits: a review *Trends Food Sci. Technol.* **20** 10 pp 438–447

[7] M Vargas, C Pastor, A Chiralt, DJ McClements, and C González-Martínez 2008 Recent advances in edible coatings for fresh and minimally processed fruits *Crit. Rev. Food Sci. Nutr.* **48** 6, pp 496–511

[8] W Pavlath, AE, and Orts 2009 Edible films and coatings: why, what and how? in *Edible Films and Coatings for Food Applications*, K Embuscado, ME, Huber, Ed Springer pp 1–23

[9] M Lacroix and C Le Tien 2005 Edible films and coat ~ ngs trom non- starch polysaccharides *Innov. Food Packag. Overiew* pp 0–12

[10] V Morillon, F Debeaufort, and G Blond 2012 Critical Reviews in Food Science and Nutrition Factors Affecting the Moisture Permeability of Lipid- Based Edible Films : A Review Factors Affecting the Moisture Permeability of Lipid-Based Edible Films : A Review *Crit. Rev. Food Sci. Nutr.* pp 37–41

[11] Y Zhang and JH Han 2006 Mechanical and Thermal Characteristics of Pea Starch Films Plasticized with Monosaccharides and Polyols *J. Food Sci.* **71** 2 pp E109–E118

[12] S Kokoszka and A. Lenart 2007 Edible coatings – formation, characteristics and use – a review," *Polish J. Food Nutr. Sci.* **57** 4 pp 399–404

[13] B Adhikari, DS Chaudhary, and E Clerfeuille 2010 Effect of plasticizers on the moisture migration behavior of low-amylose starch films during drying *Dry. Technol* **28** 4 pp 468–480

[14] K Krogars, J Heinämäki, M Karjalainen, A Niskanen, M Leskelä, and J Yliruusi 2003 Enhanced stability of rubbery amylose-rich maize starch films plasticized with a combination of sorbitol and glycerol *Int. J. Pharm.* **251** 1–2 pp 205–208

[15] S Nongtaodum and A Jangchud 2009 Effects of edible chitosan coating on quality of fresh-cut mangoes (Fa-lun) during storage *Kasetsart J.- Nat. Sci* **43** 5 SUPPL pp 282–289

[16] A Plotto, JA Narciso, N Rattanapanoneb, and EA Baldwina 2010 Surface treatments and coatings to maintain fresh-cut mango quality in storage *J. Sci. Food Agric.* **90** 13 pp 2333–2341

[17] HM. Hamzah, A Osman, CP Tan, and F Mohamad Ghazali 2013 Carrageenan as an alternative coating for papaya (Carica papaya L. cv. Eksotika) *Postharvest Biol. Technol.* **75** pp 142–146

[18] N Afifah, E Sholichah, N Indrianti, and DA Darmajana 2018 Pengaruh kombinasi plasticizer terhadap karakteristik edible film dari karagenan dan lilin lebah *Biopropal Ind.* **9** 1 pp 49–60

[19] JW Rhim and LF Wang 2013 Mechanical and water barrier properties of agar/κ-carrageenan/konjac glucomannan ternary blend biohydrogel films *Carbohydr. Polym.* **96** 1 pp. 71–81

[20] E Colla, PJ do A Sobral, and FC Menegalli 2006 Amaranthus cruentus flour edible films: influence of stearic acid addition, emulsification stirring speed and plasticizer concentration in amaranth flour based film forming solutions on barrier and mechanical properties of its films *J Agric Food Chem* **54** pp 6645–6653

[21] E Velickova, E Winkelhausen, S Kuzmanova, VD Alves, and M Moldão-Martins 2013 Impact of chitosan-beeswax edible coatings on the quality of fresh strawberries (Fragaria ananassa cv Camarosa) under commercial storage conditions *LWT -Food Sci. Technol.* **52** 2 pp 80–92

[22] BSN 2006 *SNI 01-2332.3-2006* Badan Standarisasi Nasional Indonesia

[23] A Farhan and NM Hani 2017 Characterization of edible packaging films based on semi-refined kappa-carrageenan plasticized with glycerol and sorbitol," *Food Hydrocoll.* **64** pp 48–58

[24] P Veiga-Santos, LM Oliveira, MP Cereda, AJ Alves, and ARP Scamparini 2005 Mechanical properties, hydrophilicity and water activity of starch-gum films: Effect of additives and deacetylated xanthan gum," *Food Hydrocoll.* **19** 2 pp 341–349

[25] ML Sanyang *et al.* 2015 Effect of plasticizer type and concentration on tensile, thermal and barrier properties of biodegradable films based on sugar palm (arenga pinnata) starch **7** pp. 1106–1124

[26] W Tongdeesoontorn, LJ Mauer, S Wongruong, P Sriburi, and P Rachtanapun 2011 Effect of carboxymethyl cellulose concentration on physical properties of biodegradable cassava starch-based films," *Chem. Cent. J.* **5** 1 pp 1–8

[27] M Chiumarelli and MD Hubinger 2014 Evaluation of edible films and coatings formulated with cassava starch, glycerol, carnauba wax and stearic acid *Food Hydrocoll.* **38**, pp 20–27

[28] M Freire, M Lebrun, MN Ducamp, and M Reynes 2005 Evaluation of edible coatings in fresh cuts mango fruits *Inf. Technol. Sustain. Fruit Veg. Prod.* **5** September pp 12–16

[29] SL Marpudi, LSS Abirami, R Pushkala, and N Srividya 2011 Enhancement of storage life and quality maintenance of papaya fruits using Aloe vera based antimicrobial coating *Indian J. Biotechnol.* **10** 1 pp 83–89

[30] P Hernández-Muñoz, E Almenar, V Del Valle, D Velez, and R Gavara 2008 Effect of chitosan coating combined with postharvest calcium treatment on strawberry (Fragaria × ananassa) quality during refrigerated storage *Food Chem.* **110** no pp 428–435

[31] H Dong, L Cheng, J Tan, K Zheng, and Y Jiang 2004 Effects of chitosan coating on quality and shelf life of peeled litchi fruit *J. Food Eng.* **64** 3 pp 355–358

IOP Conf. Series: Earth and Environmental Science **251** (2019) 012030 doi:10.1088/1755-1315/251/1/012030

The Effects of Solvents and Maltodextrin on the Characteristics of *Physalis angulata* L. Leaf Extract

A Triyono[1], R Luthfiyanti[1], T Rahman, N Y Pamungkas[2]

[1]Development Center for Appropriate Technology, Indonesian Institute of Sciences K.S Tubun Street No. 5 Subang, West Java, Indonesia
[2]Food Technology Program, Faculty of Engineering, Pasundan University, Jl. Dr. Setiabudi No. 9, Bandung, 40153. Indonesia

Email: ateluyono@gmail.com

Abstract. The research aimed to study the influence of the use of solvent type with the addition of maltodextrin on various concentration on the physalis angulata leaves to obtain the maximal amount of flavonoid, polyphenol, and alkaloid. The Physalis angulata L leaf was extracted using a solvent of 70% ethanol, 100% distilled water and maltodextrin with concentrations of 0%, 3%, 6%, and 9%. The UV-VIS spectrophotometric method was used to analyze the quality of flavonoid and polyphenol compounds. The gravimetric method was used to analyze the alkaloid compounds. Results of analysis of Physalis angulata leaf extract showed that the treatment of 70% ethanol added with 9% maltodextrin obtained the highest average value of flavonoid. The average amount of flavonoid was 3268.20 mg GAE/g extract. The treatment of 70% ethanol added with 6% maltodextrin produced the highest average value of alkaloid. The average amount of alkaloid was 53.16%. Results of advanced Duncan's test showed that the highest average value of all parameters analyzed appeared that the best treatment was a solvent of 70% ethanol added with 9% maltodextrin.

Keywords: Alkaloid, *Ciplukan*, Flavonoid, Polyphenol, Spectrophotometry

1. Introduction

Indonesia is one of the second largest biodiversity countries in the world after Brazil which consists of tropical plants and marine biota, with as many as 30,000 species of plants, 2500 of which are medicinal [6]. Herbal medicinal products and herbal supplements derived from medicinal plants over the last three decades are increasing rapidly. From this case, it is evident that Indonesia has the potential to develop products from herbs based on functional foods. One of the herbaceous plants that have begun to be developed is *Ciplukan* [6]. *Ciplukan* (Latin: *Physalis angulata L.*) is the name of small fruit which is covered by the enlarged petals of flowers when it is ripe. This plant is usually used as a traditional medicine in which all the parts of the plant, i.e., roots, stems, leaves, and fruit can be used as a traditional herb that is able to treat degenerative diseases such as diabetes mellitus [3]. Diabetes is a metabolic disorder caused by the deficiency of insulin hormones produced by pancreatic beta cells. Lack of insulin causes the glucose not to be stored in the blood and not processed by the cells to become energy [5].

Physalis angulata L plants have the efficacy of treating diabetes mellitus because it contains secondary metabolite compounds including flavonoids, polyphenols, and alkaloids. Flavonoids and polyphenols are antioxidant compounds that function to overcome or neutralize free radicals. Alkaloids also have the ability to regenerate damaged pancreatic beta cells and the ability to inhibit the action of the α-glucosidase enzyme [8] [10] [11].

These secondary metabolites can be obtained by extracting materials from *Physalis angulata L* plants. Many factors can affect the content of extracted compounds, i.e., the type of solvent, and the process of evaporation. Therefore, this research aimed to study and determine the influence of the type of solvent and the addition of maltodextrin at different concentrations up to the level that would obtain maximal flavonoid, polyphenol, and alkaloid compounds from the *Physalis angulata L* leaf extract.

2. Methodology

2.1. Material and Tools

The materials used were *Physalis angulata L* leaves, 70% ethanol, 30% methanol, 10% NaOH, quercetin standard solution, methanol p.a, 10% $AlCl_3$, aquadest, 1M sodium acetate, ammonia, 1% ammonium hydroxide, standard solution of gallic acid, reagents Folin-Ciocalteau (1:10), and 7.5% Na_2CO_3. The equipment used are knives, analytical balance, 100 mL and 250 mL Erlenmeyer flasks, 100 mL glass beaker, filter paper, funnel, vacuum rotary, evaporator, refractometer, pH meter, Chromameter and UV-Vis spectrophotometry.

2.2. Research Method

The fresh *Physalis angulata L* leaves were obtained from Jabong area (Latitude $6^0$31'07.80"S, Longitude $107^0$46'41.07" E, and elevation 67 MAMSL), Subang, West Java. The leaves were washed and chopped into the same small size and extracted by maceration using 70% ethanol and water solvent with leaf and solvent ratio of 1:10. The mixture of chopped leaves and solvent was poured into an Erlenmeyer covered with aluminium foil stored for 24 hours, shaken occasionally, then filtered. The filtrate from the maceration was then added with maltodextrin in different concentrations of 0%, 3%, 6%, and 9%. The extract of the *Physalis angulata L* leaf was obtained by evaporating the mixture. There was two analysis applied to characterize the extract of *Physalis anggulata leaf*, i.e. qualitative and quantitative.

The qualitative analysis was applied to characterized the parameters of TPT, pH and colour of the extract. Further analysis of the quantitative test was applied to investigate the content of flavonoid, polyphenol, and alkaloid compounds. The levels of flavonoids were obtained by a UV-VIS spectrophotometric method at a 431 nm wavelength using quercetin as the standard solvent. The levels of polyphenols were determined using the UV-Vis spectrophotometry at wavelengths of 765 nm by using gallic acid as the standard solvent. The alkaloid content was tested using a gravimetric method in which the alkaloid compounds in the sample were precipitated using ammonium hydroxide then filtered and dried.

The experiment design used was a factorial in the interaction of randomized block design with a treatment design consisting of two factors. The first factor was the type of solvent (A) which consisted of water (a_1) and 70% of ethanol solvent (a_2). The second factor was the addition of maltodextrin (B) with concentration levels of 0 % (b_1), 3 % (b_2), 6 % (b_3), and 9 % (b_4). The factorial pattern used was 2 x 4 with 3 repetitions. The response designs were physical-chemical responses, consisted of TPT and pH. Physical responses were colour, consisted of brightness, redness and yellowness (L*, a*, b*), and chemical responses were content of flavonoids, polyphenols, and alkaloids.

3. Results and Discussion

The type of solvent and the addition of maltodextrin at different concentrations in the leaf extraction process affected physical-chemical, physical, and chemical responses. Each solvent had a different polarity that affected the amount of active compounds dissolved in the process of extraction and addition of maltodextrin at different concentrations. This also affected the level of content of metabolite compounds obtained. The results of the research appeared that the best solvent to extract the *Physalis angulata L* leaves was the treatment of 70% ethanol added with 9% maltodextrin.

3.1. *Total Dissolved Solids*

The result of variation analysis showed that the solvent type (A) treatment using 70% ethanol and water solvent as well as the treatment of maltodextrin concentration (B) had significant effects (F count > F table 5%). The interaction of both treatments (AxB) also had a significant effect (F count > F table 5%). Table 1 showed the average value of total dissolved solids of the *Physalis angulata L* leaf extract.

Table 1. Average value of total dissolved solids in *Physalis angulata L* leaf extract

	Brix Percentage		
Solvent	**Maltodextrin**	**Total**	**Average**
Water (a_1)	Malto 0% (b_1)	43.00	14.33 [a]
	Malto 3% (b_2)	97.50	32.50 [c]
	Malto 6% (b_3)	116.70	38.90 [e]
	Malto 9% (b_4)	128.80	**42.93** [g]
70% Ethanol (a_2)	Malto 0% (b_1)	56.20	18.73 [b]
	Malto 3% (b_2)	105.80	35.27 [d]
	Malto 6% (b_3)	126.20	42.07 [f]
	Malto 9% (b_4)	139.20	**46.40** [h]

Note: The same letters at the end of each average value do not show actual differences

The Results showed that the average total of dissolved solids (brix) in the thick *Physalis angulata L* leaf extract on each treatment showed a significant difference. The highest average treatment by water was the addition of 9% maltodextrin with a total of dissolved solids (brix) 42.93%. The highest average treatment by 70% ethanol solvent was the addition of 9% maltodextrin with a total of dissolved solids (brix) of 46.40%. Overall, it could be concluded that the highest average of the total dissolved solids was 46.40%, which was shown by the treatment of 70% ethanol solvent with the addition of 9% maltodextrin, The lowest average of that was 14.33% shown by the water solvent with 0% maltodextrin concentration. This indicated that the increasing level of maltodextrin concentration caused the total percentage of dissolved solids increased. It was expected that the content of phytochemical compounds in the extract was also at a high level.

Water was more polar than 70% ethanol. Water could extract polar compounds, whereas 70% of ethanol could extract both polar and nonpolar compounds because it had nonpolar alkyl groups. Therefore, total dissolved solids with 70% ethanol solvent had a higher average value than water. It could be postulated that the higher the total value of dissolved solids, the better the quality of the extract. The compounds measured as total dissolved solids included complex and simple sugars, organic acids, several classes of pigments, dilutable vitamin, and other organic compounds. Maltodextrin was a solid dye in the form of fine white power without taste. Maltodextrin could be obtained by partially hydrolyzing cassava starch with the α-amylase enzyme at a temperature of 85°C for 65 minutes. Maltodextrin is a mixture of maltose, maltotriose, and maltotetraose. Therefore, the addition of maltodextrin can increase the total dissolved solids [13].

3.2. *Value of pH*

The result of variation analysis showed that the solvent type (A) treatment of 70% ethanol and water solvent, and the treatment of maltodextrin concentration (B) had a significant effect (F count > F table 5%). The interaction of both treatments (AxB) also had a very significant effect (F count > F table 5%). Table 2 showed the average pH value of the *Physalis angulata L* leaf extract.

Results of the analysis showed that the lowest average pH was 5.0l which had the highest acid content occurred in the treatment of water solvent with the addition of 9% maltodextrin. The highest average pH was 5.62 which had the lowest acid content occurred in the treatment of ethanol with 0% maltodextrin. The characteristics of water solvents were more polar than solvents of 70% ethanol; this resulted in the lowest pH of 5.01 occurred in the treatment of water solvent with the addition of 9% maltodextrin. From the average value of treatments, it could be concluded that the higher the amount of maltodextrin addition, the lower the pH value. In other words, the content of the *Physalis angulata* leaf extract had more acid. The pH obtained in this study was classified as a weak acid, which in water it produced H^+ ion imperfectly, and would result in a pH value ranged from 3-5. Acid strength is determined by the ability to produce H^+ ions. The more H^+ ions produced, the stronger the acid [14].

Table 2. Average pH value of *Physalis angulata L* leaf extract

pH			
Solvent	**Maltodextrin**	**Total**	**Average**
Water (a_1)	Malto 0% (b_1)	16.01	5.34^{d}
	Malto 3% (b_2)	16.06	5.35^{d}
	Malto 6% (b_3)	15.11	5.04 ^{a}
	Malto 9% (b_4)	15.03	**5.01** ^{a}
70% Ethanol (a_2)	Malto 0% (b_1)	16.85	5.62 ^{e}
	Malto 3% (b_2)	15.76	5.25 b^{c}
	Malto 6% (b_3)	15.68	**5.23** ^{b}
	Malto 9% (b_4)	15.91	5.30 ^{cd}

Note: The same letters at the end of each average value do not show actual differences

3.3. Brightness Level (L)*

The result of variation analysis showed that the solvent type (A) treatment using 70% ethanol solvent and water solvent and the treatment of maltodextrin concentration (B) had a significant effect (F count > F table 5%). The interaction of both treatments (AxB) also had a significant effect (F count > F table 5%). The average level of brightness (L*) of the *Physalis angulata L* leaf extract was shown in Table 3.

Table 3. The average level of brightness (L*), redness (a*) and yellowness (b*) of the physalis angulata leaves extract

Solvent	Maltodextrin	L*	a*	b*
Water (a_1)	Malto 0% (b_1)	30.035 ^{b}	0.692 ^{d}	3.637 ^{c}
	Malto 3% (b_2)	30.695 ^{cd}	0.126 ^{a}	**2.652** ^{b}
	Malto 6% (b_3)	31.052 ^{def}	**0.184** ^{bc}	2.085 ^{a}
	Malto 9% (b_4)	**31.343** ^{f}	0.093 ^{a}	2.215 ^{a}
70% Ethanol (a_2)	Malto 0% (b_1)	28.738 ^{a}	0.714 ^{d}	4.614 ^{d}
	Malto 3% (b_2)	30.403 ^{bc}	**0.222** ^{c}	**3.577** ^{c}
	Malto 6% (b_3)	30.866 ^{de}	0.192 ^{c}	2.596 ^{b}
	Malto 9% (b_4)	**31.114** ^{ef}	0.157 ^{bc}	2.618 ^{b}

Note: The same letters at the end of each average value do not show actual differences

The L* value indicates the brightness level from dark to light with a range value of 0-100. Results showed that the highest average of brightness was 31.343 occurred in the treatment of the water solvent with an addition of 9% maltodextrin. The lowest average value of that was 31.114 occurred in the treatment of 70% ethanol solvent with an addition of 9% maltodextrin. The overall treatment appeared that the highest value was caused by the chlorophyll content that was extracted. The extracts that used water solvent had a brighter green colour compared to that of the 70% ethanol solvent extracts which had a darker green colour. It meant that the 70% ethanol solvent could extract the *physalis angulata* leaf better than the water solvent. This results were in accordance with a study conducted by Widya, et al. [15], which stated that 85% alcohol and 85% acetone solvent could extract the suji leaf greater than water solvent. Results of this study were also in accordance with the researched done by De Mann [12] which stated that Chlorophyll was easily extracted using organic solvents such as acetone, alcohol, methanol, ethyl acetate, pyridine, and dimethylformamide. Chlorophyll contains phytol groups that have low solubility towards the water.

The addition of maltodextrin affected the brightness of the Physalis angulata L extract, in which the higher the concentration of maltodextrin, the higher the brightness. This was due to the dissolved chlorophyll content in the extract would be bounded by the maltodextrin which caused the level of brightness to be higher.

3.4. Redness Level (a)*

The result of variation analysis showed that the solvent type (A) treatment of using 70% ethanol solvent and water solvent and the treatment of maltodextrin concentration (B) had a significant effect (F count > F table 5%). The interaction of both treatments (AxB) also had a very significant effect (F count > F table 5%) on the treatment of the study. The average redness level (a*) of the *Physalis angulata L* leaf extract was shown in Table 3.

The a* values show that the trend of colour from green to red with a range value of -100 to +100. The greater value of a* shows the trend of an increasingly red colour. Results of the analysis showed that the highest average level of redness in the treatment of water solvent with the addition of 6% maltodextrin was 0,184. The highest average level of redness in the treatment of 70% ethanol solvent with the addition 3% maltodextrin was 0,222. The highest average level of redness from both solvents was 0,222 which occurred in the treatment of 70% ethanol solvent with the addition of 3% maltodextrin. It was suspected that the reddish colour indicated the content of alkaloid and polyphenols. Alkaloid and polyphenols are phytochemical compounds that present in plants as pigment of orange, red, brownish red, and brown. It was expected that the addition of maltodextrin as a binder retained the flavonoid compound in the *Physalis angulata L* extract which could be lost due to processing and evaporation.

3.5. Yellowness Level (b)*

The result of variation analysis showed that the solvent type (A) treatment using 70% ethanol solvent and water solvent, and the treatment of maltodextrin concentration (B) had a significant effect (F count > F table 5%). The interaction of both treatments (AxB) also had a very significant effect (F count > F table 5%). The average yellowness level (b*) of the *Physalis angulata L* leaf extract was shown in Table 3.

The b* values show the trend of colour from blue to yellow with a range value of -100 to +100. The greater the value of b* shows a trend of colour increases to yellow. Results showed that the highest average of yellowness level was 3,577, obtained by the treatment of using the 70% ethanol solvent with an addition of 3% maltodextrin. Meanwhile, the lowest value was 2.652, obtained by the treatment of using the water solvent with an addition of 6% maltodextrin. The overall highest average value occurred in the treatment of using the 70% ethanol solvent. It was believed that the yellowness indicated the presence of flavonoid compounds in the extract. Flavonoids are one of the pigments found in plants as the pigment of yellow, orange, and yellowish orange colours. It was expected that the addition of maltodextrin as a binder could retain the flavonoid compounds in the *Physalis angulata L* extract which could be lost due to processing and evaporation.

3.6. Level of Flavonoid Compounds

The result of variation analysis showed that the solvent type (A) treatment using 70% ethanol solvent and water solvent, and the treatment of maltodextrin concentration (B) had a significant effect (F count > F table 5%). The interaction of both treatments (AxB) also had a significant effect (F count > F table 5%). The value of flavonoid compounds level in the *Physalis angulata L* leaf extract was shown in Table 4.

Table 4 showed that the average level of flavonoid compounds in the *Physalis angulata L* leaf extract were significant difference in each treatment, except in treatment a_1b_3 and a_2b_1 which showed similarity. The highest average of flavonoid in the treatment of water solvent with the addition of 9%

maltodextrin was 4,594 ppm. The highest average of that in the treatment using 70% ethanol with the addition of 9% maltodextrin was 8,439 ppm. Overall, it could be concluded that the highest average level of the flavonoid compounds in the *Physalis angulata L* leaf extract was 1,421 ppm, obtained by the treatment of 70% ethanol solvent with an addition of 9% maltodextrin, Otherwise, the lowest value was 1,421 ppm, obtained by the treatment of the water solvent and with addition of 0% maltodextrin. This showed that the higher amount of maltodextrin added, the higher the level of flavonoid compounds contained in the Physalis angulata L leaf which was extracted using the ethanol solvent.

Table 4. Average value of flavonoid compounds of the *Physalis angulata L* leaf extract

Flavonoid Level			
Solvent	**Maltodextrin**	**Total**	**Average**
Water (a_1)	Malto 0% (b_1)	4.26	1.421 [a]
	Malto 3% (b_2)	7.48	2.492 [b]
	Malto 6% (b_3)	10.00	3.335 [c]
	Malto 9% (b_4)	13.78	**4.594** [d]
70% Ethanol (a_2)	Malto 0% (b_1)	9.98	3.328 [c]
	Malto 3% (b_2)	15.11	5.037 [e]
	Malto 6% (b_3)	23.48	7.826 [f]
	Malto 9% (b_4)	25.32	**8.439** [g]

Note: The same letters at the end of each average value do not show actual differences

Flavonoids play an essential role in capturing free radicals or functioning as natural antioxidants so that they can prevent complications or progression of diabetes mellitus by clearing excessive free radicals and sever the chain of free radical reactions. These antioxidant activities allow flavonoids to capture or neutralize free radicals, such as ROS or RNS, associated with phenolic OH groups in order to improve the state of damaged tissue. This meant that the inflammatory process could be inhibited, chelating metal could be bound, and the polyol pathway could be blocked by inhibiting the aldose reductase enzyme. Flavonoids also have an inhibitory effect on alpha-glucosidase enzyme by the hydroxylation bond and substitution on the β ring. The principle of inhibition is similar to that of acarbose which has been used as a drug for the treatment of diabetes mellitus, by producing delays of carbohydrate and disaccharide hydrolysis and glucose absorption as well as inhibiting the metabolism of sucrose into glucose and fructose [2] [7].

3.7. Level of Alkaloid Compounds

The result of variation analysis showed that the solvent type (A) treatment using 70% ethanol solvent and water solvent and the treatment of maltodextrin concentration (B) had a significant effect (F count > F table 5%). The interaction of both treatments (AxB) also had a significant effect (F count > F table 5%). The average value of the levels of alkaloid compounds could be seen in Table 5.

Table 5. Average value of alkaloid compounds of the *Physalis angulata L* leaf extract

Alkaloid Level			
Solvent	**Maltodextrin**	**Total**	**Average**
Water (a1)	Malto 0% (b1)	74.73	24.91 [a]
	Malto 3% (b2)	103.06	34.35 [d]
	Malto 6% (b3)	127.42	**42.47** [f]
	Malto 9% (b4)	85.29	28.43 [c]
70% Ethanol (a2)	Malto 0% (b1)	107.33	35.78 [e]
	Malto 3% (b2)	141.84	47.28 [g]
	Malto 6% (b3)	159.48	**53.16** [h]
	Malto 9% (b4)	80.78	26.93 [b]

Note: The same letters at the end of each average value do not show actual differences

Results showed that the average content of alkaloid compounds in the *Physalis angulata L* leaf extract on each treatment showed a significant difference. The highest average of alkaloid in the treatment of water solvent with the addition of 6% maltodextrin was 42.47%. The highest average of that using the 70% ethanol solvent with addition of 6% maltodextrin was 53.16%. Overall, it can be concluded that the highest average value of alkaloid content was 53.16% which occurred in the treatment of 70% ethanol solvent with addition of 6% maltodextrin. Otherwise the lowest average value of alkaloid was 24.91%, obtained by the treatment of water solvent with addition of 0% maltodextrin. In each addition of 9% maltodextrin, there was a decrease in the alkaloid compound which made the addition no longer effective. However, it could be concluded that the higher the concentration of maltodextrin (up to 6%), the higher the alkaloid compounds contained in the leaf which was extracted using the ethanol solvent.

The alkaloid is widely used in medicine. One of the benefits is to treat diabetes mellitus. Alkaloids are shown to have properties that can regenerate the pancreatic β cell. With the improvement of the pancreas tissue, there is an increase in the amount of insulin in the body so that blood glucose can enter the sell, resulting in decreased blood glucose in the body. Alkaloids work by stimulating the hypothalamus to increase the secretion of Growth Hormone Releasing Hormone (GHRH), so the secretion of Growth Hormone (GH) in the hypophysis increases. High GH levels will stimulate the liver to secrete Insulin-like Growth Factor-1 (IGF-1). IGF-1 affects inducing hypoglycemia and lowering gluconeogenesis, so that blood glucose levels and insulin requirements decrease. The IGF-1 through a negative feedback system will normalize GH levels again [4].

3.8. Level of Polyphenol Compounds

The result of variation analysis showed that the solvent type (A) treatment using 70% ethanol solvent and water solvent and the treatment of maltodextrin concentration (B) had a significant effect (F count > F table 5%). The interaction of both treatments (AxB) also had a significant effect (F count > F table 5%) on the treatment of the study. The average value of the polyphenol compounds level could be seen in Table 6.

Table 6. Average value of polyphenol compounds of the *Physalis angulata L* leaf extract

	Polyphenol Level		
Solvent	**Maltodextrin**	**Total**	**Average**
Water (a_1)	Malto 0% (b_1)	8381.33	**2793.78** d
	Malto 3% (b_2)	7175.72	2391.91 b
	Malto 6% (b_3)	7527.00	2509.00 c
	Malto 9% (b_4)	6323.46	2107.82 a
70% Ethanol (a_2)	Malto 0% (b_1)	11286.00	**3762.00** h
	Malto 3% (b_2)	8659.29	2886.43 e
	Malto 6% (b_3)	9270.85	3090.28 f
	Malto 9% (b_4)	9804.59	**3268.20** g

Note: The same letters at the end of each average value do not show actual differences

The findings showed that the average content of polyphenol compounds in the *Physalis angulata L* leaf extract on each treatment showed the significant difference. The highest content of polyphenol was 3762 mg GAE/g extract obtained by the treatment of 70% ethanol solvent with no addition of maltodextrin (0%). In the treatments of 70% ethanol solvent with an addition of different maltodextrin concentrations, appeared that there was an increase in the polyphenol content as the percentage of addition increase. The highest average value of polyphenol content was 3268.20 mg GAE/g extract which was obtained by the addition of 9% maltodextrin,

In the treatment of water solvent, the highest value was 2793.78 mg GAE/g extract which was obtained without the addition of maltodextrin. It could be concluded that the highest polyphenol levels were obtained in the use of 70 % ethanol solvent without the addition of maltodextrin. However, the addition of maltodextrin in the ethanol solvent proves that the higher the concentration of maltodextrin, the higher the polyphenol content even though the highest remained to be obtained without any addition of maltodextrin.

Polyphenols play a role in colouring a plant, for example, a leaf colour during autumn. Polyphenols are found in fruits, vegetables, and grains. The average human consumes up to 23 mg of polyphenols a day. Polyphenols are often present in polar glycosides and are easily soluble in polar solvents. The benefits of polyphenols include lowering blood glucose levels and protecting against various diseases such as cancer. Polyphenols help counter the formation of free radicals. Polyphenol antioxidants are able to reduce oxidative stress by preventing the chain reaction of superoxide to superoxide hydrogen by donating hydrogen atoms from the aromatic hydroxyl (-OH) group of polyphenols to bind free radicals and discharge them from the body through excretion systems. The role of polyphenols as antioxidants is thought to be capable of protecting pancreatic β cells from toxic effects of free radicals produced under conditions of chronic hyperglycemia [9] [1].

4. Conclusion

The results of the variation analysis of the parameters of brix, pH, brightness, redness, yellowness, levels of flavonoids, polyphenols, and alkaloids indicated that the solvent type (A) treatment using 70% ethanol and water, and the treatment of maltodextrin concentration (B) had a significant effect (F count > F table 5%). The interaction of both treatments (AxB) also had a very significant effect (F count > F table 5%). TPT observation showed that the highest average value was 46.40% obtained by the 70% ethanol treatment with an addition of 9% maltodextrin. The pH observation showed that the lowest average value, indicating the highest acid content, was 5.01, obtained in the treatment of water solvent with addition 9% maltodextrin. The brightness level (L*) observation showed that the highest average value was 31.343 obtained in the treatment of the water solvent with the addition of 9% maltodextrin. The redness level (a*) observation showed that the highest value was 0.222, obtained in the treatment of 70% ethanol with addition of 3% maltodextrin.

The yellowness level (b*) observation showed that the highest value was 3.577, obtained in the treatment of 70% ethanol with the addition of 3% maltodextrin solvent. The higher the concentration of maltodextrin added, the higher the average value of flavonoid compounds in the *Physalis angulata L* leaf extract. The highest average of flavonoid was 8,439 ppm, obtained in the treatment of 70% ethanol with the addition of 9% maltodextrin. The higher the maltodextrin concentration added (up to 6%), the higher the alkaloid compounds in the *Physalis angulata L* leaf extract. Otherwise, there was a decrease in the average of maltodextrin addition. The highest value of alkaloid was 53.16 % obtained in the treatment of 70% ethanol solvent with an addition of 6%.

In the use of 70% ethanol solvent, the addition of maltodextrins in different concentrations might increase the content of polyphenol compounds. However, the highest average value was 3762 mg GAE/g extract, obtained in the treatment of 70% ethanol solvent without any addition of maltodextrin. Overall, the 70% ethanol solvent had higher average value compared to the water solvent. Based on the Duncan advanced test table, the treatment based on the type of solvent and with the addition of maltodextrin as well as the combination of treatments had different effects. The highest value of all the analysis parameters was obtained by the treatment of 70% ethanol solvent and 9% maltodextrin.

5. Acknowledgements

Our gratitude goes out to Insinas Research Grant from The Ministries of Research, Technology, and the Higher Education Republic of Indonesia and Indonesian Institute of Sciences for funding this study. We also are grateful to Prof. Dr. Ir. Wisnu Cahyadi, M.Si from Pasundan University- Bandung for his support in carrying out this research.

6. References

[1] Barbosa D S 2007 Green Tea Polyphenolic Compounds and Human Health *J. Consumer Protection and Food Safety* **2** 407-413

[2] Botutihe 2010 *Efek Ekstak Rumput Laut Coklat (Sargasum duplicatum Bory) Terhadap Profil Radikal Bebas dan Protein Kinase C Paru Tikus (Rattus novegicus) yang Dipapar Benzo[A]piren* Tesis Universitas Brawijaya Malang: Indonesia

[3] BPOM-RI 2007 Acuan Sediaan Herba Volume Ketiga Edisi Pertama Badan POM Jakarta
[4] Bunting K, J K Wang and M F Shannan 2006 *Control of Interleukin-2-gene Transcription: a Paradigm For Inducible, Tissue Specific Gene Expressions. Interleukins*, eds G Litwack 74 Elsevier Academic Press Inc p 105-145
[5] Clark M 2008 *Diabetes Self-Management Education: a Review of Published Studi* Published Studies Primary Care Diabetes 2 p 113-120
[6] Kemendag RI 2014 *Warta Ekspor Obat Herbal Tradisional* Kementerian Kesehatan Republik Indonesia: Jakarta
[7] Mills S and K Bone 2002 *Principles and Practice of Phytotherapy:Modern Herbal Medicine* Edinburgh: Scotland Churral Livingstone
[8] Murali Krishna T 2013 In Vitro Determination Of Antioxidant Activity Of Physalis Angulata Lnn. *Int J. of Pharma and Bio Sciences* **4** p 541 – 549
[9] Ni Luh and Razimin 2013 *Bawang Dayak Si Umbi Ajaib Penakluk Aneka Penyakit* PT Agromedia Pustaka Jakarta
[10] Prameswari O, Meidiana and Widjanarko S B 2014 Uji Efek Ekstrak Daun Pandan Wangi Terhadap Penurunan Kadar Glukosa Darah Dan Histloplatologi Tikus Diabetes Melitus *J. Pangan dan Argoindustri* **2** p 16-27
[11] Winarsi H 2007 *Antioksidan Alami Dan Radikal Bebas* Kanisius Yogyakarta Indonesia
[12] De man J M 1980 *Principles of Food Chemistry* The AVI Publishing CO Inc Westport Connecticut
[13] Griffin V K and Brooks J R 1989 Poduction and Size Distribution of Rice Maltodekxtrin Hydrolyzed from Milled Riced Flour Using Heat Stabile α-Amylase *J Food Science* vol 54 issue 1 p 190-193
[14] Sutresna N 2007 *Cerdas Belajar Kimia Untuk Kelas XI Sekolah Menengah Atas Program Ilmu Pengetahuan Alam* Grafindo Media Pratama Bandung:Indonesia
[15] Putri W D R, Zubaidah E and Sholahudin N 2018 Ekstraksi Pewarna Alami Daun Suji, Kajian Pengaruh Blanching dan Jenis Bahan Pengekstrak *J. Tek Pert* **4** (1) 13 - 24

IOP Conf. Series: Earth and Environmental Science **251** (2019) 012031 doi:10.1088/1755-1315/251/1/012031

Optimization of Gluten Free Spaghetti Products from Local Food with the Taguchi Method Approach

N K I Mayasti[1], M Ushada[2], and M Ainuri[2]

[1]Development Center for Appropriate Technology, Indonesian Institute of Sciences Subang, Indonesia
[2]Agricultural Technology Faculty, Gadjah Mada University, Yogyakarta

Email: indahmayasti@gmail.com

Abstract. Pasta product, especially spaghetti, is a very potential product to be developed. As the market growth, nutritious gluten-free spaghetti product is now being developed (as a response from consumer needs. The production process of gluten-free spaghetti relies on the gelation process and retrogradation mechanism which requires an extruder machine. The objectives of this research are to optimize the production of gluten-free spaghetti by combining four factors and three levels based on controlled quality. Taguchi method will be applied in this research. Taguchi data tabulation includes SNR, ANOVA, contribution percentage, multiresponse loss function, and a confirmation test. The result of this optimization research is the gluten-free spaghetti made by local flour (60 mesh) with composition of 30% mocaf, 30% corn flour, 30% rice flour, 10% soybean flour, 1.5% guar gum with confirmed quality characteristics (within confidence interval) : moisture content 9.94±0.57%, protein content 10.33±0.55%, fat content 1.11±0.67 %, carbohydrate content 76.51±2.32%, cooking time 10.65±0.25 minutes, cooking loss 13.00±2.76, elongation 181.35±25.72, hardness 2069.03±453.07 gf, adhesiveness -32.75±3.41 g.sec, color 76.51±0.63 dL and sensory of taste 4.80±0.16.

1. Introduction

Taguchi method is one of the methods which applies to produce a robust product towards external noise. Robust design is intended to optimize experimental results and based on quality improvement by decreasing the additive effect [1]. Taguchi method is chosen for this study as it helps to include several factors and levels, as well as, to conclude response from factors and controls which results in optimum response. Product development with Taguchi is expected to create an optimum product with efficient time and cost.

The development of pasta product, especially spaghetti, is potentially advanced. The product has turned into a global commodity and becomes a staple food in western countries and many other countries. The total number of pasta production is increasing every year [2]. In Indonesia, the development of pasta product was previously limited to the high-class society, and nowadays, its consumption level is rising from 5% to 30% [3]. As the market growth, nutritious gluten-free spaghetti product with high protein and high fiber is now being developed as a response to the increasing criteria from the consumers [4].

Spaghetti is defined as a dry product with a massive stick-*ish* shape which has 1.4-2.5 mm diameter. It is produced through a mixture extrusion process of wheat, with or without additional ingredients and other allowed substances [5]. The availability of local Indonesian food is considered as a solution to dependency of imported wheat as the main ingredients in spaghetti production. Rice, cassava, and corn were the three most produced crops in Indonesia in 2016. The total number of rice production is 79,354,767 ton, the total of cassava production is 20,260,675 ton, and corn production is 23,578,413 ton [6]. The challenge in developing non-wheat spaghetti from a local product is the characteristics of the ingredients which must be gluten-free and low protein content.

The production of gluten-free spaghetti requires different process from the conventional spaghetti since the non-wheat spaghetti does not have gluten protein which has elastic characteristic on the mixture. Without the modification process, the mixture of gluten-free spaghetti will be very easily ruined. Gluten-free spaghetti counts on the gelation process of starch and the retrogradation process to form sturdy tissue structure. Therefore, extrusion system through extruder machine is required. Through the extrusion process with high compression and enough shear stress, the flour will rupture, in which most of the starch will be detached from the granola flour and will experience gelation [7]. Afifah and Ratnawati developed curly noodles made by mocaf (modified cassava) flour, rice flour, and corn flour through extrusion technique [8]. The curly noodles resulted in 10.6-14.3% cooking loss, 276-374% elongation, 4.09%-5.58% protein content. The resulted non-wheat noodles have lower protein content compared to spaghetti product. The protein content of the spaghetti can be increased through adding soybean flour as the source of protein.

This research applies Taguchi method in the optimization process of gluten-free spaghetti production combination of four factors and three levels. The research started with control factors and levels, and the quality response characteristics of spaghetti. The degree of freedom will be calculated, and the orthogonal matrix will be selected. Taguchi data tabulation includes the average calculation, SNR, impact factors, ANOVA, contribution percentage and multi-response loss function to get the optimal combination based on the intended characteristics quality [1]. Thus, the objective of this optimization research is to obtain the optimal combination factors and levels based on the intended characteristics quality.

2. Materials and Methods

2.1. Materials and Tools

Materials used in this research is the ingredients used in the process of spaghetti production which are mocaf (modified cassava flour) from Cimanggu cassava fermented with starter mocaf; rice flour Ciherang varieties, P21 hybrid corn, and local soybean flour. Each flour was made with 40, 60, and 80 mesh smoothness index. The tools used in the process of spaghetti production are extruder machine, cabinet dryer, fan scale, gas stove, steaming pan, spoon, scissors, stainless table, and sealer. Physical analytic tools are texture analyzer for elongation, hardness, and adhesiveness analysis, and physical analytic tools for *cooking loss* are *colorimeter* 3nH to test the color, chemical analytic tools such as moisture content by using gravimeter method, protein content by using Kjeldahl method, ash content by using ash method, fat content by using soxhlet, and carbohydrate content by using difference method.

2.2. Place and Time Research Conduct

The research was conducted from January to May 2018 in Pilot Plant of Non-Wheat Product, Laboratorium of Food Chemistry, Laboratorium of Instrumentation, PPTTG LIPI, West Java, Subang; PT. Saraswanti, Bogor dan Analysis Services, Agricultural Technology Faculty, Gadjah Mada University, Yogyakarta.

2.3. Experiment design with Orthogonal-Taguchi Matrix

a. Factors and levels of measurement

Table 1. Factors and levels of experiment

Smoothness Index of the Flour (A)	Mocaf Flour: Corn Flour (B)	Rice flour: soybean flour (C)	Guar gum (D)
40 mesh	40% : 20%	35% : 5%	1 %
60 mesh	35% : 25%	30% : 10%	1.5 %
80 mesh	30% : 30%	25% : 15%	2 %

b. *Degree of freedom calculation and orthogonal matrix selection*
- Degree of freedom = 4 x (3 - 1) = 8
- Selection of orthogonal matrix 3 levels: ($L_9(3^4)$, L_{27} (3^{13}), $L_8 1(3^{40})$)
- Orthogonal Standard Matrix selected is $L_9(3^4)$ because of its degree of freedom is 8, similar to degree of freedom in the experiment.

c. *Plotting factors and levels in accordance with the standard orthogonal matrix.*

Table 2. Research design

Concept	Smoothness index of the flour	Mocaf flour: corn flour (%)	Rice flour: soybean flour (%)	Guar gum (%)
1	40	35 : 25	35 : 5	1
2	40	30 : 30	30 : 10	1.5
3	40	25 : 35	25 : 15	2
4	60	35 : 25	30 : 10	1.5
5	60	30 : 30	25 : 15	1
6	60	25 : 35	35 : 5	1.5
7	80	35 : 25	25 : 15	1.5
8	80	30 : 30	35 : 5	2
9	80	25 : 35	: 10	1

2.4. *Gluten-free spaghetti preparation*
Gluten-free spaghetti was prepared using the extruder machine with 16 die-designed holes and 2mm diameter, the temperature at 60 °C, and spinning speed at 40 rpm. All the ingredients are measured as mentioned in Table 2. Then, they were mixed with dough mixer and steamed in steaming pan for 30 minutes, before the mixture was placed in the extruder machine.

2.5. *Analyses of gluten-free spaghetti*
2.5.1 *Chemical analysis*. The analysis of moisture content was done by using SNI 01-2891-1992, related to an analysis method for food and beverage. Protein content is measured according to the Dumas combustion method by using DuMaster Buchi D-480, Switzerland, ash content and fat content with soxhlet, and carbohydrate content were analyzed through *by difference* method [9].

2.5.2 *Cooking quality analysis.* Cooking quality analysis includes analyzing cooking time, cooking weight and cooking loss [10]. Cooking time analysis is started with cutting 5 gram of dried spaghetti for 4-5 cm long and cooked in 200 ml boiled water in closed beaker glass. Every 30 seconds, the sample is taken and pressed with two transparent glass up to a certain period which shows that sample is well-cooked. Cooking weight and cooking loss analysis were done by cutting 1 gram of dried spaghetti for 3-5 cm long and cooked based on the determined cooking time of each sample. The sample was then filtered by monyl T13 and washed with distilled water. The sample was filtered for one minute and put on scale to measure the cooking weight. The calculation formula in cooking weight can be seen in the equation formula (1). The cooking loss is counted based on the residual solid dissolved by the boiled water which is drained by oven in 105 °C temperature until it reaches its constant weight. The cooking loss formulation can be seen in the equation formula (2).

$$Cooking\ weight\ (\%) = \frac{W_1 - W_2 \times (1 - M)}{W_2 \times (1 - M)} \times 100\% \qquad (1)$$

$$Cooking\ loss\ (\%) = \frac{W_3}{W_2 \times (1 - M)} \times 100\% \qquad (2)$$

Explanation =
W_1 = sample weight after being cooked; W_2 = sample weight before being cooked

W_3 = weight of dry solid dissolved during cooking process; M = moisture content sample

2.5.3 *Profile Texture Analysis.* Profile texture analysis includes elongation, hardness and adhesiveness are analyzed through texture analyzer (TAXt-Plus, Stable Micro Systems, Surrey, UK). A strand of spaghetti which has been cooked based on the cooking time was drained for 2 minutes at room temperature. Texture analysis is based on Stable Micro System with several modifications [11]. Elongation of cooked spaghetti is measured with spaghetti tensile grips (A/SPR) rig pretest 1 mm/s, test speed 3 mm/s, post-test speed 10 mm/s and initial distance slamps 2 cm. Solidity and adhesiveness analysis is done by placing the strands of spaghetti on the metal plate under the probe. Experiment test with cylinder P/36R probe on mode trigger type, auto 0.5 g; pre-test speed 2 mm/s, test speed 2 mm/s; post-test speed 10 mm/s and strain 95 %. Each analysis was repeated for 5 times.

2.5.4 *Sensory test of taste.* Sensory test using hedonic scale is a test conducted to 30 panelists to find out the level of taste recipient of the product. The panelist scores are as follow: 1 (very bad); 2 (bad); 3 (quite bad); 4 (neutral); 5 (not very bad); 6 (good); 7 (very good).

2.6. Taguchi data calculation

a) Calculation of the average of every experiment
b) Value calculation of SNR per experiment
Larger the better

$$\eta = -10\log\left(\frac{1}{n}\sum_{i=1}^{r}\frac{1}{y_i^2}\right) \tag{3}$$

Nominal the best

$$\eta = 10\log_{10}\left[\frac{\mu^2}{\sigma^2}\right] \tag{4}$$

$$\mu = \frac{1}{n}\sum_{i=1}^{n} y_i^2 \tag{5}$$

$$\sigma^2 = \frac{1}{n-1}\sum_{i=1}^{n}(y_i - \mu)^2 \tag{6}$$

Smaller the better

$$\eta = -10\log\left(\frac{1}{n}\sum_{i=1}^{r} y_i^2\right) \tag{7}$$

Explanation:
η = SNR value of every experiment; n = the number of repetition in every experiment
y = value of every experiment; μ = the average of each experiment; σ^2 = deviation of every experiment

c) The ANOVA average value calculation
d) ANOVA calculation of SNR value (the same method to calculate the average of ANOVA)
e) Calculation of Multi Response Loss Function
i) Calculate loss function
Larger the better

$$L_{ij} = \frac{k}{n}\sum_{p=1}^{n}\frac{1}{y_{ijp}^2} \tag{8}$$

Smaller the better

$$L_{ij} = \frac{k}{n}\sum_{p=1}^{n}(y_{ijp}^2) \tag{9}$$

Nominal the best

$$L_{ij} = \frac{k}{n}\sum_{p=1}^{n}(y_{ijp} - m) \tag{10}$$

Explanation:
L_{ij} = loss function ; k = coefficient loss; n = number of data repetition;
y_{ijp} = result of experiment ; m = intended value

ii) Calculate normalization

$$N_{ij} = \frac{L_{ij}}{\bar{L}_{ij}} \tag{11}$$

Explanation:

N_{ij} = normalization; $\bar{L}_{ij}$ = the average loss function

iii) Calculate of loss function

$$TL_j = \sum_{i=1}^{r} w_i \, x N_{ij} \tag{12}$$

Explanation:

TL_j = total loss function; w_i = total variable response

iv) Transformation to SNR

$$\eta = -10\log(TL_j) \tag{13}$$

Creating factor effect table

v) Optimization of level combination and optimal factor

f) Confidence interval calculation (confirmation test)

i) Estimation of average optimal condition

$\mu_{prediction}$ = sum of the average from optimum factor level + y

ii) Confidence interval calculation

$$Cl_{mean} = \pm\sqrt{F_{\alpha;v1;v2} x MS_e x \frac{1}{neff}} \tag{14}$$

$$neff = \frac{number\ of\ experimental\ \ x\ replication}{number\ of\ degree\ of\ freedom\ \text{degree of freedom for measuring the mean}} \tag{15}$$

Explanation:

Cl = confidence interval ; Fα; v_1; v_2 = F-ratio from table ; α = Risk, confidence level = 1–risk ; v_1= degree of freedom numerator; v_2= degree of freedom denominator; *Mse* = the average of sum of squares polled error

iii) Average confirmation experiment

Average value calculation

$$\mu = \frac{1}{n}\sum_{i=1}^{n} y_i \tag{16}$$

Confidence interval calculation

$$Cl_{mean} = \pm\sqrt{F_{\alpha;v1;v2} x MS_e x \left(\frac{1}{neff} + \frac{1}{r}\right)}$$

Explanation: r = The number of observation used to calculate the average

3. Result and Discussion

3.1. Characteristics of the Spaghetti Quality

Table 3 presents the characteristics of the spaghetti response quality.

Table 3. Characteristic determination

Spaghetti Quality	Characteristic	Spaghetti Quality	Characteristic
Moisture content	Smaller the better	Elongation	Nominal the best
Protein content	Nominal the best	Hardness	Nominal the best
Fat level	Nominal the best	Adhesiveness	Nominal the best
Carbohydrate	Higher the better	Color	Nominal the best
Cooking time	Smaller the better	Sensory of taste	Higher the better
Cooking loss	Nominal the best		

The quality response of moisture content in this study is smaller the better. The amount of moisture content affects the shelf life of the product. The low water content of spaghetti products shows the product has good quality

3.2. Data tabulation and interpretation
This scientific paper illustrates data tabulation in one of the quality characteristics of the gluten-free spaghetti, in this case, is the moisture content.

3.2.1. The average calculation and SNR of moisture content
The average of moisture content from 9 concepts of gluten-free spaghetti product is around 9.54% to 10.77%. Based on significant difference test, the highest average of moisture content is concept 1 which has 10.77% differentiation compared to the lowest average of moisture content in product 7 with 9.54%.

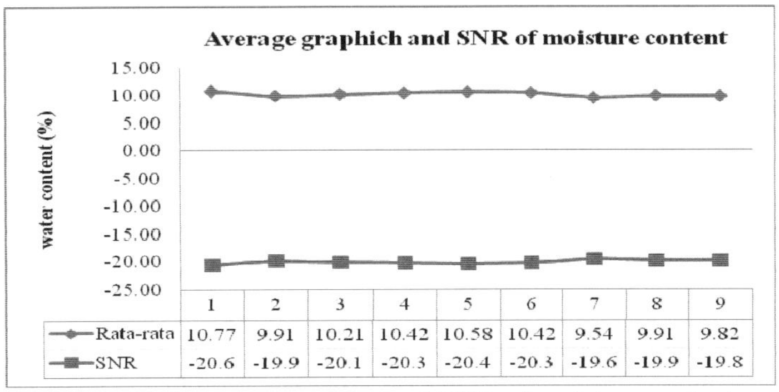

Figure 1. Average graphics and SNR of moisture content

Gluten-free spaghetti in concept 1 is made from 35% of mocaf (modified cassava flour), 25% of corn flour, 35% of rice flour, 5% of soybean flour, and 1% of guar gum. The whole flours are used to produce spaghetti concept 1 with the level of smoothness index of 80 mesh. From SNR graphic response of moisture content with smaller the better quality characteristic from 9 concepts ranged from -19.85 – (-20.65). The SNR highest value of smaller the better moisture content is 7, and the lowest is 1. This proves that the combination factor level of concept 7 has the biggest effect on the variety of moisture content response and must be controlled to meet the expected quality.

3.2.2. The average response of moisture content
Besides the average calculation and SNR of moisture content, the calculation of factor effect and SNR were also conducted to identify the effect level from factor towards the average of observed quality. The average effect of every level from each factor has resulted from the calculation of the average quality value from the contained product concept. The average effect of level from each factor then subtracted with the highest score with the lowest score to find the deviation as the average effect value from the factor level. The calculation of factor effect from the moisture content is as illustrated in Table 4.

Table 4. The calculation of factor effect from the moisture content

	A	B	C	D
Level 1	10.30	10.24	10.37	10.39
Level 2	10.47	10.13	10.05	9.96
Level 3	9.76	10.15	10.11	10.18
Deviation	0.72	0.11	0.31	0.43
Ranking	1	4	3	2

From the average of factor effect calculation of moisture content, it shows that the order of factors ranking which has the highest number of deviation in a row is factor A, D, C, B, while the combination of factor level which resulting average response of smaller the better moisture content can be seen in Figure 2.

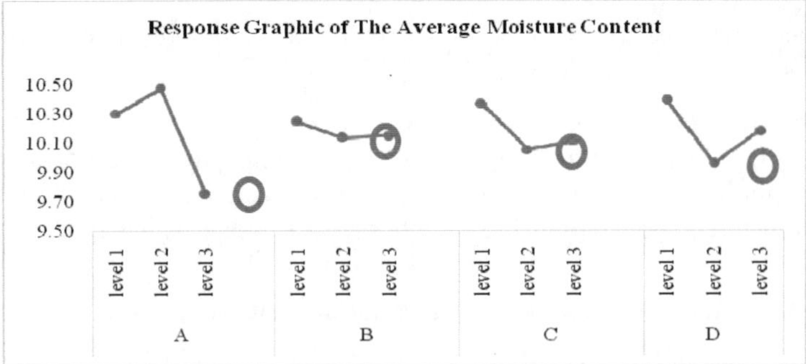

Figure 2. Response Graphic of the Average Moisture Content

From the graph of the average response of moisture content, it is known that factor A level 3 is the most influencing factor and level to the average for the moisture content response variable because it produces the lowest average moisture content response. The fineness of 80 mesh flour can be used to control the mean for the moisture content response variable if desired the lower, the better. The best combination of factors and levels to control the average response is A3, B2, C2, D2.

3.2.3. Effect of SNR moisture content

SNR effect is used to determine the effect of factors on noise. SNR effect of each level of each factor is obtained by calculating the SNR value of the quality of the product concept in which there is a combination of these levels. The SNR effect level of each factor is obtained from the highest value difference with the lowest value. From these values obtained a rating of experimental factors to the noise of each quality parameter. From the calculation of SNR factor effect of moisture content, it is known that the factor rank of the largest SNR difference is A, D, C, B, whereas the combination of factor level which produces SNR response of smaller the better moisture content can be seen in Figure 3.

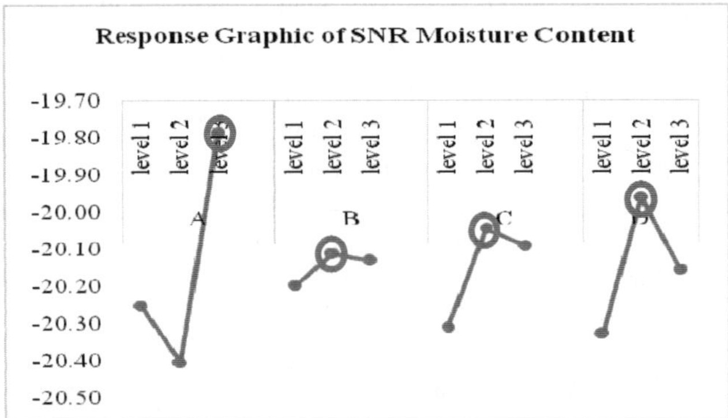

Figure 3. Response Graphic of SNR Moisture Content

From the graph of SNR response, it is known that factor A has the largest SNR response. Factor A is the most influential on the variation (noise) for the moisture content response variable. These factors should be controlled to reduce variations in the moisture content of gluten-free spaghetti products. The combinations of levels of the factors that produce the highest moisture content SNR response are A3, B2, C2, D2.

3.2.4. ANOVA of the average and SNR of Moisture Content

From the mean and SNR factor effect analysis, it is known that the A factor has the highest rank on the moisture content of spaghetti, followed by other factors. The average ANOVA calculation of moisture content shows that not all experimental factors affect the average water content of gluten-free spaghetti. Only factor A effect with F ratio 7.89, greater than F table (0.05%) 3.55. Factor A contributes 32.39% to the quality characteristics of gluten-free spaghetti moisture content. The other factor has an F ratio lower than F table. This is because of the smaller the starch particles; the starch granules will be more ruptured. Amylose that has come out of the starch granule (but is still trapped in a flour granule) can spread throughout the dough and form a strong matrix structure through hydrogen bonds when retrograde [12].

3.2.5. Loss Function of Moisture Content

Because the response observed in this study is more than one (multi-response), the approach used is loss function. The loss function is a function of loss due to the quality produced. Taguchi adopts a loss function for the optimization process. Calculation of loss function of moisture content can be seen in table 5.

Table 5. Calculation of loss function for moisture content

Concept	Lij	Nij	Wi	Wi x Nij
1	116.07	1.12	0.08	0.0932
2	98.42	0.95	0.08	0.0790
3	104.37	1.01	0.08	0.0838
4	108.61	1.05	0.08	0.0872
5	111.96	1.08	0.08	0.0899
6	108.73	1.05	0.08	0.0873
7	91.12	0.88	0.08	0.0732
8	98.21	0.95	0.08	0.0789
9	96.64	0.93	0.08	0.0776

Calculation of loss function is continued for all the characteristics of the quality of gluten-free spaghetti (protein content, fat content, carbohydrate content, cooking time, cooking loss, elongation, hardness, adhesiveness, color, and sensory of taste).

3.3 Multiresponse Loss Function Calculation

Every optimized characteristic has a different value. Therefore, in order to equalize, it needs to be normalized on each characteristic and weighted. Calculation of total loss function is then transformed in a multiresponse signal to noise ratio. Calculation of multiresponse loss function can be seen in table 6.

Table 6. Calculation of multiresponse loss function

Concept	TLj	Hj	Concept	TLj	Hj
1	1.206	-0.813	6	0.724	1.403
2	0.861	0.650	7	1.041	-0.175
3	1.968	-2.941	8	0.883	0.543
4	0.758	1.203	9	0.731	1.362
5	0.858	0.665			

3.4 Optimum selection of the optimum combination of factors and levels

Based on the calculation of the effect of the multi-factor response on the quality characteristics of gluten-free spaghetti products, among others moisture content, protein content, fat content, carbohydrate content, cooking time, cooking loss, elongation, hardness, adhesiveness, color, and sensory taste were obtained optimum combination A2, B2, C2, D2 which can be seen in figure 4.

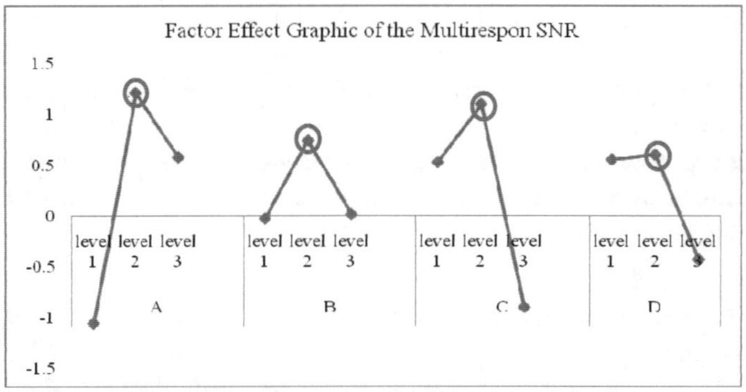

Figure 4. Factor Effect Graphic of the Multirespon SNR

3.5 Confirmation test

The confirmation test is based on reassessing product quality. Confirmation test is to find out how far the selected product can get the product with the best quality attributes. From the confirmation experiment, it is known that the optimum product selected has been in the confidence interval and has confirmed its quality.

4. Conclusion

In conclusion, we have shown that the optimum gluten-free spaghetti products made from 60 mesh local flour with a composition of 30% mocaf flour, 30% corn flour, 30% rice flour, 10% soybean flour, guar gum 1.5% with quality characteristics have been confirmed within its confidence interval. Quality analysis of this product has moisture content 9.94 ± 0.57%, protein content 10.33 ± 0.55%, fat content 1.11 ± 0.67%, carbohydrate content 76.51 ± 2.32%, cooking time 10.65 ± 0.25 minutes, cooking loss 13.00 ± 2.76, elongation 181.35 ± 25.72, hardness 2069.03 ± 453.07 gf, adhesiveness -32.75 ± 3.41 g.sec, color 76.51 ± 0.63 dL and sensory score of taste 4.80 ± 0.16. This research is expected to produce alternative spaghetti products from local Indonesian food

5. Acknowledgment

We would like to express our gratitude to the DIPA of the Center for Appropriate Technology-LIPI Subang Development and the Ministry of Research and Technology which has funded this research, as well as the research technicians who have helped this research.

6. Reference

[1] Soejanto I 2009 *Experimental Design with Taguchi Method* (Yogyakarta: Graha Ilmu)

[2] Sereewat P, Suthipinittham C, Sumathaluk S, Puttanlek C, Uttapap D, Rungsardthong V 2015 Cooking properties and sensory acceptability of spaghetti made from rice flour and defatted soy flour *Food Science and Technology* **60** 1061-1067

[3]Arif 2010 *Successful Pasta Products* [*Accessed on 22 April 2017.* http://arifh.blogdetik.com/2010/12/13/produk-pasta-sukses-di-negeri-orang]

[4] Mintel 2013 *Consumer Trends, Pasta and Couscous in The United States* Market Indicator Report (Minister of Agriculture and Agri-Food Canada)

[5] Anonim[a] 1998 *Spaghetti Quality Requirements SNI 01-4454-1998* National Standardization Agency, Indonesia

[6] Anonim[b] 2017. *Rice Production, Corn, Cassava, Soybeans by Province. Indonesian Ministry of Agriculture. [Accessed on November 20, 2017,* http://www.pertanian.go.id/ap_pages/mod/datatp]

[7] Muhandri T 2012 Mechanism of making noodles made from corn flour. *buletin teknologi pascapanen pertanian* **8 (20)** 71-79

[8] Afifah N and Ratnawati L 2017 Quality assessment of dry noodles made from blend of mocaf, rice flour and corn flour *IOP Conference Series: Earth and Environmental Science* **101**1-9

[9] Anonim[c] 2005 *Official Method of Analysis of The Association of Offical Agricultural Chemistry.* (Washington D C : Association of Official Agriculture Chemistry)
[10] Wandee Y, Uttapap D, Puncha-arnon S, Puttanlek C, Rungsardthong V, Wetprasit N 2015 Quality assessment of noodles made from blends of rice flour and canna starch *Food Chem.* **179** 85-93
[11] Impaprasert R, Piyarat S, Sophontanakij N, Sakulnate N, Paengkanya S, Borompichaichartkul C, Srzednicki G 2017 Rehydration and textural properties of dried konjac noodles : effect of alkaline and some gelling agents *Horticulture* **3** **20** 1-10
[12] Chen J S, Fei M J, Shi C L, Tian J C, Sun C L, Zhang H, Ma Z, Dong H X 2011 Effect of particle size and addition level of wheat bran on quality of dry white chinese noodles *Journal of Cereal Science* **53** 217-224

IOP Conf. Series: Earth and Environmental Science **251** (2019) 012032 doi:10.1088/1755-1315/251/1/012032

Characteristics of physical, chemical, and organoleptic properties of inulin-enriched pudding as a complementary food

D Kristanti[1] and A Herminiati[1]

[1] Development Center for Appropriate Technology, Indonesian Institute of Sciences, Jl. Aipda KS. Tubun No. 5 Subang 41213

Corresponding author, e-mail: dita.kristanti@gmail.com

Abstract. The products for complementary feeding circulating in the market are in the form of baby porridge, biscuits, and powdered milk. The functional food product to support the growth and development of children under five in the pudding form, which is made from skim milk as the raw ingredient and modified cassava flour (mocaf) as the supporting ingredient and enriched with inulin that acts to increase calcium absorption, is not yet on the market. The development of this product is expected to meet the nutritional requirements of infants aged 12-36 months. This study aimed to analyze the characteristics (chemical, physical, and organoleptic properties) of the inulin-enriched pudding. The pudding was made using four formulas; i.e., F1 (10% db mocaf and 0.8 g/100 ml inulin), F2 (10% db mocaf and 1.6 g/100 ml inulin), F3 (15% db mocaf and 0.8 g/100 ml inulin), and F4 (15% db mocaf and 1.6 g/100 ml inulin). The results of this study showed that the best pudding for complementary feeding based on the characteristics of physical properties was F2 which had a gel strength of 843.03 g Force and a syneresis of 7.78%. The F2's organoleptic properties were as follows: milky white color, distinctive aroma of milk, a rather sweet taste, and soft textured. Besides that, the contribution of one serving of F2 pudding to the recommended dietary allowance (RDA) of children aged 1-3 years was as follows: 2.46 g of protein, 1.67 g of fat, 8.11 g of carbohydrates, and 57.32 kcal of energy.

1. Introduction

The products for complementary feeding currently circulating in the market are in the form of baby porridge, biscuits, and powdered milk. One of the complementary foods to support the growth and development of children under five that can be developed is pudding. It is a dessert that uses milk as the main ingredient which contains protein and calcium. The protein functions as a bone growth regulator while calcium plays a role in bone formation and height growth. The calcium intake in children under five needs to be optimally given because it will affect their growth and development. The complementary foods containing calcium can be given to the infants aged 12-36 months.

The use of mocaf is expected to increase the competitiveness of local resources. Mocaf is flour made from cassava which is produced by modifying the cassava flour with fermentation using lactic acid bacteria (LAB). The cassava flour modification technology using LAB fermentation caused the changes in the characteristics of the flour produced such as the increased viscosity, gelation ability, rehydration capacity, and ease of dissolution [1].

In the formulation of food products, mocaf has the properties as the binder, thickener, and filler that can be applied when making the pudding. The additional ingredient that has good physical functional properties as thickener, gelling agent, and stabilizer is needed when making pudding. The additional ingredient can be obtained from carrageenan, and the sweet taste is obtained from the addition of sugar [2].

The protein intake in children under five must be optimally given because it will affect their growth and development. The calcium intake of the Indonesian people is still low, one of which is due to the low consumption of milk as a source of calcium. Each person only consumes about half a glass of milk in a week [3]. The calcium intake deficiency or impaired intestinal calcium absorption may have different effects at various age levels. This condition may cause rickets if it occurs in children under five, and it may lead to osteomalacia if it occurs in adults [4], [5].

The complementary food given to the infants aged 12-36 months must contain carbohydrates and protein according to RDA, and it should be enriched with inulin. The functional food that has a physiological activity to increase calcium absorption is needed to increase the absorption of this mineral [6]. The identification of the component and the composition of functional food that can positively affect the calcium absorption also needs to be performed [7]. One of the components that can increase the calcium absorption is inulin. The study results showed that inulin could increase calcium solubility in the lumen of the small intestine due to the short-chain fatty acids (SCFAs) formation which can decrease the pH of the small intestine resulting in the increased absorptive surface area and the increased calbindin production as the calcium-transport protein [8].

Based on the above description, the pudding formulation used the skim milk as the raw ingredient and mocaf as the supporting ingredient which was enriched with inulin. This product is expected to play a role in increasing the calcium absorption and fulfilling the nutritional requirement of the infants aged 12-36 months. This study aimed to analyze the characteristics (chemical, physical, and organoleptic properties) of the inulin-enriched pudding.

2. Materials and Methods

2.1 The materials

The raw and supporting ingredients used in this study were skim milk from Ny. Liem store in Bandung, mocaf from "Tanjung Siang" Small and Medium Enterprises (SME) in Subang, Orafti® inulin (Synergy1-Oligofructose 112), fine granulated sugar, carrageenan, and water.

2.2 Research Methods

Four pudding formulations were made in this study. The formulations are presented in Table 1, and the flow diagram of the pudding production process can be seen in Figure 1.

Table 1. Four pudding formulations.

Materials	Formulations			
	F_1	F_2	F_3	F_4
Skim milk (g)	9	9	9	9
Mocaf flour (g)	2	2	3	3
Inulin (g)	0.8	1.6	0.8	1.6
Fine granulated sugar (g)	8	8	7	7
Carrageenan (g)	1	1	1	1
Water (ml)	100	100	100	100

The pudding was made through several stages as follows: 1) weighing the raw and supporting ingredients; 2) dry-mixing the raw and supporting ingredients by using a mixer; 3) cooking the dried pudding ingredients by adding the water into the pan; 4) stirring the ingredients until dissolved (± 1 minute); 5) heating on a stove over medium heat for six minutes while stirring regularly until the final temperature of the pudding solution reached 85-95 °C; 6) pouring the pudding solution into the mold and then cooling it to room temperature; 7) cooling the pudding again in the refrigerator at 5 °C for ±12 hours.

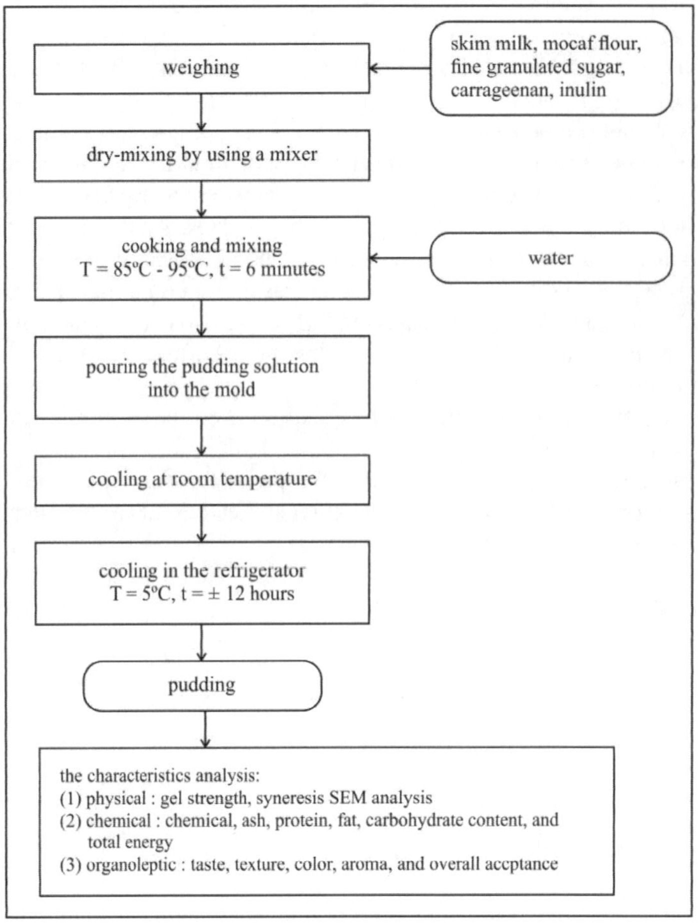

Figure 1. The flow diagram of the pudding production process.

2.2.1 The characteristics of physical analysis. The characteristics of physical properties were determined by performing gel strength analysis using an Stable Micro System TA.Xtplus which settings as follows: P 36R probe model; 0.50 mm/s of test speed; 5 g of force target; 10 mm of distance; and 30 of retrun distance and syneresis analysis using an modification of the [9] method.

2.2.2 The characteristics of chemical analysis. To determine the characteristics of chemical properties, a proximate analysis was performed which included: analysis of the moisture content by the gravimetric method [10], ash content by the gravimetric method [10], protein content by the Dumas combustion method by using DuMAster Buchi D-480, Switzerland, and fat content by Soxhlet method [11]. The carbohydrate content was determined using by difference method [10], and the total energy was determined based on the conversion formula of carbohydrates, fat, and protein contents.

2.2.3 The organoleptic test. The organoleptic test used the scoring tests on taste, texture, color, and aroma while the overall acceptance was determined by the hedonic test [12]. The organoleptic assessment was carried out by 33 mothers who had children under five in Posyandu Blok Kopti Cigadung Subang using the attribute rating scale presented in Table 2 below.

Table 2. The attribute rating scale of organoleptic.

Scale	Color	Aroma	taste	Texture	overall acceptance
5	milky white	very typical of milk	very sweet	very soft	very like
4	slightly yellowish white	typical of milk	sweet	soft	like
3	yellowish white	rather distinctive of milk	rather sweet	rather soft	rather like
2	very yellowish white	distinctive of milk	tasteless	chewy	dislike
1	Yellow	very distinctive of milk	very tasteless	very chewy	very dislike

The assessment of the best pudding for complementary feeding organoleptic was carried out by 10 childrens (1-5 years old). The assessment using the attribute hedonic scale with 1 (super bad), 2 (really bad), 3 (bad), 4 (maybe good or maybe bad), 4 (good), 5 (really good), 6 (super good) to assess the children's facial expressions based on the [13] method.

2.2.4 Characteristics of physical and chemical properties of the F2 CF pudding flour. The microstructure of the F2 CF pudding flour was conducted using Scanning Electron Microscope (SEM) JSM-6510 series. The Ca, Fe, and Zn content of of F2 CF pudding flour was determined in duplo by flame atomic absorption spectrometry

3. Results and Discussion

The development of inulin-enriched pudding product as a complementary food and the regulation regarding the use of inulin refer to the Food Standards Australia New Zealand (FSANZ) that allows the addition of inulin-derived and GOS-derived substances for single or combination use in the foods formulated for children under five up to a total maximum of 0.6 g/100 g and 1.6 g/serving. Furthermore, the European Union's Novel Food Regulation (EC 258/97) recommends that the content found in inulin (a combination of 90% oligalactosyl-lactose and 10 oligofructosyl-saccharose) should not exceed 0.8 g/100 ml for the formulation of the food products for children under five that add inulin. According to the Food and Drug Administration (2003), the use of inulin at a level of 90% is as follows: 6 g/day for infants under 12 months of age, 15 g/day for infants aged 13-24 months, and 20 g/day for infants over 25 months of age [14].

Inulin can function as a prebiotic in the complementary foods. The study results showed that inulin as a prebiotic could produce SCFAs which included propionate, butyrate, acetate, and lactate. The lactic acid can make the intestinal pH become acidic. This condition causes the calcium ions to become more soluble, thereby increasing the calcium absorption in the large intestine [15].

3.1 Characteristics of physical properties of the pudding as a complementary food

The results of the analysis on the physical properties of the pudding as a complementary food (CF pudding) can be seen in Table 3 below.

3.1.1 The gel strength of the CF pudding. The CF pudding is a food product for the children under five that still has no Indonesian National Standards and regulations from the Ministry of Health for the determination of the pudding's gel strength. The commercial CF pudding products that are already on the market are used as a comparison for the pudding's gel strength. Of the four formulas, only F2 and F3 puddings were within the range of gel strength of the commercial CF pudding products. The comparison of the CF pudding's gel strength between the formulas can be seen in Table 3. The

IOP Conf. Series: Earth and Environmental Science **251** (2019) 012032 doi:10.1088/1755-1315/251/1/012032

differences in gel strength values can influence the food texture. The pudding with low gel strength will produce a soft-textured pudding. The soft texture causes a decrease in the consumer's acceptance level. The study results showed that the difference in gel strength in each CF pudding formula did not affect the consumer's acceptance level [2]. It was evident from the results of the organoleptic test on the texture and overall acceptance of the CF puddings that were not significantly different. The consumers were more tolerant of organoleptic changes in milk pudding which was added with a functional food (i.e., HAMS) because they were more interested in the product claims that had the potential to improve health [16].

Table 3. The physical properties of the CF pudding.

CF pudding	Kekuatan gel (g Force)	Sineresis (%)
F_1	848.02±11.77^c	6.16±1.36^b
F_2	1009.04±14.73^a	7.78±0.43^b
F_3	921.92±15.60^b	7.93±1.49^b
F_4	601.49±29.40^d	13.01±1.15^a
Commercial	870.32 - 1096.74	8.74 – 10.06

Values are expressed as mean±standard deviation. Means in the same column with different letters were significantly different at $p < 0.05$

3.1.2 The syneresis of the CF pudding. Syneresis is the extraction of a liquid from a gel that occurs due to slow gel contraction, and it is affected by time. The study results showed that the greater the syneresis value, the easier the gel formed to release water. Therefore, it is not preferred by the consumers [9]. The results of CF pudding's syneresis can be seen in Table 3. The analysis results indicated that the syneresis of F1, F2, and F3 CF puddings was better than the F4. According to Subaryono et al. (2003), the product stored at low temperature (e.g., ice cream or pudding) must have high gel strength and low syneresis [2]. Based on the results of gel strength and syneresis analyses, it is known that the F2 formula is the best CF pudding compared to other formulas.

3.1.3 Characteristics of chemical properties of the CF pudding

The chemical analysis performed in this study was the analysis of moisture, ash, protein, fat, and carbohydrate contents. This analysis aimed to determine the nutritional content of the CF pudding. The results of the chemical analysis are presented in Table 4.

Table 4. The chemical properties of the CF pudding.

CF pudding	Moisture (%)	Ash (%)	Protein (%)	Fat (%)	Carbohydrate (%)
F_1	80.16±0.73^a	4.99±0.18^a	3.09±0.30^a	2.56±0.18^a	9.19±0.60^a
F_2	79.97±0.17^a	4.73±0.38^a	3.07±0.39^a	2.09±0.19^b	10.14±0.26^a
F_3	80.32±0.57^a	4.77±0.61^a	3.27±0.39^a	2.86±0.19^a	8.78±1.27^a
F_4	79.92±0.17^a	4.72±0.41^a	3.12±0.29^a	2.59±0.09^a	9.64±0.58^a

Values are expressed as mean ± standard deviation. Means in the same column with different letters were significantly different at $p < 0.05$

The results indicate that there are no significant differences in the moisture, ash, protein, fat, and carbohydrate contents between the four CF pudding formulas. The protein content of the CF pudding was influenced by the addition of skim milk. The concentration of the skim milk added to the four formulas was the same. Meanwhile, the carbohydrate content was influenced by the addition of mocaf. Although the concentration of the mocaf added to the F3 and F4 formulas was different, the analysis results of the carbohydrate content showed no significant difference. However, the addition of a higher amount of inulin to the F2 and F4 formulas showed a relatively higher difference in carbohydrate content.

3.2 Characteristics of organoleptic properties of the CF pudding

The results of an organoleptic test of the CF pudding formulas can be seen in Table 5. The results showed that the addition of mocaf and inulin did not have significant effects on the organoleptic properties (i.e., color, aroma, taste, and texture).

Table 5. The organoleptic properties of the CF pudding.

CF pudding	Color	Aroma	Taste	Texture	Overall acceptability
F1	4.39±1.04[a]	3.00±0.93[a]	3.39±0.66[a]	3.85±0.51[a]	3.37±1.45a
F2	4.61±0.66[a]	3.09±0.88[a]	3.42±0.87[a]	3.79±0.60[a]	3.69±0.79[a]
F3	4.30±0.88[a]	3.12±0.96[a]	3.18±0.92[a]	3.88±0.60[a]	3.56±1.15[a]
F4	4.36±0.86[a]	3.24±1.00[a]	3.21±0.74[a]	3.76±0.66[a]	3.63±0.72[a]

Values are expressed as mean ± standard deviation. Means in the same column with different letters were significantly different at $p < 0.05$

Based on the score of the organoleptic test on the CF pudding's colors, the F1, F3, and F4 CF puddings had a slightly yellowish white color while the F2 CF pudding had milky white color. These colors were influenced by the skim milk as the raw ingredient. Meanwhile, the addition of inulin did not affect the pudding's color. In line with the results of a previous study, the addition of 90 g/kg inulin did not affect the color of the strawberry-flavored gummy jellies [17].

The results of the organoleptic test score of the CF pudding's aroma indicated that the F1, F2, F3, and F4 CF puddings had a rather distinctive aroma of milk. It was influenced by the addition of inulin and mocaf which made the aroma of milk to be reduced.

Based on the score of the organoleptic test on the CF pudding's taste, all the CF puddings had a rather sweet taste. The results of a previous study indicated that the addition of inulin to pudding formulation could reduce the sweet taste of the pudding [16]. The results of the organoleptic test on the taste showed that the four CF pudding formulations could be given to the children under five because the taste was relatively not too sweet. However, the analysis of the total sugar content of the CF pudding needs to be performed, with the requirement not to exceed 30 g/100 g of product [18].

The results of the organoleptic test on the CF pudding's texture showed that the four CF pudding formulas had a soft texture. These results were consistent with the previous study results, the inulin addition of 90 g/kg in strawberry-flavored gummy jellies made the texture softer [17]. The inulin extract had the ability to bind water with high solubility which would affect the product's texture [19]. The results of the organoleptic test on the texture indicate that the four CF pudding formulas can be given to the infants aged 9-59 months as a snack. Soft-textured foods can already be given to the 9-month-old infants, and family foods can begin to be introduced to 10-month-old infants as a snack once a day [20]. Based on the scores of the organoleptic test, the four CF pudding formulas could be accepted by the panelists. The scores belonged to the "like" category. The highest score was obtained by the F2 CF pudding. The results of the overall assessment on the CF pudding indicated that the addition of mocaf and inulin with different concentrations was proven not to change the organoleptic properties of the CF pudding. Figure 2 and Figure 3 showed that a good response from consumers sensory perception of CF pudding based on F2.

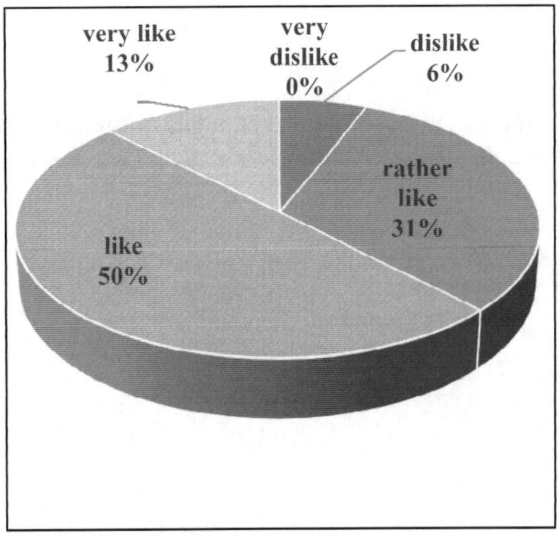

Figure 2. Consumers sensory perception of CF pudding based on F2 (mother assessment).

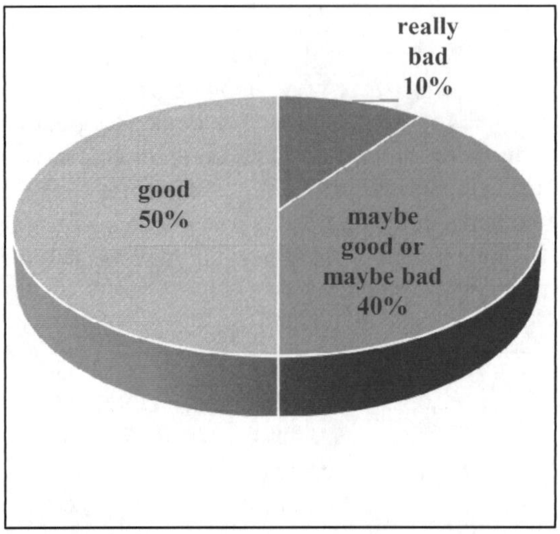

Figure 3. Consumers sensory perception of CF pudding based on F2 (childrens 1-5 years old assessment).

The best CF pudding based on the organoleptic and physical properties was the F2 CF pudding. To determine the nutrient adequacy of the CF pudding with the target of children aged 1-3 years, it is necessary to calculate the contribution of nutrients from one serving of CF pudding. Based on the results of chemical analysis on the protein, fat, carbohydrates, and energy contents, the nutrient content in one serving of the F2 CF pudding was 80 g. The nutrient contribution from one serving of the F2 CF pudding is presented in Table 6 below.

Table 6. The nutrient contribution from one serving of the F_2 CF pudding.

Nutrient content	Amount per serving	RDA for 1-3 years old	Nutrient contribution (%)
Protein (%)	2.46	26	9.45
Fat (%)	1.67	44	3.80
Carbohydrates (%)	8.11	155	5.23
Total Energy (kcal)	57.32	1125	5.09

Based on the data presented in Table 6, the F2 CF pudding gave a higher contribution to the RDA for protein than the RDAs for carbohydrates and fat because the largest composition of pudding ingredients was skim milk. This product is expected to be an alternative to complementary foods which contributes to the nutritional fulfillment of the children under five and is preferred by the children under five as the consumers. The suggestions from the mothers of the children under five who became panelists in this study need to be considered so that this product can be accepted by the consumers; i.e., the use of fruit extracts or chocolate as an addition to the aroma and color of the CF pudding product. The claims for functional food products that can improve health will only have a small impact on the customer acceptance, and the organoleptic properties of the product still have to be considered [21].

3.3 Characteristics of physical and chemical properties of the F2 CF pudding flour

The microstructure of the F2 CF pudding flour presented in Figure 4 and the minerals (Ca, Fe, and Zn) content of the F2 CF pudding flour are presented in Table 7. The varying particle size of the F2 CF pudding flour was indicated homogeneously mixed. The morphology surface of the F2 CF pudding flour was same as the inulin-enriched instant yogurt [22].The inulin in the flour mixture shown by arrows, spherical and amorphous.

The microstructure of inulin powder from chicory roots showed sperichal with a size of 50-100 μm and amorphous surface [23], [24].

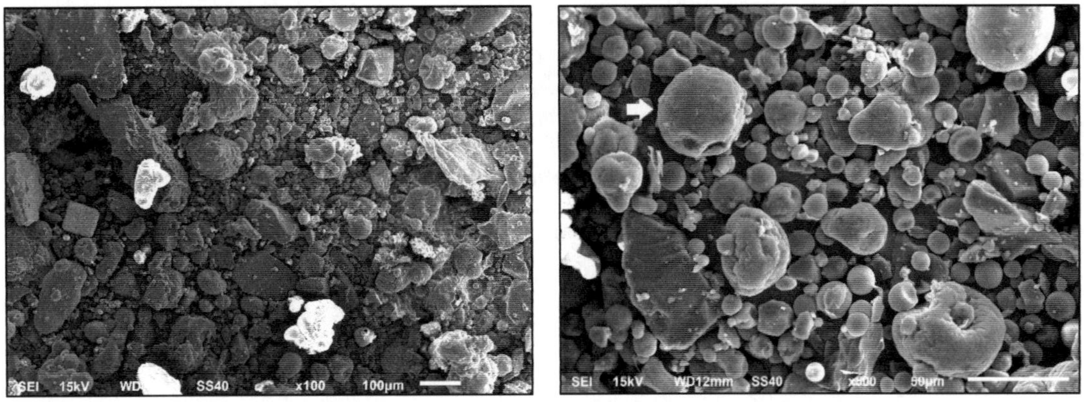

Figure 4. Microstructure of the F2 CF pudding flour.

Table 7. The minerals content of the F2 CP pudding flour.

Nutrient content	F2 CF pudding flour (mg/100 g)
Calcium (Ca)	385.11±4.04
Iron (Fe)	0.58±0.02
Zinc (Zn)	2.09 ±0.02

The calcium content of the F2 CP pudding flour was in accordance with the Indonesian National Standard (SNI) for CF instant powder, but the iron and zinc contents were not appropriate. Base on the SNI, CF instant powder must contain more than 200 mg/100 g of calcium, 5 mg/100 g of iron, and 2.5 mg/100 g zinc [18]. The iron and zinc contents of F2 CF pudding flour can be increased by the addition of micronutrient mix powder. The fortification of micronutrient powders containing iron, vitamin A, zinc and other vitamins and minerals to complementary food was beneficial for male and female children under 2 years old [25].

4. Conclusion

The F2 CF pudding has the best characteristics based on the physical, chemical, and organoleptic properties with the formulation of 10% db mocaf and 0.8 g/100 g inulin. The contribution of one serving of the F2 CF pudding to the RDA of children aged 1-3 years is as follows: 9.45% protein, 3.8% fat, 5.23% carbohydrates, and 57.32 kcal energy.

This study needs to be continued towards the analysis of the effectiveness of the product using animal models, with the aim of obtaining the evidence of scientific claims regarding the addition of inulin that acts as a prebiotic and can increase calcium absorption.

IOP Conf. Series: Earth and Environmental Science **251** (2019) 012032 doi:10.1088/1755-1315/251/1/012032

5. References

[1] Subagio A, W SW, Witono Y and Fahmi F 2008 *Rusnas Diversifikasi Pangan Pokok Prosedur Operasi Standar (POS) Produksi Mocal Berbasis Klaster* (Bogor: Southest Asian Food & Agricultural Science & Technology (SEAFAST) Center - Institut Pertanian Bogor)

[2] Darmawan M, Peranginangin R, Syarief R, Kusumaningsih I and Fransiska D 2014 Pengaruh Penambahan Karaginan untuk Formulasi Tepung Puding Instan *JPB Perikanan* **9**

[3] Khomsan A 2004 *Peran Pangan dan Gizi untuk Kualitas Hidup* (Jakarta: Gramedia Widiasarana Indonesia)

[4] Parfitt AM 2005 Vitamin D and The Pathogenesis of Rickets and Osteomalacia in *Vitamin D 2nd Edition* pp 1029-48 (San Diego: Elsevier)

[5] Anderson JJ 2008 Nutritions and Bone Health in *Food & Nutrition Therapy 12th Edition* pp 614-35 (Missouri: Saunders Elsevier)

[6] van den Heuvel EGHM, Schoterman MHC and Muijs T 2000 Transgalactooligosaccharides Stimulate Calcium Absorption in Postmenopausal Women *Journal Nutrition* pp 2938-42

[7] Kennefick S and Cashman KD 2000 Investigation of an in vitro model for predicting the effect of food components on calcium availability from meals *Journal Food Science Nutritions* **51** pp. 45-54

[8] Roberfroid M B 2005 Introducing Inulin-Type Fructans *British Journal of Nutrition* **93**

[9] Draget K.I, Gaserod O, Anue I, Andersen PO, Strobakken B, Stokke BT and Smidsrod O 2001 Effects of molecular weight and elastic segmen flexibility on syneresis in Ca-alginate gels *Food Hydrocolloids* **15**

[10] AOAC 1995 *Official Methods of Analysis of The Association of Official Analytical Chemist, 14 ed.* (Airilington: AOAC Inc)

[11] Badan Standardisasi Nasional 1992 *SNI 01-2891-1992: Cara Uji Makanan dan Minuman* (Jakarta: Badan Standardisasi Nasional)

[12] Soekarto ST and Hubeis M 1992 *Metodologi Penelitian Organoleptik* (Bogor: Pusat Antar Universitas Pangan dan Gizi, Institut Pertanian Bogor)

[13] Guinard J 2001 Sensory and Consumer Testing with Children *Trends Food Sci. Technol.* **11**, pp 273-83

[14] Soedarto and Nugrahanti Y 2008 *Kajian Regulasi Pangan Fungsional: Studi Kasus Prebiotik, Probiotik, dan Simbiotik* (Bogor: Institut Pertanian Bogor)

[15] Roberfroid MB 2007 Inulin-Type Fructans: Functional Food Ingredients *The Journal Of Nutrition* **137** pp 2493S-502S

[16] Ares G, Baixauli R, Sanz T, Varela P and Salvador A 2009 New Functional Fiber in Milk Puddings: Effect on Sensory Properties and Consumers, Acceptabilty *Food Science and Technology* **42** pp 710-6

[17] Delgado P and Banon S 2018 Effects of Replacing Starch by Inulin on The Physicochemicals of Gummy Jellies *CyTA - J. Food* **16** pp 1-10

[18] Badan Standardisasi Nasional 2005 *Makanan Pendamping Air Susu Ibu (MP-ASI) Bagian 1: Bubuk Instan* (Jakarta: Badan Standardisasi Nasional)

[19] Anan'ina NA, Andreeva OA, Mycots LP and Oganesyan ET 2009 Standardization of Inulin Extracted From Dahlia Single Tubers and Some Physicochemical Properties of Inulin *Pharmaceutical Chemistry Journal.* **43** pp 35-7

[20] Mufida L, Widyaningsih TD and Maligan JM 2015 Prinsip Dasar Makanan Pendamping Air Susu Ibu (MP-ASI) untuk Bayi 6-24 Bulan: Kajian Pustaka *Jurnal Pangan dan Agroindustri* **3** pp 1646-51

[21] Tuorila H and Cardello AV 2002 Consumer Responses to an Off-Flavor in Juice in The Presence of Specific Health Claims *Food Quality and Preference* **13** pp 561-9

[22] Herminiati A 2015 *Aplikasi dan Efektivitas Difructose Anhydride III dari Umbi Dahlia untuk Meningkatkan Penyerapan Kalsium pada Model Tikus Defisien Kalsium* (Bogor: Institut Pertaian Bogor)

[23] Ronkart S, Blecker C, Fougnies C, Van Herck JC, Wouters J and Paquot M 2006 Determination of physical changes of inulin related to sorption isotherms: An X-ray diffraction, modulated differential scanning calorimetry and environmental scanning electron microscopy study *Carbohydrate Polymers* **63** pp 210–7

[24] Zarroug Y, Abdelkarim A, Dorra ST, Hamdaoui G and Felah MEL 2016 Biochemical Characterization of Tunisian *Cichorium Intybus L.* Roots and Optimization of Ultrasonic Inulin Extraction *Mediterranean Journal of Chemistry* **6** pp 674–85

[25] Vist G, Suchdev P, De-Regil L, Walleser S and Peña-Rosas J 2011 Home fortification of foods with multiple micronutrient powders for health and nutrition in children under 2 years of age (Protocol) *Cochrane Database Systematic Reviews*

Studying the Effects Of Inhibitor Solution Soaking Time On Guava *(Psidiumguajava L.)* Storability

A Triyono[1a], R C Erwan Ardiansyah[1], M C Hapsari [2]

[1]Development Center for Appropriate Technology, Indonesian Institute of Sciences, Jl. K.S Tubun No. 5, Subang, Jawa Barat 41213, Indonesia

[2]Food Technology Department, Faculty of Engineering, Pasundan University, Jl. Dr. Setiabudi No. 9 Bandung 40153 Indonesia

[a)]Corresponding author:ateluyono@gmail.com

Abstract. This research is conducted to study the effects of the concentration and soaking time of inhibitor solution on the storability of guava using various concentrations (0, 10, 30, and 50 ppm) and soaking time (4, 8, and 12 hours). The estimation of guava storability uses simple linear regression. Results show that an $AgNO_3$ 10 ppm solution with 4 to 12 hours of soaking time can reduce the effect of storability with the following parameters: weight decrease, hardness level, vitamin C content, and total dissolved solids (TDS). Guava that is stored at room temperature is expected to have a storability that lasts for 29 days for the responses of weight decrease, hardness level, vitamin C, and TPI or it can reach around 2.5 to 3 times longer compared to the controls only stored for 9 days at room temperature. The treatment using $AgNO_3$ solution for 4 hours can suppress the effect of storage on the response parameters; the hardness level with storability reaching 28 days, and vitamin C content with storability reaching 12 days. Meanwhile, the treatment using the $AgNO_3$ solution for 12 hours can suppress the effects of storage on the response parameters; weight decrease with storability that reaches 12 days, TDS with storability that reaches 63 days.

Keywords: *guava, storability, inhibitor solution, soaking time*

1. Introduction

Fruits generally have a certain season and time to distribute, so it is necessary to extend a fruit's storability. Changes in quality during the time of storage occur because the fruits still do respiration, ripening, and then followed by the process of withering and decaying. According to Kader (1992) cited in [3], the range of loss after harvesting fresh fruits and vegetables is estimated at 5-25% in developed countries and 20-50% in developing countries.

Guava is one of the climacteric fruit groups which has a respiration rate that will continue to increase as the fruit grows. The decrease in weight and other changes might occur, especially when the fruit reaches its climatic peak. Changes occur due to external factors such as ethylene, O_2, CO_2 and temperature, and mechanical damage as well as microbiological damage resulting in a shorter shelf life [20]

Inhibition of the most important biological factors in fruits is the formation of ethylene, respiration, transpiration, and morphological factors [6]. Ethylene is a compound that functions to stimulate fruit ripening, a flower blooming, and leaf and flower shedding. Ethylene formation in the tissue, from ACC (1-*aminocyclopropane*-1-*carboxylic acid*), is converted to ethylene with the ACC oxidase catalyst. According to [10], silver nitrate($AgNO_3$) can function as an anti-ethylene. In silver nitrate solutions, Ag^+ ions may inhibit ethylene formation in tissues and negative feedback from endogenous ethylene.

Based on the explanation above, it is necessary to study the storability of Guava (*Psidiumguajava L.*) to inhibit ripening to fruit by soaking in inhibitors. The purpose and objective of this research are to study the effects of soaking treatment using an inhibitor solution on the shelf life of fresh guava (*Psidiumguajava L.*) stored at room temperature.

2. Materials and Methodology

2.1 Materials and Tools

The main material is Citayam variety grade guavas (*Psidiumguajava L.*) with 6-9 fruits/kg, fruit age at 4 months and ripeness level at 60% in Garut Regency. The research materials include solid $AgNO_3$, 1% acetate acid from PT Bratachem, Citric acid 1% and glacial acetate from PT. Bratachem and solid chitosan obtained via Tokopedia. The analysis material for vitamin C is iodine 0.01 N and 1% starch, sodium thiosulfate 0.01 N, and solid potassium iodate from PT. Merck. The volume measuring include plastic measuring (200 mL, 1000 mL), and 2000 mL glass measuring, 100 mL and 1000 mL measuring flask, and tool such as stirring rod, watch glass, glass funnel, spatula, 10 mL volumetric pipette, 25 mL burette, 250 mL Iwaki Pyrex Erlenmeyer, and a 500 mL spray bottle.

The tools used to assist analysis (physical, chemical, and physicochemical) are heaters, magnetic stirrers, Quattro Macs series digital balance, Mettler Toledo AL 204 digital analytical balance, Memmert VO200 vacuum oven, Penetrometer, and Atago PAL 1 digital refractometer.

2.2 Research Method

The estimation of storability uses the experimental design is a simple linear regression with the regression equation model [17] and the research stages are: The first stage of research to determine the concentration and soaking time of the inhibitor solution ($AgNO_3$) with a variety of concentrations (0, 10, 30, and 50 ppm) and soaking time (4, 8, 12 hours). The experimental design as follows: Y= a + bX, Y= dependent variable (response X), X= independent variable, a= intercept, b= regression coefficient (slope). The values of a and b for linear regression are calculated by the following formula:

$$a = \frac{(\sum Yi)(\sum Xi^2) - (\sum Xi)(\sum XiYi)}{n\sum Xi^2 - (\sum X\,i)^2} \quad b = \frac{n\sum XiYi - (\sum Xi)\,(\sum Yi)}{n\sum Xi^2 - (\sum X\,i)^2}$$

The response design/parameter(y) consists of: (1) physical response; weight decrease and hardness level. (2) Chemical response; water and vitamin C levels [16], and (3) physicochemical response; total dissolved solids [7].

3. Results and Discussion

Results of the analysis of the effect of soaking treatment using inhibitor solution on the response or parameters for the change of guava characteristics are as follows:

3.1. Weight Decrease

According to Royana (2012) quoted in [8], the increase of respiration rate will cause changes in the compounds and the release of water (transpiration) so that the fruit shrinks in weight. The results of the correlation of storage time with the percentage of weight decrease in the guava fruit can be seen in Figure 1 and Table 1.

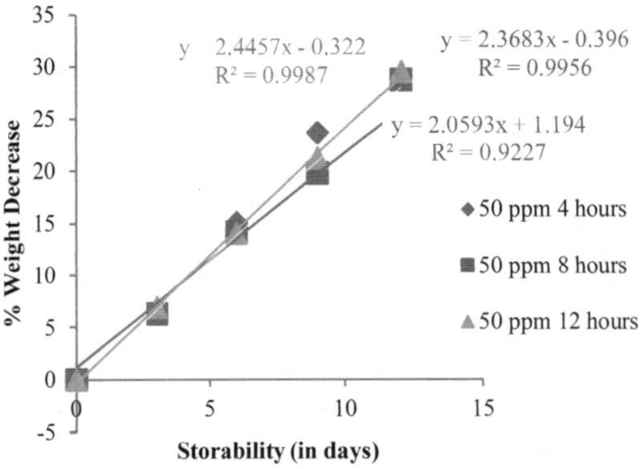

Figure1. The curve of the correlation of storability(x) with fruit weight decrease(y) on the treatment of $AgNO_3$ 50 ppm and soaking time of 4,8, and 12 hours

Based on Figure 1 and Table 1, the soaking of the fruit stem in the $AgNO_3$ 50 ppm solution for 4 hours can suppress the effect of storability (x) on the weight decrease percentage with a coefficient of determination that is the lowest among the other treatments. This treatment has a formula of y = 2.0593x + 1.194 with r^2= 0.9227, which shows that the storability has affected about 92.27% of the weight decrease with an estimation of storability reaching 12.29 days. The second treatment is the soaking using AgNO3 10 ppm solution for 12 hours, resulting in the storability effect of 97.33%. The treatment uses a formula of y = 2.139x + 0.912 with r^2 = 0.9733 and a storability estimation of 11.96 days.The third treatment uses $AgNO_3$ 10 ppm for 4 hours which resulted in a 97.66% in weight decrease. The treatment uses a formula of y = 2.403x + 0.3 with r^2 = 0.9766 % and an estimation of storability reaching 10.90 days.

Table 1. Correlation of storability(x) with fruit weight decrease (y) from the $AgNO_3$soaking treatment on the concentration and soaking time and the coefficient of determination from the lowest value

$AgNO_3$ Treatment	Formula	Coef. of Determination (r^2)*	Effect of X on Y (%)	Storability Estimation (days)
50 ppm, 4 hours	y = 2.0593x + 1.194	0.9227 (1) *	92.27%	12.29
10 ppm,12 hours	y = 2.139x + 0.912	0.9733 (2)	97.33%	11.96
10 ppm, 4 hours	y = 2.403x + 0.3	0.9766 (3)	97.66%	10.90

*) Selected starting from the smallest r^2 value from the effect of x (storage)on the response parameters

Soaking the pieces of tomato pericarp for 2 minutes into 0.3 to 5 millimolar (mM) $AgNO_3$ solution which is equivalent to 51 ppm to 850 ppm can inhibit ethylene and significant work and reduce endogenous ethylene production [10]. Silver metal (Ag^+) can also inhibit the ripening of whole tomatoes when silver thiosulfate enters the fruit's vascular system through a cut fruit stem. Tomatoes treated with soaking in a solution containing silver ions had a smaller amount of 1-*aminocyclopropane*-1-*carboxylic acid* (ACC) than the fruit that did not receive the treatment with solutions containing silver ions. The fruit treated with silver ion solutions had an ACC content of 1.8 nmol/g of fresh weight. Meanwhile, the fruit that was not soaked into a solution containing Ag^+ ions had an ACC content of 2.6 nmol/g fresh weight.

3.2. Hardness Level

Measuring the hardness level of fruit is one way to find out the characteristics of fruit ripeness, the fruit hardness will decrease or become soft. Fruit softness that occurs during the ripening process will result in a decrease in the quality of the fruit. According to Nasution (2012) quoted in [8], the value of fruit hardness shows the freshness level of fruit. The hardness level is considered good not when the value is too high or too low, but depending on the physical condition of the fruit. Fruit stored for a certain period will experience a decrease in the level of hardness. The results of the observation of the correlation of storability on the hardness level of guava can be seen in Figure 2 and Table 2.

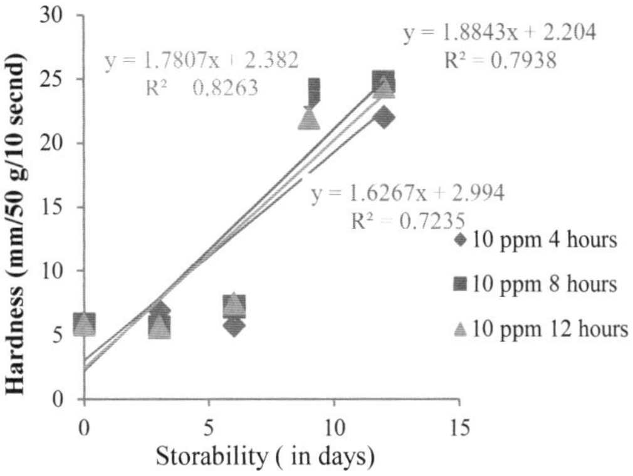

Figure 2. The curve of the correlation of storability (x) with guava hardness level(y) on the treatment of $AgNO_3$ 10 ppm and soaking time of4, 8, and 12 hours

Based on Figure 2 and Table 2, the treatment of soaking the fruit stem in $AgNO_3$ 10 ppm solution for 4 hours have the formula of **y = 1.6267x + 2.994** with $\mathbf{r^2= 0.7235}$. The $r^2 = 0.7235$ value shows that the storage time has an effect of 72.35% on the level of fruit hardness with an estimated storability of 27.91 days. The second treatment is the soaking using $AgNO_3$ 50 ppm solution for 12 hours which shows a storability effect of 72.77% on the level of hardness. The treatment uses a formula of y = 1.6033x + 2.692, with $r^2 = 0.7277$ and a storability estimation of 28.50 days. The third treatment uses $AgNO_3$ 30 ppm for 4 hours using a formula of y = 1.64x + 1.972 with $r^2 = 0.7495$ and an estimated storability of 28.30 days.

Table 2. Correlation of storability (x) with hardness level (y) from the $AgNO_3$ soaking treatment on the concentration and soaking time and the coefficient of determination from the lowest value

Inhibitor Treatment	Formula	Coef.of Determination (r^2)*	Effects of X on Y(%)	Storability Estimation (days)
10 ppm, 4 hours	y = 1.6267x + 2.994	0.7235 (1)	72.35%	27.91
50 ppm, 12 hours	y = 1.6033x + 2.692	0.7277 (2)	72.77%	28.50
30 ppm, 4 hours	y = 1.64x + 1.972	0.7495 (3)	74.95%	28.30

*) Selected starting from the smallest r^2 value from the effect of x (storage)on the response parameters

Sampebatu (2006) cited in [11] states that there is a change in the composition of the cell wall constituent due to the breakdown of insoluble protopectin into soluble pectin so that softening occurs. Afrazak (2014) cited in [4] states that during the fruit ripening process, pectin compounds resulting from protopectin degradation increase. The increase of pectin will reduce the cell wall cohesion power that binds cells to one another. As a result, the fruit's hardness will decrease, and the fruit becomes soft. According to [4] hardness is an important indicator in determining the ripeness level of an agricultural product, especially fruits. Fruits that experience the ripening process tend to have a low level of hardness than before the ripening process.

The guava fruit has increased the value of hardness (getting softer) in each treatment, because the respiration that occurs is suspected to run normally, given the respiration process needs simple organic compounds and oxygen from the air. Ag+ ions from $AgNO_3$ solution which inhibits the formation of ACC will inhibit the formation of ethylene. Respiration of guava fruit runs normally, but the ethylene hormone which can accelerate the ripening process is inhibited. When the ripening process is inhibited, then the degradation of protopectin into pectin which dissolves in water will also be inhibited so that the increase in the hardness value will be more stable where the r^2 value is near 0%. The soaking treatment of guava stalks into $AgNO_3$ 10 ppm solution for 4 hours can reduce the effect of storage time (lower r^2 value) on the response of the parameters of the hardness level that is greater than other treatments. The r^2 value of the soaking treatment of $AgNO_3$ solution was more than 10 ppm and more than 4 hours compared to the selected treatment; this could be due to a surge in the increase or decrease in the level of hardness on the guava.

3.3. Vitamin C Content

Vitamin C or ascorbic acid has the molecular formula $C_6H_8O_6$. Vitamin C is soluble in water and slightly soluble in acetone or alcohol which has a low molecular weight. Vitamin C is easily oxidized and even more so if there is a Cu, Fe catalyst, ascorbate oxidase enzyme, light, and high temperature [16]. The content of vitamin C in guava increases after being harvested. According to Purwatiningsih (2012) cited in [14], the increase of vitamin C in the handling of post-harvest fruits during storage for a certain period of time is due to the enzymes that work actively to convert simple sugars into vitamin C. Results from the observation of the correlation of storage time for vitamin C content of guava can be seen in Figure 3, and Table 3.

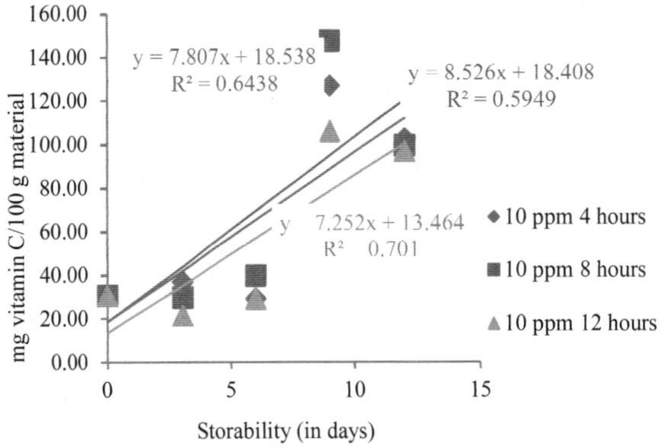

Figure 3. The Curve of the correlation of storability (x) with vitamin C level (y) on a treatment of $AgNO_3$ 10 ppm and soaking time of 4, 8, and 12 hours

Based on Figure 3 and Table 3, the treatment of soaking the fruit stem in $AgNO_3$ 10 ppm solution for 8 hours can suppress the effect of storability on the level of vitamin C in guava better than the other treatments. This treatment use the formula of **y = 8.526x + 18.408** with $\mathbf{r^2 = 0.5949}$. The r^2 = 0.5949 value shows that the storage time has an effect of 59.49% on the level of vitamin C with an estimated storability of 12,26 days. The second treatment is the soaking using $AgNO_3$10 ppm solution for 4 hours which shows a storability effect of 64,38% on the level of vitamin C. The treatment uses a formula of y = 7.807x + 18.538 with r^2 = 0,6438 and a storability estimation of 11.25 days. The third treatment uses $AgNO_3$ 50 ppm for 12 hours using a formula of y = 4.7367x + 20,21 with r^2 = 0.6507 and an estimated storability of 19.86 days.

In each treatment, the content of vitamin C tends to increase until the 9th day and then decrease on the 12th day. This increase in vitamin C will reach its peak when the fruit is ripe. The synthesis of vitamin C will be followed by a decrease in the content of vitamin C. According to, the decrease occurs of vitamin C because of the enzyme formation of vitamin C is no longer active due to lack of supply of nutrients and minerals as the fruit has been picked from the tree [14]. In addition, it can also be caused by oxidation with air which causes vitamin C to decrease. Storage of guava at room temperature can also cause damage to vitamin C in the fruit. The oxidation of vitamin C will produce L-dehydroascorbic acid which is highly labile and changes to L-diketogulonic acid which no longer has the activity as vitamin C.

Table 3. Correlation of storability (x) with vitamin C level (y) from the $AgNO_3$ soaking treatment on the concentration and soaking time and the coefficient of determination from the lowest value

Inhibitor Treatment	Formula	Coef. of Determination (r^2) *	Effects of X on Y (%)	Storability Estimation (days)
10 ppm, 8 hours	y = 8.526x + 18.408	0.5949 (1)	59.49%	12.26
10 ppm, 4 hours	y = 7.807x + 18.538	0.6438 (2)	64.38%	11.25
50 ppm,12 hours	y = 4.7367x + 20.21	0.6507 (3)	65.07%	19.86

*) Selected starting from the smallest r^2 value from the effect of x (storage)on the response parameters

When the ripening process is inhibited, vitamin C synthesis can also be inhibited. When Ag^+ ion from $AgNO_3$ solution inhibits ACC synthase enzyme work, the resulting ethylene product will be reduced to a smaller amount. ACC compounds are intermediate compounds in ethylene production, so ripening hormones can be inhibited. Thus, the ripening process is inhibited and the synthesis of vitamin C will be suppressed.

Based on the explanation above, the $AgNO_3$ 10 ppm concentration with 8 hours soaking time can withstand the effect of storage time on the value of the increase in vitamin C content of guava up to **40.51%** (100%-59.49%). Soaking time for more than 8 hours in $AgNO_3$ 10 ppm concentration did not affect because the 12-hour soaking time had a greater r^2 value than the 8-hour soaking time.

3.4. Total Dissolved Solids (TDS) Contents

Total dissolved solids are compounds in fruit that dissolve in water. In most fruits, the total dissolved solids contained are sugar. This means that the TDS can be used as a parameter of sugar content in fruit. The sugar found in fruit is sucrose, glucose, and fructose [18].

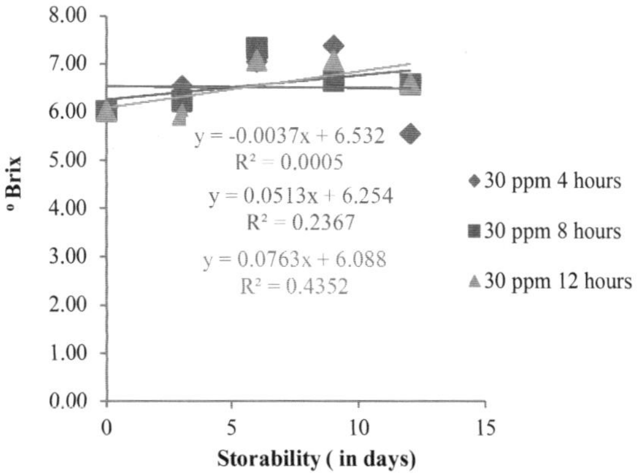

Figure 4. Curve of the correlation of storability (x) with TDS (y) on treatment of $AgNO_3$ 30 ppm and soaking time of 4, 8, and 12 hours

Based on Figure 4 and Table 4, the treatment of soaking the fruit stem in $AgNO_3$ 30 ppm solution for 4 hours can suppress the effects of storability on the TDS in guava better than the other treatments. This treatment uses the formula of **y = -0.0037x + 6.532** with $\mathbf{r^2 = 0.0005}$. The r^2 = 0.0005 value shows that the storage time has an effect of 0.05% on the TDS with an estimated storability of 58.01 days. The second treatment is the soaking using $AgNO_3$ 10 ppm solution for 12 hours which uses a formula of $y = 0.0407x + 6.638$ with $r^2 = 0.0830$. This r^2 value shows an effect of 8.30% towards the TDS and an estimated storability of 63.69 days. The third treatment uses $AgNO_3$ 10 ppm for 8 hours using a formula of $y = 0.031x + 6.33$ with $r^2 = 0.1237$ and an estimated storability of 93.55 days.

Table 4. Correlation of storability (x) with TDS (y) from the $AgNO_3$ soaking treatment on the concentration and soaking time and the coefficient of determination from the lowest value

Inhibitor Treatment	Formula	Coef. of Determination (r^2)*	Effects of X on Y %)*	Storability Estimation (days)
30 ppm, 4 hours	$y = -0.0037x + 6.532$	0.0005 (1)	0.05%	58.01
10 ppm,12 hours	$y = 0.0407x + 6.638$	0.0830 (2)	8.30%	63.69
10 ppm, 8 hours	$y = 0.031x + 6.33$	0.1237 (3)	12.37%	93.55

*) Selected starting from the smallest r^2 value from the effect of x (storage)on the response parameters

The TDS content of guava in each treatment tends to rise to a certain point and then decrease. This change is in accordance with the statement of Hakim, et al. (2012) quoted in [4] that the total sugar during storage tends to increase due to organic acids being converted into sugar. According to Wills, et al. (1981) cited in [4], the increase in total dissolved solids that occur at room temperature is caused by the accelerated reaction due to high temperatures resulting in carbohydrate breakdown by enzyme activity. This can be used as an indicator of the process of ripening and respiration.

The soaking treatment of guava fruit stalks into $AgNO_3$ solution can reduce the effect of storage time on total dissolved solids due to the Ag^+ ions from $AgNO_3$ solution which can optimally inhibit intermediate compounds (ACC) in the formation of ethylene hormone. The results of the ANOVA analysis of the response parameters from the effect of soaking treatment of $AgNO_3$ inhibitor solution and storage factor with estimated storability are recapitulated in Table 5.

Table 5. Results of ANOVA Analysis of the Effect of inhibitor solution and storage factor of Guava (at trust 95%)

No	Parameter	Treatment Effect			Explanation
		Inhibitor Factor (A)	Storage Factor (B)	AB Interaction	
1	Weight decrease	significant	significant	Significant	significant
2	Hardness	not significant	significant	not significant	not significant
3	Vitamin C	significant	significant	Significant	significant
4	Water content	not significant	significant	not significant	not significant
5	TDS	not significant	significant	Significant	significant

Table 6. Conclusions from the ANOVA analysis results of the response parameters and storability from the effects of soaking in $AgNO_3$ inhibitor solution

No	Response / Parameter	Concentration and Soaking Time of $AgNO_3$	Coefficient of Determination (%)	Storability Estimation (days)
1	Weight decrease	10 ppm 12hours	97.33 %	11.96
2	Hardness	10 ppm 4 hours	72.35 %	27.91
3	Vitamin C	10 ppm 4 hours	64.38 %	11.26
4	TDS	10 ppm 12 hours	8.30 %	63.69
				$\bar{x}= 29$

Based on Table 5 and Table 6, it can be concluded in general that the soaking of guava stems $AgNO_3$ 10 ppm solution with soaking time of 4 to 12 hours can reduce the effect of storage time on the response of the analyzed parameters; weight loss, level of hardness, vitamin C content, and total dissolved solids (TDS) with an estimated shelf life reaching an average of 29 days (Table 6). Soaking treatment in $AgNO_3$ solution with a concentration of 10 ppm and a soaking time of 4 hours can reduce the effect of storage time on the response/parameters; the level of hardness with an estimated storability of 28 days, and vitamin C with an estimated storability of 12 days. Meanwhile, the soaking treatment of AgNO3 solution with a concentration of 10 ppm and a soaking time of 12 hours can reduce the effect of storage time on the response/parameters; weight decrease with an estimated storability of up to 12 days, and TDS with an estimated storability of 63 days.

The estimation of storability/shelf life based on simple linear regression method shows that the effect of soaking in an $AgNO_3$ inhibitor solution can increase the average storability of 29 days compared to control without $AgNO_3$ soaking, which only has a storability of up to 9 days stored at room temperature or around 2.5-3 times. The $AgNO_3$ 10 ppm soaking treatment for 4 hours on the guava fruit stem can increase the fruit's storability up to 2.5 times when stored at room temperature. The $AgNO_3$ 10 ppm concentration is based on the calculation of r^2 values of linear regression where the $AgNO_3$ 10 ppm solution can suppress the effect of storage time on response parameters analyzed.

An increase in the concentration of $AgNO_3$ solution can inhibit the formation of ACC compounds, but an increase in the concentration of $AgNO_3$ solution can trigger the production of ethylene in tomatoes [10]. This is why the $AgNO_3$ with a 10 ppm concentration was chosen.

4. Conclusion

Based on the research, it can be concluded that increasing the concentration of inhibitor solution ($AgNO_3$) can reduce the effect of storage time on the responses; weight decrease, Vitamin C content, and total dissolved solids (TDS) of the guava. The soaking treatment of inhibitor solutions($AgNO_3$) and the concentration, as well as time, can suppress the effects of storability (low value of r^2) of the guava. The soaking treatment of $AgNO_3$ 10 ppm solution for 4 hours can reduce the storage effects (low value of r^2) on the response parameters; the level of hardness with a storability estimation of 28 days and vitamin C with an estimation of 12 days. Meanwhile, the treatment using $AgNO_3$ 10 ppm solution for 12 hours can reduce the storage effects (low value of r^2) on the weight decrease parameter with an estimated storability of 12 days and TDS with an estimated storability of 63 days.

Based on the results of the evaluation of the research that has been carried out, the researcher submits several suggestions including an analysis on silver metal residues on the skin and flesh of the guava fruit should be done. Further research is needed on coating the seed of guava with chitosan to suppress the effect of exogeneous factors combined with the soaking of fruit stem into the $AgNO_3$ to suppress the influence of endogenus during storage at room temperature.

5. Acknowledgments

The author would like to thank the PPTTG-LIPI Subang institution for the implementation of this research program. We would also like to express our gratitude towards Dr. Dede Zainal Arief from Universitas Pasundan Bandung for the help and guidance which assisted the research and report writing very well.

6. References

[1] Amiarsi and Pudji KU 2011 Peranan Larutan Pengawet terhadap Mutu Bunga Potong *Alpinia* selama Peragaan. *J. Hortikultura* **21** 2 185-190 Balai Besar Penelitian dan Pengembangan Pascapanen Pertanian Bogor dan Balai Penelitian Tanaman Hias Cianjur

[2] Desrosier and Norman W 1988 *Teknologi Pengawetan Pangan* UI Press Jakarta

[3] Fitrianti and Junita 2006 *Kajian Teknik Penyimpanan dan Pengemasan Jambu Biji (Psidium guajavaL.) dalam Kemasan Transportasi* Skripsi Departemen Teknik Pertanian Fakultas Teknologi Pertanian Institut Pertanian Bogor

[4] Ifmalinda 2017 Pengaruh Jenis Kemasan Pada Penyimpanan Atmosfir Termodifikasi Buah Tomat. *J Teknologi Pertanian* Andalas **21** no 1 Program Studi Teknik Pertanian Fakultas Teknologi Pertanian, Universitas Andalas

[5] Indriani and Nidahanifah 2012 *Pengaruh Komposisi Larutan Pulsing dan Chitosan sebagai Anti Transpiran dalam Meningkatkan Vase Life Mawar Potong (Rosa hybridavar 'Grand Gala')* Skripsi Departemen Agronomi dan Hortikultura Fakultas Pertanian Institut Pertanian Bogor.

[6] Kristianingrum and Susila 2007 *Beberapa Metode Pengawetan Buah-buahan* Jurusan Pendidikan Kimia Fakultas Matematika dan Ilmu Pengetahuan Alam Universitas Negeri Yogyakarta

[7] Muchtadi, Tien R, Sugiyono and Fitriyono A 2015 *Ilmu Pengetahuan Bahan Pangan* Penerbit Alfabeta Bandung

[8] Mudawamah and Nur 2014 *Pengaruh Konsentrasi dan Lama Perendaman dalam Kalsium Klorida ($CaCl_2$) terhadap Kualitas dan Kuantitas Pascapanen Buah Jambu Biji Merah (Psidiumguajava Linn)* Skripsi Jurusan Biologi Fakultas Sains dan Teknologi Universitas Islam Negeri Maulana Malik Ibrahim Malang

[9] Mukdisari and Yurisqi 2015 *Penggunaan Kitosan dan Lilin Lebah sebagai Bahan Pelapis untuk Meningkatkan Masa Simpan dan Mempertahankan Kualitas Buah Pepaya* Skripsi Departemen Agronomi dan Hortikultura Fakultas Pertanian Institut Pertanian Bogor

[10] Atta-Aly MA, Saltveit ME and Hobson GE 1987 Effect of Silver Ions on Ethylene Biosynthesis by Tomato Fruit Tissue *Plant Physiol* **83** Department of Horticulture Ain Shams University Cairo Egypt

[11] Naibaho J, Elisa J and Era Y 2013 Penyimpanan Buah Terung Belanda dengan Kemasan Aktif Menggunakan Bahan Penjerap Oksigen, Karbondioksida, Uap Air dan Etilen *J Rekayasa Pengan dan Pertanian* **1** No 3 Program Studi Ilmu dan Teknologi Pangan Fakultas Pertanian Universitas Sumatera Utara Medan

[12] Nur'aini, Hesti and Siska A 2015 Penggunaan Kitosan untuk Memperpanjang Umur Simpan Buah Duku (*Lansium Domesticum* Corr) *AGRITEPA (ISSN: 2407-1315)* **1** No 2 Program Studi Teknologi Pertanian dan Program Studi Teknologi Pangan Fakultas Pertanian Universitas Dehasen Bengkulu

[13] Rezeki and Trisma 2016 *Pelapisan Lilin Karnau badan Kitosan untuk Mempertahankan Mutu Wortel Kupas* Tesis Program Studi Teknologi Pascapanen Institut Pertanian Bogor

[14] Rohim A, Alimuddin and Erwin 2016 Analisis Kandungan Asam Askorbat dalam Buah Naga Merah (*Hylocereuspolyrhizus*) dengan Iodimetri *J Kimia Mulawarman* **14** No. 1 Jurusan Kimia FMIPA Universitas Mulawarman Samarinda

[15] Sitorus RF, Karo-Karo T and Lubis Z 2014 Pengaruh Konsentrasi Kitosan sebagai Edible Coating dan Lama Penyimpanan terhadap Mutu Buah Jambu Biji Merah *J. Rekayasa Pangan dan Pertanian* **2** No.1 Program Studi Ilmu dan Teknologi Pangan, Fakultas Pertanian, Universitas Sumatera Utara

[16] Sudarmadji S 2010 *Analisa Bahan Makanan dan Pertanian* Penerbit Liberty Yogyakarta

[17] Sudjana 2005 *Metode Statistika* Tarsito Bandung

[18] Syaefullah and Enrico 2008 *Optimasi Keadaan Penyimpanan Buah Pepaya Sebelum Pemeraman Dengan Algoritma Genetika*. Disertasi. Program Studi Ilmu Keteknikan Pertanian, Sekolah Pascasarjana, Institut Petanian Bogor

[19] Triyanto and Cahya I 2000 *Pengaruh Penggunaan Germisida Perak Nitrat dan Tetrasiklin, dan Lama Perendaman Larutan Pulsing Terhadap Mutu Bunga Mawar Potong (Rosa hybrida)* Skripsi Jurusan Teknologi Industri Pertanian FakultasTeknologi Pertanian Institut Pertanian Bogor

[20] Wardhani HK, Santoso S, Rahmawati W 2013 Pengaruh Penggunaan Modified Atmosphere Storage (MAS) dalam Kemasan Polietilen Suhu Ruang dan Suhu Chiller terhadap Kadar Vitamin C dan Kadar ß Karoten pada Jambu Biji *J. Mahasiswa Program Studi Ilmu Gizi* Fakultas Kedokteran Universitas Brawijaya

[21] Winarno FG 1992 *Kimia Pangan dan Gizi* PT Gramedia

IOP Conf. Series: Earth and Environmental Science **251** (2019) 012034 doi:10.1088/1755-1315/251/1/012034

Formulation and Shelf Life Prediction of Cookies From Modified Cassava Flour (Mocaf) In Flexible Packaging

M Kurniadi[1*], Y Khasanah[1], A Kusumaningrum[1], M Angwar[1], D Rachmawanti[2], N H R Parnanto[2] and L D Pratiwi[2]

[1]Research Unit for Natural Products Technology, Indonesian Institute of Sciences (LIPI), Yogyakarta, Indonesia
[2]Department of Science and Food Technology, Universitas Sebelas Maret, Surakarta, Indonesia

*Email: hm_kur@yahoo.com

Abstract. The product development from modified cassava flour (mocaf) cookies has not been done much. This research was done to find out the best formulation on the cookies from mocaf and to predict its shelf life on flexible packaging. Formula cookies using mocaf flour, egg - tempe flour, green bean flour and catfish flour, each with a ratio: F1 (63%: 5%: 12%: 15%); F2 (63%: 5%: 12%: 20%); F3 (58%: 5%: 12%: 25%). Shelf life prediction using Accelerated Self Life Time (ASLT) method based on critical moisture content. The packaging that used is polyethylene, aluminum foil and metalizing. The results of this research showed the best formula was F2 containing protein 12.06 %, and energy 511.56 kcal / 100 g. The critical water content of cookies 6.89 %. Permeability of polyethylene, metalizing and aluminum foil were 0.05 gH2O / day.m2.mmHg, 0.006 gH_2O / day.m^2.mmHg and 0.006 gH2O / day.m^2.mmHg respectively. That packaging was able to maintain the cookies quality for 1.5 months (polyethylene). 12 months (metalizing), and 19 months (aluminum foil).
Keyword: Shelf life, cookies, mocaf, flexible packaging

1. Introduction

Today, stunting is a problem must be solved immediately in Indonesia. In 2013, stunting was approximately 37 % (9 million) Indonesian children in various income groups, predominantly having fifth rank in the world [1]. Stunting was caused by less household access (prospective parents, pregnant mothers or babies under 5 years old) to nutritious food. Based on the case, we need to find out alternatives food with high calorie, protein, and micro-nutritional substances at a low price. Cookies are an alternative product, ready to eat and accept by people of all ages [2]. The main problem to make cookies is wheat flour, where this material is imported from other countries, with gluten that can be consumed by autism. Therefore, we need usable alternatives to wheat flour, such as modified cassava flour (mocaf). Mocaf has similar characteristics and quality to wheat flour but contains low protein, so we need to add other food materials as protein sources. Some of the food materials, which may be used to increase protein content, are egg-tempe flour, green bean flour, and catfish flour [3-5].

Cookies quality is speedily reduced because it easily absorbs water content [6]. Shelf life is a very important indicator to specify quality of a food product in a specific time. Some factors that affecting shelf life are a type of product, type of packaging, a technique of packaging, temperature, and moisture

IOP Conf. Series: Earth and Environmental Science **251** (2019) 012034 doi:10.1088/1755-1315/251/1/012034

[7]. The shelf life of cookies may be specified using Accelerated Shelf Life Time (ASLT), applying environmental condition parameters accelerating the process of reduction in a quality product, giving higher shelf life temperature [8]. Shelf life prediction using ASLT method is specified by critical water content method using Labuza equation and semi-empirical approach using Arrhenius equation [6][9]. This research was conducted to find out the best formula of cookies using basic materials with mixed mocaf flour, egg-tempe flour, green bean flour, and catfish flour as well as hypothesizing shelf life using ASLT method with critical water point approach in some flexible packages.

2. Materials and Methods

2.1. Materials and tools

Materials used in product formulation are mocaf, green bean flour, catfish flour, egg-tempe flour, sugar, butter, skim milk, baking powder, mocha pasta, and water. Materials for analysis are LiCl, KCH_3CO_2, $MgCl_2$, K_2CO_3, $Mg(NO_3)_2$, KI, NaCl, KCl, and K_2SO_4. Packaging was used are aluminum foil, metalized, and polyethylene (PE). The tools are oven, mixer, analytical scale, cup, desiccators, clamper, mortar, and laboratory glass tools, glass jars, aluminum cup, thermometer.

2.2. Methods

This research consists of some stages : (1) making of cookies, (2) analysis: organoleptic test, chemical analysis, and shelf life predictions. The formula of cookies was in Table 1. The organoleptic test includes parameters of color, texture, aroma, taste, and overall preference with the scoring method in scale from 1 to 5, using 25 semi-trained panelists [10]. Chemical analysis of selected cookies formula (results of preference test) consists of water content (Thermogravimetry)[11], ash [11], protein (Micro Kjeldhal)[11], fat (Soxhlet) [11], carbohydrate (by different), and total calories.

Table 1. Formula of cookies

Formula	*Mocaf*	Egg-tempe flour	Green bean flour	Catfish flour	Butter	Suga r	York	Mocha pasta
F1	68 g	5 g	12 g	15 g	50 g	60 g	32 g	9 g
F2	63 g	5 g	12 g	20 g	50 g	60 g	32 g	9 g
F3	58 g	5 g	12 g	25 g	50 g	60 g	32 g	9 g

Shelf life was hypothesized using ASLT method based on critical water content. The stages include: (1) determination of initial moisture content based on Thermogravimetry method (Mi)[11]; (2) determination of critical moisture content (Mc); (3) determination of equilibrium moisture content (Me); (4) determination of Packaging Permeability on Water Vapour (ASTM E96/E96M-10); and (5) determination of shelf life (t) [9].

$$\ln\left[\frac{Me-Mi}{Me-Mc}\right] = (k / x)\left[\frac{A}{Ws}\right]\left[\frac{Po}{b}\right]\Theta \qquad (1)$$

Where :

M_e : equilibrium moisture content (gH20/ 100 g solid)
M_i: initial moisture content (g H_2O/ 100 g solid)
M_c: critical moisture content (g H_2O/ 100 g solid)
k/x: moisture permeability through packaging material (g H_2O (m^2 day mm Hg)$^{-1}$)
A: Packaging area (m^2)
Ws: weight of dry food solids (g)
Po: water vapor pressure at the storage temperature (mmHg)
b: slope of the linearized isotherm portion (i.e., from Mi to Mc)

3. Results and Discussion

3.1. Preference Test

Results of cookies preference test are shown in Table 2. The parameter of the most preferred color, aroma, texture and overall preference is formula F2. However, Formula F1 has a preference score equal to parameters of overall aroma and preference. Formula with the lowest preference rate is Formula F3. Preferred overall Formula is Formula F1 and F2. However, if seen from some of the other parameters, chosen Formula is Formula F2. Thus, furthermore, the Formula F2 was tested chemically, and shelf life was specified.

Table 2. Preference Scores of Cookies

Formula	**Parameter**				
Cookies	**Color**	**Aroma**	**Taste**	**Texture**	***Overall***
F1	3.64^a	3.64^a	3.80^a	3.84^a	3.92^b
F2	3.72^a	3.64^a	3.80^a	4.00^a	3.92^b
F3	3.32^a	3.28^a	3.36^a	3.56^a	3.40^a

Where: Different notations in the same column show real differences in α 5%

3.2. Chemical Analysis

Results of selected Formula (F2) analysis indicate that water content 6.01%; ash content 1.77%; protein content 12.06%; fat content 20.99%, carbohydrate content, 59.17%; energy 511.56 kcal/100 g. The total energy of Formula F2 meets minimal quality requirement of total calories with SNI 01-2992, minimal 400 kcal/100 g [12].

3.3. Shelf life prediction

3.3.1. Initial moisture Content (Mi)

The initial moisture content of cookies was 5.02%. The low water content of cookies is caused by baking process at high temperature (108 °C) making sucrose re-crystallization discharge water steam [13].

3.3.2. Critical moisture content (Mc)

The crunchiness of cookies is stated as a critical parameter of damage to cookies because it is most easily impaired. This is caused by the absorption of moisture from the environment which can increase the moisture content of the product. Critical moisture content is expressed as water content where the crispness of the product can no longer be accepted organoleptically by consumers [6]. Scores of water content and organoleptic score during storage are shown in Figure 1.

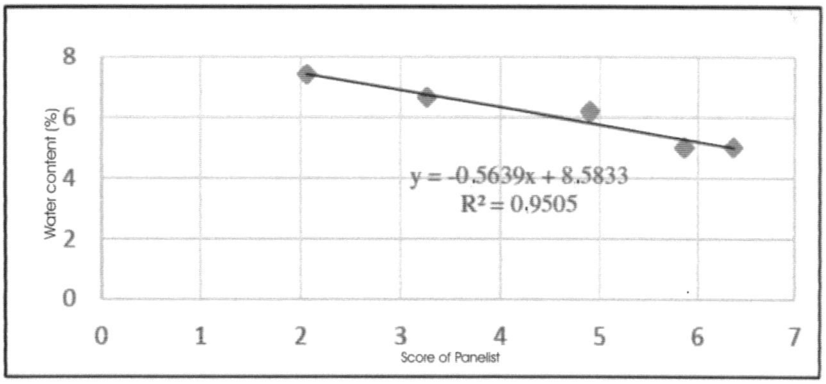

Figure 1. Correlation between water content and an organoleptic score of cookies

The graph shows that increasing the water content of the cookies panelists score is getting smaller, meaning that the panelists' preference has decreased. This shows a decrease in quality due to the increasingly soft texture due to absorption of moisture from the environment [6]. The linear equation obtained is $Y = -0.5639X + 8.5833$ with R^2 of 0.95. From these linear equations can be determined the value of critical water content by plotting a score of 3 on the x value so that the critical water content of cookies is protein and energy source of 6.8916%. The initial moisture content of this product determines critical moisture content and will affect the material adsorption process to the environment to achieve equilibrium conditions [14].

3.3.3. Equilibrium moisture content (Me)

The equilibrium moisture content of various levels of a_w based on the results of the research is shown in Table 3. The higher value of water activity (a_w), value of balanced water content of cookies is also higher. The difference in the level of a_w value from various salts is affected by the reaction between water and ionic bonds between the atoms that make up the salt. The ionic bonds between atoms the greater the activity of the salt water. Water activity is expressed as the chemical potential of water whose value varies from 0 to 1. If the water activity value is equal to 0 then the water molecule concerned does not carry out any activity in the chemical process, or the material is dry. If the water activity value is equal to 1 then the water potential in the chemical process at maximum conditions or material is pure water [6].

Table 3. Equilibrium moisture content Cookies at Various RH Storage

Salt	Water activities (a_w)	Balanced Water Content (Me) (%)
LiCl	0.112	2.3197
KCH_3CO_2	0.222	3.7545
$MgCl_2$	0.325	4.9368
K_2CO_3	0.437	6.6954
$Mg(NO_3)_2$	0.519	8.2006
KI	0.682	12.7303
NaCl	0.752	14.9064
KCl	0.838	22.7530
K_2SO_4	0.967	41.4999
LiCl	0.112	2.3197
KCH_3CO_2	0.222	3.7545

Table 3 shows the level of water absorbed by cookies to achieve equilibrium is strongly influenced by relative humidity (RH). The higher the RH, the greater the amount of water vapor absorbed by the material to achieve balance [15]. This equilibrium moisture content is then plotted with relative humidity or water activity (aw) and forms a curve called the moist sorption isotherm curve (ISL). Moist sorption isotherm curves relationship between (aw) and balanced water content are shown in Figure 2.

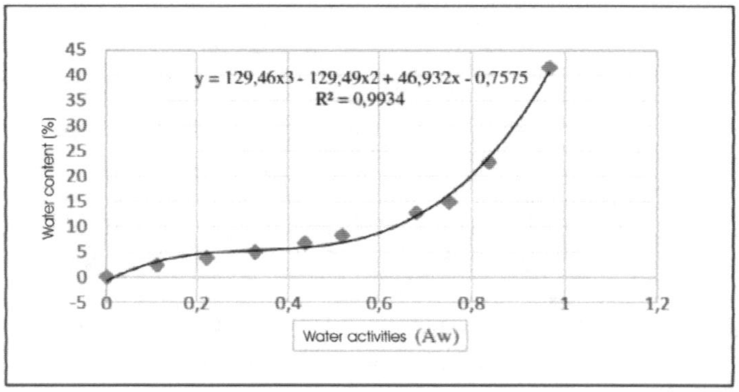

Figure 2. ISL Curve of Cookies

The relationship between water activity (a_w) and balanced water content of cookies in Figure 2 is made using the degree three polynomial equation. The constant equation produced A = 129.46; B = - 129.49; C = 46.93; D = - 0.75 with a value of R^2 = 0.99. Thus the equation of Isotherm Sorpsi Moist

Cookies is a source of protein and energy as follows: $Y = 129.46x^3\ 129.49x^2 + 46.93x - 0.75$, ($y$ = equilibrium moisture content (%); $x = a_w$).

The general form of the GAB equation is used to calculate the a_w value of the initial moisture content (M_i) and critical water content (M_c) of cookies. The initial water content, critical water content, and water activity are plotted into the GAB equation to obtain an equilibrium moisture content (Me). Data on determining the GAB curve are presented in Table 4.

Table 4. Data of GAB Equation Specification of Cookies

Salt	A_w	a_w Salt/Balanced Water Content
LiCl	0.112	0.0483
KCH_3CO_2	0.222	0.0591
$MgCl_2$	0.325	0.0657
K_2CO_3	0.437	0.0653
$Mg(NO_2)_2$	0.519	0.0633
KI	0.682	0.0520
NaCl	0.752	0.0463
KCl	0.838	0.0368
K_2SO_4	0.967	0.0233

Based on Table 4, it is furthermore plotted in GAB curve using degree 2 polynomial Equation. Water activity (a_w) as x-axis and comparison of a_w with equilibrium water content as the y-axis. The GAB curve formed can be seen in Figure 3.

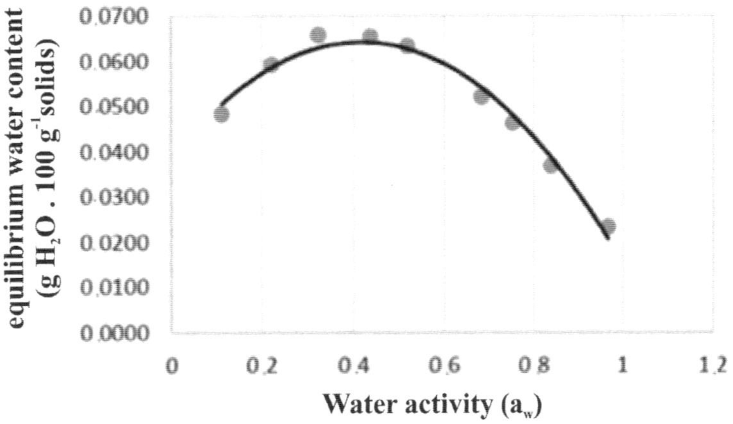

Figure 3. GAB Equation Curve of Cookies

The equation for GAB cookies is shown in Figure 3, $Y = -0.1502\ x^2 + 0.1286\ x + 0.0374$ with R^2 0.984. The K, C and Mo values obtained based on the Bizot [16] procedure in this study were 0.92; 5.73; and 5.06. C and K values show monolayer and multilayer adsorption energies. The value of C is adsorption energy associated with the water binding energy in the monolayer layer is quite low. The low energy of the adsorption is related to the type of bond involved in the binding of water. The greater the value of C, the stronger bond between water molecule of monolayer and material surface layer is stronger in bonding [17].

The sigmoid sorption type isotherm is well described if the values are $0.24 < K \leq 1$ and $5.67 \leq C \leq \infty$. While value of the monolayer water content is very necessary to determine the storage conditions and control of the damage reactions that may occur and determine the physical and chemical conditions of the dry material [18]. The water content of this monolayer layer is primarily bounded water, which cannot function as a solvent. The primarily bonded water is part of the solid because the water is absorbed on the active side of the polar solids [19]. The general form of GAB cookies is:

$$M = \frac{26{,}738\ aw}{(1-0{,}9212\ aw)(1-0{,}9212\ aw+5{,}2809\ aw} \quad (2)$$

Values of Mi, Mc, and a_w can be understood by the general form of GAB Equation. Values of Mi, Mc and a_w are shown in Table 5.

Table 5. Values of a_w of initial material (Mi), critical material (Mc), and balance (Me)

Parameter		Value
Determination a_w of initial material	Mi	5.0231
	a_w Mi	0.3168
Determination a_w of Critical moisture content (Mc)	Mc	6.8916
	a_w Mc	0.557
Determination of Me	Regressive Equation of	Y= 7.780x + 2.5586
	a_w Me	0.75
	Me	8.3934

From the Equation, value of a_w is plotted so that value of *Equilibrium moisture content* (Me) 8.3934%.

3.3.4. Moisture Permeability

Dry food such as cookies is hygroscopic so that these need packaging process to prevent water steam from entering into the product. Data for packaging permeability test results for water steam are shown in Table 6.

Table 6. moisture permeability through packaging material & Slope

Packaging	Slope	k/x ($gH_2O/day.m^2.mmHg$)	Mean
Aluminum Foil	0.00027	0.0039	0.004
	0.00024	0.0035	
	0.00026	0.0038	
Metalizing	0.00042	0.0061	0.006
	0.00045	0.0066	
	0.00044	0.0064	
PE	0.00349	0.0508	0.050
	0.00315	0.0459	
	0.00364	0.0531	

Based on Table 6, it can be understood that value of slope is directly proportional to packaging permeability. The higher value of packaging permeability, the water steam will be easier to penetrate the packaging, so that it will be easier to achieve critical water content and shorten shelf life [19]. Packaging permeability is found good and quality if it has low value. The lower value of packaging permeability, the shelf life is longer [6].

3.3.5. Shelf Life

Based on Equation of GAB curve (Figure 3), value of Balanced Water Content with RH = 75% (RH of place in which cookies is stored) is 8.3934%. Slope (b) or slope of the isothermal curve is obtained by plotting data of initial water content and critical water content in linear Equation. Linear Equation passing Me and Mc is shown in Figure 4, where value of slope is 7.7798.

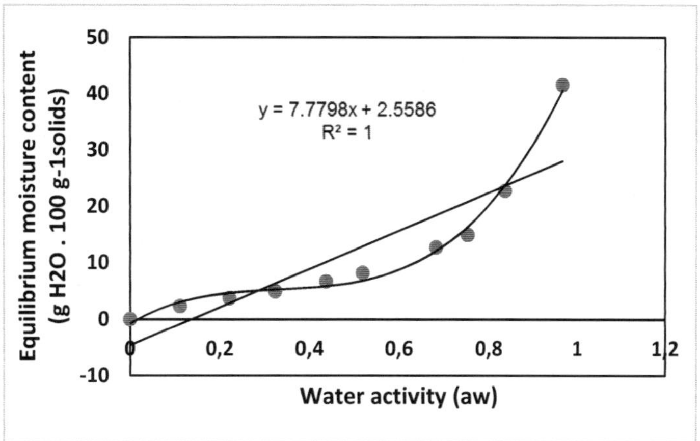

Figure 4. Linear Equation passing initial water content (Me) and critical water content (Mc)

The measure of packaging used in calculating the shelf life is 10 cm x 12 cm where packaging area is 0.024^2 in both sides. This measure is adapted to packaging used to package cookies in markets. Available data are plotted in Labuza Equation, so that hypothesis of shelf life of cookies can be seen in Table 7.

Table 7. Results of Shelf life Hypothesis of Cookies

Pack-aging	Mi	Mc	Me	k/x	A (m^2)	Ws (gram)	Po (mmHg)	Slope ISL	Shelf life	
									Day	Month
A	5.0231	6.8916	8.3934	0.004	0.024	23.80445	28.349	0.0778	588,05	19,6
B	5.0231	6.8916	8.3934	0.006	0.024	23.80445	28.349	0.0778	366.70	12,2
C	5.0231	6.8916	8.3934	0.050	0.024	23.80445	28.349	0.0	44,05	1.5

A : Aluminum foil; B: Metalizing; C: polyethylene.

Shelf lives of cookies use polyethylene packaging, metalizing, and aluminum foil are, respectively, 1.5 months; 12 months; and 19 months. Based on Table 7, it can be understood that the higher packaging permeability rate, shelf life is shorter. It is concordant to the theory of shelf life specification principle suggested by [19] indicating that the higher packaging permeability, the water steam is easier to penetrate the packaging so that it will be easier to achieve critical water content and shorten shelf life.

Packaging area and solid weight of packaging affect the shelf life of the product. The wider area of packaging, the entering water steam is wider to distribute to packaging and prevent achievement of critical water content, and total solidity is less in packaging so that shelf life will be shorter. It is caused by water steam entering into the packaging is distributed widely and accelerates achievement of critical water content [19]. This research used the packaging area of 0.024 m^2 and same solid weight, 23.8 g so that effect on shelf life is not seen clearly.

4. Conclusions

This research indicates that cookies with the best formula are formula F2 (mocaf = 63%; egg-tempe flour, 5%; green bean flour, 12%; catfish flour, 20%), contain 12.06% of protein, and energy of 511.56 kcal/100 g. The critical water content of cookies is 6.8916%. Values of polyethylene packaging permeability, metalizing, and aluminum foil are, respectively, 0.05 $gH_2O/day.m^2.mmHg$, 0.006 $gH_2O/day.m^2.mmHg$, and 0.006 $gH_2O/day.m^2.mmHg$. The three packagings are able to remain qualities respectively during 1.5 months, 12 months, and 19 months.

5. References

[1] Balitbangkes 2013 *Riset Kesehatan Dasar 2013* (Kementerian Kesehatan Republik Indonesia: Badan Penelitian dan Pengembangan Kesehatan)

[2] Iwegbue, Chukwujindu M.A. 2012. *Metal Contents in Some Brands of Biscuits Consumed in South Nigeria.* American Journal of Food Technology **7** 3 160-167

[3] Angwar M, Ariani D and Khasanah Y 2007 *Optimasi Pembuatan Tepung Tempe Telur sebagai Salah Satu Bahan Penyusun Tepung BMC Tempe* Prosiding Seminar PATPI Bandung 17-18 Juli 2007 ISBN 978-979-16436-0-7

[4] Ladamay, Arfa N and Yuwono S S 2014 Pemanfaatan Bahan Lokal dalam Pembuaan Foodbars (Kajian Rasio Tapioka:Tepung Kacang Hijau dan Proporsi CMC) *Jurnal Pangan dan Agroindustri* **2** 1

[5] Mervina C, Kusharto M and Marliyati S A 2012 Formulasi Biskuit dengan Substitusi Tepung Ikan Lele Dumbo *(Clarias gariepinus)* dan Isolat Protein Kedelai (Glycine max) sebagai Makanan Potensial untuk Anak Balita Gizi Kurang *Jurnal Teknologi dan Industri Pangan* **23** 1

[6] Kusnandar F, Dedde R, Adawiyah and Fitria M 2010 *Pendugaan Umur Simpan Produk Biskuit dengan Metode Akselerasi Berdasarkan Pendekatan Air Kritis* Jurnal Teknologi dan Industri Pangan **21** 2

[7] Ahmad U E, Darmawati and Refilia N R 2014 Kajian Metode Pelilinan Terhadap Umur Simpan Buah Manggis (Garcinia mangostana) Semi Cutting dalam Penyimpanan Dingin *Jurnal Ilmu Pertanian Indonesia* **19** 2 104-110

[8] Herawati H 2008 Penentuan Umur Simpan pada Produk Pangan *Jurnal Litbang Pertanian* **27** 4

[9] Labuza TP 1984 Moisture Sorption:Practical Asepticts of Isotherm Measurement and Use *American Association of Cereal Chemists* (Minnesota : St. Paul)

[10] Setyaningsih D, Apriyantono A and Sari MP 2010 *Sensory Analysis for Food Industry and Agro [Analisis Sensori untuk Industri Pangan dan Agro]* (Bogor : IPB Press)

[11] Association of Official Analytical Chemists (AOAC) 2002 *Official Methods of Analysis* (Washington DC : AOAC Publisher)

[12] BSN 1992 *Standar Nasional Indonesia untuk Syarat Mutu Cookies* (Jakarta : SNI 01-2973-1992)

[13] Romeo F V, Luca S D, Piscopo A, Santisi V and Poiana M 2010 Shelf-Life of Almond Pastry Cookies with Different Types of Packaging and Levels of Temperature *Journal Food Sciences and Technology International*

[14] Budijanto S, Sitanggang A B and Kartika D W 2010 Penentuan Umur Simpan Tortilla dengan Metode Akselerasi Berdasarkan Kadar Air Kritis serta Pemodelan Ketepatan Sorpsi Isothermnya *Jurnal Teknologi dan Industri Pangan* **21** 2

[15] Mustafida C and Widjanarko S B 2015 Umur Simpan Minuman Serbuk Berserat dari Tepung Porang (Amorpophallus onccophillus) dan Karagenan melalui Pendekatan Kadar Air Kritis *Jurnal Pangan dan Agroindustri* **3** 2

[16] Bizot H 1983 *Using The GAB Model To Construct Sorption Isotherms. Dalam:* Jowitt R, Escher F, and Vos G *Physichal Properties of Food* (London:Applied Science Publishers)

[17] Wijaya IMA, Suter IK and Yusa NM 2014 Karakteristik Isotermis Sorpsi Air dan Umur Simpan Ledok Instan *Agritech* **34** 1

[18] Purwanti A 2012 Evaluasi Pengaruh Lama Penyimpanan Terhadap Water Sorption Lembaran Plastis dari Kitosan *Prosiding Seminar Nasional Aplikasi Sains dan Teknologi Periode* III. Yogyakarta

[19] Cahyanti M N, Hindarto J and Lestario L N 2016 *Pemodelan Isoterm Sorpsi Air Biskuit Coklat Menggunakan Persamaan Caurie Jurnal Aplikasi Teknologi Pangan* **5** 2

[20] Hariyadi P 2008 *Prinsip-Prinsip Penetapan dan Pendugaan Masa Kadaluarsa Produk Pangan. Di dalam: Modul Pelatihan Pendugaan dan Pengendalian Masa Kadaluarsa Bahan dalam Produk Pangan* (Bogor:Departemen Ilmu dan Teknologi Pangan. Fakultas Teknologi Pertanian IPB)

IOP Conf. Series: Earth and Environmental Science **251** (2019) 012035 doi:10.1088/1755-1315/251/1/012035

Shelflife prediction of biscuits prepared from modified suweg (*Amorphophallus campanulatus B*) flour using Arrhenius model

T Rahman[1], N F Sulaiman[2], E Turmala[2], R C E Andriansyah[1], R Luthfiyanti[1] and A Triyono[1]

[1]Development Center For Appropriate Technology, Indonesian Institute of Sciences, Subang, Indonesia
[2]Food Technology Department, Faculty of Engineering, Pasundan University, Bandung, Indonesia.

Email: tkr.ttg@gmail.com

Abstract. Shelf life is very crucial information mentioned such as in food packages due to safety of food products when they reach their valid period to consume. The objectives of this study were to know shelf life of biscuit made of wheat flour and modified cassava flour with and without modified suweg flour and to know suitable packages for Shelf life prediction was based on moisture content and peroxide number in different packages: aluminium foil and metalized. This study showed that suitable packages for modified suweg biscuit were metalized package. According to the study, prediction of shelf life using the Arrhenius method with different shelf temperature resulted in modified suweg biscuit and blank biscuit (without modified flour suweg). To conclude, shelf life was determined by peroxide value for modified suweg biscuit in the temperature of 15 °C, 30 °C and 45 °C with metalized package resulting in 217, 172, and 137 days, respectively. While shelf life study of blank biscuit was done at 15 °C, 30 °C and 45 °C in a metalized package and resulted in 333, 250, and 192 days of shelf life, respectively.

Keywords: Arrhenius, biscuit, modified suweg flour, shelf life

1. Introduction

Biscuit is a dried food product made by baking the dough made of wheat flour or with its substitutes, oil or fat with or without other additional ingredients and allowed food materials [1]. Biscuit is a snack that is favored by all. Therefore, there are many kinds of biscuit in the market. In addition, there are countless modifications in the making. Basically, biscuit is made of wheat flour, but during its development, there have been many modifications by adding or replacing it with other ingredients, such as tapioca starch, modified cassava flour, sweet potato flour, and suweg flour.

Shelf life is the time interval between the time of production and consumption in which the product is in a satisfactory condition based on the characteristics of appearance, taste, aroma, texture, and nutritional value [2]. Shelf life is one type of information that food manufacturers must mention in food packages. Mentioning shelf life in the packages is very crucial information for the safety of food products and to avoid their consumption when they reach their valid period [3]. Syarief [4] stated that food quality changes mainly could be determined by the changes of quality, so determining the shelf life of a product needed measurement towards the product quality attributes.

Valid period or shelf life of biscuit can be determined by using a conventional method of *Extended Storage Studies* (ESS) and a non-conventional method of *Accelerate Storage Studies* (ASS), which is an accelerated method of *Accelerated Shelf Life Testing* (ASLT). The conventional method takes a long time and is costly since shelf life prediction is done in daily condition. It is different from the accelerated method that needs a relatively shorter period [5].

Suweg is a wild plant and grows well in places that are humid and away from sunshine [6]. Suweg has a prospect for potato flour or tapioca starch. Suweg has low amylose (24.5%) and high amylopectin (75.5%) [7]. According to [8], suweg flour has several advantages, i.e., high fiber and high protein – 13.71% and 7.20%, respectively, with low fat (0.28%). In addition, the glycemic index of suweg flour is relatively low. The study by Ekawati [9] describes that heat treatment towards suweg flour shows a significant effect towards *swelling power*, water absorption index (WAI), water solubility index (WSI), and water holding capacity (WHC). One type of biscuit that is being developed now is biscuit which was prepared for modified suweg flour, wheat flour, and mocaf. The objectives of this study were to determine the shelf life of biscuits made of modified suweg flour, wheat flour, and mocaf, and to know suitable packages for biscuit products.

2. Materials and Methods

2.1. Materials and Equipment

The materials used in this study were ingredients for making biscuit, and the materials to analyze. The ingredients for making biscuit were *autoclaving-cooling cycles* modified suweg flour, wheat flour, modified cassava flour, egg yolks, margarine, baking powder, powdered sugar (sorbitol), and chocolate powder.

The materials used for analysis are as follow: aquadest, KI, starch indicator, $Na_2S_2O_4$ 0.05 N, NaCl 0.85%, *Plate Count Agar* (PCA), N-hexane, selenium mixture, high concentration of H_2SO_4, NaOH 30%, HCl 0.01 N, boric acid, PP indicator, mixed indicator, H_2SO_4 1.25 %, NaOH 3.25 %, acetic acid glacial, alcohol 95%, and chloroform.

The equipment used in the study were a gas cooker, a kettle, a spatula, a digital scale, a mixer, cups, an exicator, an oven, 250 mL Erlenmeyer, 50mL buret, 1 mL pipette, Pasteur pipette, Petri dish, an incubator, aluminium foil, and metalized film.

2.2. The process of Making Modified Suweg Flour

The making of modified suweg flour used *autoclaving-cooling cycle* method. Suweg flour was mixed with 16% of water to make a suspension. The suspension was made into gelatin at the temperature of 70-85 °C. Then, it was heated in an autoclave at a temperature of 121 °C for 41 minutes. After that, it was cooled down in room temperature for 4-5 hours. Then it was put in the freezer of 4 °C for 12-16 hours. Then, it was dried in the drying cabinet until the water level reached 12%. Finally, it was milled and shifted through 70 mesh screens [10].

2.3. The process of Making Biscuit

The first step was to scale the main ingredients and other materials based on a predefined formula. The first mixing was to mix wheat flour, margarine, chocolate powder, and yolks by using a mixer until it created homogeneous cream. The next step was to mix modified suweg flour, wheat flour, modified cassava flour, milk, and baking powder. Then, the dough was molded into similar shape and size for the best baking result. The baking was done at around 120 °C for 45 minutes.

2.4. Method

Arrhenius model was used to determine the shelf life of biscuit prepared with and without modified suweg flour. The steps to determine shelf life using the *Arrhenius* model included determining valid period criteria (water level and peroxide value), determining types of packaging (metallized film dan aluminium foil), determining the temperature for the testing (15 ° C, 30 °C dan 45 °C), predicting the time and frequency of sampling, plotting data according to reaction order, analysing according to preservation temperature, and analysing shelf life prediction according to the lowest tolerable quality decrease [11].

2.5. Shelf Life Prediction

Arrhenius equation for shelf life prediction is as follow :

$$k = ko.\ Ae^{-Ea/RT}$$

Where: k: the rate constant; T: the absolute temperature (in Kelvin); A: the pre-exponential factor; Ea: the activation energy for the reaction; R: the universal gas constant

Shelf life was calculated by kinetic reaction equation according to predefined reaction order, whether it followed ordo 0 or ordo 1. Shelf life prediction using the equation for ordo 0 is as follow :

$$t = (A_t - A_0)/k$$

And for ordo 1 is as follow:

$$t = (\ln(A_0 / A_t))/k$$

Where : t = predicted shelf life of product; A_t = end product quality; A_0= initial product quality; k = constant

3. Results and Discussion

3.1. Chemical Properties of Suweg Biscuit

Chemical properties of suweg biscuit are presented in Table 1. Suweg biscuit has a moisture content of 4.89% and protein content 5.55 %, which complied to the national standard for biscuit (SNI 2973-2011) with a maximum moisture content of 5.55 % and minimum protein content 5 %. The ash content in suweg biscuit is 2.73 %. According to [12], ash content showed the total of minerals in a food. Mineral and an inorganic salt in a small amount of ash are declared as food residue after the combustion process into carbon-free. Table 1 below shows the requirements of biscuit quality according to SNI 2973-2011 [1].

Table 1. Indonesian national standard for biscuit (SNI 2973-2011)

No	Parameter (b/b)	Unit	Quality Requirements
1.	Moisture content	%	Max 5
2.	Protein (N x 6.25)	%	Min. 5
3.	Free fatty acids	%	Max 1

Fat content, carbohydrate and dietary fiber content of suweg biscuit are 34.65 %, 51.05 % and 15.77 %. Carbohydrate, fat, and protein are very important to the formation of the mixture, which will affect the texture of the food produced [13]. Suweg biscuit has a starch content 36.59 %. Starch is composed of two main fractions of amylose and amylopectin [14].

Table 2. Chemical properties of suweg biscuit

Parameter	Value (%)
Moisture content	4.89
Ash content	2.73
Protein content	5.55
Fat content	34.65
Carbohydrate content	51.01
Dietary fiber content	15.77
Starch content	36.59

3.2. Moisture Content

Moisture content is very important in determining the shelf life of a food product, especially biscuits because this factor will affect the physicochemical properties, chemical changes, and microbiological damage [15]. Biscuit is a type of food that goes through baking process, so it is hygroscopic. This makes biscuit easily absorb moist from its surroundings until the balance between moisture and its holding water capacity is reached. This is assumed to increase biscuit moisture content [16].

The result of moisture content analysis of modified suweg biscuit and the blank (biscuit without modified suweg flour) in the aluminium foil and metalized can be seen in Table 3 and Table 4.

Table 3. Value of moisture content of sample biscuits and the blanks on the aluminium foil packaging

Days	Moisture Content (%)					
	15 °C		30 °C		45 °C	
	Treatment	Blank Biscuit	Treatment	Blank Biscuit	Treatment	Blank Biscuit
0	4.89	4.99	4.89	4.99	4.89	4.99
7	3.58	7.18	4.54	4.94	3.38	9.76
14	4.93	11.32	4.33	6.13	4.67	5.54
21	4.63	6.84	4.62	8.37	4.12	9.65
28	2.76	5.49	3.42	10.97	3.77	5.15
35	4.74	9.52	4.59	8.19	6.55	4.43
42	2.73	5.44	5.78	6.95	5.31	4.85
49	4.38	4.1	6.63	10.8	6.65	3.24

Table 4. Value of moisture content of sample biscuit and blanks on the metalized packaging

Days	Moisture Content (%)					
	15 °C		30 °C		45 °C	
	Treatment	Blank	Treatment	Blank	Treatment	Blank
0	4.89	4.99	4.89	4.99	4.89	4.99
7	4.17	6.46	5.63	5.32	3.51	9.78
14	3.97	9.39	3.39	6.67	3.65	4.34
21	5.69	5.36	5.38	11.41	3.58	7.2
28	4.18	5.2	4.38	2.86	4.45	10.12
35	2.86	6.7	5.35	5.5	4.65	3.75
42	3.6	5.67	5.53	12.07	4.89	8.4
49	5.19	7.6	5.9	8.2	4.8	8.01

Table 5. Selection of ordo for sample of moisture content of aluminium foil packaging

Temperature (°C)	Regression Equation (Ordo 0)	Regression Equation (Ordo 1)	R^2 Ordo 0	R^2 Ordo 1
15	-0.034x + 4.912	-0.004x + 1.495	0.394	0.103
30	-0.016x + 5.244	0.005x + 1.428	0.081	0.226
45	0.011x + 4.625	0.009x + 1.344	0.028	0.404

Table 6. Selection of ordo for sample of moisture content of metalized packaging

Temperature (°C)	Regression Equation (Ordo 0)	Regression Equation (Ordo 1)	R^2 Ordo 0	R^2 Ordo 1
15	-0.009x + 4.551	-0.002x + 1.509	0.032	0.046
30	0.019x + 4.579	0.004x + 1.507	0.165	0.147
45	0.017x + 3.88	0.004x + 1.346	0.230	0.236

Table 7. Selection of ordo for blanks of moisture content of aluminium foil packaging

Temperature (°C)	Regression Equation (Ordo 0)	Regression Equation (Ordo 1)	R^2 Ordo 0	R^2 Ordo 1
15	-0.036x + 7.763	-0.006x + 2.019	0.066	0.090
30	0.101x + 5.188	0.014x + 1.651	0.540	0.595
45	-0.075x + 7.810	-0.013x + 2.044	0.290	0.363

The R^2 value in moisture content of all temperatures and packages (Table 5 and Table 6) had a lower value, since the result of the moisture content analysis was fluctuating. In addition, the R^2 value shows that the data is not linear, so it is considered inaccurate. While selecting order for determining shelf life was seen from the most linear or accurate value from the second order. For order 0, the R^2 value was taken from the original data result, whereas for order 1, the value was taken from the original data that was already changed into form. After that, the R^2 value was compared between ordo 0 and ordo 1, to see the R^2 value that was higher or more linear. The R^2 value that was higher or more linear in the chosen order was considered as a more accurate value. From Table 5 and 6, it can be seen that the correlation coefficient of order 1 is higher than correlation coefficient of ordo 0 (R^2 ordo 1 > R^2 ordo 0). Therefore, the water level rate of decline follows the reaction of ordo 1.

Table 8. Selection of ordo for blanks of moisture content of metalized packaging

Temperature (°C)	Regression Equation (Ordo 0)	Regression Equation (Ordo 1)	R^2 Ordo 0	R^2 Ordo 1
15	-0.036x + 7.763	-0.006x + 2.019	0.066	0.090
30	0.101x + 5.188	0.014x + 1.651	0.540	0.595
45	-0.075x + 7.810	-0.013x + 2.044	0.290	0.363

The calculation referring to the critical level of SNI value resulted in constant of quality decline rate (k) of the moisture content of modified suweg flour and blank biscuit in each of the packages, based on Arrhenius approach, which is as following Table 9.

Shelf life of the tested biscuit was longer than that blank biscuit because the tested biscuit absorbed less water than the blank biscuit. When they were compared, the second package type of biscuit that used *metalized* package had a longer shelf life than the one with the aluminium foil. This was probably due to the difference in the value of WVTR (*Water Vapor Transmission Rate*) in both packages. The lower the value of WVTR, the lower water vapor transmission into the packages.

Shelf life of biscuit with the moisture content parameter had a much shorter valid period than the shelf life with the water parameter. This was because the original quality of both biscuits was already close to the SNI maximum value of 5%. When it was compared with Garut starch cookies biscuit, Tahudi (2012) stated that the shelf life based on critical moisture content using metalized package ranged from 24 to 157 days. While the shelf life of composite biscuit of Banten taro tuber-based on critical moisture content using metalized package ranged from 6 to 12 months [17]. [3] stated that biscuit products usually had shelf life of 12-18 months.

Table 9. Prediction of the shelf life of sample biscuit and blanks on aluminium foil and metalized packaging

Temperature (°C)	Product	Packaging	Ea (kal/mol)	k_0	Constant (k)(/Hari)	Shelf Life (Day)
15	Sample biscuit	Aluminium Foil	4875.63	18.89	0.0037	6
30					0.0057	4
45					0.0084	3
15		Metalized	4268,32	3.87	0.0022	10
30					0.0032	7
45					0.0045	5
15	Blank biscuit	Aluminium Foil	4775.93	29.19	0.0069	0.29
30					0.01	0.2
45					0.015	0.13
15		Metalized	4430.37	6.64	0.0029	0.7
30					0.0042	0.4
45					0.006	0.3

3.3. *Peroxide Number*

Fat breakdown is indicated by bad smell and rancid taste. This process is called rancidity. Rancidity process is caused by oxidation and hydrolysis both in enzymatic and non-enzymatic reactions. The parameter that can be used to determine cookies quality decrease caused by oxidative fat breakdown is by determining peroxide value [18].

The analysis result of peroxide number of modified suweg biscuit and blank (biscuit without modified suweg flour) can be seen in Table 10 and Table 11.

Table 10. Value of peroxide number of sample biscuits and the blanks on the aluminium foil packaging

Days	Peroxide Number (mgEq/Kg)					
	15 °C		30 °C		45 °C	
	Treatment	Blank	Treatment	Blank	Treatment	Blank
0	0	0	0	0	0	0
7	0	0	0	0	0	0
14	1.37	0.59	0	0.6	0.88	0.9
21	0	0.22	0.89	0	0	0
28	1.62	0.79	0	0	1.04	0.24
35	1.35	1.95	2.69	1.06	1.9	1.47
42	2.06	0.55	2.09	2.67	1.46	1.72
49	2.11	3	4.34	5.77	3.76	4.99

Table 11. Value of peroxide number of sample biscuits and the blanks on the metalized packaging

Days	Peroxide Number (mgEq/Kg)					
	15 °C		30 °C		45 °C	
	Treatment	Blank	Treatment	Blank	Treatment	Blank
0	0	0	0	0	0	0
7	0	0	0	0	0	0
14	0.87	1.18	2.41	0.15	1.86	2
21	0.98	0	0.9	0	0	0
28	0.67	0	0	0.36	1.58	0.48
35	1.45	4.45	1.95	0.43	1.06	0
42	2.06	0	0.86	1.49	2.41	1.55
49	1.95	1.32	5.31	1.83	4.17	4.36

According to the test of peroxide number, it shows fluctuating value in all types of biscuit in both packages. However, during preservation, peroxide number breakdown was still acceptable because it was not higher than 10 mgEq/kg, whereas that is standard value stated in SNI 01-2347-1991.

In the tested biscuit with aluminium foil package, the highest value only reached 4.34 mgEq/Kg at the temperature of 30 °C, while at the temperature of 15 °C and 45 °C, it reached 2.11 and 3.76 mgEq/Kg. In the blank biscuit with the same package, the highest value reached 5.77 mgEq/kg, similar to the one at a temperature of 30 °C. Whereas the one at a temperature of 15 °C and 45 °C only reached 3 and 4.99 mgEq/Kg.

In the tested biscuit with metalized package, the highest value reached 5.31 mgEq/Kg at the temperature of 30 °C, while the one at the temperature of 15 °C and 45 °C only reached 2.06 and 4.17 mgEq/Kg. In the blank biscuit with the same package, the highest value reached 4.45 mgEq/kg, similar to the one at a temperature of 15 °C. While the value at a temperature of 30 °C and 45°C only reached 1.83 and 4.36 mgEq/Kg.

Table 12. Selection of ordo for sample of peroxide number of aluminium foil packaging

Temperature (°C)	Regression Equation (Ordo 0)	Regression Equation (Ordo 1)	R^2 Ordo 0	R^2 Ordo 1
15	0.064x - 0.441	0.023x - 0.281	0.735	0.642
30	0.007x + 1.059	0.004x + 0.286	0.006	0.013
45	0.041x + 0.055	0.013x - 0.002	0.584	0.528

The peroxide number of cookies shows an increase and a decrease, and probably it was caused by the fast oxidation reaction period influenced by humidity, air, oxygen, and light, as stated [18] in his study that oxidation level would increase adjacent to oxidation process until it reached its peak, and then it would decrease. This decrease was degradation of peroxide level that would be discussed. Fatty acids would be dissolved along with hydroperoxide conversion into aldehyde and ketone, and free fatty acids. The oxidation result would change smell and taste. From Table 12 and Table 13, it can be seen that correlation coefficient of ordo 0 is higher than correlation coefficient of ordo 1 (R2 ordo 0> R2 ordo 1). Therefore, the decrease in water level followed the reaction of ordo 0.

Table 13. Selection of ordo for sample of peroxide number of metalized packaging

Temperature (°C)	Regression Equation (Ordo 0)	Regression Equation (Ordo 1)	R^2 Ordo 0	R^2 Ordo 1
15	0.043x - 0.06	0.016x - 0.243	0.880	0.479
30	0.066x - 0.204	0.017x - 0.063	0.397	0.218
45	0.068x - 0.299	0.022x - 0.118	0.651	0.535

Table 14. Selection of ordo for blanks of peroxide number of aluminium foil packaging

Temperature (°C)	Regression Equation (Ordo 0)	Regression Equation (Ordo 1)	R^2 Ordo 0	R^2 Ordo 1
15	0.048x - 0.295	0.016x - 0.537	0.613	0.121
30	0.093x - 1.034	0.032x - 0.501	0.620	0.581
45	0.077x - 0.73	0.023x - 0.458	0.614	0.236

Table 15. Selection of ordo for blanks of peroxide number of metalized packaging

Temperature (°C)	Regression Equation (Ordo 0)	Regression Equation (Ordo 1)	R^2 Ordo 0	R^2 Ordo 1
15	0.032x + 0.075	0.010x - 0.004	0.127	0.112
30	0.036x - 0.361	0.014x - 0.693	0.753	0.084
45	0.055x - 0.315	0.016x - 0.169	0.379	0.189

The calculation referring to the critical level of SNI quality resulted in constant of decline rate (k) of peroxide number of modified suweg biscuit and blank biscuit in each package, which is as follow, based on Arrhenius approach.

Based on the calculation of shelf life of peroxide value using Arrhenius approach, it can be seen that the tested biscuit had a shorter shelf life than blank biscuit in both packages. In the tested biscuit and blank biscuit, the shelf life in the metalized package was longer than the shelf life in different packages.

Table 16. Prediction of shelf life of sample biscuit and blanks on aluminium foil and metalized packaging

Temperature (°C)	Product	Packaging	Ea (kal/mol)	k_0	Constant (k)(/Hari)	Shelf Life (Day)
15	Sample biscuit	Aluminium Foil	2223.9	2.49	0.051	196
30					0.062	161
45					0.074	135
15		Metalized	2822.9	6.35	0.046	217
30					0.058	172
45					0.073	137
15	Blank biscuit	Aluminium Foil	2957.1	9.62	0.055	182
30					0.071	141
45					0.089	112
15		Metalized	3254.3	8.97	0.03	333
30					0.04	250
45					0.052	192

One at the factors of oxidation causing rancidity is oxygen, so packaging significantly influences *Oxygen Transmission Rate* (O_2TR). It was assumed that metalized package had a smaller O_2TR value than aluminium foil package. In a study by [19], the O_2 TR value for aluminium foil was 0.8492-0.3199 cc/m^2/24 hours.

3.4. Determining Shelf Life
Based on determining suitable packages, the metalized package was selected as a suitable package for modified suweg biscuit. This was due to a longer shelf life using metalized package than using aluminium foil package. Somethings to be aware of during the packing process are the contact time between freshly processed and the packaging environment, packaging conditions that must meet safety standards [20]. Based on determining shelf life using moisture content and peroxide number, the result was that there were several different valid periods. The shelf life chosen was based peroxide number because peroxide number demonstrated a fat breakdown in the food products. Therefore, shelf life of modified suweg biscuit ranged from 217 days to 137 days, or around 7 - 4 months. The recommended shelf life mentioned in the package is at the temperature around 30 °C, because normally biscuit is preserved in room temperature at a similar temperature, around 30 °C. Therefore, the shelf life in the package is 172 days or around 5 months

4. Conclusion
Based on the study, the suitable package for modified suweg biscuit was metalized package. Referring to the study, determining shelf life using Arrhenius method with different temperatures concluded that the shelf lives determined from peroxide number for modified suweg biscuit at the temperatures of 15 °C, 30 °C and 45 °C using metalized package were 217 days, 172 days, and 137 days, respectively. The shelf life recommended mentioning in the package was 172 days or around 5 months.

5. Acknowledgment
Our deepest gratitude is sent to Insinas Research Grant from The Ministries of Research, Technology, and Higher Education Republic of Indonesia and Indonesian Institute of Sciences for funding this study. We also would like to thank Neneng Komalasari, Sutrisno, Sukwati, and Siti Khudaifanny for as well as their support and encouragement.

6. References
[1] BSN 2011 SNI 2973-2011 Biskuit", *Badan Standarisasi Nasional*, Jakarta
[2] Institute of Food Science and Technology 1974 Shelf Life of Food *J. Food Sci.* 39: 861−865
[3] Fitria M 2007 *Pendugaan Umur Simpan Produk Biskuit Dengan Metode Akselerasi Berdasarkan Pendekatan Kadar Air Kritis* (Bogor : Skripsi Fakultas Teknologi Pertanian Institut Pertanian Bogor)
[4] Syarief R dan Halid H 1993 *Teknologi Penyimpanan Pangan* (Jakarta : Arcan)
[5] Riyadi I A F 2015 *Pendugaan Umur Simpan Biskuit Berbasis Konsentrat Protein Ikan dan Spirulina platensis Berdasarkan Metode Akselerasi Dengan Pendekatan Kadar Air Kritis* (Bogor : Skripsi Teknologi Hasil Perikanan Institut Pertanian Bogor)
[6] Purwanto R 2012 Umbi suweg sebagai pangan fungsional untuk mendukung diversifikasi dan ketahanan pangan *Buletin IKATAN BPTP* Banten **2** pp 52
[7] Septiani D, Hendrawan Y, dan Yulianingsih R 2015 Uji Karakteristik Fisik, Kimia, dan Organoleptik Pembuatan Tepung Umbi Suweg *(Amorphophalluscampanulatus B)* Sebagai Bahan Pangan Alternatif *Jurnal Bioproses Komoditas Tropis* **3** 1
[8] Faridah D N 2005 Sifat fisiko kimia tepung suweg dan indeks glikemiknya *Jurnal Teknologi dan Industri Pangan* **16** 3 254-259
[9] Ekawati I G A, Ina P T, Kartika P, IDP 2015 Karakterisasi sifat fungsional tepung suweg (*Amorphophallus campanulatus B*) termodifikasi dengan metode pregelatinisasi *Seminar Nasional Sains dan Teknologi*
[10] Andriansyah R C E, Rahman T, Herminiati A, Nurhaidar, Luthfiyanti R 2017 Characteristics of chemical and functional properties of modified cassava flour (Manihot esculenta) by autoclaving-cooling cycles method, *IOP Conf. Series: Earth and Environmental Science* **101** 012023
[11] Herawati H 2008 Penentuan umur simpan pada produk pangan *Jurnal Litbang Pertanian* **27** (4) 124-130
[12] Elliason A C 2004 *Starch in food: Structure, function, and applications* (Boca Raton (US): CRC

Press, Taylor & Francis Group)

[13] Marco C and Rosell C M 2008 Functional and rheological properties of protein enriched gluten free composite flours *Journal of Food Engineering* **88** 94-103

[14] Abera S and Rakshit S K 1987 Comparison of physicochemical and functional properties of cassava starch extracted from fresh root and dry chips *Starch/Starke* **55** 287-296

[15] Buckle K A, Edwards R A, Fleet G H, Wootton M 1987 *Ilmu Pangan* (Jakarta : Universitas Indonesia Press)

[16] Castro I A, Motizuki M, Murai H, Chiu M C and Silva R S S F 2005 Effect of Anticaking Agent Addition and Headspace Reduction in The Powdered-Drink Mix Sensory Stability *J. of Food Quality* **29** 203 – 215

[17] Novita D 2011 *Evaluasi Mutu Gizi Dan Pendugaan Umur Simpan Cookies Tepung Komposit Berbasis Talas Banten (Xanthosoma undipes K. Koch) Sebagai Makanan Tambahan Ibu Hamil* (Bogor : Skripsi Fakultas Ekologi Manusia Institut Pertanian Bogor)

[18] Tahudi P A B 2011 *Pendugaan Umur simpan dan Analisis Keamanan Cookies Berbasis Pati Garut (Maranta arundinaceae L) Dengan Penambahan Torbangun (Coleus amboinicus Lour)* (Bogor : Skripsi Fakultas Ekologi Manusia Institut Pertanian Bogor)

[19] Putra M R P 2010 *Pendugaan Umur Simpan Keripik Wortel (Daucus carota L.) Dalam Kemasan Alumunium Foil Dengan Metode Akselerasi* (Bogor : Skripsi Intitut Pertanian Bogor)

[20] Eskin N A M. and Robinson D S 2001 *Shelf Life Stability: Chemical, Biochemical and Microbiological Changes* (Florida : CRC Press LLC inc)

IOP Conf. Series: Earth and Environmental Science **251** (2019) 012036 doi:10.1088/1755-1315/251/1/012036

Effect of Mocaf and Sugar Addition on the Quality and Preference Level of Pineapple Dodol

L Ratnawati[1] and N K I Mayasti[1]

[1]Development Centre for Appropriate Technology, Indonesian Institute of Sciences Subang, Indonesia

Email: lia.romeo@gmail.com

Abstract. Pineapple dodol is a traditional food of Subang; West Java made from glutinous rice flour as raw material. The aim of this study was to determine the effect of mocaf and sugar addition on the quality and level of preference for pineapple dodol. A completely randomized design in factorial experiment was used in this study with mocaf and sugar addition as treatments. There are three levels in the first factor (mocaf addition), i.e. 0, 50 and 100% mocaf addition. While, for the second factor (sugar addition), i.e. 50, 55, and 60% sugar addition based on pineapple puree. The physicochemical properties and preferences of panelists' tests have been carried out on the pineapple dodol product. The result indicated that the best treatment is the pineapple dodol with 50% mocaf and 50% sugar addition. The physicochemical properties of dodol include hardness, yield, moisture content, and activity water (a_w) have the following values 121.67 mm/g/sec, 40.96%, 11.21% and 0.74, respectively. The panelist preference score of taste, flavor, color, hardness, stickiness and overall liking were 5.5 (like moderately), 5.6 (like moderately), 5.3 (like moderately), 5.2 (like moderately), 5.8 (like moderately), and 5.54 (like moderately), respectively.

Keywords: mocaf, sugar, pineapple dodol

1. Introduction

Pineapple is one of the many cultivated fruit plants in Indonesia. One of the regions producing pineapple is Subang district, West Java. In 2014, pineapple production in Subang reached 136,567 tons with production centers located in the sub-district Jalancagak, Ciater and Cijambe [1]. Pineapple is included as perishable food because of its high moisture content, i.e., 84% [2]. Therefore, proper post-harvest handling is necessary in order to reduce the risk of fruit being wasted due to decaying in processed food products that have high economic value. One form of post-harvest handling is processing the pineapples into dodol [3].

Dodol is a processed food made from glutinous rice flour, sugar and coconut milk which is boiled until it becomes thick, oily and non-sticky. When cooled, the paste will become solid, chewy and can be sliced. The types of dodol vary greatly depending on the ingredients and how they are made [4]. Dodol includes intermediate moisture food that has moisture content 10-40 %; a_w 0.70-0.85; so that it is not effective for microbial growth, not easily damaged and resistant to storage without the use of any preservatives [5]. Good quality dodol has a good texture that is not soft, has a glossy surface, distinctive taste and not rancid. Consumers prefer dodol that has a dense texture and non-sticky [6].

Glutinous rice flour is the main raw material in the dodol production process which serves as a binding material that can increase the number of carbohydrates in fruit dodol, while sugar serves as a sweetener and to extend the shelf life of the product. Glutinous rice flour is one of the factors that influence the taste, color, texture and chemical properties of dodol [7]. Several previous studies have been carried out in order to reduce the use of glutinous rice flour that is through substitution using other types of flour. Murtiningrum and Cepeda [8] stated that substitution of tapioca flour 15% by weight of glutinous rice flour and red fruit pasta is the best formulation based on organoleptic and physicochemical properties of red fruit dodol. Meanwhile, Supiani [9] stated that Raja banana dodol made with substitute wikau maombo flour 75% and glutinous rice flour 25% had a good effect on organoleptic assessment of panelists.

However, the addition of mocaf (modified cassava flour) flour to the production of dodol has not been widely done. The selection of mocaf flour as a substitute material is based on the price of which is cheaper than glutinous rice flour and as a form of efforts to realize food security through local resource-based food diversification. The aim of this study was to determine the effect of mocaf and sugar addition on the quality and level of preference for pineapple dodol.

2. Materials and Methods

2.1. Raw materials

Pineapple dodol ingredients used in this study were pineapple puree, glutinous rice flour, sugar, mocaf, coconut milk, salt, and butter. Pineapple, glutinous rice flour, coconut milk, sugar, salt, and butter were obtained from the local market in Subang. While mocaf flour made by fermenting cassava chips for 12 hours using a starter mocaf.

2.2. Pineapple dodol preparation

Production of dodol pineapple begins with making pineapple puree. Then, pineapple puree is poured into a pan containing flour, sugar, coconut milk, and salt, and was stirred until the dough became homogenous. Dodol is cooked using medium heat (<80 °C) for 13 min. During cooking, the dough was stirred continuously in order to prevent it from burning. It is then removed from heat after the mixture began to thicken and become non-sticky. Butter is added into the cooked dough and stirred until it is smooth, before spreading it into trays. The pineapple dodol is left at room temperature for 12 h, then cut and packed using *polypropylene (PP)* plastic with a thickness of 0.8 mm.

2.3. Analysis of pineapple dodol

2.3.1 Hardness and yield determination

Force is required to compress food between the molars in order to attain a given deformation. The higher the number obtained, the softer is the product [10]. Hardness was determined using Universal Penetrometer (merk Humboldt, Type H 1200) and the spindle (H1270).

The yield was calculated by comparing the weight of the final results with an initial weight of the material that was used to make the dodol.

2.3.2 Moisture Content and Water Activity Determination

Moisture content was determined using the gravimetric method [11]. In this method, the samples in the coded, dry and clean crucible were heated in the oven (Memmert) at 105 °C until successive weighing. The moisture content of the samples is calculated based on the loss in weight of the samples on drying process. Water activity was determined using a_w meter (Smart Water Activity Meter Model no.HD-3A).

2.3.3 Best Treatment Determination (Effectiveness Index)

Best treatment was determined using Effectiveness Index [12] and were done before sensory analysis. Effectiveness index calculation procedure is to weigh the value of each variable on the relative numbers 0-1.

2.3.4 Sensory Analysis
Several samples selected based on effectiveness index were tested for sensory analysis. It was performed by 30 untrained panelists in order to evaluate taste, flavor, color, hardness, stickiness, and overall liking of dodol, using a 7-point hedonic scale (1-dislike very much, 2-dislike fairly, 3-dislike moderately, 4-neither like nor dislike, 5-like moderately, 6-like good, 7-like very good).

2.4. Experimental Design and Statistical Analysis

This study uses a completely randomized design with two factors. There are three levels in the first factor (mocaf addition), i.e. 0, 50 and 100% mocaf addition. While, for the second factor (sugar addition), i.e. 50, 55, and 60% sugar addition based on pineapple puree.

Data were analyzed and means were compared by Duncan's multiple range test with significance determined at the 0.05 level. The data analysis was assisted by statistical software (SPSS for Windows Ver.13.0, SPSS Inc.).

3. Result and Discussion

3.1. Hardness and Yield of Pineapple Dodol
The result of the hardness analysis is shown in Fig. 1, which is the effect of mocaf treatment and sugar addition to the hardness of pineapple dodol. Mocaf addition has shown to be significantly different ($p<0.05$) to the hardness of pineapple dodol. On the 50% sugar addition, dodol is made with the addition of mocaf 0, 50, and 100% that results in the hardness of 27.30, 24.50, and 31.20 mm/g/sec, respectively.

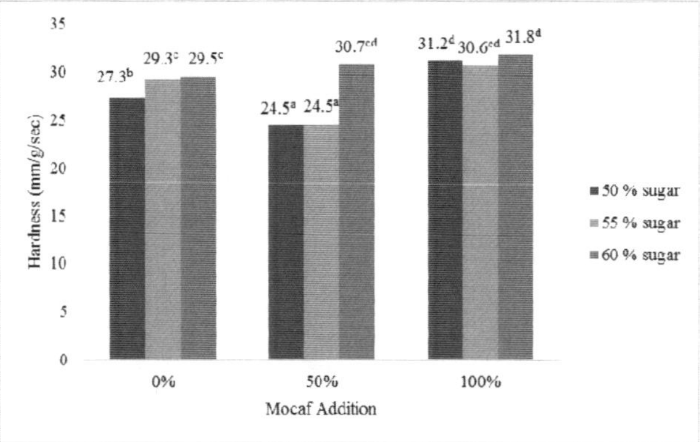

Figure 1. Effect of mocaf and sugar addition to hardness of pineapple dodol

Dodol made from 100% mocaf has a higher hardness value than dodol made without the addition of mocaf (dodol made from 100% glutinous rice flour). High hardness value indicates soft texture of the pineapple dodol. Hence, pineapple dodol made from 100% mocaf has a softer texture than dodol made from 100% glutinous rice flour. This occurs due to the influence of the ratio of amylose and amylopectin in the starch. A higher amylopectin ratio will cause a harder texture in the resulting product. Glutinous rice flour contains higher amylopectin than amylose, which is about 98% of the total starch [13,14]. Amylopectin in glutinous rice flour is also higher than in mocaf flour, resulting in lower hardness level than the latter.

The level of hardness dodol at different sugar addition is influenced by the type of flour used. Glutinous rice flour and mocaf have different gelatinization profile and water binding capabilities. Meanwhile, sugar also has the ability to bind water. Sugar has a high water binding capacity, where at high sugar concentrations, it can cause to bind water out of the material [15]. This affects the availability of water in dodol, hence affecting its texture. The more water that is bound, the harder will be its texture. Sugar is the main ingredient of dodol which is known to be able to give effect to the texture of dodol, as well as, the taste [6,16].

Fig.2 shows the yield of pineapple dodol which is in the range of 37.84 - 46.67%. The higher the mocaf and sugar addition increases the yield of dodol. This is directly proportional to the moisture content of dodol that is made from 100% mocaf flour and has the highest moisture content among other treatments.

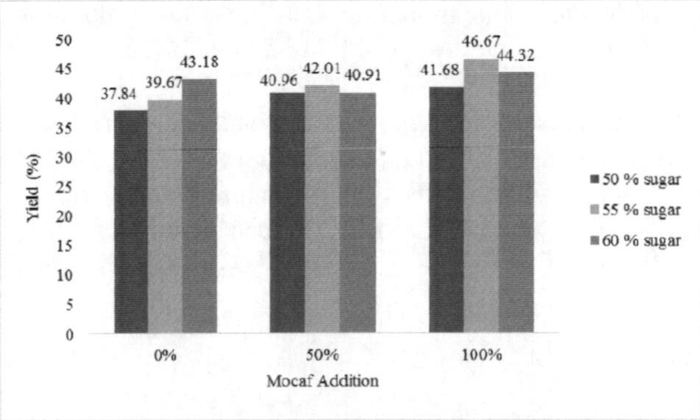

Figure 2. Effect of mocaf and sugar added to yield of pineapple dodol

3.2 Moisture Content and Water Activity (a_w)

Fig. 3 shows that sugar addition was not significantly different ($p>0.05$) on moisture content of pineapple dodol, whereas mocaf addition was significantly different ($p<0.05$) on moisture content of pineapple dodol. Pineapple dodol made from 50% sugar and 0% mocaf addition has the lowest moisture content than other treatments.

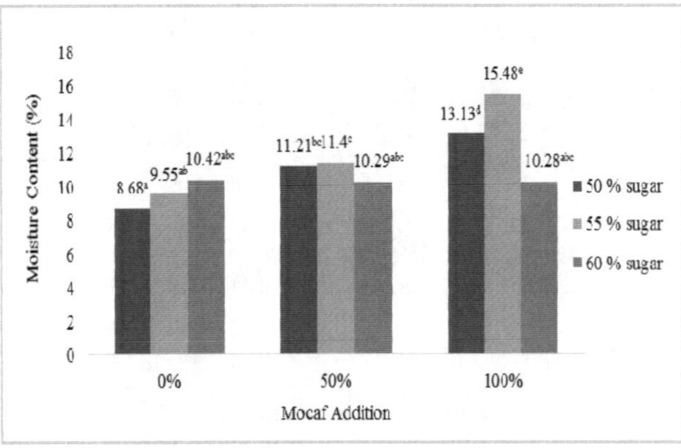

Figure 3. Effect of mocaf and sugar addition to moisture content of pineapple dodol

The treatment of mocaf addition affects the moisture content of pineapple dodol. Pineapple dodol made from mocaf has a moisture content greater than dodol that is made from 100% glutinous rice flour. The concoction of pineapple dodol was controlled at the same time. At the same cooking time, dodol made from glutinous rice flour and mocaf flour have different moisture content. This is due to both flour has a different gelatinization profile. Gelatinization profile of glutinous rice flour at a peak time, peak temperature and peak viscosity are 5.87 minutes, 67.47 °C 3996.25 mPas, respectively. While gelatinization profile of mocaf at a peak time, peak temperature, peak viscosity is 7.74 minutes, 90.6 °C, 5384.00 mPas, respectively [13]. From this profile, gelatinization can be seen that at the same cooking time, dodol made from mocaf has a moisture content greater than dodol made from glutinous rice flour due to peak temperature glutinous rice flour is lower than mocaf flour.

Fig.4 shows that mocaf addition was significantly different ($p<0.05$) on water activity (a_w) of pineapple dodol. The higher addition of mocaf flour will increase water activity of dodol. Pineapple dodol that made from 0, 50 and 100% mocaf addition have a_w 0.73; 0,74; and 0.75, respectively. Water activity (a_w) value is directly proportional to the moisture content of dodol. The water content in food ingredients will affect the resistance to microbial attack, expressed as a_w, i.e., the amount of free water that can be used by microorganisms for growth. Various types of microorganisms have a minimum a_w in order to grow properly, e.g., minimum a_w bacteria, yeasts and fungi are 0.90: 0.80-0.90; 0.60-0.70 [17]. The result of a_w in this study ranged between 0.71-0.78. a_w in this range of the possible types of microorganisms that can grow are fungi and yeast. Treatment of sugar addition affects the shelf life of the product. High sugar concentration can inhibit the growth of microbial spoilage. Sugar that is added to the food would bind with the water contained in the material so that the value of water activity is lower [18].

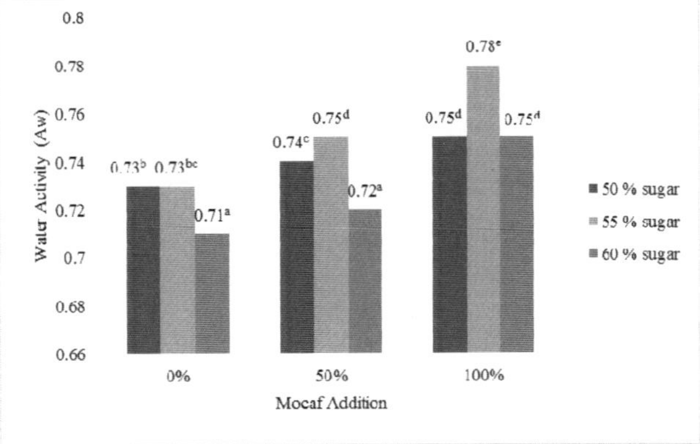

Figure 4. Effect of mocaf and sugar addition to water activity (a_w) of pineapple dodol

3.3 Best Treatment Selection (Effectiveness index)

From all treatments, four of the best treatment were selected for using effectiveness index. Parameter assessment was moisture content, a_w, hardness, and the yield of each treatment. Four best treatments from de Garmo analysis, e.g 0% mocaf addition and 60% sugar; 50% mocaf addition and 55% sugar; 50% mocaf addition and 50% sugar; and 100% mocaf addition and 60% sugar with effectiveness index 0.32; 0.32; 0.31 and 0.29, respectively.

3.4 Sensory Evaluation

Sensory test (organoleptic) was conducted to determine consumer acceptance of a product. According to Soekarto [19], physical, chemical, and nutritional testing can show a high-quality food product, but that does not mean the product cannot be consumed because it is not tasty or not acceptable organoleptically. Sensory data for pineapple dodol were presented in Table 1. Six characteristics of dodol (taste, color, flavor, hardness, stickiness and overall liking) were analyzed for acceptability.

Table 1. Effect of mocaf and sugar addition on sensory quality of pineapple dodol

Treatments	Parameter					
	Taste	*Color*	*Flavor*	*Hardness*	*Stickiness*	*Overall*
0% mocaf :60% sugar	5.3^{ab}	5.00^{ab}	4.77^{a}	4.63^{a}	4.40^{a}	4.93^{a}
50% mocaf :55% sugar	4.73^{a}	4.77^{a}	4.77^{a}	5.10^{a}	4.83^{a}	5.04^{ab}
50% mocaf :50% sugar	5.50^{b}	5.60^{b}	5.30^{a}	5.20^{a}	5.80^{b}	5.54^{b}
100% mocaf:60% sugar	5.53^{b}	5.43^{b}	5.20^{a}	4.80^{a}	4.77^{a}	5.29^{ab}

Taste is a very important attribute in determining the consumer decides to accept or reject a food product. The acceptance of taste ranged from 4.73 to 5.53 (Table 1), which means the preference level is neither like nor dislike to like moderately. Statistical analysis showed that the preference level for the taste of dodol made with 50% mocaf addition and 55% sugar was significantly different ($p<0.05$) from the other treatments. Table 1 shows that the panelists like dodol made of mocaf flour and 60% sugar with the highest score are 5.53, which means that the panelists like dodol with a sweet taste.

According to Fennema [20], color is the most important quality attributes. The acceptance of color ranged from 4.77 to 5.60 (Table 1), which means the preference level is neither like nor dislike to like moderately. Statistical analysis showed that the preference level for color of dodol made with 50%

mocaf addition and 50% sugar was significantly different from the preference level for color of dodol made 50% mocaf addition and 55% sugar ($p<0.05$). High sugar addition can also cause the color difference in dodol. In production of dodol, there is a heating process that can cause the caramelization process in the sugar, which can affect the color of dodol. Based on color acceptance, it is known that color preferred by panelist is pineapple dodol made from 50% mocaf and 50% sugar.

Flavor is a subjective sensation produced by the olfactory (smelling). In the food industry, testing of the flavor is considered important as it can quickly provide an assessment of product acceptance [19]. Panelists' preference on flavor were ranged between 4.77 to 5.30, which means that the level of preference is neither like nor dislike to like moderately. Statistical analysis of the four treatments showed that the level of preference for the flavor was not significantly different ($p>0.05$). Table 1 shows that the dodol made from 50% mocaf addition with 50% sugar has the highest preference level for flavor parameter, with a score of 5.30. The flavor of pineapple dodol has a distinctive smell that comes from its main ingredient, which is the ripe pineapple. At the time of fruit, ripening occurs, there is a decreased level of phenolic compounds that lead to reduced astringent taste due to decreased in organic acids and higher volatile substances affecting the typical taste and flavor of the fruit [21]. In the ripening fruit process, starch hydrolysis and increasing of sugar occur. Panelists tend to prefer the refreshing taste and flavor of pineapple.

Hardness and stickiness are the two parameters used in testing the sensory properties by using the senses of touch that are expressed in either hard or soft and sticky. The mean score for hardness varied from 4.63 to 5.20 (Table 1). It means the level of preference for pineapple dodol is neither like nor dislike to like moderately. Statistical analysis showed that the preference level for hardness was not significantly different ($p>0.05$). The result shows that the maximum value corresponded to the sample prepared with 50% mocaf and 50% sugar addition which has a score of 5.20. The texture is acceptable when the product is under normal condition and depending on the specific ingredient [22].

The acceptance of dodol stickiness ranged between 4.40-5.80 (Table 1), which means the level of preference neither like nor dislike to like moderately. Statistical analysis showed that the level of preference for dodol stickiness that made from 50% mocaf addition with 50% sugar was significantly different from the other treatment ($p<0.05$). It is due to the low concentration of sugar that produces dodol which is not sticky. Panelists prefer pineapple dodol with low stickiness as it is more convenient to hold by hand without the packaging

Testing the overall quality of all factors which include taste, color, flavor, hardness, and stickiness, aims to determine the level of acceptance of a product. The mean score for overall acceptability is from 4.93 to 5.54 (Table 1). It means the preference level is neutral to like slightly. Statistical analysis showed that the level of preference for overall acceptability of dodol made from 0% mocaf:60% sugar was significantly different with 50% mocaf:50% sugar ($p<0.05$). The maximum and the minimum value corresponded to the samples of dodol prepared by adding 50 % mocaf:50% sugar and 0% mocaf: 60% sugar is 5.54 and 4.93, respectively.

4. Conclusion

Mocaf addition significantly affects the hardness, yield, moisture content and water activity (Aw) of pineapple dodol. The result indicated that the best treatment is the pineapple dodol with 50% mocaf and 50% sugar addition. The physicochemical properties of dodol include hardness, yield, moisture content, and activity water (a_w) have the following values 121.67 mm/g/sec, 40.96%, 11.21% and 0.74, respectively. The panelist preference score of taste, flavor, color, hardness, stickiness and overall liking were 5.50 (like moderately), 5.60 (like moderately), 5.30 (like moderately), 5.20 (like moderately), 5.80 (like moderately), and 5.54 (like moderately), respectively.

5. Acknowledgment

The authors are grateful to Development Centre for Appropriate Technology for providing the experimental facilities and to the Indonesian Institute of Sciences for financial support.

6. References

[1] Anonim 2014 Potensi pertanian. http://www.subang.go.id/potensi_pertanian.php

[2] Agustina 2004 Pengaruh berbagai varietas nanas dalam pembuatan kripik nanas. *Jurnal Pangan dan Agroindutri* **2** (2). Hal 24-30. Institut Pertanian Bogor. Bogor.

[3] Tresnawati D 2010 Analisis pengembangan agroindustri dodol nanas di Kabupaten Subang. Skripsi. Fakultas Pertanian-Universitas Sebelas Maret Surakarta

[4] Haryadi 2006 Teknologi pengolahan beras. Gadjah Mada University Press. Yogyakarta

[5] Musaddad D dan Hartuti N 2003 Produk olahan tomat. Penebar Swadaya. Jakarta

[6] Nasaruddin F, Chin N L and Yusof Y A 2012 Effect of processing on instrumental textural properties of traditional dodol using back extrusion. *International Journal of Food Properties*, **15**:495–506

[7] Satuhu S dan Sunarmani 2002 Membuat aneka dodol buah. Penebar Swadaya. Jakarta

[8] Murtiningrum dan Cepeda G N 2011 Penggunaan bahan pengisi dalam perbaikan sifat fisikokimia dan organoleptik dodol buah merah (*Pandanus conoideus* L.) sebagai sumber β-karoten. *AGRITECH*, Vol.31, No. 1 Februari 2011 pp.14-20

[9] Supiani, Wahyuni S, Syukri M 2016 Analisis organoleptik dodol pisang raja *(Musa paradisiaca L.)* dengan substitusi tepung wikau maombo. *J. Sains dan Teknologi Pangan* Vol. 1, No.1, pp. 24-30

[10] Hubeis M 1985 Penuntun praktikum pengawasan mutu pangan. Jurusan teknologi pangan dan gizi. Fateta IPB Bogor

[11] Analysis of The Association of Official Agriculture Chemistry 1995 Official methods of analysis, 16th Edition. Gaithersburg, Maryland: AOAC International

[12] De Garmo E D, Sullivan W G and Canada J R 1984 Engineering economy. MacMillan Publishing Company. New York

[13] Imanningsih 2012 Profil gelatinisasi beberapa formula tepung-tepungan. *Penel Gizi Makan.* **35** (1) : 13-22

[14] Luh B S, Ang C Y W, Liu K, Huang Y W 1999 Rice products. In Asian Foods : Science and Technology ; Eds.; Technomic Publishing Company, Inc.: Lancaster, PA, USA, 1999; 6

[15] Astawan 2006 Pemanfaatan rumput laut untuk meningkatkan iodium dan serat pangan dalam selai dodol. *Junal Teknologi dan Industri Pangan.* **15** (1):61-69

[16] Esteller M S, Amaral R L, Da Silva Lannes S C 2004 Effect of sugar and fat replacers on the texture of baked goods. *Journal of Texture Studies* **35**, 383–393.

[17] Winarno F G 1992 Kimia Pangan dan Gizi. Gramedia Pustaka Utama: Jakarta

[18] Zakaria M H 2012 Penerapan konsep pengendalian mutu dan hazard analysis critical control points (HACCP) di usaha kecil menengah "Wajik Harso Mulyono". Skripsi S1. Fakultas Pertanian. Universitas Sebelas Maret. Surakarta

[19] Soekarto S T 1990 *Dasar-dasar pengawasan mutu dan standarisasi mutu pangan.* Bogor. IPB Press

[20] Fennema R O 1985 *Food Chemistry* 2nd Edition. Revised and Expanded. New York. Academic Press

[21] Winarno F G 2008 Kimia pangan dan gizi. M-Brio Press, Bogor

[22] Kusmawati A, Ujang H dan Evi E 2000 Dasar-dasar pengolahan hasil pertanian I. Jakarta. Central Grafika

IOP Conf. Series: Earth and Environmental Science **251** (2019) 012037 doi:10.1088/1755-1315/251/1/012037

Addition of Tempeh Flour as a Protein Source in Mixed Flour (Mocaf, Rice, and Corn) for Pasta Product

L E Yulianti[1], E Sholichah[1], N Indrianti[1]

[1] Development Centre of Appropriate Technology Indonesian Institute of Sciences, Jl. KS. Tubun No. 5 Subang West Java, Indonesia, 41213

Corresponding author: listaeka0507@gmail.com

Another author: enny.sh002@gmail.com

Abstract. Mocaf (modified cassava flour), rice flour, and corn flour were used as mixed flour to prepare pasta product. Various levels of tempeh flour (0%, 2.5%, 5%, 7.5%, 10%, 12.5% and 15%) were incorporated to enrich nutritional content of the mixed flour, primarily protein. This current work aimed to understand the effect of tempeh flour addition on physicochemical characteristics (moisture content, ash content, protein content, fat content and whiteness level) and pasting properties of the mixed flour. The results showed that addition of tempeh flour did not exhibit significant impacts to moisture (9.17-10.18%), ash (0.74-0.91%), and fat content (4.68-5.65%) of the mixed flour. On the other hand, it produced significant effects on protein level, white degree and pasting properties at a significance level of 5%. In the absence of tempeh flour, protein level and whiteness level of the mixed flour reached 5.83% and 80.96%, respectively. Meanwhile, tempeh flour could increase protein content of 6.30-9.07% and alter whiteness level of 66.08-71.97%. In addition, the presence of tempeh flour was also found to improve viscosity in comparison with control. The protein level was higher as more tempeh flour was added, while whiteness level and viscosity appeared to decrease. Addition of 5% tempeh flour increased protein content to 6.88%, and the mixed flour has the best viscosity.

Keywords: mixed flour, tempeh flour, physicochemical, pasting properties

1. Introduction

Noodle is one of the pasta products and constitutes an alternative source of carbohydrate for Indonesian people. Noodle consumed in Indonesia is mostly made of wheat flour, which provokes noticeable threats to food security considering that the wheat was not massively cultivated in Indonesia. Demand on wheat flour seems to be increased, resulting in the dependence on the importation. Therefore, numerous efforts have been made to explore the use of alternative source for substituting wheat flour; this is expected to reduce wheat dependence. Besides, there is a growing demand for gluten-free products including noodle [1].

Indonesia is endowed with local natural sources that can be developed for food ingredients. These potential sources of carbohydrate, i.e., cassava, rice, and corn, are reported to be the most abundant in Indonesia. In 2015, their production reached 21,801,415 tons, 75,397,841 tons and 19,612,435 tons, respectively [2]. They have been widely used for intermediate food products such as mocaf (modified cassava flour) through the fermentation process, rice flour, and corn flour.

The search for alternative source of wheat flour in noodle production has been investigated. Sabbatini *et al.* [3] developed noodle made of cassava starch, rice flour, and cornstarch with the addition of gum 2% as a hydrocolloid. Pato *et al.* [4] reported the production of noodle made of cassava flour (55%) and corn flour (45%) that corresponds to the national standard of instant noodle (SNI 01-3551-2000) except protein content.

The protein content of food could be improved through incorporating protein-rich components, such as tempeh which is well-known as a fermented soybean. Previous study reported that fermented soybean had better nutritional content due to enzymatic activity generated by tempeh starters (mold). The action of the enzyme could degrade protein into more digestible fragments in intestine in comparison with non-fermented soybean [5]. This research aimed to uncover the effects of tempeh flour addition on physical, chemical and pasting properties of the mixed flour (mocaf, rice, and corn flour) used for pasta noodle.

2. Materials and Methods

2.1. Raw materials

The main ingredients included tempeh flour, mocaf, rice flour, and corn flour. Mocaf was supplied from small and medium enterprise (SMEs) Tanjungsiang in Subang, while the rest (tempeh flour, rice flour, and corn flour) was prepared in Laboratory of Food, Development Center for Appropriate Technology.

To make tempeh flour, tempeh was first made through fermenting soybean for 48 h. Tempeh was sliced and dried using infra-red drier at 40-45 °C for 4-5 h. The dried tempeh was ground using disc mill and sieved to form a flour. The rice flour was made by soaking rice grains for 1 h. Afterward, the soaked rice grains were drained and then air-dried. The dried rice grains were ground using disc mill and dried to reach a moisture content of no more than 10%. Corn flour was made by using dry milling method. The milling process was carried out twice. The first milling was performed using a chopper machine to produce grits, pericarp, germ and tip cap. Only grits were used in a further step. The grits were washed and soaked in water for 3-4 h to soften them, enabling to easier milling process. The second milling (fine) was performed by using a disc mill. Corn flour and rice flour were sieved in 40-mesh.

2.2. Procedures

2.2.1. Formulation of mixed flour

The composite flour was made by mixing the raw materials as the following percentage: mocaf (40%), rice flour (35%) and corn flour (25%). The composite flour was then added with tempeh flour at 7 levels and performed at triplicate.

Table 1. Concentration levels of tempeh flour

Formula code	**Tempeh flour concentrations (%)**
K (control)	0
M1	2.5
M2	5
M3	7.5
M4	10
M5	12.5
M6	15

2.2.2. Analysis of mixed flour

The chemical analysis was carried out, including moisture, ash, protein, and fat. Determination of moisture and ash content was performed using an approved method by national standard (SNI 01-2891-1992). Meanwhile, protein content was determined using Dumas combustion method using DuMaster Buchi D-480, Switzerland. The fat content was analyzed using Soxhlet with hexane as a solvent.

Physical characteristics of the flour (L^*, a^*, b^*) were determined using NH_3 colorimeter. The whiteness level was obtained through calculation using the following formula: $100 - [(100 - L^*)^2 + a^{*2} + b^{*2}]^{½}$ [6]. Pasting properties was performed using Rapid Viscosity Analyzer (RVA-TecMaster).

Data were evaluated using One-way Analysis of Variance (ANOVA) in SPSS software. Significant difference between means was compared using Duncan Multiple Range Test (DMRT) at significance level of 5% ($p \leq 0.05$).

3. Result and Discussion

3.1. *Chemical characteristics of mocaf, rice flour, and corn flour*

Raw materials for mixed flour included mocaf, rice flour, and corn flour. Tempeh flour was added to enrich protein level of mixed flour. The chemical profile of each flour is presented in Table 2.

Table 2. Chemical profile of mixed flour raw materials

Parameter	Mocaf*	Rice Flour*	Corn Flour*	Tempeh Flour
Moisture Content (%)	10.62	5.05	6.05	6.50
Ash Content (%)	1.29	0.27	0.48	2.20
Protein Content (%)	1.77	10.07	7.07	45.43

(*): source by Afifah and Ratnawati [7]

As a result, moisture content of each was less than 11%, in which mocaf was observed to have the highest percentage. According to SNI, the maximum level of rice flour moisture and ash is 13% and 1.0%, respectively (SNI 3549-2009), as also attributed to mocaf, i.e., 13% and 1.5% (SNI 7622-2011), and corn flour, i.e., 10% and 1.5% (SNI 01-3727-1995). This suggests that the moisture and ash content of those flours is acceptable based on SNI.

Tempeh flour was found to have higher ash content compared to other flours, and it is also rich in protein. Tempeh is a processed soybean product after fermented using *Rhizopus oligosporus*. Wang *et al.* [8] asserted that total amino acids in tempeh were drastically increased up to 3-10 times higher compared to soybean. This is due to roles of *R. oligosporus* capable of hydrolyzing protein to form amino acids and peptides, leading to enhancement of nutritional content of tempeh.

3.2. *Chemical characteristics of mixed flour*

Mocaf, rice flour, corn flour and tempeh flour at various concentrations were mixed. The chemical characteristics of mixed flour are presented in Table 3.

Table 3. Chemical characteristics of mixed flour with the addition of tempeh flour

Formula Code	Moisture Content (%)	Ash Content (%)	Protein Content (%)	Fat Content (%)
K	$10.18^{a} \pm 0.24$	$0.91^{a} \pm 0.05$	$5.83^{a} \pm 0.05$	$5.22^{ab} \pm 0.11$
M1	$9.89^{ab} \pm 0.11$	$0.89^{a} \pm 0.05$	$6.30^{b} \pm 0.03$	$5.19^{ab} \pm 0.15$
M2	$9.95^{ab} \pm 0.51$	$0.82^{a} \pm 0.05$	$6.88^{c} \pm 0.05$	$5.65^{a} \pm 0.11$
M3	$9.17^{b} \pm 0.19$	$0.87^{a} \pm 0.07$	$7.24^{d} \pm 0.05$	$4.91^{ab} \pm 0.52$
M4	$9.85^{ab} \pm 0.89$	$0.81^{a} \pm 0.01$	$8.03^{e} \pm 0.11$	$4.68^{b} \pm 0.17$
M5	$9.41^{ab} \pm 0.14$	$0.83^{a} \pm 0.01$	$8.43^{f} \pm 0.11$	$4.85^{ab} \pm 0.04$
M6	$9.93^{ab} \pm 0.14$	$0.74^{a} \pm 0.03$	$9.07^{g} \pm 0.17$	$5.33^{ab} \pm 0.24$

Means with different letters in the same column differ significantly at $p \leq 0.05$

3.2.1. *Moisture content*

Moisture content is defined as the amount of water present in a material and expressed as a percentage. The moisture content of flour constitutes a considerable parameter in relation to flour storability. As presented in Table 3, the moisture content of sample K (without the addition of tempeh flour, 0%) was significantly different from sample M3 (tempeh flour 7.5%) but not different in comparison with other samples.

The moisture content of sample K was 10.18%, while other samples contained moisture content ranging from 9.17% to 9.93%. This suggests that addition of tempeh flour could reduce the moisture content of composite flour since it possessed lower content of water than other flours (mocaf, rice flour, corn flour). As reported by Murni [9], the addition of tempeh flour attenuated moisture content of chicken nugget.

3.2.2. Ash content

Ash content is one of important parameters capable of representing the level of inorganic components (mineral) in a product. It positively correlates to the abundance of inorganic components [10]. Tempe is found to contain both micro and macro minerals such as Ca, P, Fe, Mg, Cu and Zn [5].

As presented in Table 3, the highest ash content was attributed to the sample K, while the lowest one was found in composite flour added with tempeh flour 15%. This is not different statistically at $p \leq 0.05$.

3.2.3. Protein content

An increase in protein content is expected after incorporating tempeh flour, considering that protein is considered as important nutrition for infants [11-12]. The results exhibited that protein level of variously mixed flours ranged from 5.83% to 9.07%, which was statistically significant at $p \leq 0.05$.

Sample K, without the addition of tempeh flour, was observed to have the lowest protein level, while the highest one was attributed to mixed flour added with tempeh flour 15%. Protein level was higher as more tempeh flour was incorporated. This is in accordance with Aini *et al.*[13], finding that higher protein level of white corn noodle occurred along with an increase in substitution by tempeh flour. Lailiyati *et al.* [14] also reported that addition of Tempeh flour 20% could enhance protein content of corn rice.

Addition of tempeh flour appears to elevate protein content of mixed flour due to a high protein content of tempeh flour reaching 45%. Babu *et al.* [15] found that the abundance of amino acids in Tempeh which is comparable to either red meat or poultry meat.

3.2.4. Fat content

Fat has a pivotal role as an energy source and also affects texture of food as well as its organoleptic performance. Fat content in mixed flour ranged from 4.68% to 5.65%. Enrichment using tempeh flour seems to unaffect fat content of the mixed flour, according to statistical evaluation at $p \leq 0.05$, except between sample M2 (tempeh flour 5%) and sample M4 (tempeh flour 10%).

The fat content in tempeh is less than soybean since lipase hydrolyzes triacylglycerol into fatty acids during fermentation process. The fatty acid, a hydrolytic product of lipase activity, is utilized by molds to generate energy, thus lowering fat content in tempeh. As reported by a previous study, fermentation accounts for the reduction of fat content up to 26% [5].

3.3. Physical characteristics of mixed flour

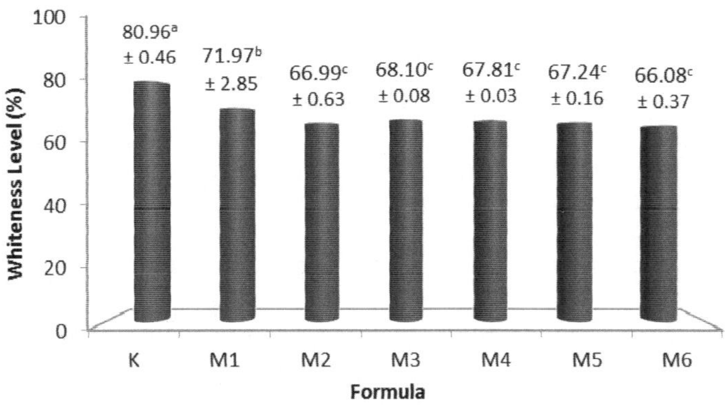

Figure 1. Whiteness level of mixed flour with the addition of tempeh flour

In this experiment, color is evaluated since it serves a noticeable attribute in food industry. Whiteness level refers to color indices commonly used to evaluate flour quality and is considered a major parameter for flour-based ingredients. It may also represent the ability of flour to reflect light coming to its surface [16]. The higher degree of whiteness is more desirable in flour.

The whiteness level of mixed flour ranged from 66.08% to 80.96%. The highest whiteness level was found in sample K, while the lowest one was attributed to the sample M6 (tempeh flour 15%). The results showed a significant difference between sample K, M1, and from sample M2 to sample M6. In contrast, the whiteness level of the sample from M2 to M6 was not significantly different.

These results suggested that incorporation of tempeh flour negatively related to whiteness. Aini *et al.* [13] reported that an increase in tempeh flour proportion could increase brown color intensity of corn noodle. Tempeh flour was also reported to decrease lightness of crackers [17].

3.4. Pasting properties of mixed flour

Paste profile of the flour was evaluated regarding the viscosity increment during heat exposure due to irreversible granular swelling in water [18]. The pasting properties of flour included pasting temperature (PT), peak viscosity (PV), hot paste viscosity (HPV), breakdown viscosity (BV), final viscosity (FV), and setback viscosity (SV). Table 4 indicates that there is a significant difference in paste profile ($p \leq 0.05$). In general, the addition of tempeh flour affected pasting properties of the composite flour.

Table 4. Pasting properties of mixed flour

Formula Code	Pasting Temp / PT (°C)	Peak Viscosity / PV (cP)	Hot Paste Viscosity / HPV (cP)	Breakdown Viscosity / BDV (cP)	Final Viscosity / FV (cP)	Setback Viscosity / SBV (cP)
K	$75.0^{d} \pm 0.0$	$2997^{ab} \pm 58$	$2753^{bc} \pm 99$	$262^{ab} \pm 6$	$3477^{c} \pm 86$	$742^{c} \pm 28$
M1	$75.5^{d} \pm 0.3$	$3237^{a} \pm 219$	$2934^{ab} \pm 206$	$303^{a} \pm 41$	$4158^{b} \pm 250$	$1224^{b} \pm 143$
M2	$76.7^{c} \pm 0.2$	$3257^{a} \pm 214$	$3053^{a} \pm 174$	$222^{b} \pm 26$	$4737^{a} \pm 157$	$1702^{a} \pm 68$
M3	$79.4^{ab} \pm 0.4$	$2862^{bc} \pm 110$	$2580^{cd} \pm 96$	$282^{ab} \pm 12$	$4110^{b} \pm 60$	$1530^{a} \pm 14$
M4	$78.9^{b} \pm 0.5$	$2915^{bc} \pm 154$	$2608^{cd} \pm 186$	$307^{a} \pm 18$	$3895^{bc} \pm 190$	$1287^{b} \pm 88$
M5	$80.1^{a} \pm 0.4$	$2699^{c} \pm 80$	$2395^{d} \pm 107$	$304^{a} \pm 23$	$3539^{c} \pm 53$	$1144^{b} \pm 76$
M6	$80.4^{a} \pm 0.2$	$2828^{bc} \pm 73$	$2529^{cd} \pm 67$	$299^{a} \pm 7$	$3596^{c} \pm 51$	$1067^{b} \pm 12$

Means with different letters in the same column differ significantly at $p \leq 0.05$

Based on Table 4, a remarkable change of pasting properties was observed after treatment of tempeh flour. Pasting temperature (PT) appeared to be higher with an increasing tempeh flour level. PT refers to a minimum temperature required to induce gelatinization. The lowest PT was observed in sample K, while the highest one was attributed to the sample M6. Richana and Sunarti [18] stated that PT was influenced by some factors such as the molecular size of amylose and amylopectin and heating condition. The protein in tempeh flour could form complex with amylose, thus generating insoluble sediment and inhibiting amylose release from granule. Presence of tempeh flour leads to a higher PT since more energy required to release amylose (gelatinization process).

The results showed that a higher proportion of tempeh flour seemed to decrease the value of pasting properties as observed in PV, HPV, BDV, FV, and SBV. As reported by Julianti *et al.* [19] and Tharise *et al.* [20], the greater proportion of tempeh flour reduced PV, HPV, BDV, FV, and SBV. Jane *et al.* [21] found that content of amylose, protein, fat negatively correlated to viscosity. Protein content of mixed flour was higher due to greater proportion of tempeh flour. The increasing protein level caused a reduction of viscosity of mixed flour.

Table 4 exhibits that addition of tempeh flour induced higher PV, HPV, BDV, FV, and SBV than sample K. Sample M2 (tempeh flour 5%) was found to have the highest PV, HPV, FV and SBV in comparison with other samples. The high PV, HPV, FV, and SBV of the samples indicate that they could induce proper gelling formation and retain it during cooling stage. The high HPV could indicate low cooking loss and superior eating quality [22].

4. Conclusion

The addition of tempeh flour to mixed flours had effects on their protein content, whiteness, and pasting properties. The greater proportion of tempeh flour increased protein level, but decreased whiteness level and viscosity of the mixed flour. The protein content of sample K was 5.83%, while the presence of tempeh flour could increase protein level up to 6.30% - 9.07%. The whiteness level of sample K reached 80.96%, which was higher than that of mixed flour added with tempeh flour, ranging from 66.08% to 71.97%. The mixed flour enriched with tempeh flour has a better viscosity compared to sample K. The viscosity of mixed flour was higher as more tempeh flour was added. Addition of 5% tempeh flour increased protein content to 6.88%, and the mixed flour has the best viscosity.

5. Acknowledgment

The authors greatly thank those have supported this research, especially to The Ministry of Research Technology and Higher Education of the Republic of Indonesia for funding this research.

6. References

[1] Jeong S, Kim HW and Lee S 2016 Rheological and secondary structural characterization of rice flour-zein composites for noodles slit from gluten-free sheeted dough *Food. Chem.*
[2] Badan Pusat Statistik 2017 https://www.bps.go.id/site/resultTab
[3] Sabbatini SB, Sanchez HD, Torre MA and Osella CA 2014 Design of a premix for making gluten free noodles *Int. J. Nutr. Food. Sci.* **3 (5)** 488-92
[4] Pato U, Yusuf Y, Isnaini RF, Dira DM 2016 The quality of instant noodle made from local corn flour and tapioca flour *J. Adv. Agr. Tech.* **3(2)** 118-23
[5] Astuti M, Meliala A, Dalais FS and Wahlqvist ML 2000 Tempe , a nutritious and healthy food from Indonesia *Asia. Pas. J. Clin. Nutr* **9** 322-25
[6] Mawarni RT, Widjanarko SB 2015 Penggilingan metode ball mill dengan pemurnian kimia terhadap penurunan oksalat tepung porang *J. Pangan Agr.* **3(2)** 571-81
[7] Afifah N and Ratnawati L 2017 Quality assessment of dry noodles made from blend of mocaf flour, rice flour and corn flour *Proc. IOP Conf. Ser.: Earth Environ. Sci.*
[8] Wang HL, Ruttle DL and Hesseltine CW 1968 Protein quality of wheat and soybeans after *Rhizopus oligosporus* fermentation *J. Nutr.* **96** 109-14
[9] Murni M 2014 Pengaruh penambahan tepung tempe terhadap kualitas dan citarasa naget ayam *Berita Litbang Industri* **3 (2)** 117-23
[10] Bastian F, Ishak E, Tawali Ab and Bilang M 2012 Daya terima dan kandungan zat gizi formula tepung tempe dnegan penambahan semi refined carrageenan (SRC) dan bubuk kakao *J. Apl. Tek. Pangan* **2(1)** 5-8
[11] Ernawati F, Rosmalina Y and Permanasari Y 2013 Pengaruh asupan protein ibu hamil dan panjang badan bayi lahir terhadap kejadian stunting pada anak usia 12 bulan di Kabupaten Bogor *Penelitian Gizi dan Makanan* **36 (1)** 1-11
[12] Anindita P 2012 Hubungan tingkat pendidikan ibu, pendapatan keluarga, kecukupan protein dan zinc dengan stunting pada balita usia 6-35 bulan di Kecamatan Tembalang Kota Semarang *J. Kes. Masy* **1(2)** 617-26
[13] Aini N, Prihananto V dan Munarso J 2012 Characteristics of white corn noodle substitued by tempeh flour *J. Teknol. Ind. Pangan* **23 (2)** 179-85
[14] Lailiyati SN, Rahmawanti D and Andriani MAM 2014 Formulasi dan kajian karakteristik nasi jagung (*Zea mays* L) instan dengan penambahan tepung tempe *J. Teknosains Pangan* **3 (1)** 155-63
[15] Babu PD, Bhakyaraj R and Vidhyalakshmi R 2009 A low cost nutritious food 'Tempeh' - a review *World J. Dairy Food Sci.* **4 (1)** 22-27
[16] Gilang R, Affandi DR and Ishartani D 2013 Karakteristik fisik dan kimia tepung koro pedang (*Canavalia ensiformis*) dengan variasi perlakuan pendahuluan *J. Teknosains Pangan* **2(3)** 34-42
[17] Mustakim, Yusmarini and Herawati N 2016 Pemanfaatan tepung jagung dan tepung tempe dalam pembuatan kerupuk *J. Faperta* **3 (2)**
[18] Richana N and Sunarti TC 2004 Karakterisasi sifat fisikokimia tepung umbi dan tepung pati dari umbi ganyong, suweg, ubi kelapa dan gembili *J. Pascapanen* **1 (1)** 29-37

[19] Julianti E, Rusmarilin H, Ridwansyah and Yusraini E 2015 Functional and rheological properties of composite flour from sweet potato, maize, soybean and xanthan gum *J. Saudi Soc. Agric. Sci.*
[20] Tharise N, Julianti E and Nurminah M Evaluation of physicochemical and functional properties of composite flour from cassava, rice, potato, soybean and xanthan gum as alternative of wheat flour *Int. Food Research J.* **21 (4)** 1641-49
[21] Jane J, Chen YY, Lee LF, McPherson AE, Wong KS, Radosavljevic M and Kasemsuwan T 1999 Effects of amylopectin branch chain length and amylose content on the gelatinization and pasting properties of starch *Cereal Chem.* **76 (5)** 629-37
[22] Bhattacharya M, Zee Sy and Corke H 1999 Physicochemical properties related to quality of rice noodles *Cereal Chem.* **76 (6)** 861-67

IOP Conf. Series: Earth and Environmental Science **251** (2019) 012038 doi:10.1088/1755-1315/251/1/012038

Dynamic characteristics analysis of agitator design for soy sauce cooking process

A Haryanto[1] and B Budiwantoro[2]

[1] Development Center for Appropriate Technology, Indonesian Institute of Sciences, Indonesia
[2] Faculty of Mechanical and Aerospace Engineering, Institut Teknologi Bandung, Indonesia

Email: aharyant03@gmail.com

Abstract. Soy sauce is one type of dark brown condiment, distinctive smell, salty or sweet taste, lumpy, and contains protein produced from fermented. One of the machines that can be used for cooking soy sauce is a pressure vessel as cooking pan with an agitator. Agitation is a process of mixing and stirring and are carried out by heat transfer and mass inter-phases or with external surfaces (due to outside influences). The agitator is a system used for mixing and stirring accompanied by a phase change. In the design of rotating machinery, it is necessary to predict the dynamic characteristic in bending and in torsion to avoid failure. Dynamic characteristics analysis of agitator design for soy sauce cooking process consists of mechanical vibration analysis and mass unbalance response. The stiffness method for the agitator shaft by dividing the shaft element into two elements based on the bearing position. The bearing is assumed to be roller supports and only moves in the direction of translation and rotation. Based on the dynamic characteristics of rotordynamics prediction by using finite element method both theoretical and software, the agitator in operating conditions with the rotation speed of 5 RPM according to Campbell diagram will not fail. The highest amplitude of the mass unbalance response is less than 2.5×10^{-2} mm.

Keywords: Soy Sauce, Agitator, Finite Element Methods, Campbell Diagram, Rotordynamics

1. Introduction

Soy sauce is one type of dark brown food product or condiment, distinctive smell, salty or sweet taste, lumpy, and contains protein produced from fermented [1] and [2]. The basic ingredients in making Indonesian soy sauce is soybeans. Soybeans are the protein source with the highest content of about 40% compared to other types of nuts [3]. Soy sauce consumption needs are increasing every year, requiring the companies to increase production capacity and quality of soy sauce products. To increase production capacity, it is necessary to design a soy sauce production process that is able to meet the needs. The flow chart of the cooking process for soy sauce, shown in Figure 1. Based on Figure 1, one step in the process of producing soy sauce is the cooking process [4]. The machine used for soy sauce cooking process is using a pressure vessel as storage and an agitator as a mixer. The function of the agitator for soy sauce cooking process is as a stirrer, transfer and evenly distribute heat to the fluid (soy sauce) of the pressure vessel wall. An agitator is machines used in industries that process products in the chemical, food, pharmaceutical and cosmetic industries [5]. If the result of the design is not appropriate with the function of the agitator, the agitator performance does not optimal. Another result is the effect of uneven heat distribution give damage to the taste of soy sauce [4].

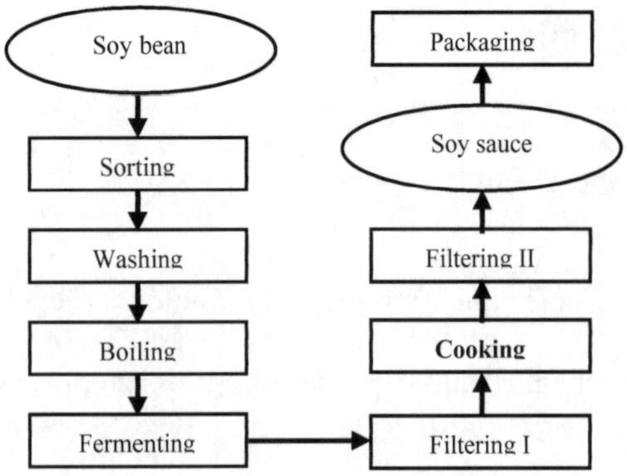

Figure 1. Flowchart of soy sauce making process [6]

The dynamic characteristics of rotordynamics in agitator must be analyzed to get satisfactory design performance and to minimize mechanical vibration that can cause failure in the planned operating conditions. The frequency, mass unbalance response, and mode shapes can be analyzed theoretically or numerically [7]. In this paper, the finite element method is employed for dynamic characteristics analysis of agitator. Comparisons are made between finite element analysis (FEA) theoretical and FEA software to vibration frequency in Campbell diagram and mass unbalance response.

2. Material and methods

2.1. Agitator design data

The agitator impeller type that has been designed based on the operating conditions in the soy sauce cooking process is anchor impeller. The viscosity of soy sauce 1900 cP, volume 6000 Liter, and density of soy sauce 1300 kg/m^3. The anchor impeller geometry adjusted with the pressure vessel. The agitator design concept is shown in Figure 2, and the impeller geometry is shown in Table 1.

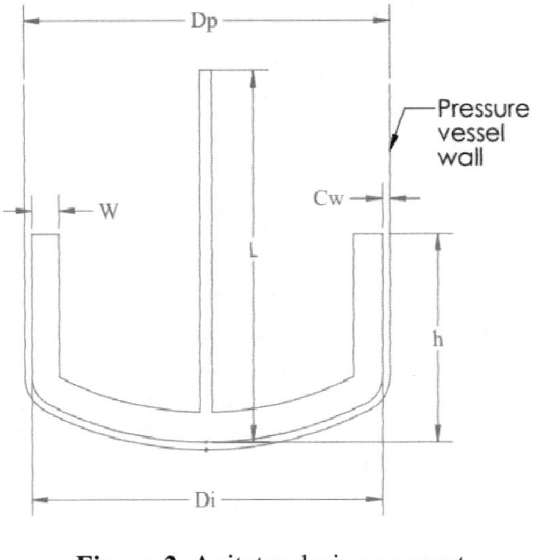

Figure 2. Agitator design concept

Table 1. Anchor impeller geometry

Geometry	Symbol	Value	
Pressure vessel diameter	D_p	2450	mm
Impeller diameter	D_i	2352	mm
Clearance	C_w	47	mm
Impeller width	W	193	mm
Impeller height	h	1344	mm
Shaft length	L	2360	mm
Shaft diameter	D	70	mm

The agitator rotation speed for soy sauce cooking process of 5 RPM, the 5052-H38 aluminum alloy is the material used with the mechanical properties as yield strength of 255 MPa, density of 2680 kg/m^3, and based on ANSI/NSF 51-1997 is a food grade material. The agitator components are shown in Figure 3.

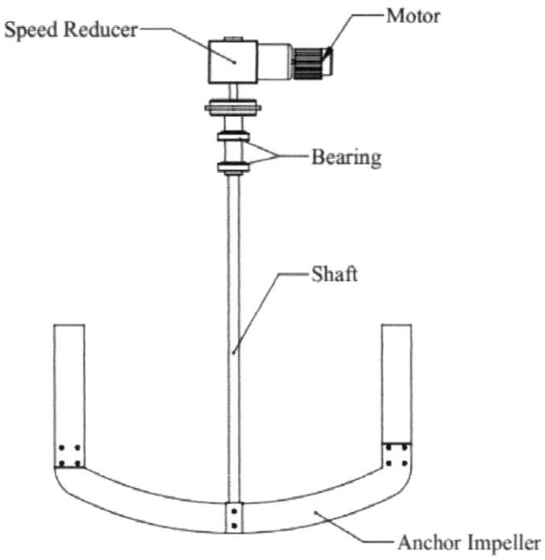

Figure 3. Agitator

2.2. Dynamic characteristics analysis

Based on the rotordynamics prediction, the rotor is part of rotating machinery including the shaft, disk, impeller, bearing, gear, coupling, and other rotating elements [8]. In this theory, the disk is assumed to be rigid and is then characterized solely by its kinetic energy, and there is no strain energy. The dynamic characteristics analysis is used to determine the ability of the agitator shaft at operation rotating speed based on natural frequency, Campbell diagram analysis, and mass unbalance response. Finite element methods both theoretical and software for dynamic characteristics analysis as follows:

2.2.1. Finite element equation for rotor disk

The agitator for soy sauce cooking process has one rotor so that the dynamic behavior of the monorotor. The equation of motion with mode shapes in Figure 4 for rotor disk as follows:

$$\frac{d}{dt}\left(\frac{\partial T}{\partial \dot{\delta}}\right) - \frac{\partial T}{\partial \delta} = \begin{bmatrix} M_d & 0 & 0 & 0 \\ 0 & M_d & 0 & 0 \\ 0 & 0 & I_{dx} & 0 \\ 0 & 0 & 0 & I_{dz} \end{bmatrix} \begin{bmatrix} \ddot{u} \\ \ddot{w} \\ \ddot{\theta} \\ \ddot{\psi} \end{bmatrix} + \Omega \begin{bmatrix} 0 & 0 & 0 & 0 \\ 0 & 0 & 0 & 0 \\ 0 & 0 & 0 & -I_{dy} \\ 0 & 0 & I_{dy} & 0 \end{bmatrix} \begin{bmatrix} \dot{u} \\ \dot{w} \\ \dot{\theta} \\ \dot{\psi} \end{bmatrix} \quad (1)$$

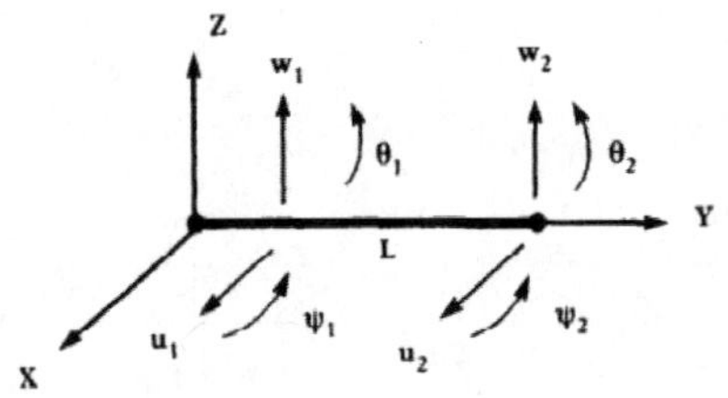

Figure 4. Mode shapes for one element and two nodes [8].

2.2.2. Finite element equation for shaft

Mass matrix, stiffness matrix, and force vector for bar element. The bar element is shown in Figure 5. For this one dimensional element, the two endpoints form the joints (nodes). This element type is affected by the axial movement or axial displacement of the shaft [9].

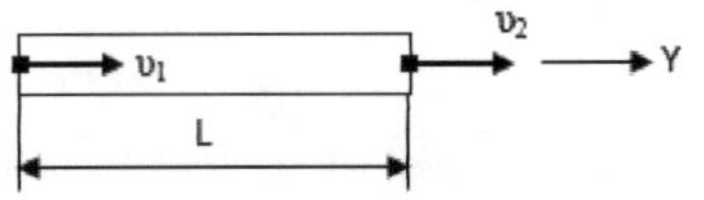

Figure 5. Bar element mode shapes

The equation of stiffness matrix for bar element as follows: [9]

$$[K] = \frac{AE}{L}\begin{bmatrix} 1 & -1 \\ -1 & 1 \end{bmatrix} \tag{2}$$

For the equation of mass matrix as follow: [8]

$$[M] = \frac{\rho AL}{6}\begin{bmatrix} 2 & 1 \\ 1 & 2 \end{bmatrix} \tag{3}$$

To calculate mass matrix, stiffness matrix in the torsion element shown in Figure 6 as follows: [9]

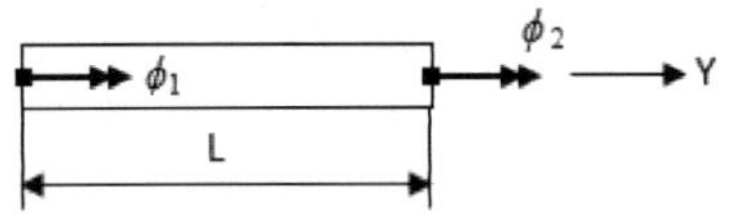

Figure 6. Torsion element mode shapes

$$[K] = \frac{JG}{L}\begin{bmatrix} 1 & -1 \\ -1 & 1 \end{bmatrix} \tag{4}$$

$$[M] = \frac{\rho JL}{6}\begin{bmatrix} 2 & 1 \\ 1 & 2 \end{bmatrix} \tag{5}$$

Figure 7 shows a uniform beam element subjected to the transverse force distribution. In this case, the joint undergo both translational and rotational displacements.

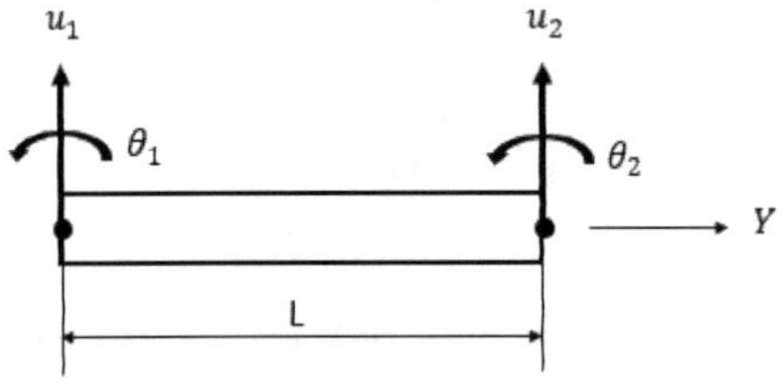

Figure 7. Beam element mode shapes

The equation of stiffness matrix and mass matrix for beam element as follow: [9]

$$[K] = \frac{EI}{L^3}\begin{bmatrix} 12 & 6L & -12 & 6L \\ 6L & 4L^2 & -6L & 2L^2 \\ -12 & -6L & 12 & -6L \\ 6L & 2L^2 & -6L & 4L^2 \end{bmatrix} \quad (6)$$

$$[M] = \frac{\rho AL}{420}\begin{bmatrix} 156 & 22L & 54 & -13L \\ 22L & 4L^2 & 13L & -3L^2 \\ 54 & 13L & 156 & -22L \\ -13L & -3L^2 & -22L & 4L^2 \end{bmatrix} \quad (7)$$

The mode shapes for three-dimensional beam element with gyroscopic matrix shown in Figure 8 and the equation as follow: [10]

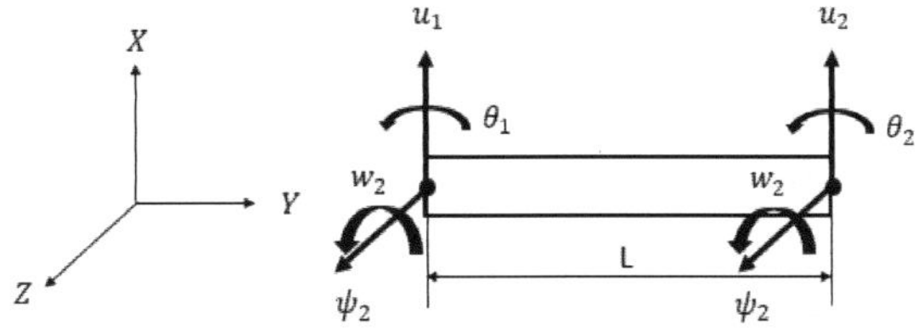

Figure 8. Three-dimensional beam element mode shapes

$$[C] = \frac{\rho I\Omega}{15L}\begin{bmatrix} 0 & -3L & 0 & -3L & 36 & 0 & -36 & 0 \\ 3L & 0 & -3L & 0 & 0 & 4L^2 & 0 & L^2 \\ 0 & 3L & 0 & 3L & -36 & 0 & 36 & 0 \\ 3L & 0 & -3L & 0 & 0 & -L^2 & 0 & 4L^2 \\ -36 & 0 & 36 & 0 & 0 & -3L & 0 & -3L \\ 0 & 4L^2 & 0 & L^2 & 3L & 0 & -3L & 0 \\ 36 & 0 & -36 & 0 & 0 & 3L & 0 & 3L \\ 0 & L^2 & 0 & -4L^2 & 3L & 0 & -3L & 0 \end{bmatrix} \quad (8)$$

2.2.3. Finite element equation for bearing
The equation of motion for rotor bearing as follows: [8]

$$\begin{bmatrix} F_u \\ F_\theta \\ F_w \\ F_\psi \end{bmatrix} = -\begin{bmatrix} k_{xx} & 0 & k_{xz} & 0 \\ 0 & 0 & 0 & 0 \\ k_{zx} & 0 & k_{zz} & 0 \\ 0 & 0 & 0 & 0 \end{bmatrix}\begin{bmatrix} u \\ \theta \\ w \\ \psi \end{bmatrix} - \begin{bmatrix} c_{xx} & 0 & c_{xz} & 0 \\ 0 & 0 & 0 & 0 \\ c_{zx} & 0 & c_{zz} & 0 \\ 0 & 0 & 0 & 0 \end{bmatrix}\begin{bmatrix} \dot{u} \\ \dot{\theta} \\ \dot{w} \\ \dot{\psi} \end{bmatrix} \quad (9)$$

The first matrix in the equation above is the stiffness matrix and the second matrix are the damping matrix. The matrix depends on the specifications of the bearing used.

2.2.4. Mass unbalance equation
The mass unbalances in rotordynamics prediction is shown in Figure 9 below. The general expression for the kinetic energy of mass unbalance and based on Lagrange's equations for finite element method was given in equation: [11]

$$\frac{d}{dt}\left(\frac{\partial T}{\partial \dot{\delta}}\right) - \frac{\partial T}{\partial \delta} = \begin{bmatrix} F_u \\ F_w \end{bmatrix} = -m_u d\Omega^2 \begin{bmatrix} sin\Omega t \\ cos\Omega t \end{bmatrix} \quad (10)$$

IOP Conf. Series: Earth and Environmental Science **251** (2019) 012038 doi:10.1088/1755-1315/251/1/012038

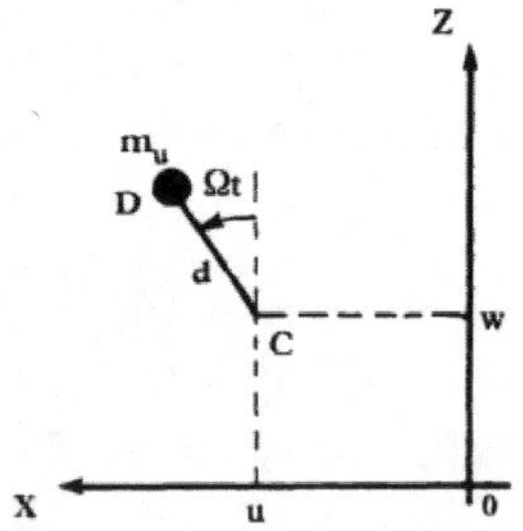

Figure 9. Mass unbalance

Where: M_d = Mass of disk (kg), I_d = Inertial mass of disk (kgm^2), A = Cross section area (m^2), E = Young modulus (N/m^2), ρ = Density (kg/m^3), J = Polar moment of inertia (m^4), G = Shear modulus (N/m^2), I = Momen of inertia (m^4), Ω = Rotation speed (rad/s), and m_u = Mass unbalance (kg).

2.2.5. General equation of motion
The most general case of the rotordynamics was enunciated, and it was briefly mentioned how to in general terms it should be solved. A solution of the equation to rotordynamics problems using finite elements will produce natural frequency and mode shapes. Two methods of solving this problem are the pseudo-modal method with quicker and more efficient, and the second methods are the direct method, more precise but more demanding of computational effort [11]. The general equation to calculate natural frequency and mode shapes as follow:

$$[M]\ddot{x} + [C](\Omega)\dot{x} + [K]x = 0 \tag{11}$$

The solution pseudo-modal method as follow: [11].

$$\begin{bmatrix} 0 & I \\ m^{-1}k & m^{-1}c \end{bmatrix} \begin{bmatrix} rP \\ P \end{bmatrix} = \frac{1}{r} \begin{bmatrix} rP \\ P \end{bmatrix} \tag{12}$$

3. Result and discussion

Dynamic characteristics analysis in rotordynamics is assumed free motion with vibration frequency and force response analysis. The impeller is assumed rigid disk, and the material properties with the shaft are same. The disk parameter to vibration frequency analysis consists of the mass of disk (Md), and inertia mass of disk of X, Y, Z axis. Mass of disk as 48.145 kg, inertia mass Idx = 7.249 kgm^2, Idy = 22.196 kgm^2, Idz = 19.415 kgm^2. The bearing as roller supports [12]. The translation and the rotation displacement of the shaft axis occur on the roller supports. To force response analysis caused by a mass unbalance when the rotor rotates then eight M14 bolts is modeled as a connector between impeller. The length of bolt 60 mm and total weight 1.6 kg. The mass unbalances position from the rotor axis is radius impeller 1176 mm. The shaft parameters needed to analyze dynamic characteristics as in Table 2.

Table 2. Shaft Parameters

Parameters	Symbol	Value
Shaft length between two bearings	L1	200 mm
Overhang shaft length	L2	2260 mm
Radius	R1	35 mm
Density	ρ shaft	2680 Kg/m^3
Young modulus	E	70 GPa
Cross section area	A	3.847e3 mm^2
Inertia	I	1.178e6 mm^4
Polar moment of inertia	J	2.356e6 mm^4
Poisson Ratio	υ	0.33

IOP Conf. Series: Earth and Environmental Science **251** (2019) 012038 doi:10.1088/1755-1315/251/1/012038

Parameters	Symbol	Value
Shear modulus	G	26 GPa

3.1. Frequency analysis of vibration

The main characteristic of the agitator for the cooking process is that it has intermediate support. For illustration, the shaft could be divided into two elements with three nodes as shown in Figure 10. The first node in the top bearing, the second node in the bottom bearing, and the last node in the disk. The model analysis uses FEA theoretical and FEA software. The stiffness method free body diagram for this model as shown in Figure 11 and Figure 12. The frequency analysis of vibration with FEA theoretical uses the equation of motion above.

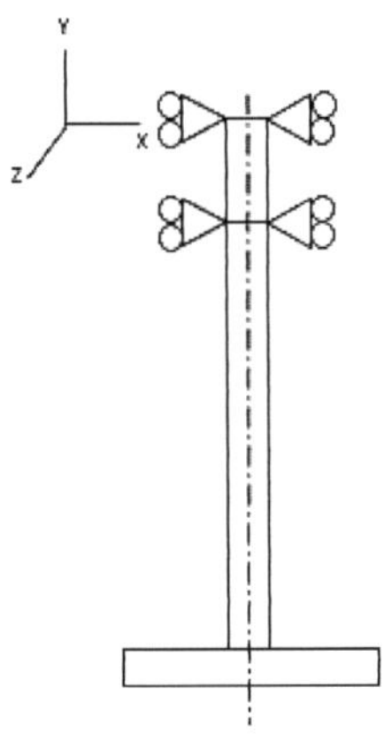

Figure 10. Roller support model

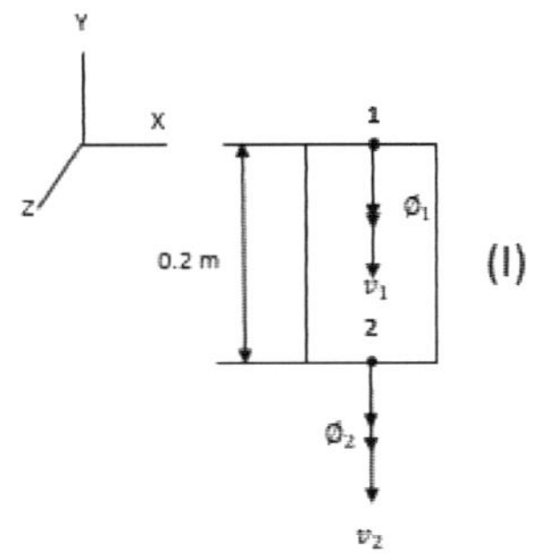

Figure 11. The stiffness method free body diagram for element I

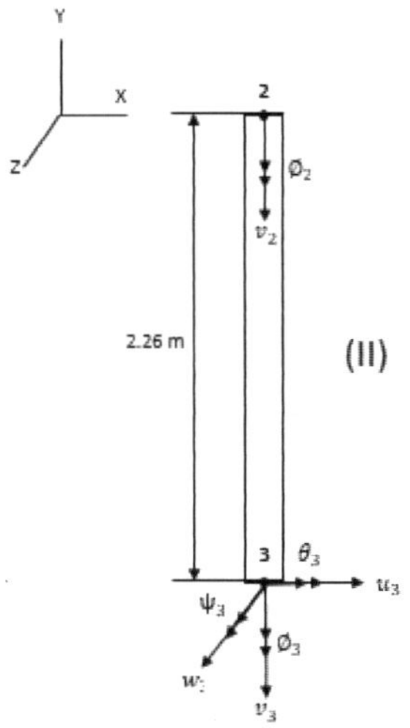

Figure 12. The stiffness method free body diagram for element II

The frequency analysis uses FEA software, four mode shapes selected based on FEA theoretical. In this FEA software problem, the rigid disk modeled in modal analysis can use *a mass point* feature, by entering data in the form of mass and inertia on all axis. Roller supports as boundary condition use *remote displacement* feature. Element type used is *tetrahedron quadratic* with the number of element 14888 or mesh size 8 mm.

Figure 13 is a Campbell diagram for two bearings model as roller support based on FEA theoretical and software. Points A, C, D, and F correspond to the intersection with N/60 and 0.5 N/60 as BW (Backward Whirl). Points B and E with FW (Forward Whirl). In points, A, B, and C is the intersections between lines of F2, F3, and F4 with N/60. At the corresponding points, a frequency of the rotor equals the frequency of the rotation. Points D, E, dan F is intersection points between lines of F2, F3, F4 with 0.5 N/60. At the corresponding points, a frequency of the rotor equals half of the frequency of rotation. These six points in the agitator design will occur resonance that is affected by vibration so that the application needs to be considered not to use rotational speed around the six points. The operation speed of agitator 5 RPM based on the Campbell diagram is no resonance. That, the design of an agitator with frequency analysis is still safe. Campbell diagram around operation speed is the plot in Figure 14.

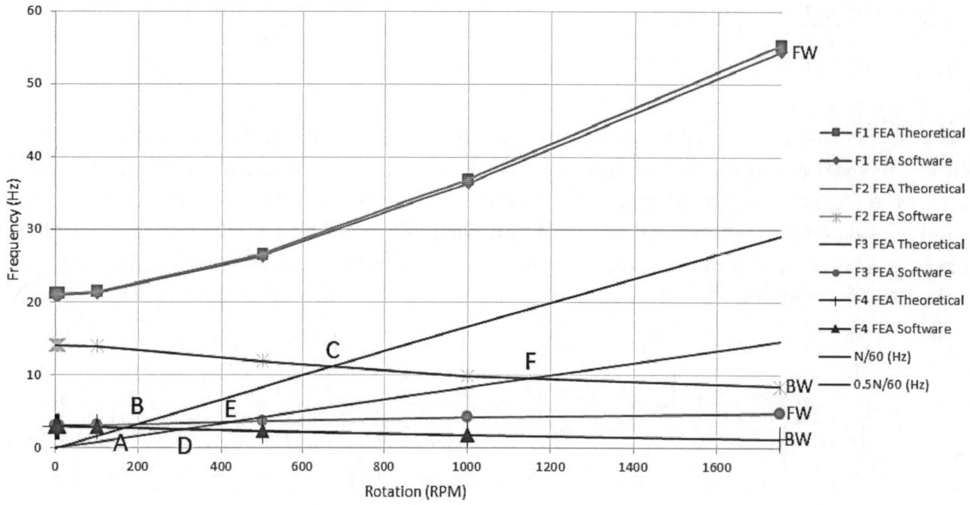

Figure 13. Campbell diagram

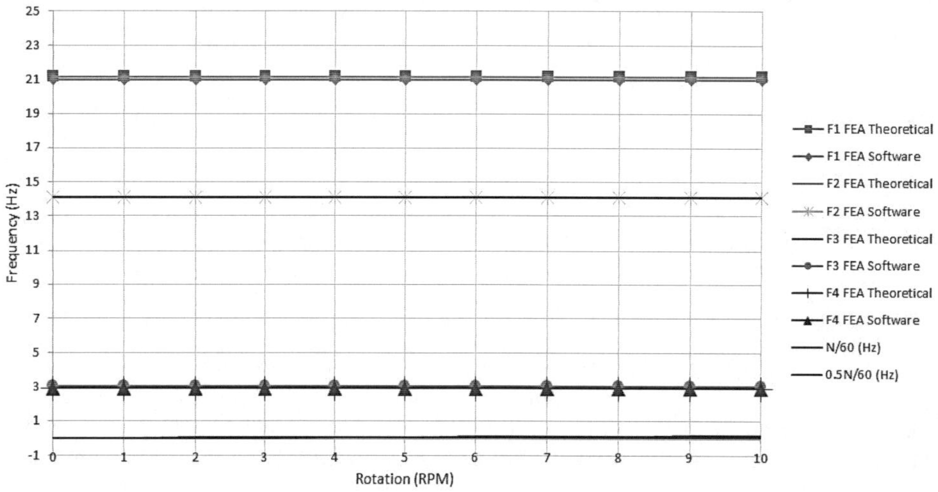

Figure 14. Campbell diagram around operation speed

Comparisons of frequency in Table 3 show good accuracy FEA theoretical and software in beam element for rotation speed is 0 and 5 RPM.

Table 3. The frequency associated with the previous Campbell diagram

		F1	F2	F3	F4
0 rotation speed	FEA Theoretical (Hz)	21.1911	14.1213	3.0916	2.9490
	FEA Software (Hz)	21.0130	14.0950	3.0908	2.9485
	Error (%)	0.84	0.19	0.03	0.02
5 RPM rotation speed	FEA Theoretical (Hz)	21.1919	14.1209	3.0920	2.9487
	FEA Software (Hz)	21.0140	14.0950	3.0911	2.9481
	Error (%)	0.84	0.18	0.03	0.02

3.2. Mass unbalance response

Using equation 10, to analyze the effect amplitude of mass unbalances with assuming one bearing model as fixed support. Comparison of force response analysis for FEA theoretical and software with up to a maximum speed of 1750 RPM motor is shown in Figure 15. The maximum amplitude occurs at point B, mode shape 3 or FW in operation speeds 190 RPM is 45.5 mm.

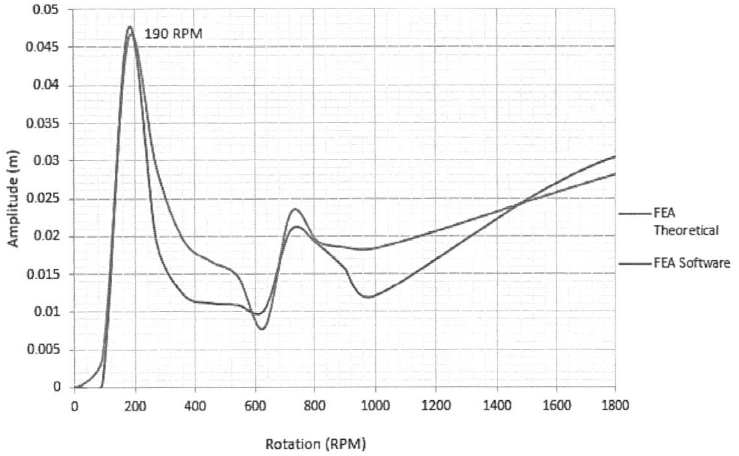

Figure 15. Mass unbalance response

Figure 16 shows the amplitude of mass unbalances response 5 RPM for agitator operation condition. The maximum amplitude in mode shape 3 is less then 2.5×10^{-2} mm and can be considered very small.

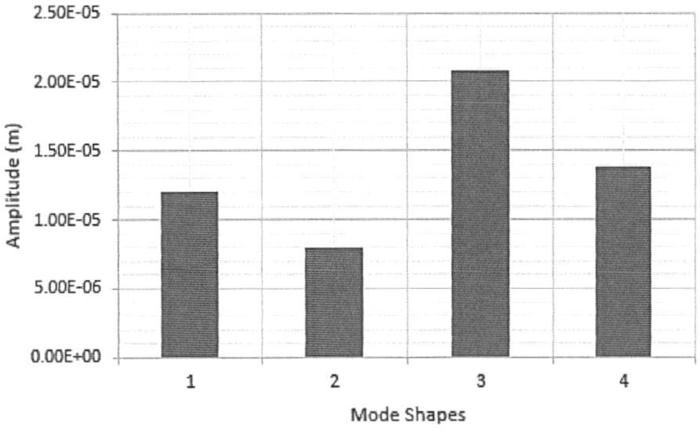

Figure 16. Mass unbalance response for 5 RPM

The rotation speed and frequency range that must be avoided when the agitator operates is based on the Campbell diagram as in Table 4.

Table 4. The rotation speed and frequency range that must be avoided

Points	The rotation speed range (RPM)	The frequency range (Hz)
A	± 160	± 2.5
B	± 190	± 3
C	± 680	± 11
D	± 310	± 2.5
E	± 430	± 3
F	± 1150	± 9

4. Conclusion

Dynamic characteristics analysis of agitator for soy sauce cooking process with FEA both theoretical and software has a similar result. The stiffness method approach for the agitator shaft is carried out by dividing the shaft element into two elements based on the bearing position. The bearing is assumed as roller support. The critical speed is not in the range of the operating speed. The agitator in operating conditions with the operation speed of 5 RPM based on Campbell diagram will not fail with the highest amplitude occurring in mode shape 3.

5. Acknowledgments

Thanks are due to Faculty of Mechanical and Aerospace Engineering, Institut Teknologi Bandung and Development Center for Appropriate Technology - Indonesian Institute of Sciences

6. References

[1] Kasmidjo R B 1990 Tempe: *Mikrobiologi dan Biokimia Pengolahan* serta *Pemanfaatannya* (Yogyakarta : PAU Pangan dan Gizi)

[2] Andiana D 2014 Pengaruh Subtitursi Kacang Gude (Cajanus Cajan) Terhadap Kadar Protein Dan Daya Terima Kecap Kedelai *Unnes Journal of Public Health 3* **(3)** ISSN 2252-6528 (Semarang)

[3] Septiani Y 2004 Kadar Karbohidrat, Lemak, Dan Protein Pada Kecap Dari Tempe (Carbohydrate, Lipid, And Protein Contents In Soy Sauce Of Tempe) *Bioteknologi* **1** (2) Nopember 2004 ISSN: 0216-6887 DOI: 10.13057/biotek/c010204 pp 48-53.

[4] Nur M A 2017 *Perancangan Bejana Tekan Dengan Jacket Untuk Proses Pemasakan Kecap Skripsi* (Bandung : Program Studi Teknik Mesin Institut Teknologi Bandung)

[5] Desai S R, Gajjal S Y and Kulloli 2016 Redesign and Structural analysis of agitator shaft for reactor pressure vessel *International Journal of Current Engineering and Technology* E-ISSN 2277 – 4106 P-ISSN 2347 – 5161 MIT College of Engineering Pune India AMET 2016 INPRESSCO IJCET Special Issue-4 (India) pp 268-273

[6] Putri M A 2009 *Analisis Usaha Pembuatan Kecap Kedelai Pada Industri"Kecap Lele" Di Kabupaten Pati Skripsi* (Surabaya : Fakultas Pertanian Universitas Sebelas Maret)

[7] Agostini C E and Sauza E A C 2010 Complex Modal Analysis Of A Vertical Rotor By Finite Elements Method *Proceedings of the 9th Brazilian Conference on Dynamics Control and their Applications Serra Negra* SP ISSN 2178-3667 (Brazil) pp 449-457

[8] Lalanne M and Guy F 1998 *Rotordynamics Prediction in Engineering* Second Edition (New Jersey : John Wiley & Sons)

[9] Rao S S 2011 *Mechanical Vibrations* Fifth Edition (Miami: Pearson Education)

[10] Jones S 2005 *Finite Elements For The Analysis Of Rotor-Dynamic Systems That Include Gyroscopic Effects* A thesis submitted for the degree of Doctor of Philosophy Department Of Mechanical Engineering Brunel University (UK)

[11] Paulo P 2011 *A time-domain methodology for rotor dynamics: analysis and force identification* Disertasi Instituto Superior Tecnico Universidade Tecnica De Lisboa

[12] Tiwari R 2010 *Finite Element Analysis of Simple Rotor System* Associate Professor Dept of Mechanical Engg IIT Guwahati (rtiwari@iitg.ernet.in). Available at: https://www.coursehero.com/file/9921790/rtiwari-rd-book-04/.

The potency of cassava starch (var. kristal merah and var. revita) for bio-succinic acid production using indigenous lactic acid bacteria (Leuconostoc sp)

A Andriani[1], I Nuryana[1], N Rahmani[1], S Hartati[1], P Lisdiyanti[1], Yopi[1]

[1]Research Center for Biotechnology, Indonesian Institute of Sciences, Jl. Raya Bogor Km 46, Cibinong-Bogor 16911, Indonesia

Email: andriani_chemie@yahoo.com

Abstract. Cassava of local strain, a plentiful agricultural product in Indonesia was evaluated for biosuccinic acid production via saccharification and fermentation using indigenous lactic acid bacteria (*Leuconostoc sp*). Cassava starch is an ideal biosuccinic acid feedstock due to its low price and high sugar content. Conversion of cassava starch into succinic acid is an effort to obtain added-value of the local product. Succinic acid has been well-known as an important platform chemical and being top 12 building block in the chemical industries. In the current study, saccharification of starch was carried out using available glucoamylase enzymes produced by *Aspergillus awamori* KT 11 (*Glutech*) in the temperature 60°C and 200 rpm. Results revealed the best yield of total reducing sugar accumulation (62.3 g L^{-1}) was obtained when the ratio of starch solution of var. kristal merah (25% w/v) and glucoamylase was kept to 3:1. HPLC analysis of sugar also showed three different sugar were produced during saccharification: glucose, maltose, and maltotriose. Fermentation of the liquid sugar as a carbon source in various media showed that *Leuconostoc sp* could produce highest succinic acid in medium SPM, with concentration 3.4 g L^{-1} in 72 h for var. kristal merah and 3.6 g L^{-1} g in 48 h for var. revita. The other valuable organic acids, such as lactic acid, citric acid, formic acid, and acetic acid were also formed during fermentation process, with concentration less than 3 g L^{-1}. The results suggested that cassava starch of 2 local strains (var. kristal merah and var. revita) has the potential to be an alternative substrate for efficient and economic valuable organic acids production such as succinic acid.

1. Introduction

Succinic acid, known as butanedioic acid, is one of the most important platform compounds that can be used as a precursor for industrial chemicals, such as 1, 4-butanediol, tetrahydrofuran, maleic anhydride, g-butyrolactone, adipic acid, and polymers through esterification process [1-3]. For sustainability purpose in this era of petroleum shortages, the production of succinic acid via the microbial conversion of renewable feedstock has attracted great interest. The production of succinic acid can be performed by fermentation using anaerobic or facultative anaerobic microorganisms. Succinic acid is a common intermediate in the metabolic pathway of the microorganisms [3]. The wild-type rumen bacterial strains such as *Actinobacillus succinogenes* and *Basfia succiniciproducens* are two of the most promising strains for industrial application [4]. However, the utilization of anaerobic microorganisms needs extra treatment to create a suitable condition for growing. Among succinic acid-producing strains, Lactic Acid Bacteria (LAB) has great potential because they can use different carbon sources to produce various valuable organic acids such as succinic acid in both aerobic and anaerobic condition. In the present study, *Leuconostoc sp*, was selected after an extensive screening assay of hundreds of LAB carried out in our laboratory (*unpublished data*).

The other important aspect in the production of bio-based chemicals such as biosuccinic acid is that low-cost raw materials with high supply should be used as fermentation feedstocks. Some researcher has reported the utilization of low-cost carbon sources such as sweet potato granular [5], spent sulphite liquor (SSL) [4], fresh cassava root [6], citrus peel [7], cane molasses, hydrolysate of carbon fibre, and wood hydrolysate [1] for fermentation to produce succinic acid. In the present study, a hydrolysate of cassava starch from 2 different local strains was used as a low-cost carbon source for fermentation by selected LAB, *Leuconostoc sp*. Enzymatic hydrolysis using glucoamylase was performed as saccharification method to obtain liquid sugar (glucose). The liquid was then used as carbon source in the fermentation process by *Leuconostoc sp*. In this study, we also determined the most suitable mineral medium for the microbe of organic acids production. Utilization of low-cost starchy materials could be another effective approach for reducing petroleum-based chemicals and also the cost of succinic acid production.

2. Methods

2.1. Chemicals

MRS agar and MRS broth were purchased from Wako Co., Ltd. (Osaka, Japan). Starch from 2 strains of cassava (kristal merah and revita) was obtained from Laboratory Molecular Genetics and Biosynthesis Pathway Modification of Plant, Research Centre for Biotechnology, Indonesian Institute of Sciences (LIPI).

2.2. Microorganisms

The lactic acid bacteria (LAB) strain used in this study was *Leuconostoc mesenteroides* (obtained from InaCC with code LAB100). The fungus was cultivated at 25 °C on MRS agar medium for 3 days in a disposable plastic Petri dish and maintained at 4 °C prior to use. The roadmap oh this research was shown in Figure 1.

2.3 Enzymatic hydrolysis of starch by crude glucoamylase produced by Aspergillus awamori KT 11

Experiments using various combination enzyme and substrate (E/S) was performed in 100-mL Erlenmeyer flask to identify the most suitable combination E/S for liquid sugar production. Gelatinization of starch (25% w/v) at 100 °C for 10 minutes at a stirring water bath was conducted prior to reaction with glucoamylase. The hydrolysis was conducted for 24, 48, and 72 h at 60 °C and agitation 200 rpm. The composition reaction for enzyme and substrate as shown in Table 1.

Table 1. Composition of E/S during enzymatic hydrolysis of cassava starch (kristal merah)

Volume glucoamylase (mL)	Starch (gram)	Aquades (mL)	E/S
10	5	10	1/1
5	5	15	1/3
4	5	16	1/5
2	5	18	1/9

Table 2. Four various of culture medium for fermentation *Leuconostoc sp*

Component	MRS (gram/L)	Medium 2*	Medium 3**	Medium 4*** (SPM)
D (+) Glucose	20	-	-	-
Enzymatic digest of Casein	10	-	-	-
Peptone	-	4	-	4
Tripton	-	-	1.1	-
Beef/Meat Extract	10	-	-	-
Yeast Extract	4	3.1	0.5	4
Tween 80	1 ml	0.2 ml	-	1 ml
Disodium Phosphate (Na_2HPO_4)	-	-	-	2
Potassium Phosphate (K_2HPO_4)	2	4	0.3	2
Magnesium Sulphate ($MgSO_4.7H_2O$)	2	-	-	0.1
Manganese Sulphate ($MnSO_4.5H_2O$)	0,04	0.04	-	0.05
Tri-Sodium Citrate	-	0.01	-	-
NaCl	-	-	0.1	-
$(NH_4)_2SO_4$	-	-	0.1	-
$CaCl.H_2O$	-	-	0.02	-
$MgCl_2.5H_2O$	-	-	0.02	-
Na_2CO_3	-	-	0.1	-
$(NH_4)_2HPO_4$	-	-	0.55	-
pH	6.5	6.5	6.5	6.5

References: *[8] **[3] ***[9]

2.4 Reducing sugar and glucose content analysis
Reducing sugar was analyzed using DNS (dinitrosalicylic acid) method by Miller [10]. 500 μL sugar solution (dilution factor = 2000) and 750 μL DNS were mixed and incubated at 100 °C for 15 minutes. The mixed solution then was precooled in the ice water bath for 10 minutes prior to absorbance measurement by spectrophotometer at wavelength 540 nm.

2.5 Selection of culture medium for fermentation by Leuconostoc sp
To obtain a suitable medium for fermentation of liquid sugar by *Leconostoc sp*, 4 type media were selected. The content of each medium was shown in Table1. The strain was inoculated for preculture in each medium (@ 5 mL) in a capped tube to minimize air circulation. The inoculated medium was incubated for 24 hours with agitation 150 rpm and temperature 37 °C. Ten percent of preculture was then inoculated into a fresh medium containing 20% of sugar liquid of cassava. Fermentation was conducted for 24, 48, 72, 120, and 168 h.

2.6 HPLC analysis of glucose and succinic acid content
The concentration of succinic acid, glucose, and other organic acids was analyzed by an HPLC system. Isocratic system using H_2SO_4 5 mM (flow rate 0.6 μL) as mobile phase and Aminex column as stationary phase were used in the analysis. The temperature of the column was set at 60 °C with the injection volume was 10 μL. RID was used as detector for analysis of these samples.

3. Result and Discussion
3.1 Enzymatic hydrolysis of starch by crude glucoamylase produced by Aspergillus awamori KT11
In the present study, non-simultaneous (separation) saccharification fermentation was used for biosuccinic acid production. Starch, a polysaccharide material, was breakdown into more digestible sugar for microorganisms (glucose and/or oligosaccharide) using glucoamylase prior to use in the fermentation step. Several combinations of enzyme and substrate (E/S) treatment were selected for saccharification to obtain the most effective and efficient method for sugar production. Based on enzyme activity analysis, glucoamylase powder from *Glutech* has activity 16.9 U/mL when diluted in 9 mL aquadest (1:9). The enzyme has moderate thermophilic characteristics with the optimum reaction temperature at 60 °C. Initial high concentration of starch (25% w/v) was used to achieve high yield and

full utilization of substrate [2]. Among the combination treatment (Figure 1) the two highest total reducing sugar revealed at combination E/S (1/1) and (1/3), with the concentration 61.3 and 62.3 g L^{-1}, respectively at 48 hours. Under these conditions, a saccharification rate of approximately 25% was obtained. Many factors can cause this low conversion of starch into glucose. Unsuitable gelatinization process, such as temperature, and utilization of unpurified enzymes are two factors that cause low yield on enzymatic hydrolysis of starchy materials [5]. The reducing sugar concentration tends to decrease after 48 h, except for treatment E/S (1/9) which gradually increased up to 43.9 g L^{-1} until 72 hours.

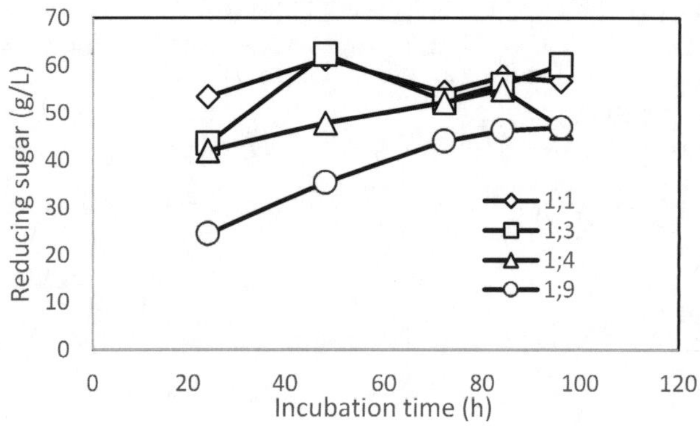

Figure 1. Total reducing sugar of hydrolysate of cassava starch (var. Kristal merah)

Based on HPLC analysis of sugar content in the hydrolysate, three different sugars were found as products: glucose, maltose, and maltotriose (Table 3). In all treatment, glucose revealed as a major product of the enzymatic hydrolysis of starch by glucoamylase. In this study, since glucoamylase, *Glutech* is an unpurified enzyme product, various products of sugar were produced. The glucose accumulation in hydrolysate (1/3) after 96 h reached 47.71 g L^{-1} which was slightly higher compared with treatment (1/1), that used the highest enzyme loading (50%). The chromatogram of hydrolysate in various treatment can be seen in Figure 3.

Some study has reported the utilization of low-cost carbon sources for fermentation of succinic acid. Akhtar and Idris [11] evaluated for the production of succinic acid via simultaneous saccharification and fermentation (SSF) using an anaerobic microbe, *A. Succinogenes* ATCC 55618 and oil palm empty fruit bunch (EFB) as a carbon source. Dessie et al., [12] also reported the utilization of fruit and vegetable wastes for saccharification in the solid-state fermentation. Chen et al., [2] reported the production of succinic acid from cassava starch and raw cassava instead of glucose by *Escherichia coli* NZN111. However, the utilization of local cassava starch for succinic acid production by LAB has not been reported to date.

Table 3. Sugar content of hydrolysate of cassava starch (var. kristal merah)

		Sugar content (g/L)		
Treatment	Incubation time (h)	glucose	maltose	maltotriose
E/S (1:1)	48 h	35.22	4.30	2.94
	96 h	46.77	4.87	3.30
E/S (1:3)	48 h	43.59	3.44	1.83
	96 h	47.71	4.30	2.94
E/S (1:4)	48 h	35.14	4.15	2.18
	96 h	37.06	4.28	1.50
E/S (1:9)	48 h	23.01	5.99	3.01
	96 h	32.02	2.87	1.45

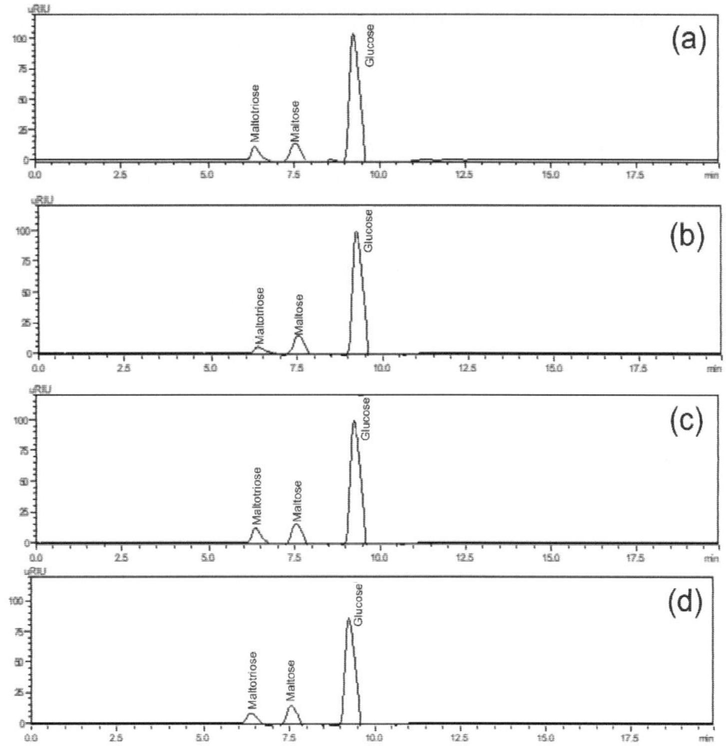

Figure 2. HPLC Chromatogram of hydrolisate of cassava starch (var. kristal merah) in various treatment E/S: 1/1 (a); 1/3 (b); 1/4 (c); and 1/9 (d).

3.2. Fermentation of starch hydrolysate by Leuconostoc sp for succinic acid production

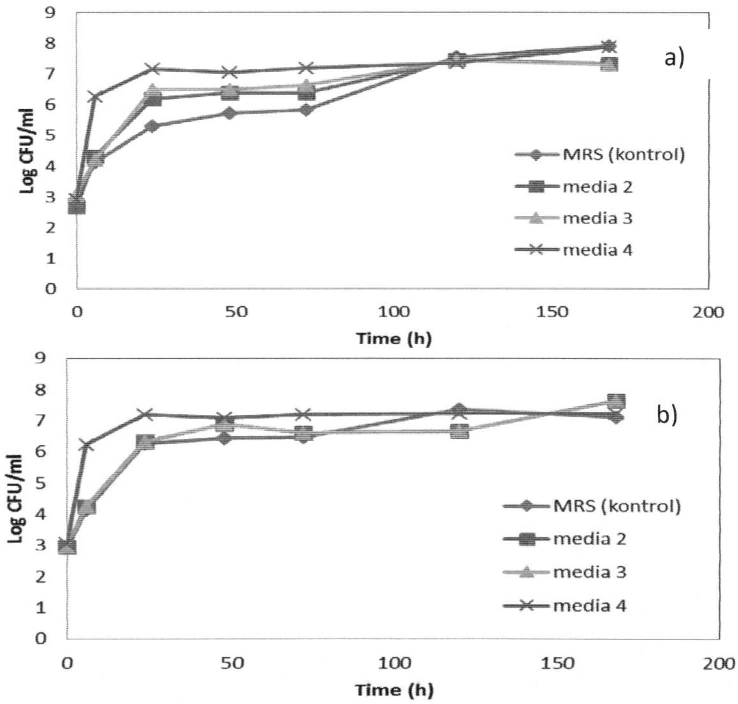

Figure 3 Microbial growth during fermentation of *Leuconostoc sp;* a) kristal merah; b) revita

After completion of enzymatic hydrolysis of starch on the optimum incubation time (48 hours), the hydrolyzed products were added into fermentation medium for further process. Culture medium is known to play an important role in maintaining cellular metabolism and enzyme activities during fermentation [3]. In the present study, we selected three types of mineral medium which specific for lactic acid bacteria, and also MRS medium as a positive control for bacterial growth. Medium 2 [8] has been reported as a suitable medium for *Lactobacillus salivarius* L29 and allows the attainment of high levels of cell mass and lactic acid production. Medium 3 [3] has been used as the growing medium for *Enterococcus flavescen* on succinic acid production). Medium 4 supported with sweet potatoes as carbon sources [9] has been reported as the developed medium for *Lactobacilli.* Twenty percent of sugar liquid from the previous study was added into each medium and inoculated with 10% bacterial inoculum (24 h). As shown in Figure 3, all medium was a suitable medium for *Leuconostoc* to grow. When using hydrolysate of var. kristal merah and var. revita, the growth rate in all three media was higher compared with MRS medium. The optimum growth rate was achieved at 24 h for medium 4 and start to enter the stationary phase after 24 h.

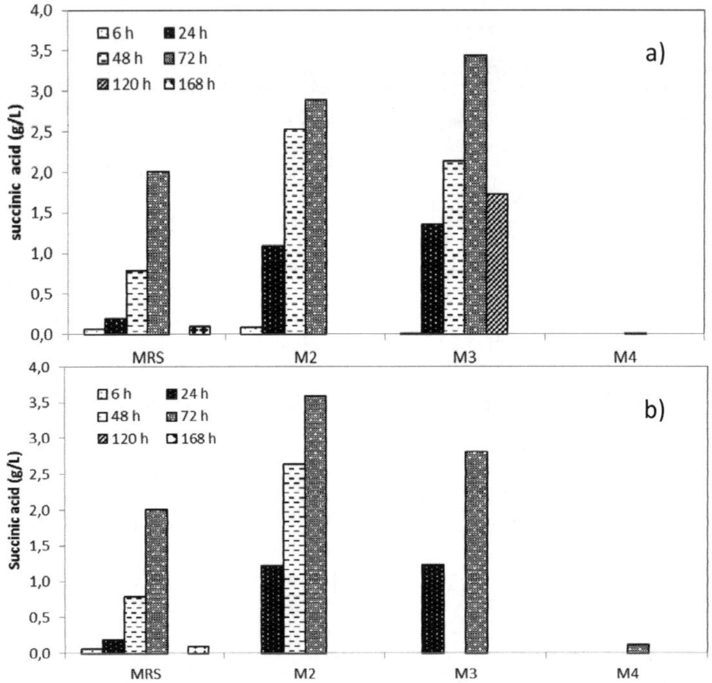

Figure 4. Total succinic acid production during fermentation by *Leuconostoc sp*: a.) var. kristal merah, and b.) var. Revita

Among four fermentation medium tested, the highest production of 3.4 g L^{-1} succinic acid was obtained in 72 h using medium 3 (M3) for var. kristal merah and 3.6 g L^{-1} succinic acid in 48 h using medium 2 (M2) for var. revita (Figure 4). There was no increase in amount of succinic acid produced on subsequent increase in incubation time. The content of mineral medium M3 is more complex than M2, with only contain minimum media such as yeast extract and peptone as protein sources, potassium phosphate and manganese sulfate as mineral sources [5] [8].

Metabolically engineered *E. coli* NZN111, constructed by disruption of ldhA and pflB encoding lactate dehydrogenase (LDH) and pyruvate-formatelyase (PFL), has been reported as a potential succinate producer with formation 106.17 g L^{-1} of succinic acid when the liquefied crude cassava powder was used directly in SSF, 106.17 g L^{-1} [2]. *Actinobacillus succinogenes* also has been reported on succinic acid production with maximum concentration, (33.4 g L^{-1}), yield (30.47 g g^{-1} substrate) and productivity 0.69 g L^{-1} h-^{1} when using pretreated empty fruit bunch sample as a carbon source [11].

Succinic acid production by *Leuconostoc sp* (only 3.6 g L^{-1}) is lower than 2 recent studies above. Many anaerobic and facultative anaerobic microorganisms such as *Leuconostoc sp.* ferment carbohydrates to a mixture of acids, e.g., formate, acetate, lactate, and succinate as end products. Phosphoenolpyruvate (PEP) is one of the central intermediates during the mixed acid fermentation. It is either converted into pyruvate resulting acetate, formate, ethanol, and lactate or it is converted into oxaloacetate resulting succinate and propionate as end products via the reversible arm of tricarboxylic acid (TCA) cycle depend on environmental factors [3] To boost succinic acid production by *Leuconostoc sp*, the effect of various environmental factors should be conducted. Beside succinic acid, the microbe also produced organic acids such as citric acid, formic acid, and acetic acid (less than 3 g L^{-1}). Lactic acid was produced in the highest concentration, up to 10.62 g L^{-1}.

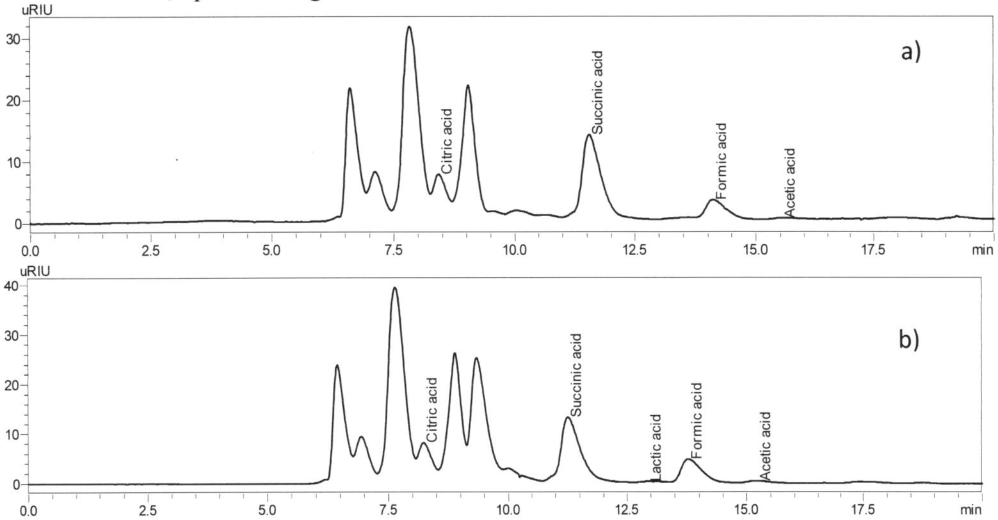

Figure 5 HPLC chromatogram of fermentation product by *Leuconostoc sp* : a.) var. kristal merah, and b.) var. revita

This is first study regarding succinic acid production by *Leuconostoc sp.* Many studies reported anaerobic microorganisms such as succinic acid production. *Actinobacillus succinogenes*, *Mannheimia succiniciproducens* (Lee et al., 2003), *Anaerobiospirillum succiniciproducens* (Meynial-Salles, et al. 2008) and *Corynebacterium crenatum* (Chen et al. 2013). Facultative microorganisms can produce organic acids such as succinic acid under aerobic and anaerobic condition.

4. Conclusion

Fermentation of the liquid sugar as a carbon source in various media showed that *Leuconostoc sp* could produce highest succinic acid in medium SPM, with concentration 3.4 g L^{-1} in 72 h for var. kristal merah and 3.6 g L^{-1} g in 48 h for var. revita. The other valuable organic acids, such as lactic acid, citric acid, formic acid, and acetic acid were also formed during fermentation process, with concentration less than 3 g L-1. The results suggested that cassava starch of 2 local strains (var. kristal merah and var. revita) has the potential to be an alternative substrate for efficient and economic valuable organic acids production such as succinic acid.

5. References

[1] Zhao Y, Cao W, Wang Z, Zhang B, Chen K, Ouyang P 2016 *Biores. Tech.* **202** 152–7
[2] Chen C, Ding S, Wang D, Li Z, Ye Q 2014 *Biores. Tech.* **163** 100–5
[3] Agarwal A, Isar J, Meghwanshi GK, Saxena RK 2007 *Enzyme. Microb. Tech* **40** 629–36
[4] Alexandri M, Papapostolou H, Stragier L, Verstraete W, Papanikolaou S, Koutinas A A 2016 *Biores. Tech.* 1–38
[5] Pereira C R, Resende J T V, Guerra E P, Lima V A, Martins M D, Knob A 2017 *Biocat. Agr. Biotech.* 1–28

[6] Thuy N T H, Kongkaew A, Flood A, Boontawan A 2017 *Biores. Tech* 1-40
[7] Patsalou M, Menikea K K, Makri E, Vasquez M, Drouza C, Koutinas M 2017 *Cleaner Prod* 1–44
[8] Lee K, Kang S K and Choi Y J 2013 *African Journal of Biotechnology* **12** (16)
[9] Hayek S A, Shahbazi A, Awaisheh S S, Shah N P and Ibrahim S A, 2013 *Bioscience, biotechnology, and biochemistry* **77** (11) 2248-2254
[10] Miller G L 1959 *Anal. Chem.* **31** 426–428
[11] Akhtar J, Idris A 2017 Renewable Energy **114** 917–923
[12] Dessie W, Zhang W, Xin F, Dong W, Zhang M, Ma J, Jiang M 2017 *Biores. Tech.* 1–44
[13] Lee P C, Lee S Y, Hong S H., Chang H N 2003 *Bioproc. Biosyst. Eng* **26** 63-7
[14] Meynial-Salles I, Dorotyn S, Soucaille P 2008 Biotechnol. Bioeng **99** 129-135
[15] Chen X J, Jiang S T, Li X J, Pan L J, Zheng Z, Luo S Z 2013 *Ann. Microbiol.* **63** 39-44

Characterization and Evaluation of Physical and Mechanical Properties of Un hulled Arabica Coffee Bean

S I Kuala, D D Hidayat, C E W Anggara and R Saparita

Development Center for Appropriate Technology, Indonesian Institute of Sciences, KS Tubun Street 5 Subang 41213, West Java, Indonesia

Email: seri.kuala.sk@gmail.com

Abstract. This work was carried out to characterize the physical and mechanical properties of unhulled arabica coffee bean. The results of this work found that at moisture content of 18.76 ± 0.19 % wb, the average length, width, thickness, geometric diameter and surface area of unhulled coffee bean ranged from 12.04 ± 1.06 mm, 8.24 ± 0.45 mm, 4.99± 0.34 mm, 7.90 ± 0.38 mm and 196.44 ± 19.09 mm^2 respectively. The average mass and volume ranged from 0.21 ± 0.06 g and 0.21 ± 0.06 cm^3 respectively. The average particle density, bulk density and sphericity ranged from 1.00 ± 0.02 g/cm^3, 0.40 ± 0.01 g/cm^3 and 0.66 ± 0.04% respectively. The average porosity, aspect ratio, and shell content ranged from 60.16 ± 1.09 %, 68.88 ± 6.38% and 16.55 ± 5.70% respectively. The average hardness and fracturability ranged from 12.34 ± 1.28 gf and 2.45 ± 0.08 gf respectively. Regarding the angle of repose on the surface of stainless steel, aluminum, acrylic, and plywood results showed that the average emptying angle of repose ranged from 26.88 ± 3.67°, 29.44 ± 3.35°, 30.01 ± 4.83° and 40.0° ± 7.00° respectively, and the average filling angle of repose ranged from 32.02 ± 2.880°. Concerning friction, results of measurement indicated that average static friction of stainless steel, aluminum, acrylic, and plywood ranged from 0.36 ± 0.02, 0.40 ± 0.03, 0.38 ± 0.04 and 0.45 ±0.03 respectively. The highest emptying angle of repose occurred on the surface of plywood inversely of that was occurring on the surface of stainless steel. There was a positive relationship between the angle of repose and the static friction.

Keywords: Un hulled coffee bean, physical properties, mechanical properties, texture profile, angle of repose

1. Introduction

Among the ASEAN countries, Indonesia has known as a producer and second largest coffee exporter after Vietnam [1]. In the world, Indonesia was the fourth largest coffee producer after Brazil, Vietnam, and Colombia [2]. In the case of coffee exports, Indonesia was the fourth largest coffee exporter in the world after Brazil, Vietnam, and Colombia [2]. In the world, Indonesia has known as specialty coffee through coffee variants. With its unique taste and aroma, Indonesia has an excellent opportunity to increase trade coffee in the world.

In Indonesia, coffee is one of the commodities that play an essential role in supporting efforts to increase non-oil exports. Foreign exchange earnings from commodity coffee in 2014 generated export value of 10.1 percent of the entire export value of agricultural commodities, or 0.5 percent of non-oil exports, or 0.4 percent of the total export value [3]. In recent years, the amount of coffee exports has declined. The decline in the value of commodities, aside from the decrease in prices of coffee in the international market and has also suspected due to the declining quality of Indonesian coffee.

As much as 66% of world coffee production was an arabica coffee, and another 34% was robusta coffee [4]. It mentioned before that in Indonesia the production of coffee was 73.57 % robusta and 26.43 % Arabicas. To increase involvement in the world coffee market related to the increase of export value, Indonesia had to be able to increase the production of Arabica coffee with quality that meets the world coffee standards.

The physical and mechanical properties of coffee beans, the same as other grains and seeds are indispensable in processing, designing, and manufacturing of equipment. Many researchers carried out research on the physical and mechanical properties of agricultural products among others were: [4] carried research on physical, mechanical and chemical properties of Jatropha curcas; [5] carried out research on physical properties of sugar beet seed, [6], [7], [8], [9], [10] and [11] carried out research on physical and mechanical properties of onion, [12] carried out research on physical and mechanical properties of wild medlar. Many other researched done on physical and mechanical properties of agricultural products were: sweet cherry [13], kiwifruit [14], jujube fruit [15], Simarouba fruit and kernel [16], date fruit [17], goldenberry [18], kumquat [19], cocoa bean were [20] and [21], [22], and [23]. Relevant parameters to be known to increase the quality of coffee bean among others are physical and mechanical properties of the coffee bean. The data of physical and mechanical properties of unhulled coffee bean has not been covering yet on those studies mentioned above. Therefore this study focused on assessing the physical and mechanical properties of unhulled arabica coffee bean

2. Materials and methods

Unhulled arabica coffee bean samples collected from Cipunagara village (6°46'57.71 "S 107°41'37.15" E, 1153 MAMSL), Cisalak district, Subang regency, Indonesia.

2.1. Measurement of physical properties

Moisture content determined by using indirect moisture methods indirect moisture methods. The instrument used was Moisture Bulk Density Tester AMTAST JV-010S (P 30) with the specification of test range: 0 – 40 %, repeat error ≤ 0.2%, test error ≤ 0.5 %, environment temperature 0-40 °C, relative humidity ≤ 80 %, amount of sample 150 - 200 grams.

Size and dimension of the 50 samples of the unhulled arabica coffee bean measured by digital vernier caliper with the accuracy of 0.01 mm. Mass (M) of samples measured by using an analytical balance. Volume (V) of samples measured by using a graduated cylinder. The volume quantity determined by counting the number of displaced water. The geometric diameter (Dgm), sphericity (ψ), surface area (Sa), particle density (ρp), bulk density (ρb), porosity (ε), aspect ratio (Ra) and shell /silverskin (Cs) content calculated from equation (1), equation (2), equation (3), equation (4), equation (5), equation (6), equation (7), and equation (8) respectively.

$$D_{gm} = \sqrt[3]{w \times l \times t} \quad (1)$$

$$\psi = \frac{D_{gm}}{l} \quad (2)$$

$$A_s = \pi (D_{gm})^2 \quad (3)$$

$$\rho_p = \frac{M}{V} \text{ gr/cm}^3 \quad (4)$$

$$\rho_b = \frac{M_{v500}}{V_{500}} \text{ gr/cm}^3 \quad (5)$$

$$\varepsilon\,(\%) = \frac{(\rho_p - \rho_b)}{\rho_p} \times 100\% \quad (6)$$

$$R_a = \frac{W}{L} \times 100\,\% \tag{7}$$

$$C_s = \frac{W_{us} - W_s}{W_{us}} \times 100\,\% \tag{8}$$

The determination of geometric diameter (Dgm), sphericity (ψ), particle density (ρp), bulk density (ρb) referred to [24], [25], [26] and [27]. Sphericity and aspect ratio were calculated regarding [28]. Surface area calculated regarding [29], [30], [31] and [32]. Porosity and shell content calculated regarding [21].

2.2 Measurement of mechanical properties

Hardness and fracturability were measured using TA X plus Texture Analyser Stable Micro System. The operative condition of the TA was: test mode of compression, the test speed of 0.1 mm/sec, distance of 2.5 mm, data acquisition of 200 pps and probe of P/2. Samples randomly chose and positioned horizontally on the platform of the Texture Analyser instrument.

Samples randomly chose and positioned horizontally on the platform of the Instron testing machine (Instron, 5566, USA). Berries placed with its front side facing upward while the beans placed with their flat sides facing upward. Uniaxial compression conducted at a rate of 0.83 cm/s. The maximum force recorded during the test before the sample fracture taken as the fracture force of the sample. Deformation determined from the fracture force. Ten samples were recorded for their values [33]. There are two types of the angle of repose, i.e., emptying and filling angle of repose. The emptying angle of repose (θe) measured by an apparatus consisting of a sloping platform with adjustable angles. The individual of coffee bean sample was dropped from the distance of 10 cm above the device of the plane board to get a natural position; then the plane board was raised gradually until the coffee bean just started to slide down. The angle of repose determined by measuring the height of the seeds at two points on the sloping platform and the horizontal distance between the points as in equation (9) [34], [10]. There were four types of materials used in determining the angle of repose, i.e., stainless steel 1 mm, aluminum 1 mm, acrylic 3 mm and plywood 18 mm.

In determining the coefficient of friction, the sample filled into a PVC cylinder (diameter 10 cm, height 5 cm) which placed on the friction surface. The PVC cylinder was then lifted 3 mm from the friction surface so that the barrel did not touch the friction surface. The friction surface was elevated gradually until the barrel started to slide down. The angle (θ) between the lifted friction surface and the horizontal surface is known as the static angle of friction between the sample and the friction surface. The coefficient of static friction obtained from equation (10) [21] and [35]. There were four types of material surfaces used to determine the static friction, i.e., stainless steel 1 mm, aluminum 1 mm, acrylic 3 mm and plywood 18 mm.

$$\theta_s = \frac{\tan^{-1}(h_2 - h_1)}{x_2 - x_1} \tag{9}$$

$$\mu_s = \tan\theta \tag{10}$$

The filling angle of repose (θf) measured by PVC cylinder (10 cm x 10 cm) in which this cylinder filled with the sample, then the barrel was lifted gradually until the example formed a cone. The diameter and height of the cone were measured, and the angle of repose was calculated using the equation as follows [36] and [35] :

$$\theta_f = \tan^{-1}\left[\frac{2h}{D}\right] \tag{11}$$

3. Results and discussion

All the properties assessed at an average moisture content of 18.76 ± 0.19 % wb.

3.1. Physical properties

Table 1 showed the value of the sample size, minimum, maximum, mean and standard deviation of the measured physical properties. The measured physical properties comprised of length (L), width (W), thickness (T), mass (M), volume (V), geometric diameter (Dgm), surface area (As), particle density (ρp), bulk density (ρb), sphericity (ψ), porosity (ε), aspect ratio (Ra), shell content (Cs) and moisture content (Mc).

Table 1. Descriptive statistics of coffee bean sample dimensions

	N	Minimum	Maximum	Mean	Std. Deviation
L (mm)	50	9.51	14.54	12.0406	1.0623
W (mm)	50	7.45	9.37	8.2386	0.4473
T (mm)	50	4.21	5.84	4.9906	0.3357
M (g)	50	0.08	0.44	0.2076	0.0566
V (cm^3)	50	0.08	0.45	0.2084	0.0574
D_{gm} (mm)	50	7.08	8.74	7.8998	0.3844
A_s (mm^2)	50	157.2	239.8	196.437	19.092
ρ_p (gr/cm^3)	50	0.96	1.04	0.9968	0.0168
ρ_b (gr/cm^3)	50	0.38	0.42	0.3968	0.0098
Ψ (%)	50	0.58	0.75	0.6588	0.0394
ε (%)	50	58.20	62.37	60.1612	1.0874
R_a (%)	50	57.26	82.12	68.8758	6.3824
C_s (%)	50	4.72	50.06	16.5474	5.7012
M_c (%)	50	18.10	19.10	18.7580	0.1939

Sahin and Sumnu [37] stated that dimension and size were essential parameters in the screening as well as grading and evaluating the quality of the products. In the hulling mechanism, the size of the coffee bean has also been significant to set the steel huller component clearance.

Table 2. Correlation coefficients among measures of unhulled coffee bean samples

		L	W	T	M	V	Dgm	Sa	Dp	Db	Sph	E	Ra	Cs
L	Pearson Correlation	1												
	Sig. (2-tailed)													
W	Pearson Correlation	.213	1											
	Sig. (2-tailed)	.137												
T	Pearson Correlation	.202	.249	1										
	Sig. (2-tailed)	.159	.081											
M	Pearson Correlation	.050	.108	.162	1									
	Sig. (2-tailed)	.728	.457	.262										
V	Pearson Correlation	.053	.099	.178	.998**	1								
	Sig. (2-tailed)	.714	.494	.216	.000									
Dgm	Pearson Correlation	.773**	.606**	.677**	.146	.152	1							
	Sig. (2-tailed)	.000	.000	.000	.312	.293								
Sa	Pearson Correlation	.772**	.603**	.680**	.150	.156	.999**	1						
	Sig. (2-tailed)	.000	.000	.000	.299	.279	.000							
Dp	Pearson Correlation	-.197	.146	-.095	-.047	-.092	-.108	-.111	1					
	Sig. (2-tailed)	.171	.311	.513	.747	.525	.455	.444						
Db	Pearson Correlation	.189	.072	-.265	.152	.143	.013	.008	.073	1				
	Sig. (2-tailed)	.188	.618	.063	.292	.321	.929	.955	.615					
Sph	Pearson Correlation	-.853**	.171	.244	.015	.019	-.340*	-.337*	.194	-.275	1			
	Sig. (2-tailed)	.000	.236	.087	.917	.896	.016	.017	.176	.053				
E	Pearson Correlation	-.272	.071	.192	-.122	-.145	-.045	-.043	.560**	-.749**	.366**	1		
	Sig. (2-tailed)	.056	.625	.182	.401	.314	.759	.765	.000	.000	.009			
Ra	Pearson Correlation	-.820**	.375**	-.026	.000	-.005	-.371**	-.371**	.266	-.152	.921**	.306*	1	
	Sig. (2-tailed)	.000	.007	.858	1.000	.975	.008	.008	.061	.293	.000	.031		
Cs	Pearson Correlation	.109	-.080	.015	-.587**	-.581**	.043	.040	-.064	-.184	-.121	.075	-.146	1
	Sig. (2-tailed)	.452	.582	.920	.000	.000	.768	.784	.659	.202	.403	.604	.313	

Lang and Sokhansanj [38] reported that mass was an essential parameter in the calculation of densities and it could determine the quality of materials. Since coffee was processed using the dry method, the volume of coffee bean was essential to consider in drying simulation models where the changed of the amount will lead to a significant error in the modeling. The volume quantity also played a necessary role in the roasting process where roasting made the beans expanded [39].

Surface area and geometric diameter related to length, width and thickness played a significant role in the drying process and were also essential for the determination of terminal velocity, the coefficient of drag and Reynolds number [24].

Mohsenin NN. [24] stated that in a separation process such as hulling and in quality evaluation particle density played a significant role and bulk density data required in handling and packing process. It can observe in Table 1 that the sphericity of coffee bean was 0.66 ± 0.04%; it meant that most of the coffee bean samples had a nonspherical shape. The result of the observation found that only 12% of the coffee bean samples had a spherical shape. The coffee bean can be defined as a spherical if the sphericity value ≥ 0.70 [40].

Porosity is one of the most critical indicators of food microstructure and texture. Accurate prediction of porosity changes during drying had great importance for the optimization of the drying process and control of food quality [41].

Table 2 showed that volume had a relationship with mass. Geometric diameter had a relationship with length feature, width, and thickness feature. The surface area had a relationship with length, width, thickness, and geometric diameter. Sphericity had contact with length, geometric diameter, and surface area. Porosity had a relationship with particle density, bulk density, and sphericity.

Aspect ratio had a relationship with length feature, width, geometric diameter, surface area, and sphericity. Moreover, shell content had a relationship with mass and volume. To obtain the best statistical model for predicting the relationship between the measures of the sample the stepwise multiple regression analysis was applied. The results of stepwise multiple regression analysis among the standards of the example presented in Table 3.

Table 3. Regression equation among the physical measures of the unhulled coffee bean sample

Equation	R^2	SEE
$V = 1.013\,M - 0.002$	0.995	0.0039
$D_{gm} = 0.218\,L + 0.522\,T + 0.311\,W + 0.058$	0.999	0.0116
$S_a = 49.64\,D_{gm} - 195.74$	0.999	0.6544
$\Psi = 0.002\,S_a - 0.054\,L + 0.991$	0.984	0.0051
$\varepsilon = 39.93\,\rho_p - 88.27\,\rho_b + 55.39$	0.941	0.2700
$R_a = 107.38\,\Psi + 7.92\,W - 8.02\,D_{gm} - 3.76$	0.998	0.3091

3.2. *Mechanical properties*

Table 4. Descriptive statistics of hardness and fracturability

	N	Minimum (g-force)	Maximum (g-force)	Mean (g-force)	Std. Deviation (g-force)
Hardness, H	15	10.53	15.13	12.3419	1.27725
Fracturability, F_r	15	2.26	2.50	2.4517	0.07635

Table 4 shows the value of the samples size, minimum, maximum, mean and standard deviation of hardness and fracturability. The fracturability is the typical characteristic of the high degree of hardness with a low degree of cohesion materials [42]. Therefore the unhulled coffee bean can be defined as the product that had high hardness and low cohesiveness.

Fracture force was necessary for equipment design such as the crusher. The fracture force will determine the maximum strength which applied to a material without damaging the processed product.

Table 5 shows the value of emptying angle of repose on various surface materials and filling angle of repose. Results show that from 50 samples the emptying angle of repose on surface of stainless steel, aluminum, acrylic and plywood ranged from 26.88 ± 3.670, 29.44 ± 3.350, 30.01 ± 4.830 and 40.00 ± 7.000, with minimum value of 15.260, 19.200, 16.040 and 20.010 and with the maximum amount of 32.650, 35.380, 43.170 and 57.360. The average filling angle of reposed ranged from 32.02 ± 2.880 with a minimum value of 20.320 and with the maximum amount of 37.450.

Table 5. Descriptive statistics of emptying and filling angle of repose

	N	Minimum (°)	Maximum (°)	Mean (°)	Std. Deviation (°)
$EAOR_{ss}$	50	15.26	32.65	26.8838	3.6691
$EAOR_{al}$	50	19.20	35.38	29.4442	3.3508
$EAOR_{acry}$	50	16.04	43.17	30.0092	4.8323
$EAOR_{plyw}$	50	20.01	57.36	40.0024	6.9950
FAoR	50	20.32	37.45	32.0158	2.8793

The highest emptying angle of repose occurred on the surface of plywood inversely of that was happening on the surface of stainless steel. In general, both emptying and filling angle of repose can determine the behavior of materials flow [43]. Therefore, the angle of repose has also used in equipment design and storage structures [24].

Table 6 shows that average static friction of stainless steel, aluminum, acrylic and plywood ranged from 0.36 ± 0.02, 0.40 ± 0.03, 0.38 ± 0.04 and 0.45 ±0.03 with minimum value of 0.32, 0.35, 0.33 and 0.40 and with the maximum value of 0.40, 0.45, 0.52 and 0.52 respectively.

The highest static friction occurred on the surface of plywood; the opposite happened on that of stainless steel. Static friction showed a positive correlation with the emptying angle of repose. The higher emptying angle of repose the higher static friction will occur. The coefficient of friction used in designing agricultural equipment, bins, silos, and other storage and handling structures [24].

Table 6. Descriptive statistics of static friction on various surfaces

	N	Minimum	Maximum	Mean	Std. Deviation
$\mu_{statis}SS$	20	0.32	0.40	0.3565	0.0211
$\mu_{statis}AL$	20	0.35	0.45	0.3990	0.0255
$\mu_{statis}ACRY$	20	0.33	0.52	0.3765	0.0404
$\mu_{statis}PLYW$	20	0.40	0.52	0.4505	0.0347

3.3. Correlation between physical and mechanical properties

Best relationship between them obtained by using the stepwise multiple regression analysis. In general emptying and filling angle of repose, tangentially were dependent much on the length, width and aspect ratio. In more detail, the relationship between the emptying and filling angle of repose was presented in Table 7.

Table 7. The regression equation of the relationship between an emptying angle of repose and the physical measures

Equation	R^2	SEE
$\theta_f = (3.14\ L - 0.175\ W + 0.005\ R_a - 0.409\ \varepsilon + 0.10\ T)\ tg\theta_f + 16.707$	0.995	0.2124
$\theta_{e\text{-}ss} = (3.919\ L - 0.045\ W + 0.002\ R_a - 0.008\ S_a + 0.253\ T)\ tg\ \theta_{e\text{-}ss} + 2.268$	0.988	0.4207
$\theta_{e\text{-}Al} = (3.605\ L - 0.117\ W + 0.003\ R_a - 0.163\ \varepsilon + 17.685\ \psi)\ tg\ \theta_{e\text{-}al} + 4.785$	0.991	0.3426
$\theta_{e\text{-}Acry} = (3.413\ L - 0.134\ W + 0.003\ R_a)\ tg\ \theta_{e\text{-}Acry} + 8.562$	0.991	0.4765
$\theta_{e\text{-}Plyw} = (2.567\ L - 0.135\ W + 0.003\ R_a) tg\ \theta_{e\text{-}Plyw} + 12.919$	0.962	1.4052

Regarding the correlation between the physical properties and mechanical properties which consisted of hardness and fracturability, results of multiple regression analysis showed that there was not found the relationship between both of them.

4. Conclusion

The results of this work found that the analyzed unhulled coffee bean, physically had an average length, width, thickness, geometric diameter, and surface area ranged from 12.04 ± 1.06 mm, 8.24 ± 0.45 mm, 4.99± 0.34 mm, 7.90 ± 0.38 mm and 196.44 ± 19.09 mm^2 respectively. The average mass and volume ranged from 0.21 ± 0.06 grams and 0.21 ± 0.06 cm^3 respectively. The average particle density, bulk density and sphericity ranged from 1.00 ± 0.02 gr/cm3, 0.40 ± 0.01 gr/cm3 and 0.66 ± 0.04% respectively. The average porosity, aspect ratio, and shell content ranged from 60.16 ± 1.09%, 68.88 ± 6.38% and 16.55 ± 5.70% respectively.

Concerning the relationship between the physical properties of unhulled coffee bean analyzed, stepwise multiple regression analysis indicated that the volume was dependent on mass. The geometric diameter was dependent on length, thickness and width, the surface area, was dependent on geometric diameter, sphericity was dependent on surface area and length, porosity was dependent on the particle, and bulk densities and aspect ratio was dependent on sphericity, width, and geometric diameter.

Results of mechanical properties measurement indicated that the average hardness and fracturability ranged from 12.34 ±1.28 g-force and 2.45±0.08 g-force respectively. Based on the value of hardness and fracturability, the unhulled coffee bean samples defined as a product that had high hardness and low cohesiveness. Regarding the angle of repose on the surface of stainless steel, aluminum, acrylic, and plywood results showed that the average emptying angle of repose ranged from 26.88±3.670°, 29.44±3.350°, 30.01±4.830° and 40.00±7.000° respectively, and the average filling angle of repose ranged from 32.02 ±2.880. Concerning friction, results of measurement indicated that average static friction of stainless steel, aluminum, acrylic, and plywood ranged from 0.36 ± 0.02, 0.40 ±0.03, 0.38±0.04 and 0.45±0.03 respectively. The highest emptying angle of repose occurred on the surface of plywood inversely of that happened on the surface of stainless steel. There was a positive association between the quantity of repose angle and the static friction. The bigger the angle of repose was the higher static friction occurred. There was not found any relationship between the physical properties and hardness and fracturability of the unhulled coffee bean.

5. Acknowledgment

This work is part of community empowerment concerning improving the quality of community coffee processing in Alor district, National Priority Program conducted at the Center for Appropriate Technology Development - Indonesian Institute of Sciences - Indonesian Institute of Sciences. We thank the members of the team responsible for this activity on their cooperation. We would like to extend my most profound appreciation to all who gave us the possibility to complete this report. And we would also like to thank Teguh Santoso, Maulana Furqon, Taufik Yudhi, and S. Khudaifanny for their help in carrying out this work.

Appendices

Dp	:	Polar diameter, mm	*Ra*	:	Aspect ratio, %
De	:	Equatorial diameter, mm	*w*	:	Width, mm
Dgm	:	Geometric diameter, mm	*l*	:	Length, mm
As	:	Surface area, mm^2	*t*	:	Tick, mm
Π	:	3,14	θ_e	:	Emptying angle of repose (°)
M	:	Mass, gr	θf	:	Filling angle of repose (°)
V	:	Volume, ml	h_1	:	High 2, mm
ρp	:	Particle density, gr, ml	h_2	:	High 2, mm
ρb	:	Bulk density, gr/ml	x_1	:	Length 1, mm
ε	:	Porosity, %	x_2	:	Length 2, mm
ψ	:	Sphericity, %	μ_s	:	Coeficient static friction

6. References

[1] United States Department of Agriculture 2016 *Indonesia: Coffee Annual*

[2] Food and Agriculture Organization 2016 *United Nations*
[3] Association of Indonesian Coffee Exporters 2005 *Statistik Kopi 1980-2005*
[4] Karaj S, Huaitalla R M and Müller J 2008 Physical, mechanical and chemical properties of Jatropha curcas L. seeds and kernels *Proceedings of the Conference on International Agricultural Research for Development. Stuttgrant, Germany* 7-9
[5] Dursun I, Tugrul K M, and Dursun E 2007 Some physical properties of sugarbeet seed *J. Stored Prod. Res.* **43** 2 149–155
[6] Yalcin H and Kavuncuoglu H 2014 Physical, chemical and bioactive properties of onion (Allium cepa L.) seed and seed oil *J. Appl. Bot. Food Qual.* **87** 87–92
[7] Ghaffari H, Marghoub N, Sheikhdarabadi M S, Hakimi A, and Abbasi F 2013 Physical properties of three Iranian onion varieties *Int. Res. J. Appl. Basic Sci.* **7** 9 pp. 587–593
[8] Eissa A H and Gamea G R 2003 Physical and mechanical properties of bulb onions. *Miser Journal of Agricultural Engineering* **20** (3) 661-676
[9] Rathinakumari A C and Jesudas D M 2015 Physical and mechanical properties of onion sets (Allium cepa, L.) *Int. J. Trop. Agric.* **33** 2 817–823
[10] Sunitha B, Sujatha G, Girish B, Kiran B R, and Ramana MV 2016 Determination of physical properties of onions (Arka kirthiman) *Internat. J. Proc. Post Harvest Technol* **7** 1 36–41
[11] Shoba H, Rajeshwari N, and Nagaraja G 2017 A Study on Physico-Mechanical Properties of Onion Varieties Under Koppal District, (Karnataka) Article History *Curr. Agric. Res. J.* **5** 381–386
[12] Haciseferoğullari H, Özcan M, Sonmete M H, and Özbek O 2005 Some physical and chemical parameters of wild medlar *(Mespilus germanica L.)* fruit grown in Turkey *J. Food Eng.*, **69** 1–7
[13] Vursavuş K, Kelebek H, and Selli S 2006 A study on some chemical and physical-mechanic properties of three sweet cherry varieties *(Prunus avium L.)* in Turkey *J. Food Eng.* **74** 568–575
[14] Celik A, Ercisli S, and Turgut N 2007 Some physical, pomological and nutritional properties of kiwifruit cv. Hayward *Int. J. Food Sci. Nutr.* **58** 6 411–418
[15] Akbolat D, Ertekin C, Menges H O, Ekinci K, and Erdal I 2008 Physical and nutritional properties of jujube *(Zizyphus jujuba Mill.)* growing in Turkey *Asian J. Chem.* **20** 1 757–766
[16] Dash A K, Pradhan R C, Das L M, and Naik S N 2008 Some physical properties of simarouba fruit and kernel Pressurized liquid extraction of bioactives from green tea and seabuckthorn for food applications View project Utilization of gaseous fuels in IC engines View project," *Int. Agrophysics* **22** 111–116
[17] Jahromi M K, Rafiee S, Jafari A, Bousejin M R G, Mirasheh R, and Mohtasebi S S 2008 Some physical properties of date fruit (cv. Dairi)," *J. Agrophysics* **22** 221–224
[18] Ersoy N and Bagci Y 2011 Altın Çilek *(Physalis peruviana L.)*, Pepino (Solanum muricatum Ait.) ve Passiflora (Passiflora edulis Sims) Tropikal Meyvelerinin Bazı Fizikokimyasal Özellikleri ve Antioksidan Aktiviteleri *J. Agric. Food Sci.* **25** 3 67–72
[19] aliliantabar F J, Lorestani A N, and Gholami R 2013 Physical properties of kumquat fruit *Int. Agrophysics*
[20] Niemenak N, Eyamo J A, Onomo P E, and Youmbi E 2014 Physical and chemical assessment quality of cocoa beans in south and center regions of Cameroon *Agric. Food Sci.* **5** 27–33,
[21] Bart-Plange A and Baryeh E A 2003 The physical properties of Category B cocoa beans *J. Food Eng.* **60** 219–227
[22] Bart-Plange A, Dzisi A K, Addo A, Teye E, and Kumi F 2012 A Comparative Study of Some Physical Properties of Large and Medium Size Cocoa Beans from Ghana *ARPN J. Sci. Technol.* **2** 3 135–141
[23] Ha L T V, Bangun N, Philip R A, Anh P H, Messens K, and Dewettinck K 2016 Physico-chemical properties of fourteen popular cocoa bean varieties in Dongnai – highland Vietnam *Can Tho Univ. J. Sci.* **4** 81–86

[24] Mohsenin N N 1986 *Physical properties of plant and animal materials: Structure, Physical Characteristics, and Mechanical Properties* New York: Gordon and Breach Science Publishers

[25] Bahnasawy A H, El-Haddad Z A, El-Ansary M Y, and Sorour H M 2004 Physical and mechanical properties of some Egyptian onion cultivars *J. Food Eng.* **62** 255–261

[26] Ismail I, Anuar M S, and Shamsudin R 2014 Physical properties of liberica coffee (Coffea liberica) berries and beans *Pertanika J. Sci. Technol* **22** 1 65–79

[27] Kaveri G and Thirupathi V 2015 Studies on geometrical and physical properties of CO 4 onion bulb (Allium cepa lvar. aggregatum don.) *Int. J. Rec. Sci. Res* **6** 2897-2902.

[28] Loghavi M, Souri S, and Khorsandi F 2011 Physical and Mechanical Properties of Almond (Prunus dulcis) in *Annual International Meeting ASABE*

[29] McCabe W L, Smith J C, and Harriott P 1986 *Unit operations of chemical engineering*, 4th Edition (New York: McGraw-Hill)

[30] Sacilik K, Öztürk R, and Keskin R 2003 Some Physical Properties of Hemp Seed *Biosyst. Eng.* **86** 2 191–198

[31] Tunde-Akintunde T Y and Akintunde B O 2004 Some physical properties of sesame seed *Biosyst. Eng.* **88** 1 127–129

[32] Altuntaş E, Özgöz E, and Taşer Ö F 2005 Some physical properties of fenugreek (Trigonella foenum-graceum L.) seeds *J. Food Eng.* **71** 1 37–43

[33] Pittia P, Dalla Rosa M, and Lerici C R 2001 Textural Changes of Coffee Beans as Affected by Roasting Conditions *LWT - Food Sci. Technol.* **34** 168–175

[34] Chandrasekar V and Viswanathan R 1999 Physical and Thermal Properties of Coffee *J. Agric. Eng. Res.* **73** 3 227–234

[35] Pradhan R C, Naik S N, Bhatnagar N, and Vijay V K 2009 Moisture-dependent physical properties of jatropha fruit," *Ind. Crops Prod.* **29** 341–347

[36] Razavi S M A, Amini M A, Rafe A, and Emadzadeh B 2007 The physical properties of pistachio nut and its kernel as a function of moisture content and variety: Part III: Frictional properties *J. Food Eng.* **81** 1 226–235

[37] Sahin S and Sumnu S G 2006 *Physical Properties of Foods* (New York: Springer Science+Business Media)

[38] Lang W and Sokhansanj S 1993 Bulk volume shrinkage during drying of wheat and canola *J. Food Process Eng.* **16** 4 305–314

[39] Franca A S, Mendonça J C F, and Oliveira S D 2005 Composition of green and roasted coffees of different cup qualities *LWT - Food Sci. Technol.* **38** 709–715

[40] Eke C N U, Asoegwu S N, and Nwandikom G I 2007 Some Physical Properties of Jackbean Seed (Canavalia ensiformis) modeling the energy requirement for cutting selected fruits and vegetables View project Biogas Technology View project Some Physical Properties of Jackbean Seed (Canavalia ensiformis) *CIGR Ejournal Manuscr.* **9** 1–11

[41] Wang D 2016 *Prediction of Texture Characteristics in Apple Drying Using Computer Vision* Dalhousie University (Doctoral dissertation)

[42] Szczesniak A S 2002 Texture is a sensory property *Food Qual. Prefer.* **13** 215–225

[43] Ileleji K E and Zhou B 2008 The angle of repose of bulk corn stover particles *Powder Technol.* **187** 110–118

Statistical Analysis on The Geometric, Physical and Mechanical Properties of Dried Robusta Coffee Cherry Resulting From Natural System Processing

R Saparita[1], D D Hidajat[1], S I Kuala[1]

[1]Development Center For Applied Technology, Indonesian Institute of Sciences

Email: rsaparita@gmail.com; zehidayat@gmail.com; seri.kuala.sk@gmail.com

Abstract. The study conducted to evaluate the in-between correlation parameters of size, shape, physical and mechanical properties to provide information of physical and mechanical properties of the dried Robusta coffee cherry in relation with the development of coffee processing equipment. The physical and mechanical properties of the dried cherry were measured. The relationship in-between the physical and mechanical properties was analyzed using the stepwise regression analysis. The results of the study showed that polar diameter, equatorial diameter, thickness, geometrical diameter, arithmetic diameter, and surface area were dependent on volume with the correlation coefficient of 0.738, 0,761, 0.775, 0.994, 0.999 and 0.999 respectively. The sphericity was dependent on bulk density with a correlation coefficient of 0.437. The shape index, cross-sectional area and frontal surface area were dependent on particle density with the correlation coefficient of 0.959, 1.00 and 0.967 respectively. The correlation coefficient of the relationship between static friction and angle of repose on acrylic, stainless steel, and plywood were 0.772, 0.700, 1.000 and 0.930 respectively; otherwise, there was no correlation between the static friction and the filling angle of repose. The hardness of the dried cherry had a weak correlation with the fracturability.

Keywords: coffee, natural, physical, mechanical, properties

1. Introduction

In Indonesia, coffee was one of the commodities that played an essential role in supporting efforts to increase non-oil exports. Coffee contributed around 10.1 percent of the entire export value of agricultural products, or 0.5 percent of non-oil exports, or 0.4 percent of the total export value [1]. Besides that, coffee was also a source of employment and income sources for farmers or other economic actors involved in both on-farm and off-farm activities.

The coffee development area reached 1.2 million hectares which consisted of 958 thousand hectares of Robusta coffee (77.77%) and 251 thousand hectares of Arabica coffee (22.23%). From that area, the coffee plantation was almost 96.15% cultivated by smallholders, involving around 1.9 million households, 1.99% grown by large private estates, and the rest of 1.82% improved by government estate.

Concerning the process of coffee cherry processing, farmers generally applied dry system processing, because it was relatively more straightforward than that of another process dry coffee bean processing also was known as a natural process. This process included the oldest technique available in the history of coffee processing. In the natural process of coffee, cherries harvested will then be spread over the surface of the plastic bases and dried in the sun. Some coffee producers sometimes left out in the brick terrace or tables that had an airflow on the bottom. When dried in the sun, these coffee beans should be turned back periodically so that the coffee beans drained evenly, and avoided decay. In the natural process, dried coffee fruit is still in the form of cherry, complete with all of the layers. The dried coffee cherry produced from this natural process then processed into the green bean and ready for roasting.

Knowledge of size, shape, physical and mechanical properties of coffee are essential in processing, designing, and manufacturing of equipment. Many researchers carried out a study on the physical and mechanical properties, and the correlation between the particular features of agricultural products among others was: Bart P. et al.[2], Nicolas N. et al.[3], and Ha et al.[4] studied physical and mechanical properties of cocoa bean; Karaj et al.[5] carried research on physical, mechanical and chemical properties of *Jatropha curcas*; Keramat et al.[6] studied on physical properties of date fruit; Jaliliantabar et al.[7] studied on physical properties of Kumquat; Zare, D. et al. [8] studied on some physical and mechanical properties of Russian olive fruit; Loghavi, M. et al.[9] studied physical and mechanical properties of Almond; Tunde-Akintunde, T.Y. et al.[10] studied some physical properties of Sesame Seed; Altuntas, E. et al.[11] studied some physical properties of Fenugreek; Celik A et al.[12] studied some physical, pomological and nutritional properties of Kiwifruit; Dursun, I. et al.[13] studied some physical properties of sugarbeet seed; Eke, C. N. U., et al. [14] studied some physical properties of Jackbean seed, and Sessiz, A. et al.[15], Amer E. A.H., et al.[16], Yalcin, H. [17], and Kaveri, G., et al.,[18] studied physical and mechanical properties of Onion.

Some researchers who had researched the physical and mechanical properties of coffee bean were Iván Darío, I.A.T., et al.[19], Couto, S.M., et al.[20], Marija R. J., et al.[21], and Coradi, P.C. et al.[22]. Most of the coffee bean studies related to fresh cherry, hulled and unhulled coffee bean. The reviews on dried coffee cherry resulting from the natural process were relatively rare.

According to the previous study, physical and mechanical properties were useful parameters for the development of the design and manufacture of coffee cherry processing equipment, i.e., shape and size were essential parameters in the screening as well as grading and evaluating the quality of the products. In the hulling mechanism, the size of the dried cherry resulting from the natural system was essential to set the steel huller component clearance [23]. The Surface area and geometric diameter played a significant role in the drying process and were also essential for the determination of terminal velocity, the coefficient of drag and Reynolds number [24]. The mass was a necessary parameter in the calculation of densities, and it could determine the quality of materials; the volume also played a role required in the roasting process where roasting made the beans expanded [25]. The hardness was essential in evaluating their feeding value as well as their size reduction and milling characteristics, and Fracture force was also necessary for equipment design such as the crusher. Both emptying and filling angle of repose can determine the behavior of materials flow [26], the angle of repose has also used in equipment design and storage structures [24].

The Study aimed to analyze the physical and mechanical properties, to determine the relationship in-between the properties component of the dried cherry.

2. Materials and methods

2.1. Material Preparation

The fresh Robusta cherry sample collected from Cupunagara village (Latitude 6°46'57.71 "S, Longitude 107°41'37.15" E, and elevation 1153 MAMSL), Cisalak sub-district, Subang district, West Java province. The sample then dried using electrical drier at temperature 50 °C for 72 hours.

The size features in this study consisted of polar diameter (Dp), equatorial diameter (De), thickness (t), arithmetic (Dam), and geometric diameter (Dgm). The shape consisted of shape index (SI) and sphericity (ψ). The area features consisted of a cross-sectional of t (Acs), frontal surface area (Afs), and total surface area (As). The physical properties included mass (m), volume (v), particle (ρp) and bulk (ρb) densities. The mechanical properties consisted of emptying (θe) and filling (θf) angles of repose, friction (μ), Hardness (H) and fracturablity (frac).

Instruments used for measuring the sample consisted of digital vernier caliper with an accuracy 0.01 mm, an electronic balance with an accuracy 0.1 gram, an analytical balance with an accuracy 0.01 mg, an apparatus to determine angle of repose and static friction, a texture analyser, a beaker glass and a graduated cylinder and high-quality colorimeter NH 310 and cabinet electric drier.

The statistical package programmes, i.e., descriptive statistics, One-Sample Kolmogorov-Smirnov Test, matrix correlation, stepwise regression analysis, and pair sample test performed to assess the characteristic and relationship among the size, shape, area, physical and mechanical properties.

2.2. Moisture content

The water content was calculated by the following equation [27]:

$$M_{wb} = \frac{W_1 - W_2}{W_1} \times 100\% \qquad (1)$$

Where: M_{wb}: Moisture content, % wet basis; W_1: The initial weight of the sample, gram
W_2: The final weight of the sample, gram

2.3. Geometric

2.3.1. Size

Size comprised of polar diameter, equatorial diameter, thickness, arithmetic diameter and geometric diameter, polar and equatorial diameter, and thick of samples measured by using a digital vernier caliper with an accuracy of 0.01 mm. There were many researchers had measured these dimensions of agricultural products similarly, among them were Shoba H., et al.[28], Mukesh D et al.[29], Kaveri[18], Vijaya&Srivastava[30], and Bahnasawy et al. [31], had measured these dimensions for onion bulbs, Karaj et al.[5] had measured these dimensions for Jatropha curcas, Dursun I., et al.[13] had measured these dimensions for sugar beet seed, Despande S.D. et al. [32]had mastered these dimensions for soybean, Mohsenin [24] and Shepherd &Bhardwaj [33] had measured these dimensions for other grains and seeds.

Geometric diameter is the cube root of the product of linear dimensions of polar diameter, equatorial diameter, and thickness. Referring to Mohsenin[24], Bahnasawy et al., [31] and Kaveri et al.[18], The geometric mean diameter determined using equation(2). The arithmetic diameter is the sum of all the three linear dimensions, i.e., polar diameter, equatorial diameter and thickness of the sample divided by the total number of linear dimensions [31]. The arithmetic diameter calculated using equation (3).

$$D_{gm} = \sqrt[3]{(D_p \times D_e \times t)} \qquad (2)$$

$$D_{am} = \frac{D_e + D_p + t}{3} \qquad (3)$$

Where : D_{gm}: Geometric diameter, mm; D_{am}: Arithmetic diameter, mm; D_p: Polar diameter, mm; D_e: Equatorial, mm; t: Thickness, mm

2.3.2. Shape

Shape and size were essential parameters in the screening as well as grading and evaluating the quality of the products. In the hulling mechanism, the size of the dried cherry resulting from the natural system was essential to set the steel huller component clearance [23]. The shape which comprised of shape index and sphericity is an essential parameter in grading to develop the quality of the product such as coffee cherry, or coffee bean Shape consisted of the shape index and sphericity. The shape index is the ratio of the equatorial diameter and root square of the product of polar and thickness of samples[31]. The shape index used to evaluate the shape of the dried cherry and calculated according to equation (4).

$$SI = \frac{D_e}{\sqrt[2]{D_p \times t}} \qquad (4)$$

Where :SI: Shape index, Dp; Polar diameter, mm; De: Equatorial, mm; t: Thickness, mm

The sphericity of the dried cherry is the ratio of the root of the product of equatorial diameter, polar diameter, and thickness of dried cherry and its equatorial diameter[9]. The sphericity calculated using equation (5) as follows :

$$\psi = \frac{\sqrt[3]{D_e \times D_p \times t}}{D_e} \qquad (5)$$

Where : ψ: Shape index; Dp: Polar diameter, mm; De: Equatorial, mm; t: Thickness, mm

2.3.3. Area of dried Robusta cherry

Surface area and geometric diameter played a significant role in the drying process and were also essential for the determination of terminal velocity, the coefficient of drag and Reynolds [24]. The area of the dried Robusta cherry determined through the frontal surface, cross-section and total surface areas of the dried Robusta cherry[31]. The frontal surface area is the representation of a solid object appeared if cut by a plane, while the cross-sectional area is the area of the section made by a plane cutting an object transversely to the longest axis.

The frontal surface area (Afs), cross-sectional area (ACS) and total surface area (As) calculated using equation (6), (7), and(8) respectively.

$$A_{fs} = \frac{\pi}{4} D_e D_p \qquad (6)$$

$$A_{cs} = \frac{\pi}{4} \text{ x } \frac{(D_e + D_p + t)^2}{9} \qquad (7)$$

$$A_s = \pi (D_{gm})^2 \qquad (8)$$

Where : Afs: Frontal surface area,mm^2; Acs: Cross-sectional area, mm^2; As: Total surface area, mm^2; D_e: Equatorial diameter, mm; D_p: Polar diameter, mm; t: Thickness, mm; D_{gm}: Geometric mean diameter, mm; π: Constant, 3,14

2.4. Physical properties

The physical properties included four features viz. mass, volume, particle and bulk densities[31]. The mass was an essential parameter in the calculation of densities, and it could determine the quality of materials. The volume also played a necessary role in the roasting process where roasting made the beans expanded [25]. The weight (m) of dried cherry samples measured by using an electronic balance with an accuracy of 0.01 gram. The volume (v) of dried cherry samples determined by the water displacement method. Particle density is the ratio of mass in grams and the amount of displacement water in cm^3. The particle density calculated according to equation (9). The bulk density is the proportion of weight in a specific volume, calculated based on formula (10).

$$\rho_p = \frac{m}{v} \text{gr/cm}^3 \qquad (9)$$

$$\rho_b = \frac{m_{v500}}{V_{500}} \text{gr/cm}^3 \qquad (10)$$

Where: ρ_p: Particle density, gr/cm^3, ρ_p: Bulk density, gr/cm^3, m: Mass, gram, v: Volume, cm^3, m_{500}: Mass of 500 cm^3 sample, gram, v_{500}: The Volume of 500 cm^3

2.5. Mechanical properties

The mechanical properties measured comprised of five features viz. hardness, fracturability, emptying angle of repose, filling angle of repose and friction. Hardness value is the peak force that occurred during the first compression; therefore the hardness of agricultural product such as cherry or bean coffee is essential in evaluating their feeding value as well as their size reduction and milling characteristics. Fracturability is the force at the first high spot which will determine the maximum strength which applied to a material without damaging the product. Fracture force is also necessary for equipment design such as the crusher. Hardness and fracturability measured using TA XT plus Texture Analyser Stable Micro System. In general, both emptying and filling angle of repose can determine the behavior of materials flow [26]. Therefore, the angle of repose has also used in equipment design and storage structures [24]. The emptying angle of repose (θe) measured by an apparatus consisting of a sloping platform with adjustable angles[34]. The filling angle of repose (θf) measured by PVC cylinder (10 cm x 10 cm) in which this cylinder filled with the cherry sample, and then the barrel was lifted gradually until the cherry formed a cone[35] and Pradhan, R.C. et al.[36].

Practically coefficient of friction used in designing agricultural equipment, bins, silos, and other storage and handling structures [24]. The coefficient of friction measured by using a PVC cylinder (diameter 10 cm, height 5 cm) which placed on the friction surface. The PVC cylinder filled with the cherry sample then lifted around 3 mm from the friction surface so that the barrel did not touch the friction surface. The friction surface lifted gradually until the cylinder started to slide down. The angle (θ) between the elevated friction surface and the horizontal surface is known as the angle of static friction between the sample and the friction surface[2],[36]. Measurement of emptying and filling angle of repose and friction performed on various surfaces of materials, i.e., stainless steel 1 mm, aluminum 1 mm, acrylic 3 mm, and plywood 18 mm. The emptying angle of repose, filling angle of repose and friction calculated according to equation (11), (12) and (13) respectively.

$$\theta_e = \frac{tan^{-1}(h_2 - h_1)}{x_2} \quad (11)$$

$$\theta_f = \tan^{-1}\left[\frac{2h}{D}\right] \quad (12)$$

$$\mu_s = \tan\theta \quad (13)$$

Where: θ_e: Emptying angle of repose; θ_f: Filling angle of repose; h_1: The Initial height of the platform; h_2: The Final height of the platform; h: Height of cone; x_1: Distance of h1; x_2: Distance of h2; D: The diameter of the cone; μ_s: Static Friction

3. Results And Discussion

All the properties of the dried robusta coffee cherry assessed at an average moisture content of 14.80±2.29 %. Results of measurements of size, shape, area and physical properties presented in Table 1.

Table 1. Descriptive Statistics of size, shape, area and physical properties

Parameters	Minimum	Maximum	Mean	Std. Deviation
Dp	11.10	15.15	12.87	0.96
De	9.92	12.41	11.04	0.65
t	8.81	11.46	9.75	0.60
Dam	10.36	12.57	11.22	0.56
Dgm	10.35	12.50	11.14	0.55
ψ	0.77	0.93	0.87	0.04
shape	0.52	1.07	0.78	0.15
Acs	45.02	91.99	67.92	12.95
Afs	39.10	79.24	58.98	11.89
As	336.34	490.57	390.47	39.00
m	0.38	0.56	0.48	0.05
v	0.58	1.02	0.73	0.11
ρp	0.45	0.92	0.68	0.13
ρb	0.42	0.48	0.45	0.016

The One-Sample Kolmogorov-Smirnov test of the geometric, physic and mechanic of the sample indicated that the data frequently scattered. The normal distribution of geometrical, physical and mechanical properties indicated by the value of asymptotic significance 2-tailed more than 0.05 ($p>0.05$). The result of matrix correlation analysis of the sample showed that polar diameter, equatorial diameter, thickness, arithmetic diameter, geometric diameter, and total surface area were dependent on volume. The cross-section area, frontal surface area, and shape index were dependent on particle density. The sphericity was dependent on bulk density. The relationship between the geometric and physical properties of the sample in detail presented as follows :

$$D_p = 6.931\,v + 8.215 \quad R= 0.738 \quad SEE= 0.6576$$
$$D_e = 4.497\,v + 7.762 \quad R=0.761 \quad SEE= 0.4313$$
$$t = 4.196\,v + 6.690 \quad R=0.775 \quad SEE= 0.3849$$
$$D_{am} = 5.025\,v + 7.558 \quad R=0.994 \quad SEE= 0.0639$$
$$D_{gm} = 4.969\,v + 7.518 \quad R=0.999 \quad SEE= 0.0306$$
$$\Psi = 1.153\rho_b + 0.350 \quad R=0.437 \quad SEE= 0.0378$$
$$S_i = 1{,}124\rho_p + 0.021 \quad R=0.959 \quad SEE= 0.4397$$
$$A_{cs} = 99.374\,\rho_p + 0.477 \quad R=1.000 \quad SEE= 0.2852$$
$$A_{fs} = 88.230\,\rho_p - 0.903 \quad R=0.967 \quad SEE= 3.0929$$
$$A_s = 352.666\,v + 133.609 \quad R=0.999 \quad SEE= 1.3856$$

Table 2 showed the texture profile, i.e., hardness and fracturability of the sample. The result of the matrix correlation showed that there was a weak relationship between the hardness and fracturabily.

Table 2. Descriptive Statistics of hardness and fracturability

	Minimum	Maximum	Mean	Std. Deviation
H	2.41	5.54	3.88	0.98
F_{rac}	0.76	2.00	1.23	0.44

The relationship between the hardness and fracturability was as follow :

$$H = 0.208\,F_{rac} + 3.623 \quad R = 0.094 \quad SEE= 1.0362$$

Table 3 showed the value of emptying and filling repose of angle, and the static friction of the sample on various surfaces. The emptying angle of repose on multiple surfaces showed a similar pattern with filling angle of repose and static friction, the more significant of the emptying angle of repose the more considerable filling angle of repose and static friction.

Table 3. Descriptive Statistics of Emptying Angle of Repose and Static Friction on various surfaces

	Minimum	Maximum	Mean	Std. Deviation
$\theta_{e\text{-}ACRY}$	15.10	18.41	16.64	0.83
$\theta_{e\text{-}SS}$	13.86	19.52	17.03	1.43
$\theta_{e\text{-}AL}$	14.55	22.76	18.46	2.26
$\theta_{e\text{-}PLYW}$	16.83	20.97	18.76	1.23
μ_{ACRY}	0.22	0.39	0.28	0.04
μ_{SS}	0.21	0.33	0.28	0.03
μ_{AL}	0.25	0.38	0.31	0.04
μ_{PLYW}	0.29	0.47	0.36	0.05
$\theta_{f\text{-}ACRY}$	12.52	19.15	15.83	1.67
$\theta_{f\text{-}SS}$	13.09	26.81	17.89	3.03
$\theta_{f\text{-}AL}$	15.05	22.72	19.24	1.65
$\theta_{f\text{-}PLYW}$	12.32	26.48	19.91	3.15

The biggest emptying angle of repose, filling angle of repose and static friction occurred on the surface plywood followed by aluminum, stainless steel, and acrylic. Results of matrix correlation analysis showed that there was a relationship between static friction of the sample on a certain surface and its emptying angle of repose; otherwise, there was no correlation between that and the filling angle of repose. The regression equations between the static friction and emptying angle of repose of each surface were as follows:

$$\mu_{ACRY} = 0.034\,\theta_{e\text{-}ACRY} - 0.280 \quad R = 0.772 \quad SEE = 0.0233$$
$$\mu_{SS} = 0.016\,\theta_{e\text{-}SS} + 0.016 \quad R= 0.700 \quad SEE = 0.0232$$
$$\mu_{AL} = 0.016\,\theta_{e\text{-}AL} + 0.017 \quad R = 1.000 \quad SEE = 0.0003$$
$$\mu_{PLYW} = 0.035\,\theta_{e\text{-}PLYW} - 0.035 \quad R= 0.930 \quad SEE = 0.0175$$

4. Conclusion

The results of this work found that there was a robust correlation between polar diameter, equatorial diameter, thickness, geometrical diameter, arithmetic diameter surface area and the volume with a correlation coefficient of 0.738, 0,761, 0.775, 0.994, 0.999 and 0.999 respectively. The strong correlation also occurred in the relationship between the sphericity, shape index, cross-sectional area, frontal surface area and the particle density with the correlation coefficient of 0.959, 1.00 and 0.967 respectively. The sphericity had a moderate correlation with bulk density; its correlation coefficient was 0.437. The hardness of the sample showed a weak correlation with the fracturability; its correlation coefficient was 0.094. The static friction had a strong association with the emptying angle of repose but not with filling angle of repose. The determination value of the relationship between static friction and emptying angle of repose of acrylic, stainless steel, aluminum, and plywood were 0.772, 0.700, 1.000 and 0.930 respectively.

5. Acknowledgment

This study was a part of coffee -based agro-industry development activities in Belu and Alor districts, National Priority Programme undertaken at Development Center For Appropriate Technology – Indonesian Institute of Sciences. We would like to appreciate all those who offered us the possibility to complete this report. We also would like to thank Teguh, S., Dadang, G., Iman R., Maulana F., Neneng, K., Sutrisno, and Sukwati for their help in carrying out this study.

6. References

[1] Association of Indonesian Coffee Exporters 2005 *Statistic of Coffee 1980-2005* Jakarta Association of Indonesian Coffee Exporters

[2] Bart-Plange A and Baryeh E A 2003 The physical properties of category B cocoa beans *Journal of Food Engineering* **60** 219-227

[3] Niemenak N, Eyamo J, Al Onomo P E and Youmbi E 2014 Physical and chemical assessment quality of cocoa beans in south and centre regions of Cameroon *Syllabus Review* on *Research Article Agricultural & Food Science* **5** 27-33

[4] Ha L T V, Nusantoro B, Phillip R A and Anh P H 2016 Physico-Chemical Properties Of Fourteen Popular Cocoa Bean Varieties In Dongnai - Highland Vietnam *Journal of Science* **4** 81 - 86

[5] Karaj S 2008 Physical mechanical and chemical properties of Jatropha curcas L seeds and kernels in *International Agricultural Research for Development* Stuttgart-Hohenheim

[6] Keramat J M, Rafiee S, Jafari A, Ghasemi B M R, Mirasheh R and Mohtasebi S S 2008 Some physical properties of date fruit (cvDairi) *International Agrophysic* **22** 221 - 224

[7] Jaliliantabar F, Lorestani A N and Gholami R 2013 Physical properties of kumquat fruit *International Agrophysics* **27** 1 107 - 109 January 2013

[8] Zare D, Salmanizade F and Safiyari 2012 Some Physical and mechanical properties of Russian olive fruit International *Journal and biological Biomolecular Agricultural Food and Biotechnologic* **6** 9 668 – 671

[9] Loghavi M, Souri S and Khorsandi F 2011 Physical and Mechanical Properties of Almond (Prunus dulcis l. cv. 7Shahrood) in *2011 Louisville, Kentucky, August 7-10, 2011* (p. 1). American Society of Agricultural and Biological Engineers

[10] Tunde-Akintunde T Y and Akintunde B O Some 2004 Physical Properties of Sesame Seed *Biosystems Engineering* 127 - 129

[11] Altuntas E, Ozgoz E and Taser O F 2005 Some physical properties of fenugreek (Trigonella foenum-graceum L) seeds *Journal of Food Engineering* **71** 37 - 43

[12] Celik A, Ercisli S and Turgut N 2007 Some physical pomological and nutritional properties of kiwifruit cv Hayward *International Journal of Food Science Nutrition* **58** 6 411-418

[13] Dursun I, Tuğrul K M and Dursun E 2007 Some physical properties of sugarbeet seed *Journal of Stored Products Research 43*(2) 149-155

[14] Eke C N U, Asoegwu S N and Nwandikom G I 2007 Some physical properties of jackbean seed (Canavalia ensiformis) *CIGR e-journal of Agricultural Engineering International* vol IX

[15] Sessiz A, Elicin A K, Esgici R., Ozdemir G and Nozdrovický L 2013 Cutting properties of olive sucker *Acta technologica agriculturae* **16** (3) 82-86

[16] Eissa A H and Gamea G R 2003 Physical and mechanical properties of bulb onions *Miser Journal of Agricultural Engineering* **20** (3) 661-676

[17] Yalcin H 2014 Physical chemical and bioactive properties of onion *(Allium cepa L)* seed and seed oil *Journal of Applied Botany and Food Quality* **87** 87-92

[18] Kaveri G and Thirupathi V 2015 Studies On Geometrical And Physical Properties Of Co 4 Onion Bulb (Allium Cepa Lvar Aggregatum Don) *International Journal of Recent Scientific Research* **6** 3 2897-2902

[19] Iván Darío I A T, José Jaime C H and Carlos E O T 2012 Physical And Mechanical Properties Correlation Of Coffee Fruit (Coffea Arabica) During Its Ripening Dyna *Dyna* **79** 172 148-155

[20] Couto S M, Batista C, Devilla I A E and Paim V T 2002 Characteristics of coffee fruits under compression *Brazilian Journal of Agricultural and Environmental Engineering* **6** 1 117-122

[21] Jokanović M R., Džinić N R, Cvetković B R, Grujić S and Odžaković B, 2012 Changes of physical properties of coffee beans during roasting *Acta Periodica Technologica* (43) 21-31

[22] Coradi P C, Flavio M B and Henrique R C 2014 Coffee Cherries Drying Process And The Influence Of Environment Relative Humidity *The Mathematical Modeling Moisture Content And Enthalpy Of Vaporization* **29** 2 148-157

[23] Sahin S and Sumnu S G 2006 *Physical properties of foods* (New York : Springer Science+Business Media)

[24] Mohsenin NN 1986 *Physical properties of plant and animal materials* Gordon and Breach

[25] Franca A S, Mendonca J C F and Oliveira S D 2005 The composition of green and roasted coffees of different cup qualities *Lebensmittel-Wissenschaft und –Technologie* **38** 709-715

[26] Ileleji K E and Zhou B 2008 The angle of repose of bulk corn stover particles *Powder Technology* **187** 110-118

[27] Association of Analytical Communities Official 1995 *methods of analysis* 16th Ed (Washington DC: Association of official analytical chemists)

[28] Shoba H, Rajeshwari N and Nagaraja G A 2017 Study on Physico-Mechanical Properties of Onion Varieties Under Koppal District (Karnataka) *Current Agriculture Research Journal* **5** 3

[29] Dabhi M and Patel N 2017 Physical and Mechanical Properties of Talaja Red Onion Cultivar *Bioprocess Engineering* **1** 4 110-114

[30] Vijaya R and Srivastava AP 2006 Physical and Mechanical Properties of Onion (Allium Cepa L) Crop Relevent to Mechanical Detopping *Journal of Agricultural Engineering* **43** 3 83-86

[31] Bahnasawy A H, El-Haddad Z A, El-Ansary M Y and Sorour HM 2004 Physical and mechanical properties of some Egyptian onion cultivars *Journal of Food Engineering* **62** 255-261

[32] Deshpande S D, Bal S and Ojha T P 1993 Physical properties of soybean *Journal of Agricultural Engineering Research* **56** 89-98

[33] Shepherd H and Bhardwaj R K 1986 Moisture dependent physical properties of pigeon pea *Journal of Agricultural Engineering Research* **35** 227-234

[34] Sunitha B, Sujatha G, Girish B, Kiran B R and Ramana MV 2016 Determination of Physical Properties of Onions (Arkakirthiman) *International Journal of Process and Post Harvest Technology* **7** 1 36-41

IOP Conf. Series: Earth and Environmental Science **251** (2019) 012041 doi:10.1088/1755-1315/251/1/012041

[35] Razavi S M A, Amini A M, Rafe A and Emadzadeh B 2007 The physical properties of pistachio nut and its kernel as a function of moisture content and variety *Journal of Food Engineering* **81** on Part III Frictional properties 226-235

[36] Pradhan R C, Naik S N, Bhatnagar N and Vijay V K, 2009 Moisture-dependent physical properties of jatropha fruit *Industrial crops and products* **29** (2-3) 341-347

[37] Silvia E and Sidebang B 2015 Engineering Properties of Coffee Beans from Various Colors of Coffee Cherries *Agriculture and Agricultural Science Procedia* **3** 274-277

[38] Zapata L, Gerard L, Davies C, Oliva L and Schvab M 2007 Mathematical correlation of tomato color indexes with textural parameters and carotenoid concentration *Science Teaching and Technology* **34** 207-226

IOP Conf. Series: Earth and Environmental Science **251** (2019) 012042 doi:10.1088/1755-1315/251/1/012042

Fruit nutrition content, hedonic test, and processed products of pidada (*Sonneratia caseolaris*)

M Basyuni[1,2*], Y S Siagian[1], R Wati[1], L A P Putri[3], E Yusraini[4], I Lesmana[5]

[1]Department of Forestry, Faculty of Forestry, Universitas Sumatera Utara, Jl. Tri Dharma Ujung No. 1 Medan, North Sumatera 20155, Indonesia
[2]Mangrove and Bio-Resources Group, Center of Excellence for Natural Resources Based Technology, Universitas Sumatera Utara, Medan, North Sumatera 20155, Indonesia
[3]Department of Agroecotechnology, Faculty of Agriculture, Universitas Sumatera Utara, Medan, North Sumatera 20155, Indonesia
[4]Department Food Science and Technology, Faculty of Agriculture, University of Sumatera Utara, Medan, Indonesia
[5]Department of Aquatic Resource Management, Faculty of Agriculture, Universitas Sumatera Utara, Medan, North Sumatera 20155, Indonesia

*Email: m.basyuni@usu.ac.id

Abstract. Exploitation of mangrove forest products still lacks consent, it is indispensable to apply appropriate technology for the sustainable use of mangrove forest products to improve the income of coastal communities, without damaging to the mangrove forest. This study aims to investigate the fruit nutrient content, hedonic test, and processed product of pidada (*Sonneratia caseolaris*) along with examining the storage period of pidada syrup. The nutritional content and the storage period of the product for seven months were performed using proximate analysis and organoleptic test. Results showed that nutritional content in pidada fruit decreased levels of vitamin C, protein, total acid compare with pidada syrup, while total sugar and total dissolved solids rather enhanced than processed syrup product. The nutritional content of pidada fruit and syrup is favourable to improve the quality and the selling value of the mangrove processed products. The present study suggested that processing affected the nutrients content and the storage time of the product. The process of processing affects the nutrients contained in the food, and also affects the storage life of the product.

1. Introduction

Pidada (*Sonneratia caseolaris* (L.) Engl., Sonneratiaceae) is a mangrove plant that widely distributed in areas or edges of river estuaries, especially in low salinity areas with freshwater mixtures [1]. Tree height reaches 16 meters, fruit with a diameter of 6-8 cm, with a shiny surface, red flower crown, and red and white stamens. The fruit of S. caseolaris is round, made up of a pompom of white stamens in a cup-shaped calyx. [1]. *S. caseolaris* is known as a traditional medicinal plant with several biological and pharmacological properties have been reported [2-3]. *S. caseoalaris* fruits have been long consumed by community living in Indonesia and other coastal areas as a food material or ingredient in various processed food products, including syrup [4], cakes and steamed pudding [5].

Pidada extract is traditionally used as an antiseptic, treats sprains, and prevents bleeding [6]. Pidada fruit has long been known to have efficacy as traditional medicines to treat several diseases, but its use is still not widely known to the public. Therefore this study aims to investigate the fruit nutrient content, hedonic test, and processed product of pidada (*S. caseolaris*) along with an evaluation of pidada syrup shelf life.

2. Materials and method

2.1. Materials
A fresh of *Sonneratia caseolaris* fruits and syrups (Figure 1A and 1B) were collected from Lubuk Kertang mangrove forest, Langkat, North Sumatra. The plant species are determined in Indonesian Institute of Science (LIPI), Research Center for Biology Bogor. The specimen vouchers have been deposited there.

2.2. Nutritional parameters of fruits
The dietary parameters the edible fruit was analyzed for attributes such as total sugars, vitamin C, protein, total acid, total soluble solids (TSS), and total microbe using standard analytical procedures [7-8].

2.3. Organoleptic test on the processed product (pidada syrup)
Pidada syrup as the processed product was evaluated according to organoleptic score for variable parameters in 5 points hedonic scale as previously reported [8]. The hedonic test was performed as treatment time (1-7 months) storage periods. The criteria for a hedonic analysis such as 5 (5 – like very much, 4—like, 3 as either like or dislike, 2 as dislike, and 1 showing dislike very much as earlier reported [9].

2.4. Shelf life evaluation of pidada syrup
Shelf life evaluation of pidada syrup was carried out through a continuous assessment method. The sample was examined for physicochemical properties, colour, flavour, taste, viscosity, product label, product label, and product seal for a seven month of storage period as previously described [10].

A B

Figure 1. Change in the level of preference of panellists in pidada syrup

3. Results and Discussions

3.1. The morphological character of pidada fruit and pidada syrup
The composition of pidada syrup processing consists of pidada juice, water, and sugar. Pidada fruit juice and the addition of granulated sugar make the characteristics of syrup to be thick, brownish in color, and have a distinctive flavor from pidada fruit. The fruit used for making syrup is chilled and cooked, which is characterized by yellowish green fruit, and textured [1].

3.2. Nutritional parameters of fruits

Testing of pidada fruit nutrition and pidupa syrup was carried out using a proximate analysis which included parameters of total sugar, vitamin C, protein, total acid, total soluble solids (TSS), and total microbes. Table 1 shows the nutritional content and changes in nutrient content found in pidada fruit with its processed product in the form of pidada syrup. Table 1 displays the nutritional content in pidada fruit decreased levels of vitamin C, protein, total acid compares with pidada syrup, while total sugar and total dissolved solids rather enhanced than processed syrup product.

Table 1. Nutritional content test on pidada fruit and syrup

	Total sugars (mg/100 g)	Vitamin C (Mg/100 g)	Protein (%)	Total acid (%)	TSS (° brix)	Total microbe (cfu/mL)
Pidada fruit	46.58 ± 11.22	187.46 ± 20.29	52.78 ± 1.07	0.2697 ± 1E-04	9.16 ± 1.44	$25x10^4$ ± 45,82
Pidada syrup	106.77 ± 23.62	151.92 ± 10.42	38.63 ± 0.61	0.2 ± 0.05	26.66 ± 2.89	$29x10^4$ ± 41.63

TSS: total soluble solid. Data was expressed as triplicate analyses ± SD (*n*= 3)

It was obtained that the average total pidada fruit sugar was 46.58 mg/100 g, while pidada syrup was 106.77 mg/100 g. Vitamin C of pidada fruit is earned on average 187.46 mg / 100 grams of ingredients, while in syrup is 151.92 mg / 100 grams of ingredients. Pidada fruit protein obtained an average of 52.78%, while pidada syrup protein was 38.63%. The total acid in pidada fruit was acquired by an average of 0.26%, whereas in pidada syrup obtained a total acid of 0.2%. The TSS in the pidada fruit was acquired by an average of 9.16° brix, while in pidada syrup was obtained an average of 26.66° brix. And the total microbes in pidada fruit were obtained on average by 25x104 cfu/mL, while in pidada syrup was obtained an average of 29x104 cfu/mL.

The highest total sugar was found in pidada syrup with a total of 106.77 mg/100 g while in fruit it was 46.58 mg/100 g. Because pidada fruit syrup has gone through a process of processing, with boiling and adding sugar, which causes the increase in the total sugar in the syrup. According to SNI number 01-3544-1994, the minimum sugar content in syrup is 65%, our results met the criterion. The sugar serves to give flavor to the product and preserve the product by inhibiting bacteria that cause decay. Total sugar content is affected by the amount of sugar added to the product. The more added sugar in the syrup, the greater the percentage of total sugar. Addition of sugar in addition to providing sweetness also functions and is involved in preservation. If sugar is added to high concentrations (at least 40% dissolved solids), then some of the water that is bound is available for the growth of microorganisms and water activity from food is reduced, whereas microorganisms have minimum water requirements for growth [11].

The ability of sugar to bind water is what causes sugar to function as a preservative. This is in line with the study [11] that the addition of high levels of sugar (minimum 40%) will bind the free water contained in food. Furthermore, the addition of sugar in large quantities will speed up the heating process, and the water will be confined in the gel. This condition happens because the fruit has gone through the boiling process [7].

The addition of sugar with longer heating will cause the caramelization process to be non-enzymatic browning reactions. Caramel formed during heating gives a brown color to food products. After testing the highest vitamin C was found in pidada fruit as much as 187.46 (mg/100 g ingredients) while pidada syrup was 151.92 (mg/100 g ingredients), because the pidada fruit had gone through a boiling process with high temperature, heating process high temperatures can cause a decrease in vitamins [12], where the vitamin C of pidada fruit was 6.74 mg. Good product handling can help stabilize vitamin content in pidada syrup. The same circumstance is true for protein in fruit and pidada syrup. The highest protein content was found in pidada fruit at 52.78%, suggested that the processing of foodstuffs through the boiling process with high temperatures can damage and reduce the levels of vitamins and proteins in these foods. The total level of dissolved solids tends to increase with increasing sugar concentration. This is in accordance with previous results on the higher the addition of sugar can produce a higher total dissolved solids [12].

3.3. Evaluation of processed product (pidada syrup)

Pidada syrup is carried out by the same test which is a hedonic test which includes the parameters of colour, flavour, taste, viscosity, product label, product packaging, and product seal. The results of pidada syrup based on hedonic testing can be seen in Figure 2. Tests were carried out 3 times on the same panellists for 7 months. After testing, the hedonic test method showed the level of people's preference for the colour of the syrup in the first month to the seventh month had not changed with a scale of 4 (like). The aroma of syrup in the first month of the fifth month also did not change, but in the sixth and seventh months decreased with scale (very dislike). Syrup taste in the first month to the fifth month did not change, but in the sixth and seventh months decreased, in the first to sixth month with a scale (neither like or dislike), while in the seventh month with a scale of 2 (disliked).

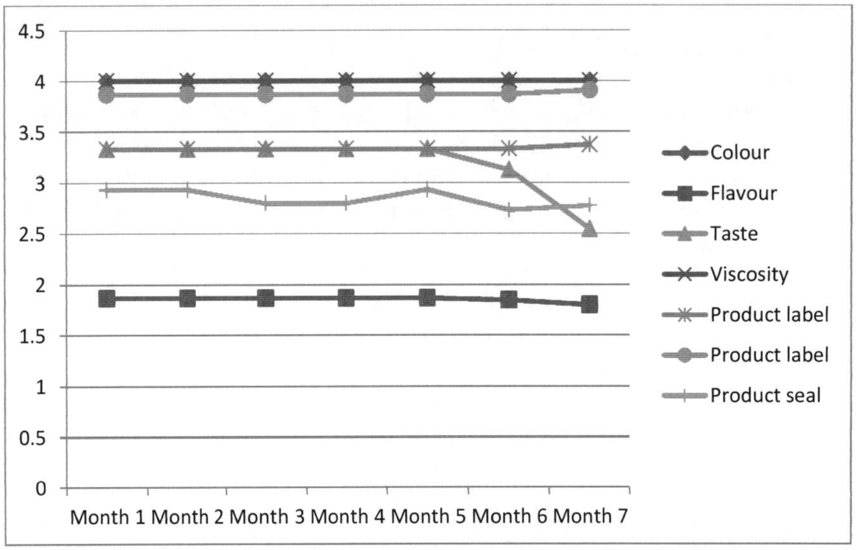

Figure 2. Change in the level of preference of panellists in pidada syrup

The thickness of the syrup in the first month of the seventh month does not change, with a scale of 4 (like). Pidada syrup product labels are obtained on a scale of 3 (either like or dislike). Pidada syrup product packaging obtained a scale of 3, while the product seal obtained a scale of 2. The changes in the level of panellists' preference for pidada syrup can be seen in Figure 2. As for the nutritional values for the mangrove species, a study [13] reported that fruits of *S. caseolaris* contain about 15.95% carbohydrate, moisture 77.10%, fat 0.86%, ash 3.85%, and protein 2.24%.

4. Conclusions

The present study confirmed the present study suggested that processing affected the nutrients content and the storage time of the product. The process of processing affects the nutrients contained in the food, and also affects the storage life of the product.

5. Acknowledgment

This study was partly supported by an International Research Collaboration and Scientific Publication Grant (No. 003/SP2H/LT/DRPM/IV/2017 to MB) from the Directorate for Research and Community Service, Ministry of Research, Technology and Higher Education, Republic of Indonesia.

6. References

[1] Kitamura S, Anwar C, Chaniago A, 1997 Baba S Handbook of Mangroves in Indonesia –Bali & Lombok- Japan International Cooperation Agency and International Society for Mangrove Ecosystems 119 p.

[2] Simlai A, Rai A, Mishra S, Mukherjee K, Roy A 2014 Antimicrobial and antioxidative activities in the bark extracts of *Sonneratia caseolaris*, a mangrove plant *EXCLI J* **13** 997-1010.
[3] Patra JK, Thatoi HN 2011 Metabolic diversity and bioactivity screening of mangrove plants: a review *Acta Physiol. Plant.* **33** 1051-1061.
[4] Abeywickrama WSS, Jayasooriya MCN 2011 Formulation and quality evaluation of cordial based on kirala (*Sonneratia caseolaris*) fruit *Trop. Agric. Res. Extension* **13** 16-18.
[5] Brown B 2006 Cooking with mangroves 36 Indonesian mangrove recipes. Mangrove Action Project Indonesia p 26-34.
[6] Tian, M., Dai, H., Li, X., & Wang, B. (2009). Chemical constituents of marine medicinal mangrove plant *Sonneratia caseolaris*. *Chin. J. Oceanol. Limnol.* 27: 288.
[7] Jariyah, Widjanarko, SB, Yunianta, Estiasih T, Sopade PA 2014 Pasting properties mixtures of mangrove fruit flour (*Sonneratia caseolaris*) and starches *Int. Food Res. J.* **21** 2161-2167.
[8] Sumardi, Basyuni M, Wati R 2018 Antimicrobial activity of polyisoprenoids of sixteen mangrove species from North Sumatra, Indonesia *Biodiversitas* **19** 1243-1248.
[9] Hersleth M, Ilseng MA, Martens M, Naes T 2005 Perception of cheese: a comparison of quality scoring, descriptive analysis and consumer responses *J. Food Qual.* **28** 333-349.
[10] Basyuni M, Ginting PY, Lesmana I 2017 Phytochemical analysis of binahong (*Anredera cordifolia*) leaves extract to inhibit in vitro growth of *Aeromonas hydrophila* AIP Conf. Proceed. **1904** 020072.
[11] Lievens B, Hallsworth JE, Pozo MI, Belgacem ZB, Stevenson A, Willems KA, Jacquemyn H 2015 Microbiology of sugar‐rich environments: diversity, ecology and system constraints. *Environ. Microbiol.* 17 278-198.
[12] Fernández-Merino MJ, Guardia L, Paredes JI, Villar-Rodil S, Solís-Fernández P, Martínez-Alonso A, Tascon JM 2010 Vitamin C is an ideal substitute for hydrazine in the reduction of graphene oxide suspensions *J. Phys. Chem. C.* **114** 6426-3642.
[13] Patil PD, Chavan NS 2013 A need conservation of mangrove genus Bruguiera as a famine food *Annals Food Sci. Technol.* **14** 294-297

IOP Conf. Series: Earth and Environmental Science **251** (2019) 012043 doi:10.1088/1755-1315/251/1/012043

Physicochemical Properties and its Relations of Beef Rendang inside Retort Pouch Packaging in Various Temperature Storage Conditions

A S Praharasti[1*], A Kusumaningrum[1], Y Khasanah[1], R Nurhayati[1], A Nurhikmat[1], A Susanto[1], Suprapedi[2]

[1]Research Unit for Natural Product Technology–Indonesian Institute of Sciences
Jln. Jogjakarta – Wonosari km, 31.5 Yogyakarta, Indonesia

[2]Research Center of Physics– Indonesian Institute of Sciences
Kawasan PUSPIPTEK Serpong, Muncul, Setu, Tangerang, Banten
**e-mail: praharasti@gmail.com*

Abstract. Physicochemical characterization for beef Rendang inside retort pouch packaging had been evaluated. The objective of this study was to know the effect of time, and temperature storage to the physical properties of beef Rendang includes interactions between them so it could be evaluated what properties were changed along storage. Commercially sterile beef Rendang products using sterilization process at a pressure of 1.3 bar for 40 minutes holding time with the lethality value (Fo) of 4.67-7.49 minutes were stored at three different temperatures (35, 45, and 55 °C). Moisture content, pH, and rancidity value (TBA) as physicochemical properties examined were observed for 6 weeks of storage. The results showed that the effects of three different temperatures and times storage had significant difference value of moisture content, pH, and TBA value. Moisture content, pH, and TBA value of beef Rendang were approximately in the amount of 38-42%; 5.15 – 5.38; and 0.12 – 0.27 mg malonaldehyde/kg sample respectively, along 6 weeks of storage.

1. Introduction

Rendang is a traditional food from West Sumatera Indonesia with generous amounts of different kinds of spices and ingredients including meat and coconut milk. The dried meat *rendang* cooking technique could prolong the food's shelf life and also might be caused by the contribution of the spices used during the cooking process[1]. *Rendang* usually was packaged in a flexible pouch and sealed uses vacuum processing. In the present study, beef rendang in pouch packaging was sterilized after the vacuum sealed processing for extending the shelf life of the product [2–4]. During the storage process, the temperature is one of the primary factors that effect in physicochemical properties of products. Lipid oxidation reaction effected in viscosity of oils rancid off flavors while cross-linking of polymers effected in loss of water-holding capacity [5]. The integrated effects of time and temperature on food products affect their color, flavor, and texture commonly referred to as quality.

Commercially sterile retort foods for ready to eat meal had been studied by many researchers such as pepper chicken [6] soy peas curry [7], milkfish [8], tuna in oil [9], black clam product [10], mushroom in brine [11], dairy dessert (kheer) [12], freshwater prawn in curry medium [13], egg-based product [14], Chettinad goat meat product [15], chopped mussel meat [16], and also samgyetang [17]. The effect of retort pouch processing to the physicochemical parameters had also examined by [18,19].

Many researchers were studied especially about physic-chemical properties of meat and retorted foods. Bindu et al. [10] were also studied about TBA value in *black clam* product in indigenous retort pouches. They found that TBA during the storage period, TBA value tends to increase and it was be expected that it caused by a product used which contained only solids, without a liquid medium. Processing of mushrooms in brine was also studied by Chandrasekar et al. [11], and they found that the pH value of the product was 5.75 and remained unchanged during 12 months storage time. Diaz et al. [18] evaluated the chemical composition, pH, water holding capacity, drip loss, color and texture of the meat of meat from capons and they found that those factors were significantly influenced by the age of the capons. The meat from the youngest animal showed higher pH than the older ones. pH value from muscle tissue in accordance with meat quality was in the range of 5.5-7.4 [20]. Ukekpe et al. [21] were also evaluated TBA value in selected fish harvested from Nigeria, and they reported that the Tilapia and catfishes with the value TBA of 6.60 and 8.00 mgMA/kg respectively have better quality within twelve hours of keeping, than elephant snout, Mud, and butterfishes. *Kheer,* as a traditional milk product of South East Asia, were examined the product quality such as F_0 value, TBA value, reflectance value, and pH by Jha et al. [12]. According to their findings, for the product with F_0 value of 12.4-14.8 minutes, the TBA value ranged 0.073-0.081 and the pH value reported were 6.04-6.10. Based on a study which had been reported by Karel [5] the remaining cause of deteriorations of foods stored at room temperature besides the microbial growth were physical changes and chemical reactions. Ma et al. [22] studied the effect of pressure and heat treatment on beef proteins, and they found that the concentration of soluble myofibrillar proteins increased from 0.273 mg/mL to 0.747 mg/mL of the meat proteins. According to Pasha et al. [23] the TBA value of rice polishing was affected ($p<0.05$) by storage period, the increasing value were reported when the storage period increased from 4 to 16 weeks. Pearson [24] found that most beef can be considered to be acceptable if the total volatile nitrogen (TVN) and TVN/FF (fat-free) figures were not exceeded 16.5 and 19.7 mg N/100g respectively. In his further study [25], peroxide values of most fresh samples were found in the range of 0.0-1.0 mequiv./kg (on the extracted fat). Study about the physicochemical deterioration and lipid oxidation were also done by Rahman et al. [26] which reported that as the increase of the freeze-thaw's number, pH value of beef muscle affected by freeze-thaw cycles was decreased ($p<0.05$), free fatty acids value were increased ($p<0.05$) and TBARS value were also increased ($p<0.05$) significantly. The increase of TBA value inside the retort-pouched sauces was also reported to increase during storage according to study had been done by Lee et al. [27], whereas the overall pH reported decreased.

2. Methods

2.1. Materials
Beef Rendang was one of traditional food from Indonesia. Beef meat, coconut milk, and traditional Indonesian spices and herbs were heated together while stirred until they got dried continuously. In this research, beef Rendang was bought from Uni Upik Restaurant in Yogyakarta.

2.2. Preparation of beef Rendang retort pouch
Beef Rendang as the sample of this research then packaged inside retort pouch by weighing preparation which had been developed by previous research [2] through preparing 60 grams of meats and 60 grams of dressing. The size of the pouch was 18 cm x 13 cm. Rendang inside retort pouch then was thermally processed using Zon Gon Horizontal Retort Machine according to sterilization method for flexible packaging which had been registered in Indonesian patent document number of P00201708373. In this research, it was chosen the retort pressure of 1,3 bar hold for 40 minutes to sterilize the samples. When it had sterilized, the samples were quarantine for 2 weeks to assure samples in stable condition achieving commercially sterile. In order to study the reaction rate, samples were stored at three different temperatures inside incubators, which were 35, 45, and 55 °C.

2.3. Commercially sterile measurement in the slowest heating zone
The requirement of a commercially sterile product was stated by US-FDA to accomplish the 12D process [28]. To check the adequacy of the thermal process, Fo meter probe was mounted with the system of the sample and pouch exactly on its slowest heating zone (SHZ). The vessel was then turned on together with the data logger. Fo result then was printed on thermo-paper.

2.4. Physicochemical properties measurements
Physicochemical properties parameters which observed in this study were moisture content, pH, and rancidity value based on TBA (thiobarbituric acid reactive substance). Those three parameters were examined every week for 6 weeks and the initial week was also measured. Moisture content determined gravimetrically; pH value was measured using pH meter brand Eutech PC 700, Thermo Scientific, II, USA; whether the TBA value was derived using a spectrophotometer and Tarladgis's method [29].

2.5. Statistical Analysis
Value of moisture content, pH, and TBA was evaluated statistically using SPSS 16.0 version. Univariate Analysis of Variance test was used in this study, accompanied by Duncan Multiple Range Test for Post-Hoc when the value was significantly different.

3. Result and Discussions

3.1. Fo value of Beef Rendang
According to the previous study which had been registered in the patent document mentioned above, sterilization process using a pressure of 1.3 bar for 40 minutes holding time was chosen for the thermal process of flexible package pouch, especially for beef Rendang. The sterilization Fo value was found

approximately in the range of 4.225-8.795 minutes; the value was varied sometimes in large value range according to quality control or might be caused by different scenarios occurs along the process. Fo value was insufficient to an amount close to 2.54 minutes as the requirement of 12D thermally processes for low acid food to reach commercially sterile product by the thermal death time of *Clostridium botulinum* [28].

3.2. Statistical analysis for moisture content, pH, and TBA value

To study the behavior of each parameter and interactions between them, the univariate analysis of variance study for moisture content, pH, and TBA value was displayed in Table 1.

Table 1. Univariate analysis of variance result for moisture content, pH, and TBA value

Properties Factors	Sig. (Moisture content)	Sig. (pH value)	Sig. (TBA value)
Temperature storage	*	*	0.011*
Time storage	*	*	0.04*
Temperature*Time storage	*	*	0.825 ns

(*) differ significantly, (ns) not significant
(Significance level used: 0.05)

Post-Hoc test for each factor was displayed in Table 2, 3, and 4 and explained below in sections.

Table 2. Moisture content in various time and temperature storages

Time storage (weeks)	Moisture content (%)	Temperature storage (°C)	Moisture content (%)
1	39.116±0.923^{a}	35	39.798±1.104^{a}
2	40.335 ±0.973^{bc}		
3	40.237 ±0.920^{b}	45	40.251±0.741^{b}
4	40.946 ±0.851^{d}		
5	39.951 ±0.642^{b}	55	40.648±1.156^{c}
6	40.807 ±1.086^{cd}		

*Same column having similar superscript did not differ significantly
(Significance level used: 0.05)

Table 3. pH value in various time and temperature storages

Time storage (weeks)	pH value	Temperature storage (°C)	pH value
1	5.303±0.017^{d}	35	5.331±0.029^{c}
2	5.312±0.046^{d}		
3	5.279±0.040^{bc}	45	5.282±0.028^{b}
4	5.290±0.050^{c}		
5	5.226±0.060^{a}	55	5.230±0.052^{a}
6	5.276±0.073^{b}		

*Same column having similar superscript did not differ significantly
(Significance level used: 0.05)

Table 4. TBA value in various time and temperature storages

Time storage (weeks)	TBA value (mg malonaldehyde/kg sample)	Temperature storage (°C)	TBA value
1	0.240±0.070[c]	35	0.233±0.071[a]
2	0.186±0.056[abc]		
3	0.219±0.098[bc]	45	0.172±0.057[a]
4	0.201±0.060[abc]		
5	0.145±0.044[a]	55	0.176±0.068[b]
6	0.172±0.049[ab]		

*Same column having similar superscript did not differ significantly
(Significance level used: 0.05)

3.3. *Moisture content*

Freshly beef Rendang in this research was in the amount of 39% (wb). Changes in moisture content were shown in Figure 1 compared with the initial condition. It could be seen that the moisture content was decreased as the sterilization effect, but during storage time the values were increased.

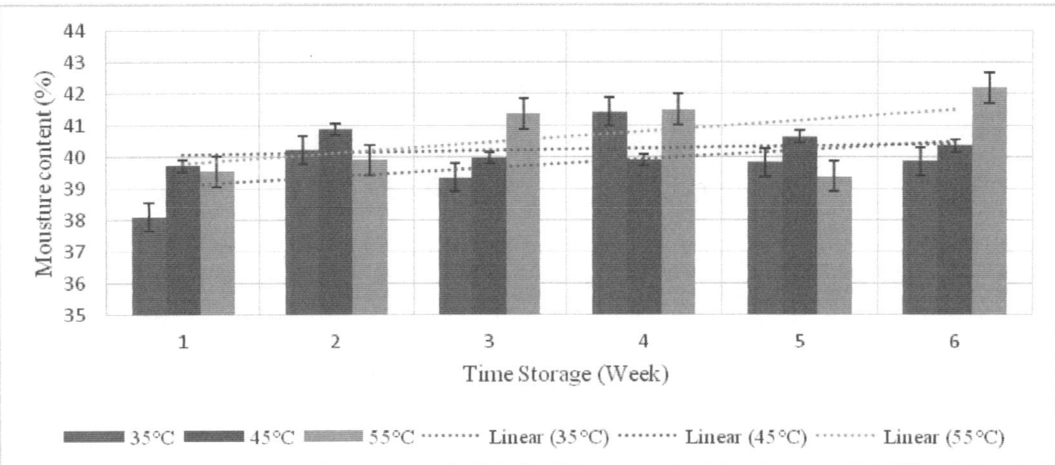

Figure 1. Moisture content during storage

According to a statistic test, it showed that the probability was <0.05. It can be concluded that the three temperature storage had significant difference value of moisture content. The higher temperature storage, moisture content value tends to increase. For storage, the time parameter could be seen that the probability value was <0.05. The different time of storage gave a different value of moisture content, significantly. Moisture content value of the first week was the lowest of all. In the interaction test between temperature and time of storage, the probability was <0.05, means that those many interactions gave significant different moisture content value.

3.4. *pH value*

Statistic test for pH value showed that the probability was <0.05. It can be concluded that the three times storage had significant difference value of pH value. The longer storage time pH value tends to decrease. For storage, the temperature parameter could be seen that the probability value was <0.05. The different temperature of storage gave a different value of pH, significantly. The higher temperature storage. pH value tends to decrease. In the interaction test between temperature and time of storage, the probability was >0.05; it means that those many interactions gave significant different pH value.

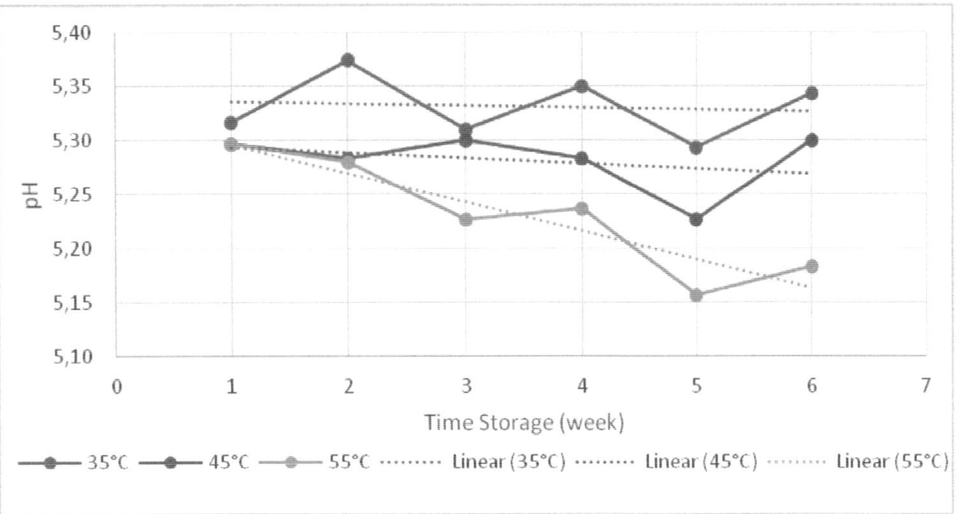

Figure 2. pH value during storage

From Figure 2 above, it is shown that the pH value a ranged between 5.15-5.38 during 6 weeks of storage. Those results were in line with a study from Methayasa [30] which found that the pH value of Bali Beef and Wagyu Beef were in the range between 5.41-5.85. It also had a similar result which was specifically mentioned also by Weglarz that meat obtained from cows had pH values above 5.8 [31]. According to G. Feiner, pH values of meat and meat product were generally between 4.6 (raw fermented salami) and 6.4 [32].

3.5. TBA value

In this research, thiobarbituric acid (TBA) was used to measure rancid level from week to week using Tarladgis equation [29] and to investigate the effect of storage temperature of Rendang. Figure 3 shows the rancidity value of ready-to-eat Rendang in flexible retort pouch during storage time.

According to a statistic test, it showed that the probability was 0.011 (<0.05). It can be concluded that the three-temperature storage had significant difference value of TBA. The higher the temperature, the lower the TBA value will be. For storage, time parameter could be seen that the probability value was 0.04 (<0.05). The different time of storage gave a different value of TBA, significantly. The more time it stored, TBA value tends to decrease. In the interaction test between temperature and time of storage, the probability was 0.825 (>0.05). It means that those many interactions did not give significant different TBA value. Meat products contain the high level of unsaturated fatty acids which can easily become rancid over time in storage due to lipid oxidation reaction [5,33].

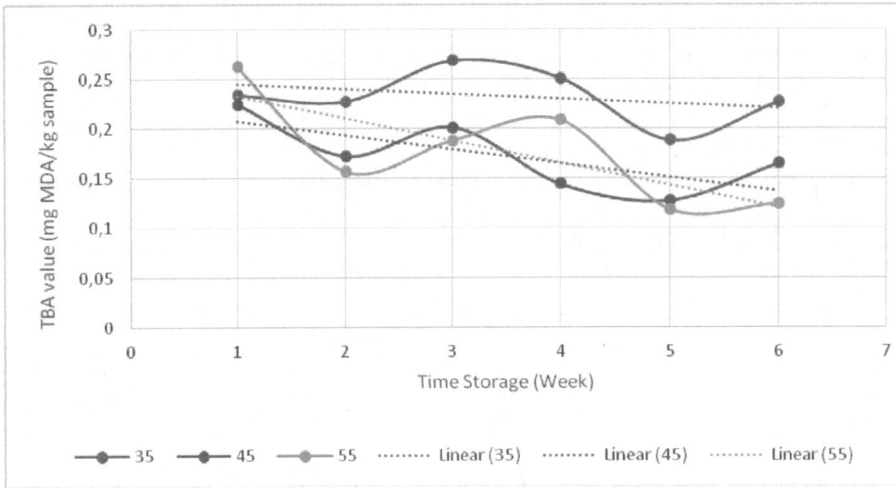

Figure 3. TBA value during storage

From Figure 3, we could see that TBA value has fluctuated. One of the reason had been stated before that the red color formed in TBA test might also be part of the reaction from the reagent with other aldehydes which formed by the breakdown of the non-fat components in food, and this may interfere with the result [34]. When TBA test been done, all samples were centrifuged together but their red colors were not read at one time. The lag time of reading between samples made the visually intensities red color changed, different from the sample which read at the same time. According to the result studied by [34], the TBA result values of beef meat ranged between 0.1 to 2.2 depending on the storage condition. For spoiled beef, TBA amount was 1.8 according to the study reported by [24]. In this research TBA values varied from 0.12-0.27, close to those value reported by [34].

4. Conclusions

Temperature difference gave significant difference moisture content. The higher the temperature, moisture content tends to increase. Different time storage gave significant different moisture content. The longer storage time, the higher moisture content will be. The interaction between temperature and time storage was significantly different. Moisture content values for Beef Rendang was approximately in the amount of 38-42%

Different time storage gave a significant difference in pH value. The longer storage times. pH value tends to decrease. The different temperature of storage gave significant different pH value. The higher temperature storage, the lower pH value. The interaction between time and temperature storage to pH value was significantly different, pH values for Beef Rendang were approximately in the amount of 5.15-5.38.

The three temperatures storage gave significant difference value of TBA (sig. 0.011). The higher the temperature, the lower the TBA value will be. Different storage time gave significant different TBA value (sig. 0.04). The longer it stored, TBA value tends to decrease. The interaction between temperature and time to the value of TBA had no significant difference (sig. 0.825). TBA values for Beef Rendang were approximately in the amount of 0.12-0.27 mg malonaldehyde/kg sample.

5. Acknowledgments

The authors would like to express their sincere gratitude to Kegiatan Unggulan LIPI 2017.

6. References

[1] Nurmufida M, Wangrimen G H, Reinalta R and Leonardi K 2017 Rendang : The treasure of Minangkabau *J. Ethn. Foods* **4** 232–5

[2] Praharasti A S, Kusumaningrum A, Frediansyah A, Nurhikmat A, Khasanah Y and Suprapedi 2017 Lethality of Rendang Packaged in Multilayer Retortable Pouch with Sterilization Process *International Conference on Engineering, Science, and Nanotechnology* vol 030079 (Solo, Indonesia: American Institute of Physics) pp 1–7

[3] Frediansyah A, Praharasti A S, Kusumaningrum A, Nurhikmat A, Susanto A, Khasanah Y and Nurhayati R 2017 Application of Static Retort Thermal Processing Technology for Dried Beef Rendang Production : Evaluation of its Post- Processing on Microbiological and Physicochemical Properties *International Conference on Engineering, Science, and Nanotechnology* vol 030079 (Solo, Indonesia) p 7

[4] Praharasti A S, Nugroho J, Rahardjo B and Nurhikmat A 2016 Evaluation for Sterilization of Rendang Packaged by Flexible Retort Pouch using Finite Difference Method *Proceeding of The International Symposium in Agricultural and Biosystem Engineering*

[5] Karel M 1984 Chemical Effects in Food Stored at Room Temperature *J. Chem. Educ.* **61** 335–9

[6] Nalini P, Abraham R J J, Rao V A, Babu R N, Rajkumar T N, Rajkumar R and Kathiravan R S 2018 Shelf-Life of Ready-To-Eat Retort Processed Pepper Chicken *Int. J. Curr. Microbiol. Appl. Sci.* **7** 832–40

[7] Abhishek V, R K, George J, S N, J.H L, T K, N M and S N 2014 Development of retort process for ready-to-eat (RTE) soy-peas curry as a meat alternative in multilayer flexible retort pouches *Int. Food Res. J.* **21** 1553–8

[8] Adepoju M A, Omitoyin B O, Mohan C O and Zynudheen A A 2016 Heat penetration attributes of milkfish (Chanos chanos) thermal processed in flexible pouches : a comparative study between steam application and water immersion *Food Sci. Nutr.* 1–4

[9] Ali A A, Sudhir B, Krishnaswamy T and Gopal S 2006 Effect of rotation on the heat penetration characteristics of thermally processed tuna in oil in retort pouches *Int. J. Food Sci. Technol.* **41** 215–9

[10] Bindu J, Ravishankar C N and Srinivasa Gopal T K 2007 Shelf life evaluation of a ready-to-eat black clam (villorita cyprinoides) product in indigenous retort pouches *J. Food Eng.* **78** 995–1000

[11] Chandrasekar V, Gopal T K S and Rai R D 2004 Heat Penetration Characteristics and Shelf-life Studies of Mushrooms in Brine Processed in Retort Pouches and Science *Packag. Technol. Sci.* **17** 213–7

[12] Jha A, Patel A A, Gopal T K S and Ravishankar C N 2012 Heat penetration characteristics and physico-chemical properties of in-pouch processed dairy dessert (kheer) *J. Food Sci. Technol.* 1–8

[13] Majumdar R K, Roy D and Saha A 2017 Textural and sensory characteristics of retort- processed freshwater prawn (Macrobrachium rosenbergii) in curry medium *Int. J. Food Prop.* **20** 2487–98

[14] R. K, George J, Kumar D, C. J, S. N, J.H. L, M.R. K, T. K, R. R, N. M and S. N 2015 Development and evaluation of egg based ready-to-eat (RTE) products in flexible retort pouches *African J. Food Sci.* **9** 243–51

[15] Rajkumar V, Dushyanthan K and Arun K D 2010 Retort pouch processing of Chettinad style

goat meat curry – a heritage meat product *J. Food Sci. Technol.* **47** 372–9
[16] Tribuzi G, Gláucia Maria Falcão de A and Laurindo J B 2015 Processing of chopped mussel meat in retort pouch *Food Sci. Technol.* **35** 612–9
[17] Triyannanto E and Lee K T 2015 Effect of Pre-cooking Conditions on the Quality Characteristics of Ready-To-Eat Samgyetang *Korean J. Food Sci.* **35** 494–501
[18] Díaz O, Rodríguez L, Torres A and Cobos A 2010 Chemical composition and physico-chemical properties of meat from capons as affected by breed and age *Spanish J. Agric. Res.* **8** 91–9
[19] Cheftel J C and Culioli J 1997 Effects of High Pressure on Meat: A Review *Meat Sci.* **46** 211–36
[20] Dutson T R 1983 The Measurement of pH in Muscle and its Importance to Meat Quality *Reciprocal Meat Conference Proceedings* vol 36 (Texas) pp 92–7
[21] Ukekpe U S, Gashua I B and Okoye U J 2014 Evaluation of rancidity rate of oil in selected fish species harvested from Hadejia-Nguru Wetlands , Nigeria *Int. J. Curr.Microbiology Appl. Sci.* **3** 122–8
[22] Ma H and Ledward D A Effect of Pressure and Heat Treatment on Beef Proteins
[23] Pasha T N, Khattak F, Khan D and Jabbar M 2018 Effect of Storage Time on the Rancidity and Metabolizable Energy of Rice Polishing in Poultry *Asian-Australian J. Anim. Sci.* **21** 420–5
[24] Pearson D 1968 APPLICATION OF CHEMICAL METHODS FOR THE ASSESSMENT OF BEEF QUALITY III .* -Methods related to fat spoilage *J. Sci. Food Agric.* **19** 553–6
[25] Pearson D 1968 Application of Chemical Methods for the Assessment of Beef Quality II* Methods related to protein breakdown *J. Sci. Food Agric.* **19** 366–9
[26] Rahman M H, Hossain M M, Rahman S M E, Amin M R and Oh D 2015 Evaluation of Physicochemical Deterioration and Lipid Oxidation of Beef Muscle Affected by Freeze-thaw Cycles **35** 772–82
[27] Shin E L, Park S Y, Jeong Y-G, Jo B-C, Kim M and Ha S-D 2015 Quality evaluation and estimation of shelf life of retort-pouched tomato- based and Korean traditional fermented food-based sauces Quality evaluation and estimation of shelf life of retort-pouched tomato-based and Korean traditional fermented food-based s *Korean Soc Appl. Biol. Chem.* **58** 229–36
[28] FDA 2015 Subpart A — General Provisions PART 113 — THERMALLY PROCESSED LOW-ACID FOODS PACKAGED IN HERMETICALLY SEALED CON- *Authenticated U.S. Gov. Inf.* 296–331
[29] Tarladgis B G, Watts B M and Margaret T Y 1960 A Distillation Method for the Quantitative Determination of Malonaldehyde in Rancid Foods **1029** 44–8
[30] Merthayasa J D, Suada I K and Agustina K K 2015 Water Holding Capacity, pH, Color, Odor, and Texture of Bali Beef ang Wagyu Beef *Indones. Med. Veterinus* **4** 16–24
[31] Węglarz A 2010 Meat quality defined based on pH and colour depending on cattle category and slaughter season *Czech J. Anim. Sci.* **55** 548–56
[32] Feiner G 2006 *Meat products handbook, practical science and technology* (Abington, Cambridge, England: Woodhead Publishing Ltd)
[33] Verma D K and Srivastav P P 2017 Proximate Composition , Mineral Content and Fatty Acids Analyses of Aromatic and Non-Aromatic Indian Rice *Rice Sci.* **24** 21–31
[34] Adhikary M 2001 *Oxidative Rancidity of Peanut Butter Slices under Different Packaging and Atmospheric Conditions* (Indian Institute of Technology Kharagpur, India)

IOP Conf. Series: Earth and Environmental Science **251** (2019) 012044 doi:10.1088/1755-1315/251/1/012044

The Alteration of Physicochemical and Thermal Characteristics of Peanut by Salting-Roasting Process

U Laila[1], R Nurhayati[1], Y Khasanah[1], D Ariani[1] and M Kurniadi[1]

[1]Research Unit for Natural Product Technology, Indonesian Institute of Sciences, Jalan Jogja-Wonosari km 31.5 Desa Gading Kecamatan Playen, Kabupaten Gunungkidul, Yogyakarta 55861, Indonesia

Email: umilaila38@gmail.com

Abstract. Peanut as nutritious food containing high protein, high fat, carbohydrate, and bioactive substance is usually processed in order to enhance palatability and consumer preference. One of the thermal processes, a traditional salting-roasting process used by BIMARAM SME in Gunungkidul to add the value of peanut, was predicted to be able to alter the properties of peanut, i.e., physicochemical and thermal characteristics. This research investigated the alteration, and the comparison of salted-roasted peanut and raw peanut properties was essentially needed to be conducted. Physicochemical characteristics were represented by parameters such as moisture content, ash content, fat content, protein content, antioxidant activity, color, mean geometric peanut kernel diameter, kernel size distribution, hardness, and fracture parameter. Meanwhile, thermogravimetric analysis (TGA) result described thermal characteristic. The salting-roasting process produced the decreasing of moisture, protein, fat, antioxidant activity expressed by antiradical power, and CIE Lab color values (lightness (L^*), redness (a^*), and yellowness (b^*)), compared with that of raw peanut. Meanwhile, the ash content of the peanut was increased by the process. All of the changes were significant ($p<0.05$) statistically. The process reduced the mean geometric peanut kernel diameter ($p<0.05$) and yielded a narrower kernel size distribution. According to TGA result, the roasted peanut was more stable thermally, which was shown by the shifting of endothermic peak to the higher temperature. The enhancing of palatability of peanut by salting-roasting was proven by hardness decreasing ($p<0.05$), although the fracture was insignificantly increased ($p>0.05$).

1. Introduction

Peanut or groundnut classified as *Arachis hypogaea* was dominantly produced by Asia's countries. China and India respectively occupy the first and second of overall world peanut production [1]. Indonesia with other southeast countries like Myanmar and Vietnam also produce peanut substantially [2]. Even, fifth-ranking position of world peanut production is occupied by Indonesia [3]. Indonesia's productivity of peanut is in the sixth position after cassava, sweet potato, paddy, maize, and soybean and the area of peanut production was dominated by Jawa Island's province [4]. DI Yogyakarta province is third peanut crops area which the contribution to national crops total is 12.46%. Based on BPS-Statistics of D.I. Yogyakarta Province 2018, Gunungkidul regency leads in harvester area of peanut which the contribution is 90.44% [5].

According to USDA National Nutrient database, peanut is nutritious food that contains protein (25.80%), fat (49.24%), fiber (8.5%), vitamin, mineral [1]. It also consists of bioactive components like arginine, resveratrol, phytosterol, phenolic acid, and flavonoid that involve with coronary heart disease risk decreasing and chemopreventive activity [6]. Peanut can be served and consumed in raw or processed form such as boiled, roasted, fried peanut and also other added value products such as peanut butter and peanut candy [7]. Peanut processing, one of them is roasting, is needed to enhance palatability and consumer preference including color, flavor, and texture attribute. Roasting was also aimed to destroy microorganism, food contaminant, and other toxins [8]. However, thermal processing including roasting can alter the properties of peanut because of its high heat penetration [9].

Besides roasting, salting process is used for taste and flavor food enhancing. The process plays a role in food properties changes because mass transfer occurs by the process. Water is released out through semi-permeable food membrane because of the osmotic pressure gradient. Meanwhile, the salt substance can be lost from water, penetrate, and enter into food matrix through the membrane. This salt entering will influence the taste and other food properties [10].

BIMARAM as one Small Medium Enterprise (SME) in Gunungkidul regency produces various peanut-based food, one of them is traditional salted-roasted peanut snack. The snack consumption can fulfill daily nutrient adequacy. However, the peanut processing firmly involved with thermal processing can alter the physicochemical and thermal characteristics properties including the nutrient. By this research, the physicochemical and thermal characteristics of salted-roasted peanut were investigated. Moisture content, ash content, fat content, protein content, and also antioxidant activity, color, hardness, fracture, peanut size, and peanut size distribution were the parameters which describe physicochemical properties. Meanwhile, the thermal properties were represented by a thermogravimetric property. Regarding with macronutrient (recommended dietary allowance) fulfillment, it is imperative to consider moisture, fat, protein content of processed peanut. The ash content interpreting mineral along antioxidant activity had to be studied because they refer to human body micronutrient. Meanwhile, the hardness, fracture, peanut size and size distribution are involved with sensory palatability and visual preference [3]. The effect of roasting on peanut had been studied by many researchers. However, the simultaneous salting-roasting have not been enormously investigated yet. The properties of salted-roasted and raw peanut are needed to be compared and discussed for investigating how far the salting-roasting process can alter the peanut properties.

2. Material and methods

2.1. Material

Raw peanut seed used by BIMARAM SME was obtained from Rongkop district harvesting. The raw peanut had been dried traditionally by sun drying, and the peanut kernel had been separated from the shell. For peanut properties analysis, some chemical was used. Petroleum ether, H_2SO_4, Na_2SO_4, HgO, NaOH, $Na_2S_2O_3$, HCl 37%, and HBO_3 were obtained from Merck (USA). 1,1-Diphenyl-2-picrylhydrazyl (DPPH) was obtained from Sigma-Aldrich (USA). Ethanol was purchased from Merck (USA).

2.2. Methods

2.2.1. Salted-roasted peanut preparation. Salted-roasted peanut was prepared by BIMARAM SME. Briefly, the raw peanut was soaked with boiling water for a moment, afterward, the peanut was cleaned by tap water. The seasoning dominated by salt was given to cleaned peanut so that the step can be mentioned as salting. The salting peanut was dried under sunlight until its moisture was reduced. Dry roasting was traditionally conducted by using a white sand medium. The roasted peanut was packaged with an air-tight plastic bag until it would be analyzed.

2.2.2. Physicochemical properties analysis. The physicochemical properties were represented by moisture content, ash content, fat content, protein content, antioxidant activity, color, mean geometric peanut kernel diameter, kernel size distribution, hardness, and fracture. The proximate (chemical compositions) of peanut consisting of moisture, protein, fat, and ash content were determined by reference based on the Association of Official Analytical (AOAC, 1995) [11]. Sample preparation before analysis was conducted by grinding peanut kernel with mortar and pestle by a method of Abdulsalami and Sheriff, 2010 [12]. Crude protein was determined by using the micro-Kjeldahl method (AOAC, 928.08). Soxhlet extraction with petroleum ether was used for determining crude fat (AOAC, 920.39c). Moisture content was analyzed by thermogravimetric method (AOAC, 950.46). The thermogravimetric method was also used for determining crude ash (AOAC, 923.03).

The antioxidant activity expressed by antiradical power (ARP) was determined by DPPH antiradical procedure [13]. However, the procedure was modified for microplate method. Before DPPH antiradical analysis, fine powder peanut defatted by using petroleum ether was extracted by ethanol 80% at room temperature for overnight. The suspension was centrifuged to get clear supernatant. The supernatant was hereinafter mentioned as a sample solution. The dilution of the sample solution and DPPH was carried out by using ethanol 80% for creating them into various concentration. The principal of DPPH antiradical method is the reaction between sample solution containing antiradical activity and purple-colored radical DPPH solution to form yellow-colored non-radical (reduced-radical) solution. For the initial step, DPPH calibration curve was essentially made by plotting some concentration DPPH (0-100 μM) versus their absorbance. The standard curve could be utilized for looking for the DPPH concentration for both initial and remaining DPPH concentration which the terms will be further explained. Afterward, about 100 μl sample solution and its diluted solution with various concentration were poured to 96-well microplate. Then, about 100 μl DPPH solution 75 μM was added to wells containing sample solution. The mixing was called a DPPH-sample mixed solution. The mixing of 100 μl DPPH solution 75 μM and 100 μl ethanol 80% was made as control depicting unreacted DPPH solution by any antiradical sample solution, and that was called as a control solution. The blank sample solution had to be made for correcting sample color (if sample color could not be ignored) by mixing of 100 μl sample solution and 100 μl ethanol 80%. Meanwhile, the blank solution used in this analysis was 200 μl ethanol 80%. After the pouring to microplate was finished, the microplate was shaken in Elisa reader for 10 seconds (Thermo Scientific Multiskan GO Type 1510, Thermo Fisher Scientific, Finland) and the absorbance was measured at 515 nm by using the reader. The absorbance of control solution of this step was utilized to look for initial DPPH concentration by referring absorbance value to the standard DPPH curve. Meanwhile, the step was resumed by incubating microplate for 30 minutes at the dark condition to allow the reaction of DPPH and antiradical sample. The absorbance was measured at the same wavelength. The further step was the finding of initial and remaining DPPH concentration for calculating remaining DPPH percentage according to the following equation:

$$\%\ \text{remaining DPPH} = \frac{[\text{remaining DPPH}]}{[\text{initial DPPH}]} \times 100\%$$

The initial DPPH determination was conducted by subtracting blank solution absorbance from control solution absorbance. On the other hand, remaining DPPH concentration was obtained by subtracting blank sample solution absorbance from DPPH-sample mixed solution absorbance. The plot of sample solution concentration versus remaining DPPH percentage was used to determine the half maximal effective concentration (EC50) figuring sample concentration that causes a 50% DPPH decrease. For clarity, another term used for expressing antioxidant activity is antiradical power (ARP = 1/EC50), in which higher ARP value represents higher antioxidant activity [14].

The measurement of peanut seed diameter was adopted from Kurt and Arioglu, 2018 [15]. The peanut kernel diameter was measured by using digital caliper with the accuracy of 0.01 mm. Principal dimension on peanut size measurement was described as following: major axis was expressed as Length (L), the intermediate axis was expressed as Width (W), the minor axis was expressed as Thickness (T). The geometric diameter (*De*) was determined according to the following equations developed by Polat et al., 2007 [16]:

$$D_e = (LWT)^{1/3}$$

IOP Conf. Series: Earth and Environmental Science **251** (2019) 012044 doi:10.1088/1755-1315/251/1/012044

Texture profile analysis was performed by using TA1 texture analyzer (Llyod Instruments) with 10 mm/s test speed and 1N preload/stress. Hardness and fracture were chosen as texture attributes because they represented palatability parameters of peanut.

Peanut color was identified colorimetry by chromameter (Konica Minolta CR-80) according to Priatková method [17]. The color is expressed as CIE L*a*b*. Value of L* indicates lightness with 0 refers to dark, and 100 refers to white, redness/greenness was indicated by a* value, in which positive (+) value refers to redness and negative (-) value refers to greenness. Meanwhile, b* indicates yellowness/blueness in which yellowness and blueness are interpreted from positive (+) and negative (-) value respectively.

2.2.3. Thermogravimetry analysis. The thermal profile of peanut was obtained by using thermogravimetric analysis (TGA) analyzer (DTG-60 Shimadzu) in which the heating temperature was of 10°C/min and the nitrogen was used as purging gas with the flow rate 30 ml/min. The sample was heated at a temperature range of 30-300 °C.

2.2.4. Statistical analysis. Collected data were statistically analyzed by SPSS 12. The independent sample t-test was used for obtaining significance of the difference among means of raw and salted-roasted peanut. Confidence interval used in this analysis was 95%.

3. Results and discussion

3.1. Physicochemical properties

A salting-roasting process involving with salt (sodium mainly) adding and heat treatment affect the proximate properties and antioxidant activity of peanut, it can be seen in Table 1. The table shows that salting and roasting caused decreasing in moisture, protein content, and also anti-radical power of peanut. However, ash content increased by the salting-roasting process.

Table 1. The proximate properties and antioxidant activity of raw and salted-roasted peanut.

Parameters	Raw peanut	Salted-roasted peanut
Moisture (% WMB)	8.68 ± 0.06^{a}	4.31 ± 0.28^{b}
Ash content(% DMB)	2.17 ± 0.17^{a}	3.87 ± 0.07^{b}
Protein content (% DMB)	31.83 ± 0.46^{a}	30.66 ± 0.51^{b}
Fat content (% DMB)	49.55 ± 0.06^{a}	45.21 ± 1.64^{b}
Antiradical Power (mg DPPH/mg dried peanut)	1.06 ± 0.13^{a}	0.44 ± 0.06^{b}

Same superscript letters at same row indicate insignificant differences

[1] WMB = wet mass basis

[2] DMB = dry mass basis

Moisture of raw peanut was higher than that of salted-roasted peanut. It is clear that high temperature (exactly more than 100°C) produced by roasting can cause dehydration (water releasing). Water was released from the food system by vaporization. In addition, at the high-temperature roasting process, dehydration is not only involved with free water but also bound water which will be released by roasting. The higher the roasting temperature, the more water can be released. Salt (mainly sodium and chloride) substances applied for peanut salting can also induce water-out of food matrix through osmotic dehydration process [10]. The result is useful for food preservation because the microorganism usually lives in a moist environment.

The salting-roasting process improved the ash content of peanut because of mineral substances mass transfer. Mineral substances composing salt (sodium and chloride) can enter into peanut matrix. Moreover, the media used for dry roasting was sand obtained Gunungkidul's beach, in which calcium, magnesium, and other mineral substances compose the sand [18]. It is predicted that there was mass transfer the substance to the food matrix.

The decrease in protein and fat mainly was induced by high thermal roasting. Our finding agreed with an investigation about peanut soluble protein decreased by air-oven roasting [19]. Non-enzymatic protein-carbohydrate reaction and denaturation of protein strengthened by high temperature were involved in the phenomenon [20]. Especially in this research, the protein was also reduced by treatment before the roasting process (peanut soaking with boiling water). Although the process took place for a moment, it allowed the soluble protein to be released from the food matrix. Meanwhile, the declining peanut fat by roasting suited the oven-air roasting of peanut conducted by Makeri et al., 2011 [21]. The roasting process disrupts peanut tissue so the fat (oil) can be liberated from it [9]. If roasting temperature reaches certain peak, the fat (oil) can permeate towards at surface and will be separated from food system [21].

Regarding peanut antioxidant activity, the antiradical power of roasted peanut was less than that of raw peanut. Phytochemicals, such as polyphenol and flavonoid, contributing peanut antioxidant activity are concentrated in peanut skin [22]. Phytochemical can be degraded easily by many external factors and the skin is the early part of peanut exposed by heat processing, so the finding that raw peanut had more efficient antioxidant activity is proper.

Salting-roasting influenced the kernel peanut geometry by reducing diameter. The difference between roasted-peanut diameter and raw peanut diameter was significant ($p<0.05$). It can be seen from Table 2.

Table 2. Mean geometric diameter of raw and salted-roasted peanut.

Parameters	Raw peanut	Salted-roasted peanut
Mean geometric diameter	9.62 ± 0.65^a	9.24 ± 0.53^b

Same superscript letters at same row indicate insignificant differences

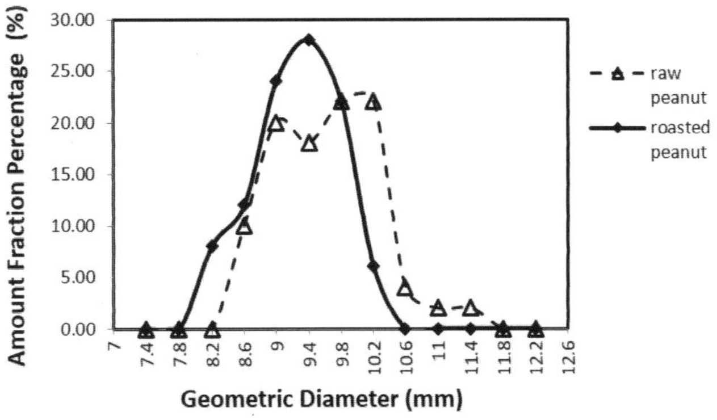

Figure 1. Peanut kernel size distribution

According to Figure 1, salting-roasting brought the kernel size distribution to be narrower than the raw peanut. It might be caused by simultaneous moisture, fat, and protein component lost. The phenomenon may be a trigger in structure modification; furthermore, it can affect geometric properties. The differences among means of raw peanut and salted-roasted peanut were significant ($p<0.05$). This implies that the influence of salting-roasting process has to be considered in peanut-based nutrient adequacy.

According to color data of Table 3, salting-roasting influenced significantly for all color properties. Lower L* value of salted-roasted peanut indicated that it was darker than the raw peanut. The darkness of roasted peanut resulted from a non-enzymatic browning reaction. The lowering a* and b* value of salted-roasted peanut indicated phytochemical degradation. The value of a* and b* of peanut contribute to polyphenol and proanthocyanidin respectively [22].

Table 3. Color properties of raw and salted-roasted peanut.

Parameters	Raw peanut	Salted-roasted peanut
L*	46.93 ± 0.12^{a}	46.60 ± 0.86^{b}
a*	11.83 ± 0.74^{a}	7.43 ± 0.47^{b}
b*	14.63 ± 0.42^{a}	12.87 ± 0.59^{b}

Same superscript letters at same row indicate insignificant differences

Salting-roasting could change textural properties of peanut by decreasing hardness of raw peanut ($p<0.05$), although the fracture value of salted-roasted peanut was higher. Fortunately, the fracture difference was not significant ($p>0.05$). This result gave an advantage because the harder snack is usually avoided by the consumer.

Table 4. Textural properties of raw and salted-roasted peanut.

Parameters	Raw peanut	Salted-roasted peanut
Hardness (N)	71.57 ± 7.11^{a}	20.97 ± 2.31^{b}
Fracture (N)	8.73 ± 2.96^{a}	9.58 ± 2.10^{a}

Same superscript letters at same row indicate insignificant differences

3.2. Thermal Properties

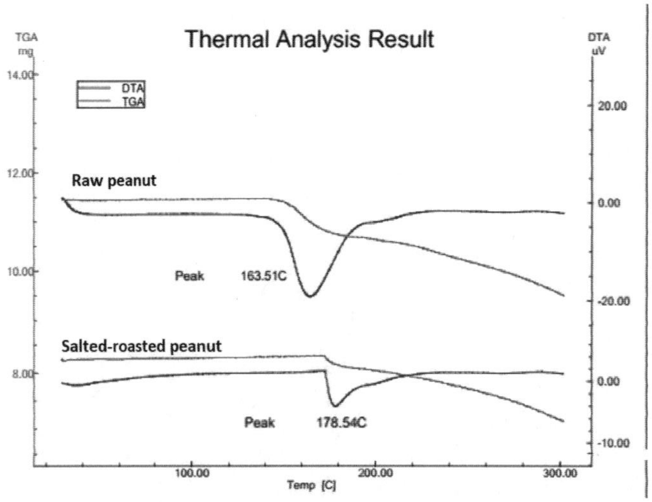

Figure 2. Thermal analysis result of raw and salted-roasted peanut.

Thermogravimetric analysis (TGA) profile of both raw peanut and salted-roasted peanut was depicted in Figure 2. The endothermic peak of both raw and roasted peanut from 100-200 °C indicated dehydration. It can be seen that the roasting process shifted the endothermic peak of peanut (properly 163.51) to the higher temperature (178.54°C). The roasting process can modify food structure so that it can influence the releasing of water (both free and bound water) from the food matrix.

4. Conclusions

Salting-roasting process alters the physicochemical and thermal characteristics of the peanut. Roasting especially brought peanut to be more stable thermally, because the free water vaporization was accomplished by roasting. The salting-roasting also made geometrically, and texture properties of peanut met palatability and visual consumer preference by improving size distribution and reducing the hardness. Generally, physicochemical properties (moisture, protein, fat, and antiradical power) was decreased by roasting. Even though, increasing in ash content happened. It was involved with osmotic dehydration and mineral mass transfer during salting, and also high temperature treating to the peanut during the roasting process. Moisture decline by the roasting process is useful because it aids food preservation. However, regarding with macronutrient (protein, ash, fat) and antiradical power uptake for the human body, the research is needed to be reinvestigated to search available traditional salting-roasting condition, including roasting temperature or time factor by using more standard roasting equipment.

5. Acknowledgments

This research was supported by PRIORITAS NASIONAL 2018 program of Indonesian Institute of Sciences. The authors also thank Ms. Arbiyanti Restuti for helping the chemical properties analysis and also to BIMARAM SME for providing the raw and salted-roasted peanut.

6. References

[1] Arya S S, Salve A R and Chauhan S 2016 Peanuts as functional food: a review *J. Food. Sci. Technol.* **53** (1) 31–41
[2] Kole C 2014 *Genetics, Genomics and Breeding of Peanuts* (Boca Raton: CRC, Taylor & Francis Group)
[3] Rozalli N H M , Chin N L and Yusof Y A 2012 Simultaneous multiple responses modelling, optimisation and correlation of asian type peanuts (arachis Hypogaea L.) roasting using response surface methodology *Acta Alimentaria* **43**(1) 142-157
[4] BPS-Statistics Indonesia 2017 *Statistical Yearbook of Indonesia* (Yogyakarta: CV. Dharmaputra) p 223
[5] BPS-Statistics of D.I. Yogyakarta Province 2018 *Daerah Istimewa Yogyakarta Province in Figures 2018* (Yogyakart: Sinar Baru Offset) p 215
[6] Gama A P, Adhikari K and Hoisington D A 2018 Peanut consumption in malawi : an opportunity for innovation *Foods* **7** (112) 1-13
[7] Varshney R K, Pandey M K, and Puppala N 2017 *The Peanut Genome* (Springer International Publishing) p 165
[8] Beuchat L R and Mann D A 2011 Inactivation of salmonella on pecan nutmeats by hot air treatment and oil roasting *Journal of Food Protection* **74** (9) 1441–1450
[9] Kita A and Figel A 2007 Effect of parameters of thermal process on the properties of peanuts *Pol.J.Food Nutr.Sci.* **57** (4B) 285-289
[10] Antonio G C Azoubel P M, Murr F E X and Park K J 2008 Osmotic dehydration of sweet potato (Ipomoea batatas) in ternary solutions *Ciênc. Tecnol. Aliment., Campinas* **28** (3) 696-701
[11] AOAC Methods: 965.33, 940.28. Official Methods of Analysis of AOAC International, 16th edn., 1995. The Association of Analytical Chemists
[12] Abdulsalami M S and Sheriff H B 2010 Effect of processing on the proximate composition and mineral content of bambara groundnut (voandezeia subterranean) *Bayero Journal of Pure and Applied Sciences* **3** (1) 188-190
[13] Brand-Williams W Cuvelier M E and Berst C 1995 Use of a free radical method to evaluate antioxidant activity *Lebensm.-Wiss.u.-Technol* **28** 25-30
[14] Mishra M, Ojha H and Chaudury N K 2012 Estimation of antiradical properties of antioxidants using dpph assay: a critical review and results *Food Chemistry* **130** 1036-1043
[15] Kurt C and Arioglu H 2018 Physical and Mechanical Properties of Some Peanut Varieties Grown in Mediterranean Environment *Cercetari Agronomice in Moldova* **51**(2) 27-34
[16] Polat R, Aydin C and Erol K 2007 Some physical and mechanical properties of pistachio nut *Bulgarian Journal of Agricultural Science* **13** 237-246
[17] Priatková L, Hlaváčová Z and Kertész Á 2010 Hunter colour determination of blueberry cultivars *Journal of Food Physics ÉLELMISZERFIZIKAI KÖZLEMÉNYEK* **XXIII** 18-21
[18] Wildan D M, Affandi R, Pratiwi N T M, Krisanti M, Ayu I P and Iswantari A 2017 *IOP Conf. Series: Earth and Environmental Science* **54** 012070

[19] McDaniel K A, White B L, Dean L L, Sanders T H and Davis J P 2012 Compositional and mechanical properties of peanuts roasted to equivalent colors using different time/temperature combinations *Journal of food science* **77** (12) C1293-C1299

[20] El-Beltagi H E S 2011 Effect of roasting treatments on protein fraction profiles, some enzyme activities of Egyptian peanuts, *International Journal of Food Sciences and Nutrition* **62** 453–456

[21] Makeri M U, Bala S. M and Kassum A S 2011 The effects of roasting temperatures on the rate of extraction and quality of locally-processed oil from two Nigerian peanut (Arachis hypogea L.) cultivars *African Journal of Food Science* **5** (4) 194-199

[22] Chukwumah Y, Walker L T and Verghese M 2009 Peanut skin color : a biomarker for total polyphenolic content and antioxidative capacities of peanut cultivars *Int.J.Mol.Sci.* **10** 4941-4952

Lactobacillus pentosus isolated from *Muntingia calabura* shows inhibition activity toward alpha-glucosidase and alpha-amylase in intra and extracellular level

A Frediansyah[1], Suryani, R Nurhayati[1], Miftakhussolikhah[1], J Sholihah

[1]Research Unit for Natural Product Technology, Indonesian Institute of Sciences Jl. Yogya-Wonosari km 31.5 Gading, Playen, Gunungkidul, Yogyakarta, Indonesia

Email: andri.frediansyah89@gmail.com

Abstract. *Lactobacillus pentosus* is one of the most prevalence lactic acid bacterium which available in plant-based fermentation. Here, we reported that the cell extract and cell-free supernatant of *L. pentosus* strain TL 2.7, TL 5.8 and TL 7.8 isolated from Kersen (*Muntingiacalabura* L.) showed potential anti-diabetic properties by inhibition of alpha-glucosidase and alpha-amylase activity *in vitro*. All isolated also enhanced the antioxidant capacity through DPPH *in vitro* test. This initial study will be beneficial for screening and application of functional food with anti-diabetic properties.

Keywords. *Lactobacillus pentosus*, inhibitor alpha-glucosidase, *Muntingia calabura,* Kersen, alpha amylase

1. Introduction

Diabetes mellitus type 2 is one of the significant contributions to the world's mortality rate. The report by the International Diabetes Federation in 2015 showed that about 415 million adults have diabetes[1]. Based on its prediction, the number of diabetes cases will increase 1.5 times higher by 2040. People with diabetes have poor insulin performance; this will lead to the elevation of blood glucose after food consumption. This stage is called postprandial hyperglycemia that results in numerous complication including diabetic foot ulcer, nephropathy, retinopathy, and neuropathy [2]. During the digestion in gastrointestinal (GI) tract, the 1,4 alpha glycosidic bond from the polysaccharide is cleaved by alpha-amylase to produce disaccharide or oligosaccharide, which later then hydrolyzed to monomers such as glucose in the presence of alpha-glucosidase. The resulting product then absorbed through the small intestine to the portal vein of the liver [3, 4]. Therefore, the presence of these inhibitors can inhibit or prevent the increasing level of blood glucose in our body.

Acarbose, voglibose, and miglitol are common synthetically drug for people with diabetes who are easily found in the market [5]. However, miglitol and voglibose remain not effective on the inhibition toward alpha-amylase. The use of synthetically material as a drug is often connected to the number of GI side effects. Moreover, about the three-quarters population with diabetes are living in non-developed countries and the high price of drugs often being one of the major problems [1]. Recently, bioactive compound is demanding to a massive group of people due to the high number of the positive outcome. This material also gives fewer side effects on the human body. Up to now, scientists have put their effort to find a way out of the risk of diabetes mellitus type 2 by finding alpha-amylase and alpha-glucosidase inhibitor from many sources such as fungi [6, 7], plants [8], marine creatures [9], and bacteria [10]. However, the research related bacteria with its anti-diabetic activity are still limited.

While bacteria especially lactic acid bacteria can be employed as a part of a functional food with anti-diabetic properties.

Muntingia calabura L, known as Kersen, is easily found in the land of Indonesia. The leaf reputed to possess anti-diabetic effect by *in vivo* study using alloxan-induced diabetic rats (*Rattus norvegicus*) [11]. Flavanoid content on the Kersen's leaf has a good activity toward alpha-glucosidase inhibition [12]. On the fruit part, *M. calabura* contain diabetic substance such as beta-carotene, riboflavin, niacin, thiamine, and ascorbic acid. Its anti-diabetic properties have been demonstrated by *in vivo* experiment using rats induced by streptozotocin [13]. Therefore, this study was conducted to isolate and characterize some potential bacteria with anti-diabetic activity from the fruit part of *M. calabura*.

2. Material and Methods

2.1. Materials
Kersen fruits (*M. calabura*) was obtained from the garden of the Research Unit for Natural Product Technology - Indonesian Institute of Sciences (BPTBA-LIPI), Yogyakarta, Indonesia in 2016.

2.2. Isolation of bacteria
The fruit sample was cleaned using tap water and sterilized by consecutive washes in 75% methanol for about 1 minute and rinsed with sterilized distilled water three times. About 20 grams of samples were taken and fermented with MRSB medium 100 ml at 37 °C for 48 hours. The isolation procedure was done using MRSA medium mixed with calcium carbonate by pour plate method. Colonies with halo zone were selected and purified on the MRSA medium by streak plate method. Several colonies with various shapes and size were picked and purified using MRSA medium by streak plate method and re-purified in the same medium for three times. The incubation was employed at 37 °C for 48 hours.

2.3. Characterization and molecular identification using 16S RNA
The single colony was then characterized based on physical and chemical properties including the shape of the cell, gram staining, motility, catalase test, and endospore test. For molecular identification, the bacterium was extracted using a modification method by Giraffa et al. 2000 [14]. The extracted DNA was used as a template for PCR amplification. DNA was then used as a template for PCR to amply the 27F (5'-AGAGTTTGATCCTGGCTCAG-3') and 1492R (5'-GGTTACCTTGTTACGACTT-3') from Promega. The result of sequencing was then compared with the National Center for Biotechnology Information (NCBI) GenBank database. The sequence was aligned with CLUSTALW parameter which includes pairwise and multiple alignments. Gap opening penalty and gap extension penalty were 2 and 0.8, respectively. A final phylogenetic tree was constructed using MEGA software (version 7.0.0).

2.4. Preparation of cell extract and cell-free supernatant
The strains of *L. pentosus* were inoculated in 5 ml MRSB medium, then incubated for 24 h at 37 °C. The medium contained cell then centrifuged at 4 °C, 10000 g, for 15 min. The biomass cell will occur at the bottom of the tube and resulting cell-free supernatant. The biomass cell was then extracted by ultra-sonication with methanol and subsequently filtrated with 0.22 μm membrane filter (Millipore). The supernatant and extracted cell were used for the activity assay.

2.5. Measurement of pH
The pH of bacterial activities during incubation was measured at 24, 48, 72 and 96 hours. The test was calculated with three independent replications.

2.6. Antioxidant scavenging activity
The 1.1-diphenyl-2-picrylhydrazyl called as DPPH; scavenging assay was performed using the 96-well plate as the previous report in Frediansyah et al. 2017 [15]. Briefly, a filtered cell-free supernatant was mixed with methanol, and the absorbance of the pre-plate reading was recorded at 517 nm. DPPH solution in methanol was then added to the well. The changing color from purple to yellow visualized the scavenging efficiency of the cell-free supernatant. After the incubation period of 30 min at room

temperature in the darkness, the change of the absorbance was monitored at 517 nm in the Multiskan® Go micro-plate spectrophotometer (Thermo Scientific, Vantaa, Finland). The reduction of absorbance in the reaction showed a higher free radical-scavenging activity. The scavenging activity against DPPH was calculated using the equation (a):

$$\text{Scavenging rate (\%)} = [1 - (\text{AbsB} - \text{AbsA})] \times 100\ \% \qquad \text{(a)}$$

Where AbsA was the absorbance of control and AbsB was the absorbance in the presence of cell-free supernatant.

2.7. Alpha-glucosidase inhibition activity
The inhibition activity toward alpha-glucosidase from cell extract and cell-free supernatant were analyzed according to previous reports [15]. In brief, the mixture contained the sample, alpha-glucosidase, and sodium phosphate buffer. The mixture was then incubated at 37 °C for 15 min. After pre-incubation, the enzymatic reaction was initiated by implemented *p*-nitrophenyl-α-D-glucopyranoside solution in the phosphate buffer. The resulting reaction mixture was incubated for 20 min at 37 °C. The absorbance was subsequently measured at 405 nm in the Multiscan® Go microplate spectrophotometer (Thermo Scientific, Vantaa, Finland). Percent inhibition was calculated by a formula (b):

$$\text{\% inhibitor} = [(\text{Abs1} - \text{Abs2})/\ \text{Abs1}] \times 100\ \% \qquad \text{(b)}$$

Where Abs1 is the absorbance of the control and Abs2 is absorbance in the presence of the sample.

2.8. Alpha-amylase inhibition activity
The inhibition activity toward alpha-amylase from cell extract and cell-free supernatant were measured according to Telagari and Hullatti, 2015 [16]. The test system was performed in a 96-well plate. Briefly, the mixture contained the sample, alpha-amylase, and sodium phosphate buffer. The mixture was then incubated at 37 °C for 20 min. After pre-incubation, the soluble starch was then added to a sodium phosphate buffer as a substrate. Then, the aliquot was incubated further at 37 °C for 30 min, followed by adding 3.5-di-nitrosalicylic acid (DNS) solution and heated at 100 °C for 10 min. The absorbance was subsequently measured at 540 nm on Multiskan® Go microplate spectrophotometer (Thermo Scientific, Vantaa, Finland). The percentage of α-amylase inhibition was calculated by the same formula in 2.7.

3. Result and Discussion

3.1. Strain characterization and taxonomical position
Three strains of *L. pentosus* from *M. calabura* fruit have been isolated and characterized based on physical and chemical properties, and also 16S rRNA. Based on physical and chemical properties as shown in Table 1, these three strains of *L. pentosus* belongs to lactic acid bacteria. This bacterial clade gives the positive reaction to gram staining, negatively react to catalase, non-motile and not forming spore [17]. Based on information from 16S rRNA gene sequencing, TL 1.7, TL 5.8 and TL 7.8 have 100%, 99% and 99 % similarity to *L. pentosus* strain 124-2, respectively.

Table 1. Physical and chemical properties of three strains of *L. pentosus* isolated from *M. calabura*

Strains	Cell form	Gram staining	Catalase test	Motility	Endospore	Identity
TL 1.7	Rod	+	-	-	-	Lactic acid bacteria
TL 5.8	Rod	+	-	-	-	Lactic acid bacteria
TL 7.8	Rod	+	-	-	-	Lactic acid bacteria

The taxonomical tree, as shown in Figure 1, showing the relative position of *L. pentosus* strain TL 2.7, TL 5.8 and TL 7.8 as inferred by the neighbor-joining method with the 16S rRNA gene sequence

IOP Conf. Series: Earth and Environmental Science **251** (2019) 012045 doi:10.1088/1755-1315/251/1/012045

from *L. pentosus* strain 124-2 as obtained from National Center for Biotechnology Information (NCBI) as the in-group sequence. The optimal tree has branch length 0.2269 and bootstrap values for a total 1,000 replicates. The evolutionary distances were computed using the Kimura 2-parameter method and in the units of the number of base substitutions per site.

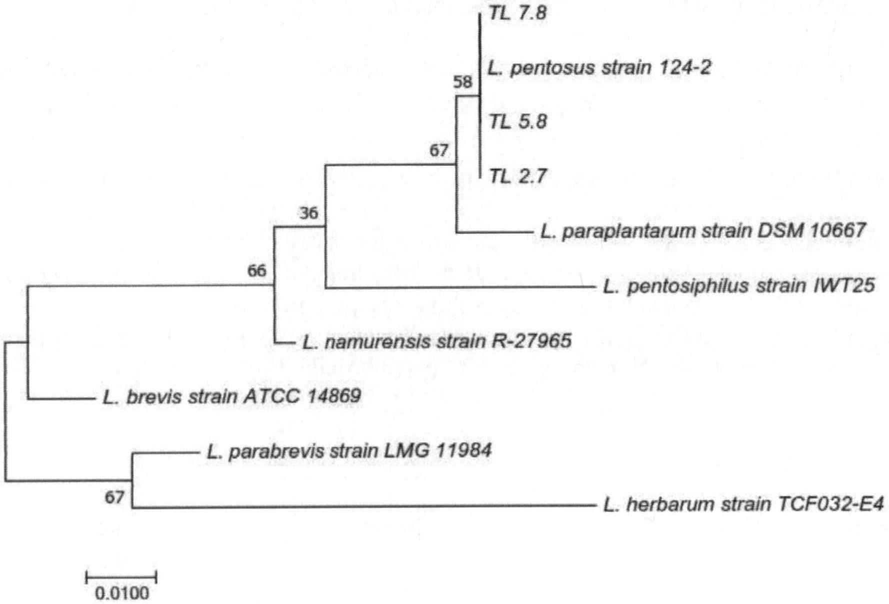

Figure 1. Phylogenetic relationship of *L. pentosus* TL 2.7, TL 5.8 and TL 7.8 based on 16S rRNA gene sequence.

3.2. Antioxidant activity and pH

The antioxidant activity from the supernatant of MRSB medium containing *L. pentosus* was investigated using DPPH free radical scavenging activity assay as shown in Figure 1. The reducing capacity of antioxidant was measured spectrophotometrically based on the color change from dark violet to yellow. The presence of an antioxidant in the bacterial medium after incubation can decrease the free radical by electron donor or hydrogen atom transfer.

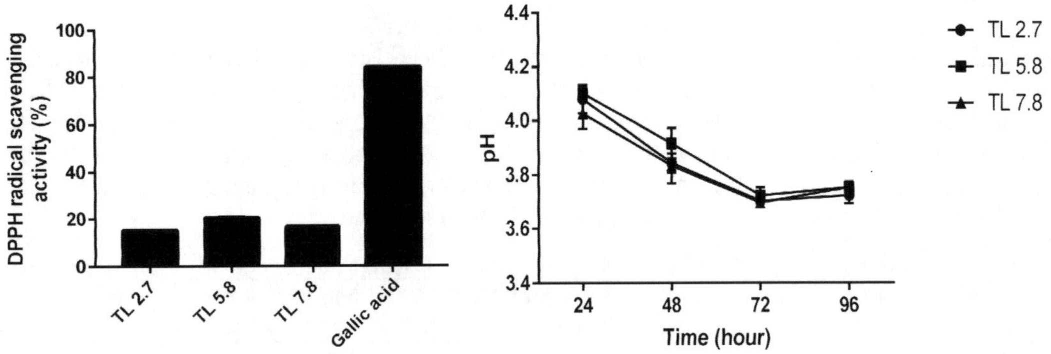

Figure 2. Antioxidant level of *L. pentosus* in MRSB using DPPH radical scavenging activity assay (left) and pH level of MRSB medium containing *L. pentosus* during 96h incubation (right).

All strains of *L. pentosus* showed scavenging activity (Figure 1). However, these activities only show about 4.12 to 5.53 times lower compare with Gallic acid as a positive control. Each result was significantly different at $p \leq 0.01$ using ANOVA followed by Tukey's HSD test. The value for TL 2.7, TL 5.8 and TL 7.8 are 15.21%, 20.44% and 16.68%, respectively. This finding revealed that *L. pentosus* is valuable in enhancing the antioxidant level in MRSB medium. Nevertheless, this result showed lower scavenging activity compared with *L. plantarum* in the medium containing black grape juice [16]. Additionally, the pH level decreased during incubation. Hachemet al 2005 [18] stated that the pH could enhance the activity of an enzyme, while it also could improve the bioavailability of compounds. The elevation of antioxidant level due to the presence of bacteria in this experiment is in agreement with many reports.

3.3. Inhibition of alpha-glucosidase and alpha amylase

Postprandial hyperglycemia is the valuable target as first-line therapy for people with diabetes. One of the ways to reduce that step is to retard the monosaccharide absorption by inhibiting alpha-glucosidase and alpha-amylase. Many studies have been conducted to inhibit these enzymes by the natural compound from a plant, including seeds, roots or fruits. Limited research has been employed on the anti-diabetic potential of lactic acid bacteria.

Cell-free supernatant and cell extract of each strain of *L. pentosus* showed good inhibition against two enzymes related to diabetes (Figure 2). Among them, cell extract showed higher restraint to alpha-glucosidase compare to the cell-free supernatant group. However, this finding is not applicable to the inhibition of alpha-amylase. Among three strains of *L. pentosus,* TL 2.7 showed the highest inhibition activity again α-glucosidase and α-amylase to both from its cell-free supernatant and cell extract. This result is not far different with these activities from lactic acid bacteria isolated from *Canna edulis* and *Xanthosoma sagittifolium* [19] and *Ganoderma lucidum* [20]. *L. pentosus* has a good potential as inhibitor alpha-amylase which is opposite to *Paenibacillus spp.* [21],*L. plantarum* [22, 23], and *L. acidophilus* [24] which are good as amylase producer. However, acarbose, a positive control in this experiment, showed higher inhibitor activity with the value of 88.22% compared with those three strains, but only restricted to inhibition of alpha-glucosidase. In detail, the inhibition of alpha-glucosidase by cell extract of TL 2.7, TL 5.8 and TL 7.8 are 1.04, 1.070 and 1.27 times lower, respectively. From the cell-free supernatant with similar order showed 1.26, 1.29 and 1.33, respectively. While for α-amylase inhibitor showed the opposite trend, all those results both from the cell-free supernatant and cell extract gave higher inhibition compare to Acarbose 1 mg.

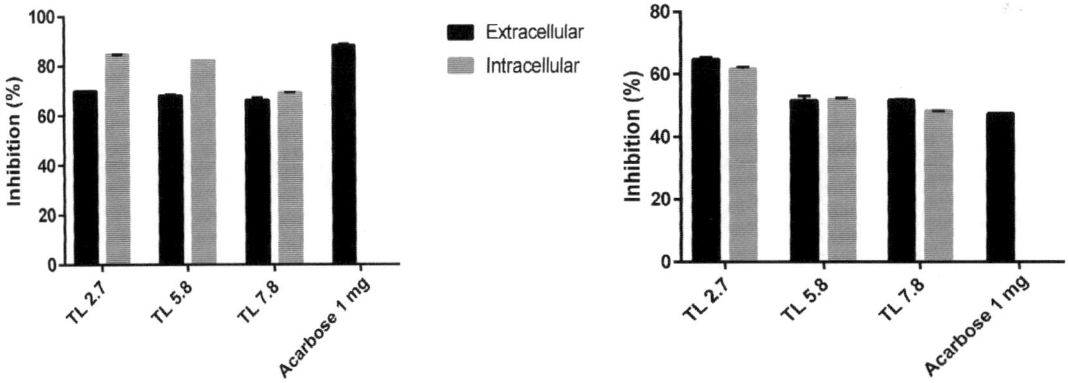

Figure 3. Alpha-glucosidase and alpha-amylase inhibition activities of *L. pentosus* strain TL 2.7, TL 5.8 and TL. 78. The left graph (A) represented inhibition α- glucosidase from an extract of the cell and cell-free supernatant from various strains of *L. pentosus.*

4. Conclusion

The current *in vitro* study showed that *L. pentosus* (TL 2.7, TL 5.8 and TL 7.8) has good potential as anti-diabetic properties by inhibition of alpha-glucosidase and alpha-amylase activity. This bacteria is also valuable to enhance antioxidant capacity. This initial study will be beneficial for screening and application of functional food with anti-diabetic properties.

5. Acknowledgment

We gratefully acknowledge the financial support of internal sources funding of DIPA LIPI. We also thank lab members in Food and Microbiology Laboratory for their valuable contributions during preparation.

6. References

[1] IDF 2015 *IDF Diabetes Atlas* Sixth Edition, International Diabetes Federation 2015
[2] American Diabetes Association 2014 *Diabetes care* **37** Supplement 1 S81-S90
[3] Lordan S, Smyth T J, Soler-Vila A, Stanton C, Ross R P 2013 *Food chemistry* **141** 3 2170-6
[4] Docherty N G, le Roux C W 2016 *Current opinion in gastroenterology* **32** 2 61-6
[5] DiNicolantonio J J, Bhutani J, O'Keefe J H 2015 *Open heart* **2** 1 e000327
[6] Artanti N, Tachibana S, Kardono L B, Sukiman H 2011 *Pakistan journal of biological sciences PJBS* **14** 22 1019-23
[7] Wang C, Guo L, Hao J, Wang L, Zhu W 2016 *Journal of natural products* **11** 2977-81
[8] Kumar S, Narwal S, Kumar V, Prakash O 2011 *Pharmacognosy reviews* **5** 9 19
[9] Lee S H, Park M H, Heo S J, Kang S M, Ko S C, Han J S, Jeon Y J 2010 *Food and Chemical Toxicology* **48** 10 2633-7
[10] Zhu Y P, Yin L J, Cheng Y Q, Yamaki K, Mori Y, Su Y C, Li L T 2008 *Food chemistry* **109** 4 737-42
[11] Herlina H, Amriani A, Solihah I, Sintya R. 2018 *Science and Technology Indonesia* **3** 1 7-13
[12] Hong H C, Li S L, Zhang X Q, Ye W C, Zhang Q W 2013 *Chinese medicine* **8** 1 19
[13] Pramono V J, Santosa R 2015 *Jurnal Sain Veteriner* **32** 2
[14] Giraffa G, Rossetti L, Neviani E 2000 *Journal of Microbiological Methods* **42** 2 175-84
[15] Frediansyah A, Nurhayati R, Romadhoni F 2017 *AIP Conference Proceedings* **1803** 1 020022 AIP Publishing
[16] Telagari M, Hullatti K 2015 *Indian journal of pharmacology* **47** 4 425
[17] Lin W H, Hwang C F, Chen L W, Tsen H Y 2006 *Food Microbiology* **23** 1 74-81
[18] Hachem J P, Man M Q, Crumrine D, Uchida Y, Brown B E, Rogiers V, Roseeuw D, Feingold K R, Elias P M 2005 *Journal of investigative dermatology* **125** (3) 510-20
[19] Nurhayati R, Frediansyah A, Rachmah D L *IOP Conference Series Earth and Environmental Science* **101** 1 012009 IOP Publishing
[20] Nurhayati R, Frediansyah A, Rahmawati F, Retnaningrum E, Sembiring L 2015 *Proceeding of Internationalon Appropriate Technology Development* **1**
[21] Frediansyah A, Sudiana, I M 2013 *Widyariset* **16** 2 201-210.
[22] Frediansyah A, Kurniadi M 2017 *AIP Conference Proceedings* **1788** 1 030111 AIP Publishing
[23] Frediansyah A 2018 *Cassava* InTech **1** 123-138
[24] Frediansyah A, Kurniadi M 2016 *Nusantara Bioscience* **8** 2 207-2014

Quality of Chocolate Bar from Fermented Cocoa Beans from Lombok, West Nusa Tenggara

Fahrurrozi[1*], S Fauziyyah[1], M N Sari[1], S Ratnakomala[1], P Lisdiyanti[1]

[1]Research Center for Biotechnology, Indonesian Institute of Sciences (LIPI)

[*]e-mail: fahrurrozi@lipi.go.id

Abstract. About 500 kgs cocoa beans have been fermented using box fermentation. The fermentation experiments were conducted at the Gangga district North Lombok, West Nusa Tenggara - Indonesia. The starter culture, which contains yeast, lactic acid bacteria, and acetic acid bacteria (10^7/ mL) was added to cocoa beans mass in box fermentation. The fermentation was conducted for 5 days and every 24 hours were mixed. After fermentation, the cocoa beans were dried for 5 days to reduce the moisture content using a sun drying. Dried fermented cocoa beans were processed to be a chocolate bar. The quality of cocoa liquor was determined using organoleptic score. Chocolate aroma, chocolate flavour, and caramelly score were 8.25. The texture and colour score were 8.00 and off flavours score was 0.00. Overall quality of cocoa liquor score was excellent. Analysis of proximate and microbiological properties was carried out to obtain the chocolate bar qualities. The result showed that quality of chocolate bar from fermented cocoa beans appropriate with SNI 7934:2014 about chocolate and chocolate products.

Keywords: Theobroma cacao L, box fermentation, starter culture, chocolate bar.

1. Introduction

Cocoa is a fruit seed from the tree of *Theobroma cacao* L., which can grow in tropical regions such as Indonesia, Central, and South America and West Africa [1,2,3]. One of the most preferred cocoa beans products is chocolate because it has a unique taste and has a positive effect on health due to it contains a number of antioxidants.

The fermentation of cocoa beans is still traditional and uncontrolled and an adequate process together with an appropriate selection and handling of raw materials. Drying and roasting can influence the character of chocolate [4]. Fermentation process involves complex microbial activities and biochemical changes that have been recently deeply reviewed by Schwan and Fleet [5]. The microorganisms responsible for the fermentation are yeast, lactic acid bacteria (LAB) and acetic acid bacteria (AAB). Recently, Ho et al. [6] elucidated the importance of yeast growth and activity for successful cocoa beans fermentation and their involvement in the development of chocolate aroma.

The fermentation process of cocoa beans is important to phase in producing aromatic compounds as well as bioactive components in cocoa and chocolate products [2]. The fermentation of cocoa beans in the process of reforming sugar and citric acid in pulp into organic acids by a consortium of microbial [1]. The resulting organic acids will induce the enzymatic reactions present in the seeds resulting in biochemical changes that will form the compounds that give rise to the smell, taste, and colour of the cocoa [2]. Without fermentation process, cocoa beans will taste bitter and astringent, so it cannot produce a distinctive aroma of chocolate when processed into chocolate products such as brown powder or brown paste [2].

The fermentation process is divided into 3 stages; the initial stage (0-36 h) is an anaerobic stage, in which yeast will convert sugars to alcohol in low oxygen and pH conditions below 4. The second stage is the dominant role of lactic acid bacteria to convert sugar and some organic acids to lactic acid. The third stage is the role of acetic acid bacteria in converting alcohol to acetic acid. During the fermentation process, cocoa pulp as a substrate containing sugar and citric acid will be converted into organic acids. The resulting organic acids will diffuse into the seeds and induce enzymatic reactions to form the candidate compounds of flavour, aroma, and colour [2].

The quality of fermented cocoa beans is determined by the acidity (pH). Processing cocoa requires a pH of seeds between 5.2-5.8 to produce quality cocoa butter [7]. In the cocoa fermentation process, the pH and total acid values are closely related to the process of seed death followed by acid diffusion into the seeds and the chemical reactions that affect the quality of the cocoa beans.

After the fermentation process, the cocoa beans are dried until the moisture content is below 7%. During the drying process, microorganism activity is still in progress in the formation of specific aroma, texture and colour characteristics [1, 3]. This study aims to determine the quality of chocolate stem produced from fermented cocoa beans from Lombok, West Nusa Tenggara.

2. Material and methods

2.1. Fermentation of cocoa beans

The raw materials used are 500 kgs of Forastero cocoa beans from Gangga district North Lombok, West Nusa Tenggara. The culture starter used for the fermentation process consists of yeast, lactic acid bacteria (LAB) and acetic acid bacteria (AAB) (10^7 CFU/mL). The yeast was grown in Malt Extract Broth medium (peptone 3 g/L, malt extract 30 g/L), LAB was grown in DeMan Rogosa Sharpe Broth medium (52.2 g/L), AAB were grown in medium Glucose, Yeast Extract, Peptone Broth (D-glucose 20 g/L, yeast extract 5 g/L, peptone 5 g/L). Each isolate was inoculated into the medium, and incubated for 48 hours at 30°C, with shaking 150 rpm.

Fermentation was conducted for five days, using box fermentation. The starter culture was added by spraying into cocoa beans mass, and then the fermentation box is covered with banana leaves and stacked with another fermentation box. Every 24 hours cocoa beans mass were mixed to get O_2 and to uniform, the cocoa beans mass. After five days of fermentation, cocoa beans were dried in the sun for five days until the water content decreased to 7%.

2.2. Chocolate preparation and analysis

The dried beans were processed to milk and dark chocolate using the facilities from Indonesian Coffee and Cocoa Research Institute (ICCRI) – Jember, East Java, Indonesia.

Sensory analysis of milk and dark chocolate were performed using a consumer acceptance test. The tests were conducted with 20 expert panelists. Samples were taken as much as 120 g of roasted beans, roosting with a temperature of 116°C for 25 minutes. The cocoa beans are peeled and then the seeds are milled to a fine paste. The fine cocoa paste is then heated in 50°C heater and tested organoleptically by a trained panelist.

The taste of chocolate paste was done organoleptically using hedonic test by 10 trained panelists from ICCRI. Characteristics of chocolate flavour tested include flavour, taste, acidity, caramel qualities, bitterness, texture, colour, and overall appearance. Scale for taste organoleptic test using 5 scale that is: 0 = zero, 1-2 = weak, 3-4 = weak enough, 5-6 = strong enough and 7-8 = strong. Scale for scent assessment, acidity, caramel nature, bitterness, texture, colour and overall appearance using 6 levels of scale i.e. 0 = unfit for consumption, 1-2 = very bad, 3-4 = bad, 5-6 = neutral, 7-8 = good and 9-10 = very good.

2.3. Analysis of chocolate bar quality

Analysis quality of chocolate bars (milk and dark chocolate) includes chemical and microbiological analysis. This analysis aims to determine the suitability quality of chocolate product with the standard quality of chocolate based on SNI 7934: 2014 [8]. The chemical analysis included proximate content in the form of fat content, non-fat cocoa solids, total cocoa solids, and metal contaminants. The microbiological analysis included i.e. the total plate count (TPC), the amount of contamination *E. coli*, *Salmonella* sp., mold, and yeast.

2.3.1. Fat content
Fat content was analyzed based on the test method contained in SNI. 01-2891-1992 concerning how to test foods and beverages [9].

2.3.2. Total cocoa solids
Cocoa solids in chocolate are cocoa products used in the manufacture of chocolate i.e. cocoa mass, cocoa powder, and cocoa fat. Total cocoa solids are calculated using the formula:

$$[(CM-(ka_{CM} \times CM) + (CP-(ka_{CP} \times CP) + (CB-(ka_{CB} \times CB)]/[(S-(ka_S \times S)) + (CB-(ka_{CB} \times CB)) + (CM- (ka_{CM} \times CM)) + (MP - (ka_{MP} \times MP)) + (CP-(ka_{CP} \times CP)) + (MF-(ka_{MF} \times MF))] \times 100\ \%$$

Where : CM = mass cocoa content in chocolate formula (%); KaCM = moisture content of cocoa mass used (%); CP = powdered cocoa content in chocolate formula (%)\; KaCP = moisture content of cocoa powder used (%); CB = cocoa fat content in brown formula (%); KaCB = moisture content of cocoa fat used (%); S = sugar content in brown formula (%); KaS = moisture content of the sugar used (%); MP = powdered milk content in brown formula (%); KaMP = moisture content of the milk powder used; MF = milk fat content used in brown formula (%); KaMF = moisture content of milk fat used (%)

The fatty cocoa solid is calculated using the formula :

$$[(CM-(ka_{CM} \times CM) + (CP-(ka_{CP} \times CP) + (CB-(ka_{CB} \times CB)]/[(S-(ka_S \times S)) + (CB-(ka_{CB} \times CB)) + (CM- (ka_{CM} \times CM)) + (MP - (ka_{MP} \times MP)) + (CP-(ka_{CP} \times CP)) + (MF-(ka_{MF} \times MF))] \times 100\ \%$$

Total non-fat cocoa solids (% b / b, bk) :

$$[(CM-(f_{CM} \times CM)- (ka_{CM} \times CM)) + (CP-(f_{CMP} \times CP) - (ka_{CP} \times CP))]/[(S-(ka_S \times S)) + (CB-(ka_{CB} \times CB)) + (CM-(ka_{CM} \times CM)) + (MP-(ka_{MP} \times MP)) + (CP-(ka_{CP} \times CP)) + (MF-(ka_{MF} \times MF))] \times 100\ \%$$

Where : CM = mass cocoa content in brown formula (%); fCM = fat content of cocoa mass used (%); kaCM = moisture content of cocoa mass used (%); CP = powdered cocoa content in brown formula (%); FCP = fat content of cocoa powder used; KaCP = moisture content of cocoa powder used (%); CB = cocoa fat content in brown formula (%); KaCB = moisture content of cocoa fat used (%); S = sugar content in brown formula (%); KaS = moisture content of the sugar used (%); MP = powdered milk content in brown formula (%); KaMP = moisture content of the milk powder used; MF = milk fat content used in brown formula (%); KaMF = moisture content of milk fat used (%)

2.3.3. Microbial contamination analysis

2.3.3.1. Total plate count
The sample was weighed 25 g and put into an Erlenmeyer containing 225 mL diluent solution. The mixture is then shaken to homogeneous and diluted to 10^{-5}. Each dilution level was taken 1 mL and inserted into a sterile petri dish by duplo and then added 12-15 mL of PCA liquid medium. The plate was then incubated for 72 hours at 30°C. Total plate count is calculated by the formula:

$$TPC = \Sigma C/\ [(1xn1)+(0.1xn2)xd]$$

Where : C = number of colonies of each petri dish; n1 = number of Petri dishes of the first dilution calculated; n2 = number of Petri dishes from the second dilution; d = first dilution calculated

2.3.3.2. Analysis of Salmonella sp.
Samples weighed 25 g and 225 mL of sterile BPW were then shaken for 2 min and incubated at 37°C for 18 h. A total of 0.1 mL of pre-enrichment culture was piped into 10 mL of RVS medium and 1 mL of other pre-enrichment cultures into 10 mL MKTT broth then vortexed to homogeneous. The RVS medium was incubated at 41.50 °C for 24 hours in a circulating water bath while MKTTn broth medium was incubated at 37 °C for 24 hours.

The incubated samples were shaken using a loop. The TT broth enrichment culture was streaked into a petri dish containing agar medium for XLD, HE, BS and RV, then incubated for 24 hours at 35 °C. The presence of *Salmonella* sp. characterized by the presence of pink colonies with or without black core on XLD media, blue to blue colored colonies with or without black core on HE medium, brown colonies, grey to black sometimes metal luster on BS media.

2.3.4. Metal contamination analysis
The analysis of metal contamination includes analysis of lead pollution (Pb), cadmium (Cd), tin (Sn), mercury (Hg) and arsenic (As) was performed by SNI 7934: 2014 methods on analysis of metal contamination on chocolate products [8].

3. Result and Discussion

3.1. The water content of cocoa beans
Fermented cocoa beans from Lombok, containing content of 8.2% water content, this value is higher than dry water content of dried cocoa beans according to SNI 2323: 2008 standard that is a maximum of 7.5%. Water content of cocoa more than 8% cause seeds susceptible to fungi and insects, thus increasing the risk of damage to the seeds, but if the seed moisture content of less than 5% will cause the seeds to break easily. Non-fermented cocoa beans will cause a sense of sprinkling and bitter and less sharp aroma in cocoa processed products thereby degrading the quality of cocoa. Quality of fermented cocoa beans from Lombok has a grade A.

Table 1. Results of analysis of fermented cocoa beans from Lombok

Parameter	**Analysis result**	**SNI 2323:2008**
Moisture content	8.2 (N.C)	Max 7.5
Life insect	Absent	Absent
Bean's smell smoky/abnormal/other taints	Absent	Absent
Number of beans / 100 g	96 beans	Grade A
Foreign matters	Absent	Absent
Mouldy beans content	0.3 %	Max.2
Slaty beans content	15 %	Max. 3
Insect damaged beans content	0 %	Max. 1
Waste contents	1.3 %	Max. 1.5
Sprouted beans content	1 %	Max. 2

3.2 Analysis of cocoa liquor
The sensory analysis resulted that aroma, flavours, caramelly and creamy of cocoa liquor score were 8.25. Beckett [10] reported that the presence of milk content in chocolate would affect the properties of cream and soft texture on chocolate products.

Table 2. Results of analysis of cocoa liquor from fermented cocoa beans from Lombok

Characteristic	**Organoleptic Score***
Aroma	8.25
Flavour	8.25
Acidity	7.25
Bitterness	7.75
Astringent	7.25
Caramelly	8.25
Creamy	8.25
Sweetness	7.00
Texture/colour	8.00
Taints/Off flavours	0.00
Preference	8.25
Quality	Excellent

Score notes :
Notation for taste: 0 = nil; 1-2 = weak; 3-4 = moderately weak; 5-6 = moderately strong; 7-8 = strong.
Notation for preference and quality 0 = inconsumable; 1-2 = very bad; 3-4 = bad; 5-6 = neutral; 7-8= good; 9-10 = excellent.
Notation for taints: 0 = nil; X = weak; XX = strong

IOP Conf. Series: Earth and Environmental Science **251** (2019) 012046 doi:10.1088/1755-1315/251/1/012046

The astringent and acidity of chocolate liquor score were 7.25. The presence of astringent and bitterness in the chocolate product caused by the presence of tannin and polyphenol in cocoa beans. Group of polyphenol compounds found in cocoa is flavonoids [11]. The bitterness of chocolate liquor has a score of 7.75. Cocoa has bitterness due to the presence of a natural substance such as alkaloid from theobromine types. The level of sweetness of chocolate liquor from fermented cocoa beans has a score of 7.00.

Texture and colour of chocolate liquor from fermented beans have score 8.00 and the off flavours have a score of 0.00. These results indicate that brown stems produced from fermented cocoa beans from Lombok has quite good quality. According to Pangabean et al. 2008, high-quality chocolate has texture smooth and buttery, it can melt with softly in a mouth [12].

3.3 Analysis of chocolate bar quality

Cocoa fat is dominated by triglycerides consisting of stearic acid (34%), palmitate (27%) and oleic (34%) which is solid at room temperature, melts at 37°C and provides a soft texture when chewed in the mouth [13]. Cocoa fat is a very important vegetable fat in the chocolate and candy industry because it has the unique physical characteristics of the triglyceride composition (TG) composed mainly of 55% POS, 5% POP and 20% SOS and has a melting point range of 32-35°C [14].

Table 3. Results of analysis of chocolate bar quality from fermented cocoa beans from Lombok

Parameter	**Unit**	**Milk chocolate**	**Dark chocolate**	**SNI 7934:2014**	
				Milk chocolate	**Dark chocolate**
Smell	-	normal	normal	Normal	normal
Flavour	-	normal	normal	Normal	normal
Colour	-	normal	normal	Normal	normal
Fat	%	54.3	48.4	≥ 18	≥ 15
Solid content non fat	%	36.5	33.5	≥ 14	≥ 2.5
Solid content	%	80.3	73.8	≥ 35	≥ 25
Metal contamination					
Lead (Pb)	mg/kg	< 0.031	< 0.031	max.1	max.1
Cadmium (Cd)	mg/kg	0.04	0.03	max.0.5	max.0.5
Lead (Sn)	mg/kg	< 0.8	< 0.8	max.40.0	max.40.0
Mercury (Hg)	mg/kg	< 0.005	< 0.005	max.0.03	max.0.03
Arsenic (As)	mg/kg	< 0.013	< 0.013	max.1	max.1
Microbial Analysis					
Total Plate Count	colony/g	< 10	< 10	Max. 1 x 10^4	Max. 1 x 10^4
E.coli	APM/g	< 3	< 3	< 3	< 3
Salmonella sp.	/25 g	Negative	Negative	Negative	Negative
Mold	colocy/g	< 10	< 10	Max. 1 x 10^2	Max. 1 x 10^2
Yeast	colony/g	< 10	< 10	Max. 1 x 10^2	Max. 1 x 10^2

Solid fat content (SFC) is one of the typical parameters needed in the cocoa fat business. Chocolate industry requires this parameter as an indication of the nature of cocoa fat liquefaction in fat processing and its use in the food industry (confectionary) [15]. Table 3. Showed that non-fat cocoa solids and total cocoa milk solids are higher than dark chocolate. This is due to milk chocolate containing more milk than dark chocolate causing solid components derived from milk to affect the number of solids in chocolate produced.

In general, the quality of the chocolate stems produced from fermented cocoa beans from Lombok fulfills the criteria for chocolate bars according to SNI 7934: 2014 regarding chocolate and chocolate products. This was shown that chocolate stems produced from fermented cocoa beans from Lombok have good quality.

4. Conclusion

The fermented cocoa beans have a water content of 8.2%. The quality of cocoa beans is included in grade A (96 seeds / 100 g). The quality of the chocolate stems produced from fermented cocoa beans is very good and in accordance with SNI 7934: 2014.

5. References

[1] Ardhana MM and Fleet GH 2003 The microbial ecology of cocoa bean fermentations in Indonesia *International journal of food microbiology* **86** 1 87-99

[2] Schwan RF and Wheals AE 2004 The microbiology of cocoa fermentation and its role in chocolate quality *Critical reviews in food science and nutrition* **44** 4 205-221

[3] Nielsen DS, Teniola OD, Ban-Koffi L, Owusu M, Andersson TS and Holzapfel WH 2007 The microbiology of Ghanaian cocoa fermentations analysed using culture-dependent and culture-independent methods *International journal of food microbiology* **114** 2 168-186

[4] Ramos CL, Dias, DR, Miguel MGDCP and Schwan RF 2014 Impact of different cocoa hybrids (*Theobroma cacao* L.) and *S. cerevisiae* UFLA CA11 inoculation on microbial communities and volatile compounds of cocoa fermentation *Food research international* **64** 908-918

[5] Schwan RF, Pereira GVM and Fleet GH 2014 Microbial activities during cocoa fermentation In *Cocoa and coffee fermentations* (London: Taylor & Francis) pp 125–135

[6] Zhao J and Fleet G 2015 The effect of lactic acid bacteria on cocoa bean fermentation *International Journal of food microbiology* **205** 54-67

[7] Wood GAR and Lass RA 2001 *Cocoa* 4th ed. Longman, London

[8] BSN 2014 *SNI Cokelat dan Produk Cokelat* 7934:2014 Jakarta

[9] BSN 1992 *SNI Cara Uji Makanan dan Minuman* 01-2891-1992 Jakarta

[10] Beckett ST 2015 *The science of chocolate* Royal Society of Chemistry

[11] Dhonsi D, and Stapley AGF 2006 The effect of shear rate, temperature, sugar, and emulsifier on the tempering of cocoa butter *Journal of Food Engineering* **77** 4 936-942

[12] Pangabean TR, Pujianto and Wahyudi T 2008 *Kakao Manajemen Bisnis dari Hulu Hingga Hilir* Penerbit Penebar Swadaya Jakarta

[13] Bucket ST 1999 *Industrial chocolate manufactures and use* (3rd ed.) Oxford: Blackwell Science pp. 153-181, 201-230, 405-428, 460-465

[14] Liu KJ, Chang HM, and Liu KM 2007 Enzymatic synthesis of cocoa butter analog through interesterification of lard and tristearin in supercritical carbon dioxide by lipase *Food chemistry* **100** 4 1303-1311

[15] Indarti E, Arpi N, and Budijanto S 2013 Kajian pembuatan cokelat batang dengan metode tempering dan tanpa tempering *J. Teknologi dan Industri Pertanian Indonesia* **5** 1:1-6

The effect of refining time to the antioxidant capacity, phenolic content, sensory and physical properties of dark chocolate couverture

R Nurhayati[1]*, E R N Herawati[1], B Oktaviani[2], I D Rachmawati[2]

[1]Research Unit for Natural Product Technology, Indonesian Institute of Sciences Jl. Yogya-Wonosari km 31.5 Gading, Playen, Gunungkidul, Yogyakarta, Indonesia.
[2]Faculty of Agriculture, Sebelas Maret University, Surakarta, Indonesia

*Corresponding author: rifa.lipi@gmail.com

Abstract. Dark chocolate is one of the chocolate products which has the highest cocoa mass content around 60-70%. Dark chocolate has health benefits because of the high antioxidant content of phenol and flavonoids. The principle of making a chocolate bar is mixing and refining cocoa butter, liquor, and sugar. This research aimed to determine the effect of refining time on the characteristics of dark chocolate couverture. Cleared tests were carried out, consisting of the descriptive sensory test, preference test, texture using "Llyod" Testing Machine, color test using chromameter, and test of antioxidant activity and phenolic content. The test results state that the refining time process affects the organoleptic properties, color, and flavors. Refining time also affects for some parameters in texture, antioxidant capacity, and phenolic content antioxidant. Dark chocolate couverture that most preferred by the panelists is chocolate with 18 hours refining time, with antioxidant activity 68,96 ± 1,482 % and phenolic content 165,280 ± 1,558 mg GAE/mg.

Keywords: Dark chocolate, antioxidant capacity, phenolic content, sensory, physical properties.

1. Introduction

Cacao (*Theobroma cacao*) is a strategic role in the national economy which is the third largest contributor of foreign exchange for the plantation sector. In 2012, cocoa contributed to the foreign exchange reserves of US$ 1,053 billion, and the value increased from 2013 until now. Currently, Indonesia is the third sequence as world Cacao beans producer after Ivory Coast and Ghana. Indonesia has a target to become the largest producer in the next decade because there is still a wide land in Eastern Indonesia and reliable human resources. According to data from the Ministry of Agriculture 2017, the cacao productivity increased from the year 2016 of 785 kg/Ha, while 2017 of 787 kg/Ha. The cocoa production of 656.817 tons in 2016 increased to 688.345 tons in 2017. Therefore we need diversification to increased the added value of cocoa beans.

Chocolate is a product which is formulated from cocoa mass, cocoa butter, and sugar. Everyone consumed chocolate because it can generate sensory and positive emotions. Chocolate is categorized into 3 types, namely dark chocolate, milk chocolate, and white chocolate. Dark chocolate is one of the products of the chocolate couverture with a cacao content of more than 60%. Dark chocolate has benefits for health because the chocolate is rich in antioxidants namely phenol and flavonoid which are capable of capturing free radicals in the body. Antioxidant content on chocolate three times more than green tea.

These antioxidants are able to reduce the blood cholesterol so it can prevent the risk of heart attacks, cancer, high blood pressure, and stroke. Besides the cocoa butter is known as cholesterol free and does not obstruct the blood vessels [1].

There are five steps of dark chocolate processing, i.e., mixing, refining, tempering, molding, and packaging. The refining process is one of the critical points to obtain the fine and smooth chocolate. This research aims to find out the influence of the refining time on the quality of the dark chocolate couverture, namely sensory characteristic, texture, color, antioxidant activity, and phenolic content.

2. Material and Methods

2.1. *Materials*

The material used in this research is cocoa liquor, cocoa butter, and sugar. Cocoa liquor and cocoa butter obtained from farmers in the village of Nglanggeran, Gunungkidul, Yogyakarta. While sugar was obtained from the supermarket with the brand of Gulaku.

2.2. *Dark Chocolate Processing*

The making of dark chocolate covers 5 stages, i.e., mixing, refining, tempering, molding, and packaging. There were 3 different refining time formula, i.e., 16 hours, 18 hours and 20 hours. The analysis for each formula were organoleptic tests, antioxidant activity, total phenol test, texture, and color.

2.3. *Sensory Evaluation*

Hedonic tests were used to evaluate the sensory parameters. There were 30 panelists. The hedonic test was used to know the level of panelist preference with 7 scales. These scale from 7 to 1 were very like, like, quite like, just normal, quite not like, does not like, very does not like [2].

2.4. *Texture and Color Measurement*

The measurement of the dark chocolate texture by using the appliance Llyod texture analyzer which aims to know the value of violence chocolate on each treatment. While the color tests using Chromameter Konica Minolta CR 400.

2.5. *Phenolic Content*

Further samples according to Ruzaidi *et al.* [3] and Othman *et al.* [4] about 5-gram samples diluted with 25 ml alcohol 70%. Extraction is done at a temperature of 50 °C for 2 hours and then filtered to taken the filtrate. This Filtrate stored in a dark container at a temperature of 20 °C. Samples solution 0.2 ml mixed with 0.4 ml Folin-Cioalteuau for 5 minutes. Added 4 ml Na_2CO_3 7 % and aquadest until 10 ml total volume.The solution incubated at 23 °C for 90 minutes then the absorbance measured at λ 750 nm. Galic acid was used as a standard

2.6. *Antioxidant Capacity*

Analysis of antioxidant activity according to Ortega *et al.* [5] by mixing 250 µl sample solution with 5 ml 2.2-diphenyl 1-picrylhydrazyl (DPPH). The solution was homogenized and incubated for 20 minutes in the dark room. The absorbance was measured in λ 515 nm. Butylated hydroxytoluene (BHT) was used as a standard.

3. Result and Discussion

3.1. *Sensory Evaluation*

Three formula of dark chocolate couverture were tested organoleptically by 30 panelists. Based data on Table 1 in accordance with the parameters of color, flavors, taste, violence, tenderness and *overall*, then samples are the most preferred panelists is dark chocolate couverture with long mixing 18 hours compared with long mixing 16 and 20 hours.

Based on the parameters of colors and flavors, 16 hours refining time has a real difference with 18 and 20 hours, but 18 and 20 hours refining time is not different from each other. Panelist likes the color and aroma of dark chocolate couverture with 18 hours and 20 hours refining time (sig<0.05). This is due to the longer refining process, the color of dark chocolate more shiny and darker. The longer refining or conching process will produce the finer dough, the color will even be distributed on the chocolate dough, and the surface area becomes larger. This allows the volatile escaping components that will provide a strong chocolate aroma impression. Aroma is one of the organoleptic characteristics which determines the level of consumer acceptance on a product. The delicious aroma can attract customers because it is a representation of a delicious taste [6].

Table 1 Panelists Preference against the Dark Chocolate Couverture

Refining time of dark chocolate	The Parameters					
	Colors	Aroma	Taste	Hardness	Smoothness	Overall
16 hours	4.20[a]	4.89[a]	4.51[a]	4.86[a]	4.24[a]	4.51[a]
18 hours	5.54[b]	5.58[b]	4.51[a]	4.35[a]	4.60[ab]	4.77[a]
20 hours	5.30[b]	5.56[b]	4.50[a]	4.86[a]	5.03[b]	4.56[a]

Description: letter a and b on each number showed that the sample is located on a different subset (p=0.05).

The parameter of taste, hardness and overall, there were no real difference in refining treatment time 16, 18 and 20 hours. Taste is the most important thing in the organoleptic characteristic of a product. The bitter taste is the natural taste of the chocolate. The bitter taste is derived from the alkaloid components such as *theobromine.*

The parameter of hardness and smoothness, there was no real difference in refining treatment time 16, 18 and 20 hours (<0.05). Refining time does not affect hardness and smoothness of chocolate. The hardness of 18 hours refining time has the lowest score; this indicates that this chocolate was not so hard. While the highest smoothness score found in dark chocolate which processed in 20 hours refining time. Good chocolate must have a smooth texture that can melt softly and gently in the mouth. Mixing and stirring in a long time can reduce the size of the chocolate particles. This affects the texture of chocolate. The large size of sugar particles is one of the causes of sandy and not a smooth texture. Therefore, we need refining or conching treatment process in a long time to reduce the size of the particles to the produced smooth texture of chocolate.

3.2. Physical Characteristics

The texture is part of the organoleptic characteristic of the product. The material which is not smooth and does not evenly mix will cause the rude texture of chocolate. This is the cause of the chocolate did not melt perfected in the mouth [7]. Tempering method will change the microstructures in the particle size of chocolate products. Particle size is inverse with texture and color, the bigger the size of particles, the more hardness of chocolate [8].

Table 2. The texture of dark chocolate Couverture

Refining time of chocolate	hardness bite 1 (N)	hardness bite 2 (N)	Cohesive-ness (N)	Adhesive-ness (Nmm)	Gummi-ness (N)	Fracture (N)	Springi-ness	Crispy-ness	Crunchi-ness (Nmm)
16 hours	338.68[a]	237.72[a]	0.4945[a]	61.91[a]	163.60[a]	129.75[a]	0.0060[a]	1.585.7[a]	413.415[a]
18 hours	583.76[a]	390.53[a]	0.3610[a]	79.24[a]	214.02[a]	416.85[b]	10.6050[a]	2.591.9 [b]	1.088.65[a]
20 hours	606.25[a]	440.42[a]	0.4250[a]	61.91[a]	260.83[a]	496.54[b]	8.3350[a]	2.602.3[b]	963.245[a]

Description: letter a and b on the number shows that the sample is located on a different subset (p=0.05).

Hardness is the peak maximum at the first pressure on the bite. Based on the data Table 2, 16, 18 and 20 hours refining time does not provide the real influence on the value of the hardness bite 1 and hardness bite 2 (sig>0.05). Cohesiveness is defined as the ratio of the pressure area during the second compression until the first compression. Cohesiveness can be measured as the level where the materials were destroyed by mechanical. secondary parameters from the cohesiveness, among others brittleness, chewiness, and gumminess. Based on the data 16, 18 and 20 hours refining time does not provide the real influence on the value of the cohesiveness and adhesiveness (sig>0.05).

Gumminess is defined as the result of the hardness value multiplied by the value of the cohesiveness. Gumminess does not have the units. Based on the data 16, 18 and 20 hours refining time does not provide the real influence on the value of the gumminess (sig>0.05). Crispiness intensity is defined as a sound when the first bite with the front teeth. Based on the data on the crispiness parameters and fracture, 16 hour refining time was different with 18 and 20 hours refining time, but 18 and 20 hours refining time were not different from each other. Springiness or elasticity can be interpreted as a recovery time between the end of the first bite and the beginning of the second bite. There are no units used because on this parameter calculating the difference in the area of time. Based on the data 16, 18 and 20 hours refining time does not provide the real influence on the value of the springiness (sig>0.05). Crunchiness is defined as the relation between fracture that has occurred on the next layer in the structure of the cells that give the impression of the expansion of the duration of the voice in time, 16, 18 and 20 hours refining time do not provide the real influence on the value of the crunchiness.

Table 3 The color of dark chocolate Couverture

Refining time of samples	The average value			ΔE
	L	A	b	
16 hours	29.31^{c}	8.83^{c}	5.55^{c}	31.11^{c}
18 hours	27.94^{a}	7.03^{b}	3.69^{b}	29.05^{a}
20 hours	28.63^{b}	6.67^{a}	3.21^{a}	29.57^{b}

Description: Letter a, b and c on the numbers shows that the sample is located on a different subset (p=0.05).
The code: ΔL = difference between light and dark, A = Δ differences in red and green, Δb = difference between yellow and blue, ΔE = total difference colors

Consumers prefer chocolate with a dark color, shiny, not fade and not blooming in its surface [9]. Table 3 showed the results of the color test using chromameter that produces the value of the L, a and b. The ΔE value of samples 16, 18 and 20 hours refining time were 31,10; 29,05 and 29.57. The value of ΔE on each sample different from each other. The results of the test statistic obtained that there was a real difference to the value of ΔE on each treatment (sig<0.05).

In Table 4. known that total phenolic content of dark chocolate couverture with refining time 16, 18 and 20 hours in a row of 105.924 ± 4.337 mg GAE/gram samples, 165.280 ± 1.558 mg GAE/gram samples and 154.900 ± 5.022 mg GAE/gram samples. While Natsume *et al.* [10] reported that the average of total phenol dark chocolate in all countries was 578.64 mg CAE/100 g (CAE = Catechin Equivalent). The biggest total phenolic content on 18 hours refining time chocolate, while the smallest total phenolic content on 16 hours refining time chocolate. The statistics result obtained that there was a difference between the total phenol on each chocolate formula (Sig.≤0.05). The total phenolic content of the beans and cocoa products is very influenced by the interaction of genetic factors with their surroundings. In the period before the harvest (pre-harvest) total phenolic content influenced by the interaction of genetic factors (genotype/varieties/clone) with environmental factors biophysics or agronomist (aquaculture). Later in the period of post-harvest, the components of the post-harvest interact each other with processing factors (the process of fruit storage, fermentation, drying, and roasting) to influence polyphenol content [11].

Table 4 Total Phenolic Content of Dark chocolate Couverture

Refining time	The total content of GAE phenol (mg/gr samples)
16 hours	105.924 ± 4.337^{a}
18 hours	165.280 ± 1.558^{c}
20 hours	154.900 ± 5.022^{b}

Description: letters a, b and c on the numbers shows that the sample is located on a different subset (p=0.05)

According to Harrington [12] the polyphenol content in chocolate products will provide benefits for the improvement of the quality of the product. This due to the polyphenol compounds has antioxidant capabilities that can prevent food oxidation of cacao butter that can cause rancidity. This will increase the shelf life of chocolate products.

Table 5 Antioxidant Activity Dark chocolate Couverture

Refining time	%Inhibition of radical DPPH
16 hours	55.01 ± 0.703a
18 hours	68.96 ± 1.482b
20 hours	60.87 ± 1.482c
BHT solution	53.77 ± 0.458a

Description: letters a, b and c on the numbers shows that the sample is located on a different subset (p=0.05)

Analysis of the antioxidant activity of three formula *dark chocolate* using radical scavenging DPPH methods. Data on Table 5 showed that the antioxidant activity of dark chocolate couverture with 16, 18, 20 hours refining time in a row of 55.01 ± 0.703 %; 68.96 ± 1.482 %; and 60.87 ± 1.482 %. While BHT solution as standard showed the antioxidant activity of 53.77 ± 0.458 %. The biggest antioxidant activity on 18 hours refining time dark chocolate. While the smallest antioxidant activity was 16 hours refining time dark chocolate. The statistics result obtained that there were very real differences in the antioxidant activity of each formula (Sig.≤0.05). According to Genovese [13], antioxidant activity on the dark chocolate in Brazil by 7.8 ± 0.4 %. Antioxidant activity and total phenol in this study different from previous research because of the difference of cocoa bean. The different varieties, environment plant, and cacao processing will affect the level of antioxidants and phenol on chocolate products.

There is a positive correlation between the antioxidant capacity of cacao and its products with the number of total polyphenols. The lower of polyphenol content, the lower the value antioxidant capacity [14]. The heating process high tends to decrease the polyphenol content in the samples. Some polyphenol compounds found in cacao beans are :

(1) hydroxybenzoic acid(gallic god syringic, protocatechuic, vanillic acid),
(2) hydroxycinnamic acid (caffeic, ferulic, p-coumaric, phloretic acids, clovamide, dideoxyclovamide),
(3) flavonols (quercetin),
(4) flavones (luteolin. apigenin),
(5) flavanones (naringenin), and
(6) flavanols (catechin, epicatechin, procyanidins/oligomers, and polymers)

4. Conclusion

Based on this research data, dark chocolate couverture with 18 hours refining time become the most preferred formula. Its characteristic was the activity of antioxidant 68.96 ± 1.482%, total phenolic content 165.280 ± 1.558 mg GAE/mg, hardness bite 1 583.76 N, hardness bite 2 390.53 N, cohesiveness 0.3610 N, adhesiveness 79.24 Nmm, gumminess 214.02 N, fracture 416.85 N and the value of the L, a and b respectively 27.94; 7.03; and 3.69.

5. Acknowledgment

We gratefully acknowledgment for Ministry of Research and Technology, Republic of Indonesia who has research funding with the 2018 National Research Incentive Program. We also thank you for Nisa Sepdifa, Arbianti Restuti, Wiwin Widiastuti and Sri Endartini for all their support.

6. References

[1] Abbas A, Ariesusanti L. 2006. *Peluang pengembangan kakao dan produk olahannya dalam upaya pemulihan ekonomi, masyarakat Kabupaten Poso: dinamika sosial dan pembangunan di Kabupaten Poso*. Jakarta (ID): LIPI Press.
[2] Soekarto E 1985 *Penilaian Organoleptik untuk Industri Pangan Dan Hasil Pertanian* (Jakarta: Penerbit Bhatara Karya Aksara)
[3] Ruzaidi A et al 2008 Hypoglycemic Properties of Malaysian Cocoa Polyphenol Rich Extract Journal International Research Food. Vol **15** 3 1-8
[4] Othman A, Ismail A, Ghani N A and Adenan I 2010 *Antioxidant capacity and phenolic content of cocoa beans* Food Chemistry **100** 1523-1530

[5] Ortega N, Romero M and Macia A 2008 Obtention and Characterization of Phenolic Extract from Different Cocoa Sources *Journal of Agricultural Food Chemistry* **56** 1 9621-9627
[6] Wahyudi T, Pangabean and Pujiyanto 2008 *Panduan Lengkap Kakao* (Jakarta : Penebar Swadaya)
[7] Minifie B W 1999 *Chocolate, Cocoa and Confectionary* : Sains and Technology. An Aspen Publication Aspen Publishers, Inc Gaithersburg, Maryland
[8] Afoakwa E O 2010 *Chocolate Science and Technology* (Ist ed) (Blackwell West Sussex: Wiley)
[9] Jinap S, Thien J and Yap T N 1994*Effect of Drying on Acidity and Volatile Fatty-Acids Content of Cocoabeans* Journal of the Science of Food and Agriculture **65** (1) 67-75
[10] Natsume M, Osakabe N, Yamagishi M, Takizawa T, Nakamura T, Miyatake H, Hatano T and Yoshida T 2000 Analysis of polyphenols in cacao liquor. cocoa and chocolate by normal phase and reversed phase HPLC *Bioscience Biotechnology and Biochemistry* **64** 12 2581-2587
[11] Wardiana E 2015 Faktor-Faktor Yang Berpengaruh Terhadap Kandungan Polifenol Pada Biji Dan Produk Berbasis Kakao *Indonesian Industrial and Beverage Crops Research Institute Jurnal Nasional* **1** 1 31-33
[12] Harrington W L 2011 The Effects of Roasting Time and Temperature on The Antioxidant Capacity of Cocoa from Dominican Republic, Equador, Haiti, Indonesia and Ivory Coast *Thesis of Master of Science* The University of Tennessee Knoxville USA **66**
[13] Genovese, Caetano M S dan Lannes S 2009 Comparison of total phenolic content and antiradical capacity of powders and "chocolates" from cocoa and cupuassu *Ciênc Tecnol Aliment Campinas* **29** 4 810-814
[14] Towaha J 2014 Kandungan Senyawa Polifenol pada Biji Kakao dan Kontribusinya terhadap Kesehatan *Sirinov* **2** 1 2-3

Oligosaccharides Profile and Prebiotic Potential of Gembolo Tuber (Dioscorea bulbifera)

E R N Herawati[1]*, M Miftakhussolikhah[1], R Nurhayati[1], K W Sari[2] and Y Pranoto[2]

[1] Research Unit for Natural Product Technology, Indonesian Institute of Sciences, Jl. Jogja-Wonosari km 31,5 Gading, Playen, Gunungkidul, Yogyakarta- 55861, Indonesia
[2] Faculty of Agricultural Technology, Gadjah Mada University, Jln. Sosio-Yustisia, Bulaksumur, Yogyakarta, Indonesia

Email: ervika.rnh@gmail.com

Abstract. Indonesia has many potential tubers that not yet utilized, whereas the tubers can be utilized in functional food. *Dioscorea bulbifera* contains oligosaccharides which can be used as functional food prebiotics based. This research aim was to analyze the content of oligosaccharides from *Dioscorea bulbifera* tuber flour and its prebiotics potential. The extraction of oligosaccharides from *Dioscorea bulbifera* flour was done using ethanol 70% and analyze the content of oligosaccharides using HPLC (High-Performance Liquid Chromatography). The inoculum of colonic microbial used from volunteer infants aged less than 6 months and only consume breast milk. The prebiotic index was determined by enumerating the growth of total bacteria, microbial probiotics (*Bifidobacteria* and *Lactobacillus*) and microbial pathogens (*Bacteroids* and *Clostridium*) in anaerobic conditions at 0, 24 and 48 hours. The results showed that oligosaccharides profile of *Dioscorea bulbifera* flour were Inulin 0.76; Raffinose 0.07; and Lactulose 0.09% b/v. The prebiotics index value of oligosaccharide of extracts of *Dioscorea bulbifera* flour at 48th hour is 1.2045. It can be concluded that the oligosaccharide from *Dioscorea bulbifera* tuber flour has potential as a prebiotic.

Keywords: functional food, *Dioscorea bulbifera*, oligosaccharides, prebiotic, probiotic

1. Introduction

Currently, food not only serves to meet the needs but also to maintain health known as functional food. Functional food could be described as any food or ingredient that may provide a health benefit beyond the traditional functions hitherto known[1]. One of the food ingredient that beneficially affects the body is prebiotic. In recent years, increasing attention has been focussed on the possible beneficial effects of prebiotics [2]. Prebiotics has been proved for their effect in increasing absorption and bioavailability of calcium, magnesium, zinc, and iron [3]. Prebiotics are selectively fermented materials that cause specific changes in both the composition and / or microbiota activity in the colon that provide benefits to host health [4]. Prebiotics are generally carbohydrates (poly- and oligosaccharides) that can not be digested in the host channel. Some types of oligosaccharides that have been known as prebiotics are fructooligosaccharides (FOS), glucooligosaccharides (GOS), inulin, and Raffinose. Oligosaccharides are commonly found in grains, beans, and tubers. Gembolo (*Dioscorea bulbifera L.*) is a kind of tubers from *Dioscoreaceae* family that grows much as wild plants in Indonesia but not yet widely used.

Many researchers have investigated the medical potency of gembolo. Gembolo, which rich in diosgenin, a steroid saponin which believed to possess preventive and therapeutic properties against several ailments including arthritis, cancer, diabetes, gastrointestinal disorders, high cholesterol and inflammation [5]. Copper nanoparticles synthesized by Dioscorea bulbifera has α-amylase and α-glucosidase inhibitory activity [6]. Gembolo tubers contain high carbohydrates (19.8%), and glucomannan as the main polysaccharide content. Among the carbohydrate content, the oligosaccharide compound in the gembolo tuber is potentially prebiotic. At the moments, there is little data research on the oligosaccharide profile and prebiotic potential of gembolo tuber. Therefore, this research aim was to analyze the content of oligosaccharides from *Dioscorea bulbifera* tuber flour and its prebiotics potential.

2. Methodology

2.1. Materials

The material used was gembolo (*Dioscorea bulbifera*) tuber from local area (Gunungkidul, Yogyakarta), 70% ethanol, sugar standard (raffinose, inulin, lactulose, glucose, galactose, and fructose), feces from baby volunteers younger than 6 months and still consuming breast milk, Phosphate Buffer Saline (Sigma), basal media, CO_2 gas, media plate count agar (Oxoid), agar Columbia media (Oxoid), rogossa agar medium (Oxoid), reinforced clostridial agar medium (Oxoid), tryptone soya agar medium (glucose), glucose, heamin (Sigma), glacial acetic acid (Merck), L-cysteine HCl (Merck), lached horse blood (Oxoid), brucella selective supplement (Sigma), antibiotics include kanamycin, vancomycin, and colistin (Sigma). The tools used were a grinder, sieve, oven (Memmert), Soxhlet tools, rotary evaporator, HPLC (High-Performance Liquid Chromatography), magnetic stirrer, hand counter, hot plate (Thermo scientific), autoclave (TOMY SX-700), pH meter.

Research methodologies consisted of (i) production of gembolo flour; (ii) extraction and analysis oligosaccharides profile of gembolo flour; and (iii) characterization prebiotic potency and determine prebiotic index.

2.2. Production of Gembolo flour

Making of gembolo flour consisted of some procedure: peeled Gembolo tuber skin and washed with clean water, sliced Gembolo in small part, dried with an oven at 50-60 °C, then ground using a grinder and sieved using a 60 mesh sieve to obtain the Gembolo flour.

2.3. Extraction and analysis oligosaccharides profile of Gembolo flour

The oligosaccharide extraction of Gembolo powder refers to reference [7]. Analysis of total dissolved solids in Gembolo extract was referred to as reference [8]. Analysis oligosaccharides profile of gembolo flour using HPLC method [9]. The column used was Metacharb 87 °C with RID detector, eluent H_2O, the flow rate of 0.6 ml/min and temperature about 85 °C. Sugar standards used were inulin, raffinose, lactulose, glucose, galactose, and fructose with each concentration was 2500 ppm. The sample was centrifuged at 12000 rpm for 5 minutes. Then diluted 5 times with aquabidest and filtered with millex 0.45 μm. A sample of 20 μl was injected into HPLC.

2.4. Characterization prebiotic potency and determine prebiotic index

2.4.1. Preparation of fecal fluid

Analysis of prebiotic potency initiated with the preparation of colonic microbes. Fecal fluid preparation refers to the reference [10]. Feces samples obtained from infants younger than 6 months and consuming only breast milk were collected and dissolved in 0.1 M phosphate buffer saline (PBS) sterile, pH 7 with a ratio of 1:10 (w/v). Then homogenized with a magnetic stirrer (200 rpm, 2 min). The liquid at the top of the precipitate was taken and used as an inoculum for further testing.

2.4.2. Preparation of colonic microbes
Preparation of the culture of colonic microbial was referred to the reference [10], incubated at 37 °C and samples taken at certain time intervals (0, 24, and 48 hours). Parameters analyzed for each sampling were bacterial growth, prebiotic index, and pH decreased. Isolation and analysis of total growth of bacteria, microbial probiotics *(Bifidobacteria and Lactobacillus)* and microbial pathogens *(Clostridium and Bacteroides)* referred to the reference [11-12] with modification. Total bacteria isolation was done using medium plate count agar. Isolation of microbial probiotics *(Bifidobacteria)* was done using Columbia agar medium supplemented with glucose (5g/l), L-cysteine.HCl (0.5 g/l), propionic acid (0.5 ml/m), or adjusted to pH 5 and isolation Lactobacilli microbial probiotics use rogosa agar supplemented with glacial acetic acid (1.32 ml/l). The medium used for microbial isolation of Clostridium pathogens was reinforced clostridial agar supplemented with vancomycin (8 mg/l), colistin (8 mg/l). Isolation of Bacteroides pathogenic microbials was done using tryptone soya supplemented with brucella selective supplement (5 ml/l), kanamycin 975 mg/l), haemin (5 mg/l), vancomycin (75 mg/l), laced horse blood (50 ml/l). Enumeration of the total microbial colon then calculated by measuring the growth of Total Bacteria, *Bifidobacteria, Lactobacillus, Clostridium,* and *Bacteroides*. Each time interval of the 0, 24, and 48 hours was done to sampling the colonic microbial culture. Prebiotic Index was a measure to the reference [13]. Prebiotic index of oligosaccharide from Gembolo can be determined by using the equation:

$$PI = (Bif / Total) - (Bac / Total) + (Lac / Total) - (Clos / Total)$$

Where: Bif is the amount of Bifidobacteria, Bac is the amount of Bacteroides, Lac is the amount of Lactobacilli, and Clos is the amount of Clostridia.

The sampling time used were 0, 24 and 48 hours. In this equation, it is assumed that the population increase of *Bifidobacteria* and / or *Lactobacilli* is a positive effect while the increase of Bacteroides and Clostridia is negatively affected. Data analysis was done by analyzing the obtained data and present it in the form of tables and graphs.

3. Results and discussion

3.1. Analysis Oligosaccharide Profile of Gembolo Flour
Analysis of oligosaccharide content in Gembolo flour extract was performed using High-Performance Liquid Chromatography (HPLC). HPLC test results for the analysis of oligosaccharide content of flour Gembolo can be seen in Table 1.

Table 1. Oligosaccharide Profile of Gembolo Flour

Parameter	(% b/v)
Inulin	0.76
Raffinose	0.07
Lactulose	0.09
Galactose	0.22
Glucose	2.53
Fructose	2.98

As shown in Table 1, the highest oligosaccharide content in Gembolo extract was inulin which contains about 0.76%. Inulin is a polysaccharide consisting of fructose units with a glycosidic bond β- (2-1) and a glucose terminal at the end [14]. Inulin is not digested by enzymes in the small intestine due to the presence of the β-(2,1) linkages, contributing to its functional properties as prebiotic [15]. Inulin and galactooligosaccharides (GOS) are two major carbohydrates that fulfill the prebiotics criteria [16]. Many research report about the prebiotic activity of inulin from tuber [17-20]. The high content of inulin in the Gembolo extract shows the potential of prebiotics. The lowest oligosaccharide content in Gembolo was raffinose which contains about 0.07%. Raffinose is a disaccharide consisting of glucose, galactose and fructose monomers. Raffinose can be metabolized by intestinal microflora resulting in lactic acid, acetic acid, butyric acid, hydrogen peroxide, bacteriocin and other metabolites [21]. Raffinose can stimulate the growth of *Bifidobacteria.* Raffinose consumption in humans about 15 g/day can significantly increase the number of fecal *Bifidobacteria* and decrease the amount of *Clostridium sp.* and *Bacteroides* due to a decrease in fecal pH. As described in Table 1, the largest monosaccharide content was fructose which contains about 2.98%. The fructose content may show the inulin content present in the extract because inulin is a polymer of the fructose units associated with the β- (2-1) glycosidic bond with the glucose terminal group. The content of glucose and galactose in Gembolo tuber was 2.53 and 0.22%.

3.2. Prebiotic Analysis of Gembolo Flour Oligosaccharide

The growth of microbial on the media added gembolo flour oligosaccharide extract could be seen in Table 2.

Table 2. The growth of Total Bacteria, Lactobacillus, Bifidobacteria, Bacteroides, Clostridium in the extract of gembolo flour oligosaccharide

Observation time (h)	Log Cfu/ml				
	Lactobacillus	*Bifidobacteria*	*Bacteroides*	*Clostridium*	*Total Bacteria*
0	3.46	6.95	1.30	4.52	5.62
24	4.88	7.35	1.89	5.25	6.90
48	6.03	7.44	2.84	4.57	6.84

As shown in Table 2, at 0 hour, the total bacteria amount was 5.62 log cfu/ml, then increased at 24 hours to 6.90 log cfu/ml, and at 48 hours it decreased to 6.84 log cfu/ml. Total bacteria are the total amount of microbes present in the sample, so the amount should be higher than the number of other microbes. Increasing the total number of bacteria indicates that microbes are able to grow on a medium containing Gembolo oligosaccharide extract. At 0 hour, the number of Lactobacillus was 3.46 log cfu/ml, then at the 24th hour it increased to 4.88 log cfu/ml, and at the 48th hour, it increased to 6.03 log cfu/ml. The presence of oligosaccharides in the media causes Lactobacillus growth well. At first sampling, the number of Bifidobacteria was 6.95 log cfu/ml, then at 24 hours it increased to 7.35 log cfu/ml, and at 48 hours it increased slightly to 7.44 log cfu/ml. Oligosaccharide present in the media is utilized by Bifidobacteria to metabolism and growth well. Colonic microbes, such as Lactobacillus and Bifidobacteria, have increased in each time interval of observation. This suggests that the microbial colon is able to grow on a medium containing Gembolo oligosaccharide extracts. Oligosaccharides stimulate the growth of colonic microbes in the presence of inulin and glucose. Glucose is included in the monosaccharide group, so bacteria can directly use it without having to chop it first into smaller monomers.

At the 0 hour, the number of Bacteroides was 1.30 log cfu/ml, then at 24 hours it increased to 1.89 log cfu/ml, and at 48 hours it increased to 2.84 log cfu/ml. The increasing number of Bacteroides at 24 hours may be due to the pH of the medium that is still the pH optimum for Bacteroides growth. At first sampling, the Clostridium count was 4.52 log cfu/ml; then at 24 hours, it increased to 5.25 log cfu/ml and the 48th hour it slightly decreased to 4,57 log cfu/ml. The increasing number of Clostridium in the 24th hour can be due to the high glucose content in the medium can still be utilized by Clostridium to metabolism and growth. Pathogenic microbes, such as Bacteroides, increase in the 24th hour and 48th time, unlike Clostridium which decreases at 48 hours. This suggests that pathogenic microbes can still grow on a medium although the growth rate is different. Microbial pathogens can still grow because pathogenic microbes still use the existing glucose to grow. The growth pattern of Total Bacteria, Lactobacillus, Bifidobacteria, Bacteroides, and Clostridium in Gembolo oligosaccharides extract can be seen in Figure 1.

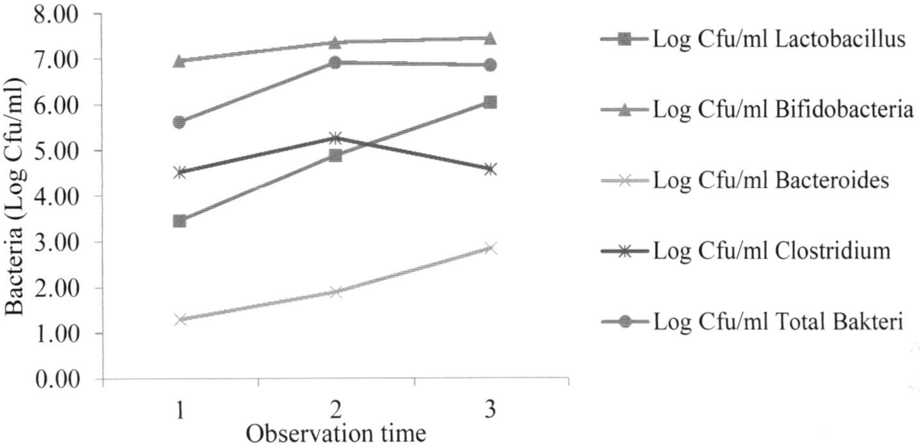

Figure 1. The growth pattern of Total Bacteria, Lactobacillus, Bifidobacteria, Bacteroides, and Clostridium in gembolo oligosaccharides extract

The growth of all colon bacterium genus will be used to calculate the value of the prebiotic index. The prebiotic index values describe the prebiotic properties of oligosaccharides that can spur the growth of probiotic bacteria compared with pathogenic microbes [22]. The oligosaccharide index of the gembolo flour is determined based on microbial growth from 0 hour to 48th hours incubated at 37 °C under anaerobic conditions. The value of the prebiotic index can be seen in Table 3.

Table 3. Prebiotic Index of Gembolo Flour Oligosaccharide Extract

Time (hours)	Prebiotic Index
24	0.3876
48	1.2045

As shown in Table 3, the value of the prebiotic index at the 24th hour was 0.3876, and at the 48th hour, it increased to 1.2045. The increase in prebiotic index values indicates that Gembolo flour oligosaccharide extract is able to support the growth of colonic microbes (Lactobacillus and Bifidobacteria) well, although the growth of pathogenic microbes (Bacteroides and Clostridium) also increases. An increase in the number of pathogenic microbes can be caused by pathogenic microbes still using glucose present in the media to perform their metabolism, thus increasing their growth. Prebiotic index values obtained are quite low when compared with other prebiotic index values of oligosaccharides such as inulin which has a prebiotic index of 1.82 and lactulose which has a prebiotic index of 4.90 [13]. Oligosaccharides present in the extract are expected to inhibit the growth of pathogenic microbes such as Clostridium and Bacteroides, but the extract also contains high monosaccharide that can be used the microbial pathogens to metabolism and growth. In addition, the pH of the media is still not able to inhibit the growth of pathogenic microbes because the pH is still not low enough so that pathogenic microbes can still grow.

4. Conclusion

Oligosaccharides profile in gembolo flour extract were 0.76% Inulin; 0.07% Raffinose; and 0.09% Lactulose w/v. The highest oligosaccharide content in gembolo extract was inulin, and the largest monosaccharide content was fructose. Gembolo flour oligosaccharides extract has prebiotic potency related to the value of the prebiotic index at 48 hours which about 1.2045.

5. References

[1] Hasler C M 1998 "Functional Foods: Their Role in Disease in: Developing New Food Products for a Changing Prevention and Health Promotion" *Food Technology* **52** 2 57-62
[2] Supriya A, Yadav, Snehal S G, Vikram B L, Smita S N and Vaishali V A 2014 In vitro screening of indigenous plant materials for prebiotic potential *Int.J.Curr.Microbiol.App.Sci* **3** 11 137-150
[3] Miyazato S, Nakagawa C, Kishimoto Y, Tagami H, Hara H 2010 Promotive effects of resistant maltodextrin on apparent absorption of calcium, magnesium, iron and zinc in rats *Eur. J. Clin. Nutr.* **49** 165-171
[4] Roberfroid M B 2001 Prebiotics: preferential substrates for specific germs? *Am. J. Clin. Nutr.* **73** 406-409
[5] Ghosh S, More P, Derle A, Patil A B, Markad P, Asok A 2014 Diosgenin from *Dioscorea bulbifera*: Novel Hit for Treatment of Type II Diabetes Mellitus with Inhibitory Activity against α-Amylase and α-Glucosidase *PLoS ONE* **9** 9 e106039
[6] Ghosh S, More P, Nitnavare R, Jagtap S, Chippalkatti R, Derle A, Kitture R, Asok A, Kale S, Singh S, Shaikh M, Ramanamurthy B, Bellare J, Chopade 2015 Antidiabetic and Antioxidant Properties of Copper Nanoparticles Synthesized by Medicinal Plant Dioscorea bulbifera *Journal of Nanomedicine & Nanotechnology* **S6** 007
[7] Muchtadi, D 1989 *Laboratorium Procedure of Nutrient Evaluation* (Bogor: PAU Pangan dan Gizi IPB)
[8] AOAC (Association of Official Analytical Chemist) 2005 *Official Methods of Analysis of the Association of Official Analytical Chemists* (New York : Chemist Inc.)
[9] Apriyantono A, Fardiaz D, Puspitasari N, Yasni S and Budiyanto 1989 *Laboratorium Procedure of Food Analysis* (Bogor : IPB Press)
[10] Vulevic J, Rastall R A, Gibson G R 2004 Developing a quantitative approach for determining the *in vitro* prebiotic potential of dietary oligosaccharides *FEMS Microbiol Lett* **236** 153–159
[11] Vardakou M, Palop C N, Christakupoulos P, Faulds C B, Gasson M A, Nabad A 2008 Evaluation of the prebiotic properties of wheat arabinoxylan fraction and induction of hydrolase activity in gut microflora *International Journal Food Microbiology* **123** 166-170.
[12] Polnaya, F J 2013 Pyhsicochemistry Characteristic and Prebiotic Potency of Phosphate Sago Starch *(Metroxylon rumphii) Disertation* Faculty of Agricultural Technology Universitas Gadjah Mada
[13] Palframan R, Gibson G R, Rastall R A 2003 Development of a quantitative tool for the comparison of the prebiotics effect of dietary oligosaccharides *LettnAppl Microbial* **37** 281-284
[14] Niness KR 1999 Inulin and oligofructose: what are they? *J. Nut.* **129** 1402S–6S
[15] Kalyani N K, Kharb S, Thompkinson D K 2010 Inulin Dietary Fiber with Functional and Health Attributes— A Review *Food Rev. Intern.* **26** 189–203
[16] Wilson B, Whelan K 2017 Prebiotic inulin type fructans and galacto oligosaccharides: definition, specificity, function, and application in gastrointestinal disorders *Journal of Gastroenterology and Hepatology* **32** 64-68
[17] Rubel I, Perez E, Genovese D, Manrique G 2014 *In vitro* prebiotic activity of inulin-rich carbohydrates extracted from Jerusalem artichoke (*Helianthus tuberosus* L.) tubers at different storage times by *Lactobacillus paracasei* *Food Research International* **62** 59-65

[18] Caleffi E, Krausova G, Hyrslova I, Paredes L, Santos M, Sassaki G, Goncalves R, Oliveira A 2015 Isolation and prebiotic activity of inulin-type fructan extracted from *Pfaffia glomerata* (Spreng) Pedersen roots *International Journal of Biological Macromolecules* **80** 392-399.

[19] Samal L, Chaturvedi V B, Pattanaik A K 2017 Effects of dietary supplementation with Jerusalem artichoke (*Helianthus tuberosus* L.) tubers on growth performance, nutrient digestibility as well as activity and composition of large intestinal microbiota in rats *J. Anim. Feed Sci.* **26** (1) 50–58

[20] Miyaguchi Y, Tomatsuri T, Toyoda A, Inoue E, Ogawa Y 2015 Effect of Yacon Tuber (*Smallanthus sonchifolius*)-derived Fructooligosaccharides on the Intestinal Flora and Immune System of OVA-sensitized BALB/c Mice *Food Science and Technology Research* **21** 2

[21] Mishra C & Lambert J 1996 Production of antimicrobial substances by probiotics *Asia Pac J Clin Nutr* **5** 20–24

[22] Mandalari G, Nueno-Palop C, Bisignano G 2008 Potential prebiotic properties of almond (Amygdalus communis L.) seeds *Appl Environ Microbiol* **74** 4264–4270

IOP Conf. Series: Earth and Environmental Science **251** (2019) 012049 doi:10.1088/1755-1315/251/1/012049

In vitro rumen fermentation characteristics of ammoniated stover of Samurai 2 sorghum fertilized with different level of Urea

A S Anggraeni[1], H Herdian[1], A Sofyan[1], L Istiqomah[1], M Anwar[1], M F Karimy[1], M R Afany[2], and M Ekaningrum[1]

[1]Research Unit for Natural Products Technology (BPTBA), Indonesian Institute of Sciences (LIPI), Jl. Jogja-Wonosari Km. 31.5, Gading, Playen, Gunungkidul, DI Yogyakarta, Indonesia.
[2]Faculty of Agriculture, Universitas Pembangunan Nasional (UPN) "Veteran" Yogyakarta, Jl. SWK 104 (Lingkar Utara), Condongcatur, Yogyakarta, Indonesia.

Email: ayus001@lipi.go.id / ayu.anggraeni07@gmail.com

Abstract. The limitation stock of forages in dry season cause serious problem for ruminant productivity. Various efforts have been made to obtain an alternative feed. Samurai 2 has a high productivity, immune to leaf rust disease and midrange rot and suitable for animal feed. However, sorghum stover is a poor-quality fodder, low in protein and digestibility. It is needing to be processing to increase nutritive value by providing a source of protein and or non-protein and certain minerals through ammoniation. The effect of ammoniation treatments on a sorghum stover on *in vitro* feed fermentation and gas production on ruminal fluid was the main objective of this study. P12 treatment gave highest gas production than other treatment. Total gas production 48 hours, gas production from soluble fraction (*a*) and insoluble fraction (b) also total fraction (*a*+*b*) of the sorghum stover was significantly affected by treatments ($P<0.05$). Volatile Fatty acid (VFA), *in vitro* degradability, NH_3, and biomass microbial rumen indicated no significant difference among treatments. In this study sorghum stover using basic fertilizer 90 kg/ha with ammoniation treatment has the best degradability on rumen fermentation characteristics *in vitro* of sorghum stover.

1. Introduction

The limitation stock of forages in dry season cause serious problem for ruminant productivity. Various efforts have been made to obtain an alternative feed, which is easy to obtain, cheap and contains nutrients to overcome scarcity of forages in dry season [1]. Sorghum is an upright growing plant that has a wide tolerance of soil fertility and waterlogging as well as drought and is relatively resistant to pest and diseases, high production, also requires less input [1]. Sorghum has a high nutritional content, 332 calories and 11.0 g protein / 100 g seeds in the seeds, and vegetative / forage 12.8% crude protein, so it can be cultivated intensively as a source of forage for ruminants mainly in dry season [2]. Pahat, Samurai 1 and Samurai 2 have produced by National Nuclear Energy Agency of Indonesia (BATAN) as sorghum mutant varieties to be developed in dry and ideal areas in the food industry, bioethanol and has a good palatability for animal [3]. Samurai 2 sorghum varieties is the result of breeding with radiation mutations derived from the Pahat varieties. It is has a high productivity, biomass of 47 ton/ha, immune to leaf rust disease and midrange rot [4]. This type of sorghum is suitable for animal feed [3,4]. *In vitro* research showed that leaves, stems, and bagasse Samurai 2 sorghum varieties can be used for animal feed [4], while *in vivo* testing using sweet sorghum bagasse can improve buffalo productivity [5].

Balanced fertilization can increase yields [6;7]. Nitrogen fertilizer promotes sucrose content, protein percentage, growth rate, production and quality of forage crops also can slow maturation of sorghum seeds (extending vegetative period) [8;9]. Nitrogen has an important role in growth of plant through cell division [10;11;12]. Addition dose of urea fertilization leads more N, will be available as raw material of protein formation so that the higher protein content of sorghum [9]. Added information by [9] explains that the end product of urea is in the form of the NH4 + and NO3 ions to be absorbed by the plant. K fertilizers also increase yield of sorghum responses from increasing levels of nitrogen fertilizer alone [13]. K is needful for efficient transformation of solar energy into chemical energy [14]. Biomass and carbohydrate of sweet sorghum can be stimulated with accurate application of N and K fertilizers which is considered as an important factor in food and industrial usage.

Sorghum stover is a poor-quality fodder, low in protein and digestibility. It is needing to be processing to increase nutritive value. Ruminant productivity can be improved by providing a source of protein and or non-protein and certain minerals. Nutritive value of rice straw could be increased through urea treatment [15], [16]. Alkali source like urea has been reported to be effective in improving fibrous feed quality [17], [18]. The urea also supplies N for rumen microbial growth besides improves fiber utilization. Volatile fatty acid (VFA), gas production and microbial biomass are the main product of feed of organic matter fermentation in ruminant [19]. The effect of ammoniation treatments on a sorghum stover on *in vitro* feed fermentation and gas production on ruminal fluid was the main objective of this study. However, the study of rumen fermentation characteristics *in vitro* of sorghum roughage is still limited in scientific publication. The comparative study of rumen fermentation characteristics *in vitro* of sorghum with different levels of fertilization and ammoniation process is needed, to determine a good product quality of sorghum ammoniation, and to evaluate the effect of different levels of fertilization and ammoniation treatment on rumen fermentation characteristics *in vitro*.

2. Materials and methods

2.1. Sample preparation

Samurai 2 sorghum varieties was planted in field laboratory of BPTBA LIPI using basic fertilizer urea. Sorghum plant was harvested at 150 days after harvesting the seeds. The sorghum straw used includes the leaves and stems. After this process leaves and stems were weighed and chopped manually around ± 5 cm. Ammoniation treatment using urea as much as 5% with simple open clip (SIMPOC) method refers to [20]. All treatment kept for one month at room temperature. Each treatment was conducted in three replicates. The treated sorghum analyzed for dry matter (DM) by following the procedure from [21].

2.2. Treatment and experimental design

This research was conducted based on Factorial Block Randomized Design (FBRD) with eight treatments and three replications based on [22]. Main factor is ammoniation (treated and untreated) and second factor is fertilizer used (0 kg/ha, 30 kg/ha, 60 kg/ha, 90 kg/ha). Treatments described as follows:

P3	:	0 kg/ha urea fertilizer SIMPOC
P3.o	:	0 kg/ha urea fertilizer SIMPOC.o
P6	:	30 kg/ha urea fertilizer SIMPOC
P6.o	:	30 kg/ha urea fertilizer SIMPOC.o
P9	:	60 kg/ha urea fertilizer SIMPOC
P9.o	:	60 kg/ha urea fertilizer SIMPOC.o
P12	:	90 kg/ha urea fertilizer SIMPOC
P12.o	:	90 kg/ha urea fertilizer SIMPOC.o

Information: "o" it is mean without ammoniation (untreated)

Nutrient value of Samurai 2 Sorghum stover after ammoniation process conducted based on paired t-test [22].

Table 1. Nutrient value of Samurai 2 Sorghum stover after ammoniation treatment.

Treatment	Dry Matter (%)	Ether Extract (%)	Crude Protein (%)	Crude Fiber (%)	Ash (%)	NFE (%)
Non-amoniation	90.74±0.33	0.74±0.38	5.45±5.62	35.59±52.61^{a}	8.19±2.58	50.02±28.05^{a}
Ammoniation	91.43±0.44	0.75±0.07	4.40±0.97	45.56±6.47^{b}	6.57±2.39	42.72±1.56^{b}

2.3. Fermentability and in vitro degradability assessment

The sample of sorghum roughage each treatment and rumen liquid were prepared before *in vitro* assessment. Two fistulated Ongole crossbreed cattle used as rumen liquor donor, those adapted by feeding consisted of forage (*P.* hybrid) and concentrated (80:20 in dry matter basis). Rumen fluid was taken using aspirator, and immediately transported in pre-warmed vacuum flask (39 ºC water temperature) then filtered.

In vitro gas production technique according to [23] used to evaluate *in vitro* fermentability. The exponential equation according to [24] was used to calculate gas production kinetics. The exponential function is P = a+b (1-e-ct) with describing P is cumulative total gas production, a is shared gas production from soluble fraction, b is the gas production from insoluble fraction, c is the rate of gas production, t is the time of incubation and e is Euler's constant (2.7183…). Fitting curve method using Neway Software [25] apply to the estimated value of *a, b* and *c*. Fermentation evaluation used 100 mL syringe glass (Fortuna model, Poulten and Graft Gmbh Germany). Three syringes are containing rumen-buffer without sample (blank) used in the experiment. All of the syringes consisted of blank and samples were randomly incubated for 48 hours in an incubator at 39 °C [26].

Gas production cumulative recorded at 0, 2, 4, 8, 12, 24, 36 and 48 hours incubation. Gas was released after 48 h of incubation and the fluid contained in the syringe taken for analysis of VFA, and *in vitro* dry matter degradability. Substrate from each syringe measured according to [21] method to get dry matter (DM) value. Percentage of DM differences between initial and after incubation and then corrected with blank were calculated as *in vitro* degradability as followed the formulas previously described by [27]:

$$IVDMD = \frac{[DMf-(DMr-DMb)]}{DMf}$$

$$IVOMD = \frac{[OMf-(OMr-OMb)]}{OMf}$$

Where: IVDMD *in vitro* dry matter degradability. DMf: a dry matter of feed, DMb: a dry matter of blank, DMr: a dry matter of residue.
IVOMD in vitro organic matter degradability. OMf: an organic matter of feed, OMb: an organic matter of blank, OMr: an organic matter of residue.

Volatile fatty acid (VFA) product from fermentation measured according to [28]. Meta-phosphoric acid added to the sample and stored at -20 °C before analysis. Gas chromatography (Shimadzu type 8A) was used to analyzed VFA content using GP10% SP-1200/1% H_3PO_4 column with 80/100 Chromosorb WAW (Supelco, Bellefonte, PA). NH_3 measured using [29] method. Measurement microbial cell protein using principle gradually modified centrifugation from [30].

2.4. Data Analysis

Variables measured were *in vitro* degradability (IVDMD & IVOMD) fermentability (gas production kinetics, a, b and c), NH_3 and biomass microbial rumen and individual volatile fatty acids (VFA). Analysis of covariance (ANCOVA) was used to evaluated data, and post hoc test of Duncan's Multiple Range Test was used to analyzed differences among mean treatments that performed by the CoSTAT statistical software [31].

3. Result and discussion

Sorghum with basic fertilizer urea (urea) 90 kg/ha with ammoniation treatment (P 12) was generated the highest gas production compared to the others (Figure 1).

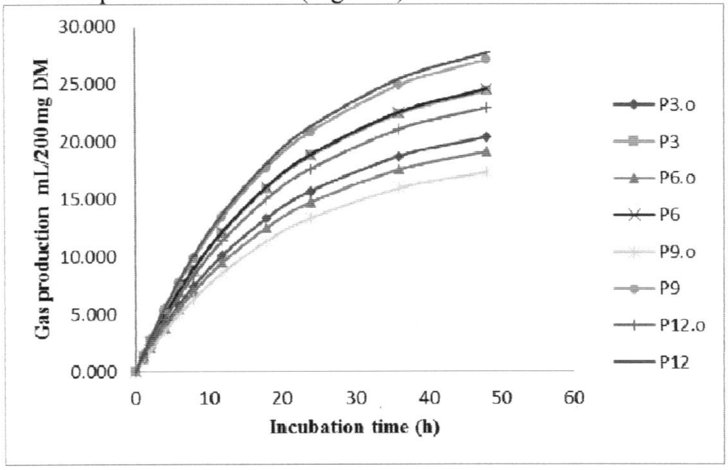

Figure 1. Cumulative gas production of sorghum ammoniation and without ammoniation treatment used different level of fertilizer incubated during 48 hours.

Figure 1. showed that P12 treatment (fertilizer 90 kg/ha with ammoniation treatment) gives the highest gas production than other treatment. Kinetic gas production parameters used as an indication of fermentability evaluation by *in vitro* gas production [28]. Gas production is the result of fermentation caused by the amount of microbial activity occurring in the rumen, as well as showing the amount of digested organic material [26]. The gas production kinetics continue to increase from hours 0 to 48 hours. The gas production curve continued to increase at 24 hours incubation time, while at 48 hours the incubation curve started to look linear although there was still no slowing. This condition may be due to the incubation time of 24 hours the presence of readily available carbohydrate sources (Readily Available Carbohydrate) in sufficient quantities to produce gas [20]. The highest cumulative gas production found at P12 treatment followed by P9, P6, P3, P12.o, P3.o, P6.o and P9.o. It can say that ammoniation treatments in sorghum stover with fertilizer 90kg/ha produce the best sorghum stover degradability than other treatment. Urea treatment with the highest fertilizer used showed to be effective in improving the nutritive value of sorghum stover. The reason is that urea supplement may act as nonprotein nitrogen source that rumen microbes could utilize it and then passed as microbial protein for the host animal [32].

Table 2. Gas production of sorghum stover with different level of fertilizer and ammoniation treatment

Variables	Level of Fertilizer (Urea)				Average
	0 kg/ha	30 kg/ha	60 kg/ha	90 kg/ha	
Gas production kinetic parameters					
a (mL)					
Ammoniation	-0.27 ± 1.17	-0.17 ± 1.29	-0.16 ± 1.23	0.20 ± 1.11	-0.10 ± 0.18[a]
Non-ammoniation	-0.82 ± 0.38	-0.91 ± 0.24	-0.87 ± 0.04	-0.78 ± 0.28	-0.85 ± 0.05[b]
Average	-0.55 ± 0.77	-0.54 ± 0.77	-0.52 ± 0.64	-0.29 ± 0.69	
b (mL)					
Ammoniation	30.73 ± 11.78	37.33 ± 14.64	40.06 ± 14.49	39.78 ± 13.22	36.98 ± 13.53[a]
Non-ammoniation	41.33 ± 20.56	30.79 ± 13.84	30.50 ± 16.16	33.26 ± 11.40	33.97 ± 15.49[b]
Average	36.03 ± 16.17	34.06 ± 14.24	35.28 ± 15.33	36.52 ± 12.31	
c (mL/h)					
Ammoniation	0.03 ± 0.02	0.04 ± 0.02	0.04 ± 0.02	0.04 ± 0.02	0.04 ± 0.02
Non-ammoniation	0.04 ± 0.02	0.04 ± 0.02	0.03 ± 0.02	0.03 ± 0.02	0.03 ± 0.02
Average	0.04 ± 0.02	0.04 ± 0.02	0.03 ± 0.02	0.04 ± 0.02	
a+b (mL)					
Ammoniation	40.50 ± 19.39	36.42 ± 13.35	39.19 ± 13.26	39.00 ± 12.11	38.78 ± 14.53[a]
Non-ammoniation	30.46 ± 11.40	30.62 ± 13.60	30.44 ± 16.07	33.45 ± 11.68	31.24 ± 13.19[b]
Average	35.48 ± 15.40	33.52 ± 13.47	34.81 ± 14.66	36.23 ± 11.90	
Gas (48 h) (mL)					
Ammoniation	27.31 ± 4.17	27.12 ± 3.65	28.12 ± 1.44	29.62 ± 2.75	28.04 ± 3.00[a]
Non-ammoniation	22.96 ± 3.67	21.76 ± 3.76	19.43 ± 2.96	22.55 ± 0.52	21.68 ± 2.72[b]
Average	25.14 ± 3.92	24.44 ± 3.71	23.78 ± 2.20	26.08 ± 1.64	

Table 3. *In vitro* digestibility and ruminal fermentation profiles of sorghum stover with different level of fertilizer and ammoniation treatment

Variables	Level of Fertilizer (Urea)				Average
	0 kg/ha	30 kg/ha	60 kg/ha	90 kg/ha	
In vitro digestibility					
IVDMD (%)					
Ammoniation	42.33 ± 8.27	39.99 ± 11.14	42.48 ± 9.27	39.02 ± 10.56	40.96 ± 9.81
Non-ammoniation	37.45 ± 9.95	39.45 ± 9.59	35.78 ± 16.69	39.41 ± 3.11	38.02 ± 9.83
Average	39.89 ± 9.11	39.72 ± 10.37	39.13 ± 12.98	39.21 ± 6.84	
IVOMD (%)					
Ammoniation	87.71 ± 1.67	83.98 ± 0.76	89.34 ± 1.02	85.87 ± 1.87	86.72 ± 1.33
Non-ammoniation	88.96 ± 0.49	95.74 ± 10.82	85.05 ± 1.27	86.63 ± 1.97	89.09 ± 3.64
Average	88.33 ± 1.08	89.86 ± 5.79	87.19 ± 1.15	86.25 ± 1.92	
Ruminal Fermentation					
MCP					
Ammoniation	5.69 ± 1.33	4.08 ± 1.18	4.09 ± 0.73	3.88 ± 1.03	4.43 ± 1.07
Non-ammoniation	3.98 ± 1.57	3.61 ± 1.42	4.26 ± 0.81	4.00 ± 0.60	3.96 ± 1.10
Average	4.83 ± 1.45	3.84 ± 1.30	4.18 ± 0.77	3.94 ± 0.82	
Ammonia					
Ammoniation	128.66 ± 3.40	105.52 ± 4.50	107.99 ± 14.88	103.69 ± 7.61	111.47 ± 7.60
Non-ammoniation	92.56 ± 16.46	78.07 ± 19.56	99.49 ± 4.84	52.73 ± 74.58	80.71 ± 23.09
Average	110.61 ± 9.93	91.80 ± 12.03	103.74 ± 9.86	78.21 ± 41.10	

IVDMD = *in vitro* dry matter digestibility, IVOMD = *in vitro* organic matter digestibility, MCP = microbial cell protein

IOP Conf. Series: Earth and Environmental Science **251** (2019) 012049 doi:10.1088/1755-1315/251/1/012049

Table 4. Volatile fatty acids characterictics of sorghume stover with different level of fertilizer and ammoniation treatment

Variables	Level of Fertilizer (Urea)				Average
	0 kg/ha	30 kg/ha	60 kg/ha	90 kg/ha	
VFA total (mM)					
Ammoniation	168.46 ± 56.81	135.10 ± 58.18	316.96 ± 180.89	205.79 ± 84.21	206.58 ± 95.02
Non-ammoniation	208.29 ± 138.58	299.79 ± 293.62	163.28 ± 52.69	218.08 ± 105.71	222.36 ± 147.65
Average	188.37 ± 97.70	217.44 ± 175.90	240.12 ± 116.79	211.93 ± 94.96	
Acetate (%)					
Ammoniation	55.94 ± 1.29	56.91 ± 0.73	56.61 ± 4.14	57.57 ± 3.77	56.76 ± 0.74
Non-ammoniation	59.53 ± 4.96	55.43 ± 1.06	56.11 ± 1.74	58.04 ± 2.77	57.28 ± 0.71
Average	57.74 ± 3.86	56.17 ± 1.15	56.32 ± 2.70	57.80 ± 3.13	
Propionate (%)					
Ammoniation	28.53 ± 0.64	28.34 ± 15.26	27.73 ± 1.64	27.32 ± 1.47	27.98 ± 0.35
Non-ammoniation	26.51 ± 2.68	28.83 ± 83.46	28.54 ± 0.55	27.12 ± 1.17	27.75 ± 0.33
Average	27.52 ± 2.10	28.58 ± 49.36	28.19 ± 1.11	27.22 ± 1.26	
Butyrate (%)					
Ammoniation	15.54 ± 0.10	14.76 ± 0.38	15.66 ± 2.49	15.11 ± 2.60	15.27 ± 0.46
Non-ammoniation	13.97 ± 2.38	15.74 ± 1.35	15.35 ± 1.27	14.84 ± 1.65	14.98 ± 0.44
Average	14.75 ± 1.89	15.25 ± 1.06	15.48 ± 1.70	14.98 ± 2.06	
A/P ratio					
Ammoniation	1.96 ± 0.08	2.01 ± 0.11	2.05 ± 0.28	2.12 ± 0.25	2.04 0.18
Non-ammoniation	2.28 ± 0.44	1.92 ± 0.02	1.97 ± 0.10	2.15 ± 0.52	2.08 0.27
Average	2.12 ± 0.26	1.97 ± 0.06	2.01 ± 0.19	2.13 ± 0.38	

Ruminal fermentability characteristic productions were shown in Table 2. Total gas production 48 hours, gas production from soluble fraction (a) and insoluble fraction (b) also total fraction (a+b) of the sorghum stover was significantly affected by ammoniation treatments ($P<0.05$), but were not affected by the level of fertilizer treatment ($P>0.05$). The production of gas formed will increase with the length of the incubation time. The cumulative gas production increased during 48 hours incubation period [20]. The highest cumulative gas production found at P12 during 48 hours incubation. This result may be due to the higher digestibility of ammoniation treatment of sorghum stover with 90 kg/ha fertilizer. This study has a similar outcome with [33] that urea ammoniated sugarcane trash (USCT) more digestible than sugarcane trash (SCT) that similarly, concentrates are highly edible hence higher gas production recorded. The kinetics of gas production appear to be determined by two distinct phases; the first one insoluble but potentially fermentable fraction and second suitable to the degradation of the soluble fraction of the tested mixtures. ‘a’ is nutritionally represented by soluble carbohydrates of rapid ruminal degradation. The fraction ‘b’ is represented by cell wall components, of slow ruminal degradation, due to structural arrangements of these components with lignin [34]. Ruminal microbes need an adaptation time (*lag* phase) before degrading the soluble particle this condition reflected in negative results on gas production produced from the soluble fraction (*a*) this results in accordance with the previous study reported by [26;34; 35], that possibility there was negative "a" value arising from lag time of soluble fraction degradation activity by ruminal microbes and then to adhere to cellulosic fraction.

In vitro dry matter degradability (IVDMD) and *In vitro* organic matter degradability (IVOMD), Microbial cell protein (MCP) and ammonia (NH_3) shown in Table 3. IVDMD had a tendency ($P{\sim}0.11$) affected by ammoniation treatments, but IVOMD was not influenced by ammoniation treatment ($P>0.05$). Moreover, both of IVDMD and IVOMD were not influenced by the level of fertilizer treatment ($P>0.05$). Biomass microbial rumen was not influenced by ammoniation and level of fertilizer treatment ($P>0.05$). NH3 had a tendency ($P{\sim}0.11$) affected by ammoniation treatments, but not controlled by the level of fertilizer treatment ($P>0.05$). Value of IVDMD booth of treated and untreated ammoniation is 40.96, and 38.02 include in the low category of IVDMD. Grain-sorghum residues, like residues from small grain crops, had a low IVDMD (37.7%) [37]. Ammoniation treatment has tendency effect on IVDMD than without ammoniation treatment even though there are no significant. IVDMD value of ammoniated sorghum stover in this study was 7,708% units higher than untreated, this result lower than the result of [37] treated ammoniation 17-18% units higher than untreated. Urea treatment showed to be effective in improving the nutritive value of sorghum stover by reducing ADF content of sorghum stover [1;32]. Urea provided ammonia after being hydrolyzed to produce stepwise ammonium carbonate and ammonium carbamate [17]. Such treatment reduces physical strength of primarily fibrous feed, disrupts the silicified cuticular barrier and cleavages of some lignin-carbohydrate bonds [17], [38].

Higher IVDMD values compared than untreated appeared as an effect of increasing ammoniation-treated sorghum stover degradability. Apart from modifying structural carbohydrate of the sorghum stover, urea addition could contribute to feeding the microbes an additional nitrogen source for microbial protein synthesis purposes and subsequent utilization of the protein by the host animal; it will give better digestibility than other treatment [1;32]. IVDMD affected by degradation rate of particle and gas production. On the other value of *in vitro,* organic matter degradability (IVOMD) for treated and untreated ammoniation were 86.72 and 89.09%. Contrast results with IVDMD, untreated ammoniation has higher IVOMD than treated. It might be caused by existence rumen microbes that can not utilize organic matter on a regular basis maximum for growth, lack of time fermentation of soluble carbohydrates, lack of nitrogen [35; 39].

Microbial cell protein (MCP) was not influenced by ammoniation and level of fertilizer treatment ($P>0.05$), while NH_3 had a tendency ($P{\sim}0.11$) affected by ammoniation treatments, but not influenced by the level of fertilizer treatment ($P>0.05$). Ammonia value of all treatment ranged between 78-111 mg/L. Ammonia concentration that can support microorganisms in synthesizing microbial proteins is between 6 - 21 mM or 85 - 300 mg / l [40]. Ammonia is one of the fermentation products in the rumen derived from proteins degradation that will used by rumen microbes for its growth [39]. The rumen microbe will be utilizing ammonia as nitrogen source used in protein synthesis. Increased synthesis of microbial proteins shows that there is the growth of rumen bacteria in utilizing a substrate. This result is proportional to the production of ammonia of each treatment. Synthesis of microbial proteins associated with ammonia concentration, where ammonia is a source of nitrogen which is used for protein synthesis. Value low ammonia indicates utilization of ammonia as the source of nitrogen for microbial growth rumen [39]. A lot of nutrient supply also provides more substrate for rumen microbes to grow and to yield more microbial protein from their biomass [41], [42], which also observed in this study.

A/P ratio, NGR, Total VFA, Acetate, Propionate, and Butyrate were not influenced by the level of fertilizer and ammoniation treatment ($P>0.05$) (Table 4). The resulting gas production shows the fermentation process of feed by rumen microbes, which hydrolyzes carbohydrates into monosaccharides and disaccharides which are further fermented into fatty acids (VFAs) such as acetic acid, propionate and butyrate and methane (CH_4) and CO_2 [40]. Total VFA varied between 135-316 mM per mL of rumen fluid. VFA range required for rumen microbial growth was 80-165.81 mM [43], added by [26] that total VFA production ranged from 202-268 mM, the VFA production was sufficient condition for optimal rumen microbial protein synthesis. Increased VFA concentration correlated with increasing microbial production [44]. An increase in the number of VFA shows quickly or not the feed is degraded by rumen microbes. The composition of VFA at rumen changes with the difference of physical form, the composition of feed, level and feeding frequency, also feed processing. Production of total VFA from non ammoniation higher 7,64% than ammoniation treatment ($P>0.05$). However, the contrast result in NH_3 also rumen microbial protein, ammoniation treatment give higher result than non ammoniation treatment (Table 2). This condition may be because low VFA production is not necessarily bad it could be that VFA will be converted into microbial protein. A/P ratio seemed constant in all treatments with value ± 2. In Table 2, ammoniation treatment using resulted in the ratio of acetate to propionate (A/P) lower than untreated. A/P ratio of ammoniation treatment was 2.04 while the A/P ratio untreated was 2.08. It has been recommended that there is an increased proportion of propionate in the rumen compared to acetate. Urea treatment plays a role in glucose metabolism affected the production of propionate that was glucogenic. This result in line with [45] that supplementation of microencapsulated flaxseed oil resulted in the lowest proportion of acetate and the ratio of A : P, and the highest proportion of propionate. This result was presumably because microencapsulated flaxseed oil supplementation can stimulate the growth of bacteria propionate producers in the rumen system so that the ruminal propionate production increased. The proportion of C2: C3: C4 was constant about 57:28:15 for all treatments. Based on the ruminal fermentation stoichiometry, individual VFA proportion consisting of C2, C3, and C4 were 55-62%, 21-33% and 8-11% respectively [45,46]. Urea treatment could improve the nutritive value of cocoa pod; it will reduce fiber content and increase the protein content of the cocoa pod [20]. Production of VFA mostly depends on the fermentation of the carbohydrate feed, a fraction of the protein [47], [48].

4. Conclusion

In vitro degradability indicated nutrient utilization in the rumen which attributed by gas production kinetics parameters, NH3, rumen microbial protein and VFA. In this study sorghum stover using basic fertilizer 90 kg/ha with ammoniation treatment has the best degradability on rumen fermentation characteristics *in vitro* of sorghum stover

5. Acknowledgment

The author would like to thank for Faculty of Agriculture Universitas Pembangunan Nasional (UPN) "Veteran" Yogyakarta as collaborator research and also for Mrs. Madina Nurrohmah, Mr. Andri Suwanto and Mr. Diding Ristino for their help in this activities and data collection.

6. References

[1] Andini L, Shinta and Suharyono 2009 Nilai biologis substitusi suplemen pakan multinutrien pada hijauan sorgum sebagai pakan ternak ruminansia secara *in vitro Seminar Nasional Teknologi Peternakan dan Veteriner 2009* 201-207

[2] Hosamani S V, Mehra U R and Dass R S 2003 Effect of different source of energy on urea mineral molasses block intake nutrient utilization, rumen fermentation pattern and blood profile in murah buffaloes (*Bubalus bubalis*) *Asian-Aust. J. Anim. Sci.* **6** 818 – 822

[3] Soeranto 2018 Pemuliaan Tanaman Sorgum http://batan.go.id/patir/pert.html[1] may 04 2018

[4] Wahyono T, Astuti D A, Wiryawan K G and Sugoro I 2014 Pengujian ransum kerbau berbahan baku sorghum sebagai sumber serat secara *in vitro* dan *in sacco J. Ilmiah Aplikasi Isotop dan Radiasi* **10** 2 98-106

[5] Seshaiah C V, Rao S J, Reddy Y R., Mahendar M and Kumar M K 2013 Effect of feeding differently processed sweet sorghum bagasse based complete rations on feeding behaviour, milk production and cost economics in graded murrah buffaloes *Buffalo Bull.* **32** 3 231-238

[6] Johnston, A E 2000 Efficient use of nutrients in agricultural production systems *Commun Soil Sci. Plant Anal.* **31** 1599-1620

[7] Rego T J, Rao V N, Seeling B, Pardhasaradhi G and Rao J V D K K 2003 Nutrient balances a guide to improving sorghum and ground based dry land cropping systems in semi-arid tropical India *Field Crop Res.* **81** 53-68

[8] Tsialtas J T and Maslaris N 2005 Effect of N fertilization rate on sugar yield and non-sugar impurities of sugar beets (*Beta vulgaris*) grown under Mediterranean conditions *J. Agron. Crop Sci.* **191** 330-339

[9] Koten B B, Soetrisno R D, Ngadiyono N and Suwignyo B 2012 Produksi tanaman sorgum (*Sorghum bicolor* (*L.) moench*) varietas lokal Rote sebagai hijauan pakan ruminansia pada umur panen dan dosis pupuk urea yang berbeda *Buletin Peternakan* **36** 3 150-155

[10] Stales H and Inze D 2001 When plant cells decide to divide *Trends Plant Sci.* **8** 359-64

[11] Zhao D K, Reddy V R, Kakani G and Reddy V R 2005 Nitrogen deficiency effects on plant growth, leaf photosynthesis and hyperspectral reflectance properties of sorghum *Europ. J. Agron.* **22** 391-403

[12] Saraswathy R, Suganya S and Singaram P 2007 Environmental impact of nitrogen fertilization in tea eco-system *J. Environ. Biol.* **28** 779-788

[13] Almodares A, Taheri R, Chung I M and Fathi M 2008 The effect of nitrogen and potassium fertilizers on growth parameters and carbohydrate contents of sweet sorghum cultivars *J. Environ. Biol.* **29** 6 849-852

[14] Mengel K and Kirkby E A 2001 Principles of plant nutrition 5th Edn. (Dordrecht: *Kluwer Academic Publisher*) p 605-650

[15] Sundstøl F, Coxworth E and Mowat D N 1978 Improving the nutritive value of wheat straw and other low quality roughages by treatment with ammonia *World Anim. Rev.* **26** 13-21

[16] Abate D and Melaku S 2009 Effect of supplementing urea-treated barley straw with lucerne or vetch hays on feed intake, digestibility and growth of Arsi Bale sheep *Trop. Anim. Health Prod.* **41** 579-86

[17] Schiere J B and Ibrahim M N M 1989 Feeding of urea-ammonia treated rice straw*: A compilation of miscellaneous reports produced by the straw utilization project* (Sri Lanka: Pudoc, Wageningen)

[18] Chenost M and Kayouli C 1997 Roughage utilization in warm climates *FAO animal production and health* (Rome)

[19] Alexander G, Singh B, Sahoo A and Bhat T K 2008 *In vitro* screening of plant extracts to enhance the efficiency of utilization of energy and nitrogen in ruminant diets *Anim. Feed Sci. Technol.* **145** 229-44

[20] Anggraeni A S, Herdian H, Sakti A A, Sofyan A and Ekaningrum M 2017 Effect of urea treatment of cocoa pod on rumen fermentation characteristics *in vitro IOP Conf. Ser: Earth and*

Environmental Science **101** 012031 1-8

[21] AOAC Association of Official Analysis Chemist 2012 *Official methods of analysis. AOAC International 19th ed.* (Washington DCUSA: Association of Official Analysis Chemist)

[22] Steel R G D and Torrie J H 1993 *Principles and procedures of statistics* 2nd ed McGraw Hill (London UK: International Book company)

[23] Menke K H and Steingass H 1988 Estimation of the energetic feed value obtained from chemical analysis and *in vitro* gas production using rumen fluid *Anim Res Dev.* **28** 7-55

[24] Ørskov E R and McDonald I. 1979 The estimation of protein degradability in the rumen from incubation measurements weighted according to rate of passage *J Agric Sci.* **92** 499-503

[25] Chen X B 1997 Neway-excel microsoft office: a utility for processing data of feed degradability and *in vitro* gas production (Aberdeen UK: Rowett Research Institute)

[26] Sofyan A, Sakti A A, Herdian H, Khairulli G, Suryani A E, Karti P D M H and Jayanegara A 2017 *In vitro* gas production kinetics and digestibility of king grass (*Pennisetum* hybrid) added by organic mineral and natural crude tannin *J Appl Anim Res* **45** 122-5

[27] Sofyan A, Sakti A A, Karimy M F, Julendra H, Istiqomah L, Herdian H, Damayanti E and Suryani A E 2015 Effectivity of probiotic, micromineral enriched yeast and their combination with *Azadirachta indica* leaves containing tannin on fermentability and digestibility of *Pennisetum* hybrid *JITV* **20** 95-104

[28] Sun P, Wang J Q and Deng L F 2013 Effects of *Bacillus subtilis* natto on milk production, rumen fermentation and ruminal microbiome of dairy cows *Animal* **7** 216-22

[29] Chaney A L and Marbach E P 1962 Modified reagents for determination of urea and ammonia *Clin Chem.* **8** 130-132

[30] Makkar H P S, Sharma O P, Dawra R K and Negi S S 1982 Simple determination of microbial protein in rumen liquor *J Dairy Sci* **65** 2170-2173

[31] Cohort 2008 CoSTAT Version 6.400. Copyright 1998-2008. *Cohort Software. 798* (USA: Lighthouse Ave, Montere, CA.93940)

[32] Laconi E B and Jayanegara A 2015 Improving nutritional quality of cocoa pod (*Theobroma cacao*) through chemical and biological treatments for ruminant feeding: *In vitro and In vivo* Evaluation *Asian Australas J. Anim. Sci.* **28** 343-50

[33] Jaishankar N, Ramachandra B, Thirumalesh T, Jag J R, Biradar U S and Suranagi M D 2018 In vitro rumen fermentation kinetics and microbial biomass synthesis of unconventional sugarcane trash and concentrates using cattle inoculums *International Journal of Agriculture Sciences* **10** 8 5859-5862

[34] Reis S T D, Marcus V G L, Eleuza C J D S, Flávio P M, João P S R and Leonardo D T S 2016 Fermentation kinetics and in vitro degradation rates of grasses of the genus Cynodon *Acta Sci. Anim Sci.* **38** 3 249-254

[35] Sandoval-Castro C A, Capetillo-Leal C, Cetina-Gongora R, Ramirez-Aviles L 2002 A mixture simplex design to study associative effects with an *in vitro* gas production technique *Anim Feed Sci Technol.* **101** 191-200

[36] Arhab R, Laadjimi K, Driss D, Djabri B and Bousseboua H 2010 Evaluation of feed mixture interactions by using in vitro gas production method *Livest Res Rural Develop.* **22** http://www.lrrd.org/lrrd22 /11/arha22217.htm

[37] Snyman L D and Joubert H W 2002 The chemical composition and in vitro dry matter digestibility of untreated and ammoniated crop residues *S. Afr. J. Anim. Sci.* **32** 2 83-87

[38] Van Soest P J 2006 Rice straw, the role of silica and treatments to improve quality *Anim. Feed Sci. Technol.* **130** 137-71

[39] Imanda S, Effendi Y, Sihono and Sugoro I 2016 Evaluasi *In Vitro* Silase Sinambung Sorgum Varietas Samurai 2 yang Mengandung Probiotic BIOS K2 dalam Cairan Rumen Kerbau *Jurnal Ilmiah Aplikasi Isotop dan Radiasi* **12** 1 1-12.

[40] McDonald P, Edwards R A and Greenhalgh J F D 2002 *Animal Nutrition 6th Edition* (London : UK Longman)

[41] Clark J H, Klusmeyer T H and Cameron M R 1992 Microbial protein synthesis and flows of nitrogen fractions to the duodenum of dairy cows *J. Dairy Sci.***75** 2304-23

[42] Bach A, Calsamiglia S and Stern M D 2005 Nitrogen metabolism in the rumen *J. Dairy Sci.* **88** 9-21

[43] Van Soest P J 1994 *Nutritional Ecology of The Ruminant 2nd ed.*(Ithaca and London: Comstock Publishing Associates a Division of Cornell University Press)

[44] Istiqomah L, Herdian H, Febrisantosa A, and Putra D 2011 Waru leaf (*Hibiscus tiliaceus*) as saponin source on *in vitro* ruminal fermentation characteristic *J.Indonesian Trop. Anim. Agric.* **36** 43-49

[45] Hidayah N, Suharti S, and Wiryawan K G 2014 In vitro rumen fermentation of ration supplemented with protected vegetable oils *Medpet* **37** 2 129-135

[46] Bo Y K, Yang H J, Wang W X, Liu H, Wang G Q and Yu X 2012 Metabolisable Energy, *In situ* Rumen Degradation and *In vitro* Fermentation Characteristics of Linted Cottonseed Hulls, Delinted Cottonseed Hulls and Cottonseed Linter Residue *Asian-Australas J Anim Sci.* **25** 2 240–247

[47] Van Houtert M F J 1993 The production and metabolism of volatile fatty acids by ruminants feed roughages: a review *Anim. Feed Sci. Technol.* **43** 189-225

[48] Herdian H, Istiqomah L, Febrisiantosa A and D Setiabudi 2011 Pengaruh penambahandaun*Morinda citrifolia* sebagai sumber saponin terhadap karakteristik fermentasi, defaunasi protozoa, produksi gas dan metana cairan rumen secara *in vitro JITV* **16** 99-104

IOP Conf. Series: Earth and Environmental Science **251** (2019) 012050 doi:10.1088/1755-1315/251/1/012050

Adhesion activity assay of *Lactobacillus plantarum* AKK30 combined with oligosaccharides

A Sophian [1,3], H Julendra[2,3], A Sofyan [2], M F Karimy [2], Abinawanto[3*]

[1]National Agency of Drug and Food Control in Gorontalo, Indonesia.
[2]Research Unit Natural Product Technology- Indonesian Institute of Sciences (LIPI), Yogyakarta, Indonesia.
[3]Department of Biology, Faculty of Mathematics and Natural Sciences (FMIPA) – the University of Indonesian (UI), Depok, Jakarta, Indonesia.
[*]Corresponding author: *abinawanto.ms@sci.ui.ac.id*

Abstract. This objective of study was to measure the activity of lactic acid bacteria (LAB) adhesions in the intestinal mucosa. Those of lactic acid bacteria used were *L. plantarum* AKK 30 and were combined either with inulin oligosaccharides (OS) or mannan oligosaccharides (MOS). The oligosaccharides concentration used in this experiment were, 0%, 0.5%, 1%, 1.5% and 2%, respectively. Data collected were total plate count (TPC) before and after adhesion assay. The microstructure of *L. plantarum* AKK 30 was observed by scanning electron microscope (SEM). The TPC results of the combination of *L. plantarum* and inulin (OS) were 18 x 10^7(0%), 48 x 10^7 (0.5%), 109 x 10^7 (1.0%), 129 x 10^7 (1.5%), and 99 x 10^7 (2.0%), respectively, while the combination of *L. plantarum* and mannan were 18 x 10^7 (0%), 53 x 10^7 (0.5%), 125 x 10^7 (1.0%), 210 x 10^7 (1.5%), and 66 x 10^7 (2.0%), respectively. The highest growth average of LAB were shown by the combination *L. plantarum* and 1.5% of MOS. In addition, the TPC results of assay adhesion showed that the combination of *L. plantarum* and inulin oligosaccharides were: 3.89 x 10^8 (0.5%), 4.39x 10^8 (1.0%), 3.19 x 10^8 (1.5%), and 5.09 x 10^8 (2.0%), respectively, whereas the combination of *L. plantarum* and mannan were: 7.29 x 10^8 (0.5%), 15.49 x 10^8 (1.0%), 2.89 x 10^8 (1.5%), and 4.49 x 10^8 (2.0%), respectively. There were significant differences in the TPC results of assay adhesion among the treatment groups and control (1.16 x 10^8). The highest activity of LAB adhesion was shown by the combination of *L. plantarum* and 1% of mannan. The TPC result was supported by the SEM observation. The combination of *L. plantarum* with oligosaccharides affects the growth and the adhesion length of lactic acid bacteria in the intestine. Further, the optimum concentration was shown by 1.0% of mannan combined with *L. plantarum*.

1. Introduction

Lactic acid bacteria live and develop in the small intestine in chickens that work by supplying protease and amylase enzymes to help digestion [1]. Besides lactic acid bacteria can hydrolyze glucose to lactic acid and volatile fatty acids also produces bacteriocin pathogenic anti-microbes which can eliminate pathogenic bacteria. The antimicrobial activity of Lactobacillus is mainly caused by the production of lactic acid, acetate, formate, caproic, propionate, butyrate and valeric acid [2]. Lactobacillus also produces other inhibitors such as H_2O_2 [3]. 2003) and bacteriocin, a protein compound produced by bacteria that has bactericidal and bacteriostatic activity [4] and then Ammor et al. [5] added that bacteriocin is a low molecular weight thermostable compound and has the ability to inhibit Gram positive or Gram negative bacteria and has a therapeutic effect.

Lactobacillus plantarum is one of the types of lactic acid bacteria that has the characteristics of a straight rod shape with a width range 0.9 - 1.2 μm and a length of 3 μm, single size or form a short chain, and is a gram-positive [6]. *Lactobacillus plantarum* colonies are white or yellow, and some strains are motile. This bacterial colony in the media has characteristics of round, smooth, solid, white, sometimes bright or dark yellow, 3 mm in diameter, facultative anaerobic. These bacteria can grow at a temperature of 15°C in general and cannot grow at a temperature of 45°C, with optimal temperatures ranging from 30-35°C [7]. According to Fraizer and Dennis [8], *Lactobacillus plantarum* is rod-shaped and immovable (non-motile), this bacterium has negative catalase, aerobic or facultative anaerobic properties, is able to melt gelatin, digest protein quickly, does not reduce nitrate, is acid tolerant, and is capable of produce lactic acid.

Oligosaccharides are one of the prebiotics that has various types [9]. Oligosaccharides are referred to as prebiotics because they are not degraded by endogenous enzymes produced by the host, so they are not digested and not absorbed [10]. For the *Lactobacillus plantarum,* adhesion test can use ingredients that are "non-digestible oligosaccharide" including mannan oligosaccharides (MOS) and inulin. Mannan and inulin play a role in improving health by modifying the balance of the intestinal microflora [11] and selectively stimulating the growth of beneficial bacteria such as lactic acid bacteria[12]. These specific carbohydrates serve as food for beneficial bacteria [13]. Mannan in feed is used as a prebiotic substitute for antibiotics because it can reduce mortality and increase antibodies [14].

MOS supplemented into poultry feed gives some influence. These effects include increasing production, in this case, body weight gain (UN) and feed conversion due to the utilization of nutrients in the gastrointestinal tract [15]. This effect is in line with the use of MOS as reported by Shashidhara [16] which explains that the use of MOS increases the antibody titer to IBD in boiler breeding chickens.

This study aims to measure the activity of lactic acid bacteria adhesions in the intestinal mucosa. The lactic acid bacteria used were *L. plantarum* AKK 30 which was combined with oligosaccharides (OS) inulin and mannan oligosaccharides (MOS) and with different levels of OS administration.

2. Materials and Methods

2.1. Preparation of suspension of bacterial isolates and chicken intestines

Preparation of L. plantarum + inulin and L. plantarum + MOS suspensions in MRSB incubated for 24 hours at 37 ° C. Biomass cells were harvested by centrifugation at a speed of 5000 rpm for 15 minutes. Pellets which had been separated from the supernatant were washed with PBS 2 times then suspended in the same solution (OD culture suspension around 0.9 to 1.0). The suspension is used as an inoculum. 30-day old chicken intestine is taken 10 cm from the duodenum. The small intestine is cut longitudinally to facilitate cleaning of dirt and put into sterile Petri dish then washed with PBS solution 3 times and cut into 1 cm^2 piece. The intestine is used as a binding medium for bacteria.

2.2 Analysis of the number of initial bacteria on the intestinal surface

Intestinal pieces are inserted into a test tube containing 10 ml of 0.85% NaCl and vortexed at high speed for 60-90 seconds to release bacteria that attached to the intestine. Suspension of intestinal bacteria was diluted in 0.85% NaCl solution, and serial dilution was performed (10^{-1}-10^{-4}). The result of dilution was platted in MRSA, and after 24 h of incubation the colony number was calculated using the Total Plate Count (TPC) method. The TPC enumeration results in the number of initial bacteria in the intestine (control).

2.3 Calculation of the total number of bacteria on the intestinal surface

The attachment test was done by incubating the intestinal sample into a petri dish containing a solution of bacterial suspension in PBS at 37 ° C for 60 minutes then rinsed with PBS. The rinsed intestine is then put into 10 ml of physiological solution and distorted at high speed for 60-90 seconds. The bacterial suspension obtained was then diluted in stages (10^{-1}-10^{-7}) and the results of dilution of the plate in MRSA were carried out by calculating the colony using the Total Plate Count (TPC) method. The number of test bacteria attached to the intestine is calculated as the difference in the number of bacteria in the control and the amount in the sample.

2.4 Qualitative analysis of adhesion with SEM

LAB colonies in each treatment and control were taken using an ose needle and placed on the carbon tape and coated using ion sputter with a 10 mA setting for 60 seconds. Coated samples were included in batches for SEM analysis. The images taken are magnification of 5,000, 10,000 and 20,000 x

3. Results and Discussion

The TPC results of the combination of *L. plantarum* and inulin (OS) were 18 x 10^7(0%), 48 x 10^7 (0.5%), 109 x 10^7 (1.0%), 129 x 10^7 (1.5%), and 99 x 10^7 (2.0%), respectively, while the combination of *L. plantarum* and mannan were 18 x 10^7 (0%), 53 x 10^7 (0.5%), 125 x 10^7 (1.0%), 210 x 10^7 (1.5%), and 66 x 10^7 (2.0%), respectively. The highest growth average of LAB were shown by the combination *L. plantarum* and 1.5% of MOS. In addition, the TPC results of assay adhesion showed that the combination of *L. plantarum* and inulin oligosaccharides were: 3.89 x 10^8 (0.5%), 4.39x 10^8 (1.0%), 3.19 x 10^8 (1.5%), and 5.09 x 10^8 (2.0%), respectively, whereas the combination of *L. plantarum* and mannan were: 7.29 x 10^8 (0.5%), 15.49 x 10^8 (1.0%), 2.89 x 10^8 (1.5%), and 4.49 x 10^8 (2.0%), respectively. There were significant different in the TPC results of assay adhesion among the treatment groups and control (1.16 x 10^8) . The highest activity of LAB adhesion was shown by the combination of *L. plantarum* and 1% of mannan.

Table 1. TPC value after adhesion assay

Oligosaccharide	Control chicken intestine without treatment(A)	treatment with chicken intestines (B) (..x 10^7)	B-A (..x 10^8)
	30 x 10^5	0.3 x 10^7	0.03 x 10^8
I0	30 x 10^5	11.6	1.157
Ia	30 x 10^5	39	3.897
Ib	30 x 10^5	44	4.397
Ic	30 x 10^5	32	3.197
Id	30 x 10^5	51	5.097
M0	30 x 10^5	0.3	0.03
Ma	30 x 10^5	73	7.297
Mb	30 x 10^5	155	15.497
Mc	30 x 10^5	29	2.897
Md	30 x 10^5	45	4.497

Figure 1. Scanning electron microscope (SEM) HITACHI SU 3500, 5K magnification. (A) Colonies of lactic acid bacteria from chicken intestinal mucosa that have not been given treatment (B) LAB-combined L. plantarum + Inulin colonies. (C) LAB-colony colonization of L. plantarum + Mannan oligosaccharides.

The TPC result was supported by the SEM observation. The combination of *L. plantarum* with oligosaccharides affects the growth and the adhesion length of lactic acid bacteria in the intestine. Further, the optimum concentration was shown by 1.0% of mannan combined with *L. plantarum*.

The attachment of probiotic bacteria (adhesion) on the surface of the intestinal wall has a beneficial effect on the health and condition of the intestinal microflora [17], attachment of *Lactobacillus sp* on the surface of the intestinal wall is influenced by the hydrophobicity of cell surfaces and bile salts [18]. Mannan in feed is used as a prebiotic substitute for antibiotics because it can reduce mortality and increase antibodies [17]. Mannan oligosaccharides are capable of binding to gram-negative bacteria, where the bacteria will come out of the intestine so that colonization does not occur [19]

4. Acknowledgments

The authors would like to thank the Universitas Indonesia, for the financial support. We are also grateful to the Head Office of Research Unit Natural Product Technology- Indonesian Institute of Sciences (LIPI), Yogjakarta.

5. References

[1] Ray R 1996 *Fundamental Food Microbiology* (New York: CRC Press Boca Raton Inc.)
[2] Corsetti A, Gobbetti M, Balestrieri F, Paoletti F, Russi L and Rossi J 1998 Sourdough lactic acid bacteria effects on bread firmness and staling. *J. Food Sci.* **63** 347–351
[3] Ito A, Sato Y, Kudo S, Sato S, Nakajima H and Toba T 2003 The screening of hydrogen peroxide-producing lactic acid bacteria and their application to inactivating psychrotrophic food-borne pathogens *Curr. Microbiol.* **47** 231-236
[4] Ogunbanwo S Sanni A and Onilude A 2003 Influence of cultural conditions on the production of bacteriocins by *Lactobacillus brevis OG1 Afr. J. Biotechnol.* **2** (7) 179-184
[5] Ammor S, Tauveron G, Dufour E and Chevallier I 2006 Antibacterial activity of lactic acid bacteria against spoilage and pathogenic bacteria isolated from the same meat small-scale facility 1-screening and characterization of the antibacterial compounds *Food Control* **17** 454-461
[6] Salminen S and Wright A V 1998 *Lactic Acid Bacteria* 2nd Ed (New York: Marcell Dekker Inc.)
[7] Haryati T 2011 Probiotics and Prebiotics as Nonruminansia Feed Feed Livestock Research *Center Wartazoa* **21** 3
[8] Frazier W B and Dennis W 1998 *Food Microbiology* 3rd ed. (New York: McGraw-Hill, Inc.)
[9] Durst L 1996 Inclusion of fructooligosaccharides in broiler diets *Archiv. Geflugelkunde* **60** 160 – 164

[10] Gilliland S E 1986 *Role of starter culture bacteria in food preservation in Bacterial Starter Cultures for Food. Gilliland*, S.E. (ed.). (Florida: CRC Press, Inc.)
[11] Cummning J H, Macfarlane G T and Englyst H N 2001 Prebiotic digestion and fermentation *Am. J. Clin. Nurt.* **73** 415-420
[12] Crittenden R G and Playne M J 1996 Production, Properties and Applications of Food-grade Oligosaccharides *Trends in Food Science and Technology* **7** 353-361
[13] Patterson J A and Burkholder K M 2003 Application of prebiotics and probiotics in poultry production *Poult. Sci.* **82** 627-631
[14] Murwani R 2008 Antibiotic Replacement Natural Additives (Semarang : Unnes Press)
[15] Ferket P L Parks C W and Grimes J L 2002 Benefits of dietary antibiotics and mannan oligosaccharides supplementation for poultry *Proc. of Poultry state meeting*
[16] Shashidhara R G and Devegowda G 2003 Effect of Dietary Mannan Oligosaccharide on Broiler Breeder Production Traits and Immunity *Poultry Science Association Inc.* **82** 1319–1325
[17] Thompson K L and Applegate T J 2006 Feed withdrawal alter small-intestinal morphology and mucus of broilers *Poultry Science Association* **85** 1535–1540.
[18] Muñoz-Provencio D, Llopis M, Antolín M, de Torres I, Guarner F, Pérez-Martínez G and Monedero V 2009 Adhesion properties of Lactobacillus casei strain to resected intestinal fragments and components of the extracellular matrix *Arch. Microbiol.* **191** 153-161
[19] Dimitroglou ADL Merrifield R Moate SJ Davies P Spring J Sweetman and Bradley G 2009 Dietary mannan oligosaccharide supplementation modulates intestinal microbial ecology and improves gut morphology of rainbow trout, Oncorhynchus mykiss (Walbaum) *J. Anim. Sci.* **87** 3226-3234

Improving antibacterial activity and viability of *Lactobacillus plantarum* AKK30 as feed additive by addition of different oligosaccharides

H Julendra[1,2], A Sofyan [1], Abinawanto[2] and Yasman[2]

[1]Research Unit Natural Product Technology- Indonesian Institute of Sciences (LIPI), Yogyakarta, Indonesia.
[2]Department of Biology, Faculty of Mathematics and Natural Sciences (FMIPA) – The University of Indonesian (UI), Depok, Jakarta, Indonesia.

Abstract. Combination of lactic acid bacteria as probiotic with oligosaccharides substances has potentially improved nutrient utilization and poultry performance. A study was conducted to evaluate-the addition of oligosaccharides inulin or mannan oligosaccharides (MOS) with different level of oligosaccharides into growth medium on antibacterial activity and viability of *Lactobacillus plantarum* AKK30 *in vitro*. The experiment was arranged on the factorial design using two factors, the A factors were oligosaccharides types (inulin or MOS), and the B factors were increased level of oligosaccharides (0, 0.5, 1, 1.5 and 2%). Each treatment was conducted in 3 replications. Data were analyzed with analysis of variance (ANOVA) among the treatments that show significant differences ($p<0.05$) followed by *Duncan* post hoc test. The results showed that the combination of *L. plantarum* and oligosaccharides ($P<0.05$) influenced *L. plantarum* growth according to the value of OD (OD-590nm). However, a combination of *L. plantarum* and oligosaccharides was not significant ($p>0.05$) for antibacterial activity. Addition of mannan oligosaccharides (MOS) 0.5 - 2% could increase growth of *L. plantarum* growth with 0.5% MOS as optimum level.

1. Introduction

Lactobacillus and *Bacillus subtilis* have been increasingly tested as alternative growth promoters [1]. *Lactobacillus acidophilus* has antibacterial activity [2], improved intestinal health [3]. It affects intestinal morphology and thickness of broiler intestine [4] and resists lincomycin antibiotics, tetracycline and enrofloxacin [5]. *L. plantarum* has potential as antimicrobial [6] candidate as a probiotic.

Viability of Lactobacillus in chicken intestines is considered important because it affected the performance of broilers and digestibility [7] and accelerated the process of absorption and cell proliferation because a Lactobacillus could hydrolyze undigested carbohydrates into monosaccharides through anaerobic fermentation in the intestines [8]. Oligosaccharides are a source of nutrients for probiotic bacteria because the growth of probiotic bacteria can be increased by oligosaccharides [9]. Oligosaccharides are substrate that was used selectively and useful for certain microorganisms including Fructo Oligosaccharide (FOS), Mannan-Oligosaccharide (MOS) and inulin [10]. The use of oligosaccharides and probiotics could modulate antibacterial activity, and specificity [9] enhanced the growth of representative probiotic strains in vitro [11]. Combination of lactic acid bacteria as probiotic with oligosaccharides substances has potentially improved nutrient utilization and poultry performance

A study was conducted to evaluate the addition of oligosaccharides (OS) consisting of inulin or mannan oligosaccharides (MOS) with different level of OS added into growth medium on in vitro antibacterial and growth activities of Lactobacillus plantarum AKK30.

2. Material and methods

2.1. *Culture and media*

The study was conducted at the Research Unit of Natural Product Technology (BPTBA)- Indonesian Institute of Sciences (LIPI) Microbiology Laboratory, from May to June 2018 using *L. plantarum* obtained from the AKK isolate belonging to the BPTBA-LIPI Microbiology Laboratory and the commercially obtained oligosaccharides consisting of Inulin [Orafti, Beneo-USA] and Mannan oligosaccharide (MOS) [Technomos, Biochem-Germany]. The factorial design experiment was arranged using two factors: the A factors were oligosaccharide types consisted of *L. plantarum* with inulin (Lp + inulin) and *L. plantarum* with MOS (Lp + MOS) and the B factors were increased level of oligosaccharides concentration (0, 0.5, 1, 1.5, and 2%; v/v) of which each treatment consisting of 3 replications. The parameters measured were optical density (OD), total lactic acid content, and antibacterial activity. Two mL of MRSB and 1% of *L. plantarum* were added into microtube and then were vortexed and incubated 24 hours at 37 °C. A series of falcon tubes filled with 10 mL of sterile MRSB were added with either inulin or MOS oligosaccharides (0%, 0.5%, 1%, 1.5%, 2%) and 1% of *L. plantarum* culture (Table 1).

Table 1. Experimental model

Oligosaccharides	treatment	Inoculum (1% 0.1 mL/ 10 mL)
MOS	M0	0 %
	Ma	0.5%
	Mb	1.0%
	Mc	1.5%
	Md	2.0%
Inulin	I0	0 %
	Ia	0.5%
	Ib	1.0%
	Ic	1.5%
	Id	2.0%

2.2. *Growth activity assay*

Growth activity assay of *Lactobacillus plantarum* was measured by turbidimetric (optical density) using microplate reader (MultiSkan GO, Thermoscientific) at a wavelength of 590 nm [9]. Each 150 µL of inoculum (1% 0.1 mL / 10 mL) was poured in a wheel tube according to the scheme as seen in Table 1. The inoculum in the wheel tube was stored in a stock collection shutter and inserted on the plate reader. The absorbance value of the inoculum was recorded every 2 hours for 24 hours.

2.3. *Analysis of lactic acid concentration*

Analysis of lactic acid content was the principle of acid-base titration according to [12]. Two or three drops of 1% phenolphthalein (PP) $C_{20}H_{14}O_4$ indicator was mixed with the samples and was titrated with 0.05 N NaOH solution until the color was pink. The total titrated solution was the total lactic acid in the sample. The formula was as follows:

$$\text{Crude Lactic acid (\%)} = \frac{(VTs - Vto) \times N \times MW \times Df}{V\ sample \times 1000} \times 100\%$$

VTs = titrant sample volume (mL)
Vto = blank (mL)
N = normality of the titrant (NaOH 0.05 N)
MW = relative molecular of lactic acids 90.0 (g/mol)
Df = dilution factor
V sample = sample volume

2.4. Antibacterial activity assay

Each isolate was grown in its media for 24 h at 37 °C and was centrifuged at *4136 x g* for 15 min at 4 °C. Twenty-five ml of cell-free supernatant was incorporated into a blank paper disk (0.6 mm) [oxoid] and applied to an agar plate containing pathogenic bacteria as microbial indicator. Inhibition activities test against *Escherichia coli* FNCC 0194, *Staphylococcus aureus* FNCC 6049, *Pseudomonas aeruginosa* FNCC 0063, and *Salmonella pullorum* ATCC 13036 in nutrient agar (NA) media [Merck] were performed using agar diffusion method [13] with the incubation time and temperature of 24 h at 37 °C respectively. The parameters measured were diameter (mm) of the visible clear zone.

2.5. Data Analysis

Quantitative data from lactic acids content and antibacterial activity were analyzed by using analysis of variance (ANOVA) and followed by Duncan's multiple range test to distinguish the effect of different treatment mean using CoSTAT statistical software [14].

3. Result and discussion

3.1. Growth activity

The growth curve of *Lactobacillus plantarum* AKK30 on different oligosaccharide media were shown in Figure 1. The average of optical density value on Lp + inulin (0.990) was significantly different ($p < 0.05$) compared to Lp + MOS (0.703) and the addition of 0.5% oligosaccharides were the optimal level of growth of *L. plantarum* (1.009). This result indicated that the addition of inulin give better effect for the growth of *L plantarum* in comparison with MOS as well as control and the optimal addition level was 0.5% which confirm the result of Setiarto [15] research who found that the addition of 0.5% inulin concentration to the MRSB media was the most effective modification in increasing *L. acidophilus* growth when compared to other treatments. It is probably because inulin is a good source of nutrition for the growth of *L. plantarum*. Oligosaccharides were prebiotic [16] and were a source of nutrition for lactic acid bacteria [17].

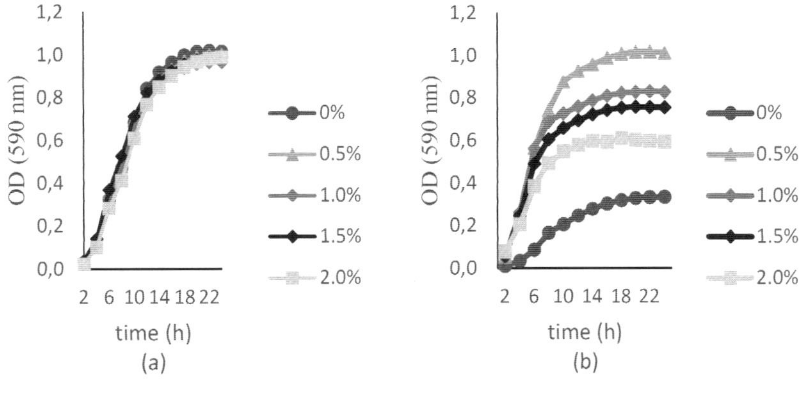

Figure 1. The growth curve of *L. plantarum* AKK30 which was supplemented with Inulin (a) and MOS (b)

From Figure 1, it can be seen that the combination of Lp + inulin started to enter and reached its stationary phase at 18 and 24 hours, respectively. Meanwhile, the optimum OD value could be found in the combination of Lp + 0.5% MOS which was significantly different in comparison with the control ($p < 0.05$) but not with other concentration treatments. The growth of the 2nd to 12th-hour incubation from all levels of MOS has entered the exponential phase but not for the control and from the 14th hour started to slow to the stationary phase 24 hours incubation. This result showed that the growth of *L. plantarum* was influenced by the addition of oligosaccharides even though the level of administration was not significantly different ($p > 0.05$). Oligosaccharides are carbohydrates that are digested into glucose [10], as a nutrient source for lactic acid bacteria growth [18], inulin and other oligosaccharides are appropriate for growth of lactic acid bacteria including *L. plantarum* [19. 20] but not all oligosaccharides are able to assimilate the lactic acid bacteria fermentation process [9]. The rate at which oligosaccharides dissolved [18]. This indicates that inulin and MOS are needed as a source of carbohydrates in the *L. plantarum* fermentation process.

3.2. Analysis of lactic acid concentration

The effectiveness of lactic acid fermentation from the combination of *L. plantarum* and oligosaccharides were shown in Table 2.

Table 2. Effect of BAL combination with oligosaccharides on crude lactic acid concentration

Oligosaccharides	Crude lactic acid concentration (%)					
	0.0%	0.5%	1.0%	1.5%	2.0%	average
	------------------ (%) ------------------					
Inulin	19.44±0.63	22.32±1.81	19.83±3.73	20.07±2.44	20.10±3.01	20.35[a]
MOS	18.15±1.53	19.20±2.76	20.82±1.68	19.02±1.05	22.65±1.15	19.97[a]
Average	18.79[a]	20.76[a]	20.33[a]	19.55[a]	21.38[a]	

MOS : Mannan oligosaccharides mean in the same column, and row differ significantly (P<0.05)

From Table 2, it can be seen that Lp + inulin from lactic acid concentration was not significantly different (p> 0.05) with Lp + MOS and but not with other concentration treatments. This result indicated that the possibility of oligosaccharides was not being used optimally by *L. plantarum* and because the type of BAL fermentation was different according to the type of BAL. *L. plantarum* is a homofermentative lactic acid bacteria that converts almost all the sugar used into lactic acid by the enzyme Lactate dehydrogenase through the reaction of glycolysis [18]. Lactic Acid Bacteria are gram-positive bacteria that cannot produce ATP so that ATP was only obtained from fermenting sugars [21] and producing lactic acid [8]. Other possibilities were the role of oligosaccharides such as inulin as prebiotics which regulate acidic conditions in lactic acid bacterial fermentation [22], prebiotics provides changes in intestinal microflora activity and regulate acidic conditions in the stomach and stimulate intestinal bacterial activity [10].

3.3. *Antibacterial Activity*

Antibacterial activity of Lp +inulin and Lp + MOS can be seen in Table 3.

Table 3. Effect of *L. plantarum* with oligosaccharides on antibacterial activity

Oligosaccharides	Clear zone diameter (mm ±SD)					
	0.0%	0.5%	1.0%	1.5%	2.0%	average
S. aerus	------------------					
Inulin	12.30±1.13	13.40±2.06	14.00±0.09	12.65±0.21	12.30±1.12	12.93[a]
MOS	12.40±0.00	13.80±1.17	13.15±0.92	12.11±0.42	13.21±1.98	12.90[a]
Average	12.35[a]	13.60[a]	13.56[a]	12.38[a]	12.76[a]	
S. pullorum	------------------					
Inulin	16.35±2.05	15.7±1.69	18.95±5.44	16.95±1.77	16.35±2.05	20.35[a]
MOS	17.10±0.00	17.55±3.89	16.50±1.69	16.55±3.46	16.00±4.22	19.97[a]
Average	16.73[a]	16.53[a]	17.73[a]	16.75[a]	16.18[a]	
E. coli	------------------					
Inulin	10.55±0.21	11.25±0.92	10.95±0.50	10.40±0.42	10.55±0.21	10.74[a]
MOS	10.20±0.00	10.80±0.14	10.75±0.07	10.70±0.10	9.90±0.57	10.47[a]
Average	10.38[a]	11.03[a]	10.85[a]	10.55[a]	10.23[a]	
P. aeruginosa	------------------					
Inulin	9.20±0.71	9.90±0.71	9.20±2.12	11.40±3.25	9.20±0.71	10.57[a]
MOS	11.1±0.00	10.10±1.28	11.15±2.33	10.65±2.90	9.85±0.92	9.78[a]
Average	10.15[a]	10.00[a]	10.18[a]	11.03[a]	9.53[a]	

MOS : Mannan oligosaccharides mean in the same column and row differ significantly (P<0.05)

Antibacterial activity of Lp + inulin and Lp + MOS were not significantly different (p>0.05) and but not with other concentration treatments. In Table 4, *L. plantarum* demonstrated antibacterial ability as evidenced by the inhibition zone on the growth of pathogenic bacteria, *Staphylococcus aureus* FNCC 6049, *Escherichia coli* FNCC 0194, *Salmonella pullorum* ATCC 13036 and *Pseudomonas aeruginosa* FNCC 0063, in the media but the anti-bacterial ability was not increased even though added oligosaccharides such as inulin and MOS. This indicated that oligosaccharides are nutrients for the growth of lactic acid bacteria including *Lactobacillus plantarum,* different from Baurhoo [23] that MOS has the ability to inhibit the growth of Gram-negative type 1 pathogenic bacteria in the digestive tract. Possibility not all types of oligosaccharides can synergize with BAL in accordance with Schrezenmeir [21] that probiotic Bifidobacteria can be synergized with oligofructose, but *Lactobacillus casei* cannot be in accordance with Figure 2.

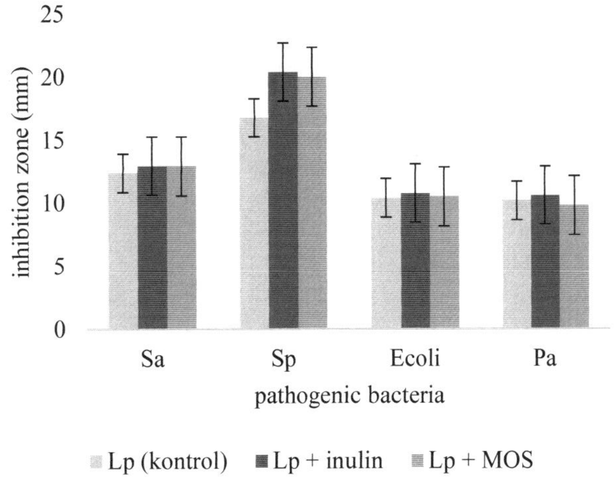

Sa : *Staphylococcus aerus*, Sp : *Salmonella pullorum,* Ec : *E. coli* dan Pa : *Pseudomonas aeruginosa*

Figure 2. Inhibitory effect of the combination of *L. plantarum* with oligosaccharides against pathogenic bacteria

The treatment of Lp + inulin could be inhibited better than that of Lp + MOS for all tested pathogenic bacteria. The biggest inhibition was demonstrated in *Salmonella pullorum* (Figure 2). The result indicated that the combination of *L. plantarum* and inulin were effective to inhibit *Salmonella pullorum* in comparison with MOS as well as control, which confirm the result of Feng [24] research which found that the antibacterial activity of the *L. plantarum* was better to inhibit the growth of *Salmonella pullorum.* Antibacterial activity of *L. plantarum* was also caused by its ability to survive longer in the digestive tract even though it was acidic and could be produced more lactic acid [25]. Bacteriocin produced by probiotic bacteria had narrow and broad-spectrum activities against pathogenic bacteria [26] so that *L. plantarum* could be used as a probiotic for animals [24].

4. Conclusion

Addition of oligosaccharides (OS) consisted of inulin or mannan oligosaccharides (MOS) in medium growth did not significantly affect lactic acid production and antibacterial activity of *L. plantarum* AKK30 against to pathogenic bacteria (*S. aureus* FNCC 6049, *S. pullorum* ATCC 13036, *E. coli* FNCC 0194 *and P. aeruginosa* FNCC 0063). However, growth of *L. plantarum* AKK30 could be increased by addition of mannan oligosaccharide (MOS) at 0.5 – 2 % (w/v) in growth-medium.

5. Reference

[1] Patterson J A and Burkholder K M 2003 Application of Prebiotics and Probiotics in Poultry Production *Poult. Sci.* **82** 627-631

[2] Kabir S M L 2009 The Role of Probiotics in the Poultry Industry *Int. J. Mol. Sci.* **10** 3531-3546

[3] Yang J, Qian K, Wu D, Zhang W, Wu Y and Xu Y 2017 Effects of different proportions of two *Bacillus sp.* on the growth performance, small intestinal morphology, caecal microbiota and plasma biochemical profile of Chinese Huainan Partridge Shank chickens *Journal of Integrative Agriculture* **16** (6) 1383–1392

[4] Tsirtsikos P, Fegeros K, Balaskas C, Kominakis A and Mountzouri K C 2012 Dietary Probiotic Inclusion Level Modulates Intestinal Mucin Composition and Mucosal Morphology in Broilers *Poultry Science* **91** 1860–1868

[5] Dec M, Renata U C, Dagmara S P and Andrzej W 2017 Assessment of antibiotic susceptibility in Lactobacillus isolates from chickens *Gut Pathog.* **9** 54

[6] Lin T H and Pan T M 2017 Characterization of an antimicrobial substance produced by *Lactobacillus plantarum* NTU 102 *Journal of Microbiology Immunology and Infection*

[7] Mountzouris K C, Tsitrsikos P, Palamidi I, Arvaniti A, Mohnl M, Schatzmayr G and Fegeros K 2010 Effects of Probiotic Inclusion Levels in Broiler Nutrition on Growth Performance, Nutrient Digestibility, Plasma Immunoglobulins, and Cecal Microflora Composition *Poult. Sci.* **89** 58–67

[8] Ohashi Y and Kazunari U 2009 Health beneficial Effects of Probiotics: Its Mode of Action *J. Anim. Sci.* **80** 361 -371

[9] Pranckute R, Arnoldas K, Nomeda K and Donaldas J C 2016 Combining Prebiotics with Probiotic Bacteria Can Enhance Bacterial Growth and Secretion of Bacteriocins *International Journal of Biological Macromolecules* **89** 669–676

[10] Gibson G R, Robert H, Mary E S, Susan L P, Raylene A R, Seppo J S, Karen S, Catherine S, Kelly S S, Patrice D C, Kristin V and Gregor R 2017 The International Scientific Association for Probiotics and Prebiotics (ISAPP) Consensus Statement on The Definition and Scope of Prebiotics *Mcmillan Publishers Limited. Part of Springer Nature, Nature Reviews Gastroenterology & Hepatology* **14** 491-502

[11] Su. P, Anders H and Hazel M 2007 Selected Probiotics Support The Growth of Probiotics Monocultures *In vitro. Anaerob* **13** 134-139

[12] AOAC 1995 Official Methods of Analisis of The Association of Official Analytical Chemists Association of Official Analytical Chemists Washington, D.C

[13] Bonev B, James H, and Judacael P 2008 Principles of Assessing Bacterial Susceptibility to Antibiotics Using the Agar Diffusion Method *J. Antimicrob. Chemotheraphy* **61** 1295-1301

[14] Cohort 2008 CoSTAT Version 6.400 Cohort Software 798 (Moneterey)

[15] Setiarto R H B, Nunuk W, Iwan S dan Rina M S 2017 Pengaruh Variasi Konsentrasi Inulin Pada Proses Fermentasi Oleh *Lactobacillus acidophilus, Lactobacillus bulgaricus* dan *Streptococcus thermophiles Biopropal Industri* **8** 1 17

[16] Machado M T C, Kaliana S E, Vieira G S, Menegalli F C, Martínez J and Hubinger M D 2015 Prebiotic oligosaccharides from artichoke industrial waste: evaluation of different extraction methods *Industrial Crops and Products* **76** 141–148

[17] Yang Y, Paul A I and Chot M 2009 Dietary Modulation of Gut Microflora in Broiler Chickens : a review of The Role of Six Kinds of Alternative to In-feed Antibiotics *World's Poultry Science Journal* **65**

[18] Khalid K 2011 An Overview of Lactic Acid Bacteria *International Journal of Biosciences* (IJB) **1** 3 1-13

[19] Brink M, Todorov S D, Martin J H, Senekal M and Dicks L M T 2006 The effect of prebiotics on production of antimicrobial compounds, resistance to growth at low pH and in the presence of bile, and adhesion of probiotic cells to intestinal mucus *Journal of Applied Microbiology* 1364-5072

[20] Roberfroid M B, Gibson G R, Hoyles L 2010 Prebiotics Effect : Metabolic and Health Benefits *British Journal of Nutrition*

[21] Schrezenmeir J and de Vrese M 2001 Probiotics, prebiotics, and synbiotics approaching a definition *Am. J. Clin. Nutr.* **73** 361S–4S.
[22] Roberfroid M B 2007 Inulin-Type Fructans: Functional Food Ingredients *J. Nutr.* **137** 2493S–2502S
[23] Baurhoo B, Letellier A, Xi Z and Ruiz- Feria C A 2007 Cecal population of Lactobacilli and Bifidobacteria and *Escherichia coli* after in vivo *Escherichia coli* Challenge in Birds Fed Diets with Purified Lignin or Mannanoligosaccharides *Poult. Sci.* **86** 2509–2516
[24] Feng Y, Lin Q, Rui L, Hongming Y, Changbin G 2017 Potential Probiotic Properties of Lactic Acid Bacteria Isolated from The Intestinal Mucosa of Healthy Piglets *Ann Microbiol* **67** 239–253
[25] Palaniyandi S A, Karthiyaini D, Joo W S, Seung H Y 2017 In Vitro Characterization of *Lactobacillus plantarum* Strains with Inhibitory Activity on Enteropathogens for Use as Potential Animal Probiotics *Indian J. Microbiol.* **57** 2 201–210
[26] Cotter P D, Ross R P, and Hill C 2013 Bacteriocins a viable alternative to antibiotics? *Nat. Rev. Microbiol.* **11** 95-105

IOP Conf. Series: Earth and Environmental Science **251** (2019) 012052 doi:10.1088/1755-1315/251/1/012052

Substitution of commercial cation with cation containing bean sprout waste on production and physical quality of egg

H S Iman Rahayu [1], Y Retnani[2], Kurniawati[2], Z Rizwantiara[2]

[1]Department of Animal Production and Technology, Faculty of Animal Science, Bogor Agricultural University, Bogor 16680, Indonesia

[2]Department of Nutrition and Feed Technology, Faculty of Animal Science, Bogor Agricultural University, Bogor 16680, Indonesia

E-mail: rahayu21@hotmail.com

Abstract. This research was aimed to evaluate substitution of commercial rations with rations contains bean sprout waste on production and physical omega 3 egg quality. Bean sprouts are a waste of green beans the remainder of beans sprout production consisting of green bean skin and shredded bean sprout. The use of bean sprout in layer diet is very limited because the bean sprouts are low palatability and high crude fiber. This experiment used eighty female birds strain of Lohman Brown were in a Completely Randomized Design allocated to five dietary treatments in four replicates. The treatment diets were R0: 100 % commercial feed, R1: 75 commercial feed + 25 % experiment feed, R2 : 50 % commercial feed + 50 % experiment feed, R3: 25 % commercial feed + 75 % experiment feed, R4: 100 % Experiment feed, respectively variable observed were egg production, feed consumption, feed conversion, Hen Day, IOFC, layer mortality, egg weight, egg mass, weight percentage of albumen, yolk, and shell, shell thickness, yolk color, egg, albumen and yolk index, and Haugh unit. The results showed there is no effect on all cumulative production (except on hen day and ration conversion) and physical quality of omega-3 eggs (except eggshell thickness). The highest IOFC value and the highest egg production were obtained at 50% substitution of commercial rations and 50% of research rations containing 2.5% of bean sprouts waste.

1. Introduction

The development of poultry farms in Indonesia is currently increasing, one of which is laying chicken farms because egg production from year to year is increasing. Egg production by province from 2009-2017 increased, which amounted to around 90,951,931 tons in 2009 and increased to 152,713,487 tons in 2017 [1]. Egg production of commercial hybrid laying hens in hot and humid area can reach 180-200 egg/year whereas in temperate area commercial can reach 250-300 egg/year [2]. The composition of eggs physically consists of 9.5% shells (eggshells or shells), 61.5% albumin, and 29% egg yolks [3]. Good quality eggs are eggs that have clean, smooth, slippery, non-cracked, and normal-shaped shell [4].

Ration is the main factor that influences the success of laying chicken farms in addition to seedlings and maintenance management. Feed must meet all the food needs of laying hens and do not contain chemicals that can endanger livestock to keep egg production stable and even increase and produce quality eggs. Commercial feed for laying hens on the market tends to have higher prices [5], so alternative feed is needed to reduce production costs and increase egg productivity and quality. One alternative feed to reduce the use of commercial feed is waste bean sprouts.

Bean sprouts waste is organic waste the residual production of bean sprouts consisting of green peanut shells and broken pieces of green bean sprout can be used animal feed [6]. Bean sprouts waste is easy to obtain, especially in the Bogor area, because the total production of bean sprouts in the Bogor area can reach 6.5 tons/day and the chance to produce sprouts waste is 1.5 tons/day. Bean sprouts waste also has good nutritional value, with active compounds, i.e. tocopherol, folic acid, vitamin c, flavonoids and phenols [7]. The use of bean sprouts waste for laying chicken feed has not been widely studied, so it is necessary to analyze the effect of its use on egg production and quality.

In addition to the addition of bean sprouts waste in the research ration, all rations were given an omega 3 supplement as a breakthrough product to produce value-added food products. Omega 3 supplementation (Patent no ID P 0023652) in commercial rations and research is intended to increase the value of functional food products in the form of omega 3 eggs. Omega 3 supplements are obtained from the mixing of lemuru fish oil processing waste with fermented tofu waste which functions as a filler. Omega 3 supplements are mixed with commercial rations and research with mixing methods.

Based on the description, substitution of commercial rations with rations containing bean sprouts waste are expected to maintain the production of omega-3 eggs and improve egg quality. This study aims to test the substitution of commercial rations with rations containing omega 3 bean sprouts waste to the production and physical quality of omega 3 eggs.

2. Materials and Methods

2.1 Birds and Cages

The birds used were 80 Lohman Brown laying eggs in the final phase of production with 98 weeks of age. The enclosure used is a battery cage made of wire with 20 cages. Each cage contains 4 chickens that are equipped with a place to eat and a drinking water container in the form of nipple. The cages are also equipped with lights and fans.

2.2 Ration

The ration used is commercial rations and research rations in the form of mash. The composition of commercial ration ingredients consists of corn, bran, corn gluten, Pollard, meat and bone flour, soybean oilcake, oil, calcium phosphate, calcium carbonate, sodium chloride, amino acids, vitamins, trace minerals, antioxidants, and omega 3 supplements. While the composition of research ration ingredients consists of corn, Pollard, soybean oilcake, coconut oilcake, bean sprouts (2.5%), DDGS, meat flour, CPO, $CaCO_3$, DCP, L-Lysin, DL-Methionine, vitamins and minerals and omega 3 supplements Nutrient content of commercial rations and research rations can be seen in Table 1.

2.3 Ration Making

The research ration used contains sprouts waste taken from Bogor market which was then dried and milled. The making of study feed was prepared based on nutrient requirements recommended by Lesson and Summers for laying hens aged more than 72 weeks (End Period) [8]. Feed formulation is carried out by trial and error while still considering the balance of protein and energy content in the ration.

2.4 Omega-3 supplementation in ration

Omega-3 supplements are made through the process of emulsifying and dispersing from the lemuru fish boiling waste material (which is known to contain lots of Omega-3, 6 and 9, and has good quality compared to the Omega-3 sources of plants) with tofu pulp (as filler) , in order to obtain an omega-3 supplement formula which is then mixed in commercial and research feed as much as 5%.

Table 1. Nutrient content of commercial and research rations (%)

Nutrient	Commercial rations[a]	Research rations[b]
Dry matter (%)	88	89.44
Crude protein (%)	18	21.61
Crude fat (%)	7	4.87
Crude fiber (%)	6	3.8
Ash (%)	13.5	16.68
Nitrogen Free Extract (%)	55.5	53.04
Calcium (%)	3.4-3.6	-
Phospor (%)	0.6-0.9	-
Metabolizable energy [1] (kcal kg)	3443.94	3198.32

[a] This value was obtained from the ration product label used

[b]This value was obtained from proximate analysis in Center for Research on Biological Resources & Biotechnology Laboratory, Bogor Agricultural University

2.5 Rearing
Chicken rear was done for 7 weeks. The chickens were divided into 5 treatments with 4 replications, each of which was cage filled with 4 chickens. The placement of chickens in a cage is done randomly. Feeding is carried out 2 times a day, namely morning and evening. Drinking water is given ad libitum.

2.6 Productivity Measurement
The number of eggs produced is recorded every morning and evening. Weighing is carried out every morning and evening and weighing the remaining food is done every morning. Calculation of food consumed is done once a week. Productivity measured include: feed consumption and conversion, egg weight, hen day production, egg mass.

2.7 Physical Quality Test of Eggs
Testing the physical quality of eggs is done 6 times in the 2nd, 3rd, 4th, 5th, 6th and 7th weeks of maintenance. The eggs tested are all eggs produced during the test day. The physical quality of the eggs measured were weight, length, and width of the egg; weight, length, height, and yolk score; weight, length, and height of albumen; thickness and weight of eggshell. Albumen and yolk indexes were calculated by dividing height with a diameter of albumen and yolks multiplied by 100%. The egg index is calculated by dividing the width by the egg length times 100%.

2.8 Statistic analysis
This study used a Completely Randomized Design (CRD) with 5 treatments of ration substitution and 4 replications. Treatment includes:

R0 = Commercial Feed 100%
R1 = Commercial Feed 75% + Experimental Feed 25%
R2 = Commercial Feed 50% + Experimental Feed 50%
R3 = Commercial Feed 25% + Experimental Feed 75%
R4 = Experimental Feed 100%

The data obtained were analyzed statistically using the SPSS Application with variance analysis (ANOVA), and if the results of the analysis showed different effects, then further testing would be carried out using the Duncan test. Data analysis was carried out based on the equations of Steel and Torrie [9] with the following mathematical formula.

$$Yij = \mu + \sigma_i + \varepsilon_{ij}$$

Note:

Y = The value of the observations in the i-th treatment and j-th replication
μ = General average value
σ_i = Effect of the i-th treatment
ε_{ij} = The i-th treatment error dan j-th replication

IOP Conf. Series: Earth and Environmental Science **251** (2019) 012052 doi:10.1088/1755-1315/251/1/012052

3. Results and Discussion

3.1 Productivity

The results of the analysis of variance (can be seen in Table 2) show that the substitution of the research ration had no effect on all parameters of productivity, except for the production of hen day eggs (P <0.01) and ration conversion (P <0.05).

Table 2. Productivity performance of research chickens for 4 weeks of maintenance [a]

Variables	Treatments				
	R0	R1	R2	R3	R4
Rations intake (g head^{-1} day^{-1})	84.16±6.97	89.89±8.35	90.13±8.00	87.15±6.90	79.71±8.13
Weight of eggs (g egg^{-1})	69.49±3.12	68.01±3.10	68.28±1.61	67.76±1.54	67.17±3.01
Hen day production (%)	46.80±10.35	39.73±10.68	55.65±15.73	51.79±13.12	43.06±12.74
Egg mass (g)	21829± 111.31	18226±118.55	25591±174.31	23541±139.35	15626 ±99.99
Rations convertions	2.56±0.83	3.40±1.07	2.49±0.62	2.70±0.81	2.70±1.23
IOFC (Rp)	434 062,50	286 481,10	549 381,10	493 766,64	234 492,56

[a] Notes are referenced using alpha superscripts. capital letters show very significantly different (P <0.01) and small letters show significantly different (P <0.05)

R0: 100% commercial rations; R1: 75% commercial rations + 25% research rations; R2: 50% commercial rations + 50% research rations; R3: 25% commercial rations + 75% research rations; R4: 100% research rations

Hen-day egg production in R2 treatment (50% commercial ration and 50% research ration (containing 2.5% bean sprouts waste)) had the highest egg production rate (60.27%) compared to R1 (75% commercial ration, 25% research ration) and R4 (100% research ration). The ration conversion value in the group of chickens fed 100% (R4) real research ration (P <0.05) had the highest conversion value compared to the other groups (R4 vs. R0, R1, R2, R3).

The consumption of research rations ranged from 84.40-92.87 g head^{-1} day^{-1} indicating that consumption did not meet needs optimally. The feed requirements of laying hens in the final phase of production are 110 g head^{-1} day^{-1} [8]. This can be caused by the content of crude fiber of bean sprouts waste contained in a high research ration, although the crude fiber content in the research ration is still within limits tolerable by chickens. The crude fiber content in commercial ration was 6% and in the research ration 3.8%.

The size/weight of eggs produced from this study is classified as large size (weight more than 60 g egg^{-1}) in accordance with the provisions of SNI (2008). Large egg weights can be produced because of the protein contained in high rations. Protein content in the commercial ration was 18%, and the research ration was 21.61%. Protein can increase egg weight because when there is no ovulation of egg yolk, protein accumulation will occur, so protein availability increases to form one egg the next day [10].

The hen-day egg production was very significantly (P <0.01) highest in R2 treatment (50% commercial ration and 50% research ration (containing 2.5% bean sprouts)). This can be due to the right mix/substitution of rations to increase the production of hen-day. In line with high egg production because feed consumption is also high in R2 treatment, although not different from the consumption of feed in other treatments.

Bean sprouts contain vitamin E which can increase egg productivity. Vitamin E is able to break various chains of free radical reactions by removing hydrogen from free radicals from polyunsaturated fatty acids that have undergone peroxidation [11]. This is what improves the quality of egg productivity.

Egg mass is obtained from the sum / total weight of eggs produced in one week. The treatment does not affect egg mass. This can be due to the right combination of feed and high hen day production. Egg mass increases when production increases [12].

The results of the analysis of variance showed that the treatment significantly affected the value of feed conversion ($P<0.05$). The ration is more efficient with a lower conversion value, shown in all four treatments (R0, R1, R2, and R3). Most inefficient in the treatment of 100% research ration (R4) with a conversion value of 5.98. This is because the treatment with 100% using the research ration of chickens undergoes a change in feed phase which can cause livestock to stress and ultimately decrease egg production. Feed conversion calculations are intended to determine the ability of chickens to convert the food consumed into eggs and see the chicken's response to the quality of feed given [13].

IOFC (Income over Feed Cost) is the result of income derived from the difference between revenue from the sale (in units) of omega-3 eggs and the cost of feed [14]. The addition of bean sprouts waste in the substitution of rations to produce omega-3 eggs as functional food has a positive effect on the IOFC value. IOFC values tend to increase as the research ration increases except for the R1 and R4 treatments because it produces low egg production, which causes the IOFC value to be low. The R2 treatment has the highest IOFC value of Rp 206,958.

3.2 Physical Quality of Omega-3 Eggs

The effect of giving substitution of commercial ration and ration containing 2.5% of bean sprouts waste gradually on the physical quality of omega-3 eggs is presented in Table 3.

The results of various analyzes of the physical quality of omega-3 eggs indicate that ration substitution does not affect on egg weight; weight and percentage of albumen, yolk, and eggshell; yolk color score; index of egg, white and yolk; Haugh units. Ration substitution of R4 (100% research ration) was significantly effected to decreasing eggshell thickness ($P <0.05$).

Table 3. Physical quality of Omega-3 eggs [a]

Variables	Treatments				
	R0	R1	R2	R3	R4
Weight of eggs (g)	70.45±4.61	67.15±5.27	69.64±3.69	67.97±3.69	68.47±4.21
Albumen					
- Weight (g)	45.59±4.21	44.01±4.34	45.46±3.68	44.25±3.51	44.55±4.18
- Percentage (%)	64.58±2.18	65.31±1.40	65.13±1.95	64.97±1.58	64.88±2.23
Yolk					
- Weight (g)	18.10±1.18	17.04±0.56	17.55±0.64	17.34±0.17	17.40±0.41
- Percentage (%)	25.81±2.32	25.57±1.24	25.34±1.65	25.62±1.37	25.58±1.88
Eggshell					
- Weight (g)	6.77±0.64	6.10±0.44	6.63±0.15	6.38±0.42	6.52±0.28
- Percentage (%)	9.61±0.41	9.12±0.35	9.53±0.36	9.42±0.48	9.54±0.40
Eggshell thickness (mm)	34.80±0.84^{a}	34.12±2.20^{a}	33.91±0.62^{a}	33.86±1.70^{a}	30.99±1.14^{b}
Yolk color	14.95±0.09	14.84±0.11	14.86±0.17	14.81±0.16	15.00±0.00
Egg index	0.72±0.02	0.72±0.02	0.73±0.01	0.73±0.01	0.72±0.03
Albumen index	0.10±0.01	0.10±0.02	0.09±0.01	0.10±0.01	0.10±0.01
Yolk index	0.37±0.00	0.38±0.02	0.37±0.00	0.38±0.02	0.36±0.02
Haugh Unit	84.30±5.05	84.12±5.39	81.78±4.70	86.24±4.58	87.69±2.33

[a] Notes are referenced using alpha superscripts show significantly different ($P <0.05$)

R0: 100% commercial rations; R1: 75% commercial rations + 25% research rations; R2: 50% commercial rations + 50% research rations; R3 : 25% commercial rations + 75% research rations; R4: 100% research rations

Albumen index is a comparison between albumen height and albumen diameter. The more protein content in the ration, it will produce thicker albumen. It means to be the higher index value of the albumen and then good to maintain during storage [4].

Yolk weight is affected by fat content because most fat deposits are in egg yolk [15]. The resulting yolk color score has the highest value in the yolk color fan scale on the substitution of commercial rations with 100% research ration (R4), which is equal to 15.00. This can be caused by the presence of 68 μg of carotenoid pigment in bean sprouts which is the main pigment to form yellow, orange, and red on yolks [7]. The yolk color score is affected by the nutrient content in the ration, such as beta-carotene and xanthophyll [16]. The yolk index is a comparison between height and diameter of yolk average. Factors that influence the yolk index include storage time, storage temperature, vitellin membrane quality, and feed nutrition.

The highest shell thickness ($P < 0.05$) was in the control treatment (R0 = commercial ration containing omega-3), while the R4 treatment produced the lowest shell thickness. This can be caused by the low calcium content found in the R4 ration compared to the commercial ration of 3.4-3.6%. Eggshells are mostly built on calcium carbonate ($CaCO_3$) so that calcium content in the ration needs to be considered to get the optimum thickness of the eggshell. Shell thickness should be noted because the thin eggshells have low ability to protect quality of inner egg [17].

The egg index is obtained from the ratio of the width and length of the egg which affects the shape of the egg. The index value of the egg will affect the appearance of the egg itself. The egg index is close to 1 if the egg is rounded, while it is close to 0 if the egg is elliptical. The egg shape is disproportionately affected by the content of Ca and P in the ration, if the content of the two nutrients is insufficient in the ration, it will also affect the shape of the egg and will greatly affect the egg index value [18].

Haugh Unit (HU) is the relationship between thick or high albumen/albumen with the overall weight of eggs, is the basis for measuring the egg quality index. The results showed that feed substitution did not affect HU values. A good HU value is affected by the freshness of the egg. The high value of HU in R4 treatment, which is equal to 87.69 can be due to the level of protein ration efficiency.

4. Conclusion
Substitution of commercial rations with rations containing 2.5% bean sprouts waste does not affect productivity (except hen-day egg production and ration conversion) and physical quality of omega-3 eggs (except eggshell thickness). The highest omega-3 egg production in R2 treatment, while the worst / highest conversion in R4. Eggshell thickness decreased with 100% substitution of commercial rations (R4). The highest IOFC value and the highest egg production were obtained at 50% substitution of commercial rations and 50% of research rations containing 2.5% of bean sprouts waste.

5. Acknowledgment
Thanks are conveyed to the Intellectual Product Unit Development Program (Program Pengembangan Unit Produk Intelektual Kampus/P2UPIK) No 1808/IT3.11/PN/2018 and No 1811/IT3.11/PN/2018 which have provided funding during the research.

6. References
[1] Badan Pusat Statistik 2017 *Produksi telur ayam petelur menurut provinsi (ton) 2009-2017* Jakarta (ID) Pusat Satistik. http://www.bps.go.id/ (5 Juni 2018)
[2] Zaheer K 2015 An updated review on chicken eggs: production, consumption, management aspects and nutritional benefits to human health *Food and Nutr Sci* **6** 1208-1220
[3] Shafey T M, Mahmoud A H, Abouheif M A 2014 Dealing with multicollinearity in predicting egg components from egg weight and egg dimension **13** 715-719
[4] Standar Nasional Indonesia 2008 *Telur Ayam Konsumsi* SNI-3926:2008 (Jakarta: Badan Standarisasi Nasional)
[5] Daud R A and Hasni A 2009 Kajian Ekonomi Wilayah dan Kelembagaan Usaha Peternakan Ayam Ras Petelur di Kabupaten Tasikmalaya *Agritech* **10** (2) 81-95

[6] Nurcholis, Raden I A and Mohamd Y 2015 Pengaruh Pakan Limbah Tauge dan Suplementasi Omega-3 Terhadap produksi Spermatozoa Domba Garut *Agricola* **5** (2) 133-142
[7] Guo X, Li T, Tang K and Liu R H 2012 Effect of germination on phytochemical profiles and antioxidant acrivity of mung bean sprouts (*Vigna radiata*) *Journal of Agricultural and Food Chemistry* **60** 11050-11055
[8] Leeson S and Summer J D 2005 *Commercial Poultry Nutrition* 3rd Edition (Nottingham (US): Nottingham University Press)
[9] Steel R G D and Torrie J H 1993 *Prinsip dan Prosedur Statistika* Edisi Ke-3 (Terjemahan. Jakarta (ID): PT Gramedia)
[10] Widjastuti T and Kartasudjana R 2006 Pengaruh pembatasan ransum dan implikasinya terhadap performa puyuh petelur pada fase produksi pertama *J. Indon. Trop. Agric.* **31** (3) 162-166
[11] Agarawal A and Alamaneni S S R 2004 Role of Free Radical in Female Reproductive Dissease and Asisted Reproduction *Reproductive BioMedicine* **9** (3) 338 – 3347
[12] Nasikin M, Nangoy F J, Sarajar C L K and Kawatu M H M 2015 Pengaruh subtitusi sebagian ransum dengan tepung tomat (*Solanum Lycopersicum L*) terhadap berat telur, berat kuning telur dan massa telur ayam ras *J. Zootek* **35** (2) 225 – 234
[13] Lokapirnasari W P, Soewarno and Dhamayanti Y 2011 Potensi crude spirulina terhadap protein efisiensi rasio pada ayam petelur *J. Ilmiah Kedokteran Hewan* **2** 5-8
[14] Mayulu H B Suryanto Sunarso M Christiyanto F I Ballo and Refa'i 2009 Feasibility of Complete feed Based on Ammonitiated Fermented Rice Straw Utilization on the Beef Cattle Farming *J. I. Tropic Anim. Agri.* **34** 74-78
[15] Argo L B, Tristiarti I and Mangisah 2013 Kualitas Fisik Telur Ayam Arab Petelur Fase I dengan Berbagai Level *Azolla microphylla Animal Agricultural Journal* **2** (2) 445-457
[16] Nuraini Sabrina and S A Latif 2008 Performa ayam dan kualitas telur dengan penggunaan ransum mengandung onggok fermentasi dengan *Neurospora crassa J Media Peternakan* **31** (3) 195-202
[17] Chang R 2005 *General Chemistry* The Essential Concepts Sixth Edition (New York: McGraw-Hill Companies)
[18] Hafez E S E 2000 *Reproduction in Farm Animals* 7 Ed (Philadelphia : Lea & Febiger)

Endo-xylanase enzyme from marine actinomycetes and its potential for xylooligosaccharide production

N Rahmani[1], P Apriliana[1], A M Jannah[1], S Ratnakomala[1], P Lisdiyanti[1], E Hermiati[2], B Prasetya[1,3] and Yopi[1]

[1]Research Center for Biotechnology, Indonesia Institute of Science, Jl. Raya Bogor KM.46 Cibinong-Bogor, West Java, Indonesia.
[2]Research Center for Biomaterial, Indonesia Institute of Science, Jl. Raya Bogor KM.46 Cibinong-Bogor, West Java, Indonesia.
[1]National Standardization Agency of Indonesia, Gedung I BPPT Jl. M.H. Thamrin No. 8 Kebun Sirih, Jakarta Pusat. 10340, Indonesia.

Corresponding author's e-mail address: yopi@bioteknologi.lipi.go.id

Abstract. Marine microorganism was emerging as the great source for the discovery of novel enzyme. The search and discovery of new enzymes derived from actinomycetes have received much attention nowadays due to a growing need for applications in various industries including food industry. Endo-β-1,4-xylanases is considered one of critical enzyme involves in the degradation of xylan that is able to cleave the β-1,4-glycosidic linkages in the xylan backbone resulting in various xylooligosaccharides and xylose. Screening and characterization of novel actinomycetes that is capable of producing high xylanase are necessary to produce xylooligosaccharides from hemicellulose hydrolysis process. The objective of this research is screening and characterizing of novel marine actinomycetes from working culture of Research Center for Biotechnology, Biotechnology Culture Collection (BTCC) that are capable of producing high-level hemicellulase-degrading enzymes, especially endo-xylanase enzyme. Approximately, 70 strains from Indonesia marine Actinomycetes have been screened by using congo red, SDS-PAGE and zymogram methods. Of these 70 strains, 3 strains were successfully identified that is capable in producing high level of xylanase enzymes based on the diameter of clear zone more than 1.5 cm on xylan plate medium pH 5.0 and 6.0. Three strains from marine Actinomycetes were identified based on analysis a 16S rRNA gene sequence revealed that all of strains belonging to the genus Streptomyces. Strain 47 is closely related with *Streptomyces variabilis* (98%). All strains had ability to produce xylanase at optimum pH 5.0 and temperature range between 50- 60 °C with range activity from 2.5 – 4.3 U/mL. Each isolate had a molecular weight from 20-50 kDa based on SDS-PAGE and zymogram analysis. Based on TLC analysis using beechwood xylan, each isolate had ability to produce xylooligosaccharides. These characterizations showed that these isolates are potentially used for xylooligosaccharides productions. Strain 47 was selected for the next analysis for molecular cloning based on the pattern and results of TLC clearer then strain 41 and strain 42.

Keywords: Endo-xylanase, marine actinomycetes, Streptomyces, xylooligosaccharide

1. Introduction

Hemicellulose is the second most abundant renewable polymer saccharide in nature after cellulose. Hemicellulose plays a role in crosslinking the other components of biomass via hydrogen and covalent bonds [1] [2] [3]. The hemicellulosic system is more complex due to the heterologous nature of hemicellulose, which is composed of various sugar units, with attached side chains and side groups. The hemicellulosic fraction represents a rich source of xylooligosaccharides (XOS) and D-xylose, a biotechnologically-relevant sugar for bioethanol production by C5-fermenting microorganisms and as precursor for synthesis of chemical compounds such as xylitol and organic acids, which are value-added products for industry. There is growing interest in production of XOS based on their beneficial health effects including prebiotic effects in gastroenteritis, functional properties as a soluble fiber and ability to reduce cholesterol blood levels [4]. It can be used as food ingredients, symbiotic food, pharmaceuticals or fertilizer [5].

β-1,4-xylan is a major component of hemicellulose and has a backbone of β-1,4-linked D-xylopyranosidase residues substituted with acetyl, arabinosyl and uronyl side chains [6]. Xylanolytic enzyme system carrying out complete hydrolysis of xylan requires the synergistic action of different xylanolytic enzymes, including endo-1,4-β-D-xylanases (EC 3.2.1.8), exo-1,4-β-D-xylosidases(EC 3.2.1.37) and accessory enzymes, such as a α-glucuronidases (EC 3.2.1.139), acetyl xylan esterases (EC 3.1.1.6), α-L-arabinofuranosidase (EC 3.2.1.55), α-galactosidase (EC 3.2.1.22) and phenolic acid esterase (ferulic and ρ-caumaric acid). All these enzymes act cooperatively to convert xylan into its constituent sugars.

Marine microorganisms are widely distributed in ocean all over the world including Indonesia. Indonesia consists of three-quarter of sea containing high biological diversity, including microorganism. The ocean floor has been recently demonstrated as an ecosystem with many unique forms of actinomycetes [7] [8]. It appears that they are widely distributed throughout the ocean and can be found in seawater [9] [10]. The presence of such a multifunctional xylanolytic enzyme system is quite widespread among the fungi, actinomycetes, and bacteria. *Streptomyces* sp. has been reported as the dominant xylanolytic species of actinomyces producing the hemicellulose degradation enzymes [11]. *Streptomycetes* sp. is model organisms for the studies of several enzymes important in the food industry and other industries. A strategy of screening and characterization of novel actinomycetes capable of producing high-level xylanase is necessary. Characterization of novel bacteria capable of producing high-level xylanase seems to be all suited starting point for obtaining improved hemicellulose hydrolysis to produce xylooligosaccharides. The purpose of this study was to screen and characterizing of novel marine actinomycetes from working culture of Research Center for Biotechnology, Biotechnology Culture Collection (BTCC) that is capable of producing high-level hemicellulase-degrading enzymes, especially for endo-xylanase enzyme and study potential the selected strains for xylooligosaccharide production.

2. Materials and Methods

2.1. Biomass preparation

The sugarcane bagasse used in this study was obtained from a sugarcane factory, PG Rajawali, at Subang, West Java, Indonesia. The sugarcane bagasse was sun-dried, ground, and sieved for material with particle sizes equal to a 40-60 mesh. The material was kept in sealed plastic bags that were stored in a container.

2.2. Strain, Materials, and Chemicals

The actinomycetes used in this study were obtained from working culture of Research Center for Biotechnology, Biotechnology Culture Collection (BTCC), Indonesian Institute of Sciences (LIPI). Xylan was purchased from Sigma-Aldrich (St. Louis, Mo. USA) and Nacalai, Japan. All other reagents were of the highest grade available.

2.3. Culture and growth conditions

Actinomycetes were grown at 28 °C in 4 mL ISP2 medium containing artificial seawater (ASW) for 3 days. The composition of the ISP2 medium (w/v) was as follows: Yeast extract 0.4 %, Malt extract 1%, Glucose 0.4%. Erlenmeyer (100 mL) containing 10 mL ISP2 medium without glucose, plus 0.5 % of sugar cane bagasse were sterilized, cooled and inoculated with pre-culture of the isolate. The fermentation was carried out on a rotary shaker at 28 °C for 7 days. Sampling was carried out every 24 hours, and the enzyme was harvested for the recovery of xylanase enzyme from the culture supernatant by centrifuge in 5000 rpm, 4 °C for 10 min. Xylanase degrading activity was conducted on xylan agar, pH 5 and 6. Clear zones were appeared around growing bacterial colonies indicating xylanase hydrolysis.

2.4. Screening for lignocelluloses degrading enzyme

Xylanase degrading activity was conducted on xylan agar. Clear zones were appeared around growing bacterial colonies indicating xylan hydrolysis. The isolates which exhibited maximal zone of degradation (> 1.5 cm) were chosen as the organism for the further study. Strains were cultured on ISP2 agar medium. A single colony was pre-cultured in ISP2 broth at 28 °C, 190 rpm for 3 days, then inoculated into 10 mL modified ISP2 broth contained 0.4% yeast extract, 1.0% malt extract, and 0.5% each of sugarcane bagasse in 100 mL Erlenmeyer flasks. The fermentation was carried out on a rotary shaker at 28 °C and 190 rpm for 7 days. Sampling was carried out every 24 h. The xylanase was harvested from the culture supernatant via centrifugation at 4 °C and 9,000 g for 10 min. The analysis of xylanase activity against each of the xylan substrates was measured on xylan agar (0.5% xylan and 1.8% agar) by individually spotting an aliquot of supernatant (5 μL) taken from the culture. The plates were then incubated at 37 °C for 3 days to ensure that a clear halo formation had occurred due to the degradation of xylan by the respective enzymes. To visualize the hydrolysis zones, the plates were flooded with an aqueous solution of 0.25% Congo red (Nacalai, Kyoto, Japan) for 30 min. Then, the plates were washed with 1M NaCl for 15 min twice, and 0.5% acetic acid to more clearly check the xylan degradation. Xylan degradation appeared as clear zones around the colonies as opposed to the unclear zones that had not degraded.

2.5. Dye staining of xylan agar

The xylanolysis basal medium pH 5, 6 containing substrate xylan 0.5% and 1.8% agar were prepared and autoclaved. The sterile media was aseptically transferred to the sterile Petri dishes and inoculated with the supernatant individually (5 μL). The plate was then incubated at 37 °C for 3 days. To visualize the hydrolysis zone, the plates were flooded with an aqueous solution of 0.25% congo red for 1 hour, washed with 1M NaCl 30 min 2 times and washed with 0.5% acetic acid to get better picture. Xylan degradation around the colonies appeared as clear zone against not clear zone upgraded. To indicate the activity of organisms, diameters of clear zone around the colonies on cellulose agar plate were measured. Colonies with the highest clear zone on the plate were selected as a potential strain and subjected to saccharification test.

2.6. Determination of molecular weight by SDS-PAGE and zymogram analysis

Samples were solubilized in sample buffer, and equal amount of protein was loaded into 12% SDS-PAGE and zymogram was carried out at constant current (100 mA) for 3.5 hours [12].

2.7. Culture of lignocelluloses producing actinomycetes

Selected isolates were cultured on Erlenmeyer flask (500 mL) containing 2.5 % of sugarcane bagasse. The fermentation was carried out in 500 mL Erlenmeyer flasks on a rotary shaker at 28 °C for 7 days. At the end of the 7 days, the enzyme was harvested for the recovery of lignocellulose enzyme from the culture supernatant by centrifuge in 9000 g, 4 °C for 10 min.

2.8. Characterization of endoxylanase from the isolated candidate

Xylanase activity was assayed using 0.5% (wt/vol) birchwood glucuronoxylan (Sigma) as a substrate according to the method of [13]. The amount of reducing sugars liberated was determined using the dinitrosalicylic acid (DNS) method with xylose used as standard [14].To determine the optimal pH of xylanase activity, evaluations were conducted between pH 3.0 and 10 under standard assay conditions using the following buffers: 50 mM sodium citrate buffer (pH 3.0-5.0), sodium phosphate buffer (pH 6.0-7.0), and Glycine-NaOH buffer (pH 7.0-10.0), respectively. The optimal temperature of the xylanase activity was evaluated by incubating the enzyme samples with the substrate at temperatures ranging from 30 to 90 °C in 50 mM sodium citrate buffer (pH 5.0).

2.9. Enzymatic hydrolysis of xylan by selected strains
The main product hydrolysis of 0.5% (w/w) xylan beechwood substrates from selected strain were carried at 45 °C in 50 mM MES buffer (pH 6.5), for various reaction time (0, 1, 2, 3, 4 and 24 h). The reaction mixture was incubated for 0, 1, 2, 3, 4 and 24 h in a shaker incubator at 30 °C. At various reaction times (0, 1, 2, 3, 4 and 24 h), the reaction mixture was withdrawn and heated at 100 °C for 5 min to terminate the reaction. The reaction mixture was spotted on TLC Silica gel $60F_{254,\ 20\text{-}20\ cm}$ (EMD/Merck, Darmstadt, Germany) and developed with a mixture of n-Butanol/Acetic Acid/water (2:1.1:1, v/v/v). Spots were staining by using DAP that contained diphenylamine, aniline, acetone, and phosphoric acid (Merck KGaA, Darmstadt, Germany), and subsequently heating at 120 °C for 15 min. Xylose (X1), xylobiose (X2), xylotriose (X3), xylotetraose (X4) and xylopentaose (X5) and xylosehexaose (X6) from Megazyme were used as a standard.

2.10. Molecular identification of the selected strain
Molecular identification of the selected strain was conducted based on the 16S rRNA gene as established by Lisdiyanti et al. [15]. A pair of 9F (5' AGRGTTTGATCMTGGCTCAG 3') and 1510R (5' TACGGYTACCTTGTTAYGACTT 3') primers were used for amplified the 16S rRNA gene via polymerase chain reaction (PCR) technique [16]. The sequence was confirmed via ABI 3130 DNA sequencer (Applied Biosystems, Foster City, CA, USA) and then compared with others available in the GenBank/DDBJ/EMBL database.

3. Results and Discussion

3.1. Screening for endoxylanase-producing strain using agar plates
Marine microorganism is not only becoming an attractive source of novel natural products but also serving as the source of valuable genes and useful industrial enzymes [17]. We have already screened marine actinomycetes in order to determine the xylanase production capacity from 70 Indonesia Actinomycetes strains, working culture of Research Center for Biotechnology, Biotechnology Culture Collection. These strains were cultivation in ISP2 medium containing ASW using sugarcane bagasse as a carbon source. The qualitative analysis for xylanolytic activity was conducted using Congo red dye. After three days of incubation on agar plate medium buffer sodium acetate pH 5.0 and sodium phosphate pH 6.0 containing xylan as the sole carbon source, the diameters of the colonies and the clear halos around them were measured. We found three selected strains have ability to produce xylanase activity with diameter clear zone more than 1.5 cm (Figure 1).

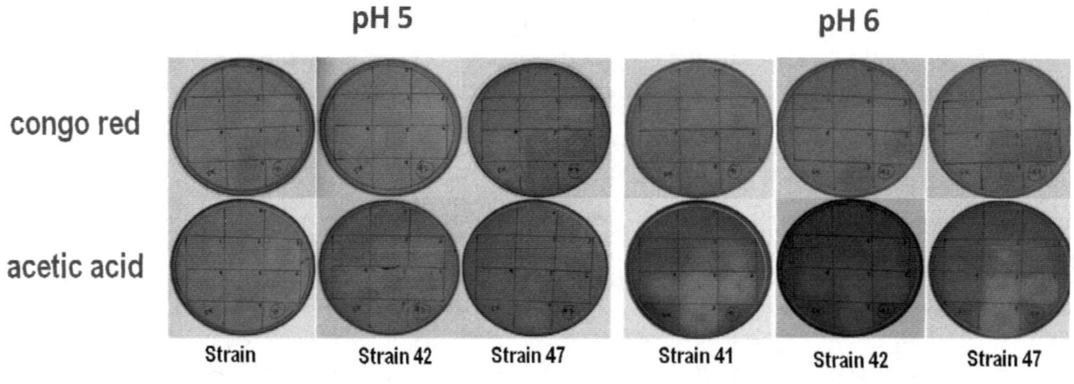

Figure 1. Clear zone in the xylan agar medium at pH 5.0 and 6.0 using the culture supernatant of the candidate strain, staining by congo red and acetic acid indicating the hydrolysis of xylanase (0, 1, 2, 3, 4, 5, 6 and 7 days cultivation).

We continue SDS-PAGE and zymogram analysis of three selected strains on xylan pH 5.0 and pH 6.0. The result of Figure 2 showed that there are four patterns of variance molecular weight range from 20-50 kDa. Strain 41 and strain 47 have similar pattern on zymogram analysis produce three thick bands predicted below 37 kDa, and one thin band predicted 50 kDa, whereas strain number 42 produces three thick bands predicted below 37 kDa.

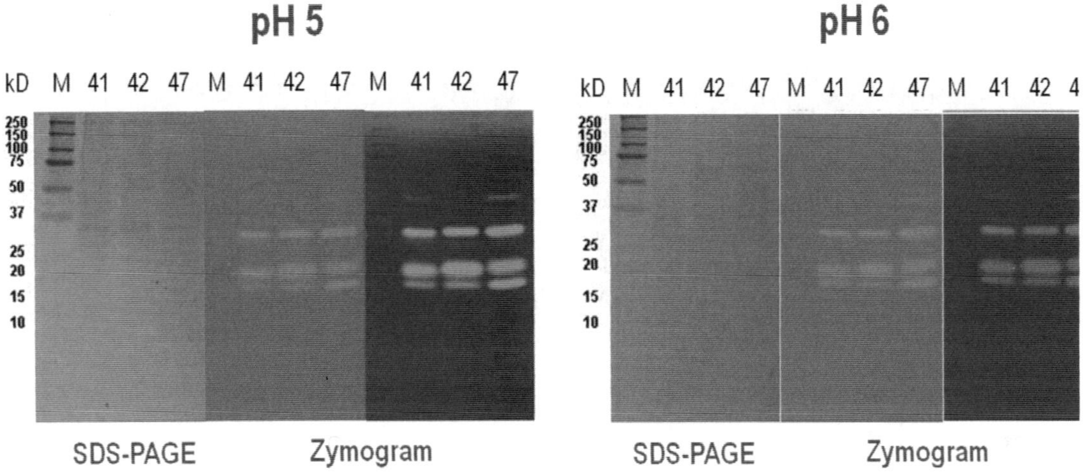

Figure 2. SDS-PAGE and zymogram analysis on xylan pH 0.05 mM sodium acetate buffer 5.0 and 0.05 mM sodium phosphate buffer 6.0 from selected strains.

3.2. Characterization of selected strains as an endoxylanase producer
Streptomyces have been found to be the most abundant hemicellulases producer among actinomycetes [18]. Several *Streptomyces* species, which are high activity as a xylanase producer have been reported from terrestrial Actinomycetes [19]. In this study, three strains from marine Actinomycetes were identified based on analysis a 16S rRNA gene sequence revealed that all of strains belonging to the genus *Streptomyces* (Table 1). Strain 47 is closely related with *Streptomyces variabilis* (98%).

The characteristic of xylanase activity was measured on various pH and temperature. The result xylanase activity on various pH was shown in Table 1. All strains were optimum at pH 5, temperature optimum range at 50 °C for strain 41 and 42 and 60 °C for strain 47 with range activity from 2.5 – 4.3 U/mL. This research has been similar to Beg et al. [20] who reported that xylanase from *Streptomyces* generally has optimal activity at 50-60 °C. Compared with other of xylanase, these xylanases were similar with xylanase from *S. megasporus* DSM 41476 which is optimum at pH 5.5 [19].

Table 1. Characterization of selected strains as a xylanase producer

Strain codes	16S rDNA analysis	pH optimum	Temperature optimum (°C)	Enzyme activity (U/mL)
Strain 41	*Streptomyces* sp.	5.0	50	2.7
Strain 42	*Streptomyces* sp.	5.0	50	2.5
Strain 47	*Streptomyces variabilis*	5.0	60	4.3

3.3. Hydrolysis of xylan commercial by selected strains
The hydrolysis of xylan substrates by the culture supernatant from three selected strains using xylan commercial was analyzed via TLC. Hydrolysis of 1% (w/v) beechwood xylan was performed containing 50 mM sodium citrate buffer (pH 5.0) in 20 mL total reaction. Each strain has ability to degrade substrate xylan to produce various xylooligosaccharide. The main hydrolysis xylooligosaccharides released using the action of endoxylanase from three strain have similar patterns from xylobiose (X2) to xylohexaose

(X6). This suggested that the xylanase from these strains belong to members of xylanase family. Kluepful et al. [21], Yan et al. [22], Li et al. [23], Shin et al. [24] reported that many members of xylanase family 11 hydrolyze xylans to xylo-oligosaccharides with the higher degree of polymerization ≥ 2. This suggested that the enzyme preferentially cleave the internal glycosidic bonds of these xylo-oligosaccharides. Strain 47 was selected for the next analysis for molecular cloning based on the pattern and results of TLC clearer then strain 41 and strain 42.

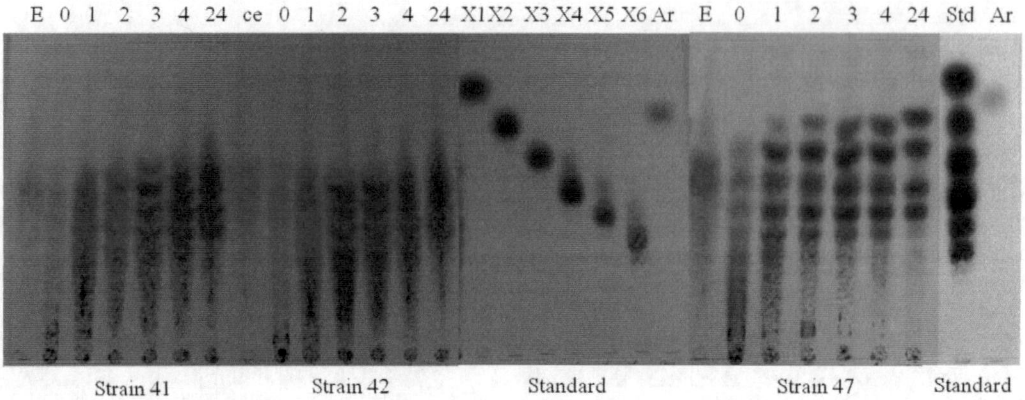

Figure 3. TLC analysis of hydrolysis products performed by three selected strains using beechwood xylan substrates. Standards (STD) : Xylose (X1), xylobiose (X2), xylotriose (X3), xylotetraose (X4) and xylopentaose (X5) and xylosehexaose (X6). Substrate concentrations 1% beechwood xylan in 50 mM sodium citrate buffer (pH 5.0), reaction time: 0, 1, 2, 3, 4, and 24 h at 30 °C, respectively.

4. Conclusion

This study report three strains were successfully identified that is capable in producing high level of xylanase enzymes based on the diameter of clear zone more than 1.5 cm on xylan plate medium pH 5.0 and 6.0. Analysis a 16S rRNA gene sequence revealed that all of strains belonging to the genus Streptomyces. All strains had ability to produce xylanase at optimum pH 5.0 and temperature range between 50- 60 °C with range activity from 2.5 – 4.3 U/mL. Each isolate had a molecular weight from 20-50 kDa based on SDS-PAGE and zymogram analysis. Based on TLC analysis using beechwood xylan, each isolate had ability to produce xylooligosaccharides. These characterizations showed that these isolates are potentially used for xylooligosaccharides productions. Strain 47 which is closely related with Streptomyces variabilis (98%) was selected for the next analysis for molecular cloning based on the pattern and results of TLC clearer then strain 41 and strain 42.

5. Acknowledgments

We thank Dr. Katsuhiko Ando and Dr. Tomohiko Tamura from National Institute and Technology Evaluation (NITE), Japan, Dr. Yantyati Widyastuti from Research Center for Biotechnology, Indonesian Institute of Sciences, Indonesia, and Biotechnology Culture Collection (BTCC) for providing the actinomycetes strains. This work was supported by Japan Science and Technology Agency (JST) and Japan International Cooperation Agency (JICA), Science and Technology Research Partnership for Sustainable Development (SATREPS) FY2014-2018, Collaboration research between Research Center for Biotechnology, Indonesian Institute of Sciences and Department of Chemical and Engineering, Graduate School of Engineering Kobe University, Japan. This work was supported by PhD Dissertation RONPAKU Fellowship from the Japan Society for the Promotion of Science (JSPS), FY2015-2017.

6. References

[1] Das N N, Das S C, Dutt A S and Roy A 1981 *Carbohydrate Res.* **94** 73–82
[2] Lai YZ 1995 *Reactivity and accessibility of cellulose, hemi-cellulose, and lignins. In: Hon DNS (ed) Chemical modification of lignocellulosic materials* (Boca Raton : CRC Press)

[3] Wong D W 2006 *Appl Biochem Biotechnol.* **133** 87–112.
[4] Palframan R J, Gibson G R, Rastall R A 2002 *Anaerobe.***8** (5) 287-92
[5] Moure A, Dominguez H and Parajo J C 2006 *Process Biochemistry* **41** (2) 447–456
[6] Biely P 1985 *Trends in Biotechnology* **3** (11)
[7] Jensen P R, Gontang E, Mafnas C, Mincer TbJ, Fenical W 2005 *Environ Microbiol.* **7** 1039–48
[8] Fenical W and Jensen P R 2006 *Nat Chem Biol.* 2006 **2** 66673
[9] Ramesh S, Jayaprakashvel M, Mathivanan N 2006 *India. Mar Ecol* **27** 198–203
[10] Ramesh S and Mathivanan N 2009 *World J Microbiol Biotechnol* **25** 2103–11
[11] Subramaniyan S and Prema P 2002 *Critical Reviews in Biotechnology* **22** 33-64
[12] Laemmli U K 1970 *Nature* **227** (5259) 680-685
[13] Bailey M J, Biely P and Poutanen K 1992 *J. Biotechnol.* **23** 257–270
[14] Miller G 1959 *Anal Chem.* **31** 426–428
[15] Lisdiyanti P, Otoguro M, Ratnakomala S, Lestari Y, Hastuti R D, Triana E, Ando K and Widyastuti Y 2010 *Int J Syst Evol Microbiol* **60** 2331–2335
[16] Burggraf S, Olsen G J, Stetter K O and Woese C R 1992 *Syst. Appl. Microbial.* **15** 352–356
[17] Liu Z, Zhao X and Bai F 2013 *Appl. Microbiol. Biotechnol.* **97** 4361–4368
[18] Saini A, Neeraj K, Aggarwal, Anuja S and Anita Y 2015 *Enzyme Research* **15** 279381
[19] Qiu Z, Shi P, Luo H, Bai Y, Yuan T, Yang P, Liu S and Yao B 2010 *Enzyme and Microbial Technology* **46** 506-512
[20] Beg Q, Kapoor M, Mahajan L and Hoondal G 2001 *Appl. Microbiol Biotechnol.* **56** (3) 326–338
[21] Kluepful D, Vats-Mehta S, Aumont F, Shareck F and Morosoli R 1990 *Biochem. J.* **267** 45-50
[22] Yan Q, Hao S, Jiang Z, Zhai Q and Chen W 2009 *J. Mol. Cat. B: Enzym.* **58** 72-77
[23] Li N, Shi P, Yang P, Wang Y, Luo H, Bai Y, Zhou Z and Yao B 2009 *Appl. Biochem. Biotechnol.* **159** 521-531
[24] Shin J H, Choi J H, Lee O S, Kim Y M, Lee D S, Kwak Y Y, Kim W C and Rhee I K 2009 *Biotechnol. Bioprocess Eng.* **14** 391-399

IOP Conf. Series: Earth and Environmental Science **251** (2019) 012054 doi:10.1088/1755-1315/251/1/012054

Analysis of organic acids produced by lactic acid bacteria

I Nuryana, A Andriani, P Lisdiyanti, and Yopi

Research Center for Biotechnology, Indonesian Institute of Sciences, Jl. Raya Bogor Km 46, Cibinong-Bogor 16911, Indonesia

E-mail: isa.nuryana@lipi.go.id; yopi@bioteknologi.lipi.go.id

Abstract. Organic acids can be useful as starting materials in multiple applications including for food supplements, bio-based material, and biodegradable polymers. The use of microorganisms for organic acids production has been promising to replace petroleum-based processes because of their abilities to utilize inexpensive and renewable feedstocks. In order to obtain potential bacteria for organic acids production, a total of 4 selected lactic acid bacteria (LAB) isolates were investigated. The production of organic acids was carried out for 5 days of incubation using a liquid fermentation process. The organic acids analyzed were acetate, citrate, formate, lactate, and succinate. The analysis of these compounds was performed using High – Performance Liquid Chromatography (HPLC). Results revealed the growth of 4 LAB isolates increased and reached about 10^7 CFU/ml during 5 days of fermentation. Lactate and acetate were found as the dominant products of organic acid fermentation. The highest amount of acetate, lactate, and formate was achieved at 44.81 g/L, 17.11 g/L, and 7.90 g/L respectively by BAL 100 isolate whereas the highest content of succinate was 20.83 g/L and produced by BAL 690 isolate. The experiments indicated that BAL 100 isolate could be the most potential for organic acid production and BAL 690 isolate is considered as promising isolate for succinic acid production.

1. Introduction

Organic acids are widely applied in many industries because of its versatile use as monomer and starting materials for food supplements, bio-based material, and biodegradable polymers [1] [2]. Those can be useful as chelating agents because of providing both protons and organic anions and become a key group among building block chemicals [3].

Organic acids can be produced by microbial processes. Most of them are natural product or intermediates in major metabolite pathways of microbes. The use of bio-based process for organic acids production can replace petroleum-based [4]. Due to the increase of crude oil price and the reduction of fossil fuel resources, organic acids production from petrochemicals are becoming costly [5]. It leads some manufacturers to find new methods which can utilize inexpensive and renewable feedstocks.

The Lactic Acid Bacteria (LAB) are popular bacteria in producing organic acids such as lactate, acetate, formate, succinate, and citrate. The LAB is Gram-positive microorganisms that produce lactic acid as a major fermentation product. It is called homofermentative if lactic acid is yielded as single fermentation products and the other is called heterofermentative if ethanol and CO_2 are mainly produced as by-products [6].

IOP Conf. Series: Earth and Environmental Science **251** (2019) 012054 doi:10.1088/1755-1315/251/1/012054

Sugar is the preferred substrates of the heterofermentative LAB. However, the homofermentative LAB can ferment hexose and produce lactic acid by the reaction of glycolysis and lactate dehydrogenase. Catabolism of LAB plays an important role to determine the fermentation products formed. At limiting catabolism, mixed acid fermentation with formate, lactic, ethanol and acetate as the products is generated. Whereas catabolism with high glycolytic flux, homofermentative metabolism with lactic acid as the sole product is produced. Under slow growth condition and low glycolytic flux rates, the homofermentative LAB change to mixed acid fermentation [7].

Development of microbial processes for organic acids production by employing LAB has been studied by many researchers. Lactate and succinate are currently produced at commercial scale; further research are required to scale up the other organic acids from laboratory to commercial scale processes [1]. Succinate, acetate, and formate can be produced by LAB through citrate metabolism. There are at least two different mechanisms of citrate utilization in the LAB. First, citrate conversion to succinate by reducing oxaloacetate using the enzymes malate dehydrogenase, fumarate reductase, and fumarase [8]. Second, citrate conversion to acetate and formate via pyruvate [9].

The analysis of organic acids in fermentative product is necessary in order to find potential Lactic Acid Bacteria. This research focuses on characterizing the selected LAB isolates and obtaining their potentiality for organic acids production. The organic acids can be determined and quantified by various methods, one of them is usually performed by high-performance liquid chromatography (HPLC). In a combination of RP-C18 columns, 5mM sulfuric acid as eluent and UV for detection are used for the separation of these compounds [10].

2. Materials and Methods

2.1. Media
The medium used for stock cultures was de Man, Rogosa and Sharpe (MRS) agar containing (g/liter): enzymatic digest of casein 10.0, meat extract 10.0, yeast extract 4.0, D(+)glucose 20.0, dipotassium hydrogen phosphate 2.0, tween 80 1.0, di-ammonium hydrogen citrate 2.0, sodium acetate 5.0, magnesium sulfate heptahydrate 0.2, manganese sulfate monohydrate 0.04 and agar 14.0. The liquid medium used for inoculums and fermentations was MRS broth.

2.2. Microorganisms
The isolates used were obtained from Culture Collection of Biocatalyst and Fermentation Laboratory, Research Center for Biotechnology, LIPI. A total of 4 selected LAB isolates were used in this work. Those were given labels as LAB 100, LAB 120, LAB 690, and LAB 761. The strains were stocked on MRS agar plate.

2.3. Inoculum Preparation
A single colony from each LAB isolate on MRS agar plate was selected and inoculated in 15 ml MRS broth medium. The inoculums were aerobically incubated at 37 ^{0}C and shaken at 150 rpm for 24 hours.

2.4. Fermentation Process
The fermentation process was carried out in a screw cap test tube containing 5 ml of fermentative medium and inoculated with 5% of the inoculum. The temperature of fermentation was set at 37 ^{0}C and constant agitation of 150 rpm for 5 days.

2.5. Sample Preparation
All the experiments were carried out in triplicates. The LAB culture in MRS media was prepared as the samples. The sampling process was performed by withdrawing each test tube periodically at an interval of 24h for 5 days of fermentation. Each sample was centrifuged at 10.000 rpm, temperature at

4^0C for 20 minutes. The supernatant was then used for HPLC analysis. A 400 µl of supernatant was filtered through syringeless filter Whatman pore size 0.45 µm.

2.6. Analytical Methods

The cell number of each LAB isolate was measured using spectrophotometer based on the optical density at λ 600 nm and converted into Colony Forming Unit using Total Plate Count method. Samples were diluted in serial dilution of 10^{-1} to 10^{-6} and poured them into 15 ml MRS agar plate. After 3 days of incubation, the colony number was enumerated. Correlation value was measured by comparing the colony number and absorbance.

The products of organic acids were analyzed using Shimadzu HPLC System equipped with Aminex column compartment, autosampler and refractive index detector (RID). A volume of 10 µl was injected with a run time of 40 minutes for each sample. The mobile phase used was 5.0 mM/L H_2SO_4 in ultrapure water (HPLC grade) at a flow rate of 0,6 mL per minute. Standard solutions were injected to obtain the retention time for each compound.

3. Results and Discussion

3.1. Lactic Acid Bacteria Growth

The growth of LAB in MRS broth medium was observed from 0 to 120 hour of the experiment (5 days of fermentation). The growth increased, starting at about 10^2 – 10^3 CFU/ml after inoculation and exceeded 10^7 CFU/ml at the end of fermentation. The exponential phase was reached at about 0 to 24h whereas, at above 24h, the growth of LAB was on the stationary phase (Figure 1). Due to the reduction of nutrients in the medium, the LAB's growth remains steady. In the stage of stationery, the LAB started to produced secondary metabolites, such as organic acids.

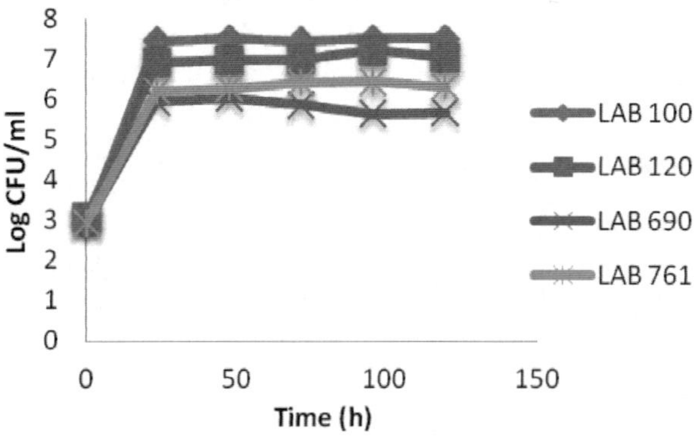

Figure 1. The growth of Lactic Acid Bacteria in MRS broth medium for 5 days of fermentation

3.2. Organic Acids Production

A total of 4 selected LAB isolates have been investigated their ability in producing 5 compounds of organic acids: acetate, citrate, formate, lactate, and succinate. The main activity of the LAB is to produce the end product lactic acid by metabolizing the sugar. In this experiment, results showed that acetate was present in large amounts in all the fermentation broths compared to the other organic acids. The highest amount of acetate was reached at 44.81 g/L whereas that of lactate was only 17.11 g/L. A small amount of lactate in fermentation broth might be caused by the presence of oxygen. The fermentation process was not completely carried out in anaerobic condition. The limited oxygen was still present in liquid medium. It could trigger the LAB to produce acetate with high content instead of lactate.

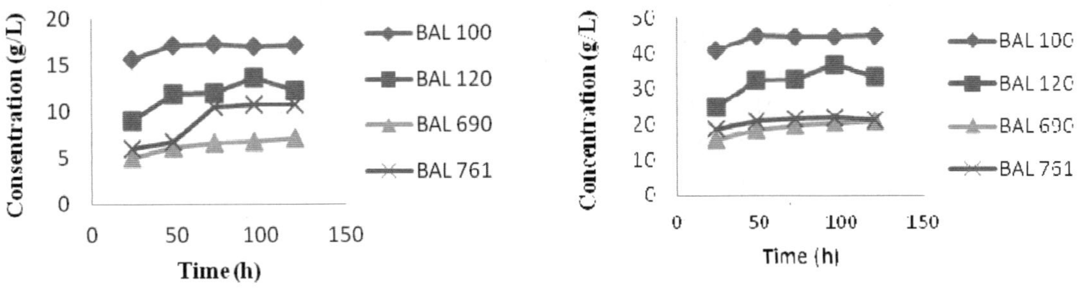

(a) (b)

Figure 2. The content of lactate (a) and acetate (b) produced by the Lactic Acid Bacteria

The BAL 100 isolate had the highest amount of lactate and acetate, followed by BAL 120, BAL 690 and BAL 761 (Figure 2). Lactate and acetate were dominant products in this work. It has been reported by Paillart et al. (2016) that the LAB was able to produce acetic and lactic acid as the dominant products in food fermentation [11].

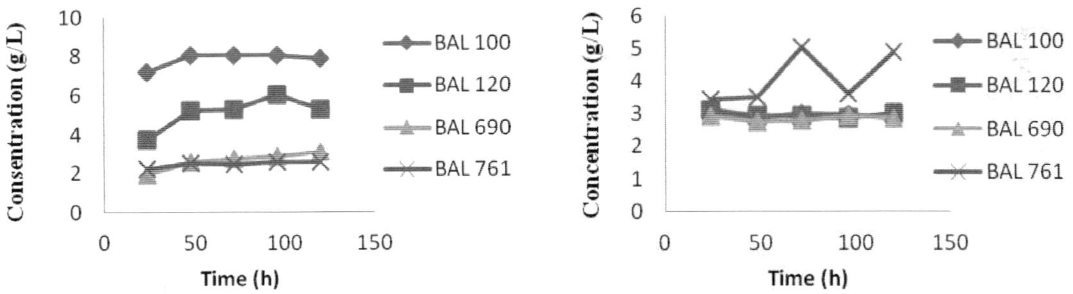

(a) (b)

Figure 3. The content of formate (a) and citrate (b) produced by the Lactic Acid Bacteria

Beside lactate, formate and citrate were also formed in the fermentation broths. The results showed formate and citrate that produced by four isolates were detected in small amounts. A fluctuate graph of citrate was shown in Fig 3 as citrate is an intermediate of the tricarboxylic acid (TCA) cycle and it is necessary to provide precursor for its cycle [12]. The limited number of LAB are able to ferment organic acids including citrate [13].

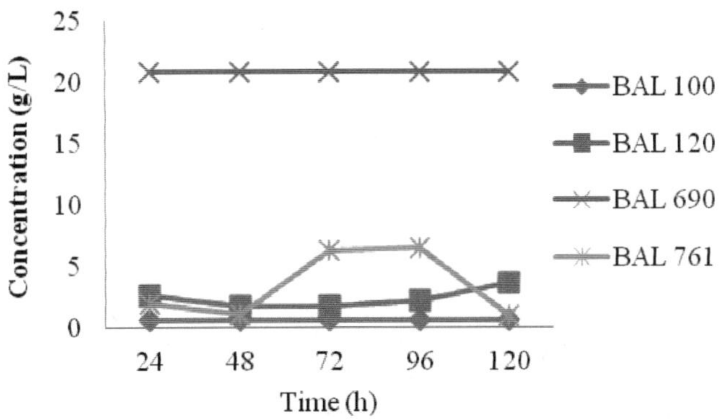

Figure 4. The content of succinate produced by Lactic Acid Bacteria

The largest content of succinate was reached by BAL 690 isolate at 20.83 g/L as depicted in Fig. 4. However, the ability of other isolates to produce succinate was low. The isolate of BAL 761 showed a fluctuate graph, and it was able to produce succinate in small amounts. Song and Lee (2006) has reported that succinate is one of the fermentation end-product of anaerobic metabolism and an intermediate of TCA cycle in aerobic condition [14]. A fluctuate graph might be caused by the use of succinate either in aerobic or anaerobic condition.

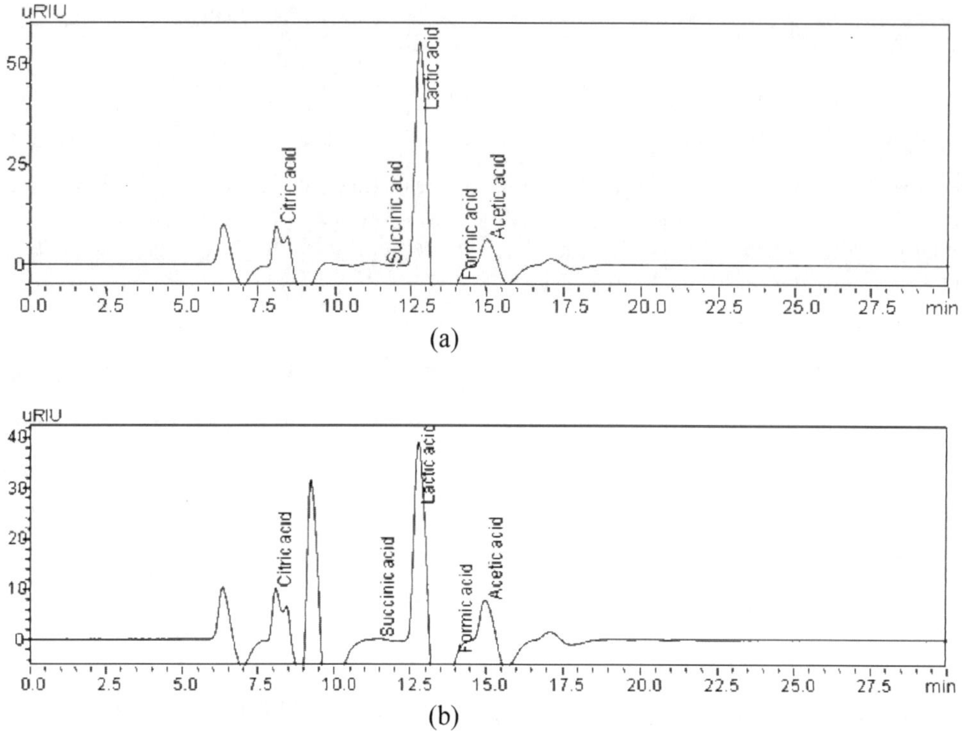

Figure 5. HPLC chromatograms of organic acids from fermentation medium for isolate: BAL 100 (a) and BAL 120 (b)

The HPLC method was able to quantify different types of organic acids (acetate, citrate, succinate, formate, and lactate) that exist in the liquid medium of fermentation. The separation for each compound of the organic acids peaks was clear. In general, the use of HPLC is an accurate and convenient analytical technique for quantification and identification of organic compounds including organic acids [15].

Table 1. The content of organic acids products

Isolate	Organic Acids (g/L)				
	Acetate	Citrate	Formate	Lactate	Succinate
BAL 100	44.81	2.87	7.90	17.11	0.65
BAL 120	33.11	2.99	5.31	12.28	3.56
BAL 690	20.75	2.85	3.11	7.20	20.83
BAL 761	21.10	4.90	2.59	10.80	6.51

The content of organic acids produced by four LAB isolates was summarized in Table 1. The potential isolates were obtained based on their abilities to produce organic acids in high amounts.

4. Conclusions
This research concluded that Lactic Acid Bacteria (LAB) were able to produce various compounds of organic acids, in particular, lactate and acetate as dominant products during the fermentation. The highest amount of acetate, lactate, and formate was produced by BAL 100 isolate whereas the highest value of succinate was produced by BAL 690 isolate. The potential isolates for organic acids production were found through this research.

5. Acknowledgment
The authors are thankful to the Indonesian research funding (Prioritas National Pangan 2018) for supporting and funding the research project.

6. References
[1] Pleissner D, Dietz D, van Duuren J B J H, Wittmann C, Yang X, Lin CSK and Venus J 2017 *Adv. Biochem. Eng. Biotechnol.* 1-38
[2] Liaud N, Ginies C, Navarro D, Fabre N, Crapart S, Gimbert I H, Levasseur A, Raouche S and Sigoillot J C 2014 *Fungal Biology and Biotechnology* 1 1
[3] Posso E J S , de Prager M S and Rojas C A C 2017 *Acta Agron.* **66** (2) pp 241-247
[4] Sauer M, Porro D, Mattanovich D and Branduardi P 2007 *Elsevier Ltd, Cell Press* pp 100-108
[5] Cao Y, Zhang R, Sun C, Cheng T, Liu Y and Xian M 2013 *Biomed Research International.*
[6] Madigan M T, Martinko J M, Stahl D A and Clark D P 2012 *Brock Biologi of Microorganisms 13th Edition* (San Francisco: Benjamin Cummings) p 375
[7] Zaunmuller T, Eichert M, Richter H and Unden G 2006 *Appl. Microbiol. Biotechnol.* **72** 421-429
[8] Cselovszky J, Wolf G and Hammes W P 1992 *Appl. Microbiol. Biotechnol.* **37** 94-97
[9] Lindgren SE, Axelsson LT and McFeeters R 1990 *FEMS Microbiol. Lett.* **66** 209-214
[10] Coelho E M, Padilha C V S, Miskinis G A, de Sa A G B, Pereira G E, de Avezedo L C and Lima M S 2017 *Journal of Food Composition and Analysis*
[11] Paillart M J M, van der Vossen J M B M, Lommen E, Levin E, Otma E C, Snels J C M A and Woltering E J 2016 *Acta Hortic.* pp 289-296
[12] Hugenholtz J 1993 *FEMS Microbiology Reviews* **12** 165 – 178
[13] Neal-McKinney J M, Lu X, Duong T, Larson C L, and Call D R 2012 *PLoS ONE* **7** (9)
[14] Song H and Lee SY 2006 *Enzyme and Microbial Technology* 39 pp 352-361
[15] Zaky A S, Pensupa N, Eiroa A A, Tucker G A and Du C 2017 *Journal of Food Composition and Analysis* **56** 25-33

Effect of Milking Frequency on Body Condition Score, Somatic Cell Count, and Reproductive Performance of Dairy Cows Milked by An Automatic Milking System

A Astuti[1], T Obitsu[2], K Taniguchi[2], and T Sugino[2]

[1]Faculty of Animal Science, Universitas Gadjah Mada. Jl. Fauna No. 3 Bulaksumur, Yogyakarta, Indonesia 55281
[2]Graduate School of Biosphere Science, Hiroshima University. 1-4-4 Kagamiyama, Higashi-Hiroshima City Hiroshima, Japan 739-8528

Corresponding author: andriyaniastuti@ugm.ac.id

Abstract. The objective of this study was to determine the effect of milking frequency using an Automatic Milking System (AMS) on body condition scores (BCS), somatic cell count (SCC), and reproductive performance of Holstein cows. The study was conducted in Hiroshima University Farm. Eight multiparous cows and four primiparous cows were divided into two treatments of groups (2 and 4 times milking frequency). Cows were fed by Partial Mixed Ration (PMR) ad libitum and concentrate diet (5-8 kg) at every milking. The PMR consisted of Italian ryegrass silage 35% dry matter (DM), oat hay (8% DM), alfalfa hay (8% DM), pellet bit pulp (8% DM), formula feed (39% DM), minerals and vitamin. The PMR nutrients contents were 13.6% crude protein (CP) and 70% total digestible nutrient (TDN), while concentrate contained 20.6% CP and 82.4% TDN. On days 20, 35, and 50 postpartum, BCS was observed, and milk samples were collected for SCC analysis. Reproductive performance data collected were S/C, calving interval, and postpartum mating. The MIXED model for repeated measures (SAS 2004) was used for testing the difference in BCS and SCC. Significance was declared at $P < 0.05$, and tendency at $P < 0.10$. The research showed that no difference on BCS and SCC between 2 and 4 times milking by AMS, but the probability of occurrence of high SCC (> 100,000) was lower for four times milking cows than 2 times milking cows (6% vs 59%). The effect of milking frequency did not show any difference on S/C, calving interval, and postpartum mating. Increasing milking frequency may positively affect on SCC without any effects on BCS and reproductive performance in this experiment.

Keywords: Automatic milking system, BCS, milking frequency, reproductive performance, SCC

1. Introduction

Somatic cell count (SCC) in milk has been known as an indicator of udder health, especially contagious pathogenic infections resulting in mastitis [6]. The SCC in milk increases when capacity of mammary cells to produce milk is damaged due to mammary gland infection [9]. When SCC exceeds 100,000 cells/ml, the infection is identified by bacteriological testing [13]. Generally, a healthy mammary gland contained less than 10,000 cells/ml milk, but SCC may also vary depending on the lactation stage, parity of cows [4], infectious status, estrus, heat stress and day-to-day variation in milk production [9]. Hale et al. [8] reported that milking frequency did not affect SCC as well as reproductive performance. The calving interval (CI) is defined as the time from one calving to the next calving. For ideal intensive production and beneficial economy, CI is targeted 12 months [10, 14]. The body condition score (BCS), udder health, and reproduction control are important factors for practical dairy production. However, the effect of milking frequency using AMS on BCS, SCC, and reproductive performance are unclear. In this study, BCS, SCC and some reproduction data obtained from cows milked by AMS were analyzed to detect the related problems.

2. Materials and methods

This study was conducted according to the Animal Care and Use Committee of Hiroshima University guidelines. Eight multiparous cows and four primiparous cows were divided into two treatments groups of milking frequency, namely, MF2 (twice milking per day) and MF4 (four times milking per day). The milking machine used was AMS (Astronaut A3 next, Lely, Drachten, the Netherlands). The cows were divided into two blocked groups based on parity (multiparous and primiparous) with the average of cow's parity for 2 times milking cows were 3.0 and for 4 times milking cows were 2.3. Cows were fed a partial mixed ration for 5 times (09.30, 11.00, 14.00, 15.30 and 17.00 hours) daily ad *libitum* and 5–8 kg/d of concentrate. Freshwater was offered at all times. The PMR consisted of Italian ryegrass silage 35% dry matter (DM), oat hay (8% DM), alfalfa hay (8% DM), pellet bit pulp (8% DM), formula feed (39% DM), minerals and vitamin. Chemical composition of the PMR were 47.9% dry matter, 92.2% organic matter, 13.6% crude protein, 3.6% ether extract, 38.5% aNDFom, and 70% estimated TDN. Concentrate composition were 88.0% dry matter, 94.3% organic matter, 20.6% crude protein, 3.8% ether extract, 16.4% aNDFom, and 82.4% estimated TDN [2].

The animals were observed for BCS at 20, 35, 50 days postpartum and milk samples for 24 hours were collected at each milking at days 20, 35, and 50 with the sampling device of AMS. Somatic cell counts were measured by a somatic cell counter (NucleoCounter SCC-100 ChemoMetec, Denmark). Reproduction data (days to first service, service per conception (S/C) and CI) for the cows in these experiments were obtained by the routine reproduction management in the experimental farm. Usually, oestrus was detected by the sensing system of AMS which automatically recorded behavior activity. The MIXED model for repeated measures [12] was used for testing the difference in SCC between treatments. Significance was declared at $P < 0.05$, and tendency at $P < 0.10$.

3. Results and discussion

Means of BCS, SCC, and reproductive performance data are shown in Table 1. In this study, BCS of cows between milking frequency treatments were similar and related to higher dry matter intake of MF4 cows in order to produce higher milk yields compared MF2 cows. Higher milking frequency produced higher milk yield (43.1 vs. 32.0 kg/d) [2]. Somatic Cell Count at day 20 was numerically lower in MF4 cows than for MF2 cows. At day 20, the proportion of cows showing more than 100,000 SCC were 60% for MF2 cows, while 0% for MF4 cows. Similar results were obtained at day 35, such that all of MF4 cows had numerically lower SCC (< 100,000, average 42,833) and MF2 cows had numerically higher SCC (> 100,000, average 146,000). At day 50, one cow from MF4 group had a high SCC due to mastitis infection, but the average of SCC in MF4 cows was numerically lower compared to MF2 cows (27,000 vs. 409,667). Overall, the probability for occurrence of high SCC (> 100,000) was lower for MF4 cows than that for MF2 cows (6% vs 59%), even though the difference in SCC between treatment groups was not significant. Previous results showed that increased milking frequency did not only increase milk yield but also decrease milk SCC [13, 15]. Calving intervals was 411 and 408 days for MF4 and MF2 cows, respectively. Days to first service were 106 and 95 days for MF4 and MF2 cows, respectively. Service per conception was 1.6 and 2.0 for MF4 and MF2 cows, respectively. Gisi et al. [7] found that milking frequency did not affect Calving Interval, days open or S/C. Barnes *et al.* (1990) also demonstrated that milking frequency did not affect days to first inseminations, S/C or days open. In contrast, a trend for increased days open with increasing milking frequency was observed by Amos et al [1]. DePeters et al [5] also reported a decreased reproductive performance in cows milked 3 times compared with cows milked 2 times daily. In this study, however, no differences in reproductive performances of cows milked twice or 4 times daily were obtained.

Table 1. Reproductive performance and somatic cell counts (SCC) on early lactation cows milked by AMS

Milking frequency	BCS	High SCC *	Low SCC**	Calving to first service (day)	S/C	Calving interval (day)
2 times	2.77±0.175	59%	41%	95±16.0	2.0±0.71	408±54.2
4 times	2.77±0.167	6%	94%	106±12.3	1.6±0.63	411±44.5

High: SCC > 100,000 cell/ml; Low: SCC < 100,000 cell/ml
*High SCC: proportion of cows which showed more than 100.000 cell/ml, on average of 20, 35 and 50 days
**Low SCC: proportion of cows which showed less than 100.000 cell/ml, on average of 20, 35 and 50 days
S/C: service per conception ratio

4. Conclusion

Increasing milking frequency may lower milk SCC, and that 4 times milking frequency has higher probability for low SCC without any effects on BCS and reproductive performance. For future research, an observation with large number of cows and long time period are needed to clarify the more precise reproductive performance in cows milked by AMS.

5. References

[1] Amos H E, Kiser T and Loewenstein M 1985 *J. of Dairy Science* **68** 732
[2] Astuti A, Obitsu T, Sugino T, Taniguchi K, Okita M and Kurokawa Y 2015 *Animal Science J.* **86** 499
[3] Barnes M A, Pearson R E and Lukes-Wilson A J 1990 *J. of Dairy Science* **73** 1603
[4] Blackburn P S 1966 *J. of Dairy Research* **33** 193
[5] DePeters E J, Smith N E and Acedo-Rico J 1985 *J. of Dairy Science* **68** 123
[6] Dufour S, Fréchette A, Barkema H W, Mussell A and Scholl D T 2011 *J. of Dairy Science* **94** 563
[7] Gisi D D, DePeters E J and Pelissier C L 1986 *J. of Dairy Science* **69** 863
[8] Hale S A, Capuco A V and Erdman R A 2003 *J. of Dairy Science* **86** 2061
[9] Harmon R J. 1994 *J. of Dairy Science* **77** 2103
[10] Holmann F J, Shumway C R, Blake R W, Schwart R B and Sudweeks E M 1984 *J. of Dairy Science* **67** 636
[11] Kelly A L, Reid S, Joyce P, Meaney W J and Foley J 1998 *J. of Dairy Research* **65** 365
[12] Statistical Analysis System (SAS) 2004 *SAS/STAT 9.1 User's Guide* (SAS Institute Inc., Cary. NC)
[13] Smith J W, Ely L O, Graves W M and Gilson W D 2002 *J. of Dairy Science* **85** 3526
[14] Strandberg E and Oltenacu P A 1989 *Acta Agriculturae Scandinavica* **39** 407
[15] Waterman D F, Harmon R J, Hemken R W and Langlois B E 1983 *J. of Dairy Science* **66** 253

IOP Conf. Series: Earth and Environmental Science **251** (2019) 012056 doi:10.1088/1755-1315/251/1/012056

In vitro study of *Nigella sativa* meal as protein source and its combination with the readily available carbohydrates for ruminant diet

N N Barkah[1], Y Retnani[1] and K G Wiryawan[1]

[1]Department Nutrition and Feed Science, Faculty of Animal Science, Jl. Agatis, Kampus IPB Dramaga, Bogor, 16680, West Java, Indonesia

E-mail: nisa_barkah@apps.ipb.ac.id

Abstract. *Nigella sativa* meal (NSM) is a waste generated from the oil processing industry of black cumin or known as habbatussauda. NSM contains high crude protein that can be used by ruminant. The purpose of this study was to evaluate the quality of crude protein of NSM in rumen and its combination with the readily available carbohydrates (RAC). This study was began by observing ammonia (NH_3) concentration from NSM and volatile fatty acid (VFA) concentration from several types of RAC that were incubated in rumen fluid in vitro. Then followed by evaluation of rumen fermentability of diets containing NSM that was combined with different types of RAC. This study used a complete randomized design with 3 treatments and 5 replications. The treatments were differentiated by the type of RAC that was used in the diet, which consisted of pollard (P), rice bran (R), and corn (C). The level of NSM is equated in each treatment. Diet consisted of forage and concentrate with the ratio of 30:70. The results showed that NSM has high NH_3 concentration in the rumen, especially in the first 4 hours and after 6 hours of incubation time in rumen fluid in vitro. The combination of NSM in rations containing corn significantly suppressed NH_3 concentration ($P<0.05$). In addition, NSM and its combination with corn in the diet produced microbial protein synthesis higher (9.66 mg $10ml^{-1}$) compared with another treatments R (9.64) and P (7.85). The use of NSM as much as 20% in rations and its combination with the RAC derived from corn can optimize the use of NSM as a source of feed protein in ruminants by maximize the utilization of NH_3 and VFA to encourage the production of microbial protein synthesis that were important for ruminants.

1. Introduction

The exploration and diversification of feed ingredients becomes ones of essential topics for substituting the main feed commonly used in animal diets. This is one of the efforts that can be taken to eliminate dependence on a feed ingredient, especially feed protein sources with highly prized. Black seed (*Nigella sativa*) or known as habbatussauda or black cumin in the group of *Ranunculaceae* family is well known as herbal plants. *N. sativa* contain pharmacological properties which has been used as a medicine for various deseases. Pharmacological activity in *N. sativa* is related with the presence of thymoquinone compounds as the main compounds contained in oil which was found in *N. sativa* seeds [1]. Habbatussauda processing industry produces waste in the form of extraction residue of habbatussauda oil known as *Nigella sativa* meal (NSM). The use of habbatussauda in large and medium industries in Indonesia reached 144 817 kg year^{-1} [2]. The potency of NSM produced is 101

372 kg year^{-1} or as much as 70% of total raw materials [3]. NSM derived from habbatussauda oil processing residues taken using pressing method contain 8.1% water content, 23.3% crude protein, and 9.6% ash [4]. NSM also contains some minerals such as Mg, Fe, Cu, Ca, and K [5].

The use of NSM waste in Indonesia as feed ingredients is still small. Though NSM has the potential to be used as a source of feed protein because of the high crude protein content in it. In ruminant, when feeding the protein source into the rumen, the protein will be degraded into ammonia (NH_3), then the ammonia is brought into the bloodstream to the liver to be converted into urea. The lack of ammonia production in the rumen can suppress the flow of microbial protein synthesis from the rumen [6]. The excess ammonia production can increase blood ammonia levels, increase urine urea synthesis, and reduce the nitrogen retention of cattle [7]. In previous study reported that the use of NSM as much as 10 and 20% in the feed significantly improved the performance of lambs [8] . This indicates that NSM has good nutrient content, so it can improve the performance of lambs. However, the utilization of 10 and 20% NSM in feed significantly increases levels of blood urea nitrogen (BUN) [9]. The high levels of BUN indicate that NSM added in the diet causes an increase in protein degradation in the rumen. The high degradation of feed protein in the rumen may result in inefficiency because of the amount of protein degraded before it reaches the intestine to be utilized by the whole body [7].

Several ways are done to maintain and optimize the quality of feed, especially sources of feed protein for its usefulness in the feed material. One way that can be done is to combine source of feed protein with readily available carbohydrate (RAC). RAC will spur the growth of microbes in the rumen which also leads to high use of ammonia for microbial growth so that the potential ammonia that is converted into urea will be reduced [10]. Ideal conditions occur when the production of feed ingredients of fermented energy sources in the rumen is as fast as the production of NH_3, so that when NH_3 was formed there is volatile fatty acid (VFA) derived from carbohydrates to be used as the source and carbon chain of the amino acid microbial protein. The purpose of this study was to evaluate the quality of the crude protein of NSM in rumen in vitro and its combination with different types of readily available carbohydrates.

2. Material and Methods

The first stage of the research was analyze the rate of degradation of single feed. The rate of degradation observed as the concentration of NH_3 and VFA which described the rate of degradation of proteins and carbohydrates in the rumen. The next stage of the research was analyze degradation of feeds that have been formulated according to nutrient requirements of 5-month-old sheep in National Research Council [11]. Four experimental diets were prepared for this study. Diet contained 20% NSM (% DM) that was combined with pollard (P), rice bran (R), and corn (C), respectively. Forage and concentrate feeds were given at a ratio of 30:70. Preparation of treatment ration starting from the preparation of raw materials consisting of pollard, corn, onggok (tapioca by-products), rice bran, NSM, molasses, premix, and $CaCO_3$. Before all the ingredients were mixed, NSM was milled using a grinder with die diameter size of 6 mm. Then, the feed material was weighed according to the formulation in Table 1 and mixed manually. Mixing begins by including large quantities of ingredients. As for premix and $CaCO_3$ mixed first separately, then mixed with the carbohydrate source in the same volume with the mixture of minerals. This was done so that the mineral mixing is evenly distributed. The napier grass used in the ration mixture was dried for 2 days in oven 60 °C.

The feed is weighed into 0.5 gram fermentor tube (Pyrex glass tube with diameter 3 cm with height 20 cm) with ratio forage and concentrate 30:70. Then the feed was incubated in rumen fluid and McDougall solution [12]. The rumen fluid comes from a 1-year-old male sheep taken by a stomach tube. The fermentor tube is incubated in a waterbath shaker with a temperature of 39 °C. After 4 hours of incubation, the fermentor tube was removed from the shaker waterbath and was measured its pH. The liquid from each tube is stored in two different places, one for microbial protein synthesis (MPS) analysis and the remainder for analysis of NH_3 and VFA concentration. The samples for NH_3 and VFA concentration analysis were given two drops of $HgCl_2$ to kill microbial activity. Then the MPS analysis is done according to [13] and the calculation of microbial N content using Lowry's method [14]. The production of NH_3 was analyzed by Conway method [15] and VFA with steam distillation method.

This experiment used a complete randomized design with 3 treatments and 5 replications. Data obtained from this study were statistically analysed by ANOVA (Analysis of Variance) by IBM SPSS Statistics. Duncan's multiple range tests was used to test the significance among means (P<0.05) [16].

Table 1. The composition of the feed ingredients in each treatment.

Ingredients	**P (%)**[a]	**R (%)**[b]	**C (%)**[c]
Napier grass	30.00	30.00	30.00
Pollard	19.60	0	0
Rice bran	0	31.50	0
Corn	0	0	24.50
Onggok (tapioca by-products)	19.95	8.05	15.05
NSM	19.60	19.60	19.60
Molasses	6.30	6.30	6.30
$CaCO_3$	0.70	0.70	0.70
Premix	0.35	0.35	0.35
CPO (crude palm oil)	3.50	3.50	3.50
Nutrient content (%DM) [d]			
Dry matter	89.63	90.46	89.71
Ash	8.72	13.12	7.87
Crude protein	15.70	15.68	15.66
Crude fat	2.80	6.26	3.90
Crude fiber	16.46	17.70	13.41
Nitrogen free extract	56.32	47.24	59.16

[a] P: NSM combined with RAC pollard
[b] R: NSM combined with RAC rice bran
[c] C: NSM combined with RAC corn
[d] Estimated results of calculations based on the nutrient content of each ingredient

3. Results and Discussion

3.1. In vitro degradation rate of NSM

Degradation is a chemical change reaction or decomposition of compound or molecule into simpler form. Degradation rate of feed ingredients in vitro describes the time spent by the rumen microbes to decompose the nutrients present in a feed ingredient in an environment that has been designed similar with the digestive process that occurs in the ruminant digestive tract. The results showed that NSM has a fluctuating rate of ammonia degradation. At the fourth hour of incubation, ammonia concentration was higher than the previous hour, because the fourth hour of incubation is the optimal time for the rumen microbes to degrade the nutrients that enter into it. The maximum production of ammonia is achieved when 2-4 hours after feeding [17], so ammonia concentration in the rumen indicate how efficiently the protein digestion process is in the rumen [18]. The presence of protein and non-protein nitrogen (NPN) that enter to rumen causes the protein and NPN to be degraded by rumen microbes to NH_3, peptides, and amino acids so that they will form microbial proteins. NH_3 is combined into amino acids and finally formed into microbial proteins by rumen microbes [19]. Ammonia levels needed to support microbial growth are between 4-12 mM [17].

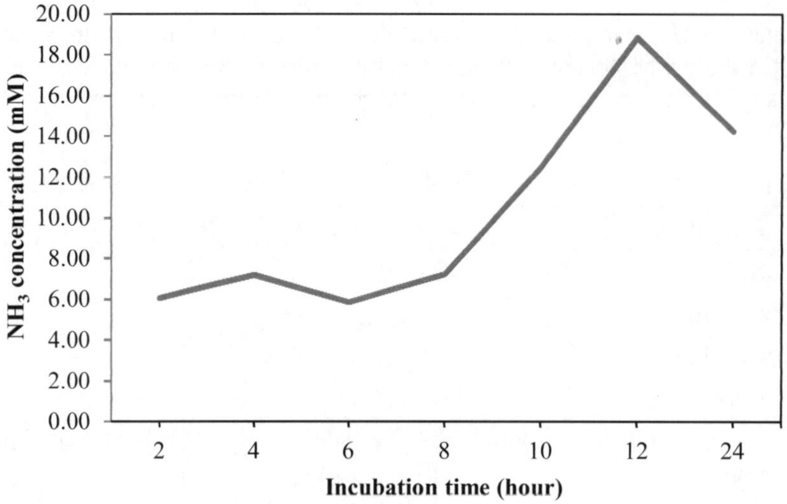

Figure 1. The ammonia concentration that measured over different incubation time of NSM.

At sixth hour incubation, NH_3 production decreased due to the use of NH_3 by rumen microbes for the purpose of microbial protein synthesis. Optimizing the growth of rumen microbes not only requires sufficient N availability, but other nutrients are also needed, such as energy, amino acids, minerals and vitamins. In the 8th to 12th hour, there is an increase in ammonia production. An increase in rumen microbial population driven by increased rumen microbial protein synthesis in the previous hours caused a comparable increase in protein degradation or NPN to ammonia. However, this ammonia cannot be used again to form microbial protein synthesis due to the lack of energy availability, so that ammonia accumulates in the Conway dish and its concentration will increase. The energy needed for microbial protein synthesis is energy in the form of ATP, while VFA which is beneficial for rumen microbes is only branching and is needed as a source of carbon framework. Branched chain amino acids strongly support microbial protein synthesis because they will undergo deamination and decarboxylation to produce branched chain fatty acids. At the 24th hour, the concentration of ammonia decreases may be caused by ammonia accumulation that has reached its maximum level, making it toxic to the rumen microbes which then suppresses the rumen microbial population, so that the process of protein or NPN degradation to ammonia also decreases. The optimal NH_3 concentration in the rumen is between 6-21 mM [20].

3.2. In vitro degradation rate of several types of RAC into VFA

Feed ingredients for carbohydrate sources in feed can increase microbial metabolic activity, both in degrading protein and carbohydrates. The end product of carbohydrate degradation will produce acidic compounds, such as acetate, propionate, and butyrate which will then reduce the rumen pH. The level of acidity that is too low will inhibit microbial growth, so that it will interfere with the food degradation process. Observations regarding the rate of carbohydrate degradation in Figure 2 show that each feed material has a rate of degradation to produce carbohydrates in different concentrations. In all types of RAC there was an increase in VFA concentration at the six hour incubation. Corn kernels consisted of starch (61 to 78%, dry matter (DM)), polysaccharides instead of starch (about 10%, DM), protein (6 to 12%, DM), and lipids (3 to 6%, DM) as a major component [21, 22]. Starch is the main component of the normal corn kernel, and is generally located in the endosperm (98% to 99% of total starch [21].

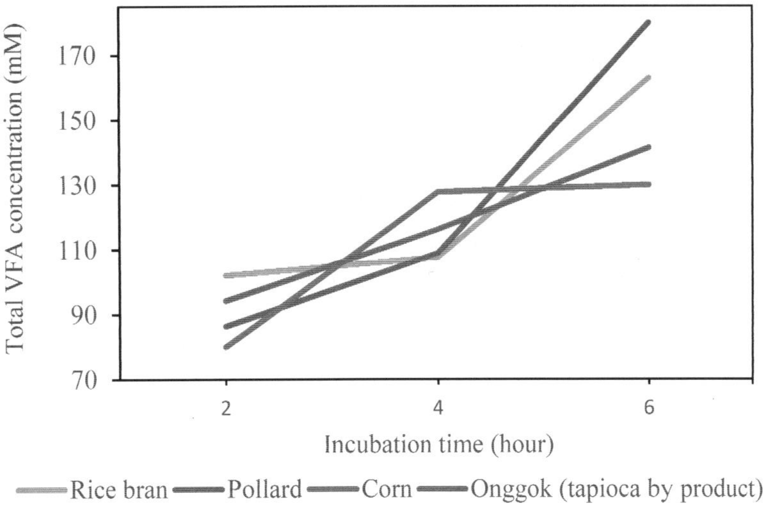

Figure 2. The VFA concentration that measured over different incubation time of several RAC.

Normal starch consists two types of α-d-glucopyranose polysaccharides: amylose and amylopectin. Amylose is basically a linear polymer that is connected by α (1 → 4) glycosidic bonds with several α (1 → 6) branch bonds, whereas amylopectin is a high branched polymer with about 5% α (1 → 6) branch relationships [23]. Amylose content and branch chain structure of amylopectin from corn starch also affect its digestion rate. The digestive level of starch, in general, decreases with increasing amylose content and the length of the branch chain of amylopectin [24]. Corn contains 71-73% amylopectin and 22-27% amylose [25]. The cassava contains 83% amylopectin and 17% amylose, wheat contains 74% amylopectin and 26% amylose. Rice bran contains 5.66% amylose [26]. The source of rice bran used in this study may come from low-amylose rice varieties in Taiwan. Other factors, such as surrounding by cellulose and hemicellulose fibers and / or protein matrix and interact with lipids, can also reduce the susceptibility of starch to enzyme hydrolysis [27].

The difference in the content of amylose and amylopectin in the four types of materials used and the different content of cellulose and hemicellulose in them causes differences in the rate of starch degradation by rumen microbes. According to [10] easily fermented carbohydrates will stimulate microbial growth in the rumen which also causes high use of ammonia for microbial growth so that the potential for ammonia to be converted into urea will be reduced. The ideal condition occurs when the production of fermented energy source feed ingredients in the rumen is as fast as NH_3 concentration, so that when NH_3 is formed there is a VFA derived from carbohydrates that will be used as a source and carbon framework of microbial protein amino acids.

3.3. In vitro study of NSM and its combination with several types of RAC

NSM is an alternative source of protein. The results of previous study about degradation rate of NSM was indicated that NSM is rapidly degraded in the rumen, especially in the first 6 hours of fermentation in the rumen. The source of protein that enters the rumen will be degraded to ammonia and if this degradation process occurs quickly, it will cause inefficiency of protein use. Efforts that can be made to protect protein for ruminant animal feed are combine it with feed ingredients with a high content of RAC, because this RAC will be an energy source for rumen microbes which causes the use of ammonia in the rumen by microbes will increase, thereby reduce accumulation ammonia in the rumen. Ammonia will be used by rumen microbes to form microbial protein synthesis. Rumen microbes contribute considerable protein to the needs of ruminant animals [28]. This MPS is also an important protein with high biological value in ruminants. The formation of the MPS also requires the supply of carbon branch chains derived from carbohydrates, and also requires vitamins and minerals.

Table 2. Production of NH_3, VFA, and microbial protein synthesis in each treatments.[a]

Variables	P [b]	R [c]	C [d]
pH	6.84±0.09	6.96±0.09	6.94±0.11
NH_3 (mM)	4.94±0.50[a]	4.02±0.71[b]	2.25±0.36[c]
VFA (mM)	128.51±20.43	128.87±17.14	146.01±15.26
Microbial protein synthesis-(mg $10mL^{-1}$)	7.85±0.97	9.64±0.67	9.66±3.20

[a] Notes are referenced using alpha superscripts show significantly different (P <0.05)
[b] P: NSM combined with RAC pollard
[c] R: NSM combined with RAC rice bran
[d] C: NSM combined with RAC corn

NSM combination with different RAC derived from pollard, rice bran, and corn showed that differences in RAC had a significant effect on Nh3 production. The highest NH_3 concentration was found in the rations with RAC pollard and lowest NH_3 concentration in rations with RAC derived from corn. NH_3 describes the degradation products of feed protein and or NPN, so that excessive amounts show inefficiency. The optimum amount of ammonia in the rumen is 6-21 mM [20]. In this study, the amount of ammonia is below the optimum limit which can indicate two things, namely the lack of ammonia or the amount of ammonia used by the rumen microbes to form MPS, so that the accumulation of ammonia in the rumen is reduced. The difference in RAC in rations containing NSM did not have significant effect on pH.

4. Conclusion

The use of NSM as much as 20% in rations combined with carbohydrate sources derived from corn can optimize the use of NSM as a source of feed protein in ruminants by encouraging the production of microbial protein synthesis that were useful for ruminants.

5. Acknowledgment

Thanks are conveyed to the Intellectual Product Unit Development Program (Program Pengembangan Unit Produk Intelektual Kampus/P2UPIK) No 1811/IT3.11/PN/2018 which have provided funding during the research.

6. References

[1] Ali B H and Blunden G 2003 Pharmacological and toxicological properties of *Nigella sativa* *Phytother. Res.* **17** 299-305
[2] Wahyuni S 2009 Opportunities for cultivation and benefits of black cumin (*Nigella sativa L.*) *Warta Penelitian dan Pengembangan Tanaman Industri* **15** 23
[3] Gokdogan O, Eryilmaz T and Yesilyurt M K 2015 Determination of energy use efficiency of *Nigella sativa* L. (Black seed) oil production *Am-Euras. J. Agric. and Environ. Sci.* **15** 1
[4] Ali M A, Sayeed M A, Alam M S, Yeasmin M S, Khan A M and Muhamad I I 2012 Characteristics of oils and nutrient contents of *Nigella sativa* Linn. and *Trigonella foenum-graecum* seeds *Bull. Chem. Soc. Ethiop.* **26** 55
[5] Cheikh-Rouhou, Besbes S, Hentati S, Blecker B, Deroanne C and Attia H 2007 *Nigella sativa* L.: Chemical composition and physicochemical characteristics of lipid fraction. *Food Chem.* **101** 673
[6] Hume I D R, Moir J and Sommers M 1970 Synthesis of microbial protein in the mmen. I. Influence of the level of nitrogen intake *Aust. J. Agric. Res.* **21** 283

[7] Satter L D and Roffler R E 1975 Nitrogen requirement and utilization in dairy cattle. *J. Dairy Sci.* **58** 1219
[8] Gustian R A 2017 Utilization of habbatusauda industry waste *(Nigella sativa)* as feed for local sheep (Bogor: Bogor Agricultural University)
[9] Barkah N N 2017 Metabolites and blood profiles of male local sheep feed with diet containing habbatussauda (*Nigella sativa*) waste processing industry (Bogor: Bogor Agricultural University)
[10] Ranjhan S K 1977 *Animal Nutrition and Feeding Practices in India* (New Delhi: Vikas Publishing House PVT Ltd)
[11] National Research Council 2006 *Nutrient Requirment of Sheep* (Washington DC: National Academy Press)
[12] Tilley J M A and Terry R A 1963 A two stage technique for the in vitro digestion of forage crops *J. British Grassl. Soc.* **18** 104
[13] Makkar H P S, Sharma, Dawra R K and Negi S S 1982 Simple determination of microbial protein in rumen liquor *J. Dairy Sci.* **65** 2170
[14] Lowry O H, Rosenbrough N J, Farr A L and Randall R J 1951 Protein measurement with the folin phenol reagent *J. Biol. Chem.* **193** 265
[15] Conway E J 1962 *Ammonia. General method, In Microdiffusion Analysis and Volumetric Error.* (London: Crosby Lockwood and Son Ltd)
[16] Steel R G D and Torrie J H 1993 *Principles and Procedures of Statistics. A Biomedical Approach*, 3rd Ed (Tokyo: Mc Graw Hill)
[17] Sutardi, Sastradipradja D T, Toharmat T, Sardiana A and Permana I G 1994 *Increased Ruminant Livestock Production Through Low-grade Fiber Feed Ammonia, Defaunation, and Protein Supplementation Resistant to Rumen Degradation* (Bogor: IPB Press)
[18] Wohlt J B, Clarj J H and Balaisdell F S 1976 Effect of sampling location, time and methode on concentration of ammonia nitrogen in rumen fluid. *J. Dairy Sci.* **59** 459
[19] Waghorn 2008 Beneficial and detrimental effects of dietary condensed tannins for sustainable sheep and goat production–Progress and challenges *Anim. Feed Sci. Tech.* **147** 116
[20] McDonald P, Edwards R A, Greenhalgh J F D, Morgan C A, Sinclair L A and Wilkinson R G 2002 *Animal Nutrition Seventh Ed* (Singapore: Pearson)
[21] Watson S A 2003. *Description, Development, Structure, and Composition of The Corn Kernel.* In: White PJ, Johnson LA, editors. *Corn: Chemistry and Technology.* 2nd ed. (St. Paul American Association Of Cereal Chemists, Inc)
[22] Sinha A K, Kumar V, Makkar H P S, De Boeck G and Becker K 2011. Non-starch polysaccharides and their role in fish nutrition-a review *Food Chem.* **127(4)** 1409–26
[23] Hizukuri S 1986 Polymodal distribution of the chain lengths of amylopectins, and its significance *Carbohyd Res.* **147** 342
[24] Li L, Jiang H X, Campbell M, Blanco M and Jane J 2008 Characterization of maize amylose-extender (ae) mutant starches. Part I: Relationship between resistant starch contents and molecular structures *Carbohyd. Polym.* **74** 396
[25] Nalin T, Sperb-Ludwig F and Venema K 2015 Determination of amylose/amylopectin ratio of starches *J. Inherit Metab. Dis.* **38** 985
[26] Fabian C, Ayucitra A, Ismadji S and Yi-Hsu Ju. 2011. Isolation and characterization of starch from defatted rice bran. *J. of The Taiwan Ins. of Chem. Engine* **42** 86
[27] Ai Y, Hasjim J and Jane J. 2013. Effects of lipids on enzymatic hydrolysis and physical properties of starch. *Carbohyd. Polym.* **92** 120
[28] Sniffen C J and Robinson P H 1987 Microbial growth and flow as influenced by dietary manipulations. *J. Dairy Sci.* **70** 425

The Effect of Probiotics on High Fiber Diet in Rumen Fermentation Characteristics

N F Sari*, R Ridwan, Rohmatussolihat, R Fidriyanto, W D Astuti and Y Widyastuti

Research Center for Biotechnology, Indonesian Institute of Sciences, Jl Raya Bogor km 46, Cibinong 16911, Indonesia

*e-mail: nurulfitrisari@gmail.com

Abstract. The productivity of cattle must be achieved through feed efficiency which has strong correlation with the rumen work. The efforts to increase productivity of cattle can use probiotics as feed additive in animal feeding. Probiotics are increasingly used in commercial animal production operations to profitably alter the gastrointestinal flora, thus improving animal health and productivity. This study aimed to investigate the effects of different type of probiotics on high fiber diet in live weight change, rumen fermentation, the total rumen of bacteria, and the total of lactic acid bacteria. Three fistulated cattle received a daily ration 2% from body weight (DM) of *Pennisetum purpureum* (70% of DM), concentrates (30% of DM), and mineral mix (1% of DM). The experiment was arranged as a crossover design with six treatments and three sampling periods as replications. The treatments were (R0) control diet, (R1) control diet + 10 mL of *Lactobacillus plantarum* TSD-10 per day, (R2) control diet + 10 mL of *L. plantarum* MX-16 per day, (R3) control diet + 10 mL of *L. brevis* SPCE-39 per day, (R4) control diet + 10 mL of mix *Lactobacillus plantarum* TSD-10, *L. plantarum* MX-16, and *L. brevis* SPCE-39 per day, (R5) control diet + 10 mL of commercial probiotics per day. Neither pH nor the total of lactic acid bacteria (LAB) were significantly affected by the treatment. Among others treatment, R3 showed a tendency body weight gain. The total of rumen bacteria was affected significantly for R2, R3, R4, and R5 compared with R0. The acetate to propionate ratio were affected for R3, R4, and R5, but were not affected significantly for R2 and R3 compared with R0. The butyrate ratio and rumen ammonia concentration were not affected by the treatment. In conclusion, probiotics change the population of the rumen bacteria. In addition, based on metabolism characteristics, particularly on propionate production and C2/C3 ratio, R3 was shown the best probiotics candidate on high fiber diet. Nevertheless, the difference pattern in volatile fatty acid indicates that different mechanisms may be involved.

Keywords: probiotics, cattle, lactic acid bacteria, rumen fermentation

1. Introduction

The growth number of the human population has increased the need to improve the productivity of the cattle. For this reason, most efforts in the field of animal production, have been directed to raise the productivity of ruminant using biotechnology approach to manipulate the ruminal ecosystem [1]. The use of probiotics can manipulate the process of rumen fermentation to maximize the production efficiency while decreasing energy loss [2]. This appears to be strongly driven by the ban of most of the antibiotic feed additives within the European Union, because of speculated risk for generating antibiotic resistance in pathogenic microbiota [3], chemical residues in animal products, release of antibiotics into the environment [4;5] and also the advent of present-day organic agriculture.

Probiotics, in general as reported by different authors, have the ability to enhance intestinal health by stimulating the development of a healthy microbial ecosystem [6;7], increase digestive capacity and their bio-availability [8], prevent enteric pathogens from colonizing the intestine [9], restore the gut microflora [7], lower pH, and improve mucosal immunity and nutrient absorption [10;6]. As a result, they should increase productivity and the general health of ruminants.

The ruminants highly rely on a symbiosis between the host and the rumen microbes, with the microorganisms supplying protein, vitamins and short-chain organic acids for the animal host [11]. Most of the energy in the ruminants are derived from microbial origins, such as the absorption of energy, the formation of glucose in the liver and the digestion of protein in the abomasum. In ruminants that function normally, in fact, as much as 90% of the protein that reaches the small intestine and up to 50% of the host's energy needs is provided by reticulorumen microbial cells [12]. Microbial communities are involved in the digestion and fermentation of plant polymers, which are very important in adult herbivores [6].

The rumen is a complicated ecosystem where nutrients consumed by the microorganisms for example bacteria, protozoa, and fungi are digested anaerobically. The main end products of fermentation are volatile fatty acids (VFA) and microbial biomass, which are used by ruminants. The interaction between microorganisms and host animals produces a symbiotic relationship that allows ruminants to digest fiber-rich and low-protein feeds. In the rumen, the environment supports microorganisms to provide the enzymes needed to digest nutrients. Ruminants have the ability to convert low-quality fibrous ingredients into products such as meat, milk, and fiber, which are useful for humans. The ability of rumen microorganisms to produce enzymes needed for the fermentation process allows ruminant animals to obtain energy contained in forages efficiently [13]. However, the rumen fermentation process is not entirely efficient because it produces several end products such as methane gas [14]. A symbiotic relationships exist in the rumen which provides the environment necessary for the formation of microorganisms and substrates needed for their maintenance. In turn, microorganisms provide nutrients to the ruminant host to produce energy [15]. The objective of this study was to investigate the effects of probiotics on a high fiber diet in rumen fermentation characteristics.

2. Material and Methods

2.1. Animal Care

Feeding trials were carried out using three fistulated Ongole cattle. The use of the cattle in this experiment was approved No. 9879/WK/HK/XI/2015 by the Ethics Clearance Committee of Indonesian Institute of Sciences.

2.2. Preparation of Probiotic Candidates

Lactobacillus plantarum TSD-10, *L. plantarum* MX-16, *L. brevis* SPCE-19 were chosen as probiotic candidates. The probiotic candidates from previous study were grown on the specific probiotic medium. The MRS is used as specific probiotic medium for the production of lactic acid bacteria [6]. The quality of candidate probiotics is done by calculating the number of colonies using TPC method.

2.3. Diet and Feeding

This experiment was used a high fiber feed formulation with a proportion of *Pennisetum purpureum* (70% of DM), concentrate (30% of DM), and mineral mix (1% of DM). *Pennisetum purpureum* were provided by the plant collection of the Research Center for Biotechnology at Indonesian Institute of Sciences. Concentrate was made from the following formulation: 30% of rice bran, 13% of pollard bran, 2% of soybean meal, 15% of coconut milk, 10% of palm kernel mill, 7% of coffee pulp, 20% of cassava flour waste, 2% of molasses, and 1% of mineral mix.

The experiment was arranged as a crossover design with six treatments and three sampling periods as replications. The treatments were :
R0: control diet
R1: control diet + 10 mL of *Lactobacillus plantarum* TSD-10 per day
R2: control diet + 10 mL of *L. plantarum* MX-16 per day
R3: control diet + 10 mL of *L. brevis* SPCE-39 per day
R4: control diet + 10 mL of mix among *Lactobacillus plantarum* TSD-10, *L. plantarum* MX-16, and *L. brevis* SPCE-39 per day
R5: control diet + 10 mL of commercial probiotics per day.

Each treatment was administered for 17 days, and rumen samples were collected on days 7, 12, and 17 as replication sampling periods. All cattle were given amounts of feed equal to 2% dry matter of their body weight (360 kg). The Nutrient and chemical composition of diets are shown in Table 1.

2.4. Chemical and pH analysis
Samples of P.purpureum and fermented concentrate were analyzed for dry matter (DM) and ash according to AOAC [17]. Crude protein (CP), crude fiber (CF), ether extract (EE) were determined followed the manufacturing procedure of FOSS. The NH_3-N concentration was determined by a Conway micro-diffusion method. Volatile Fatty Acids (VFA) were determined in centrifuged samples (1mL) by Gas Chromatography (GC) as described by Goering and Van Soest [18]. pH measurement of mix buffer rumen fluids was measured using pH meter (Jenway Model 3505, UK). Prior to the measurements, pH meter was calibrated by using pH 7 buffer solution and it was prepared carefully as the manufacturers instructions. All chemical composition analyses were conducted in triplicate.

2.5. Microbiology analysis
Total lactic acid bacteria and rumen bacteria were counted as colony forming units (CFU) per mL. The lactic acid bacteria were grown on MRS in agar medium and counted by using total plate count [19], while the rumen bacteria on 98-5 agar medium were grown anaerobically by following the method of roll tube [20]. Incubation was done at 38^0C for 3-4 days.

2.6. Statistical analysis
The experiment design was a block randomized. The statistical analysis of data was performed, using SPSS 16 for Windows. The difference among groups was determined by one-way ANOVA analysis. Significance was defined at $P<0.05$. Microbial diversity data were analyzed descriptively.

3. Results and discussions

3.1. Probiotics Production
Probiotics have been defined by the FAO/WHO as live microorganisms which in the appropriate amounts will give the health benefit on the host. In addition, it has specificity to the host, able to survive in the preparation process and the storage period can live and play in the target organ, must be consistently controlled for the observed response and can eliminate pathogenic microbes and optimize the immune system response of host [21;6]. Some of lactic acid bacteria (LAB) strains species belong to genera *Lactobacillus*, which are considered as a beneficial effect to the host. Therefore, we used three strains of *Lactobacillus plantarum* collected from biotechnology culture collection, Indonesian Institute of Sciences as candidate probiotics.

Table 1. The total of LAB Population during in vivo

Treatment	Strain	LAB Population (CFU/ml)
R0	Control	
R1	*L. plantarum* TSD-10	3.4×10^8
R2	*L. plantarum* MX-16	2.8×10^8
R3	*L. brevis* SPCE-39	7.8×10^7
R4	Mix R1-R3	5.9×10^8
R5	Commercial probiotics	1.0×10^6

The total population of probiotics candidate was shown in Table 1 which range from 1,0 x 10^6 – 5.9 x 10^8 CFU/ml. The volume of probiotics used in all treatments in PO fistulated cattle was 10 ml/head/day. This probiotic assumption is an early study related to survival or probiotic survival in rumen in vivo conditions.

3.2. Chemical and pH analysis

Chemical analysis of feed material is very important to evaluate the nutrient content of feed. proximate analysis is one of parameter to determine the quality of feed material. The nutrition composition of feed material that used in this study is shown in Table 2.

Table 2. Chemical composition (g/kg DM)

Item[a]	(%) DM	(%) ASH	(%) OM	(%) CF	(%) EE	(%) CP
Pennisetum purpureum	87.94	17.49	82.51	32.28	2.02	10.17
Concentrate	92.19	12.17	87.83	20.46	3.56	13.93
TMR (30% C, 70 *Pennisetum purpureum*)		15.89	84.11	28.74	2.48	12.10

DM= dry matter; OM = organic matter; CF= crude fiber; EE= ether extract ; CP = crude protein; TMR = total mix ratio aData are the means of three replicate analysis.

The pH rumen indicates rumen conditions that are appropriate or inappropriate for the growth of rumen microbes. This study showed that the use of different types of probiotics did not affect the pH of rumen (Table 3). The pH of the rumen is still within the normal range of 6.24 -6.58. The same pH values as control showed that the use of types of probiotics does not interfere rumen fermentation activity. According to Dehority [22], the rumen's normal pH is about 5.5-7.0. The pH value represents the amount of acid produced by microbial activity in the rumen [23;24]. The lowest pH value is generally achieved about 2-6 hours after meals, in accordance with the maximum acid production. Changes in pH are affected by time after meal, lactic acid supply, feed properties, and feeding frequency of animal [22;25].

3.3. The Total Rumen Bacteria and LAB

Provision of probiotics in fistulated cattle as a model of experimental cattle tends to show a positive effect when given on a high fiber formulation. This can be seen from the C_2/C_3 ratio R3, R4, and R5 has the lowest value compared to other treatment which can lead to increase the productivity.

The rumen microbiological profile of the probiotics treatment showed that R2, R3, R4, and R5 had significantly different from R0 and R1 (Table 3). However, the probiotics did not affect the total of LAB among all treatment. This can be explained; although the total LAB counts did not change, it does not mean that the existing population did not change. The total LAB counts might be same, but the LAB population might be different.

Table 3. The comparison among body weight, pH total rumen bacteria, and total lactic acid bacteria during in vivo in each treatment

Treatment	Body Weight	pH	The Total of Rumen Bacteria (Log 10 CFU/ml)	The Total of LAB (Log 10 CFU/ml)
R0	363.33	6.51	8.29a	7.23
R1	362.00	6.24	7.39a	6.33
R2	360.67	6.32	6.84b	6.48
R3	365.78	6.43	7.36b	6.84
R4	360.11	6.58	7.27b	6.84
R5	362.83	6.38	7.36b	6.89

LAB = lactic acid bacteria; R0; Control, R1; *L. plantarum* TSD-10, R2; *L. plantarum* MX-16, R3; *L. brevis* SPCE-39, R4; Mix R1-R3, R5; Commercial Probiotics

3.4. NH_3 and VFA Production

Ammonia is an important component for the synthesis of amino acids and microbial cell proteins. Protein from feed will be degraded by proteolytic bacteria with the help of protease enzyme into peptide then hydrolyzed to amino acid, and amino acid will go through deamination process to become ammonia. According to McDonald et al. [26] states that ammonia is produced in conjunction with several small peptides and free amino acids, then it is utilized by rumen organisms for microbial protein synthesis. In addition, some of the amino acids will be further digested into organic acids, ammonia (NH_3), and carbon dioxide (CO_2). The results of the average concentration of NH_3 are presented in Table 4. This study showed the effect of probiotics on high fiber feed formulation on N-NH_3 production gave no significant difference for all treatment of probiotic and control. This showed that rumen microbial activity tends to be stable in utilizing source N on high fiber formulations.

The effect of probiotics on high fiber feed formulation leads to the formation of high production of acetic acid. The acetate to propionate ratio was affected by R3, R4, and R5, but was not affected significantly for R2 and R3 compared with R0. The butyrate ratio and rumen ammonia concentration were not affected by the treatment. The ratio of C2/C3 shows a pattern of microbial fermentation in the rumen that has the potential to form acetic acid or propionate. The lowest C2/C3 ratio is produced by R3, R4, and R5 as shown in Table 4. The low of C2/C3 ratio is due to the high amount of fermentable organic matter enabling the formation of higher propionic acid than acetic acid. However, it is recommended to study the molecular approach to identify the rumen population. As reported by Moran [27] that the fermentative digestion of carbohydrates especially starch will produce lactic acid, this lactic acid will be converted to propionic acid by microbial lactate utilization, such as *Propionibacteria* sp, *Veillonella alkalescens,* and *Peptostreptococcus elsdeini*. The low of C2/C3 ratio will stimulate fattening and tend to body fat formation. In the treatment of R0, R1, and R2 showed higher C2/C3 ratio values than other treatments; this indicates that the microbial fermentation pattern leads to the formation of acetic acid because the carbohydrates contained in the feed are still dominated by fibrous carbohydrates and microbes that grow well are cellulolytic and hemicellulolytic microbes.

Table 4. Production of Volatile Fatty Acid (% mM) and NH_3 (mM) During Each Treatment

Variable	Treatment					
	R0	R1	R2	R3	R4	R5
C_2	63.72^a	62.85^a	66.99^a	56.40^b	56.71^b	57.46^b
C_3	21.15^b	20.61^b	18.95^b	25.27^a	23.70^a	25.21^a
C_4	10.20	11.05	9.89	10.90	11.95	10.12
Iso-C_4	2.90^c	3.19^c	2.50^c	4.27^a	4.34^a	3.77^{ab}
C_5	0.71^{ab}	0.90a^b	0.61^b	0.98^{ab}	1.00^{ab}	1.15^a
Iso-C_5	1.32^b	1.40^b	1.07^b	2.19^a	2.30^a	2.30^a
C_2/C_3	3.01^a	3.05^a	3.54^a	2.23^b	2.39^b	2.28^b
NH_3	10.77	9.68	10.32	10.86	10.01	9.03

R0; Control, R1; *L. plantarum* TSD-10, R2; *L. plantarum* MX-16, R3; *L. brevis* SPCE-39, R4; Mix R1-R3, R5; Commercial Probiotics ; C_2 = acetate ; C_3 = propionate ; C_4 = butyrate ; Iso-C_4 = Iso butyrate ; C_5 = valeric acid ; Iso-C_5 = Iso-valeric acid ; NH_3 = ammonia ; $^{a-c}$Different superscripts within the same column are significantly different at P<0.05

4. Conclusion

Based on this study, it could be concluded that probiotics in high fiber diet can change the population of the rumen bacteria. In addition, based on metabolism characteristics, particularly on propionate production and C2/C3 ratio, R3 was shown the best probiotics candidate on high fiber diet. Nevertheless, the difference pattern in volatile fatty acid indicates that different mechanisms may be involved.

5. Acknowledgment

This research was funded by Unggulan-LIPI DIPA Grant for Food Security and Drug's Discovery of the Research Center for Biotechnology, Indonesian Institute of Sciences (2015-2017). In addition, the authors would like to thank the Center of Excellence of Biotechnology Animal of Beef Cattle and Dairy Cattle (PUI Kemenristekdikti 2018) for supporting this publication.

6. References

[1] Castillo-González A R, Burrola-Barraza M E, Domínguez-Viveros J, Chávez-Martínez A 2014 *Archivos de Medicina Veterinaria* **46 (3)** 349-361
[2] Allen H K, Levine U Y, Looft T, Bandrick M, Casey T A 2013 *Trends Microbial.* **21** 114-119
[3] Windisch W, Schedle K, Plitzner C, Kroismayr A 2008 *J. Anim. Sci.* **86 (14)** Suppl E140-E148
[4] Yamamoto S, Nakano M, Kitagawa W, Tanaka M, Sone T, Hirai K, Asano K 2014 *Microbes Environ.* **29** 136-144
[5] Martinez – Vaz B M, Fink R C, Diez-Gonzalez F, Sadowsky M J 2014 *Microbes. Environ.* **29** 123 – 135
[6] Uyeno Y, Shigemori S, Shimosato T 2015 *Microbs. Environ.* **30 (2)** 126-132
[7] Musa H H, We S L, Zhu C H, Seri H I, Zhu G Q 2009 *J. Anim. Vet.* **8** 313-321
[8] Oyetayo V O, Oyetayo F L 2005 *Afr. J. Biotech.* **4** 123-127
[9] Casas I A, Dobrogosz W J 2000 *Microb. Ecol. Health. Dis.* **12** 247-285
[10] Timmerman H M, Mulder L, Everts H, van Espen D C, van der Wal E, Klaassen G, Rouwers S M G, Harteminck R, Rombouts F M, and Beynen A C 2005 *J. Dairy. Sci.* **88** 2154–2165
[11] Pinloche E, McEwan N, Marden J P, Bayourthe C, Auclair E, Newbold C J 2013 *PLoS ONE* **8(7)** e67824.
[12] Russell J B, Mantovani H C 2002 *J. Mol. Microbiol. Biotechnol.* **4** 347-355
[13] Burns J C 2008 *J. Anim. Sci.* **86** 3647-3663
[14] Kingston-Smith A H, Marshall A H, Moorby J M 2012 *Animal* **1** 1-10
[15] Russell J B, Rychlik J L 2001 *Science* **292** 1119-1122
[16] de Man J C, Rogosa M, Sharpe M E 1960 *J. Appl. Bacteriol.* **23** 130–135
[17] AOAC 2005 *Official Methods of Analysis* 18th edition (Arlington : AOAC International)
[18] Goering H K and Van Soest P J 1970 *Agriculture handbook* (Washington (DC) : Agriculture Research Service USDA) 379
[19] Cappuccino J G and Sherman N 2001 *Microbiology A Laboratory Manual* 6th Edition (San Francisco : Benjamin Cummings)
[20] Hungate R E, 1969 *A Roll Tube Method for Cultivation of Strict Anaerobes* (London : Academic Press)
[21] McAllister T A, Beauchemin K A, Alazzeh A Y, Baah J, Teather R M, and Stanford K I M 2011 *Can. J. Anim. Sci.* **91** 193-211
[22] Dehority B A and Tirabasso P A 2004 *J. Anim. Sci.* **79** 2908 - 2912
[23] Elghandour M M, Salem A Z, Castaeda J S M, Camacho L M, Kholif A E and Chagoy J C V 2015 *Journal of Integrative Agriculture* **14** 526-533
[24] Hutton P G, Durmic Z, Ghisalberti E L, Flematti G R, Duncan R M, Carson C F and Riley T V 2012 *Anim. Feed Sci. Technol.* **176** 170-177
[25] Seo J K, Kim S W, Kim M H, Upadhaya S D, Kam D K and Ha J K 2010 *Asian- Australas. J. Anim Sci.* **23** 1657-1667
[26] McDonald P, Edwards R and Greenhalgh J 2010 *Animal Nutrition* 7th Edition. (New York : Ashford Colour Press)
[27] Moran, J 2005 *Tropical Dairy Farming* (Australia : Landlinks Press)

Effect of isoamylase application on chemical characteristic of cassava root meal starch

P S Puspita[1], W Hermana[2] and Nahrowi[2]

[1]Graduate Student of Department Animal Nutrition and Feed Science, Animal Husbandry Faculty, Bogor Agricultural University, Bogor 16680, Indonesia.
[2]Lecturer of Department Animal Nutrition and Feed Science, Animal Husbandry Faculty, Bogor Agricultural University, Bogor 16680, Indonesia.

Email: pypuspita@gmail.com

Abstract. Cassava root meal (CRM) is still rarely used as an ingredient for poultry feed. The high viscosity component can be reduced by enzymatic hydrolysis. However, there is limited information about enzyme treatment on cassava root meal. The objective was to identify and evaluate the effect of isoamylase hydrolysis on the chemical characteristic of hydrolyzed cassava root meal starch which has amylose/amylopectin ratio approached the corn. The cassava root meal was weighed each 200 g containing five levels of isoamylase (CRM-A) 0.01%, (CRM-B) 0.02%, (CRM-C) 0.03%, (CRM-D) 0.04%, (CRM-E) 0.05% (w/w) were moistened with 135 ml distilled water and incubated ~~to~~ at 50 ^{0}C for 12 hours. Total starch, amylose, amylopectin, total sugar, and reducing sugar content of hydrolyzed product were measured. The result showed that there were increasing of total starch and amylose content of hydrolyzed cassava root meal and decreasing in amylopectin content and amylose/amylopectin ratio. Reducing sugar increased along with increasing enzyme addition 1.81 – 3.07% DM which corresponds to degree of hydrolysis to 54.52 – 72.99%. It can be concluded that the different levels of isoamylase addition will produce cassava root meal starch with different chemical characteristics. The addition 0.05% isoamylase is able to make amylose/amylopectin ratio of hydrolyzed cassava root meal approached the corn.

1. Introduction

Corn is one of the most important commodity also the main feed ingredient of poultry farm and become raw material of processed industry. The rapid development of the poultry industry will encourage the increasing demand for corn. Rising of raw material costs have accelerated the demand for finding alternative feed ingredients that can replace this material with lower production costs. Cassava is the highest source of carbohydrate among staple crops and can potentially replace corn as a source of energy in poultry feed. However, cassava also has a disadvantage that is high levels of resistant starch and amylopectin content. Resistant starch can not be digested and utilized by poultry. The main composition of cassava starch generally consists of amylose and amylopectin. Cassava starch contains 17% amylose and 83% amylopectin, compared to corn starch which has 28% amylose and 72% amylopectin [1-2]. High levels of starch resistance in cassava were probably 82.85% amylopectin with 5.79% of branch association and amylose 17.25% with 0.48% branch association [3]. Another possible explanation is amylopectin in cassava has a relatively long chain length. Starch is degraded by amylases, of which isoamylase (glycogen-6-glucanohydrolase, E.C.3.2.1.68) also known as debranching enzyme hydrolases 1, 6-α-D-glycosidic linkages of glycogen, amylopectin and α and β limit dextrins, producing linear malto oligosaccharides [4]. Isoamylase is used primarily in the production of food ingredients from starch like glucose, maltose, trehalose and cyclodextrins [5]. An understanding of the relationship between enzyme levels and chemical characteristic of hydrolyzed products is essential for using in the poultry feed industry. This study was designed to identify and evaluate the effect of isoamylase hydrolysis on the chemical characteristic of hydrolyzed cassava root meal starch which has amylose/amylopectin ratio approached the corn.

2. Material and method

2.1. Material

Cassava root meal was obtained from UD. Setia Flour Product (Cikarawang, Bogor). Isoamylase (Food Grade) was provided from Creative Enzymes (USA), Creative Biomart, Inc. This enzyme has optimum activity at a pH around 5.6 – 7.0 and at temperature between 50 – 55 ^{0}C. The enzyme activity ≥ 10,000 U/g.

2.2. Enzyme hydrolysis

The cassava root meal (CRM) was weighed each 200 g containing five levels isoamylase (CRM-A) 0.01%, (CRM-B) 0.02%, (CRM-C) 0.03%, (CRM-D) 0.04%, (CRM-E) 0.05% (w/w) and wetted with 135 ml of distilled water. The mixtures were incubated in waterbath at 50 ^{0}C for 12 hours. After that, the samples were dried in oven at 60 ^{0}C for 24 hours to stop the enzyme activity.

The moisture content were determined by gravimetric methods [6]. Total starch were analysed using calorimetric method [7] and amylose content were analysed by spectrophotometric [8]. Amylopectin content was calculated by difference between total starch and amylose content. The degree of hydrolysis was measured as an increase in the content of reducing sugars. It was measured using Luff Schrool method [9].

2.3. Data analysis

All chemical analyses were performed in two replicates and the result were expressed as the mean ± standard deviation (SD).

3. Result and discussion

The data presented in Table 1 showed the comparison of starch content in corn and native cassava as a control and hydrolyzed cassava root meal. All hydrolyzed cassava root meal showed increasing in total starch and amylose content and tended to decrease amylopectin content as the increasing of isoamylase addition. Cassava root meal with the addition of 0.05% isoamylase (CRM-E) produces 64.27% amylopectin content which is almost closely to corn 57.41% and has the lowest amylose / amylopectin ratio 1:2.63. Hydrolysis of starch by enzyme is influenced by enzyme diffusion rate, substrate accessibility, substrate recognition leading to the formation of enzyme substrate complex and appropriate catalytic action [10].

Similarly, the long incubation time required to achieve complete hydrolysis is an indicator for the slow rate of enzyme diffusion imposed by highly viscous cassava gel. While isoamylase capable of hydrolyzing the bond of 1, 6-α-D-glycosidic of amylopectin [11]. This can be seen with increasing addition of isoamylase tends to decrease amylopectin content in cassava root meal in order to lower amylose/amylopectin ratio.

The data presented in Table 2 revealed the degree of hydrolysis by isoamylase in cassava root meal. The total sugar of hydrolyzed cassava root meal ranged from 3.31 – 4.20% DM and reducing sugar ranged between 1.18 – 3.07% DM which correspond to degree of hydrolysis 54.52 – 72.99%. The low reducing sugar value of cassava starch implies an abundance of amylopectin and amylose polymers that are not degraded in starch [12]. On the contrary the result showed that reducing sugar value in hydrolyzed cassava root meal tended to increase with the increasing of isoamylase compared to native cassava which is the total sugar 3.28% DM and reducing sugar 1.04% DM corresponds to a degree of hydrolysis only 31.77%.

Table 1. Total starch, amylose, and amylopectin content of corn and cassava root meal

Sample	Total Starch (% DM)	Amylose (% DM)	Amylopectin (% DM)	Amylose/Amylopectin Ratio
Corn	85.30 ±0.18	27.89 ±0.03	57.41 ±0.16	1:2.05
Cassava	85.78 ±0.28	20.52 ±0.03	65.26 ±0.31	1:3.18
CRM-A	85.47 ±0.07	20.55 ±0.06	64.92 ±0.01	1:3.16
CRM-B	87.01 ±0.40	21.11 ±0.06	65.91 ±0.34	1:3.12
CRM-C	88.16 ±0.04	21.45 ±0.00	66.71 ±0.04	1:3.11
CRM-D	88.34 ±0.12	21.61 ±0.24	66.72 ±0.12	1:3.09
CRM-E	88.68 ±0.12	24.41 ±0.12	64.27 ±0.33	1:2.63

Table 2. Total sugar and reducing sugar content of cassava root meal

Sample	Total Sugar (% DM)	Reducing Sugar (% DM)	Degree of Hydrolysis (%)
Cassava	3.28	1.04	31.77
CRM-A	3.31	1.81	54.52
CRM-B	3.28	2.06	62.88
CRM-C	3.44	2.31	67.20
CRM-D	3.83	2.59	67.71
CRM-E	4.20	3.07	72.99

4. Conclusion

The different levels of isoamylase addition will produce cassava root meal starch with different chemical characteristics. The higher level of isoamylase addition can increase the total starch, amylose, total sugar, and reducing sugar content, and tend to reduce the amylopectin content as well as the amylose amylopectin ratio of hydrolyzed cassava root meal. The addition 0.05% isoamylase is able to make amylose/amylopectin ratio of hydrolyzed cassava root meal approached the corn.

5. Acknowledgments

The authors would like to thank LPDP (Indonesia Fund Endowment for Education) for the financial support.

6. References

[1] Gomes E, Souza S R, Grandi R P, Silva R D 2005 Production of thermostable glucoamylase by newly isolated Aspergillus flavus A1.1 and Tthermomyces lLanuginosus A13.37 *Braz. J. Microbiol.* **36** 75e82

[2] Kiatponglarp W, Tongta S 2007 Structural and physical properties of debranched tapioca starch. *Suranaree J. Sci. Tech.* **14** (2) 195e204

[3] Promthong S, Kanto U, TIrawattanawanich C, Tongyai S, Isariyodom S, Markvichitr K, et al. 2005 Comparison of nutrient compositions and carbohydrate fractions of corn, cassava chip and cassava pellet ingredients: animals. *In: Proceedings of 43rd Kasetsart University Annual Conference* Thailand p 211e20

[4] Fang T Y, Lin L L, Hsu W H 1994 Recovery of isoamylase from Pseudomonas amyloderamosa by adsorption-elution on raw starch *Enzyme Microbiol. Technol.* **16** 247-252

[5] Olemposka-Beer, Z 2007 Isoamylase from *Pseudomonas amyloderamosa Chem. Tech. Assess.* 1- 6

[6] AOAC, 2005 *Official Methods of Analysis* Association of Official Analytical Chemists (Washington: Benjamin Franklin Station)

[7] Dubois M, Gilles K A, Hamilton J K, Rebers P A, and Smith F 1956 Colorimetric method for determination of sugar and related substances *Anal. Chem.* **28** (3) 350-356

[8] Apriyantono A, Fardiaz D, Puspitasari N L, Sedarnawati, Budiyanto, S 1989 *Analisis Pangan: Petunjuk Laboratorium* (Bogor: PAU Pangan dan Gizi IPB)

[9] Sudarmadji S, Haryono B, Suhardi 1997 *Prosedur Analisis Untuk Bahan Makanan dan Pertanian* (Yogyakarta: Liberty) p 160

[10] Colonna P, Leloup V, and Buleon A 1992 Limiting factors of starch hydrolysis *Eur. J. Clin. Nutr.* **46** Suppl 2, S17-32

[11] Yamamoto T 1994 *Enzyme chemistry and molecular biology of amylases and related enzyme* (CRC Press Boca Raton Fl)

[12] Christopher M, Calvin O, Doris J, Thomas H, Herald R 2009 Determination of optimum condition for enzymatic debranching of cassava starch and synthesis of resistant starch type III using central composite rotatable design *Starch/Starke* **61** 367-376

Effect of Cutting Frequency of Cassava Leaves on Composition and Production during the Dry Season

R Utomo[1], B Suhartanto[1*], B Suwignyo[1], S Widodo[1] and Harimayastuti[1]

[1] Faculty of Animal Science, Universitas Gadjah Mada, Bulaksumur, Yogyakarta 55281, Indonesia

*Corresponding author: bamsuhar@ugm.ac.id

Abstract. Cassava leaves are an alternative feed derived from agricultural crops by-product. This study aimed to determine the effect of cutting frequency on the production and chemical composition of cassava leaves. This study was conducted in the Kwarasan Wetan, Kedungkeris, Nglipar, Gunungkidul during the dry season. Farmland with a size of 18 x 12 m has been planted with cassava plantation 5 months old and was divided into 4 blocks as replication, the size of each block was 18 x 3 m. Each block consisted of 3 treatments with the plot size of each treatment was 6 x 3 m which were cut 1, 2 and 4 times. Each treatment consisted of 12 plants, and each plant was conditioned to have 2 stems and cut with different intensities. Leaves and edibles stem were analyzed. The variables observed were content of dry matter (DM), organic matter (OM), crude fiber (CF), crude protein (CP), extract ether (EE), nitrogen-free extract (NFE), total digestible nutrients (TDN), and dry matter, organic matter, crude protein, and total digestible nutrients production. The data obtained were analyzed according to the completely randomized design (CRD), the differences between treatments were further tested by Duncan's new multiple range test (DMRT). The results showed that the cutting frequency had a significant effect ($P<0.05$) on dry matter and total digestible nutrients. The highest dry matter content was shown at 2 times cutting frequency (32.21%) and the lowest crude fiber at cutting 4 times frequency (17.97%). The 2 and 4 times cutting frequencies were able to increase the production of cassava leaves ($P<0.05$). Based on the results it can be concluded that the most appropriate cutting frequency is 2 times.

1. Introduction

The availability of feed that is able to meet the needs both qualitatively and quantitatively is one of the main factors that need to be considered in the selection of animal feed. In general, the availability of forage in Indonesia is influenced by the season. Feeds are available in large quantity during the rainy season however lack during the dry season.

One alternative in fulfilling feed during the dry season is to use agricultural crops by-products. Cassava is an agriculture plant that is often used as animal feed, especially its leaves. Cassava (*Manihot utilissima*) is an annual plant; this plant is a food commodity that can grow in both tropical and subtropical regions. Cassava can be planted in sandy clay types with low organic matter content, low rainfall, and high temperature. These plants in some countries are generally grown by small farmers along with other farming systems as additional sources of income [1]. Based on data, cassava production in Indonesia has decreased in the last five years from 24,044,025 tons in 2011 to only 21,801,415 tons in 2015 or decreased by 9.33% [2].

One of the problems in utilizing cassava leaves as a source of animal feed is the high crude fiber content. The high content of crude fiber is caused by the age of cutting the leaves which are carried out together with the process of harvesting cassava. The cutting age is classified as quite old because cutting is done once a year. The older of cutting age can increase the production but the inversely proportional to the quality of feed (crude fiber content increased, crude protein decreased) [3].

One alternative that can be done to reduce the crude fiber content is by plant cutting periodically. It is expected to provide optimal results and be able to maintain leaf quality. Another goal of periodic cutting is that the availability of cassava leaves is maintained even in the dry season. The quality of animal feed can be seen based on its chemical composition. The older the plant has decreased in the quality of its chemical composition. Structural carbohydrate fractions in older plants tend to be higher compared to young plants. Based on this problem, an alternative means is needed to maintain maintenance and forage quality.

2. Material and Methods

2.1. Location

This study was conducted in Kwarasan Wetan, Kedungkeris, Nglipar, Gunungkidul, Yogyakarta during dry season. The altitude of field location is at 200 meters above sea level with an average temperature of 20 to 30° C and an average rainfall of 2,000 mm per year.

2.2. Material

The material used in this study was cassava plants of 5 months old variety of Gatotkaca. Cassava planted among eucalyptus plants in the type of alfisol soil that had been fertilized with TSP and urea as much as 5 kg and 12.5 kg per plot size of 18 x12 m. Cassava was planted on flat soil with planting space of 75 cm intercropped with peanut plants. The material used in the chemical analysis consisted of 1.25% H_2SO_4, 1.25% NaOH, ethyl alcohol, concentrated H_2SO_4, H_3BO_3 0.1 N, Kjeldahl tablets (Merck's® catalyst for nitrogen detection), mix indicator, 50% NaOH and HCl 0.1 N.

The equipment used in this study include: Shanghai Yamato scales with sensitivity of 1 g, ATS scales with sensitivity of 100 g, oven with a temperature of 55 °C, Willey Mill with a screen diameter was 2 mm, silica disc, crucible, beaker glass, Soxhlet, and Erlenmeyer.

2.3. Methods

This study was conducted following a completely randomized design (CRD). Farmland with a size of 18 x 12 m was divided into 4 blocks measuring of 18 x 3 m as replication. Each block was divided into 3 plots with a range of 6 x 3 m which was determined randomly. Each plot consists of 12 plants, and each was conditioned to have 2 plant stems. As a treatment namely control (leaves cut at 9 months 15 days), cut 2 times (cut at 5 months and 9 months 15 days) and cut 4 times (cut at 5 months, 6.5 months, 8 months, and 9.5 months). Defoliation at each treatment was carried out with a distance of 20-30 cm above the soil surface and for further cutting the part was cut on the stem regrowth. The sample of cassava leaf was then weighed and dried at 55° C, after which the dried sample then ground using a Willey mill. The samples were then analyzed for chemical composition [4]. Total digestible nutrients (TDN) were calculated using a regression [5]. Significant differences between treatments were further tested by Duncan's new multiple range test (DMRT) [6].

3. Result and Discussion

3.1. Chemical Composition

The results showed there was a significant difference ($P<0.05$) in the dry matter content of cassava leaves with different cutting frequencies. The dry matter content in the control, 2, and 4 times harvesting were 28.18, 32.21 and 21.15% respectively. The highest dry matter content was obtained at cutting twice. Older plants were thickening of cell walls, which resulted in increased dry matter. The older plant contained higher cell wall content [7].

Organic matter content in cassava leaves with different cutting frequencies did not show any significant differences ($P>0.05$). The organic matter content in control, 2, and 4 times cutting were

91.22, 91.63, and 91.77%. The content of organic matter was not significantly different; it was probably due to the same soil conditions and varieties. The organic matter content of cassava leaves at harvesting is 91.58% [8].

Analysis of variance did not show any significant differences (P>0.05) on crude fiber of cassava leaves with different cutting frequencies. Crude fiber content in the control, 2, and 4 cutting times were 19.17, 19.74, and 17.94%. The crude fiber content in the plants increases with the length of life of the cutter. Therefore, the older plant had higher fiber content. In connection with the development of maturity (age of plants) forage, there will also be an increase in fiber concentration [7].

There was no significantly different (P>0.05) on crude protein content of cassava leaves cut in several cutting frequencies. Crude protein content in the control, 2, and 4 cutting times were 22.05, 22.39, and 23.56% respectively. Extended cutting interval will decrease crude protein content. Decreasing of crude protein content with addition of plant age was also caused by a decreasing in the proportion of leaves with petioles and stems, where the leaves have a higher protein content compared to the petioles and stems [9].

The extract ether content of cassava leaves at different cutting frequencies did not also show any significant differences (P>0.05). The extract ether content of cassava leaves in control, cutting twice, and cutting 4 times were respectively 7.62, 6.73, and 5.25%.

Table 1. Chemical composition of cassava leaves with different cutting frequencies

Parameter	Control	Cutting 2 times	Cutting 4 times
Dry matter (%)	28.18 ± 0.96[b]	32.21 ± 4.45[c]	21.15 ± 1.17[a]
Organic matter (%)	91.22 ± 0.41	91.63 ± 0.44	91.77 ± 0.42
Extract ether (%)	7.62 ± 2.60	6.73 ± 1.28	5.25 ± 1.80
Crude fiber (%)	19.17 ± 0.70	19.74 ± 0.62	17.97 ± 0.46
Crude protein (%)	22.05 ± 1.58	22.39 ± 2.01	23.56 ± 2.13
Nitrogen free extract (%)	42.40 ± 3.07	42.77 ± 3.39	45.00 ± 2.33
Total digestible nutrients (%)	68.22 ± 1.08[a]	68.45 ± 0.30[a]	70.45 ± 0.87[b]

[a,b,c] Different superscript in same-row shows significant difference (P<0,05)

The results of the analysis of variance showed that the nitrogen-free extract (NFE) on cassava leaves with different cutting frequencies did not show any significant differences (P>0.05). NFE in the control, 2 and 4 cutting times were 42.40, 42.77 and 45.00%. NFE content is strongly influenced by other fractions contained in plants. Increasing other fraction lowered the NFE content of the plant.

Total digestible nutrients (TDN) in cassava leaves with different cutting frequencies showed significant differences (P<0.05). The TDN in the control, 2, and 4 cutting times were 79.23, 76.81, and 74.42%. The highest TDN was shown at 4 times cutting intervals. TDN was strongly influenced by the chemical composition of plants. The higher fraction that was easily degraded such as protein and NFE can increase the TDN.

3.2. Production

Dry matter production of cassava leaves with different cutting frequencies showed significant differences (P <0.05). Dry matter production on control, 2, and 4 cutting times were 102.98, 1,333.03, and 1,332.53 kg/ha respectively. Control treatment resulted in lower dry matter production compared to cutting 2 and 4 times. The presence of cutting causes regrowth and produce more dry material production compared with only one cut. Increasing of fresh produce will increase the production of dry matter because the production of dry matter is the result of fresh produce with dry matter content [10].

The results of the analysis of variance showed a significant difference (P <0.05) on the organic matter production of cassava leaves with different cutting frequencies. Organic matter production of cassava leaves which was cut 1, 2, and 4 times were 93.99, 1,220.37, and 1,223.04 kg/ha. Organic matter production at cutting 2 and 4 times were higher than the control treatment. The high of organic matter production was in direct proportion to the high of dry matter production. The production of organic matter will depend largely on fresh and dry matter production, as well as the content of plant organic matter [11].

Table 2. Dry matter, organic matter, crude protein, and total digestible nutrients production of cassava leaves with different cutting frequency (kg/ha)

Treatment	DM Production	OM Production	CP Production	TDN Production
Control	102.98 ± 43.80[a]	93.99 ± 40.49[a]	22.89 ± 10,04[a]	63.95 ± 27.21[a]
Cutting 2 times	1,333.03 ± 394.64[b]	1,220.37 ± 357.39[b]	292.51± 60,04[b]	834.98 ± 242.72[b]
Cutting 4 times	1,332.53 ± 179.43[b]	1,223.04 ± 16.74[b]	314.85 ± 58,22[b]	862.53 ± 125.21[b]

[a,b,c] Different superscript in same column shows significant effect (P<0,05)

Crude protein production of cassava leaves with different cutting frequencies showed significant differences ($P<0.05$). The results of crude protein production in the treatment of cutting 1, 2, and 4 times were 22.89, 292.51, and 314.85 kg/ha. The cutting of 2 and 4 times resulted in an increase crude protein production of cassava leaves, but there was no significant difference between cutting 2 and 4 times. The presence of cuts affected the chemical composition and plant growth. Cutting tended to slow the maturity of the plant by forming new branches so that the protein produced was higher than the old one. Crude protein production on forage depended on dry matter production of and crude protein content of forage [12].

The results of the analysis of variance in the total digestible nutrients production (TDN) in cassava leaves with different cutting frequency showed significant difference ($P<0.05$). TDN Production of cassava leaves at 1, 2 and 4 times cuts was 63.95, 834.98 and 862.53 kg/ha. These results indicated that there was a cut of 2 and 4 times to increase TDN production in cassava leaves. Cutting caused the growth of young leaves that had higher non-structural components compared to old leaves. Non-structural components in young leaves increased TDN production in cassava leaves. Increasing in cutting age will increase the proportion of twigs and reduce the proportion of leaves. Increasing the proportion of twigs will increase the structural content, thereby reducing TDN [3].

Based on production of dry matter, organic matter, crude protein and TDN there was no difference between the cutting frequency 2 and 4 times, but the cutting frequency 4 required more time, so the cutting frequency 2 times was better.

4. Conclusion

Cassava leaves can be cut after 5 months. Cutting of cassava leaves can be done 2 until 4 times to get the highest production on crude protein and TDN. It is recommended that cassava leaves better to be cut twice on 5 and 9.5 month.

5. References

[1] Wanapat M 2003 Manipulation of cassava cultivation and utilization to improve protein to energy biomas for livestock feeding in the tropic *Asian-Aust. J. Anim. Sci.* **16** 463 – 472
[2] Badan Pusat Statistik 2016 Produksi Ubi Kayu Menurut Provinsi (ton) 1993-2015 https://www.bps.go.id/linkTableDinamis/view/id/880
[3] Savitri M V, Sudarwati H and Hermanto 2013 Pengaruh umur pemotongan terhadap produktivitas gamal (*Gliricidia sepium*) *J. Ilmu-Ilmu Peternakan* **23** (2) 25 – 35
[4] AOAC 2005 *Official Methods of Analysis Association of Official Analytical Chemists* (Washington : Benjamin Franklin Station)
[5] Harris L E, Kearl L C and Fonnesbeck P V1972 Use of regression equations in predicting availability of energy and protein *Journal of Animal Science* **35** 658-680
[6] Steel C J and Torrie J H 1995 *Prinsip dan Prosedur Statistik* (Jakarta : PT. Gramedia)
[7] Djuned H, Mansyur and Wijayanti H B 2005 Pengaruh umur pemotongan terhadap kandungan fraksi serat hijauan murbei (Morus indica L. Var. Kanva-2) *Seminar Nasional Teknologi Peternakan dan Veteriner*

[8] Fitraya R 2005 Kajian kualitas dan anti kualitas daun, ubi, dan kulit ubi ketela pohon (*Manihot esculenta* Crants) *Skripsi* (Yogyakarta : Fakultas Peternakan, Universitas Gadjah Mada)
[9] Mansyur, Djuned H, Dhalika T and Hardjosoewignyo S and Abdullah L 2005 Pengaruh interval pemotongan dan inveksi gulma *Chromolaena odorata* terhadap produksi dan kualitas rumput *Brachiaria humidicola Media Peternakan*
[10] Wong C C, Daham M D M and Abdullah O 2008 Effects of defoliation (cutting) on forage yield and quality of selected kenaf accessions *J. Trop. Agric. Feed. Sci.* **36** 21-28
[11] Abqoriyah, Utomo R, Suwignyo B 2015 Productivity of calliandra (*Calliandra calothyrsus*) as a forage in the different defoliation time *Buletin Peternakan* **39** (2) 103-108
[12] Balabanli C, Albayrak S and Yuksel O 2010 Effect of nitrogen, phosporus and potassium fertilization on the quality and yield of native rangeland *Turkish J. Field Crops* **15** 126-130

Effect of chitosan oligosaccharide (COS) and l-arginine supplementation on broiler performance

M Pramujo[1], R Mutia[2] and I Wijayanti[2]

[1]Master Program on Animal Nutrition and Feed Science, Faculty of Animal Science, Bogor Agricultural University (IPB),
[2]Department of Animal Nutrition and Feed Science , Faculty of Animal Science, Bogor Agricultural University (IPB). Jl. Agatis, Bogor, 16680, Indonesia

Email: mujokds@gmail.com

Abstract. Usage of antibiotic growth promoter is eliminated for the poultry production starting from January 2018. This policy may affect animal performance and health status in tropical countries like Indonesia. A study of Chitosan Oligosaccharide (COS) and L-arginine supplementation was carried out to evaluate the growth performance of broiler. A total of 300 DOCs (day old chickens) broilers with an initial average body weight of 46 g, were randomly treated into 5 treatments with 6 replications (10 unsexed birds in each pen) for 35 days. Dietary treatments were (P0) basal diet (BD) without supplement, (P1) BD + Antibiotic growth promoter (Zinc Bacitracine 200 mg kg $^{-1}$) of diet, (P2) BD + COS 100 mg kg $^{-1}$ of diet, (P3) BD + L-arginine 1.9% of diet, (P4) BD + COS mg kg $^{-1}$ + L-arginine 1.9% of diet. The results showed that treatments with COS and L-arginine significantly ($P < 0.05$) increased feed intake, body weight, weight gain and reduce feed conversion ratio at starter period compared with fed P0. Furthermore, feed conversion ratio and mortality were significantly ($P < 0.05$) decreases at all treatments and body weight and weight gain significantly ($P < 0.05$) increased only at COS treatment compared with fed P0, at the overall experimental periods. In conclusion, dietary addition of COS 100 mg kg-1 is able to use as alternative for antibiotic growth promotor replacement without giving effect on broiler performance

1. Introduction

Prohibition of using Antibiotic Growth Promoters (AGP) in Indonesia makes probiotics, prebiotics and enzymes as an alternative to potential AGP substitutes. Chitosan is one of the prebiotics that can be used as an alternative to AGP. Wang et al [1] reported that supplementation with 0.1% (w/w) chitosan reduced the amount of E. Coli in the chicken's digestive tract, also increased microvilli density of the small intestine, serum titer antibody to fight Newcastle disease virus , and growth performance. Chitosan in the form of Oligochitosan or called Chitosan Oligosaccharide (COS) has a smaller molecular weight of 1000-5000 Da. This makes COS is easier absorbed in the digestive tract of chicken so that it can optimize its function as a prebiotic. Huang et al [2] reported that administration of Oligochitosan supplementation (COS) can increase nutrient digestibility in ileal and improve broiler performance. Other Oligosaccharide such as Fructooligofructose (FOS) and Manna-oligosaccharide (MOS) are also currently accepted for use in livestock as an alternative to antibiotics. This substance works through many mechanisms, including by inhibiting pathogenic enteric adsorption, modulating immune function, and regulating nutrient metabolism (for example: the use of arginine in the intestine) [3-4].

One of the mechanisms of Oligosaccharide in improving livestock performance is by regulating nutrient metabolism (for example: the use of arginine in the intestine). Arginine acts to synthesize NO and polyamine , which are involved in inflammatory modulation and wound healing [5]. When there is a chicken coccidia infection requires more Arg to produce NO which has an important role in the process of killing directly or indirectly parasites (peroxynitrite , $ONOO^-$) [6]. This shows that Arg works more specifically to increase the immune system in livestock. Research on pigs and chickens shows that Arg supplementation can increase the function of the chicken immune system when in a state of immune stress [7]. Tan et al [8] reported that supplementation of L-Arginine 1.9 % increase the body weight gain and decrease the feed conversion ratio in the starter phase. L-arginine supplementation can reduce inflammation in chickens induced by lipopolysaccharide (LPS). The hypothesis of the present study was supplementation of COS, L-arginine and combination of them could replace the usage of AGP. The objective of this study was to identify the effect of oligochitosan and L-arginine supplementation on broiler performance.

2. Material and method

2.1. Experimental design, diets, and management

Three hundred unsexed DOCs (day old chickens) broilers with initial average body weight of 46 g assigned randomly to 5 dietary treatment groups (10 birds per pen with 6 pens per treatment). Five diets prepared for study. Dietary treatments were (P0) basal diet (BD) without supplement (Table 1), (P1) BD + Antibiotic growth promoter (Zinc Bacitracine 200 mg kg-1) of diet, (P2) BD + COS 100 mg kg-1 of diet, (P3) BD + L-argine 1.9% of diet, (P4) BD + COS 100 mg kg-1 + L-arginine 1.9% of diet. The COS used in this study was a mixture of oligosaccharides with molecular weights between 1000 and 5000 Da. The diets were fed in a crumble form for starter phase and pellet form for finisher phase. All groups of broiler had free access to feed and drinking water. The broiler were raised on a brooder during day 1 to 14 and then transferred into floor pens during day 15 to 35.

Table 1. Composition of the basal diets for starter and finisher broilers

Feed ingredients	*Starter* (1-21 days)	*Finisher* (22-35 days)
Corn	60.92	65.66
Soybean meal	24.82	17.89
Palm oil	1.58	2.25
Meat Bone Meal	7.00	7.00
CGM	3.45	4.95
CaCO3	0.64	0.70
Salt	0.15	0.15
Methionine	0.24	0.19
Lysin	0.21	0.20
Premix	1.00	1.00
Amount (%)	100	100
Feed Nutrient Content :		
Crude Protein (%)	22.50	20.59
ME (kcal kg^{-1})	3100	3200
Calcium (%)	0.95	0.95
Available phosphorus (%)	0.42	0.41
Methionine (%)	0.55	0.49
Methionine + cystine (%)	0.86	0.78
Lysine (%)	1.40	1.20

The composition of kg^{-1} premix : Vitamin A 11 500 000 IU, Vitamin D3 4 500 000 IU, Vitamin E 75 g, Vitamin K3 3.5 g, Vitamin B1 3 g, Vitamin B2 8 g, Vitamin B6 5.5 g, Vitamin B12 30 mg, Niacin 65 g, Panthothenic Acid 15 g, Folic Acid 1.5 g, Biotin 250 mg, Choline chloride 1.5 mg, Vitamin C 50 g, Phytase 700 000 000 FYT, Antoxidant 0.5 g, and carrier added to 1 kg.

2.2. Broiler Chicken Performance
Chickens reared for 35-day trial (5 weeks), the amount of feed intake and body weight were weighed every week.

2.2.1 Feed intake (g head $^{-1}$)
Average feed intake is calculated from the difference between the rations given and the rest of the ration divided by the number of chickens in one plot. Measurement of residual feed is done once a week in the morning.

2.2.2 Final Body Weight (g head $^{-1}$)
The final body weight is obtained by weighing all chickens at each treatment at the end of maintenance then averaged.

2.2.3 Body Weight Gain (g head $^{-1}$)
Body weight gain is obtained from the calculation of the final body weight minus the initial body weight. Body weight is measured once a week.

2.2.4 Feed conversion ratio
Ration conversion is calculated from the ratio between the average feed intake and the average body weight gain.

2.2.5 Mortality (%)
Mortality is calculated from the ration between the dead chicken and the life chicken during the maintanance.

2.2.6 Statistical analysis
The experimental design used in this study was Completely Randomized Design (CRD) 5 treatments and 6 replications. The data was analyzed by analysis of variance (ANOVA), followd by Duncan's multiple range test [9], using SPSS software (SPSS® version 16.0) with α 5%.

3. Result and discussion

COS and L-arginine supplementation and their combinations to the broilers have a significant effect (P <0.05) increasing on feed intake, body weight, body weight gain and decreasing feed conversion ratio in the starter phase (Table 2). In the finisher phase the all treatments only had a significant effect (P <0.05) on body weight. Giving Chitosan Oligosaccharide 100 mg kg^{-1} in feed rations improved the performance of broilers with the highest yield compared to other treatments.

The addition of 100 mg kg^{-1} of Chitosan Oligosaccharide had a positive effect on increasing feed intake, BW, BWG, and significantly reduced FCR (P <0.05) in starter phase and had a positive effect on increasing feed intake, BW, BWG, and decreasing FCR until end of maintenance. These results are consistent with the observations of Huang et al. [2] reported that supplementation of 100 mg kg^{-1} of COS in broilers was able to increase nutrient digestibility in the ileum and daily body weight gain which was as effective as the addition of AGP. Moreover, COS can stimulate the secretion of digestive enzymes derived from the stomach, pancreas, and intestinal mucosa [10]. In order to, this effect is expected to reduce local inflammation in the intestinal mucosa, facilitate modification of complex molecules to be simpler, and improve the integrity of enterocytes, thereby supporting nutrient digestibility and absorption [11].

The administration of L-arginine supplementation up to 1.9% gave a positive effect on feed intake BW, BWG, and significantly reduced FCR (P <0.05) in the starter phase . But it has no effect on the finisher and cumulative phases . These results are consistent with the observations of Tan et al. [8] reported that administration of L-arginine in feed with a percentage above the requirement of the 1994 NRC standard was able to increase the body weight gain and decrease the feed conversion ratio in the Starter phase. L-arginine supplementation can reduce inflammation in chickens induced by lipopolysaccharide (LPS).

The combination treatment between COS 100 mg kg-1 and L-arginine to 1.9% (P4) gave a significant effect (P <0.05) to increase BW, BWG, and reduce FCR in the starter phase and reduce FCR in the cumulative phase compared to control feed (P0). These results indicate that giving COS 100 mg kg-1 effectively improves performance both in the starter , finisher and cumulative phases , while giving L-arginine to 1.9% is only effective in the starter phase.

Table 2. Performance of broiler chickens in maintenance of various growth phases

Variable	Treatment				
	P0	P1	P2	P3	P4
DOC Body Weight (g head^{-1})	46.27	46.13	47.47	46.92	47.68
Starter (1-21 days)					
Feed intake (g head^{-1})	1003. 80^{c}	1036.26^{bc}	1102.97^{a}	1063.39^{ab}	1041.42^{bc}
Body weight (g head^{-1})	763.34^{c}	815.33^{b}	867.75^{a}	850.91^{ab}	817.67^{b}
Body weight gain (g head^{-1})	717.08^{c}	769.20^{b}	820.28^{a}	803,99^{ab}	769.98^{b}
FCR (FI / BWG)	1.4^{a}	1.35^{b}	1.34^{b}	1.32^{b}	1.35^{b}
Finisher (22-35 days)					
Feed intake (g head^{-1})	1710.29^{ab}	1741.59^{ab}	1798.7^{a}	1633.73^{b}	1668.69^{b}
Body weight (g head^{-1})	1647.28^{b}	1752.75^{ab}	1811.33^{a}	1715.29^{ab}	1740.23^{ab}
Body weight gain (g head^{-1})	883.94	937.42	943.58	864.38	922.56
FCR (FI / BWG)	1.93	1.86	1.91	1.90	1.81
Cumulative (0-35 days)					
Feed intake (g head^{-1})	2714.09^{b}	2777.85^{ab}	2901.67^{a}	2697.12^{b}	2710.11^{b}
Body weight (g head^{-1})	1647.28^{b}	1752.75^{ab}	1811.33^{a}	1715.29^{ab}	1740.23^{ab}
Body weight gain (g head^{-1})	1601.01^{b}	1706.62^{ab}	1763.87^{a}	1668.37^{ab}	1692.54^{ab}
FCR (FI / BWG)	1.70^{a}	1.63^{b}	1.64^{b}	1.62^{b}	1.60^{b}
Carcass weight (g head^{-1})	1151.33^{b}	1333.67^{a}	1312.00^{a}	1323.67^{a}	1231.00^{ab}
Mortality (%)	12	0	0	3	6

FI, feed intake; BWG, body weight gain; FCR, feed conversion ratio.
^{a,b,c}superscripts in the same row are significantly different at $P < 0.05$.

4. Conclusion

Dietary addition of COS 100 mg kg^{-1} is able to use as alternative for antibiotic growth promotor replacement without giving effect on broiler performance.

5. Acknowledgements

The authors would like to thank LPDP (Indonesia Fund Endowment for Education) for the financial support.

6. Reference

[1] Wang X W, Du Y G, Bai X F, Li S G 2003 The effect of oligochitosan on broiler gut flora, microvilli density, immune function and growth performance *Acta Zoonutr. Sin.* **15** 32–45

[2] Huang R L, Yin Y L, Wu G Y, Zhang Y G, Li T J, Li L L 2005 Effect of dietary oligochitosan supplementation on ileal digestibility of nutrients and performance in broilers *Poult. Sci.* **84**1383–1388

[3] Davis M E, Maxwell C V, Brown D C, de Rodas B Z, Johnson Z B, Kegley E B 2002 Effect of dietary mannan oligosaccharides and (or) pharmacological additions of copper sulfate on growth performance and immunocompetence of weanling and growing/finishing pigs *J. AnimSci.* **80** 2887–2894

[4] Schley P D and Field C J 2002 The immune-enhancing effects of dietary fibres and prebiotics *Br J. Nutr.* **87** S221–S230

IOP Conf. Series: Earth and Environmental Science **251** (2019) 012060 doi:10.1088/1755-1315/251/1/012060

[5] Shi H P, Efron D T, Most D 2000 Supplemental dietary arginine enhances wound healing in normal but not inducible nitric oxide synthase knockout mice *Surgery* **128** 374–378

[6] Alvarez M N, Peluffo G, Piacenza L 2011 Intraphagosomal peroxynitrite as a macrophage-derived cytotoxin against internalized Trypanosoma cruzi: consequences for oxidative killing and role of microbial peroxiredoxins in infectivity *J. Biol. Chem.* **286** 6627–6640

[7] Han J, Liu Y, Fan W 2009 Dietary L-arginine supplementation alleviates immunosuppression induced by cyclophosphamide in weaned pigs *Amino Acids* **37** 643–651

[8] Tan J, Liu S, Guo Y 2014 Dietary L-arginine supplementation attenuates lipopolysaccharide-induced inflammatory response in broiler chickens *Br. J. Nutr.* **111** 1394–1404

[9] Steel R G D, Torrie J H 1995 *Prinsip dan Prosedur Statistika. Terjemahan: B. Sumantri* (Jakarta : Gramedia)

[10] Hou Q L and Gao Q S 2001 *Chitosan and Medicine* (Shanghai : Shanghai Science Technology Press)

[11] Wu G 1998 Intestinal mucosal amino acid catabolism *J. Nutr.* **128** 1249–1252

A study on antioxidative and antimicrobial activities of saliva extract of Indonesian local leeches

B Malik[1,2,5], D A Astuti[3], D J F Arief[3] and M Rahminiwati[4]

[1] Postgraduate Student at Bogor Agricultural University, Bogor, Indonesia
[2] Faculty of Agriculture, Djuanda University, Bogor, Indonesia
[3] Faculty of Animal Science, Bogor Agricultural University, Bogor, Indonesia
[4] Faculty of Veterinary Medicine, Bogor Agricultural University, Bogor, Indonesia

[5] Corresponding Author: burhanudinmalik @gmail.com

Abstract. Antioxidative and antimicrobial activities of saliva extract of Indonesian leeches were assessed. Leeches were obtained from Situgede Bogor, Sawangan Depok, and Sijunjung Regency, West Sumatra. A completely randomized design with three treatments and five replicates was used. Treatments consisted of three places of leech origins. Measurements were taken on the amount, total protein content, IC50 value, zone of inhibition to bacterial growth, and minimum inhibitory concentration of crude leech saliva extract (CLSE). Results showed that the amount of CLSE obtained was 3.696-4.633g/g BW and total protein content of CLSE was 48.303-64.742 µg/mg. The IC50 value was 47-108 µg/mg while that of ascorbic acid was 3.312 µg/mg. Antimicrobial activity was found to the growth of S. aureus but not to the growth of E. coli with a minimum inhibitory concentration at 2.472-4.528 µg/ml.

1. Introduction

Leech, a freshwater amphibian annelid, commonly has dark-colored body of 10 cm long. There are more than 600 species of leeches belonging to the genus *Hirudo*. Traditionally, leeches have been known and used since ancient times in India and China. This leech medicinal therapy developed well in the Arabian and European regions [1]. Leeching has regained its popularity since early twentieth century. Today, studies on leeches of *Hirudo medicinalis* and *Hirudo verbana* and the active substances in their saliva are extensively done in the USA and Europe. Active substances of leech saliva and leeching have been used in medicinal field. In Asia, *Hirudinaria manillensis* is the species commonly used in leeching and studies [2].

In Indonesia, leeching practices are spreading throughout the country. However, scientific studies on the potential of leech and its active substances for human and animal health is very limited. Meanwhile, recent studies in Malaysia revealed that saliva extract of Malaysian leeches had anticoagulative [3,4], antioxidative [5], antimicrobial [6], and anticancer [7] activities. These biological activities found in leech saliva extract are needed in control of various diseases including mastitis in dairy animals. Therefore, studies on biological activities of saliva extract of Indonesian local leeches deserve to be done. This study was conducted to assess the antioxidative and antimicrobial activities of saliva extract of leeches originating from some places in Indonesia.

2. Materials and methods

2.1. Site of study

The study was done at the Biochemistry, Physiology, and Microbiology of Animal Nutrition Laboratory, Faculty of Animal Science, Bogor Agricultural University.

2.2. Leeches
Leeches were taken from three places, namely Situgede, Bogor where leeches were obtained from a river, Sawangan, Depok where leeches were obtained from a leech farm, and Sijunjung Regency, West Sumatra where leeches were obtained from a swamp area. Situgede leeches usually fed themselves by sucking blood from buffaloes which were commonly found to be bathed by their owners in the afternoon. Sawangan leeches were reared for the purpose of leeching practices. They were fed artificial feed. From each place, 1000 leeches were obtained and put in plastic ponds in the laboratory. They were left unfed for about two weeks before they were used as samples.

2.3. Leech saliva extraction
Leech saliva was extracted by using a method used by Abdualkader *et al.* [6] without killing the leech. Leeches were fed a phagostimulatory solution made of a mixture of 0.001 M arginin and 0.15 M sodium chloride solution which was heated at 37°C [8]. This solution was placed in an inverted funnel coated with a parafilm sheet. Leeches were allowed to suck the solution through the parafilm until they were satiated and detached themselves from the parafilm (Figure 1a). The weight of leeches before and after sucking the phagostimulatory solution was recorded.

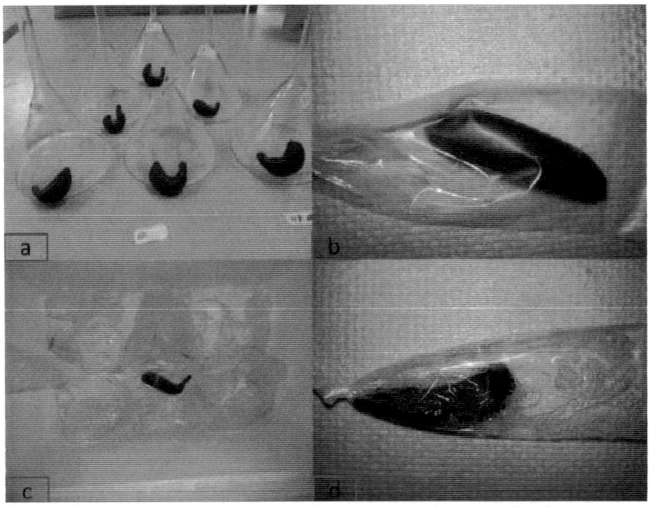

Figure 1. Leech saliva extraction method

A satiated leech which detached itself from the parafilm was then placed in a small plastic bag (Figure 1b). The plastic bag with a leech in it was closely tied and put in a box containing ice cubes for 15-20 minutes until the leech got unconscious and vomitted all the solution it sucked (Figure 1c). The leech body was then carefully squeezed from posterior to anterior to take all the remaining sucked solution out (Figure 1d). Only clear uncolored liquid was collected and put in plastic containers. The milked leech was then put into warm water (37°C) for 25-20 minutes until it got its consciousness back.

Collected liquid was centrifuged at 4°C, 9000 rpm for 10 minutes. Supernatant obtained after centrifugation was refered to as crude leech saliva extract (CLSE). Only fresh CLSE was used in this study.

2.4. Estimation of total protein
Total protein content of CLSE was estimated by using a Bradford protein assay method using *bovine serum albumin* (BSA) as a standard [9]. Phagostimulatory solution was used a blank. A series of BSA standard and CLSE solutions were prepared. Absorbances (A595) of these solutions were measured with the existence of Bradford reagent.

2.5. Antioxidative activity

Antioxidative activity of CLSE was assessed by using a DPPH method [10] as used by [5]. In this antioxidative test, 1,1-diphenyl-2-picrylhydrazyl. (DPPH) was used as a free radical. Using this method, antioxidative activity of CLSE was measured based on its ability to catch hydrogen released from DPPH.

CLSE was diluted in methanolic solvent and the solution was referred to as 3 times CLSE concentration (3 CLSE). Three two-fold serial dilutions of methanolic CLSE were prepared first by dissolving 100 µL CLSE in 100 µL methanol. 100 µL of this methanolic CLSE solution was then dissolved in another 100 µL methanol. The same step was repeated once again until methanolic CLSE solutions with concentrations of 1:2, 1:4, and 1:8 were obtained. Then, into each tube methanol solution was added until a final volume of 300 µl was reached. Simultaneously, DPPH was diluted in methanol to make 0.002 M DPPH methanolic solution. 15 µl DPPH methanolic solution was added into each tube and after 15 minutes, A_{516} absorbance reading was done. Phagostimulatory solution was used as a negative control. As a control reading, 15 µl DPPH was diluted in 300 µl methanol and its A_{516} absorbance reading was done. The same procedures were used to prepare three two-fold serial dilutions of methanolic L-ascorbic acid (50 µg/ml) as positive control. Antioxidative activity was calculated by using the following formulation.

$$\%\ \text{antioxidative activity} = \frac{Control\ absorbance - Sample\ absorbance}{Control\ absorbance} \times 100$$

The concentration of ascorbic acid and CLSE protein (µg/ml) needed to inhibit free radical activity of DPPH 50% (IC50) was estimated from the curve obtained by plotting % antioxidative activity to concentration.

2.6. Antimicrobial activity

Antimicrobial activity of CLSE was assessed by using a disc diffusion test and minimum inhibitory concentration test methods as described by [11].

2.6.1. Disc diffusion test

In this test, gram positive and gram negative bacteria were used. These gram positive and gram negative bacteria were *Staphylococcus aureus* (B2647) and *Escherichia coli* (B2743), respectively. Inocula of these bacteria were obtained from Bacteriology Laboratory of Indonesian Research Centre for Veterinary Science (IRCVS), Bogor. Tetracycline antibiotic (30 µg) and phagostimulatory solution were used as positive and negative controls.

Inoculum was prepared in a tube filled with 5 ml saline solution. The turbidity of the suspension was adjusted to McFarland standard 3.0 (9 x 10^8 CFU/ ml). This adjustment was done by placing the tubes containing inoculum and McFarland standard in front of a piece of white paper with black lines on it. Inoculum was used immediately before 18-24 hours after it was prepared.

Bacterial inoculum of 50 µl was added to a Mueller-Hinton agar (MHA) plate. The plate was left at room temperature so that the excess water favor could be absorbed by the agar medium. This was done by placing the plate in an incubator and the cover of the plate was left ajar. MHA should be 4 mm thick. After the medium was solidified, four wells with diameter of 0.5 cm were made. CLSE solution was injected into two wells and tetracycline and phagostimulatory solution were injected into the other two wells. The plate was then incubated at 37 °C for 18-24 hours. Zone of inhibition formed around each well was measured by using a calipher.

2.6.2. Minimum inhibitory concentration test.

Minimum inhibitory concentration test was performed to determine minimum concentration of CLSE which could inhibit the growth of bacterium isolate. The test was performed by referring to the procedures described by [11].

A two-fold serial dilutions of CLSE solution was prepared in 96-well microtiter plates. Mueller-Hinton Broth of 100 µl was placed into eight wells from the first to the eleventh columns of 96-well microtiter plate. CLSE solution of 100 µl was added into ten (CLSE of each treatment was added into two adjacent wells) wells containing the broth in the first row. Dilution was done by vertically transfering 100 µl aliquot from ten wells in the first row to ten wells in the second row. Similar procedures were applied for the six wells in row three to the last row. Solution containing bacterial inoculum of 10 µl (10^7 *CFU*/ml) was added into each well in 10 columns. Wells in the eleventh column was left without bacterial inoculum and used as control. Microtiter plates were covered and incubated at 37ºC for 24 hours.

After 24 hour incubation, the plates were taken out from the incubator. No contamination should be found in the purity plate and there should be a minimal growth of 2 mm in the positive control well and negative control well should be clean. Afterall, minimum inhibitory concentration of CLSE could be determined.

2.7. Statistical analysis

Measurement on antioxudative and antimicrobial activities were done to Indonesian local leeches originating from three places, namely 1) Situgede Village, Bogor, where leeches were obtained from a river, 2) Nagari Kampung Dalam, Lubuk Tarok District, Sijunjung Regency, West Sumatra where leeches were obtained from a swamp area, and 3) Bedahan Village, Sawangan District, Depok where leeches were obtained from a leech farm.

A completely randomized design was used. Places of leech origins were used as treatments and five replicates were allocated into each treatment. Data were subjected to an analysis of variance and a Duncan test by referring to the procedures of [12].

3. Results and discussion

3.1. Extraction of leech saliva and estimation of its total protein

One-hundred-fifty leeches with an average initial body weight (BW) of 1.24 ± 0.62 g/head were used. The amount of CLSE obtained and its protein content are listed in Table 1. No difference was found (P>0.05) in the amount of CLSE obtained. No difference was either found in total protein contents of CLSE.

Table 1. Amount and total protein content of CLSE (±SEM)

Leech Origin	Amount (g/g BW)	Total Protein (µg/g)
Situgede	3.854 ± 0.465	48.303 ± 4.838
Sawangan	4.633 ± 0.371	49.439 ± 7.515
West Sumatra	3.696 ± 0.458	64.742 ± 6.017

Note: P>0.05, SEM: Standard Error of Mean

Before their saliva was milked, leeches were allowed to suck phagostimulatory solution until they were satiated and voluntarily detached themselves from the parafilm. Leeches were then made unconscious and vomit the solution they sucked. The vomitted solution was mixed with saliva. This was why, as listed in Table 1, the amount of CLSE obtained was higher than their initial BW. This was in line with the finding that in nature, a leech might be able to suck blood in the amount equals to 10 times of its initial BW [13].

Total protein contents of CLSE (48.303-64.742 µg/mg) obtained in this study were lower than those obtained by [6] (62.602-79.962 µg/ml) and [5] (78.753 µg/ml). These researchers used leeches from a river in Kuantan and a natural lake in Trengganu, Malaysia. Meanwhile, leeches from Pahang region [14] were found to have saliva extract with total protein range of 39-105 µg/ml with an average of 67.918 µg/ml.

3.2. Antioxidative activity

Antioxidative activities of CLSE in this study is listed in Table 2. IC50 (Inhibitory Concentration 50) defined as the concentration of tested substance causing the loss of DPPH activity by 50% was used as a parameter. The loss of DPPH activity as a free radical occurs as the tested substance donates a hydrogen (H) atom to DPPH [15] as illustrated in Figure 2.

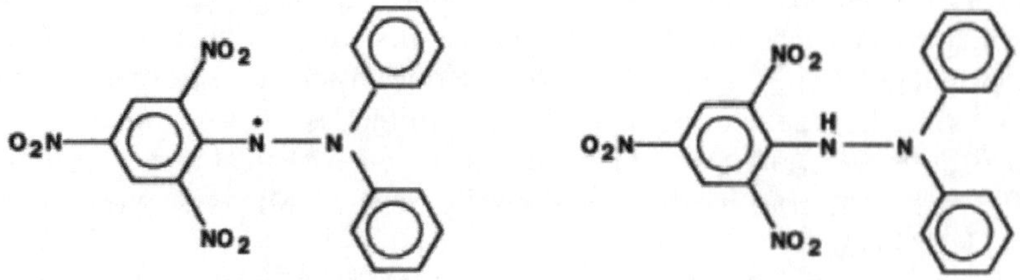

Figure 2. Molecular structure of DPPH as free radical (left) and non radical (right)

Table 2. IC50 L-ascorbic acid and CLSE

Substrate	IC50 (μg/g) ± SEM
Situgede CLSE	108.048 ± 0.465
Sawangan CLSE	47.425 ± 10.252
Sumatera Barat CLSE	88.437 ± 5.413
L-ascorbic acid	3.312

Note: P>0.05, SEM: Standard Error of Mean

Results in this study showed that CLSE obtained from leeches of several places in Indonesia had some antioxidative activities. There was no different IC50 values of CLSE. However, there was a tendency ($P<0.10$) that Sawangan leeches produced CLSE with higher antioxidative capacity (47.425 μg/mg) than those from West Sumatra (88.437 μg/mg) and Situgede (108.048 μg/mg).

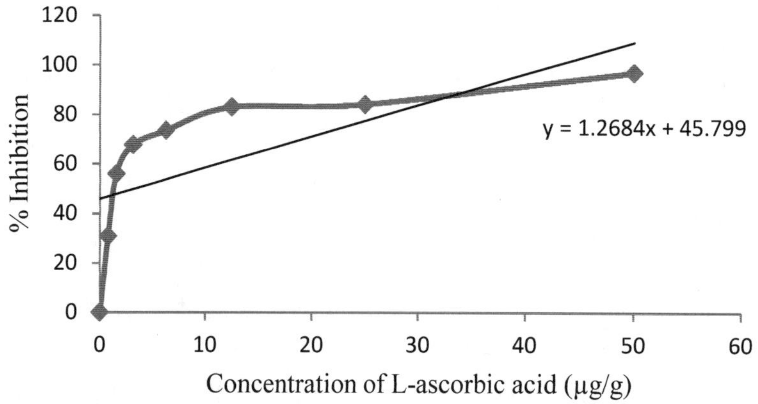

Figure 3. Curve of DPPH inhibition by L-ascorbic acid (N=8)

Vitamin C (L-ascorbic acid) was used as a positive control as this acid is a very strong alternative antioxidant. The IC50 value of ascorbic acid was needed as a comparation to the IC50 values obtained from CLSE to determine the strength of CLSE antioxidative capacity. The curve of DPPH activity inhibition is depicted in Figure 3. The regression equation obtained was used to determine the percentage of inhibition of the tested CLSE samples. Results showed that the antioxidative power of CLSE in this study was very low compared to that of the ascorbic acid. With IC50 value ranges of CLSE of 47-108 μg/mg and ascorbic acid of 3.312 μg/mg, the antioxidative activity capacities of CLSE were 3.07, 3.75, and 6.98% of that of the ascorbic acid for CLSEs of Situgede, Sumatera Barat, and Sawangan leeches, respectively.

Antioxidative activity of leech saliva extract was first reported by Malaysian researchers [5]. The CLSE they got from Malaysian local leeches had an IC50 value of 7.282 μg/ml and the IC50 value of the ascorbic acid they used was 5.803 μg/ml. A current study [16] revealed that saliva extract of

Nigerian leech, *Aliolimnatis michaelseni*, also had a high antioxidative capacity. The IC50 value of saliva extract of this leech was 8.169-8.670 µg/ml compared to 5.025 µg/ml of the ascorbic acid they used. These findings meant that saliva extract of Malaysian and Nigerian leeches respectively had about 80 and 58-62% the antioxidative capacity of ascorbic acid they used.

3.3. Antimicrobial activity

Results of disk diffusion test are listed in Table 3. There was no difference found in the inhibition zones of CLSE samples to the growth of *S. aureus*. No inhibition zone was detected to the growth of *E.coli*. These results were in contrast to the findings of [6] with their Malaysian leeches. They used fresh and lyophilized saliva extract obtained from leeches which were fasted for 14, 23, and 25 weeks. They found that only fresh saliva extract obtained from leeches fasted for 14 weeks had antimicrobial activity toward *E.coli* as indicated by an inhibition zone of 25 mm. In this study, fresh CLSE was not found to have an antimicrobial activity to the growth of *E. coli*.

Table 3. Antimicrobial activity of CLSE

Samples	Inhibition Zone (mm) ± SEM	
	*S.aureus**	*E.coli*
Situgede CLSE	2.54 ± 0.44	0
Sawangan CLSE	1.51 ± 0.41	0
West Sumatra CLSE	2.70 ± 0.76	0
Tetracycline (30 µg)	13.82	6.75
Phagostimulatory solution	0	0

Note: *: P>0.05, SEM: *Standard Error of Mean*

In this study, CLSE was found to have an antimicrobial activity to the growth of *S. aureus*. This was different from what was found in the study of Malaysian researchers. The fresh leech saliva extract they used did not show any inhibition zone to the growth of these bacteria. This difference might be caused by the different conditions of the environment from where the leeches originated.

Table 4. Minimum inhibitory concentration (MIC) of CLSE

Samples	MIC (µg/ml) ± SEM
Situgede CLSE	4.528^{A} ± 0.774
Sawangan CLSE	2.472^{B} ± 0.376
West Sumatra CLSE	2.804^{B} ± 0.189

Note: Different superscripts indicate significant difference (P<0.05), SEM: *Standard Error of Mean*

Results of minimum inhibitory concentration (MIC) test are listed in Table 4. The test was done only on S. aureus inoculum as no antimicrobial activity was detected to the growth of *E. coli*. Results showed that CLSE obtained from Situgede leeches had the highest MIC. This indicated that CLSE obtained from Sawangan and West Sumatra leeches was stronger than that obtained from Situgede leeches in inhibiting the growth of *S. aureus*. However, the MIC figures found in this study were much lower than that (78.253 µg/ml) obtained by [6] from lyophilized leech saliva extract. This difference was suspected to be caused by different concentration of microbe inoculum used in the two studies. In both studies, concentration of microbe inoculum was determined by using an estimate according to MacFarland 3.0 standard (9 x 10^8 CFU/ml).

4. Conclusions

CLSE of Indonesian local leeches had some antioxidative and antimicrobial activities The IC value was 47-108 µg/mg. Antimicrobial activity was found to the growth of *S. aureus* but not to the growth of *E. coli* with a minimum inhibitory concentration at 2.472-4.528 µg/ml.

A furher study is required to assess the effects of treatments on leeches before they are milked for their saliva. These treatments may be in the forms of leech feeding, leech starving before saliva extraction, and other kinds of treatment that might have effect on the quality of leech saliva extract, especially on its biological activities.

5. Acknowledgement

This study was funded by *Hibah Disertasi Doktor* 2015 scheme of Directorate General of Higher Education (*Dikti*) Ministry of Research and Higher Education Republic of Indonesia.

6. References

[1] Wittke-Michalsen E 2007 *The history of leech therapy* in *Medicinal Leech Therapy* (Stuttgart : Thieme) 4-11

[2] Elliot J M and Kutschera U 2011 Medicinal leeches: historical use, ecology, genetics, and conservation *Freshwater Reviews* **4** 21-41

[3] Zulhisyam A K, Ismail A A and Omar I C 2010 Preliminary study on leech crude extract as an anti-coagulant agent *Proceedings of the 9th UMT International Symposium on Sustainability Science and Management (UMTAS) 2010* 8-11 May 2010

[4] Ghawi A M, Abdualkader A M, Merzouk A and Alaama M 2012 Season variation and starvation period influence on the antithrombotic activity of leech saliva extract from the medicinal Malaysian leech *Hirudinaria manillensis Journal of Bioequivalence and Bioavailability* **S14**

[5] Ghawi A M, Abdualkader A M, Merzouk A and Alaama M 2012 Free radical scavenging activity of the medical Malaysian leech saliva extract, *Hirudinaria manillensis Journal of Bioequivalence and Bioavailability* **S14**

[6] Abdualkader A M, Merzouk A, Ghawi A M and Alaama M 2011 Some biological activities of Malaysian leech saliva extract *IIUM Engineering Journal* **12** (4) 1-9

[7] Merzouk A, Ghawi A M, Abdualkader A M, Abdullahi A D and Alaama M 2012 Anticancer effects of medical Malaysian leech saliva extract (LSE) *Pharmaceutica Analytica Acta* **S15**

[8] Rigbi M, Levy H, Iraqi F, Teitelbaum M, Orevi M, Alajoutsijärvi A, Horovitz A and Galun R 1987 The saliva of the medicinal leech Hirudo medicinalis I Biochemical characterization of the high molecular weight fraction *Comparative Biochemistry and Physiology B.* **87** (3) 567-73

[9] Bradford M M 1976 A rapid and sensitive method for the quantification of microgram quantities of protein utilizing the principle of protein-dye binding *Analytical Biochemistry* **72** 248-254

[10] Blois M S 1958 Antioxidant determinations by the use of a stable free radical *Nature* **181** 1199-1200

[11] Coyle M B 2005 Manual of Antimicrobial Succeptibility Testing *American Society for Microbiology*

[12] Gomez K A and Gomez A A 1995 *Prosedur Statistik untuk Penelitian Pertanian* Edisi Kedua (Jakarta : Universitas Indonesia Press)

[13] Roth M 2007 *The biology of leeches* in *Medicinal Leech Therapy* (Stuttgart : Thieme)

[14] Alaama M, Alnajjar M, Abdualkader A M, Mohammad A and Merzouk A 2011 Isolation and analytical characterization of local Malaysian leech saliva extracts *IIUM Engineering Journal* **12** (4) 51-59

[15] Molyneux P 2004 The use of the stable free radical diphenylpicrylhydrazyl (DPPH) for estimating antioxidant activity *Songklanakarin Journal of Science Technology* **26 (2)** 211-219

[16] Omalu I C J, Egwin, E C, Mgbemena C C, Eke S S, Ubanwa D, Busari M B and Ossai P C 2015 Free radical scavenging activity of the Nigerian leech (*Aliolimnatis michaelseni*) saliva extract *British Journal of Pharmaceutical Research* **8** (4) 1-6

IOP Conf. Series: Earth and Environmental Science **251** (2019) 012062 doi:10.1088/1755-1315/251/1/012062

Synergistic effect between *Lactobacillus plantarum* AKK30 and *Saccharomyces cerevisiae* B18 and the probiotic properties of microencapsulated cultures

L Istiqomah[1], E Damayanti[1], A E Suryani[1], M F Karimy[1], D Maknuna[2], Lutfiani[3]

[1]Research Unit for Natural Products Technology, Indonesian Institute of Sciences, (BPTBA-LIPI). Jl. Jogja-Wonosari Km. 31.5, Gading, Playen, Gunungkidul, D.I. Yogyakarta 55861, Indonesia
[2]Faculty of Science and Technology, Maulana Malik Ibrahim State Islamic University, Malang Jalan Gajayana No. 50 Malang 65144, Indonesia
[3]Faculty of Mathematics and Science, Lambung Mangkurat University. Jl. Ahmad Yani Km 36, Loktabat Selatan, Banjarbaru Selatan, Banjarbaru, Kalimantan Selatan 70714, Indonesia

Email: lusty.istiqomah@gmail.com

Abstract. The objectives of this study were to observe the microbial interactions between lactic acid bacteria (LAB - *Lactobacillus plantarum* AKK30) and yeast (*S. cerevisiae* B18) whether their effect is antagonistic or synergistic and to determine the potential of microencapsulated culture as probiotics for animal feed additive. The cell morphology of isolates were screened by macroscopic and microscopic examination. The synergistic effect of LAB and yeast was studied by disk diffusion and dilution method. The microencapsulated culture of LAB and yeast was prepared by spray drying method. Probiotic properties of microencapsulated cultures were examined for its viability on low pH, gastric juice and bile salt *in vitro*. The result of morphology cell showed that *L. plantarum* AKK30 had bacilli shaped with average length of cell was 1.35 µm at 48 h of incubation, whereas *S. cerevisiae* B18 had small budding characteristics with average length of cell was 3.19 µm at 96 h of incubation. The synergistic assay showed that *L. plantarum* AKK30 and *S. cerevisiae* B18 were alive and grew in a commensal relationship with diameter of clearing zone less than 5 mm and the population between two microorganisms was only 1 cycle log difference. Viability of microencapsulated culture was 97.25% for *L. plantarum* AKK30 and 91.66% for *S. cerevisiae* B18. The microencapsulated culture of *L. plantarum* AKK30 and *S. cerevisiae* B18 could be applied as probiotic for animal with good viability on pH 2 (95.18 and 87.95% respectively) and pH 3 (88.63 and 96.84% respectively) at 45 min of incubation, viability on gastric juice (66.58 and 92.70% respectively) at 45 min of incubation, and viability on 0.3% of bile salt (99.11 and 94.09% respectively) at 120 min of incubation. Thus, it could be concluded that LAB and yeast had synergistic effect and potentially use as a probiotic.

1. Introduction

Probiotics are defined as "live microorganisms (bacteria or yeasts), which when administered in adequate amount confer a health benefit on the host" [1]. Lactic acid bacteria (LAB) and yeast are potentially use as a probiotic agent due to its ability to survive in a variety of limiting conditions along the digestive tract, such as low pH of the stomach and the presence of bile salts in the duodenum [2,3], colonize in the digestive tract, and inhibit the growth of pathogenic bacteria. Several potential strains of LAB include Bifidobacterium, *L. reuteri*, and *L. casei* [4], while Sourabh *et al.* [5] reported that *S. cerevisiae* HM535662 had good tolerance at pH 2. This microbial survival at low pH may be due to various factors such as cell size, cell wall, pH composition, and variations in microbial strains [6-8]. The viable counts of probiotics for consumption are expected reach minimum acceptable concentration of 10^6 cfu/g or mL [9].

The antagonistic or synergistic effect is one of the most important properties in system that different microorganisms are living together [10]. The synergistic effect should be promoted, whereas the antagonistic effect is prevented to provide effective culture for probiotic purpose. Basically, different microorganisms between either different genera or species could be applied as a co-culture in the form of probiotics if they could live together and had a synergistic effect on each other. Commonly, microorganisms grown together in harsh conditions owing their resistivity and survival to the synergism between the microbial strains. Correlatively, the increase of microbial growth rate is determined in co-culture systems compared with the growth rate they experience alone in the medium [11].

Furthermore, both isolates were prepare in microencapsulated culture by spray drying method to maintain its viability for probiotic purpose. The viability was determine by comparing the cell colony before and after spray drying [12]. Viability assay of isolates was needed to determine the cell ability to grow and encapsulan ability to maintain the live cell of probiotic [13].

The objectives of this study were to determine the potential of LAB and yeast as probiotics candidate for animal feed additive and to observe the microbial interactions between LAB and yeast strains whether their effect is antagonistic or synergistic.

2. Materials and methods

2.1. Morphology observation of isolates

In this study, we used *L. plantarum* AKK30 and *S. cerevisiae* B18 with cholesterol-lowering activity that has been isolated from gastrointestinal tract (GIT) of Indonesian native chicken and Javanase duck [14,15]. *Lactobacillus plantarum* AKK30 was grown in de Man Rogosa Sharpe (MRS, Oxoid) at 37 °C for 24 h, whereas *S. cerevisiae* B18 was grown on Yeast Extract Glucose agar medium [(g/l) yeast extract, 5; glucose, 20; agar, 12.0; pH 7.0 ± 0.2] at 30 °C for 48 h. The cell morphology of isolates was screened by macroscopic and microscopic examination. Microstructure observation of *L. plantarum* AKK30 and *S. cerevisiae* B18 using scanning electron microscopy (SEM, Hitachi SU3500) were carried out by growing LAB on MRS Agar medium and incubated at 37 °C for 48 h, whereas yeast inoculated on YG Agar medium then incubated at 30 °C for 96 h. Each colony of LAB and yeast was taken with an ose needle then swab onto each carbon tape and dried, then coated with ion sputter with a setting of 10 mA for 60 seconds with magnification 20.000x and vacc (accelerating voltage) 10 kV for bacteria, whereas ion sputter with a setting of 10 mA for 50 seconds with magnification 10.000x and vacc. 5 kV for yeast.

2.2. Assay for antagonistic effect between LAB and yeast cultures

Antibacterial and antifungal effects were screened of *L. plantarum* AKK30 and *S. cerevisiae* B18 to each other. Experiments were performed by modified disk diffusion method according to Çakır [16]. Briefly, *S. cerevisiae* B18 was inoculated into YG Agar by spread over the whole surface of the medium as a thin film. On the other hand, a sterilized 6 mm diameter blank disk (Oxoid) was put on the YG agar. *Lactobacillus plantarum* AKK30 was grown in MRS Broth then centrifuged at 5000 rpm for 10 min and supernatants were transferred to another sterile tubes. Supernatants were treated into two groups. The first group was filtered using syringe filters (0.45 µm pore size), whereas the second group was not filtered. Blank disk that placed on the *S. cerevisiae* B18 film was loaded with 20 µL of filtered/non-filtered supernatants obtained from *L. plantarum* AKK30 then incubated at 37°C for 24 h. The inhibition of *L. plantarum* AKK30 growth was evaluated by measuring the diameter of the transparent inhibition zone around disk. This procedure was conducted for another group of petri dishes containing MRS agar spread with *L. plantarum* AKK30 by loading blank disks with filtered/non-filtered supernatants (20 µL) obtained from *S. cerevisiae* B18. Inhibition was scored positive if the width of the clear zone around the disc of the antimicrobial compound producer strain was 5.0 mm or larger.

2.3. Assay for synergistic effect between LAB and yeast cultures
The synergistic assay were performed for *L. plantarum* AKK30 and *S. cerevisiae* B18 according to modified dilution (the plate count) method on medium described by Korukluoglu *et al.* [10]. Two microorganism together (*L. plantarum* AKK30 and *S. cerevisiae* B18) were inoculated (2.5%, v/v respectively) into tube containing Tryptic Soy (TS) Broth. The tubes, including TS Broth and test microorganisms were incubated at 30 °C for 0 and 24 hours then after that, test microorganism was spread to the MRS and YG Agar by surface plate method to count the viable cell amounts. The counting was carried out to determine the synergistic effect between *L. plantarum* AKK30 and *S. cerevisiae* B18 cultures.

2.4. Microencapsulated culture preparation of LAB and yeast
Microencapsulated culture of *L. plantarum* AKK30 was prepared by encapsulation method using spray dryer (Buchi) according to Barbosa-Canovas [17]. Skim solution containing 20% of skim in distilled water (w/v) was prepared and sterilized (110 °C for 10 min). LAB isolate was inoculated in MRS Broth medium at 37 ºC for 18 h. The culture was centrifuged at 4500 rpm for 10 min, then the biomass was mixed with skim solution. The solution was homogenized with a magnetic stirrer during spray drying process. Spray dryer operating conditions as follows: inlet air temperature 110 °C, outlet air temperature 63 °C, and the speed of pump 3. Colony of LAB microencapsulated culture obtained from spray dryer was count by the total plate count (TPC) method.

Encapsulation of *S. cerevisiae* B18 was performed using spray dryer (Buchi) refers to Chandralekha *et al.* [18] method. Yeast isolate was inoculated in YG Broth medium at 30 ºC for 24 h. The culture was centrifuged at 4500 rpm for 10 min, then the biomass was mixed with sterile maltodextrin solution containing 20% of maltodextrin in distilled water (w/v) and stirred with a magnetic stirrer during spray drying process. The operating conditions were: inlet air temperature 105 °C, outlet air temperature 66 °C, air pressure (2.5 psi), air flow rate (0.50 m^3/min), flow rate (41 ml/min), and the speed of pump 3. Colony of yeast microencapsulated culture obtained from spray dryer was count by the total plate count (TPC) method.

2.5. Assay for probiotic properties of LAB and yeast
Microbial probiotic could applied in animal accordance to the feed retention time in each GIT. Taheri *et al.* [19] reported that retention time of feed were varied, start from crop (30 min; pH 4.5), proventriculus (15 min; pH 4.4), gizzard (90 min; pH 2.6), intestine (90 min; pH 6.2), and colon (15 min; pH 6.3). The microencapsulated culture of *L. plantarum* AKK30 and *S. cerevisiae* B18 were assayed as a probiotic candidate based on its ability to survive in acidic pH condition, gastric juice, and bile salt refers to Sabikhi *et al.* [20] method with slight modification on the incubation time according to retention time of feed in GIT.

Acid tolerance assay was observed by mix 0.5 g of the microencapsulated culture with 1 mL sterile Phosphate Buffer Saline (PBS) then 10 µL of the mixture was inoculated in 990 µL of PBS at pH 2 and 3 (pH adjusted with 1 M HCl) respectively then homogenized and incubated at 37 ºC (*L. plantarum* AKK30) and 30 °C (*S. cerevisiae* B18) for 0 and 45 min. Serial dilutions of samples was performed in PBS then each culture inoculated on MRS/YG Agar and incubated for 48 h. Each cell colony was calculated by the total plate count (TPC) method.

Gastric juice tolerance was performed by prepare the artificial gastric juice from pepsin (Sigma, 3 g/L) solution at pH 2. Microencapsulated culture of *L. plantarum* AKK30 (0.5 g) was suspended with 1 mL sterile PBS. A total of 0.2 ml suspension was taken and mixed with 1 ml of sterile PBS containing artificial gastric juice. The mixture liquid was homogenized and incubated at 37 ºC (*L. plantarum* AKK30) and 30 °C (*S. cerevisiae* B18) for 0 and 45 min. Serial dilutions of samples was made with sterile PBS and then each culture inoculated on MRS/YG Agar and incubated for 48 h for cell colony observation.

Bile salt tolerance was determined by mix 0.5 g of the microencapsulated culture with 1 mL sterile PBS then homogenized. A total 0.2 ml each pellet suspension was suspended in 1 mL PBS containing 0.3% (w/v) bile salt (Merck). The mixture liquid was homogenized and incubated at 37 ºC (*L. plantarum* AKK30) and 30 °C (*S. cerevisiae* B18) for 0 and 120 min. Serial dilutions of samples was performed then each culture inoculated on MRS/YG Agar and incubated for 48 h. Each cell colony was calculated by the total plate count (TPC) method.

2.6. Data analysis

The length of cell from SEM was analysed by ImageJ software [21]. The total of microbial cell (cfu/ml) from viability assay was converted to the logatimic value [22]. Percentage log viability = (log N / log N_0) x 100, where N – count (cfu/ml) after incubation, N_0 – count at time 0 (cfu/mL). Statistical analyses were performed with Costat [23]. The quantitative data was subjected to analysis of variance (ANOVA) and the differences among mean treatments were analysed for significance using Duncan's Multiple Range Test [24].

3. Result and discussion

3.1. Morphology of LAB and yeast

Macroscopic morphology of *L. plantarum* AKK30 colonies were milky white, smooth slippery texture, shiny colony surfaces and flat edges, while macroscopic observations of *S. cerevisiae* B18 showed round, white, gray to brownish colonies, sparkling colony surfaces until slippery, dull, with soft texture [25].

Microscopic morphological observations of *L. plantarum* AKK30 cell using SEM with ion sputter coating (setting 10 mA for 60 seconds) and *S. cerevisiae* B18 cell with ion sputter coating (setting 10 mA for 50 seconds) are shown in Figure 1.

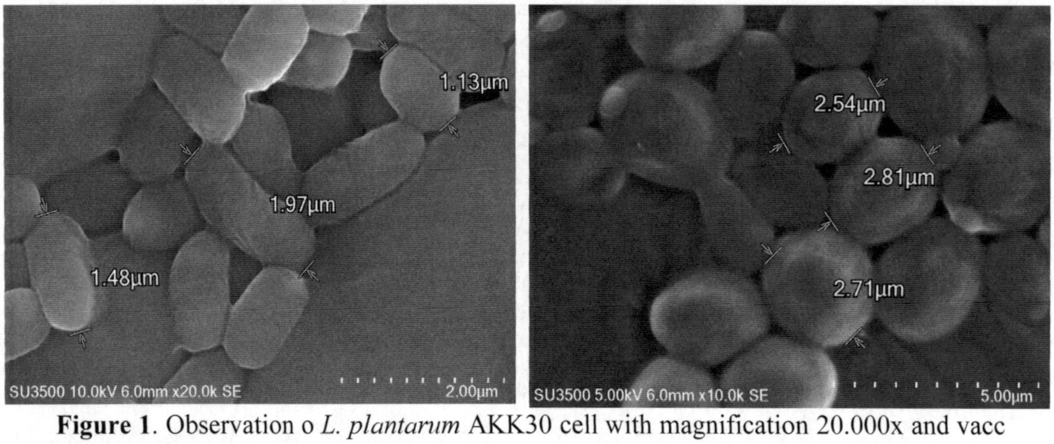

Figure 1. Observation o *L. plantarum* AKK30 cell with magnification 20.000x and vacc 10 kV (left) and *S. cerevisiae* B18 cell with magnification 10.000x and vacc 5 kV

Figure 1 showed that *L. plantarum* AKK30 had bacilli shaped and *S. cerevisiae* B18 had small budding characteristics. *Lactobacillus plantarum* tend to be short rod-shaped in suitable growth conditions and will tend to be longer under unfavourable conditions [26]. The microscopic appearance of these bacteria tends to be in the form of short rods with suitable growth conditions and tend to be longer in unfavourable conditions [27]. According to Saito *et al.* [28], the type of germination consists of no budding, small, medium, and large. The microscopic form of *S. cerevisiae* in the form of blastopores which are oval, cylindrical, oval, or ovoid short and long affected by strains. These yeast will divide and produce germinate budding multipolar at propagation time and the budding can form on the entire cell surface.

The analysis by using ImageJ software resulted that the average length of *L. plantarum* AKK30 was 1.35 µm at 48 h of incubation. Isolate could be seen with clear cell wall boundaries. The average length of *S. cerevisiae* B18 was 3.19 µm at 96 h of incubation. Isolates can be seen with clear and visible cell wall boundaries and also budding formation.

3.2. Antagonistic effect between LAB and yeast

Antifungal and antibacterial assay were carried out using the disc diffusion method. The overall results are summarized in the Table 1.

Table 1. Diameter of clearing zone between LAB and yeast

Isolate	Diameter of clearing zone (mm)	
	Filtered	Non-filtered
L. plantarum AKK30 against *S. cerevisiae* B18	0.53 ± 0.11^b	0.56 ± 0.18^b
S. cerevisiae B18 against *L. plantarum* AKK30	0.95 ± 0.00^a	3.08 ± 0.46^a

Note: Average in the same column with different superscript shown significantly differ (P<0.05)

The results obtained showed that the diameter of clearing zone in the filtered supernatant of both isolates tested were ranged from 0.53 to 0.95 mm, whereas non-filtered supernatant resulted 0.56 to 3.08 mm. This data showed that non-filtered supernatant containing some viable cells resulted larger clearing zone (P<0.05) than filtered supernatant. Non-filtered supernatant obtained from *L. plantarum* AKK30 produced larger clearing zone (0.58 mm) against *S. cerevisiae* B18 compared to the filtered supernatant (0.53 mm), whereas diameter of clearing zone obtained from *S. cerevisiae* B18 against *L. plantarum* AKK30 also resulted the similar trend where non-filtered supernatant produced the larger clearing zone (3.08 mm) than filtered supernatant (0.95 mm).

Figure 2 and 3 showed the zone formation results. Korukluoglu *et al.* [10] observed that the non-filtered supernatant obtained from LAB containing some living cells and grown around the disk did not form a zone formation around the visible colonies of *S. cerevisiae* strain. These results indicated that the culture of *S. cerevisiae* and LAB isolated from kefir were alive and grew together in a commensal relationship.

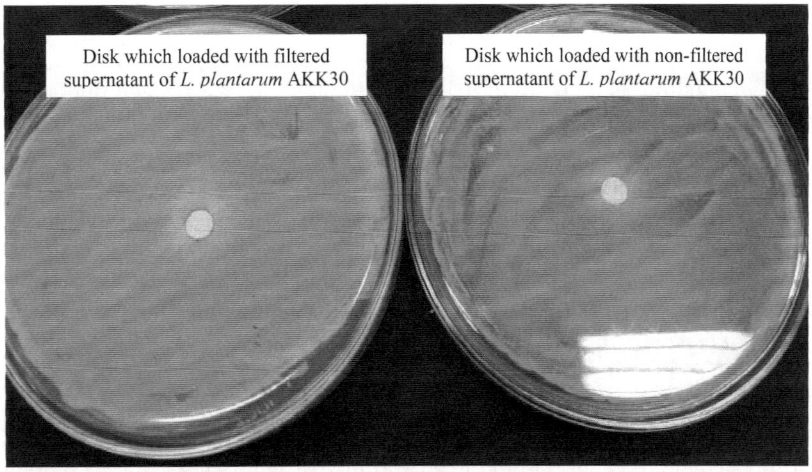

Figure 2. The inhibition effect of *L. plantarum* AKK30 against *S. cerevisiae* B18

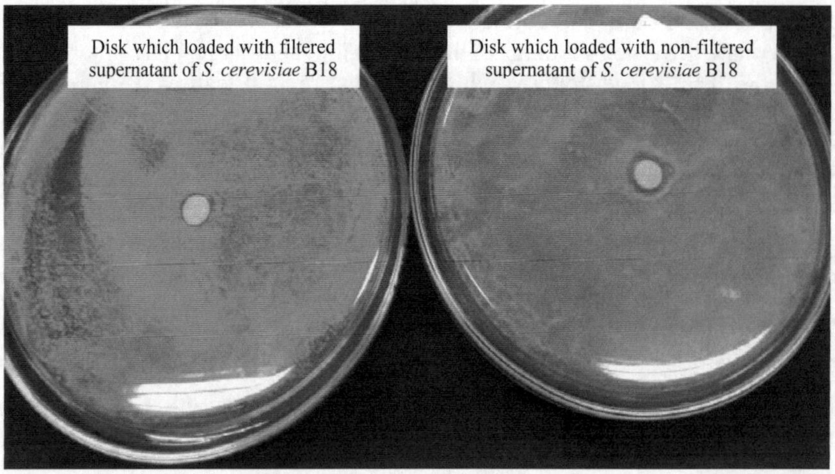

Figure 3. The inhibition effect of *S. cerevisiae* B18 against *L. plantarum* AKK30

The clearing zone values in this study were less than 5 mm, therefore *L. plantarum* AKK30 showed no antifungal activity against *S. cerevisiae* B18 and *S. cerevisiae* B18 showed no antibacterial activity against *L. plantarum* AKK30. Lactic acid is one of the volatile organic acids; thus, the results showed that the accumulation of lactic acid was not sufficient to inhibit *S. cerevisiae* strains in this study. In contrast to our results, Durlu-Özkaya *et al.* [29] reported that except *L. plantarum* Lp19, all of the *Lactobacillus* strains they studied showed strong antifungal activity against *S. cerevisiae*.

3.3. Synergistic effect between LAB and yeast

The results of the synergy effect of *L. plantarum* AKK30 isolates and *S. cerevisiae* B18 by dilution (total plate count) method are shown in Table 2.

Table 2. Population of LAB and yeast colonies

Isolate	Population cfu/mL	Population (log cfu/mL)
L. plantarum AKK30	2,3 x 10^8	8.35 ± 0.06^a
S. cerevisiae B18	1,6 x 10^7	7.20 ± 0.13^b

Note: Average in the same column with different superscript shown significantly differ (P<0.05)

Based on population data with the dilution method, it was found that both microbes (*L. plantarum* AKK30 and *S. cerevisiae*) could alive and grew together in the same medium and had a difference in population by 1 log cycle thus both isolates had a synergistic effect. A study by Soulides [30] reported that the viability of *Lactobacillus* and *Streptococcus* was increased as a result of their association with yeasts which, through synergism, either neutralized or assimilated lactic acid. This result was contrast with the research of Korukluoglu *et al.* [10] that according to the colony calculation in agar media (dilution method) showed that both non-volatile and volatile compounds produced by both LAB and *S. cerevisiae* had an inhibitory effect on each other. In addition, some of these microbes could increase the growth of other microorganisms as co-culture.

3.4. Viability of microencapsulated cultures

Population of *L. plantarum* AKK30 and *S. cerevisiae* B18 before and after spray drying were carried out to calculate the cell viability. The viability of microencapsulated cultures are shown in Table 3.

Table 3. Population of LAB and yeast cultures by spray drying

Isolate	Population (log cfu/g)		Viability (%)
	Before spray drying	After spray drying	
L. plantarum AKK30	9.28 ± 0.06[a]	9.02 ± 0.33[a]	97.25
S. cerevisiae B18	7.37 ± 0.09[a]	6.71 ± 0.13[b]	91.16

Note: Average in the same row with different superscript shown significantly differ (P<0.05)

The population of *L. plantarum* AKK30 before and after spray drying using skim solution as filler did not significantly different (P>0.05) and the viability was 97.25%. The population of *S. cerevisiae* B18 colonies before spray drying was 7.37 log cfu/g and after spray drying using maltodextrin as a filler decreased (P<0.05) to 6.71 log cfu/g or the viability was 91.66%. These results indicated that microencapsulation process using a spray dryer could maintain the LAB and yeast's survivability.

Rizqiyati *et al.* [12] stated that bacterial resistance was determined by comparing the number of cells after and before spray drying. Viability of *L. plantarum* after spray drying for encapsulation treatment in the form of skim milk and arabic gum was relatively good around 89.36% where the number of bacteria before spray drying was 9.4 log cfu/g, after spray drying decreased to 8.4 log cfu/g, meaning that the population declined only 1 log cycle. Arslan *et al.* [31] stated that the population of *S. cerevisiae* var. boulardii with spray drying method using maltodextrin solution (20%, b/v) was 8.61 log_{10} cfu/g or viability of 89.24%. Decreased viability during the spray drying process arises from dehydration and high temperatures. These two mechanisms occurred simultaneously, and had a negative effect on microbial survival [32]. The results of this study showed that the viability of both microorganism after the spray drying process was good. This good viability was the basis that the probiotic candidate microbes could be processed, packaged, and stored in dry culture. In order to provide health benefits, the number of bacteria must be a minimum of 10^6 cfu/gram or 10^7 cfu/gram in the product or be consumed in sufficient quantities to produce a daily intake of 10^8 cfu/mL [33].

3.5. Probiotic properties of LAB and yeast

Lactic acid bacteria isolated from the digestive tract of animals and humans are functional bacteria as a probiotic with the ability to survive in various limiting conditions along the digestive tract from the mouth to the intestine, such as low pH in the stomach and the presence of bile salts in the duodenum [19,20]. The results of the *L. plantarum* AKK30 and *S. cerevisiae* B18 tolerance tests on acid pH conditions (pH 2 and 3), gastric juice, and bile salts are shown in Figure 4.

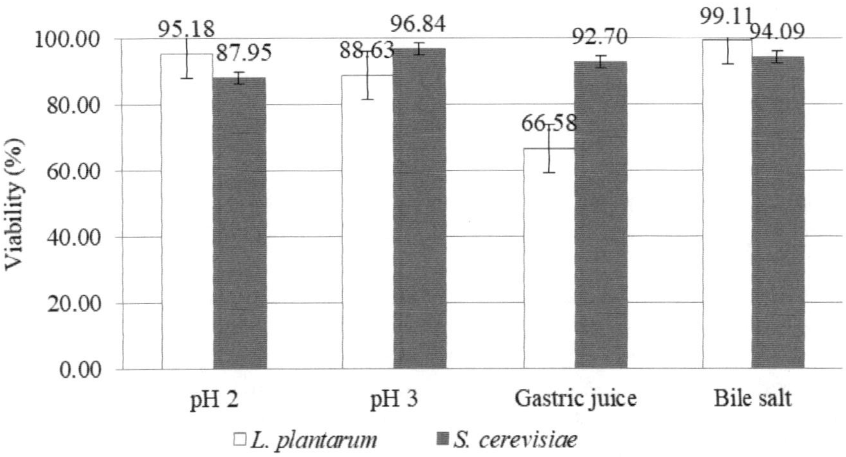

Figure 4. Viability of LAB and yeast in several assay condition

Figure 4 showed that *L. plantarum* AKK30 had good viability on pH 2 and 3 (95.18 and 88.63%) at 45 min of incubation, simulated gastric juice (66.58%) on 45 min of incubation, and 0.3% of bile salts (99.11%) at 120 min of incubation. Several studies showed that LAB had potential as a probiotic. Sabikhi *et al.* [20] reported that microencapsulation of *L. acidophilus* LA1 in sodium alginate and starch provided better protection at simulated conditions of gastric pH (1.0, 1.5, and 2.0) and at high bile salt concentrations (1.0%, 1.5%, and 2.0%). Microencapsulation process resulted in better survival of cells at bile salt concentrations and low pH as compared to free cells. Microencapsulation of *L. acidophilus* using calcium chloride coating on sodium alginate capsules increased tolerance of the bacteria against harsh acidic (pH 2) and bile (1%) conditions [34]. The ability to survive in bile salts is likely because LAB is conjugated with bile salts so it is effective in reducing serum blood cholesterol levels [35]. The high activity of bile salts which are hydrolyzed in the intestinal lumen can reduce the ability of conjugated bile salts to break down fat [36]. Lin *et al.* [37] stated that the survival of LAB ≥ 50% at pH 2, this was assumed to be due to gastric fluid, while LAB survival in bile salt stress was considered good if the viability was ≥ 60%.

The survivability of *S. cerevisiae* B18 on pH 2, pH 3, gastric juice, and bile salt were 87.95, 96.84, 92.70, and 94.09% respectively. A study by Sourabh *et al.* [5] observed that *S. cerevisiae* (HM535662) tolerant at pH 2 (2.71 to 4.12 log cfu/ml), decreased population at pH 3 (0.88 to 3.06 log cfu/mL). Arslan *et al.* [31] stated that the survivability of microencapsulated *S. boulardii* using maltodextrin against simulated gastric solution was 63.62% at pH 2. The ability to tolerate gastric juice was one of an indicator, because good probiotics must be able to pass stresses in the gastrointestinal tract such as pH stress and gastric juice [6].

Microencapsulation technology ie spray drying was efficient method to provide protection against adverse conditions during gastrointestinal passage [9] and to maintain the viability of probiotic from harsh external environments [38].

4. Conclusion

The cell morphology showed that *L. plantarum* AKK30 had bacilli shaped and *S. cerevisiae* B18 had small budding characteristics with average length of cell was 1.35 μm and 3.19 μm respectively. Both microorganism were alive and grew in a commensal relationship with a clearing zone diameter of less than 5 mm and the population between two microorganisms differs only 1 log cycle. The microencapsulation process using spray dryer could maintain the viability of *L. plantarum* AKK30 (97.25%) and *S. cerevisiae* (91.66%). Microencapsulated cultures of *L. plantarum* AKK30 and *S. cerevisiae* B18 had good viability on pH 2 and 3, simulated gastric juice, and bile salts thus it could be used as a probiotic.

5. Acknowledgement

Authors would like to thanks to Ministry of Research, Technology and Higher Education of the Republic of Indonesia for funding this study and also for technician Mrs. Madina Nurohmah and Nisa Grendpina for laboratory assistance.

6. References

[1] World Health Organization, Food and Agriculture Organization of the United Nations (WHO/FAO) 2006 Probiotics in food : Health and nutritional properties and guidelines for evaluation FAO

[2] Palacios M C, Haros M, Sanz Y and Rosell C M 2008 Selection of lactic acid bacteria with high phytate degrading activity for application in whole wheat breadmaking *LMT* **41** 82-92

[3] Korbekandi H, Mortazavian A M and Iravani S 2011 *Technology and stability of probiotic in fermented milks containing probiotics and prebiotics* In: *Probiotic and Prebiotic Foods: Technology, Stability and Benefits to the Human Health* Shah N P, A G da Cruz, J A F Faria (Eds.) (USA: Nova Science Publishers Inc.)

[4] Triana E and Yulinery T 2012 The stability assay of probiotic *Lactobacillus plantarum* Mar8

encapsulated in the preparation of ORS with a viability analysis *Journal of Environmental Technology* **13** (2) 119-24

[5] Sourabh A, Kanwar S S and Sharma O P 2012 In vitro characterization of *Saccharomyces cerevisiae* HM535662 obtained from an indigenous fermented food "Bhaturu" of Western Himalayas *Afr. J. Biotechnol.* **11** (52) 11447-54

[6] Czerucka D, Piche T and Rampal P 2007 Review article: yeast as probiotics – *Saccharomyces boulardii Aliment Pharmacol. Ther.* **26** 767-8

[7] Sofyan A, Aswari A N, Purwoko T and Damayanti E 2013 Screening of lactic acid bacteria from rumen liquor and king grass silage as well as their antibacterial activities *Media Peternakan* **36** 216-23

[8] Damayanti E, Julendra H, Sofyan A and Hayati S N 2014 Bile salt and acid tolerant of lactic acid bacteria isolated from proventriculus of broiler chicken *Media Peternakan* **37** (2) 80-6

[9] Iravani S, Korbekandi H and Mirmohammadi S V 2015 Technology and potential applications of probiotic encapsulation in fermented milk products *J. Food Sci. Technol.* **52** (8) 4679-96

[10] Korukluoglu M, Arik G, Erdogan C and Kocakoglu S 2017 Screening of antagonistic/synergistic effect between lactic acid bacteria (lab) and yeast strains isolated from kefir *International Journal of Nutrition and Food Engineering* **11** (4) 282-8

[11] Calvo M A, Arosemena E L, Shiva C and Adelantado C 2012 Antimicrobial activity of plant natural extracts and essential oils *Science against microbial pathogens: communicating current research and technological advances* 1179-1185

[12] Rizqiyati H 2006 Viablity of *Lactobacillus plantarum* encapsulated with skim milk and gom arab after drying and storage (Bogor : *Thesis* Institut Pertanian Bogor)

[13] Vinderola G, Ana B, Patricia B and Jorge R 2011 Cell viability and functional of probiotic bacteria in dairy products *Frontiers in Microbiology* **2** (1) 1-6

[14] Julendra H, Suryani A E, Istiqomah L, Damayanti E, Anwar M and Fitriani N 2017 Isolation of lactic acid bacteria with cholesterol-lowering activity from digestive tracts of Indonesian native chickens *Media Peternakan* **40** (1) 35-41

[15] Istiqomah L, Anwar M, Anggraeni A S and Damayanti E 2018 cholesterol assimilation of *Saccharomyces cerevisiae* B-18 isolated from gastrointestinal tract of Javanese duck *J. Indonesian Trop. Anim. Agric.* **43** (2) 149-58

[16] Çakır İ 2010 Antibacterial and antifungal activities of some lactic acid bacteria isolated from naturally fermented herbs *Journal of Food, Agriculture & Environment* **8** (2) 223-6

[17] Barbosa-Canovas G V, Ortega-Rivas E, Juliano P and Yan H 2005 *Food Powders: Physical Properties, Processing, and Functionality* (New York: Kluwer Academic/Plenum Publishers)

[18] Chandralekha A, Tavanandi A H, Hebbar H U, Raghavarao K S M S and Gadr R 2016 Encapsulation of yeast (*Saccharomyces cereviciae*) by spray drying for extension of shelf life *Int. J. Drying Technology. Drying Technology*

[19] Taheri H R, Moravej H, Tabandeh F, Zaghari M and Shivazad M 2009 Screening of lactic acid bacteria toward their selection as a source of chicken probiotic *Poultry Sci.* **88** 1586-93

[20] Sabikhi L, Babu R, Thompkinson D K and Kapila S 2010 Resistance of microencapsulated *Lactobacillus acidophilus* LA1 to processing treatments and simulated gut conditions *Food Bioprocess Technol.* **3** 586-93

[21] Schneider C A, Rasband W S and Eliceiri K W 2012 NIH image to ImageJ: 25 years of image analysis *Nat. Methods* **9** (7) 671-5

[22] Williamson K J and Johnson OG 1981 A bacterial assay for assessment of wastewater toxicity *Water Res.* **15** 383-90

[23] Cohort 2008 *CoSTAT Version 6.400 Cohort Software 798* (USA: Lighthouse Ave)

[24] Gomez K A and Gomez A A 2007 *Statistical Procedures for Agricultural Research* 2nd Ed (Jakarta: UI Press)

[25] Barnett J A, Payne R W and Yarrow D 1983 *Yeasts: Characteristics and Identification* (Cambridge: Cambridge University Press)

[26] Pederson C S, Breed R S and Smith N R 1957 *Bergey's Manual of Determinative Bacteriology* 7th Ed (Baltimore: The Williams and Wilkins Co.)

[27] Yulinery T and Nurhidayat N 2016 The viability analysis of probiotic *Lactobacillus* that encapsulated in coating dextrin and passion fruit juice (*Passiflora edulis*) *Jurnal Teknologi Lingkungan* **13** (1) 109-21

[28] Saito T L, Ohtani M, Sawai H, Sano F, Saka A, Watanabe D, Yukawa M, Ohya Y and Morishita S 2004 SCMD: *Saccharomyces cerevisiae* morphological database *Nucleic Acids Research* **32** D319-22

[29] Durlu-Özkaya F, Karabıçak N, Kayalı R and Esen B 2005 Inhibition of yeasts isolated from traditional Turkish cheeses by *Lactobacillus* spp *Int. J. Dairy Technol.* **58** 111-4

[30] Soulides D A 1955 A synergism between yoghurt bacteria and yeasts and the effect of their association upon the viability of the bacteria *App. Micorbiol.* **3** (3) 129-31

[31] Arslan S, Erbas M, Tontul I and Topuz A 2015 Microencapsulation of probiotic *Saccharomyces cerevisiae* var. boulardii with different wall materials by spray drying *LWT - Food Science and Technology* **63** 685-90

[32] Peighambardoust S H, Tafti A G and Hesari J 2011 Application of spray drying for preservation of lactic acid starter cultures: a review *Trends in Food Science & Technology* **22** (5) 215-24

[33] Lopez-Rubio A, Gavara L and Lagaron J M 2006 Bioactive packaging: Turning foods into healthier foods through biomaterial *Trends Foods Science Technology* **17** 567-75

[34] Chandramouli V, Kalasapathy K, Peiri P and Jones M 2004 An improved method of microencapsulation and its evaluation to protect *Lactobacillus* spp. in simulated gastric conditions *Journal of Microbiological Methods* **56** 27-35

[35] Begley M, Hill C and Gahan C G M 2006 Bile salt hydrolase cctivity in probiotics *Appl. Environ. Microbiol.* **72** 1729-38

[36] Torshizi M, Rahimi A K Sh, Mojgani N, Esmaeilkhanian S and Grimes J L 2008 Screening of indigenous strains of lactic acid bacteria for development of a probiotic for poultry *Asian-Aus. J. Anim. Sci.* **21** 1495-500

[37] Lin W H, Hwang C F, Chen L W and Tsen H Y 2006 Viable counts, characteristic evaluation for commercial lactic acid bacteria product *Food Microbiology* **23** 74-81

[38] Das A, Ray S, Raychaudhuri U and Chakraborty R 2014 Microencapsulation of probiotic bacteria and its potential application in food technology *International Journal of Agriculture, Environment & Biotechnology* **6**(1) 63-9